T5-AWF-158

PROGRESS IN CLINICAL AND BIOLOGICAL RESEARCH

1984 TITLES

Vol 137: **PCB Poisoning in Japan and Taiwan,** Masanori Kuratsune, Raymond E. Shapiro, *Editors*

Vol 138: **Chemical Taxonomy, Molecular Biology, and Function of Plant Lectins,** Irwin J. Goldstein, Marilynn E. Etzler, *Editors*

Vol 139: **Difficult Decisions in Medical Ethics: The Fourth Volume in a Series on Ethics, Humanism, and Medicine,** Doreen Ganos, Rachel E. Lipson, Gwynedd Warren, Barbara J. Weil, *Editors*

Vol 140: **Abnormal Functional Development of the Heart, Lungs, and Kidneys: Approaches to Functional Teratology,** Casimer T. Grabowski, Robert J. Kavlock, *Editors*

Vol 141: **Industrial Hazards of Plastics and Synthetic Elastomers,** Jorma Järvisalo, Pirkko Pfäffli, Harri Vainio, *Editors*

Vol 142: **Hormones and Cancer,** Erlio Gurpide, Ricardo Calandra, Carlos Levy, Roberto J. Soto, *Editors*

Vol 143: **Viral Hepatitis and Delta Infection,** Giorgio Verme, Ferruccio Bonino, Mario Rizzetto, *Editors*

Vol 144: **Chemical and Biological Aspects of Vitamin B_6 Catalysis,** A.E. Evangelopoulos, *Editor*. Published in 2 volumes. Part A: **Metabolism, Structure, and Function of Phosphorylases, Decarboxylases, and Other Enzymes.** Part B: **Metabolism, Structure, and Function of Transaminases**

Vol 145: **New Approaches to the Study of Benign Prostatic Hyperplasia,** Frances A. Kimball, Allen E. Buhl, Donald B. Carter, *Editors*

Vol 146: **Experimental Allergic Encephalomyelitis: A Useful Model for Multiple Sclerosis,** Ellsworth C. Alvord, Jr., Marian W. Kies, Anthony J. Suckling, *Editors*

Vol 147: **Genetic Epidemiology of Coronary Heart Disease: Past, Present, and Future,** D.C. Rao, Robert C. Elston, Lewis H. Kuller, Manning Feinleib, Christine Carter, Richard Havlik, *Editors*

Vol 148: **Aplastic Anemia: Stem Cell Biology and Advances in Treatment,** Neal S. Young, Alan S. Levine, R. Keith Humphries, *Editors*

Vol 149: **Advances in Immunobiology: Blood Cell Antigens and Bone Marrow Transplantation,** Jeffrey McCullough, S. Gerald Sandler, *Editors*

Vol 150: **Factor VIII Inhibitors,** Leon W. Hoyer, *Editor*

Vol. 151: **Matrices and Cell Differentiation,** R.B. Kemp, J.R. Hinchliffe, *Editors*

Vol 152: **Intrahepatic Calculi,** Kunio Okuda, Fumio Nakayama, John Wong, *Editors*

Vol 153: **Progress and Controversies in Oncological Urology,** Karl H. Kurth, Frans M.J. Debruyne, Fritz H. Schroeder, Ted A.W. Splinter, Theo D.J. Wagener, *Editors*

Vol 154: **Myelofibrosis and the Biology of Connective Tissue,** Paul D. Berk, Hugo Castro-Malaspina, Louis R. Wasserman, *Editors*

Vol 155: **Malaria and the Red Cell,** John W. Eaton, George J. Brewer, *Editors*

Vol 156: **Advances in Cancer Control: Epidemiology and Research,** Paul F. Engstrom, Paul N. Anderson, Lee E. Mortenson, *Editors*

Vol 157: **Recognition Proteins, Receptors, and Probes: Invertebrates,** Elias Cohen, *Editor*

Vol 158: **The Methylxanthine Beverages and Foods: Chemistry, Consumption, and Health Effects,** Gene A. Spiller, *Editor*

Vol 159: **Erythrocyte Membranes 3: Recent Clinical and Experimental Advances,** Walter C. Kruckeberg, John W. Eaton, Jon Aster, George J. Brewer, *Editors*

Vol 160: **Reproduction: The New Frontier in Occupational and Environmental Health Research,** James E. Lockey, Grace K. Lemasters, William R. Keye, Jr., *Editors*

Vol 161: **Chemical Regulation of Immunity in Veterinary Medicine,** Meir Kende, Joseph Gainer, Michael Chirigos, *Editors*

Vol 162: **Bladder Cancer**, René Küss, Saad Khoury, Louis J. Denis, Gerald P. Murphy, James P. Karr, *Editors.* Published in 2 volumes: Part A: **Pathology, Diagnosis, and Surgery**. Part B: **Radiation, Local and Systemic Chemotherapy, and New Treatment Modalities**

Vol 163: **Prevention of Physical and Mental Congenital Defects**, Maurice Marois, *Editor.* Published in 3 volumes: Part A: **The Scope of the Problem.** Part B: **Epidemiology, Early Detection and Therapy, and Environmental Factors.** Part C: **Basic and Medical Science, Education, and Future Strategies**

Vol 164: **Information and Energy Transduction in Biological Membranes,** C. Liana Bolis, Ernst J.M. Helmreich, Hermann Passow, *Editors*

Volume 165: **The Red Cell: Sixth Ann Arbor Conference**, George J. Brewer, *Editor*

Vol 166: **Human Alkaline Phosphatases,** Torgny Stigbrand, William H. Fishman, *Editors*

Vol 167: **Ethopharmacological Aggression Research**, Klaus A. Miczek, Menno R. Kruk, Berend Olivier, *Editors*

Vol 168: **Epithelial Calcium and Phosphate Transport: Molecular and Cellular Aspects,** Felix Bronner, Meinrad Peterlik, *Editors*

Vol 169: **Biological Perspectives on Aggression,** Kevin J. Flannelly, Robert J. Blanchard, D. Caroline Blanchard, *Editors*

Vol 170: **Porphyrin Localization and Treatment of Tumors,** Daniel R. Doiron, Charles J. Gomer, *Editors*

Vol 171: **Developmental Mechanisms: Normal and Abnormal,** James W. Lash, Lauri Saxén, *Editors*

Please contact the publisher for information about previous titles in this series.

PORPHYRIN LOCALIZATION AND TREATMENT OF TUMORS

PORPHYRIN LOCALIZATION AND TREATMENT OF TUMORS

Proceedings of the Clayton Foundation International Symposium on Porphyrin Localization and Treatment of Tumors, Santa Barbara, California
April 24–28, 1983

Editors

Daniel R. Doiron
Charles J. Gomer

Clayton Program
Childrens Hospital of Los Angeles
Los Angeles, California

ALAN R. LISS, INC., NEW YORK

Address all Inquiries to the Publisher
Alan R. Liss, Inc., 150 Fifth Avenue, New York, NY 10011

Printed in the United States of America.

Library of Congress Cataloging in Publication Data

Clayton Foundation International Symposium on Porphyrin Localization and Treatment (1983 : Santa Barbara, Calif.)
Porphyrin localization and treatment of tumors.

Includes bibliographies and index.
1. Photochemotherapy--Congresses. 2. Cancer--Radiotherapy--Congresses. 3. Porphyrin and porphyrin compounds --Therapeutic use--Congresses. 4. Cancer--Diagnosis--Congresses. 5. Porphyrin and porphyrin compounds--Diagnostic use--Congresses. 6. Porphyrin and porphyrin compounds--Physiological effect--Congresses. I. Doiron, Daniel R. II. Gomer, Charles J. III. Clayton Foundation for Research. IV. Title. [DNLM: 1. Neoplasms--Diagnosis-congresses. 2. Neoplasms--drug therapy--congresses. 3. Photochemotherapy--congresses. 4. Porphyrins--diagnostic use--congresses. 5. Porphyrins--therapeutic use--congresses. W1 PR668E v. 170 / QU 110 C622p 1983]
RC271.P43C53 1983 616.99'20631 84-21781
ISBN 0-8451-5020-0

Contents

IN VIVO PORPHYRIN PHOTOBIOLOGY

Contributors

Gabriel Adam, Department of Radiobiology, Nuclear Research Center–Negev, Beer-Sheva, Israel **[115]**

Katsuo Aizawa, Department of Physiology, Tokyo Medical College, Tokyo, Japan **[227]**

H. Ali, MRC Group in the Radiation Sciences, University of Sherbrooke Medical Center, Sherbrooke, Quebec, Canada **[315]**

Ryuta Amemiya, Department of Surgery, Tokyo Medical College, Tokyo, Japan **[747]**

C. Amsterdamsky, Department of Chemistry, Wayne State University, Detroit, MI **[563]**

Robert E. Anderson, Department of Cerebrovascular Research, Mayo Clinic, Rochester, MN **[483]**

Oscar J. Balchum, Department of Medicine, Pulmonary Disease Section, University of Southern California School of Medicine, Los Angeles, CA **[521, 727, 847]**

Pierre Band, Division of Epidemiology and Biometry, Cancer Control Agency of British Columbia, Vancouver, British Columbia, Canada **[571]**

J.L. Baumann, Department of Research, St. Vincent Hospital and Medical Center, Toledo, OH **[335]**

David A. Bellnier, Department of Urology Research, Massachusetts General Hospital, Boston, MA **[187, 361, 533]**

Ralph C. Benson, Jr., Department of Urology, Mayo Clinic, Rochester, MN **[795]**

A.E. van den Berg-Blok, Radiobiological Institute TNO, Rijswijk, The Netherlands **[637]**

Michael L. Berman, Department of Obstetrics and Gynecology, University of California, Irvine Medical Center, Orange, CA **[767]**

Michael W. Berns, Department of Surgery, Department of Developmental and Cell Biology, University of California, Irvine, Irvine, CA **[501, 681, 767]**

Jean-Guy Besner, Faculty of Pharmacy, University of Montreal, Montreal, Quebec, Canada **[571]**

A.J. Blake, Department of Physics, University of Adelaide, Adelaide, South Australia, Australia **[149, 693]**

F.P. Bolin, Division of Radiation Physics, Department of Therapeutic Radiology, Henry Ford Hospital, Detroit, MI **[211]**

Donn G. Boyle, Department of Radiation Medicine, Roswell Park Memorial Institute, Buffalo, NY **[759]**

The number in brackets is the opening page number of the contributor's article.

Robert A. Bruce, Jr., Department of Ophthalmology, The Ohio State University, Columbus, OH **[777]**

Robert G. Burns, Department of Surgery, University of California, Irvine Medical Center, Orange, CA **[767]**

B.W. Cain, Division of Radiation Physics, Department of Therapeutic Radiology, Henry Ford Hospital, Detroit, MI **[211]**

Fulvio Calzavara, Divisione di Radioterapia, Ospedale Civile, Padova, Italy **[829]**

Christian Chaussy, Urological Clinic and Policlinic, University of Munich, Klinikum Großhadern, Munich, Bavaria, Federal Republic of Germany **[249]**

M.-H. Chow, Department of Developmental and Cell Biology, University of California, Irvine, Irvine, CA **[501]**

Terje Christensen, Department of Biophysics, Norsk Hydro's Institute for Cancer Research, Oslo 3, Norway **[381, 419]**

Luigi Corti, Divisione di Radioterapia, Ospedale Civile, Padova, Italy **[829]**

Prue A. Cowled, Department of Medicine, Queen Elizabeth Hospital, Woodville; University of Adelaide, Adelaide, South Australia, Australia **[693, 709]**

Ivo Cozzani, Istituto Biologia Animale, Universita di Padova, Padova, Italy **[471]**

James J. Crute, Department of Biochemistry, University of Rochester School of Medicine, Rochester, NY **[323]**

Philip J. Disaia, Department of Obstetrics and Gynecology, University of California, Irvine Medical Center, Orange, CA **[767]**

Daniel R. Doiron, Department of Ophthalmology, Ocular Oncology Center, Childrens Hospital of Los Angeles; University of Southern California School of Medicine, Los Angeles, CA **[41, 459, 521, 613, 727, 847]**

Thomas J. Dougherty, Division of Radiation Biology, Department of Radiation Medicine, Roswell Park Memorial Institute, Buffalo, NY **[75, 301, 601, 759]**

Mohamed El-Far, Division of Gastroenterology, Department of Internal Medicine, University of California, Davis, Sacramento, CA **[657, 661]**

Reinold Ellingsen, Division of Physical Electronics, University of Trondheim, Trondheim-NTH, Norway **[647]**

Jan F. Evensen, Department of General Oncology, The Norwegian Radium Hospital, Oslo 3, Norway **[541]**

Harvey Farmer, Department of Therapeutic Radiology, Henry Ford Hospital, Detroit, MI **[583]**

G. Firnau, Department of Nuclear Medicine and Radiology, McMaster University, Hamilton, Ontario, Canada **[629]**

Christopher S. Foote, Department of Chemistry, University of California, Los Angeles, CA **[3]**

Ian J. Forbes, Department of Medicine, University of Adelaide, Woodville, South Australia, Australia **[693, 709]**

Stanley W. Fountain, Clayton Ocular Oncology Center, Childrens Hospital of Los Angeles, Los Angeles, CA **[459]**

Makoto Fujime, Department of Urology Research, Massachusetts General Hospital, Boston, MA **[187]**

Shobha N. Gandhi, Division of Gastroenterology, Department of Medicine, University of California, Davis, Sacramento, CA **[673]**

Pierre Gervais, Faculty of Pharmacy, University of Montreal, Montreal, Quebec, Canada **[571]**

Scott L. Gibson, Department of Biochemistry, University of Rochester Medical Center, Rochester, NY **[201, 323]**

P.J. Goldblatt, Department of Pathology, Medical College of Ohio at Toledo, Toledo, OH **[335]**

Charles J. Gomer, Clayton Ocular Oncology Center, Childrens Hospital of Los Angeles, Los Angeles, CA **[459, 591]**

W. Gorisch, Urological Clinic and Policlinic, University of Munich, Klinikum Großhadern, Munich, Bavaria, Federal Republic of Germany **[249]**

Leonard I. Grossweiner, Department of Physics, Biophysics Laboratory, Illinois Institute of Technology, Chicago, IL **[391]**

C. Hammer, Urological Clinic and Policlinic, University of Munich, Klinikum Großhadern, Munich, Bavaria, Federal Republic of Germany **[249]**

Marie Hammer-Wilson, Department of Surgery, University of California, Irvine, Irvine, CA **[501]**

Cheryl A. Hanzlik, Department of Physics and Astronomy, University of Rochester, Rochester, NY **[201]**

Yoshihiro Hayata, Department of Surgery, Tokyo Medical College, Tokyo, Japan **[227, 747]**

Barbara W. Henderson, Division of Radiation Biology, Roswell Park Memorial Institute, Buffalo, NY **[601]**

Fred W. Hetzel, Department of Therapeutic Radiology, Henry Ford Hospital, Detroit, MI **[563, 583]**

Russell Hilf, Department of Biochemistry, University of Rochester Medical Center, Rochester, NY **[201, 323]**

Atle H. Hindar, Department of Biophysics, Norsk Hydros Institute of Cancer Research, Oslo 3, Norway **[541]**

Haruo Hisazumi, Department of Urology, School of Medicine, Kanazawa University, Kanazawa, Japan **[239, 443, 785]**

Gerald C. Huth, Institute for Physics and Imaging Science, University of Southern California School of Medicine, Marina Del Ray, CA **[727, 847]**

Fred J. Jacka, Mawson Institute, University of Adelaide, Adelaide, South Australia, Australia **[149, 693, 709]**

Petter B. Jacobsen, Department of Biophysics, Norsk Hydro's Institute for Cancer Research, Oslo 3, Norway **[419]**

Dean B. Jacques, Department of Neurosurgery, Huntington Medical Research Institutes, Pasadena, CA **[613]**

Skip Jacques, Department of Neurosurgical Research, Huntington Medical Research Institutes, Pasadena, CA **[719]**

William P. Jeeves, Department of Medical Physics, Ontario Cancer Treatment and Research Foundation, Hamilton, Ontario, Canada **[115, 629]**

Dieter Jocham, Urological Clinic and Policlinic, University of Munich, Klinikum Großhadern, Munich, Bavaria, Federal Republic of Germany **[249]**

Charlotte A. Jones, Department of Urology Research, Mayo Clinic, Rochester, MN **[483]**

Giulio Jori, Istituto Biologia Animale, Universita di Padova, Padova, Italy **[373, 471, 829]**

Harubumi Kato, Department of Surgery, Tokyo Medical College, Tokyo, Japan **[227, 747]**

Norihiko Kawate, Department of Surgery, Tokyo Medical College, Tokyo, Japan **[227]**

R. Keck, Department of Surgery (Urology), Medical College of Ohio at Toledo, Toledo, OH **[335]**

David Kessel, Department of Medicine and Pharmacology, Wayne State University School of Medicine, Detroit, MI **[405]**

J.E. Klaunig, Department of Pathology, Medical College of Ohio at Toledo, Toledo, OH **[335]**

Wayne H. Knox, Institute of Optics, University of Rochester, Rochester, NY **[201]**

Chimori Konaka, Department of Surgery, Tokyo Medical College, Tokyo, Japan **[227]**

Martha Kreimer-Birnbaum, Department of Research, Porphyrin Laboratory, St. Vincent Hospital and Medical Center, Toledo, OH **[335]**

Edward R. Laws, Jr., Department of Neurosurgery, Mayo Clinic, Rochester, MN **[483]**

Pauline B. Leakey, Department of Biochemistry, University of Rochester School of Medicine, Rochester, NY **[323]**

Julia G. Levy, Department of Microbiology, University of British Columbia, Vancouver, British Columbia, Canada **[351]**

J.E. van Lier, MRC Group in the Radiation Sciences, University of Sherbrooke Medical Center, Sherbrooke, Quebec, Canada **[315]**

Chi-Wei Lin, Department of Urology Research, Massachusetts General Hospital, Boston, MA **[187, 361, 533]**

F.M. Little, Department of Neurosurgery, USC School of Medicine, Los Angeles, CA **[591]**

U. Löhrs, Urological Clinic and Policlinic, University of Munich, Klinikum Großhadern, Munich, Bavaria, Federal Republic of Germany **[249]**

Diane M. Lowe, Department of Medical Physics, Ontario Cancer Treatment and Research Foundation, Hamilton, Ontario, Canada **[115]**

Diane Maass, Department of Nuclear Medicine and Radiology, McMaster University, Hamilton, Ontario, Canada **[629]**

Sylvie Mailhot, Faculty of Pharmacy, University of Montreal, Montreal, Quebec, Canada **[571]**

Patrick B. Malone, Division of Radiation Biology, Roswell Park Memorial Institute, Buffalo, NY **[601]**

Giovanni Malvadi, Istituto Biologia Animale, Universita di Padova, Padova, Italy **[471]**

Giovanni Mandoliti, Divisione di Radioterapia, Ospedale Civile, Padova, Italy **[471, 829]**

Thomas S. Mang, Department of Radiation Biology, Roswell Park Memorial Institute, Buffalo, NY **[177]**

J.P.A. Marijnissen, Department of Physics, The Dr. Daniel den Hoed Cancer Center, Rotterdam Radio-Therapeutic Institute, Rotterdam, The Netherlands **[133, 637]**

G.R. Mason, Department of Surgery, University of California, Irvine, Orange, CA **[681]**

James S. McCaughan, Jr., Department of Surgery, Grant Hospital, Columbus, OH **[805]**

Glenn A.J. McCulloch, Department of Medicine, University of Adelaide, Adelaide, Australia **[709]**

Jerry L. McCullough, Department of Dermatology, University of California, Irvine Medical Center, Orange, CA **[767]**

J.D. McMahon, School of Dentistry, University of Detroit, Detroit, MI **[563]**

Daphne Mew, Department of Microbiology, University of British Columbia, Vancouver, British Columbia, Canada **[351]**

Toshimitsu Misaki, Department of Urology, School of Medicine, Kanazawa University, Kanazawa, Japan **[239, 785]**

Norio Miyoshi, Department of Urology, School of Medicine, Kanazawa University, Kanazawa, Japan **[239, 443, 785]**

Johan Moan, Department of Biophysics, Norsk Hydro's Institute for Cancer Research, Oslo 3, Norway **[381, 419, 541]**

Patrick C. Mock, Department of Dermatology, Massachusetts General Hospital, Boston, MA **[533]**

Richard Moisan, Faculty of Pharmacy, University of Montreal, Montreal, Quebec, Canada **[571]**

Maurice Nahabedian, Department of Surgery, University of California, Irvine, Irvine, CA **[501]**

Kazuyoshi Nakajima, Department of Urology, School of Medicine, Kanazawa University, Kanazawa, Japan **[443]**

Haruhiko Nakamera, Department of Surgery, Tokyo Medical College, Tokyo, Japan **[227]**

Tom M. Nordlund, Department of Physics and Astronomy, University of Rochester, Rochester, NY **[201]**

Jean Novotny, Department of Surgery, University of California, Irvine, Orange, CA **[681]**

Arne Ødegaard, Department of Pathology, University of Trondheim, Trondheim, Norway **[647]**

M.D. O'Hara, Department of Therapeutic Radiology, Henry Ford Hospital, Detroit, MI **[563]**

Elizabeth Z. Olson, Nursing Service, Grant Hospital, Columbus, OH **[841]**

Robert S. Olson, Department of Cell Biology, Huntington Medical Research Institutes, Pasadena, CA **[613]**

Jutaro Ono, Department of Surgery, Tokyo Medical College, Tokyo, Japan **[227, 747]**

Masaki Otawa, Department of Surgery, Tokyo Medical College, Tokyo, Japan **[227]**

J.G. Parker, Applied Physics Laboratory, The Johns Hopkins University, Laurel, MD **[259]**

John A. Parrish, Department of Dermatology, Massachusetts General Hospital, Boston, MA **[533]**

Victor Passy, Department of Surgery, University of California, Irvine, Orange, CA **[681]**

K.B. Patel, Pulmonary Disease Section, Los Angeles County–USC Medical Center, Los Angeles, CA **[521]**

M.J. Phillip, School of Dentistry, University of Detroit, Detroit, MI **[563]**

Neville Pimstone, Division of Gastroenterology, Department of Medicine, University of California, Davis, Sacramento, CA **[657, 661, 673]**

Cesare Polico, Divisione di Radioterapia, Ospedale Civile, Padova, Italy **[829]**

William R. Potter, Department of Radiation Biology, Roswell Park Memorial Institute, Buffalo, NY **[177, 301, 759]**

L.E. Preuss, Division of Radiation Physics, Department of Therapeutic Radiology, Henry Ford Hospital, Detroit, MI **[211]**

A.E. Profio, Department of Chemical and Nuclear Engineering, University of California at Santa Barbara, Santa Barbara, CA **[163, 847]**

George R. Prout, Jr., Department of Surgery, Urological Associates, Massachusetts General Hospital, Harvard Medical School, Boston, MA **[187]**

Y.-N. Qin, Los Angeles County–USC Medical Center, Los Angeles, CA; Associate Research Member Cancer Institute, Peking, China **[521]**

Umo Rao, Department of Pathology, Roswell Park Memorial Institute, Buffalo, NY **[759]**

Nicholas J. Razum, Clayton Ocular Oncology Center, Childrens Hospital of Los Angeles, Los Angeles, CA **[459]**

Elena Reddí, Istituto Biologia Animale, Universita di Padova, Padova, Italy **[373, 471, 829]**

H.S. Reinhold, Department of Experimental Radiotherapy, Erasmus University, Rotterdam; Radiobiological Institute, Rijswijk, The Netherlands **[637]**

Mark A. Rettenmaier, Department of Obstetrics and Gynecology, University of California, Irvine Medical Center, Orange, CA **[767].**

Michael A.J. Rodgers, Center for Fast Kinetics Research, University of Texas at Austin, Austin, TX **[373]**

Elian Rossi, Istituto Biologia Animale, Universita di Padova, Padova, Italy **[471]**

Donald E. Rounds, Department of Cell Biology, Huntington Medical Research Institutes, Pasadena, CA **[613]**

J. Rousseau, MRC Group in the Radiation Sciences, University of Sherbrooke Medical Center, Sherbrooke, Quebec, Canada **[315]**

Natalie Rucker, Clayton Ocular Oncology Center, Childrens Hospital of Los Angeles, Los Angeles, CA **[459]**

Makoto Saito, Department of Surgery, Tokyo Medical College, Tokyo, Japan **[227]**

Harumasa Sakai, Department of Surgery, Tokyo Medical College, Tokyo, Japan **[227]**

Torbjørg Sandquist, Department of Biophysics, Norsk Hydro's Institute for Cancer Research, Oslo 3, Norway **[381]**

J. Sarnaik, Department of Chemical and Nuclear Engineering, University of California, Santa Barbara, CA **[163]**

A.P. Schaap, Department of Chemistry, Wayne State University, Detroit, MI **[563]**

Kevin L. See, Department of Medicine, University of Adelaide, Woodville, South Australia, Australia **[693, 709]**

S.H. Selman, Department of Surgery (Urology), Medical College of Ohio at Toledo, Toledo, OH **[335]**

C.H. Shelden, Department of Neurosurgery, Huntington Medical Research Institutes, Pasadena, CA **[613]**

Hideki Shinohara, Department of Surgery, Tokyo Medical College, Tokyo, Japan **[227]**

Lars Smedshammer, Department of Biophysics, Norsk Hydros Institute for Cancer Research, Oslo 3, Norway **[381]**

Stein Sommer, Department of Biophysics, Norsk Hydros Institute of Cancer Research, Oslo 3, Norway **[541]**

John D. Spikes, Department of Biology, University of Utah, Salt Lake City, UT **[19]**

Gerd Staehler, Urological Clinic and Policlinic, University of Munich, Klinikum Großhadern, Munich, Bavaria, Federal Republic of Germany **[249]**

W.D. Stanbro, Applied Physics Laboratory, The Johns Hopkins University, Laurel, MD **[259]**

W.M. Star, Department of Physics, The Dr. Daniel den Hoed Cancer Center and Rotterdam Radio-Therapeutic Institute, Rotterdam, The Netherlands **[133, 637]**

Lars O. Svaasand, Division of Physical Electronics, University of Trondheim, Trondheim, Norway **[91, 647]**

A.G. Swincer, Department of Organic Chemistry, University of Adelaide, Adelaide, South Australia, Australia **[285, 693]**

Hidenobu Takahashi, Department of Surgery, Tokyo Medical College, Tokyo, Japan **[227]**

Luigi Tomio, Divisione di Radioterapia, Ospedale Civile, Padova, Italy **[471, 829]**

G.H.N. Towers, Department of Botany, University of British Columbia, Vancouver, British Columbia, Canada **[351]**

V.C. Trenerry, Department of Organic Chemistry, University of Adelaide, Adelaide, South Australia, Australia **[285]**

Osamu Ueki, Department of Urology, School of Medicine, Kanazawa University, Kanazawa, Japan **[443]**

Eberhard Unsöld, Zentrales Laserlabor, GSF Neuherberg, Neuherberg, Bavaria, Federal Republic of Germany **[249]**

Ronald G. Vincent, Director's Office, Ellis Fischel State Cancer Center, Columbia, MO **[759]**

Robert J. Walter, Department of Developmental and Cell Biology, University of California, Irvine, Irvine, CA **[501]**

A.D. Ward, Department of Organic Chemistry, University of Adelaide, Adelaide, South Australia, Australia **[285, 693, 709]**

Chi-Kit Wat, Department of Botany, University of British Columbia, Vancouver, British Columbia, Canada **[351]**

Gerald D. Weinstein, Department of Dermatology, University of California, Irvine Medical Center, Orange, CA **[767]**

K.R. Weishaupt, Oncology Research and Development, Cheektowaga, NY **[301]**

Robert E. Wharen, Jr., Department of Neurosurgery, Mayo Clinic, Rochester, MN **[483]**

Alan G. Wile, Department of Surgery, University of California, Irvine Medical Center, Orange, CA **[501, 681]**

P.A. Wilksch, Mawson Institute, University of Adelaide, Adelaide, South Australia, Australia **[149, 693]**

Brian C. Wilson, Department of Medical Physics, Ontario Cancer Treatment and Research Foundation, Hamilton, Ontario, Canada **[115, 629]**

W. Wright, Department of Surgery, University of California, Irvine, Irvine, CA **[501]**

Tetsushi Yamada, Department of Surgery, Tokyo Medical College, Tokyo, Japan **[227]**

Kazuo Yoneyama, Department of Surgery, Tokyo Medical College, Tokyo, Japan **[227]**

Pier L. Zorat, Divisione di Radioterapia, Ospedale Civile, Padova, Italy **[829]**

Preface

This book represents the proceedings of the Clayton Foundation International Symposium on Porphyrin Localization and Treatment of Tumors held April 24 through April 28, 1983 in Santa Barbara, California. The symposium included review sessions on photochemistry, photobiology, photophysics and instrumentation of porphyrins and an invited review by Thomas J. Dougherty on the current status of photoradiation therapy. Regular sessions addressed the areas of physics and instrumentation, porphyrin chemistry, basic, preclinical and clinical studies. The enclosed papers represent presentations in all these sessions and a poster session.

The symposium was supported by the Clayton Foundation for Research, Houston, Texas, and Childrens Hospital of Los Angeles. The editors take final responsibility for any editorial changes present in the enclosed papers. We would like to acknowledge and thank the staff of the Clayton Foundation Ocular Oncology Program at Childrens Hospital of Los Angeles, without whose help the symposium and these proceedings would not have been possible. A special thanks to Carla Rother, Betsy DiBella, and Ben Szirth.

Introduction

The history of porphyrins and their role as a diagnostic and therapeutic modality is of a brief duration in the history of medicine. Cruentene, a product extracted by alcohol and sulphuric acid from hemoglobin (cruorin) was referenced by Thudicum in the report of the Medical Office Privy Council in 1868. This material was named hematoporphyrin by Hoppler-Seyler in 1871.

Unique properties of the porphyrins were discovered over the next few decades. Hausmann in the early 1900's reported on murine phototoxicity following subcutaneous injection with hematorporphyrin. Meyer-Betz, in 1913, was able to relate a firsthand account of human photosensitivity after he injected himself with 200 mg of hematoporphyrin. The effects lasted for several months but eventually resolved. In 1924, A. Policard reported a spontaneous fluorescence in experimental tumors exposed to a Woods lamp. He attributed this fluorescence to an accumulation of porphyrins in neoplastic tissue. Nearly twenty years later Auler, in 1942, after administering hematoporphyrin parenterally to mice, observed that it was preferentially retained in malignant tissue. This phenomenon was also observed by Figge in 1948, who recognized the potential importance of hematoporphyrin as a tool for the diagnosis of cancer.

Throughout the 1950's, several groups, among them those of Peck, Taxdall-Rassmusen, and Mack, studied hematoporphyrin to confirm its tumor localization properties. However, an obstacle existed that had to be resolved before the diagnostic potential of the drug could be realized. Although the fluorescent properties of hematoporphyrin that contribute to its detection in neoplastic tissue had been confirmed now by several teams of investigators, the fluorescence observed had been inconsistent. Thus, reliable results could not be obtained and large amounts of hematoporphyrin were necessarily administered, increasing the risk of phototoxic reactions. Schwartz and his colleagues, in the late 1950's, pointed out that the hematoporphyrin product commercially available contained a mixture of porphyrins with varying localizing properties that in some cases exceeded those of pure hematoporphyrin. This awareness led Lipson and his co-workers, in 1960, to develop the derivative of hematoporphyrin that enhanced the amount of tumor localizing component in the clinically used drug. With the improved product, diagnostic procedures using hematoporphyrin derivative (HpD) proliferated.

While the importance of porphyrins and now HpD in cancer diagnosis was becoming accepted, it only became apparent in the late 1960's and early 1970's that the role of the drug would go beyond that of diagnostic agent. Several investigators, during the early and mid 1970's, Diamond, Kelly, Snell, and Dougherty, in particular, realized that HpD and light together had a potential capability for tumor destruction and began to conduct in vitro and in vivo studies to determine the response of tumors to this combined approach. In 1978, after extensive preclinical studies, Dougherty reported on a series of 25 patients, showing partial or complete response in 111 of 113 tumors treated with photoradiation therapy (PRT) as this modality was then called. In recent years the term photodynamic therapy (PDT) is becoming the preferred designation.

In the years since those initial studies were carried out by the few pioneering investigators working in this area, interest in the basic, preclinical and clinical aspects of phototherapy has expanded tremendously. This increased interest is evident when one observes the history of unofficial and official meetings to discuss photoradiation or photodynamic therapy. The first unofficial meeting was held as recently as 1979 at Roswell Park Memorial Institute with forty participants. The second meeting, held in Washington D.C. for two days in 1981, attracted 180 attendees and 23 presentations. Approximately 240 persons were in attendance at this four-day symposium and the program included 72 presentations and posters.

Perhaps more informative than the increasing number of persons internationally involved in phototherapy in the last few years would be a look at the constellation of disciplines represented in these proceedings. In addition to several specialties within the medical field and the field of engineering, the basic sciences—physics, chemistry, and biology—are well represented. This multidisciplinary involvement points to the dynamic state of this yet evolving therapy.

As is true with most symposia in a new field, these presentations have posed more questions than they have provided answers. It is hoped, however, that both the symposium and these proceedings will help to disseminate information on the current state of the art in phototherapy and will enhance the growth and development of an exciting and promising area of basic research and clinical medicine.

REVIEW LECTURES

Porphyrin Localization and Treatment of Tumors, pages 3–18

MECHANISMS OF PHOTOOXYGENATION

Christopher S. Foote

Department of Chemistry and Biochemistry
University of California
Los Angeles, California 90024

Photosensitizers, light and oxygen have been known for many years to be toxic to cells (Blum 1941). This toxicity is called the "photodynamic effect", and biologically important examples of it are the light sensitivity of porphyrias or photodynamic damage to animals caused by photosensitizing materials in their food.

The recent development of tumor phototherapy based on the photodynamic action of porphyrins, particularly "hematoporphyrin derivative" (Kessel, Dougherty 1983) has increased the importance of understanding the basic photochemical processes involved. This lecture provides a short review of the basic photochemistry behind photosensitized oxidation reactions responsible for photodynamic action, and suggests methods for investigating the chemical mechanisms (Foote 1982).

BASIC PHOTOCHEMISTRY OF PHOTODYNAMIC ACTION

Photodynamic action requires light, oxygen, and a photosensitizer. Only light that is absorbed can cause photochemical reactions; the sensitizer serves to absorb the light, and is converted to an excited electronic state. The electronically excited states of organic molecules are in general more reactive than their ground states, being both more readily oxidized and more readily reduced.

There are two classes of excited states, singlets (in which all electron spins are paired) and triplets (in which there are two unpaired spins). Since almost all organic molecules have singlet ground states, and since excitation usually involves conservation of spin, the first excited state to be reached is normally the singlet.

Most singlet states of organic molecules are quite short lived and either return to the ground state directly (with or without emission of light) or undergo a spin inversion ("intersystem crossing") to the triplet state. The triplet state is longer lived than the singlet, and in many cases the triplet is the reactive state involved in the action of sensitizers. In particular, in the case of most porphyrins, the triplet state is populated efficiently, with a quantum yield which is often as high as 0.9; this is true for hematoporphyrin and its derivatives of interest in phototherapy (Cannistraro, Jori, Van de Vorst 1983; Bonnett et al. 1980; Petsold, Byteva, Gurinovich 1983; Sinclair et al. 1980). We will not consider further the reactions of singlet sensitizers.

$$\text{Sens} \longrightarrow {}^1\text{Sens} \longrightarrow {}^3\text{Sens}$$

THE TYPE I REACTION

There are two fundamental types of sensitized photooxygenation (Foote 1976). In Type I, the sensitizer interacts directly with the substrate (for example, a hydrogen or electron donor) with a resulting hydrogen atom or electron transfer to produce radicals. These radicals can subsequently react with oxygen to produce oxidized products. These products can vary widely, depending on the system, but are often peroxides, which can in turn break down to induce free radical chain autoxidation, leading to further oxidation in a non-photochemical step. An example of a Type I reaction which involves free radicals is the benzophenone-sensitized photooxidation of isopropyl alcohol, shown below (Schenck et al. 1963).

$$^3Ph_2C{=}O + CH_3CH(OH)C_2H_5 \longrightarrow Ph_2\dot{C}OH + CH_3\dot{C}(OH)C_2H_5$$

$$CH_3\dot{C}(OH)C_2H_5 \xrightarrow{O_2} CH_3C(OO\cdot)(OH)C_2H_5$$

$$CH_3C(OO\cdot)(OH)C_2H_5 \xrightarrow{H\cdot\ Trans.} CH_3C(OOH)(OH)C_2H_5 + Ph_2C{=}O$$

Porphyrins are not as good hydrogen atom transfer agents as benzophenone, but they can undergo electron transfer processes to produce superoxide ion (O_2^-) by the following processes: (Cox, Whitten 1983; Jori et al. 1983).

$$^3Porph + Subs \longrightarrow Porph^- + Subs_{ox}$$

$$Porph^- + O_2 \longrightarrow Porph + O_2^-$$

$$^3Porph + \dot{O}_2 \longrightarrow Porph^+ + O_2^-$$

These reactions produce O_2^-, which can subsequently give the very reactive hydroxyl radical (OH$^\bullet$) by several pathways. These radicals can react with organic molecules in a variety of ways, or can initiate radical chain autoxidation (see below) (Singh 1982). The sorts of compounds that react in the Type I reaction are those that are electron rich or have easily abstractable hydrogens. Particularly reactive, for example, are aromatic amines, phenols, and sulfhydryl compounds.

An important characteristic of the type I reaction is that it is very concentration dependent. The reason is that there is always a competition between substrate and oxygen for the sensitizer triplet (Foote 1976). Almost all triplet molecules react with oxygen at a rate between 1 and 2×10^9 $\underline{M}^{-1}sec^{-1}$. Since the oxygen concentration in areated biological media is between 2×10^{-4} and 10^{-3} M, the product of the substrate concentration times the rate of reaction with sensitizer triplet has to exceed about 2×10^{5} to 2×10^{6} sec^{-1} if a major fraction of the reaction is to go by way of the Type I process. If the substrate concentration is too low, the type I reaction becomes very inefficient. In systems where the sensitizer is bound to an easily oxidized biomolecule, the local concentration of substrate becomes very high, and Type I reactions become highly favored. Thus it is by no means possible to dismiss this mechanism for the reactions of HPD in cells, where it is bound.

$$\text{Type I} \xleftarrow[\text{Subst.}]{} {}^{3}\text{Sens} \xrightarrow[O_2]{} \text{Type II}$$

THE TYPE II REACTION

In the Type II reaction, the sensitizer can transfer its excitation energy to a ground state oxygen molecule (Foote 1976). Since spin is conserved in this process and the ground state of oxygen is itself a triplet, singlet molecular oxygen is produced. This process occurs on every deactivation of most triplet molecules, in particular porphyrins, and occurs at the rates mentioned previously. Singlet oxygen is a metastable species with a lifetime varying from about 4 microseconds in water to 25 to 100 microseconds in nonpolar organic media which are reasonable models for lipid regions of the cell (Wilkinson, Brummer 1981). It is quite reactive, and reacts with many organic molecules to give peroxides or other oxidized products. Although it is reactive, singlet oxygen is also quite selective and fails to react with molecules which are not electron rich enough and simply returns to the ground state.

$${}^{3}O_2 \xleftarrow[\text{Decay}]{} {}^{1}O_2 \xrightarrow[\text{Acceptor}]{} AO_2$$

A number of compounds have been found to quench (i.e., deactivate without reaction) singlet oxygen efficiently (Foote 1979a). For example, β-carotene inhibits photooxidation of 2-methyl-2-pentene efficiently at 10^{-4} M without itself being appreciably oxidized. Certain amines are quenchers; e. g. DABCO (1,4-diazabicyclooctane). Other amines both quench singlet oxygen and react with it, depending on conditions. Azide ion is a somewhat better quencher; phenols also quench singlet oxygen; some also react chemically with it. α-Tocopherol is an interesting example, quenching singlet oxygen at a high rate in all solvents, but being consumed appreciably only in polar solvents.

$$^1O_2 + \text{Quencher} \longrightarrow {}^3O_2 + \text{Quencher}$$

The major classes of reaction of singlet oxygen that are important in biological systems are (Foote 1981):

1) Addition to the diene system in heterocycles to give endoperoxides. Histidine is an example of a substrate which undergoes this reaction (Ryang, Foote 1979).

R
N
H
$^1O_2$
R
N
H
O O

2) Oxidation of olefins with allylic hydrogens atoms to give hydroperoxides (Gollnick, Kuhn 1979). Cholesterol and unsaturated fatty acids are important biological substrates in this kind of reaction (Kulig, Smith 1973).

$$R{-}CH_2{-}CH{=}CH{-}R \longrightarrow RCH{=}CH{-}\underset{\displaystyle |}{\overset{H-O-O}{C}}HR$$

3) Reaction of electron rich double bonds to give a 2 + 2

addition, which eventually results in cleavage of the double bond; tryptophan is an example of a substrate for this reaction (Saito et al. 1977).

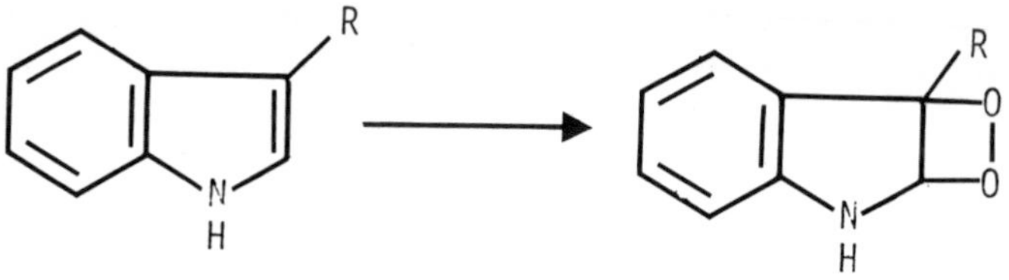

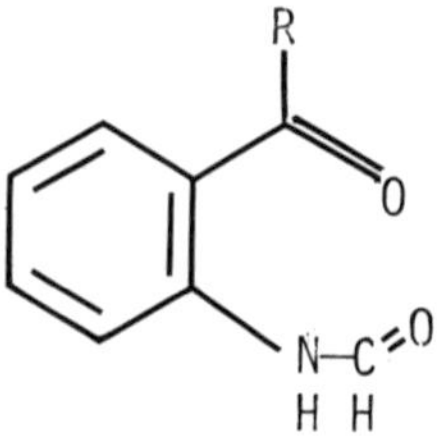

4) Oxidation of sulfides to sulfoxides. For example, methionine reacts with singlet oxygen to give the sulfoxide (Sysak, Foote, Ching 1977).

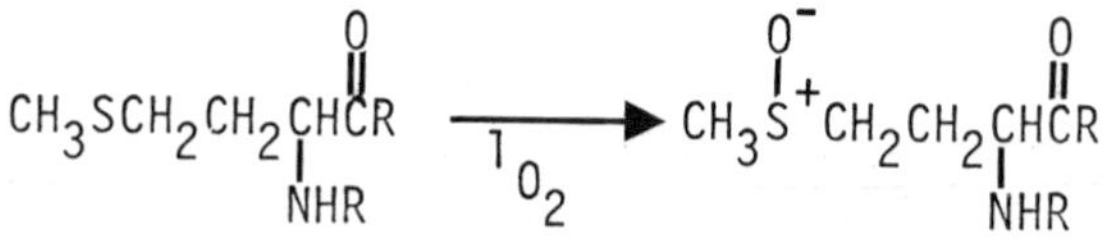

$$CH_3SCH_2CH_2CH(NHR)C(=O)R \xrightarrow{^1O_2} CH_3S^+(O^-)CH_2CH_2CH(NHR)C(=O)R$$

The major biological targets for singlet oxygen are now well known (Foote 1981; Spikes 1981). Membranes are peroxidized, leading to fragility and easy lysis. The initial hydroperoxides appear to break down in a subsequent slower step, probably involving formation of free radicals and subsequent radical chain autoxidation to cause increased degradation of unsaturated molecules in the membrane

(Lamola, Yamane, Trozzolo 1973).

Nucleic acids are also an important target for photooxidation; guanine is the major target of this reaction. The guanine ring is oxidized; intermediate products recently reported (Cadet et al. 1982) suggest the intervention of endoperoxides in this case as well.

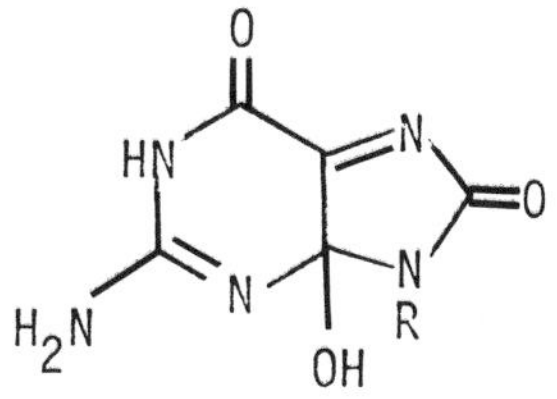

Many proteins and enzymes are damaged by photooxidation; the major targets are histidine, methionine, tryptophan (all of which are probably singlet oxygen targets, although tryptophan can probably react by a type I mechanism also). Tyrosine and cysteine are also photooxidation targets, although these appear more likely to be type I reactions.

DETERMINATION OF MECHANISM

Establishment of the mechanism by which photosensitized oxygenations occur is in general not a trivial problem, at least when the substrates are as electron rich as many biological targets (Foote 1979b). In homogeneous solution, the problem is at its easiest and a variety of kinetic and trapping techniques can be used. Of course, the detection of singlet oxygen in a system does not necessarily mean that it is the reactive species in the oxidation of the target molecules; this needs to be established separately. That is, the detection of a reactive species is a necessary but not sufficient condition for its intermediacy. Also, in general, finding that a reactive intermediate is present in a system is meaningless without quantitation. Furthermore, we need to know not only how much of the intermediate is produced, but what fraction of the actual reaction actually goes via that pathway. It is also of the utmost importance that the technique used be specific, since many chemical reagents can react with different oxidizing species to give similar products. Alkoxy and alkylperoxy radicals can also be formed by decomposition of peroxides formed in the Type I and Type II reaction, or in radical autoxidation processes. These radicals and those mentioned previously can also initiate free radical chain autoxidation. This process is a non-photochemical reaction which may occur following either the type I or type II reaction, and can lead to the oxidation of many more molecules than originally formed in the initial photochemical process. Thus the photochemical step may not even be the one which damages the majority of the molecules, although it initiates the whole process.

There are a number of standard tests for the intermediacy of the reactive oxygen species formed in the Type I reaction. Superoxide ion is not a very reactive species; its major activity is as a reducing agent and it is usually detected by its ability to reduce certain compounds such as nitroblue tetrazolium or cytochrome C (Fee and Valentine 1977). The inhibitability of this reduction by the enzyme superoxide dismutase provides a rather specific test for O_2^- (Fridovich 1982). The inhibition of damage to targets by SOD also appears to be quite specific for O_2^-, which may be a precursor of more reactive intermediates such as $OH^\bullet$.

The usual method of testing for hydroxyl radical is to inhibit its reactions with targets by adding hydrogen donors such as alcohols or traps such as benzoic acid (Singh 1982). Since hydroxyl radical is extremely reactive, these inhibitors must be present at quite high concentration and there is very little time for them to diffuse to the site of production of hydroxyl radical. Thus a more specific test is by detection of products of trapping; for example, hydroxylation of benzoic acid forms phenols which can be isolated (Richmond et al. 1981). This area is in its infancy.

Another method of trapping both O_2^- and $OH^\bullet$ is to trap them with an ESR "spin label" such as the compound DMPO (Janzen 1980). The resulting radicals can then be detected by ESR. Unfortunately, quantitation is very difficult by this technique.

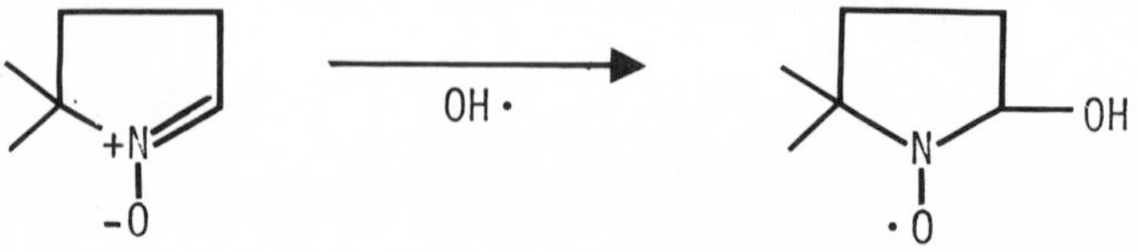

+N
-O
OH·
N
·O
OH

Radical chain oxidations are usually detected by their inhibition by a chain terminator such as a phenol, but this is often difficult to do with specificity in the presence of other oxidizing agents.

$$RO^{\bullet} + PhOH \longrightarrow ROH + PhO^{\bullet} \quad \text{(Chain Termination)}$$

SINGLET OXYGEN TESTS

Traps

Specific methods for detection of singlet oxygen include trapping, quenching and detection of luminescence (Foote 1979b). A large number of traps for singlet oxygen have been developed which give isolable products. The first class to be used in biological systems was the furans; unfortunately, these are the least specific since they are oxidized by almost all strong oxidants. For example, furans are converted to diketones by halogens, peroxy radicals and electrochemical oxidation, so that the production of diketone only serves to indicate that some strong oxidant was present in solution (Foote 1979b).

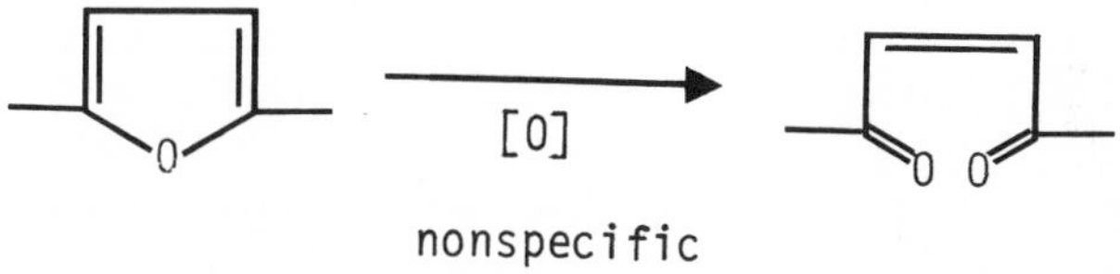

O [O] O O

nonspecific

A better group of traps is provided by substituted anthracene derivatives; these can be made water soluble with suitable substituents (S) and an isolable endoperoxide is formed (Schaap et al. 1974; Aubry et al. 1983). Unfortunately, these compounds sensitize their own photooxidation, so that the production of small amounts of product is difficult to avoid except in the complete absence of light. Also, their usefulness in complex systems may be limited because of adsorption and compartmentalization problems.

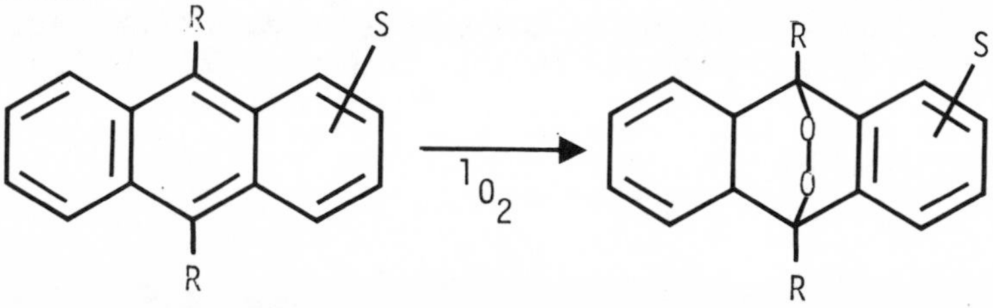

R S R $^1O_2$ R S O O R

One of the more specific traps which has been developed is cholesterol; this gives a single product (the 5-α-hydroperoxide) on reaction with singlet oxygen (Kulig, Smith 1973; Nickon, Bagli 1961). Radical oxidation gives a very complex mixture which contains none of the 5-α-product. Cholesterol is useful as a marker for the presence of singlet oxygen in biological systems; however, a major drawback is its low reactivity, which results in poor sensitivity. Singlet oxygen can also be detected by its reaction with certain amines to produce ESR active N-oxyl compounds by an obscure mechanism (Cannistraro, Jori, Van de Vorst 1983).

Kinetic Methods

The second method of detecting singlet oxygen is to inhibit its reactions by adding something which reacts with it or quenches it. This technique can be quite effective in homogeneous solution, since the rate constants for a large number of compounds are well known (Wilkinson, Brummer 1981). To be most effective, these inhibition studies should be carried out quantitatively, so that all rate constants in this system are determined. It is much less effective simply to carry out one point inhibition experiments, because all singlet oxygen quenchers and reagents are compounds of low oxidation potential and are expected to react with other strong oxidants as well. Thus the kinetic parameters provide a fingerprint for the reactive species being trapped by the inhibitor. Of course, it is much more difficult to use quantitative techniques in inhomogeneous systems, since targets and quenchers are localized, and their local concentrations are usually not known.

D_2O Effect

Another technique which has often been used is to determine the effect on the rate of the observed reaction on substituting D_2O for water. This technique is based on the fact that the lifetime of singlet oxygen in D_2O is about 14 times longer than in H_2O, so that reactions of singlet oxygen with substrates tend to proceed more efficiently since a larger fraction of the singlet oxygen often survives to react (Nilsson, Kearns 1973). This technique requires that the reaction be run in such a way that only a small fraction of the singlet oxygen is actually reacting with the substrate or with other quenchers; that is, deactivation by the solvent determines the lifetime of singlet oxygen. In the case where most of the singlet oxygen produced is already reacting with substrate or being quenched by a quencher, the lifetime is limited by these reactions, and not by the solvent, and the overall rate of the process would be the same in D_2O and water. Another problem with this technique is that other reactive species may show solvent deuterium isotope effects on their lifetimes. For example, O_2^- is known to live substantially longer in D_2O than in water (Bielski, Saito 1971); thus O_2^- reactions would be expected to show a solvent deuterium isotope effect.

Luminescence

The luminescence of singlet oxygen is the property by which it was first detected (Khan, Kasha 1964). There are two types, the direct luminescence from a single molecule of singlet oxygen, which appears in the infrared at 1.27 μ, and "dimol" luminescence which appears at 634 and 704 nm. Both are extremely inefficient in solution. The infrared luminescence is inherently weak because the lifetime of singlet oxygen in solution is short compared to the radiative lifetime. The dimol luminescence is also limited by the lifetime; in addition, since dimol luminescence requires a bimolecular collision between two short-lived species, its efficiency should be very low unless singlet oxygen is present at very high concentrations (Foote 1976).

$$^1O_2 \longrightarrow h\nu \ (1.27\,\mu)$$

$$2(^1O_2) \longrightarrow h\nu \ (634, 704 \text{ nm})$$

Dimol luminescence has often been used for the identification of singlet oxygen in biological systems. Unfortunately, this weak luminescence is usually accompanied by light of other wavelengths in complex systems. Although these emissions have been assigned to higher vibrational states of singlet oxygen (Kasha, Khan 1970), it seems more likely that the extraneous emission comes from excited carbonyls or other species in the system. Precise wavelength determination is essential if this technique is to be used for the detection of singlet oxygen; it appears to be of qualitative usefulness at best, since it seems almost hopeless to quantitate the singlet oxygen yield, since the luminescence depends on a second order process.

More hope exists for detection of 1.27μ emission. This luminescence, though weak, can be detected, both in steady state (Khan, Kasha 1979); Krasnovsky 1982) and time resolved systems (Ogilby, Foote 1982; Hurst, MacDonald, Schuster 1982; Parker, Stanbro 1982; Rodgers, Snowden 1982) and the wavelength is specific. We have used this system to detect singlet oxygen produced by porphyrins excited by a laser flash (Ogilby, Foote 1982). It seems likely that this technique could be used to detect singlet oxygen if it is produced by hematoporphyrin and its derivatives _in vivo_, although sensitivity requirements will be extreme; however the question of whether it is the reactive species in photochemotherapy will still remain.

CONCLUSION

This paper has been a short review of some of the chemical processes which can occur in biological systems exposed to photosensitizers, light, and oxygen, and of some of the techniques which are useful for characterizing reactive intermediates in these systems.

REFERENCES

Aubry JM, Schmitz C, Rigaudy J, Cuong NK (1983). Echange de substituants en serie anthracenique par cyclo-additions suivies d'eliminations spontanees. Tetrahedron 39:623.

Bielski BHJ, Saito E (1971). Deuterium isotope effect on the decay kinetics of perhydroxyl radical. J Phys Chem 75:2263.

Blum H (1941). "Photodynamic Action and Diseases Caused by Light". New York: Rhinehold, Inc.

Bonnett R, Charalambides A A, Land E J, Sinclair

R S, Tait D, Truscott G (1980). Triplet states of Porphyrin esters. J Chem Soc Farad Trans I 76:852.

Cadet J, Decarroz C, Wang SY, Midden WR (1982). Products of photosensitized degradation of deoxynucleosides. Abstr Vancouver Am Soc Photobiol Mtng:126.

Cannistraro S, Jori G, Van De Vorst A (1982). Quantum yield of electron transfer and of singlet oxygen production by porphyrins: an ESR study. Photobiochem Photobiophys 3:353.

Cox GS, Whitten DG (1983). Excited state interactions of protoporphyrin IX and related porphyrins with molecular oxygen in solutions and molecular assemblies. Advan Exptl Med and Biol 160:279.

Fee JA, Valentine JS (1977). Chemical and Physical Properties of Superoxide. In Michelson AM, McCord JT, and Fridovich I (eds): "Superoxide and Superoxide Dismutases", New York: Academic Press.

Foote CS (1976). Photosensitized oxidation and singlet oxygen: consequences in biological systems. Free Radicals in Biology 2:85.

Foote CS (1979a). Quenching of singlet oxygen. In Wasserman HH and Murray RW (eds): "Singlet Oxygen," New York: Academic Press, p 139.

Foote CS (1979b). Detection of singlet oxygen in complex systems. In Caughey WS (ed): "Biochemical and Clinical Aspects of Oxygen," New York: Academic Press, p 603.

Foote CS (1981). Photooxidation of model biological compounds. In Rodgers MAJ, Powers, EL (eds): "Oxygen and Oxy-Radicals in Chemistry and Biology" New York: Academic Press, p 425.

Foote CS (1982). Light, oxygen and toxicity. In Autor AP (ed): "Pathology of Oxygen," New York: Academic Press p 21.

Fridovich I (1982). Superoxide dismutase in biology

and medicine. In Autor AP (ed): "Pathology of Oxygen", New York: Academic Press p 1.

Gollnick K, Kuhn HJ (1979). Ene-Reactions with singlet oxygen. In Wasserman HH and Murray RW (eds): "Singlet Oxygen," New York: Academic Press p 287.

Hurst JR, MacDonald JD, Schuster GB (1982). Lifetime of singlet oxygen in solution directly determined by laser spectroscopy. J Am Chem Soc 104:2065.

Janzen EG (1980). A critical review of spin trapping in biological systems. Free Radicals in Biology 4:116.

Jori G, Reddi E, Tomio L, Calzavara F (1983). Factors governing the mechanism and efficiency of porphyrin-sensitized photooxidations in homogeneous solutions and organized media. Advan Exptl Med Biol 160:193.

Kasha M, Khan AU (1970). The physics, chemistry, and biology of singlet molecular oxygen. Ann N Y Acad Sci 171:5.

Kessel D, Dougherty TJ (eds) (1983): "Porphyrin Sensitization". (Advan Exptl Med Biol 160:) New York: Plenum Press.

Khan AU, Kasha M (1964). Rotational structure in the chemiluminescence spectrum of molecular oxygen in aqueous system. Nature 204:241.

Khan AU, Kasha M (1979). Direct spectroscopic observation of singlet oxygen emission at 1268 nm excited by sensitizing dyes of biological interest in liquid solution. Proc Nat Acad Sci 76: 6047.

Krasnovsky AA Jr (1982). Delayed fluorescence and phosphorescence of plant pigments. Photochem Photobiol 36:2069.

Kulig MJ, Smith LL (1973). Sterol metabolism. XXV. cholesterol oxidation by singlet molecular oxygen. J Org Chem 38:3639.

Lamola AA, Yamane T, Trozzolo AM (1973). Cholesterol

hydroperoxide formation in red cell membranes and photohemolysis in erythropoietic protoporphyria. Science 179:1131.

Nickon A, Bagli JF (1961). Reactivity and geometry in allylic systems. I. Stereochemistry of photosensitized oxygenation of monoolefins. J Am Chem Soc 83:1498.

Nilsson R, Kearns DR (1973). A remarkable deuterium effect on the rate of photosensitized oxidation of alcohol dehydrogenase and trypsin. Photochem Photobiol 17:65.

Ogilby PR, Foote CS (1982). Chemistry of singlet oxygen. 36. Singlet molecular oxygen ($^1\Delta_g$) luminescence in solution following pulsed laser excitation. Solvent deuterium isotope effects on the lifetime of singlet oxygen. J Am Chem Soc 104:2069.

Parker JG, Stanbro WD (1982). Optical determination of the collisional lifetime of singlet molecular oxygen ($^1\Delta_g$) in acetone and deuterated acetone. J Am Chem Soc 104: 2067.

Petsold OM, Byteva IM, Gurinovich GP (1983). Photochemical method for determining the efficiency of the conversion of molecules into the triplet state. Opt Spekt 34:343.

Richmond R, Halliwell B, Chauhan J, Darbre A (1981). Superoxide-dependent formation of hydroxyl radicals: detection of hydroxyl radicals by the hydroxylation of aromatic compounds. Anal Biochem 118:328.

Rodgers MAJ, Snowden PT (1982). Lifetime of $O_2(^1\Delta_g)$ in liquid water as determined by time-resolved infrared luminescence measurements. J Am Chem Soc 104: 5541.

Ryang HS, Foote CS (1979). Chemistry of singlet oxygen. 31. Low temperature NMR studies of dye-sensitized photooxygenation of imidazoles: direct observation of unstable 2,5-endoperoxide intermediates. J Am Chem Soc 79:6683.

Saito I, Matsuura T, Nakagawa M, Hino T, (1977). Peroxidic intermediates in the photosensitized oxygenation of tryptophan derivatives. Accounts Chem Res. 10:346.

Schaap AP, Thayer AL, Faler GR, Goda K, Kimura T (1974). Singlet molecular oxygen and superoxide dismutase. J Am Chem Soc 96:4025.

Schenck GO, Becker H-D, Schulte-Elte K-H, Krauch CH (1963). Mit Benzophenon photosensibilisierte Autoxidation von sek. Alkoholen und Aethern. Darstellung von α-Hydroperoxyden. Chem Ber 96:509.

Sinclair RS, Tait D, Truscott TG (1980). Triplet states of protoporphyrin IX and protoporphyrin IX dimethyl ester. J Chem Soc Farad Trans I 76:417.

Singh A (1982). Chemical and biochemical aspects of superoxide radicals and related species of activated oxygen. Can J Physiol Pharm 60:1330.

Spikes JD (1981). The sensitized photooxidation of biomolecules. In Rodgers MAJ and Powers EL (eds): "Oxygen and Oxy-Radicals in Chemistry and Biology", New York: Academic Press, p 421.

Sysak PK, Foote CS, Ching T-Y (1977). Chemistry of singlet oxygen. 25. Photooxidation of methionine. Photochem Photobiol 26:19.

Wilkinson F, Brummer JG (1981). Rate constants for the decay and reactions of the lowest electronically excited singlet state of molecular oxygen in solution. Phys Chem Ref Dat 10:809.

Porphyrin Localization and Treatment of Tumors, pages 19–39

PHOTOBIOLOGY OF PORPHYRINS

John D. Spikes

Department of Biology
University of Utah
Salt Lake City, Utah 84112 USA

One type of photobiological response involves photochemical reactions in which the absorption of light by one molecule leads to the chemical alteration of another molecule. Such processes, usually termed photosensitized reactions, can proceed by several alternate pathways (Foote 1976). The light absorbing molecule is termed a photosensitizer (or sensitizer) and the other molecule is called the substrate of the reaction. Most, but not all, photosensitized reactions require molecular oxygen and are often called "photodynamic" reactions by biologists (Blum 1941). The first example of a porphyrin photosensitized process in biology was the demonstration by Hausmann (1908) that hematoporphyrin efficiently sensitizes the photohemolysis of rabbit red blood cells. Immediate reactions of this type which lead to the injury or killing of cells or organisms are now usually termed phototoxic processes (Spikes 1983a). Since Hausmann's observations, a very large amount of research has been done using a wide variety of natural and synthetic porphyrins and related compounds as photosensitizers for many different kinds of biological systems (Jori, Spikes 1983). Porphyrins sensitize the photodegradation of a number of types of biomolecules which are not directly sensitive to visible radiation. In cells, this can interfere with the normal functioning of certain organelles. This can injure or kill cells and, in turn, injure or kill organisms. Thus in porphyrin photobiology we are concerned with a hierarchy of processes of increasing complexity ranging from the photochemistry of porphyrins in solution and the photooxidation of simple biomolecules to photosensitized responses in mammals, including man. Research on the photobiology of porphyrins has increased impressively in recent years. This is probably due largely to developments in the use of porphyrins in photodiagnostic and phototherapeutic techniques in medicine, especially the promising approach of using visible light plus sensitizing porphyrins for the treatment of tumors (Dougherty et al. 1978 and 1983; Kessel, Dougherty 1983). This paper will briefly review several aspects of porphyrin photobiology including the chemistry, photophysics and photochemistry of porphyrins, the porphyrin-sensitized photooxidation of biomolecules, and the phototoxic effects of porphyrins on mammalian cells and mammals. Some fundamental aspects of photochemistry and photobiology are also included. Because of space limitations, most of the references given are reviews or very recent papers. These may be consulted for more detailed bibliographies. The earlier literature has been reviewed by Blum (1941), while the reviews by Spikes (1975) and Jori and Spikes (1983) have listings of more recent references.

SOME PRINCIPLES OF PHOTOBIOLOGY (see refs. in Spikes 1983b)

It may be useful to review some basic principles of photochemistry and photobiology for those readers who do not routinely work with light. Light is an energy form propagated at very high velocity as electromagnetic waves. In some of its interactions, as with lenses and prisms, it behaves as a waveform; however, in the process of being absorbed by molecules, it delivers its energy in the form of small discrete packets termed photons. The wavelength range of light (i.e., the visible spectrum) extends from approximately 400 to 700 nm (nanometer = 10^{-9} meter). Shorter wavelength radiation (200-400 nm) is termed ultraviolet (UV) and longer wavelength the infrared (IR). The color of light in the visible spectrum changes continuously in going from 400-700 nm in the sequence violet, indigo, blue, green, yellow, orange, red. The energy associated with a photon of light (in ev = electron volts) = (1,240)/(wavelength in nm); as this equation indicates, the energy of a photon is inversely related to its wavelength. The energy of a mole of photons (an

Einstein) in kcal may be obtained by multiplying the energy/photon in ev by 23. Thus 400 nm blue light has energies of 3.1 ev/photon and 71 kcal/Einstein of photons; for red light at 700 nm the values are 1.77 and 41. Photons in the UV-visible-near IR range have sufficient energy to electronically excite molecules; the resultant excited state molecules can undergo chemical changes and/or cause chemical changes in other molecules, as in photosensitized reactions.

Photochemical and photobiological reactions occur only if a molecule absorbs a photon; such reactions cannot occur if energy is not absorbed. The wavelengths absorbed by an organic molecule depend on its structure; the longer the conjugated system (alternating double and single bonds) of a molecule, the longer the maximum wavelengths that will be absorbed. Only a very few types of biomolecules (such as anthocyanins, carotenoids, flavins, and, of course, porphyrins and related compounds such as the chlorophylls) have sufficiently long conjugated systems to absorb radiation in the visible range. Thus the process of light absorption by molecules is highly selective. We usually describe the light absorbing properties of a molecule in terms of its absorption spectrum, which is a plot of its absorbance as a function of wavelength, i.e., the probability of light absorption by the molecule at different wavelengths. Materials present as dimers and higher aggregates (as in the case of many prophyrins in aqueous media) scatter light and thus introduce errors into spectral measurements. It is very difficult to measure the spectra of molecules in cells and tissues precisely because of light scattering by the cells and certain of their components; also cells are very heterogeneous, which provides sites with many different properties, which may in turn affect the absorption spectra of molecules.

A plot of the quantitative relationship between the rate of a photobiological process (as measured at the same incident photon flux = incident photons per unit area per unit time) and the wavelength of light used is termed the action spectrum for the process. The action spectrum for a simple photodynamic system in solution usually corresponds to the absorption spectrum of the photosensitizer. However, such measurements with cells and tissues are very difficult due to light absorption by molecules not involved in the reaction of interest and to the loss of light by scattering, which increases exponentially with decreasing wavelength. The measurement of action spectra has been very useful in establishing the types of light absorbing molecules involved in various photobiological responses, e.g., the involvement of DNA in the short wavelength UV killing and production of mutations in cells, the involvement of carotenoid-protein complexes in vision, the role of chlorophyll and other pigments in photosynthesis, etc. Action spectrum measurements on photoresponses of mammals are difficult to make, especially in the quantifying of the biological responses. However, pertinent to porphyrin photobiology, the action spectrum for skin photoresponses (erythema, etc.) in patients with sensitizing types of porphyrias (in which porphyrins such as protoporphyrin, uroporphyrin and coproporphyrin accumulate in the body) correspond fairly closely to the absorption spectra of the porphyrins. In many kinds of photobiological experiments it is important to know the wavelength of the action spectrum peak for the process, since light of wavelengths corresponding to this will be the most efficient in driving the reaction. In some photobiological systems other consideration may become important. For example, in the porphyrin-sensitized phototherapy of tumors, red light is used (even though porphyrins absorb very poorly in this wavelength region) because it penetrates into tissues much more efficiently than the near UV-blue light absorbed maximally by the porphyrin.

PORPHYRINS: STRUCTURES; PHOTOPHYSICAL AND PHOTOCHEMICAL PROPERTIES (see refs. in Jori, Spikes 1983; Jori et al. 1983).

Porphyrins are tetrapyrrolic pigments based on porphin, which is made up of four pyrrole subunits joined together by four methine bridges to give a planar cyclic molecule. A very large number of different porphyrins can be made by attaching different side chains to the positions beta to the pyrrole nitrogens and to the four methine carbon atoms (meso-positions). Thus different porphyrins can vary widely in their solubility properties and in their electrical charge under physiological conditions, depending on their particular array of substituents. Porphyrins in monomeric solution have a narrow, very intense absorption band in the near UV-blue regions of the spectrum, the so-called Soret band. Free base (metal-free) porphyrins also have four weaker satellite absorption bands at longer wavelengths (500-700 nm range). Metallo-porphyrins, i.e., porphyrins with a metal ion coordinated with the pyrrole nitrogen atoms have only two satellite bands. Many porphyrins form dimers and higher aggregates in aqueous solvents as their concentrations are increased; this results in a decrease in the absorptivity, while the Soret band becomes broader and the peak shifts to shorter wavelengths.

Illuminated porphyrins fluoresce in the red region of the spectrum as a result of the radiative decay of the first excited singlet state; fluorescence lifetimes are in the nanosecond range. Singlet excited state porphyrins typically undergo so-called intersystem crossing to give high yields of long-lived triplet states (half-lives in the several hundred microsecond range). Triplet lifetimes are decreased somewhat in porphyrins containing diamagnetic metals such as Mg^{2+} or Zn^{2+}, while paramagnetic metals such as Cu^{2+} and Fe^{3+} give porphyrins with very short triplet lifetimes. Triplet porphyrins react with ground state oxygen (which is in the triplet state) by energy transfer (a Type II process) to give singlet excited oxygen, which is reactive chemically. Triplet porphyrins also react with oxygen with low efficiency by an electron transfer (Type I = free radical) process to give superoxide (O_2^-). Finally, in some cases, triplet porphyrins can react directly with certain substrates, thus initiating other free radical processes (Felix et al. 1983). Various free base porphyrins differ somewhat in their sensitizing efficiencies. For example, the relative efficiencies for sensitizing the photooxidation of histidine in aqueous buffer (a singlet oxygen reaction) by hematoporphyrin, coproporphyrin III and uroporphyrin III are 1.00, 0.71 and 3.27, respectively. Protoporphyrin, deuteroporphyrin and mesoporphyrin are typically less efficient sensitizers in aqueous solution than hematoporphyrin; also some porphyrins are photodegraded much faster than others (Jori et al. 1983). The efficiency of singlet oxygen generation by porphyrins depends to a considerable extent on their triplet lifetimes; thus metalloporphyrins tend to produce singlet oxygen with much lower efficiencies than free base porphyrins. For example, in singlet oxygen mediated photooxidation reactions in aqueous solution, the relative efficiencies of Mg^{2+}-, Zn^{2+}-, Cu^{2+}- and Fe^{3+}-substituted hematoporphyrins are approximately 0.16, 0.04, essentially zero, and zero, respectively, in comparison with free base hematoporphyrin (Jori et al. 1983).

PORPHYRIN-SENSITIZED PHOTODEGRADATION OF BIOMOLECULES (see refs. in Spikes 1975; Jori, Spikes 1983; Jori et al. 1983).

Porphyrins sensitize the photooxidation of many kinds of biologically important molecules. For example, unsaturated fatty acids and their esters are typically photooxidized to lipid peroxides via a singlet oxygen pathway. The unsaturated fatty acid side chains of lecithins are also photooxidized with porphyrins. With hematoporphyrin, cholesterol is photooxidized almost quantitatively to its 5-alpha-hydroperoxide by a singlet oxygen mechanism.

Deoxyguanosine is photooxidized under physiological conditions with porphyrins, probably by a singlet oxygen pathway; the primary photooxidation product(s) have not been established. Deoxyadenosine, deoxycytidine and thymidine are not photooxidized appreciably under the same conditions. Under physiological conditions guanine residues are selectively destroyed when thymus DNA is illuminated in the presence of porphyrins. Hematoporphyrin derivative and other porphyrins sensitize the photochemical formation of both single and double strand breaks in E. coli plasmid DNA, again, by a singlet oxygen process. Binding of the porphyrin to the DNA is not necessary for the reaction. Recent studies show that guanine is fairly rapidly photooxidized at neutrality with hematoporphyrin; thymine is photooxidized only slowly and the other nucleic acid bases not at all. With increasing pH, guanine and thymine are photooxidized progressively faster, and at pH values greater than 10, the other bases (adenine, cytosine and uracil) are photooxidized at appreciable rates. Thus the anionic forms of both purines and pyrimidines are much more susceptible to sensitized photooxidation than the neutral bases. Calf thymus DNA is slowly photooxidized with hematoporphyrin at neutrality, and much more rapidly at high pH; altering the conformation of the DNA in 8 M urea significantly increases the photooxidation rate. Illumination of guanosine in the presence of hematoporphyrin and certain proteins results in the covalent coupling of the nucleoside to the proteins (Dubbelman et al. 1982).

A rather large amount of research has been done on the porphyrin-sensitized photooxidation of amino acids, both free in solution and incorporated into the polypeptide chains of proteins (Spikes 1975; Jori, Spikes 1983). Only five of the amino acids typically present in proteins are susceptible to photooxidation: cysteine, histidine, tyrosine, methionine and tryptophan. The rates of photooxidation of the first three amino acids increase with increasing pH with dependencies indicating that only the ionized (anionic) form of cysteine, the unprotonated imidazole ring of histidine, and the ionized phenol group of tyrosine are susceptible to photooxidative attack. Cysteine is photooxidized largely to cystine by a singlet oxygen process; cystine is very resistant to photooxidation. The imidazole ring of histidine is destroyed on photooxidation, again by a singlet oxygen reaction. Little is known of the mechanism of tyrosine photooxidation. The rates of photooxidation of methionine and tryptophan increase much less with increasing pH than for the other three susceptible amino acids. Methionine is quantitatively photooxidized to the singlet oxygen product, methionine sulfoxide. Tryptophan appears to be photooxidized by competing singlet oxygen and free radical pathways, with the relative participation determined by the reaction conditions, the aggregation state of the porphyrin, etc. (Jori et al. 1983). The singlet oxygen photooxidation product of tryptophan is formylkynurenine. The photooxidation of amino acid mixtures can be more complex. For example, illumination of histidine in the presence of protoporphyrin and the non-photooxidizable amino acid, glycine, results in a photocoupling of the histidine with glycine (Verweij et al. 1981).

A number of studies have been made of the porphyrin-sensitized photooxidation of proteins. Such studies are complicated by the fact that protein molecules typically contain more than one of each of the types of photooxidizable amino acid residues located at various sites along the polypeptide chain. Further, certain of these residues may be protected from photooxidative attack to a greater or lesser extent depending on their degree of burial within the three dimensional structure of the protein molecule. Protein studies include determinations of the chemical and the physical-chemical changes which result from photooxidation, the reaction kinetics, and examinations of the changes in the biological functions. Varying patterns of amino acid destruction are observed, depending on the protein and the reaction conditions; pH is especially important because of its effect on the photooxidation rates of different amino acid residues and, possibly, on the conformation of the protein molecule, which could change the array of residues exposed. If the heme group (which is not a photosensitizer because of the iron atom present) of a hemoprotein is replaced by a sensitizing porphyrin, illumination will result in the selective photooxidation of only those susceptible amino acid residues in, or very close to, the heme binding site. Thus, in any case where a selective binding of the porphyrin occurs, sensitized photooxidation can be used as a probe of the three dimensional structure of the protein molecule. A number of proteins, including serum albumin, antibodies, fibrin, spectrin from red blood cell membranes, etc. are covalently crosslinked on illumination in the presence of porphyrins.

Proteins typically show a loss of their biological function as a result of photooxidation with porphyrins. For example, a number of enzymes are inactivated, including alcohol dehydrogenase, alkaline mesentericopeptidase, lipase, lysozyme, papain, pepsin, ribonuclease A, urease, urokinase, and several kinds of DNA polymerases; probably almost any enzyme would be susceptible. The photooxidation of some proteins, such as egg albumin, destroys their antigenic function and their ability to react with antibodies toward the native protein; the reactivity of antibodies is similarly destroyed. Illumination of porphyrin-fibrinogen mixtures results in an increase in the clotting time, and a binding of the porphyrin to the protein. In summary, although a large amount of research has been done on the porphyrin-sensitized photodegradation of amino acids and proteins, much remains to be learned about the details of the reaction mechanisms involved.

Several other types of molecules of biological importance can be photooxidized with porphyrins as sensitizers. Carbohydrates are rather resistant, although the slow porphyrin-sensitized photodegradation of glucose has been reported. Hyaluronic acid, a mucopolysaccharide, is depolymerized on illumination in the presence of hematoporphyrin. Other photooxidizable biomolecules include ascorbic acid, NADH, NADPH and carotenes (Jori, Spikes 1983).

PORPHYRIN-SENSITIZED PHOTOEFFECTS ON VIRUSES

Only a few studies on viruses have been carried out with porphyrins as sensitizers, although most viruses might be expected to be quite sensitive since they are composed of protein, nucleic acids, and, in some cases, lipids. For example, herpes simplex viruses lose the ability to form plaques when illuminated in air in the presence of hematoporphyrin derivative. The treated virus particles appear unable to adsorb to or penetrate into the host cells, suggesting that the photodynamic damage involves surface proteins and/or lipids rather than the viral DNA. Some damage to the virus is also produced under anaerobic conditions, possibly by free radical reactions (Lewin et al. 1980).

PORPHYRIN-SENSITIZED PHOTOEFFECTS ON MODEL CELL SYSTEMS (see refs. in Jori, Spikes 1981 and 1983)

Porphyrin photosensitized processes in cells are enormously more complex than the solution studies decribed above. One intermediate step in making the jump from solutions to cells involves the use of model systems, which may approximate in a crude and over-simplified fashion some aspects of phototoxic reactions in vivo. Several models used in porphyrin photosensitization studies will be described briefly including porphyrins bound to macromolecules and particles, porphyrins in micelles, and porphyrins incorporated into liposomal membranes.

Various sensitizing porphyrins bind non-covalently to specific sites on some proteins such as serum albumin and lysozyme, and to the heme-binding site of hemoproteins. This mimics the in vivo situation where it is quite likely that some porphyrins are similarly bound to cellular proteins. Such bound porphyrins are still effective sensitizers; in some cases, at least, they are more effective sensitizers for certain amino acids in the protein to which they are bound than are sensitizers free in the medium. In particular, binding probably increases the efficiency of Type I photoreactions. Further, bound porphyrins sensitize the photodegradation only of those susceptible groupings in the immediate vicinity of the binding site, since even singlet oxygen has only a very small diffusional radius because of its short lifetime in solution (Jori, Spikes 1981 and 1983; Jori et al. 1983). Uroporphyrin I covalently bound to agarose gel particles through a spacer arm is almost as effective in sensitizing the photooxidation of amino acids and other biomolecules as uroporphyrin in solution. Laser flash kinetic studies show that the triplet half-lives (under nitrogen) are essentially the same for free and bound uroporphyrin (~ 900 usec); however, the bimolecular rate constant for oxygen quenching of triplet porphyrin was significantly less for the bound uroporphyrin (Spikes JD, Burnham BF, unpublished experiments).

Surfactant micelles in aqueous systems are used as models for some properties of cell membranes since they have both hydrophobic and hydrophilic regions; lipid soluble sensitizers and substrates are both solubilized in micelles. In porphyrin photosensitization studies, the photochemical reactions of various combinations of sensitizers and substrates in solution and in the same and different micelles have been studied. Although the data from micellar systems are complex, they demonstrate that singlet-oxygen can be generated by micellar incorporated porphyrins with high efficiency and can diffuse out of the micelle and oxidize substrates in solution and in other micelles. Photogenerated electrons can also behave in the same way (Jori, Spikes 1981 and 1983; Jori et al. 1983). Liposomes (bilayered microscopic vesicles composed of phospholipids) are also used extensively as models for cell membranes. Illumination of liposomes with membrane incorporated hematoporphyrin and cholesterol results in the photooxidation of the sterol to its singlet oxygen product (Suwa et al. 1978). Egg lecithin (an unsaturated phosphatidylcholine) liposomes with membrane incorporated hematoporphyrin are lysed on illumination as a result of the sensitized photodegradation of the phospholipid. The reaction mechanism changes from a Type II singlet oxygen mediated pathway at low porphyrin concentration to a Type I process at high sensitizer concentrations (Grossweiner et al. 1982). Water insoluble porphyrin esters incorporated into the membranes of liposomes prepared from saturated phospholipids sensitize the photooxidation of amino acids, proteins, etc. in the external medium with quantum yields as high as those found with porphyrins in solution (Spikes 1983b). Studies of porphyrin-sensitized reactions in micelles and liposomes are just in their beginnings; hopefully, as more work is done with such model systems, the accumulated information will help in our understanding of the reactions occurring in the membranes of porphyrin photosensitized cells.

PORPHYRIN-SENSITIZED PHOTOEFFECTS ON CELLS (see refs. in Jori, Spikes 1983)

Although it has been shown that porphyrins are phototoxic to the cells of a few types of prokaryotic and eukaryotic microorganisms, including certain bacteria (especially Gram positive), a fungus (yeast), an alga (a dinoflagellate) and a protozoan (paramecium), most cellular level studies have been carried out with various types of mammalian cells. Much interesting work, especially on membrane effects, has been done with red blood cells; however, more has involved the use of various kinds of tissue cells in culture. This paper will consider only mammalian cells (see Ito 1978 and 1983, Pooler and Valenzeno 1981, and Spikes 1983a for general reviews of photodynamic effects on cells; the review of Jori, Spikes 1983 has an extensive listing of references on porphyrin-sensitized phototoxic effects on mammalian cells). In working in this area, it should be remembered that some porphyrins are cytotoxic in the dark; thus careful dark control measurements should always be made.

A large number of studies have been made of the kinetics of the porphyrin-sensitized photokilling of mammalian cells using various criteria for cell death. Cells of different types and cell lines, and cells from the same line but in different metabolic states can differ greatly in their sensitivity to photokilling. Cells from some lines, but not all, show different responses at different stages of the cell division cycle, with the greatest sensitivity occurring in the middle S phase (see Christensen et al., 1981; Gomer and Smith 1980). The mechanistic studies carried out so far indicate that the porphyrin-sensitized photokilling of mammalian cells is mediated by singlet oxygen. Survival curves (plots of the log of the fraction of cells surviving vs. the duration of light exposure) for photodynamically treated mammalian cells typically show an initial plateau at low light doses followed by a somewhat linear region representing exponential cell killing. Both the width of the shoulder region and the slope of the linear part of the survival curve vary with the cell type, the properties of the porphyrin used as sensitizer, the experimental conditions and the physiological state of the cells. Although kinetic studies of cell photokilling give some useful information, they do not provide much data on the specific sites of cell damage or on the molecular mechanisms involved in the generation of cell damage.

Somewhat hydrophobic porphyrins that bind to or penetrate into cells are much more phototoxic than those that remain in the external medium (see Sandberg and Romslo, 1981; Kessel, 1982; Kessel and Chou, 1983; Moan et al., 1982); the effects of porphyrin charge and aggregation state on penetration have not been examined in detail. In the one case studied, the action spectrum for porphyrin-sensitized photokilling of cells resembles the absorption spectrum of bound, rather than free porphyrin (Moan and Christensen, 1981). The mechanisms by which porphyrins enter mammalian cells are poorly understood. It is generally assumed that they diffuse into the cell; however cells take in materials by transport processes, as well as by pinocytosis and/or phagocytosis in some cases. It is probably dangerous to generalize too far about porphyrin-cell interactions on the basis of our present knowledge.

In many cases the plasma membrane of the cell appears to be the main site of porphyrin-sensitized photodamage.

Porphyrins and porphyrin mixtures such as protoporphyrin, mesoporphyrin, hematoporphyrin, hematoporphyrin derivative, etc. sensitize a number of kinds of photo-alterations in cell membranes including the oxidation of membrane components and the inactivation of membrane enzymes (Pooler and Valenzeno, 1981), the crosslinking of membrane proteins (Dubbelman et al., 1980b; Girotti and Deziel, 1983; Van Steveninck et al., 1983), alterations in membrane permeability and damage to membrane transport systems (Pooler and Valenzeno, 1981), and cell lysis (Pooler and Valenzeno, 1981, Bellnier and Dougherty, 1982).

With increasing incubation time, some porphyrins penetrate into cells and distribute into different cytoplasmic and nuclear sites, depending on their physico-chemical properties. Thus, on illlumination, somewhat selective effects on different cellular structures might be expected to occur (Jori and Spikes, 1983). For example, some porphyrins localize in mitochondria or lysosomes; subsequent illumination produces damage to these structures (Sandberg et al., 1981). Isolated organelles including mitochondria and lysosomes (Sandberg, 1981) and microsomes (Maines and Kappas, 1975) are also damaged by photodynamic treatment with porphyrins. In some cases such treatment results in morphological changes in the nucleus; single strand chromosome breaks, sister chromatid exchanges and other types of chromosomal abberations are also produced (Jori and Spikes, 1983; Evensen and Moan, 1982; Gomer et al., 1983). On this basis it might be expected that porphyrin-sensitized phototreatment could lead to the production of mutations in mammalian cells as has been found with certain other sensitizers, but this has not been reported, as yet (Gomer et al., 1983).

PORPHYRIN-SENSITIZED PHOTOEFFECTS ON MAMMALS INCLUDING MAN (see refs in Blum, 1941; Zalar et al., 1977; Valenzeno and Pooler, 1979; Jori and Spikes, 1983)

It was shown in 1911 by Hausmann (see Blum, 1941) that mice injected with hematoporphyrin show a characteristic pattern of phototoxic responses when exposed to light, including scratching the skin, hyperactivity, skin injury (erythema, edema, ulceration), generalized body damage (circulatory collapse, hemorrhaging of the intestines) and death. Other mammals such as dogs, rabbits and guinea pigs show similar responses with the severity depending on the dose of porphyrin used and the intensity and time of illumination. The first stages of the response resemble histamine-release reactions (Blum, 1941; Spikes, 1982). Other porphyrins, including hematoporphyrin derivative, tetrasulfophenylporphine and zinc tetrasulfophenylporphine injected intraperitoneally also act as phototoxic agents for mice (Grenan et al., 1980). Mice rendered protoporphyric by certain drugs become light sensitive. A hereditary porphyria (pink tooth) has been observed from time to time in cattle; such animals are light sensitive as a result of accumulated porphyrins (Ruth et al., 1977).

Photosensitivity in humans associated with the excretion of abnormal porphyrins in the urine was reported in the last century. Myer-Betz performed his famous "self experiment" in 1912 by injecting himself with 200 mg of hematoporphyrin and then exposing various parts of his body to light; erythema, edema and pain resulted. He remained light sensitive for several months (see Blum, 1941). More recently it has been shown that patients receiving hematoporphyrin derivative show the same types of responses in illuminated areas of the skin; the action spectrum peaks at 400 $\pm$ 5 nm as would be expected for a porphyrin-photosensitized process (Zalar et al., 1977). Most of our information on porphyrin photosensitization in humans comes from studies of patients with porphyrias; porphyrias are diseases, either hereditary or induced by certain chemicals, in which abnormal amounts of certain porphyrins accumulate in the body (Wintrobe, 1981). One type, erythropoietic protoporphyria, is a somewhat rare hereditary disease in which protoporphyrin IX accumulates in the plasma, erythrocytes and feces. Patients with this condition show acute skin responses on illumination including itching and burning sensations, erythema and edema; the action spectrum peaks at around 400 nm (Magnus, 1976). Chronic exposure of these individuals to sunlight results in thickening and scarring of the skin. High doses of beta-carotene, a compound which quenches triplet photosensitizers and singlet oxygen, decreases the light sensitivity in many patients (Mathews-Roth, 1982).

Probably the most common type of photosensitizing porphyria is porphyria cutanea tarda in which uroporphyrin, and to some extent, coproporphyrin, accumulate. Production of the porphyrins in this case is caused or triggered by alcohol and other drugs. Again, the action spectrum for skin photoresponses parallels the absorption spectrum of porphyrins. The morphology of skin photodamage in patients with erythropoietic protoporphyria differs somewhat from that in those with porphyria cutanea tarda as observed by electron microscopy; this may represent differences in the localization in the skin of the porphyrins in the two cases, since protoporphyrin is rather hydrophobic while uroporphyrin is highly water soluble.

One last porphyrin sensitized photoeffect should be mentioned, sensitized photocarcinogenesis. Chronic photodynamic treatment of mice with several different kinds of photosensitizers induces skin tumors (Santamaria et al., 1980). A porphyrin was used in one of the first demonstrations of this interesting phenomenon. If mice are injected subcutaneously with hematoporphyrin solution and then exposed to sunlight for extended periods, skin tumors develop; no tumors are found in sensitized animals kept in the dark or in unsensitized animals receiving the same light exposure (Bungeler, 1937). Thus porphyrins must be considered as two-edged swords. On one hand, they can sensitize the photoinduction of tumors, while on the other they can sensitize tumor destruction.

CONCLUDING COMMENTS

An enormous amount of work remains to be done in order to usefully understand the mechanisms involved in the porphyrin-sensitized phototherapy of tumors in mammals. We have a fair appreciation of the photophysical, photochemical and photosensitizing reactions of porphyrins in aqueous solution, although even here more information is needed on the effects of porphyrin structure, behavior of porphyrin aggregates and possible interactions of porphyrins in mixtures. Additional studies are desirable on the photobehavior of porphyrins in a variety of simple model systems including sensitizer bound to proteins and other macromolecules and incorporated into micelles and into liposomal membranes. Further, at the basic level, much research is needed on the relationship between the structures of porphyrins (especially steric factors), their physical-chemical properties, and their uptake, retention, localization and photosensitizing behavior in mammalian cells.

At the organismal level, not enough is known in terms of porphyrin structure of the mechanisms involved in the uptake and/or retention of porphyrins in tumors; even the site(s) of the phototherapeutic action of porphyrins in tumors is not well established, e.g., what is the relative involvement of tissue stroma, blood vessel and tumor cell damage in destroying the tumor. It was shown a number of years ago that localized illumination of a hematoporphyrin-sensitized frog leads to the rapid clumping of red blood cells in the capillaries of the illuminated area; blood flow soon slows down and stops, and hemolysis begins (Castellani et al., 1963). Similar studies in mammalian systems might be informative. Only a few of the very large number of possible porphyrins have been tested as phototherapeutic agents. Thus many more should be synthesized and examined; particular emphasis should be placed on those that absorb strongly in the red region in order to maximize photochemistry with light in this tissue penetrating region of the spectrum. Also, known porphyrin mixtures should be examined to determine possible interactions between the various components which make up hematoporphyrin derivative and the commercially available hematoporphyrins (see Kessel, 1982). Finally, at this stage, there is no reason to assume that porphyrins are the best photosensitizers for tumor phototherapy. Thus as many photodynamic sensitizers as possible, especially those absorbing in the red, should be tested. Once more information is acquired on the structure-function relationships involved, the chemists should be able to design and synthesize even more effective types of molecules.

Other site-directed phototherapeutic approaches should be explored more fully. For example, it has been shown recently (Mew et al., 1983) that hematoporphyrin covalently bound to monoclonal antibodies toward a particular strain of tumor cells still retains good photodynamic activity. Further, the conjugate still reacts specifically with the tumor cells and gives selective photokilling both in vitro and in vivo. Again, a variety of photodynamic sensitizers which absorb at longer wavelengths could probably be used in place of hematoporphyrin and thus permit therapy with light which penetrates tissues more efficiently. In conclusion, then, it is clear that a very large amount of research is urgently needed on porphyrin photobiology as well as on the photobiology of other possible phototherapeutic sensitizers.

The preparation of this paper was supported in part by NIH Grant No. GM27921, NIH Biomedical Research Support Grant No. RR07092, and the University of Utah Research Fund.

REFERENCES (in general, only reviews and recent papers have been listed; please refer to these for more extensive reference lists)

Bellnier DA, Dougherty TJ (1982). Membrane lysis in Chinese hamster ovary cells treated with hematoporphyrin derivative plus light. Photochem Photobiol 36:43.

Blum HF (1941). "Photodynamic Action and Diseases Caused by Light", New York: Reinhold.

Bungeler W (1937). Ueber den Einfluss photosensibilisierende Substanzen auf die Entstehung von Hautgeschwulsten. Z Krebsforsch 46:130.

Castellani A, Pace GP, Concioli M (1963). Photodynamic effect of hematoporphyrin on blood microcirculation. J Pathol Bacteriol 86:99.

Dougherty TJ, Kaufman JE, Goldfarb A, Weishaupt KR, Boyle DG, Mittelman A (1978). Photoradiation therapy for the treatment of malignant tumors. Cancer Res 38:2628.

Dougherty TJ, Boyle DG, Weishaupt KR, Henderson BA, Potter WR, Bellnier DA, Wityk KE (1983). Photoradiation therapy - clinical and drug advances. In Kessel D, Dougherty TJ (eds): "Porphyrin Photosensitization", New York: Plenum Press, p. 3.

Dubbelman TMAR, Haasnoot C, Van Steveninck J (1980). Temperature dependence of photodynamic red cell membrane damage. Biochim Biophys Acta 601:220.

Dubbelman TMAR, Van Steveninck AL, Van Steveninck J (1982). Hematoporphyrin-induced photo-oxidation and photodynamic cross-linking of nucleic acids and their constituents. Biochim Biophys Acta 719:47.

Evensen JF, Moan J (1982). Photodynamic actions and chromosomal damage: a comparison of haematoporphyrin derivative (HpD) and light. Br J Cancer 45:456.

Felix CC, Reszka K, Sealy RC (1983). Free radicals from photoreduction of hematoporphyrin in aqueous solution. Photochem Photobiol 37:141.

Foote CS (1976). Photosensitized oxidation and singlet oxygen: consequences in biological systems. In Pryor WA (ed) "Free Radicals in Biology", New York: Academic Press, p. 85.

Girotti AW, Deziel MR (1983). Photodynamic action of protoporphyrin on resealed erythrocyte membranes: mechanisms of release of trapped markers. In Kessel D, Dougherty TJ (eds) "Porphyrin Photosensitization", New York: Plenum Press, p. 213.

Gomer CJ, Rucker N, Banerjee A, Benedict WF (1983). Comparison of mutagenicity and induction of sister chromatid exchanges in Chinese hamster cells exposed to hematoporphyrin derivative photoradiation, ionizing radiation or ultraviolet radiation. Cancer Res (in press).

Grenan M, Tsutsui M, Wysor M (1980). Phototoxicity of the chemotherapeutic agents hematoporphyrin D, meso-tetra(p-sulfophenyl) porphine and zinc-tetra(p-sulfophenyl) porphine. Res Commun Chem Pathol Pharmacol 30:317.

Grossweiner LI, Patel AS, Grossweiner JB (1982). Type I and Type II mechanisms in the photosensitized lysis of phosphatidylcholine liposomes by hematoporphyrin. Photochem Photobiol 36:159.

Hausmann, W (1908). Ueber die sensibilisierende Wirkung tierischer Farbstoffe und ihre physiologische Bedeutung. Biochem Z 14:275.

Ito T (1978). Cellular and subcellular mechanisms of photodynamic action: $^1O_2$ hypothesis as a driving force in recent research. Photochem Photobiol 28:493.

Ito T (1983). Photodynamic agents as tools for cell biology; photodynamic action at the cellular level. In Smith KC (ed) "Topics in Photomedicine", New York: Plenum Press (in press).

Jori G, Spikes JD (1981). Photosensitized oxidations in complex biological structures. In Rodgers MAJ, Powers EL (eds) "Oxygen and Oxy-Radicals in Chemistry and Biology", New York: Academic Press, p. 441.

Jori G, Spikes JD (1983). The photobiochemistry of porphyrins. In Smith KC (ed) "Topics in Photomedicine," New York: Plenum Press (in press).

Jori G, Reddi E, Tomio L, Calzavara, F (1983). Factors governing the mechanism and efficiency of porphyrin-sensitized photooxidations in homogeneous solutions and organized media. In Kessel D, Dougherty TJ (eds) "Porphyrin Photosensitization", New York: Plenum Press, p. 193.

Kessel D (1982). Determinants of hematoporphyrin-catalyzed photosensitization. Photochem Photobiol 36:99.

Kessel D, Chou T-H (1983). Porphyrin localizing phenomena. In Kessel D, Dougherty TJ (eds) "Porphyrin Photosensitization", New York: Plenum Press, p. 115.

Kessel D, Dougherty TJ (1983). "Porphyrin Photosensitization" New York: Plenum Press.

Lewin AA, Schnipper LE, Crumpacker CS (1980). Photodynamic inactivation of herpes simplex virus by hematoporphyrin derivative and light. Proc Soc Exptl Biol Med 163:81.

Magnus IA (1976). "Dermatological Photobiology. Clinical and Experimental Aspects", Oxford: Blackwell.

Maines MD, Kappas A (1975). The degradative effects of porphyrins and heme compounds on components of the microsomal mixed function oxidase system. J Biol Chem 250:2363.

Mathews-Roth MM (1982). Beta carotene therapy for erythropoietic protoporphyria and other photosensitivity diseases. In Regan JD, Parrish JA (eds) "The Science of Photomedicine", New York: Plenum Press, p. 409.

Mew D, Wat C-K, Towers GHN, Levy JG (1983). Photoimmunotherapy: Treatment of animal tumors with tumor-specific monoclonal antibody-hematoporphyrin conjugates. J Immunol 30:1473.

Moan J, Christensen T (1981). Photodynamic effects on human cells exposed to light in the presence of hematoporphyrin: localization of the active dye. Cancer Lett 11:209.

Moan J, McGhie JB, Christensen T (1982). Hematoporphyrin derivative: photosensitizing efficiency and cellular uptake of its components. Photobiochem Photobiophys 4:337.

Pooler JP, Valenzeno DP (1981). Dye-sensitized photodynamic inactivation of cells. Med Phys 8:614.

Ruth GR, Schwartz S, Stephenson B (1977). Bovine protoporphyria: the first non-human model of this hereditary photosensitizing disease. Science 198:199.

Sandberg S (1981). Protoporphyrin-induced photodamage to mitochondria and lysosomes from rat liver. Clinica Chem Acta 111:55.

Sandberg S, Romslo I (1981). Porphyrin-induced photodamage at the cellular and subcellular level as related to the solubility of the porphyrin. Clin Chem Acta 109:193.

Sandberg S, Glette J, Hopen G, Solberg CO, Romslo I (1981). Porphyrin-induced photodamage to isolated human neutrophiles. Photochem Photobiol 34:471.

Santamaria L, Bianchi A, Arnaboldi A, Daffar P (1980). Photocarcinogenesis by methoxypsoralen, neutral red and proflavine. Possible implications in photochemotherapy. Med Biol Environ 8:179.

Spikes JD (1975). Porphyrins and related compounds as photodynamic sensitizers. Ann NY Acad Sci 244:496.

Spikes JD (1982). Photodynamic reactions in photomedicine. In Regan JD, Parrish JA (eds) "The Science of Photomedicine", New York: Plenum Press, p. 113.

Spikes JD (1983a). Photosensitization in mammalian cells. In Parrish JA, Morison WL, Kripke ML (eds) "Photoimmunology", New York: Plenum Press, p. 23.

Spikes JD (1983b). Comments on light, light sources and light measurements. In Daynes R (ed) "Experimental and Clinical Photoimmunology", Boca Raton: CRC Press (in press).

Spikes JD (1983c). A preliminary comparison of the photosensitizing properties of porphyrins in aqueous solution and liposomal systems. In Kessel D, Dougherty TJ (eds) "Porphyrin Photosensitization", New York: Plenum Press, p. 181.

Suwa K, Kimura T, Schaap AP (1978). Reaction of singlet oxygen with cholesterol in liposomal membranes. Effect of membrane fluidity on the photooxidation of cholesterol. Photochem Photobiol 28:469.

Valenzeno DP, Pooler JP (1979). Phototoxicity, the neglected factor. J Am Med Assn 242:453.

Van Steveninck J, Dubbelman TMAR, Verweij H (1983). Photodynamic membrane damage. In Kessel D, Dougherty TJ (eds) "Porphyrin Photosensitization", New York: Plenum Press, p. 227.

Verweij H, Dubbelman TMAR, Van Steveninck J (1981). Photodynamic protein cross-linking. Biochim Biophys Acta 647:87.

Wintrobe MM (1981). "Clinical Hematology", Philadelphia: Lea and Febiger, Chap. 44.

Zalar GL, Poh-Fitzpatrick M, Krohn DL, Jacobs R, Harber LC (1977). Induction of drug photosensitization in man after parental exposure to hematoporphyrin. Arch Dermatol 113:1392.

Porphyrin Localization and Treatment of Tumors, pages 41–73

PHOTOPHYSICS OF AND INSTRUMENTATION FOR PORPHYRIN DETECTION AND ACTIVATION

Daniel R. Doiron, Ph.D.

Ocular Oncology Center
Childrens Hospital of Los Angeles
University of Southern California
School of Medicine

Western Institute for Laser Treatment and
Advanced Biomedical Instrumentation
Santa Barbara, California

INTRODUCTION

Those working and studying in the area of porphyrin detection and activation should have a basic understanding of the photophysical principles underlying these techniques and the characteristics of the instrumentation available for use. Many basic errors and misinterpretation of experimental results can and have been made in these areas due to a lack of understanding of basic principles. This presentation, along with the reviews by J.S. Spikes and C.S. Foote, are meant to enable persons reading these proceedings to better understand and accurately evaluate the results presented.

This review is separated into two sections: (1) Photophysics and (2) Instrumentation. Photophysics covers the basic principle of light, its interaction with and penetration in biological systems, and light dosimetry. Instrumentation includes sources of light, light detection methods, light detection equipment and delivery systems.

The area of porphyrin detection and activation is a multidisiplinary field including physics, biology, chemistry and clinical medicine. It is hoped that this review will provide a basic background for those investigators not experienced in the physics and instrumentation areas.

PHOTOPHYSICS

The term photophysics is used in this review to cover the basic physical nature of light, its interaction with molecules, how it penetrates tissue, and the units of light dosimetry.

Nature of Light

Light is electromagnetic radiation as are radiowaves, microwaves, x-ray and gamma-rays. Light is the portion of the electromagnetic spectrum characterized as having wavelengths from 10^{-6} m to 10^{-10} m. This range is split into the ultraviolet (12×10^{-9}m to 0.4×10^{-6}m or 12nm to 400nm), visible (400nm to 750nm) and the infrared (750nm to 20 μm). Table 1 outlines the electromagnetic spectrum in terms of wavelength, frequency and energy.

Electromagnetic radiation is composed of discrete packets of energy called photons. These packets have properties of waves and discrete particles. As such, they can be described either by the wavelength and/or frequency and have discrete quantities of energy (1). The wavelength frequency and energy are related by equations 1A and 1B:

$$\lambda = c/\nu \quad (1A)$$
$$E = h\nu = hc/\lambda \quad (1B)$$

where

λ = wavelength
ν = frequency
C = speed of light
E = energy
h = Plank's constant
= 4.14×10^{-15} eV-s

TABLE 1

The Electromagnetic Spectrum in Terms of Wavelength, Frequency and Energy

Name	Wavelength, m	Frequency, Hz	Energy, eV
Radio waves	1	$3x10^{8}$	$1.24x10^{-6}$
Microwaves	10^{-3}	$3x10^{11}$	$1.24x10^{-3}$
Extreme infrared	$15x10^{-6}$	$2x10^{13}$	0.083
Far infrared	$6x10^{-6}$	$5x10^{13}$	0.207
Middle infrared	$3x10^{-6}$	$1x10^{14}$	0.414
Near infrared	$0.75x10^{-6}$	$4x10^{14}$	1.65
Visible	$0.4x10^{-6}$	$7.5x10^{14}$	3.1
Ultraviolet	$12x10^{-9}$	$2.4x10^{16}$	100
Soft X-rays	$1.25x10^{-10}$	$2.4x10^{18}$	10
Hard X-rays	$1.25x10^{-11}$	$2.4x10^{19}$	10
γ rays			

As the wavelength increases. the frequency and energy per photon decreases. The energy in a photon will determine how it interacts with its environment. High energy electromagnetic photons such as x-ray and γ-rays may cause direct ionization while extreme infrared and microwaves principly cause heating.

Visible wavelengths range from 400nm to 750nm which corresponds with what is perceived as violet to deep red (see Table 2). The visible range relates to the ability of the human eye to detect the light. Ultraviolet and infrared are also part of the light spectrum but are not perceivable by the human eye.

It must be remembered that just because light is not visible, does not mean it is not present and it may still initiate photochemical reactions. The human eye is a very sensitive detector. but only for light in the 400nm-750nm range. The eye's sensitivity is highly wavelength dependent and non-linear (Figure 1). For this reason, it is a very poor instrument for quantitative work (2).

TABLE 2

The Light Spectrum in Terms of
Color, Wavelength, Frequency and Energy

Name	Wavelength, nm	Frequency, Hz	Energy, eV
	750	4.0×10^{14}	1.65
Red			
	610	4.9×10^{14}	2.03
Orange			
	590	5.1×10^{14}	2.10
Yellow			
	570	5.26×10^{14}	2.17
Green			
	500	6.0×10^{14}	2.48
Blue			
	450	6.7×10^{14}	2.76
Violet			
	400	7.5×10^{14}	3.10

Light Interaction with Matter

A graphic representation of the interaction of light with matter is shown in Figure 2. Photons of light can scatter from the molecules within the matter, be absorbed by these molecules, or possibly not interact at all. The type of interaction that takes place depends on the wavelength, i.e. energy of the photon, and the structure of the matter.

If a photon scatters, its energy is not significantly changed (elastic scattering) but its direction may be altered. If the potential for scattering is high, the scattering will act as a diffusing process. In addition, if absorption is low and the scattering is low, the material would appear clear. Scattering does not directly add to photochemical reactions but may indirectly affect

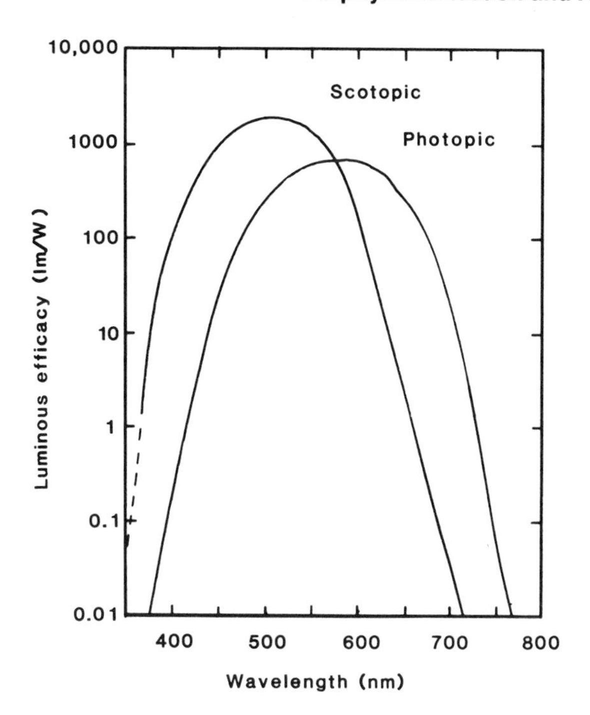

Figure 1. The spectral sensitivity (luminous efficacy) of the human eye for scotopic (color) and photopic (black and white) vision.

the amount of reaction by keeping photons within the matter for greater lengths of time; thus, increasing the probability of absorption.

If a photon is absorbed by a molecule, a variety of reactions may take place (Figure 3). In order for a photon to be absorbed by a molecule, its energy must match the energy difference of the allowable excited states of the electron bond in the molecule. For the visible and infrared light this corresponds to the electrons in the outer valence levels. It cannot be over-emphasized that in order for a photochemical reaction such as fluorescence or photosensitization to take place, the photon must first be absorbed. These are absorption initiated reactions. Once a photon is absorbed and the

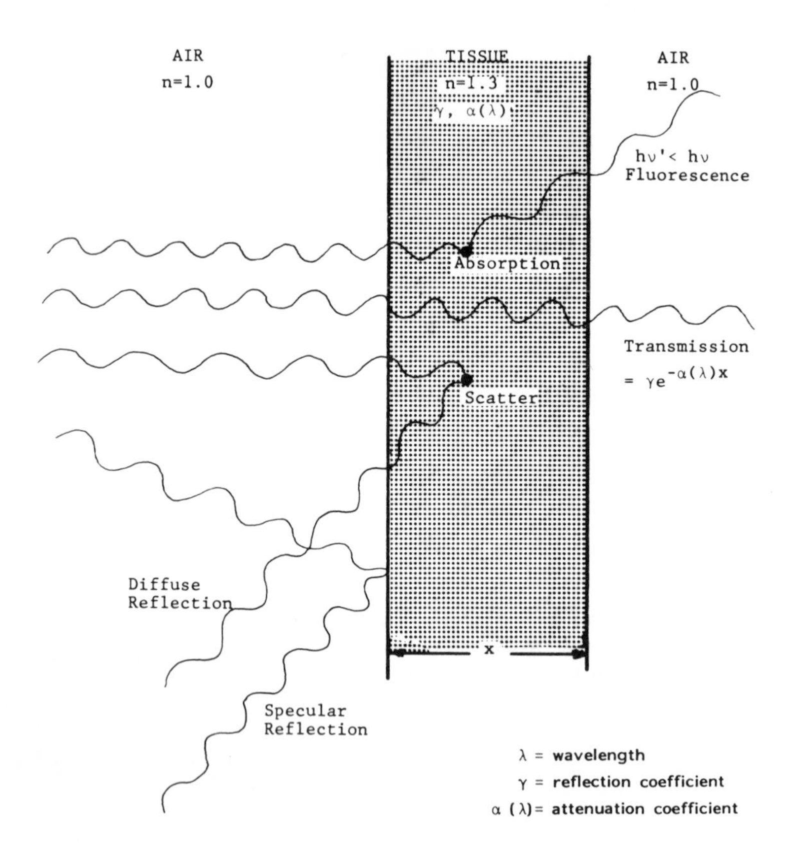

Figure 2. Possible interaction of light with tissue (matter). Multiple scattering may take place in the tissue.

molecule is raised to an excited energy state, it may de-excite by a variety of pathways. These pathways are competitive with each other. The pathway chosen is independent of the photon that excited the molecule as long as sufficient energy is absorbed to excite the molecule to an adequately excited state. This principle does not

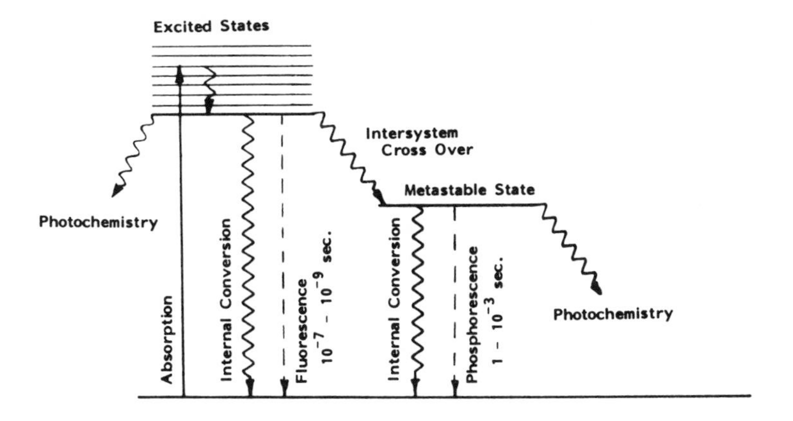

Figure 3. Schematic of possible de-excitation mechanisms.

consider the possibility of a non-linear mechanism or two photon processes when extremely short and/or high energy light pulses are used. The ability for a given wavelength of light to drive a photochemical reaction will be determined by the absorption characteristics of the molecule. In biological systems the molecules with absorption characteristics in the visible or near infrared are those containing double or conjugated bonds.

If a fluorescence photon is emitted, its energy will be less than the absorbed photon due to loss of energy in the absorption and deexcitation process. The wavelength for the fluorescence, therefore, is always longer. Also, if fluorescence deexcitation does occur, the energy of the fluorescence photon is lost and any photosensitivity reaction via the metastable triplet state is not possible.

Porphyrins have many double conjugated bonds and a resonant ring structure. As such, they exhibit characteristic absorption ranging from the near ultraviolet to the near infrared. The exact absorption structure will depend on the individual structure of the porphyrin and its associated environment. Figure 4 shows the typical aetio absorption spectrum of hematoporphyrin derivative in saline and saline with 10% fetal calf serum. Note the red shift with binding of the porphyrin to serum proteins. It must be remembered that HpD is not a pure compound and what is shown in Figure 4 is an averaging from the various components present (3). The spectrum found in-vivo may also be expected to be shifted and changed in the relative rearrangements of the peaks.

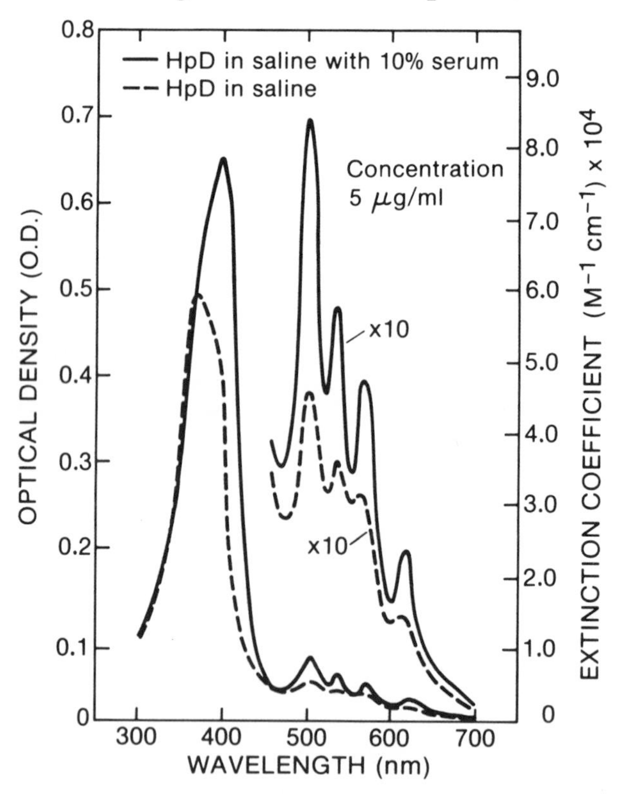

Figure 4. Absorption spectra of hematoporphyrin derivative in saline and saline with 10% serum.

The overall net effect of a photon interaction with a biological system will depend on factors other than the total number of photons absorbed. Reactions in biological systems will also depend on physical and physiological factors such as blood flow, oxygen content and temperature. The net biological effect may be extremely complicated with synergistic relationships between many factors.

Light Penetration in Tissues

Most tissues are very non-homogenous in nature and tend to exhibit a great deal of scattering and absorption for visible and infrared light. The one exception to this is the cornea, lens and vitreous of the eye. In the non-homogenous tissues, exact solution of the equations governing the propagation of light (Maxwell's Equation) is not practical. This inhomogenous condition does permit the neglect of spatial coherence and interference effects and the use of considerable approximations. Lightly pigmented tissue is highly scattering in nature and permit the use of diffusion theory in describing the propagation of light in it. Equation 2 is the solution of the light distribution in tissue at a point "r" based on diffusion theory (4).

$$F(r) = F_o(a/r)\, e^{-\alpha(r-a)} \qquad (2)$$

where α = attenuation coefficient (cm^{-1})

$= \sqrt{\beta/\xi}$

β = absorption coefficient

ξ = diffusion coefficient

$= 1/3\,\kappa$

κ = scattering coefficient

$F(r)$ = space irradiance @ point r

F_o = space irradiance @ r=a

a = source radius

The space irradiance "F" is defined in the next section. Use of the diffusion theory is only valid for highly scattering conditions, i.e. $\kappa >> \beta$, and for distances greater than a few penetration lengths ($\delta = 1/\alpha$) from the source or boundaries. The penetration length δ is the $^{1}/e$ or 37% drop-off point, and should not be mistaken for the ultimate limit of the light penetration. The light will actually go to zero only at $r = \infty$ due to the exponential decay. For applications such as photodynamic therapy (PDT), it is the space irradiance at depth that is of interest in obtaining a therapeutic response. Note that the "$^{a}/r$" term will only be dominant close to the source "$r \cong a$", and that the light distribution will be characterized by the exponential term at greater distances. This fact can be used to evaluate α for various tissues by plotting the space irradiance at depth and determining the slope of the curve. This technique is being used by a number of groups studying light penetration in tissue (5-7).

Characteristically, it is observed that tissue exhibits high attenuation (limited penetration) in the ultraviolet and violet region, slightly decreased attenuation in the blue region, increased attenuation in the green and green-yellow regions due to increased hemoglobin absorption and then significantly decreased attenuation (increased penetration) in the red and near infrared. Table 3 gives data at a variety of wavelengths for a cat brain in-vivo and in vitro that typifies this behavior. Maximum tissue penetration is in the near infrared from 700-850nm and 1000-1100nm. Between 850-1000nm and above 1100nm, penetration in tissue is greatly limited by the absorption characteristics of water.

What is desired for obtaining maximum penetration in tissue and therefore maximum photochemical reaction at depth is a photoactive drug with maximum absorption in either the 700-850nm or 950-1100nm range. A great deal is yet to be done in the area of light penetration in tissues and how it might be quantatively modeled and verified in various tissue. The work presented in these proceedings is a large move forward in this area.

TABLE 3

Attenuation Coefficients (α) and Penetration Length (δ) for a Cat Brain

Wavelength (nm)	In-Vivo		In-Vitro	
	α (cm^{-1})	δ (mm)	α (cm^{-1})	δ (mm)
632.	5.0	2.0	5.3	1.9
	9.8	1.0	8.9	1.1
577.	25.9	0.39	-	-
545.	34.4	0.29	-	-
514.5	-	-	13.3	0.75
501.7	-	-	13.2	0.76
496.5	-	-	13.2	0.76
488.	-	-	10.9	0.92
405 - 410	44.1	0.23	46.1	0.22

Light Dosimetry

Light dosimetry in the area of porphyrin detection and activation is in the realm of radiometry. Photometry deals with light detection as it relates to the spectral sensitivity of the human eye. Its application in porphyrin detection is only relevant in the area of direct visual detection of porphyrin fluorescence. The human eye response drops off considerably in the red region where most porphyrins emit fluorescences, (i.e. > 610nm), and therefore is somewhat limited as far as detection by the human eye. This is especially true for quantitative measurement of fluorescence, since the eye is only a qualitative detector. Radiometry deals with light without relationship to the spectral response of a given detector. The units used in radiometry are outlined in Table 4.

TABLE 4

Units of Radiometry

Name	Concept, Symbol	Radiometric Unit
Flux	Flux, ϕ	w (watts)
Incidance (Irradiance)	ϕ/unit area, E	$^{W}/m^2$ (watts/sq. meter)
Exitance (Radiance)	ϕ/unit area, M	$^{W}/m^2$ (watts/sq. meter)
Intensity	ϕ/unit solid angle, I	$^{W}/sr$ (watts/sterradian)
Radiance	ϕ/unit area unit solid angle, L	$^{W}/m^2$-sr (watts/ sq. meter/ sterradian)

The space irradiance is the integral of the radiance over all possible solid angles, typically 4π, and is sometimes called the power density (8). Absorption is independent of the angle of irradiance of the photon on the molecules. For isotropic scattering the space irradiance, F(r), is given by equation 3:

$$F(r) = 4\pi\phi(r) \quad (3)$$

It is the space irradiance that is important in photoactivation of porphyrins at a given spatial point within the tissue.

The unit most used in porphyrin detection and activation is milliwatt (mw = 10^{-3} W). This is a unit of power or energy per unit time. One watt equals one joule per second (J/sec). The unit giving the total energy delivered is the joule and is determined by the watts x time (sec) = joules. Note that the power must be expressed in watts to obtain joules.

Photoactivation is a discrete mechanism of a single photon absorption. The critical factor in this mechanism is the number of photons absorbed or incident. The number of photons in a joule depends on the wavelength of the light. If two different wavelengths of light are used, the number of photons per joule varies as the inverse ratio of the wavelengths. For example, if 700nm and 400nm wavelength light are compared, the number of photons in one joule of 700nm is 1.75 times greater than for 400nm. This difference must be considered when comparing different wavelengths.

All considerations up to this point have been for monochromatic light. Strongly filtered lamps or lasers are the only source for nearly monochromatic light. Use of polychromatic light requires that the individual wavelength be weighted by the corresponding absorption of the molecule being activated. For this case, the effective space irradiance (F_{eff}) for the spectrum would be obtained using equation 4:

It is this effective space irradiance that must be used in comparing various light sources for a given photoactive molecule with an excitation coefficient spectra $\varepsilon(\lambda)$. Equation 4 is valid only for point r, and does not account for differences in attenuation at various wavelengths in obtaining $F(\lambda,r)$. Equation 2 would have to be evaluated for the spectral range of the polychromatic source in order to evaluate $F(r)$.

Monochromatic lasers allow evaluation at a single wavelength, but care must be taken to assure that experiments compared are done at the same wavelength. This will be especially true for dye laser systems which can operate over a range as great as 60nm for a given dye.

$$F_{eff}(r) = \frac{\int_{\lambda L}^{\lambda up} \varepsilon(\lambda)\, F(\lambda, r) d\lambda}{\int_{\lambda L}^{\lambda up} \varepsilon(\lambda) d\lambda} \qquad (4)$$

where

$F(\lambda, x)$ = spectral space irradiance, at wavelength λ, and point r.

ε = extinction coefficient of absorbing molecule at wavelength λ.

λL = lower wavelength of the polychromatic light.

λup = upper wavelength of the polychromatic light.

Another factor that must be considered is the delivery rate of the light. If the light is delivered at a great enough rate, significant heating of a molecule and its surroundings may take place (9,10). At very high rates this can lead to burning or direct vaporization as seen in laser surgery. At medium rate, temperature increases may be great enough to cause thermal disruption of biological functions and/or possible synergistic effects in photochemical reactions (10,11). All reported light dosimetry should also state the power, so the potential for these thermal effects can be considered.

The geometry of light delivery will be significant in determining the delivered space irradiance at depth and potential dose rate (power) effects. Table 5 outlines the units appropriate for various geometries of light application.

TABLE 5

Units for Various Light Application Geometries

Geometry	Power Unit (Dose Rate)	Light Energy Unit
Surface	w/m^2 or w/cm^2	J/m^2 or J/cm^2
Interstitial:		
Point	w	J
Cylinder	w/cm	J/cm

For interstitial applications using fiberoptics, a point would be a flat cut fiber tip or a point diffusing tip. For cylindrical diffusing fibers used interstitially, the total power out of the cylinder divided by its length gives the power per unit length (i.e. W/cm). See the delivery system section of this review for further description of these types of applications.

INSTRUMENTATION

Instrumentation significant to porphyrin detection and activation in tumor includes the light source, light detectors and delivery systems.

Sources of Light:

The major types of light sources (1) incandescent, (2) gas discharge and (3) lasers.

Incandescent: Incandescent lighting is thermal black body radiation obtained by resistive heating. It represents a broad spectrum of polychromatic light whose exact shape is dependent on the temperature of the material heated. A typical spectrum of a tungsten halogen lamp is shown in Figure 5 (12). Note that the majority of light is in the infrared region which may produce significant heat in material to which it is exposed. This infrared radiation will usually not directly affect photochemical reactions but needs to be removed by filtering to minimize heating. Incandescent lamps are characterized by low brightness and are usually limited to 1 kw in size. Table 7 outlines the characteristics of the various light sources for application to hematoporphyrin derivative photodynamic therapy (HpD PDT) for activation at 630 nm.

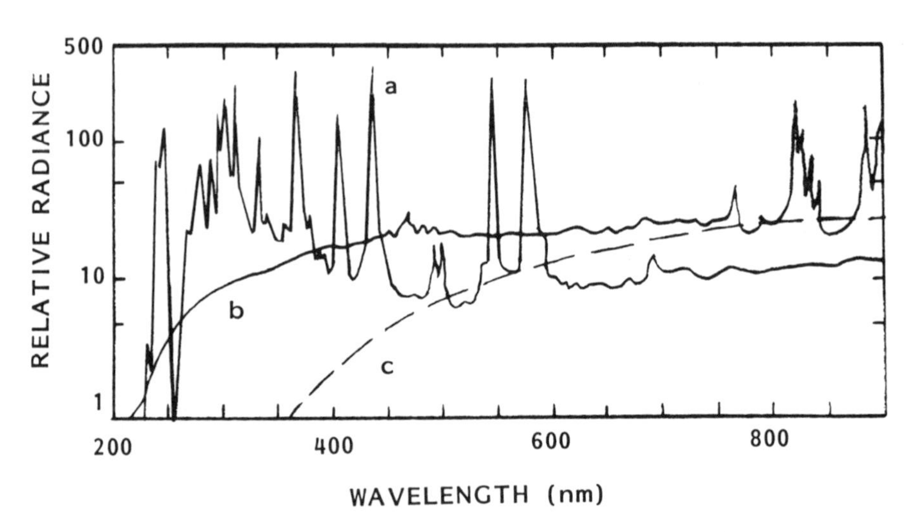

Figure 5. Relative spectral radiance of (a) short arc mercury lamp, (b) Xenon arc lamp and (c) quartz-tungsten-halogen. All for 1000 watt lamps (adapted from Oriel Corporation, Stanford, Connecticut).

<u>Gas Discharge:</u> Gas discharged light sources rely on an ionized gas to produce light emmission. The ionized gas is actually a plasma. The spectral output of the lamp is characterized by the gas and its molecular structure. The output spectrum is generally very broad with strong spectral lines. Figure 5 shows the relative spatial output for a high pressure Xe and Hg arc lamps which are the most common gas discharged sources. Due to this broad spectral shape and

large infrared component these lamps require significant filtering in photobiology and photochemical uses. The brightness of these lamps is considered medium to high with commercial systems available up to 5 kw (see Table 6).

Lasers: The laser is a light source that only came into existence as of 1958, although it was theoretically described by Einstein back in (1917). Laser is an acronym for Light Application by Stimulated Emmission of electromagnetic Radiation. A laser has properties significantly different from the incandescent and gas discharge sources that make it highly useful in porphyrin detection and activation.

A typical ion gas laser is schematically shown in Figure (6). An ionized gas is contained in a low vacuum chamber having highly reflective mirrors at each end. The gas is excited by an electrical discharge within the tube and will emit light characteristic of its electronic structure. If the energy structure of the gas is such that more molecules can be in the excited metastable state than in the ground state, an inverted population, the process of stimulated emission may take place provided other conditions are met. The metastable state is one in which the probability of decay to the ground or lower state is small. Some molecules in this state, though, can spontaneously de-excite with emission of a photon characteristic of the energy difference between the metastable and the lower state. In the cavity of Figure 6 the photons can be reflected back and forth through the gas medium by the mirrors. If this photon passes a molecule in the excited metastable state, it can stimulate it to emit a photon, i.e. stimulated emission. This stimulated emitted photon has a very special relationship to the stimulating photon; namely, it is of the same energy (wavelength) as the exciting photon, and will travel in the same direction and be in phase with the stimulating photon. Being in phase means the wave nature of each photon has a direct correlation to that of the other. This property of being in-phase is called coherence.

TABLE 6

Light Sources for HpD Photodynamic Therapy

Source	Radiant Power (620-640nm)	Percent of Spectrum (620-640nm)	Typical Delivered Irradiance	Electrical Power Requirements	Cooling Requirements	Cost
Incandescent						
Tungsten Halogen 1KW	∿ 3mw/cm^2 @10cm	1.0	2.6mw/cm^2	110VAC, 1ϕ	Air	$ 4K
Gas Discharge						
Direct 5KW-Xenon	75mw/cm^2 @10cm	1.5	30.mw/cm^2	220VAC, 1ϕ	Air, H_2O for filters	$17K
Fluorescent	∿ 1mw/cm^2	40.	0.4mw/cm^2	110VAC, 1ϕ	Air	$ 1K
Lasers						
HeNe (632.8nm)	50mw	100.	43.mw	110VAC, 1ϕ	None	$10K
Tuneable Dye (CW)						
Large	4.0w	100.	3.2w	440VDC, 3ϕ 60amp/ϕ	H_2O, 6 GPM	$65K
Small	2.0w	100.	1.6w	220VDC, 3ϕ 40amp/ϕ	H_2O, 2½ GPM	$40K
Gold Vapor 5KHz, 30nsec, (628nm)						
Large	5.0w	100.	4.0w	208VAC, 3ϕ 20amp/ϕ	H_2O, 2 GPM	$41K
Small	2.0w	100.	1.6w	208VAC, 1ϕ 25amp/	H_2O, 1 GPM	$29K

TABLE 7

Characteristics of the Four Major Light Detections (Non-Imaging)

Device Type	Possible Wavelength Range	Spectral Sensitivity	Response Time	Sensitivity/ Power Range
Thermopile	10nm-10 micron	flat	slow	low/hi
Photoelectric	Near UV-Infrared	varied	fast	med-hi/low-med
Pyroelectric	10nm-10 micron	flat	slow-med	low-med/med-hi
Photomultiplier Tube	UV-1000nm	varied	fast	hi/low

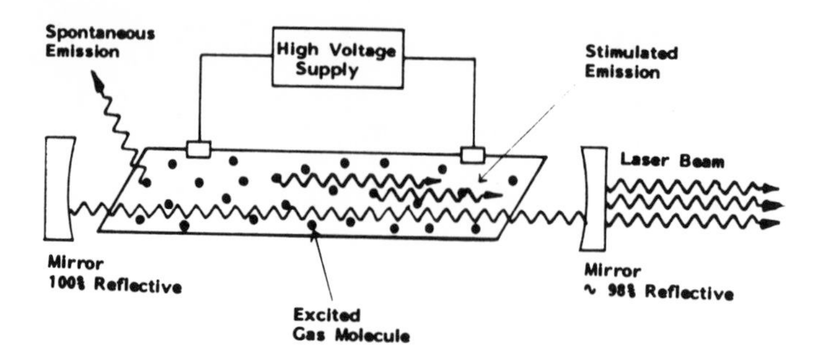

Figure 6. Schematic of ion gas laser.

By making one of the mirrors in the cavity transmit a few percent of light in the cavity, the laser beam is released from the cavity. This laser light, or beam as it is referred to, has the special properties of being highly monochromatic, directional and coherent. Because of these properties, it has a brightness much greater than the sun, and can be focussed to a very small spot of extremely high power densities. This focussing ability permits the beam to be coupled efficiently into small size optical fibers.

The wavelength of the laser is determined by the electronic structure of the medium, which can be a gas, liquid or solid. The requirement for the medium is that an inverted population density can be obtained. For a variety of organic dyes, the inverted population and stimulated transitions are possible over a large wavelength range.

The major lasers presently used in porphyrin detection and treatment are the ion laser, dye laser, and more recently the metal vapor lasers.

Ion Lasers: The most common ion lasers in porphyrin activation for tumor studies are the krypton and argon lasers. The krypton ion laser is used for the flourescence activation around 410nm due to the typical soret absorption band of porphyrins (5). This laser emits simultaneously at 406.7nm, 413.1nm and 415.4nm, with a power distribution of 36%, 60% and 4% respectively. In small size systems only the 406.7 and 413.1nm lines are operational. Figure 7 shows these wavelengths on a fluorescence, excitation efficiency curve for a spontaneous sarcoma tumor for a patient injected with HpD. Note that the spectrum does show minor peak around 510nm but it is approximately 3 times less effective than the violet light at 410nm. This laser has been used to detect early cancer in the lung (16,17).

The argon laser is the most common ion laser with multiple emission wavelengths in the blue-green region (454nm to 514.5nm). The two primary emissions are at 488nm and 514.5nm. The emission at 514.5 nm is being evaluated for HpD fluorescence detection (13) and possibly for HpD photosensitization induction in superficial tumors. Its main use in porphyrin work has been in optically pumping the continuous dye lasers. The difficulty with the ion lasers has been their relatively poor efficiency as illustrated in Table 6 by the electrical and water cooling requirements.

The helium-neon laser is also an ion laser but its power is limited to approximately 50mw due to limitations in maintaining a continuous lasing action in the larger tube bore needed for higher power output.

Dye Lasers: As mentioned previously, a variety of organic dyes can be made to lase over a broad spectrum of wavelengths. For continuous operation, these systems are generally pumped with an argon or krypton ion laser by focusing the beam on a fast flowing stream of dye. The point of the focussed ion laser beam and dye interaction is positioned in an optical cavity, usually formed by

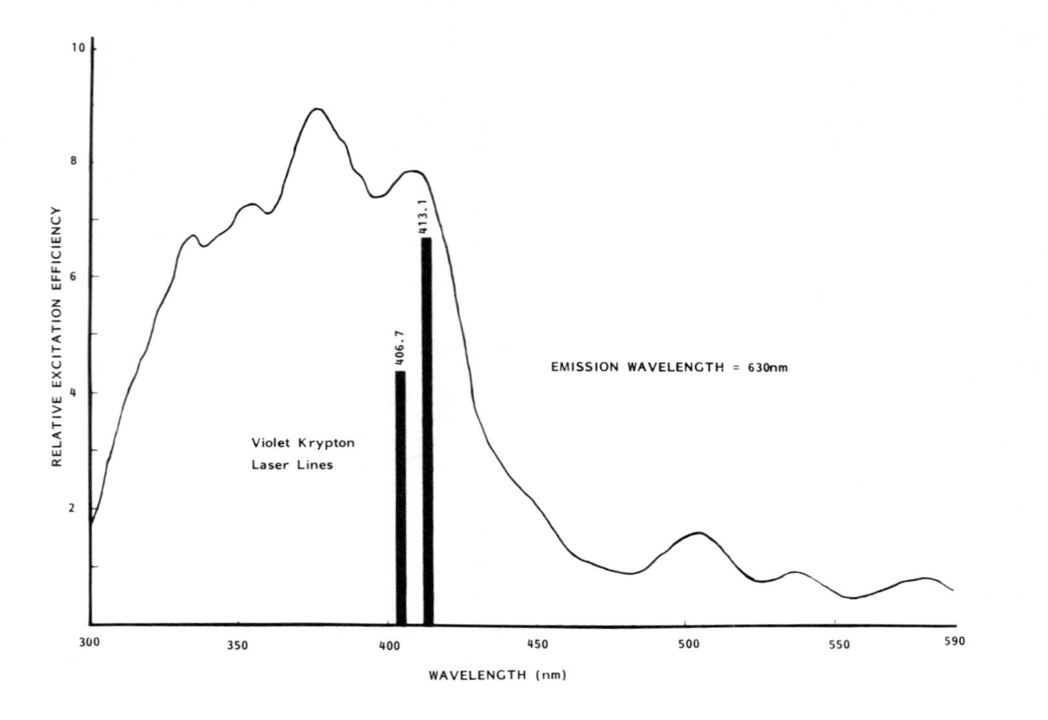

Figure 7. Relative excitation efficiency of an undifferentiated sarcoma from a patient after HpD injection.

three mirrors. This 3 mirror cavity is schematically shown in Figure 8. The optical cavity can then provide the resonator structure such as in the ion laser case. By introduction of an optical tuning element into the cavity, the system can be forced, or tuned, to operate over the possible wavelength of the specific dye used. By choosing the dye and the appropriate wavelength for pumping it, dye lasers have been made to operate from the near UV to the near infrared. The significance of the dye laser in photoactivation is its ability to be precisely tuned to the absorption spectrum of the activating molecule.

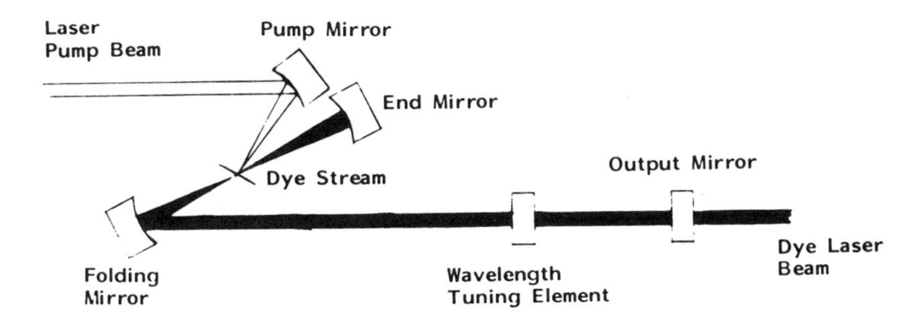

Figure 8. Schematic of a typical three mirror dye laser cavity. (Dye stream is coming out of the paper).

<u>Metal Vapor Laser</u>: As the name implies, this laser uses a metal compound in a vapor state to obtain a lasing action. It is limited to working in a pulse mode, but can be operated at a high repetition rate (5 KHz) to obtain high average powers. Presently, metal vapor lasers are available commercially for copper and gold metals. The copper laser emits at 510nm and 578nm. The gold laser emits at 628nm, which is effective in activating HpD PDT in-vivo. The conversion efficiency is higher for these lasers compared to the ion lasers, but their reliability is just now being established.

Light Detection:

The understanding of the characteristics of the various light detector will help to assure the proper selection for a given application. The four major light detectors in use are (Table 7):

1. Thermopile
2. Photoelectric
3. Pyroelectric
4. Photomultiplier tube

1. Thermopile: The thermopile, as the name implies, relies on the measurement of a temperature increase induced by light absorbed in a material. The material used has nearly perfect absorption potential and shows no spectral response (i.e. a flat response curve). Since the measurement is dependent upon significant thermal change, the thermopile's use is limited to high power application and is therefore well suited for laser operations. For low and medium power levels, the reading can be highly influenced by environmental conditions that may lead to background heat gain or loss by the detector.

2. Photodiodes: A photodiode is a solid state device that can directly convert light into an electrical current or voltage. The current device is called a photoconductive diode and the voltage device a photovoltaic diode. The spectral response will depend on the absorption characteristics of the material and its electric structure. Diodes show a large spectral sensitivity variation and must be calibrated at the wavelength of interest for quantitative measurements. The sensitivity of a diode in its maximum efficiency wavelength region can approach that of a photomultiplier tube in some applications. Its response can be very fast if configured properly and is therefore useful in pulsed light application. Use of the diode is limited to low-medium power unless substantial attenuation of the light is done.

3. Pyroelectric: This detector also relys on the measurement of a temperature change to determine incident power. It is based on the measurement of a current produced in a crystal by a change in its temperature. The current is zero for a steady state temperature, so the light source must be modulated or the detector shuttered. It uses a black absorber attached to the crystal and has a flat spectral response. Its sensitivity is in the medium-low range and is limited to uses in medium-high power application. Its response time is faster than the thermopile.

4. Photomultiplier Tube (PMT): The PMT relies on photoelectric material to absorb the light photon and emit an electron. This electron is accelerated in a vacuum by a high potential field (i.e. voltage) and cascaded through a series of dynodes for multiplication. The photoelectric material exhibits large sensitivity variation with wavelength so that PMTs must be calibrated and correction made when doing spectral measurements. The sensitive photoelectric material and high gain from the dynode array multiplication of the PMT makes them extremely useful in low light level detection. Exposure of the tubes to medium to high light levels can destroy them. Response time of PMT's can be extremely short and can be used to detect individual photons. This usually requires cooling of the PMT to reduce thermal noise and sophisticated electronics to discriminate the pulses. The PMT requires a very stable high voltage supply.

Delivery Systems:

Lamps: Light delivery from lamps is either by direct exposure of the subject to the lamp or by the use of a series of lenses and/or mirrors. Due to the broad spectral output of lamps (see light source section), most require some sort of filtering to eliminate undesireable wavelengths such as infrared and ultraviolet. These wavelengths may not directly add to photodynamic activation but may cause heating and undesirable photoactivated processes, i.e. DNA alteration. The filtering may require cooling systems in order to avoid deterioration and failure of optical elements. An optical integrator may be required to obtain a uniform illumination field.

Optical fiber bundles can be used as a lamp delivery system, but are limited to relatively large bundles, small bright light sources and have relatively poor overall transmission. Lamps are not suitable for use with single optical fibers due to the limited acceptance angle of the fibers and the relative low brightness (radiance) of lamps.

Lasers: Because of the coherent nature and intense brightness (small angle of divergence) of a laser beam, it is well suited for use with small single optical fibers. A laser beam can be focussed to a diffraction limited spot whose dimensions are defined by the focal length of the focussing lens, the wavelength of the laser light and the mode structure of the laser beam (14). Focal spots obtainable for 630nm laser light are on the order of 10 microns with a full angle convergence of approximately 6°. The typical optical fiber used in porphyrin detection and activation studies have optical transmitting cores of 200 to 400 microns and overall outer diameters of 600 and 850 microns respectively.

Fiber optics are based on the principle of total internal reflection. Light travelling from a medium with an index of refraction n to one of lower index n' will be totally reflected if the angle of incidence, normal to the interface of the two materials, is greater than the critical angle. The critical angle is given by $\sin \theta_o = n'/n$. Light entering into a transmitting optical fiber, made up of two materials of different refractive index $(n' < n)$, (Figure 9) with an angle θ, normal to the input face, will be totally internally reflected within the fiber, if $\theta \leq \theta_o$, where θ_o is determined by the refractive index of the materials and is defined by equation 5:

$$\sin \theta_o = (n^2 - n'^2)^{1/2} \tag{5}$$

This angle of acceptance is specified by the term Numerical Aperature "NA" where:

$$NA = n_o \sin \theta_o \tag{6}$$

where n_o = 1.0 for air.

For an air interface, a silica core and a silicon-resin clad, the theoretical NA is Ø.4Ø. In actuality, the NA is dependent on fiber length since the light incident closer to θ_o will be more heavily attenuated and lead to a smaller effective or steady state NA. For a silica/silicon-resin fiber, the steady state NA is typically Ø.22 for lengths greater than 5Ø meters. At 1Ø meters, it is Ø.28 N.A.

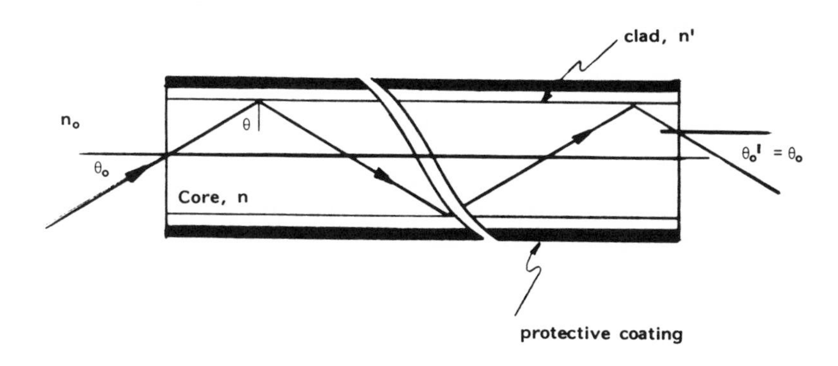

Figure 9. Schematic of a step-index optical fiber.

There are two types of optical fibers, namely step-index and graded index. The step-index fiber is shown in Figure 9. A graded-index fiber introduces the refractive index change by doping the outer region of the fiber core, giving a graduated index change and thereby total internal reflection. A graded index fiber generally has greater attenuation but a larger band width than a step-index type, and is more applicable in the telecommunications field. Optical fibers with cores greater than 1Ø microns permit multiple modes of light to be transmitted simultaneously and are referred to as "multimode fibers". Fibers with cores less than 1Ø microns will permit transmission of only one mode of light and are called "single mode fibers". Single-mode fibers are of interest in telecommunications due to their inherent greater bandwidth and smaller pulse broadening characteristics.

The spectral transmission of an optical fiber is a function of the structural materials, their purity and the wavelength of light. Figure 1Ø shows a typical attenuation curve for a silica/ silicon-resin fiber. The use of silica provides for good ultraviolet transmission. Attenuation in fiber optics is stated as decibels per kilometer (db/km) which is defined by equation 7.

$$A = 10 \log \frac{Pin}{Pout} \qquad (7)$$

A = attenuation (db/km)
Pin = power into the fiber
Pout = power out of the fiber at 1 kilometer length.

The percent transmission, less reflective losses at the faces, for a fiber of length X (in kilometers) with an attenuation of $A'(\lambda)$ at a wavelength λ is:

$$\%T = 10^{-[0.1A'(\lambda)X]} \qquad (8)$$

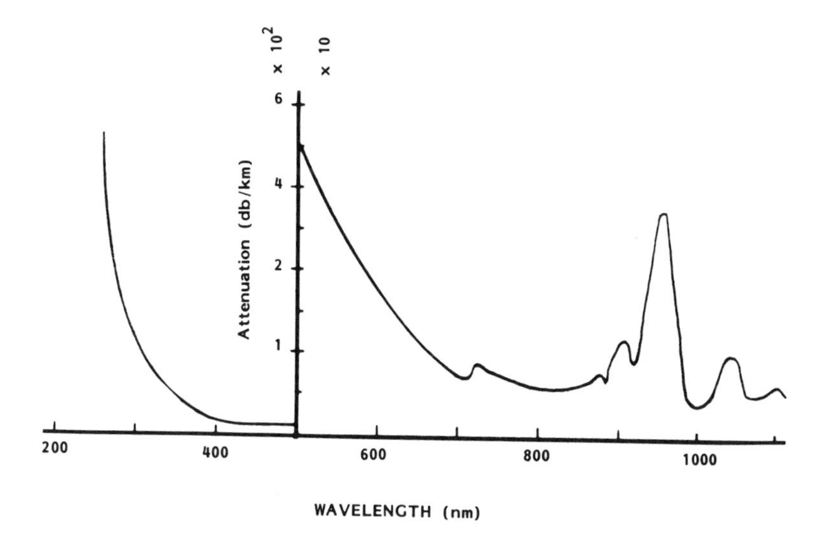

Figure 1Ø. Typical spectral attenuation of a silica/silicon-resin optical fiber. (QSF-A series, Quartz Products Corporation, Planfield, New Jersey).

The major loss in coupling a laser into a fiber at lengths less than 1Ø's of meters is the reflection losses at the air/core interface at the input and output ends. These losses are typically 4% per face. Total transmission for a 1Ø meter long silica/silicon-resin fiber including coupling losses is approximately 86%. Due to the small size of a single optical fiber, less than 1.Ømm total outer diamter, they are highly flexible and can be easily inserted through endoscopes and needles.

A variety of applicators can be attached to the distal tip of an optical fiber to distribute the laser light in a more suitable pattern. A small microlens can be used to generate an illumination field with sharp edges and uniform light distribution. This is in contrast to the non-uniform and diffuse edges obtained from a straight cut optical fiber as shown in Figure 11. To distribute the light along the inner surface of a tubular or spherical structure, optical diffusers can be attached to the fiber tip. These diffusers distribute the light circumferentially around the fiber over a given length, i.e. cylindrical diffuser, or into a spherical geometry, i.e. point diffuser. Cylindrical diffusers are especially useful for illumination within tubular structures such as the trachea, bronchus, or esophagus. Point diffusers are useful for illuminating spherical cavities such as an inflated bladder. Both types of diffusers can also be directly inserted into a mass of tissue by use of needles. This type of interstitial delivery of the light increases the irradiance to a larger volume of tissue and minimizes any induced thermal effects.

SUMMARY

In conclusion, when considering the photophysics of porphyrin detection and activation, it must be remembered that both processes are photon absorption initiated. As such, the absorption spectrum of the porphyrin, the wavelength or spectrum of the activating light, the rate of its delivery, its mode of delivery and the optical characteristics of the tissue must all be considered.

When considering instrumentation, the spectral output or spectral sensitivity of the instrumentation must be considered.

It is hoped that this review will provide basic knowledge to the new and established investigators in this area and thereby help to accelerate the field.

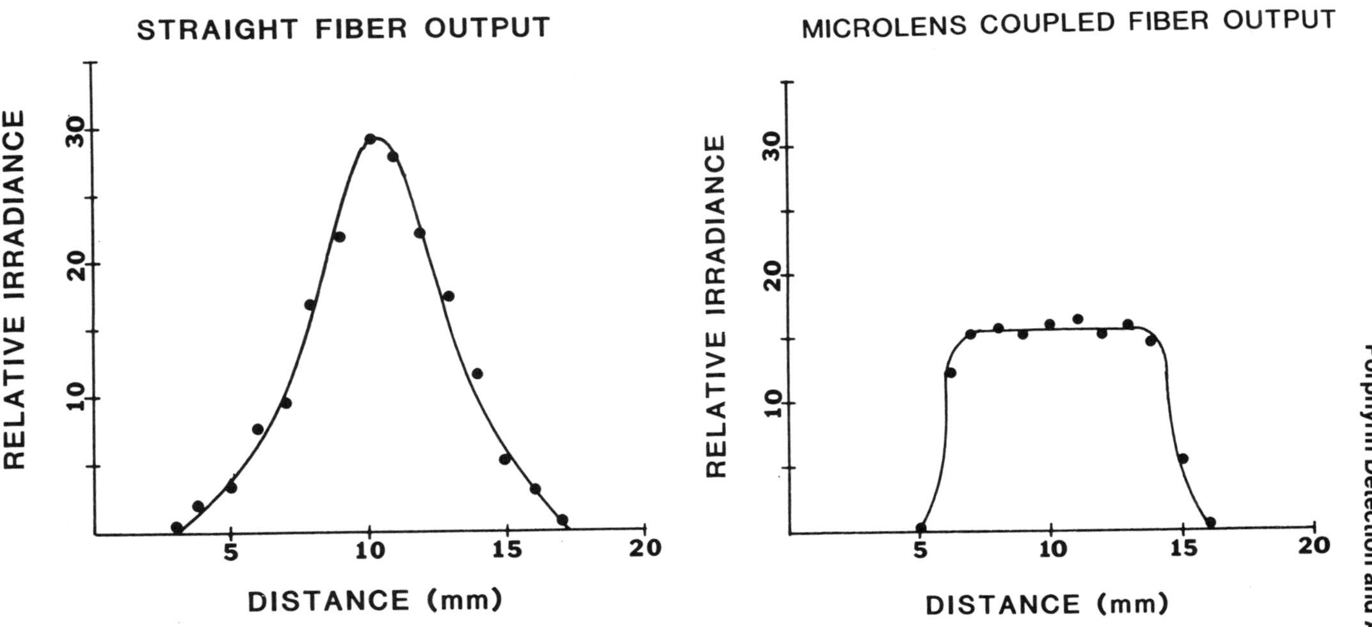

Figure 11. Relative irradiance across the illumination field of (a) straight cut optical fiber and (b) microlens coupled output of an optical fiber.

REFERENCES

1. Optics. Eugene Hecht and Alfred Zajae. Addison Wesley Publishing Company. Menlo Park, California, 1975.

2. Electro-Optics Handbook, Technical Series EOH-11, RCA Corporation, Lancaster, PA, 1974

3. Dougherty, T.J., Boyle, D.G., Weishaupt, K.R., Henderson, B.A., Potter, W.R., Bellnier, D.A., and Wityk, K.E., "Photoradiation Therapy - Clinical and Drug Advances" in Porphyrin Photosensitization. D. Kessel and T.J. Dougherty, eds., Plenum Press, N.Y., 1983.

4. Doiron, D.R., Svassand, L.O., and Profio, A.E., "Light Dosimetry in Tissue Applications to Photoradiation Therapy" in Porphyrin Photosensitization. D. Kessel and T.J. Dougherty, eds., Plenum Press, N.Y., 1983.

5. Wilson, B.C., Jeeves, WP, Lowe, D.M., Adam, G., "Light Propagation in Animal Tissues in the Wavelength Range 375-825 nanometers." Porphyrin Localization and Treatment of Tumors. IN: Clinical and Biological Research. D. R. Doiron and C.J. Gomer, eds., Alan R. Liss, N.Y., 1984.

6. Bolin, FP, Preuss L.E., Cain B.W., "A Comparison of Spectral Transmittance for Several Mammalian Tissues: Effects at PRT Frequencies." Porphyrin Localization and Treatment of Tumors. IN: Clinical and Biological Research. D. R. Doiron and C.J. Gomer, eds., Alan R. Liss, N.Y., 1984.

7. Svassand, L.O., and Ellingsen, R.: "Optical Properties of Human Brain", Photochemistry and Photobiology, 30(3):293-299, 1983.

8. Profio, A.E. and Doiron, D.R.: "Dosimetry Considerations is Phototherapy". Medical Physics, 8(2):190-196, 1981.

9. Lasers in Biology and Medicine, "F. Hellenkamp, R. Prates, and C.A. Sachi, eds., Plenum Press, New York, NY, 1980.

10. Svaasand, L.O., Doiron, D.R., and Dougherty, T.J., "Temperature Rise During Photoradiation Therapy of Malignant Tumors". Medical Physics, 10(1):10-17, 1983.

11. Dougherty, T.J., Potter W.R., Weishaupt K.R., "The Structure of the Active Component of Hematoporphyrin Derivative." Porphyrin Localization and Treatment of Tumors. IN: Clinical and Biological Research. D. R. Doiron and C.J. Gomer, eds., Alan R. Liss, N.Y., 1984.

12. Handbook of Optics. W.G. Driscoll, and W. Vaughan, eds, McGraw-Hill Book Company, New York, N.Y., 1978.

13. Van der Patten, W.J.M., and Van Gemert, M.J.C., "Hematoporphyrin-Derivative Fluorescence In-Vitro and in an Animal Model." Physics in Medicine and Biology, 28(6):633-638, 1983.

14. An Introduction to Lasers and Their Applications, by D.C. O'Shea, W.R. Cullen and W.T. Rhodes. Addison-Wesley Publishers, Menlo Park, California, 1977.

Porphyrin Localization and Treatment of Tumors, pages 75–87

AN OVERVIEW OF THE STATUS OF PHOTORADIATION THERAPY

Thomas J. Dougherty, Ph. D.

Division of Radiation Biology
Roswell Park Memorial Institute
Buffalo, New York

STATUS OF CLINICAL TRIALS

To date approximately 1,500-2,000 patients have been treated by photoradiation therapy (PRT) utilizing hematoporphyrin derivative (Hpd) for tumor photosensitization. Clinical data has been reported from 12 to 15 centers in the U. S. and abroad (Table 1). The main toxicity remains photosensitivity which persists from 3 to 4 weeks generally but much longer (up to 8 weeks) in a few patients. The major categories treated by PRT are tumors in the skin (approximately 200 patients) and bronchogenic carcinoma (approximately 120 patients). Table 2 summarizes data from several investigators in these areas with followup ranging from 3 to 4 years. While the response rate is high in both categories, these patients have been treated under a rather wide variety of conditions, e.g. metastatic breast tumors have been treated from 15 to 75 J/cm^2 at dose rates of 5 to 120 mw/cm^2. The drug dose has been 2.5 to 3.0 mg/kg given generally 3 to 5 days prior to treatment. It is difficult to assess the exact effect of light dose since tumor sizes ranged from a few mm to several cm and from cutaneous to subcutaneous. However, as a generalization, for cutaneous lesions (other than melanin-containing malignant melanoma) of less than 3 mm in size, complete responses (no palpable tumor) can be obtained at the lower end of the scale with essentially no permanent damage to adjacent skin. Exceptions are patients with lesions treated in previous ionizing radiation fields treated by PRT within a short period of also receiving adriamycin. These patients showed severe reaction in apparently non-involved

Table 1. CENTERS CARRYING OUT PRT

	Approx. No. Pts. (as of 1/83)
United States	
RPMI	150
USC	100
UC Irvine	100
Mayo Clinic	60
Grant Hospital, Columbus	40
Philadelphia Hospital	10
U. Iowa	15
Henry Ford	30
UC Davis	*
Childrens Hospital of L.A.	10
Abroad	
Japan (7-10 centers)	500
China (7-10 centers)	100
Australia	100
Italy	50
France	30
Canada	100
Germany	*
England	*
Total	1,400

*Clinical trials just starting

Table 2. PRT FOR TREATMENT OF SKIN TUMORS

Type	Approx. # Pts./ Sites	Response* (%) CR	PR	NR	Longest Followup without Recurrence (to date)
Metastatic breast cancer	120/>1000	60-70	20-30	10	4 years
Basal cell carcinoma	15/50	70-80	20-30	0	4 years
Squamous cell carcinoma	5/10	20	70-80	10	1 year
Malignant melanoma	50/>1000	50	30	20	1 year
Mycosis fungoides	5	20	60	20	1 year
Kaposi's sarcoma	5/>100	80	20	0	3 years
Bowen's disease	2/10	100	-	-	1.5 years

*Summary of data from Forbes (1980); Dougherty, et al. (1975, 1983); Kennedy (1983); Tokuda (1983); Wile, et al. (1983).

CR = Complete Response (disappearance of tumor or biopsy proven)
PR = at least 50% reduction in tumor volume
NR = less than 50% reduction in tumor volume

skin even at low light doses. Otherwise, in general, lesions can be effectively and safely treated in ionizing radiation fields.

In general, patients who benefit most are those with early recurrences treated before it has become widespread and ulcerated. The pigmented melanomas require a higher light dose (50-70 J/cm^2) generally confined to the lesion to prevent skin damage.

The patients with lung cancer have been treated by either interstitial PRT or by surface treatment (Table 3).

Table 3. PRT FOR TREATMENT OF LUNG CANCER

Stage	No. of Patients	Response CR	PR	NR	Longest Followup (to date)
Early/CIS	14	13	1	0	3 years
Advanced (non-obstructing)	52	13	36	3	1 year
Obstructing (bronchus/trachea)	47	0	36+	9*	<1 year

+Obstructions were relieved partially or completely in these cases
*Unable to evaluate

Data of Kato, et al. (1983); Vincent and Dougherty (1982); Balchum and Doiron (1983), Cortese (1982)

In about half the patients, ordinary quartz fibers with flat ends were used and directed at the lesion so as to cover the area or moved over the surface during treatment to cover the area. Treatment doses ranged from very high (~200 J/cm^2) to 30-50 J/cm^2 at dose rates of 50 to 200 mw/cm^2. Recently investigators have been using the fibers with diffusers to produce 360° lateral illumination. This greatly increases the surface area of the light emitting portion of the fiber (e.g. from 0.0013 cm^2 for a 400 μM fiber to approximately

1 cm^2 for a 3 cm cylindrical diffuser, 1000 μM in diameter). This provides a much more efficient light distribution for treatment by either surface illumination or by interstitial PRT. Svaasand has determined that a diffusing fiber requires nearly ten times more power to produce the same temperature rise as a flat cut fiber (this workshop). We have documented that such fibers emitting 400-500 mw/cm length for 20 min (i.e. 480-600 J/cm) in a non-pigmented tumor produced a radius of necrosis of 9 mm x length of cylinder. Temperature rise at the cylinder surface is estimated to be 5-10° (compared to >20° estimated at the surface of a flat cut fiber carrying 400-500 mw). These results are in accord with documented necrosis to 1 cm (i.e. to the luminal wall) in obstructing lung tumors and in esophageal tumors treated in a similar way.

Patients with lung cancer who derive the maximum benefit are those with non-obstructing tumors or early stage patients, primary or recurrent. Essentially all patients have been considered inoperable. Severe complications can be encountered in advance stage patients if edema and heavy exudate result in further obstruction. In these cases it is necessary to remove dead tissue and exudate as soon as possible. In cases where a collapsed lung has been reinflated following PRT, pneumonia can present a severe problem. Several cases of hemoptysis have been reported within a few weeks following PRT. Obstructions in remote lobes present less of a problem in general.

A few patients with esophageal tumors have been treated, most with total obstructions. The general procedure was to treat at day 3 or 4 following 3.0 mg/kg Hpd and inserting a fiber with a diffuser delivering 300-500 mw/cm to a total dose of 180-600 J/cm length of diffuser. These tumors frequently slough resulting in some temporary palliation for the patient. In several cases the tumor reobstructed within a few weeks of treatment. These tended to be the cases receiving the lower light dose. Also, in one case an esophageal-tracheal fistula was formed apparently as a result of tumor slough from an area where it had penetrated the wall.

Perhaps one of the most exciting possibilities for PRT is the treatment of bladder cancer, particularly in cases where the disease has become multi-centric. Benson has shown a high degree of selectivity of Hpd uptake in bladder cancer (1982). Approximately 50 cases have been treated to date. The few cases of CIS have all been complete responders but followup is short. Followup to two years has been obtained in non-invasive transitional cell carcinoma with the majority of patients showing complete response (Table 4). It is possible to use fibers

Table 4. PRT FOR TREATMENT OF BLADDER CANCER

Type	No. of Sites	Response CR	PR	NR	Longest Followup To Date
CIS	4	4	-	-	4 months
TCC non-invasive	47*	25	11	11	2 years
TCC invasive	1	-	1	-	10 months

*Lesions of 1-3 cm were treated; non-responders tended to be lesions >2 cm

Data of Benson, et al. (1982); Oi, et al. (1983); Hisazumi, et al. (1983)

within the bladder which produce isotropic light distribution to the whole bladder to treat superficial, multi-centric disease, a frequent finding in patients with bladder cancer. Light dose rates of up to 200 mw/cm^2 have produced no measurable temperature increases (the bladder is generally irrigated). Light doses of at least 100 J/cm^2 are recommended by Hisazumi, et al. (1983).

Recently excellent results (although early) have been obtained in various tumors of the eye. Murphree, et al. (this workshop) has shown that retinoblastoma and choroidal melanoma respond to PRT at relatively low dose rates (e.g. 50 mw/cm^2) to a total dose of up to 200 J/cm^2. Bruce and McCaughan (this workshop) have treated several choroidal malignant patients with very high dose rates (>500 mw/cm^2 in some cases). Thermal effects are likely a major contribution to response in these cases. However delivered the response appears to be confined to the treated site in the eye.

Very few gynecological tumors have been treated by PRT although some written excellent results have been reported for advanced cases (Ward, et al., 1982; Soma, et al., 1982). The same is true for tumors of the head and neck, where a few cases of good results of PRT have been reported by Wile and Berns (1982).

Application of PRT to glioblastoma is perhaps more problematical. In most instances it is used after debulking of the tumor in the expectation that residual tumor cells will be destroyed. However, the principle remains to be established, that Hpd is in fact taken up by cells in these residual areas. Nonetheless, McCulloch has obtained some intriguing results using PRT and ionizing radiation after surgical removal of glioblastoma (this workshop).

ROLE OF HYPERTHERMIA

A topic of interest is the interaction of PRT and heat generated during 630 nm illumination. Recent results on this subject have been reported by Kinsey, et al. (1983) and Svaasand, et al. (1983) clearly demonstrating that if enough power is used, considerable heat can be generated. Results in small tumors reported by Kinsey indicate temperature rises higher than those reported for larger tumors by Svaasand. This is likely due to the limited cooling capacity of tumors of small volume near the surface of the animal. Table 5 demonstrates temperature rises measured in our laboratory in the SMT-F tumors of various sizes.

Table 5. TEMPERATURE RISE IN THE SMT-F TUMOR IN DBA/2 MICE (INTERSTITIAL 630 nm LIGHT)

Tumor Size/Power	200 mw	300 mw
8 x 10 mm	10	16
11 x 12 mm	5.4 (7.9*, 11**)	8.2
15 x 17 mm	4.7	7.5

Temperature rise measured at 3 mm from flat cut quartz fiber
*2 mm distance
**1 mm distance

Since in most instances interstitial PRT is not used on small lesions, the values for the larger tumors in Table 4 are more appropriate and in fact agree with measurements in patients done at Roswell Park and data reported by Svaasand, et al. (1983).

It should be noted that for superficial illumination of 50-100 mw/cm^2 direct thermal effects are of little consequence. Perhaps a more pertinent question is the possible interaction of PRT and heat at distances comparable to the biological effects of PRT, i.e. 5 to 15 mm. Temperature rises in this region are generally below hyperthermic levels as presently treated (1 to 3°), but could enhance PRT effects. We have studied this question in the SMT-F tumor by applying heat (RF generated) to the tumor for 30 min before or after PRT at subcurative doses (i.e. 7.5 mg/kg Hpd, 24 h, 75 mw/cm^2, 630 nm, 30 min). Table 6 shows the results for several temperatures

Table 6. CURE OF SMT-F TUMOR - PRT + HEAT*

Treatment	%
PRT	3
40.5^o	0
41.5^o	0
44.5^o	20
PRT + 40.5^o	15
PRT + 41.5^o	17
**PRT + 44.5^o	50
40.5^o + PRT	
No treatment	0

*Percentage of animals free of tumor at 90 days post treatment. Minimum of 50 animals per group
**Followup to 30 days currently

applied immediately after PRT. These data indicate a possible synergistic interaction between even low levels of heat and PRT. This raises the interesting possibility of combining heat and PRT clinically. It would not be practical in most instances to attempt to raise the temperature 3 to 7° at distances of 5 to 15 mm (or greater) by using 630 nm light alone (Svaasand, et al., 1983). However, relatively low power from a Nd Yag laser emitting at 1060 nm which penetrates 2 to 3 times deeper than 630 nm may be a very practical means of doing so. This wavelength can be delivered in the same way at 630 nm. Power levels in the order of 200-300 mw/cm^2 would be required for these modest temperature increases.

PRELIMINARY EVALUATION OF THE ACTIVE HPD FRACTION (PHOTOFRIN II)

To date we have treated ten patients using Photofrin II: Four with recurrent bronchial carcinoma, four with metastatic melanoma, and two with breast cancer metastasis of the skin. Table 7 summarizes the current results.

We can make the following generalizations:

1. At drug doses of 1.5 to 2.0 mg/kg the skin can tolerate at least 108 J/cm^2 without damage. This is approximately twice the tolerable dose to the skin with 2.5-3.0 mg/kg Hpd.

2. Tumor response in patients receiving 1.5-2.0 mg/kg Photofrin II appears to be at least equivalent to that using 3.0 mg/kg Hpd although the followup time is short. Since the skin can tolerate up to 108 J/cm^2 tumors can be treated with higher light doses than those used with Hpd (30-50 J/cm^2 maximum to avoid skin damage).

3. Patients receiving 3.0 mg/kg Photofrin II have about the same skin sensitivity as those receiving 3.0 mg/kg Hpd (although the skin and tumor levels are expected to be twice that resulting from 3.0 mg/kg Hpd).

4. Photosensitivity in patients receiving 2.0 mg/kg is less than in those receiving 2.5-3.0 mg/kg Hpd but still persists for about four weeks.

It should be kept in mind that these conclusions are strictly preliminary. However, it is possible that there will be two approaches to using Photofrin II; for patients with early stage tumors where generalized photosensitivity is a problem, doses of 1.5 mg/kg (or less) can be used in conjunction with moderate to high light doses; for patients with large tumors the higher dose of 3.0 mg/kg Photofrin II can be used to provide deeper biological responses (due to higher tumor levels) without enhancing the photosensitivity now encountered with Hpd.

Table 7. SUMMARY OF RESULTS OF PHOTOFRIN II

				Tumor Response	Skin Response
LUNG CANCER (4) (RECURRENT)	3 - 2.0 mg/kg; 1 - 3.0 mg/kg				
1. SCC, RUL	480 J/cm			CR (1 mo)	
LLL	480 J/cm			?	
2. RUL (carina)	600 J/cm			?	
3. SCC, RLL	300 J/cm			PR	
4. LC - 100% obst. RMB >50% obst. LMB	600 J/cm (I)			?	
MELANOMA (4)					
1. Multiple 0.5-2 cm				No FU	
2. Multiple lesions 0.5-1.0 cm (s.c.)	1.5 mg/kg	Day 3:	54 J/cm^2	CR	0
			72 "	CR	0
			36 "	CR	0
			54 "	NR	0
		Day 4:	72 J/cm^2	CR	0
			72 "	CR	0
			72 "	PR	0
		Day 5:	72-108 "	NR	0

Table 7. (continued)

				Tumor Response		Skin Response (Time after Test)
3. Multiple lesions 1-3 cm non-pigmented	1.5 mg/kg		106 J/cm^2	PR-CR?		NA
4. Multiple lesions	3.0 mg/kg		36 J/cm^2	CR		0
			72 "	CR		eschar
		Test	108 "	-		eschar
BREAST CA (2)	2.0 mg/kg					
1. >5 cm		Test (Day 5)	36 J/cm^2			0 (1 mo)
			72 "			0
			108 "			tan
		Tumor	72 + 36	PR		
	2.0 mg/kg	Test	108 J/cm^2		Day 1	2.0 (1 day)
					2	2.0 (1 day)
					3	2.0 (1 day)
		Tumors	320 "			?
2. Multiple, deep (~0.5 cm)	2.0 mg/kg	Test	36 J/cm^2			0 (1 mo)
			72 "			sl. tan
			108 "			tan
		Tumors	72 " slight phot. at 5 wks	CR (1 mo)		0

REFERENCES

Balchum O, Doiron D (1983). Unpublished results.

Benson R, Farrow GM, Kinsey JH, Cortese DA, Zincke H, Utz DC (1982). Detection and localization of in situ carcinoma of the bladder with hematoporphyrin derivative. Mayo Clin Proc 57:548.

Cortese DA, Kinsey JH (1982). Endoscopic management of lung cancer with hematoporphyrin derivative phototherapy. Mayo Clin Proc 57:543.

Dougherty TJ (1983). Unpublished results.

Dougherty TJ, Lawrence G, Kaufman J, Boyle D, Weishaupt KR, Goldfarb A (1979). Photoradiation in the treatment of recurrent breast carcinoma. J Nat Can Inst 62(2):231.

Forbes I, Cowled PA, Leong AS-Y, Ward AD, Black RB, Blake AJ, Jacka FJ (1980). Phototherapy of human tumors using haematoporphyrin derivative. Med J Aust 2:489.

Hisazumi H, Misaki T, Miyoshi N (1983). Conditions for the treatment of bladder tumors. In Hayata Y (ed): "Laser Photoradiation for Tumor Detection and Treatment," Tokyo: Igaku-Shoin Ltd, in press.

Kato H, Konaka C, Ono J, Matsushima Y, Nishimiya K, Lay J, Sawa H, Shinohara H, Saito T, Kinoshita K, Tomono T, Aida M, Hayata Y (1983). Effectiveness of Hpd and radiation therapy in lung cancer. In Kessel D, Dougherty TJ (eds): "Porphyrin Photosensitization," New York: Plenum Publishing Corp, p 23.

Kennedy J (1983). Hpd photoradiation therapy for cancer at Kingston and Hamilton. In Kessel D, Dougherty TJ (eds): "Porphyrin Photosensitization," New York: Plenum Publishing Corp, p 53.

Kinsey JH, Cortese DA, Neel HB (1983). Thermal considerations in murine tumor killing using hematoporphyrin derivative phototherapy. Can Res 43:1562.

Oi T, Tsuchiya A (1983). Superficial bladder tumors. In Hayata Y (ed): "Laser Photoradiation for Tumor Detection and Treatment," Tokyo: Igaku-Shoin Ltd, in press.

Soma H, Akiya K, Nutahara S, Kato H, Hayata Y (1982). Treatment of vaginal carcinoma with laser photoirradiation following administration of haematoporphyrin derivative. Ann Chir Gynaec 71:133.

Svaasand LO, Doiron DR, Dougherty TJ (1983). Temperature rise during photoradiation therapy of malignant tumors. Med Phys 10:10.

Tokuda Y (1983). Primary skin cancer. In Hayata Y (ed): "Laser Photoradiation for Tumor Detection and Treatment," Tokyo: Igaku-Shoin Ltd, in press.

Vincent R, Dougherty T (1982). Unpublished results.
Ward B, Forbes IJ, Cowled PA, McEvoy MM, Cox LW (1982). The treatment of vaginal recurrences of gynecologic malignancy with phototherapy following hematoporphyrin derivative pre-treatment. Am J Obstet Gynecol 142(3):356.
Wile AG, Dahlman A, Burns R, Berns M (1982). Laser photoradiation therapy of cancer following hematoporphyrin sensitization. Lasers Surg Med 2:163.

PHOTOPHYSICS AND INSTRUMENTATION

Porphyrin Localization and Treatment of Tumors, pages 91–114

OPTICAL DOSIMETRY FOR DIRECT AND INTERSTITIAL PHOTORADIATION THERAPY OF MALIGNANT TUMORS

Lars O. Svaasand

Division of Physical Electronics
University of Trondheim
Trondheim, Norway

INTRODUCTION

Photoradiation therapy using the cytotoxic action of photoactivated hematoporphyrin derivative has been found effective against a variety of malignant tumors (Dougherty, Toma, Boyle, Weishaupt 1981; Kato, Konaka, Ono, Matsushima, Nishimiya, Lay, Sawa, Shinohara, Saito, Kinoshita, Tomono, Aider, Hayata 1983; Kennedy 1983).

The development of photoradiation therapy into an efficient clinical technique requires a precise dosimetry (Profio, Doiron 1981). This dosimetry must be based on the information of the drug concentration and the optical distribution within the neoplastic as well as the surrounding normal tissue. Detailed analysis of the exact optical distribution in as highly an inhomogeneous medium as tissue, however, is an impossible task.

It is therefore important to simplify the description to a level where it will meet the requirements for accuracy during therapeutic procedures, and simultaneously be of reasonable mathematical complexity. The regions which require the most precise dosimetry are the outer edges of the treated region, i.e. the diffuse boundaries between the normal and the invasive malignant tissue. The therapeutic depth during photoradiation therapy is several millimeters (Dougherty, Thoma, Boyle, Weishaupt 1981). Thus the relevant regions will be several millimeters below the skin during transcutaneous illumination or several millimeters from the fiber end during therapy with interstitial fibers.

This paper emphasizes a simple description of the light distribution for these regions. The various types of clinically used light applicators are discussed and tested; collimated beam irradiation, interstitial single fibers and interstitial cylindrical applicators.

The light distribution is analyzed in the brain from human cadavers and in a Lewis Lung mice tumor model in vivo.

PROPAGATION OF LIGHT IN TISSUE

The propagation of light in tissue is governed by scattering from inhomogeneities of the cellular structure and by absorption in absorbing constituents as hemoglobin and melanin. The principles for propagation of an incident collimated beam are illustrated in Fig. 1.

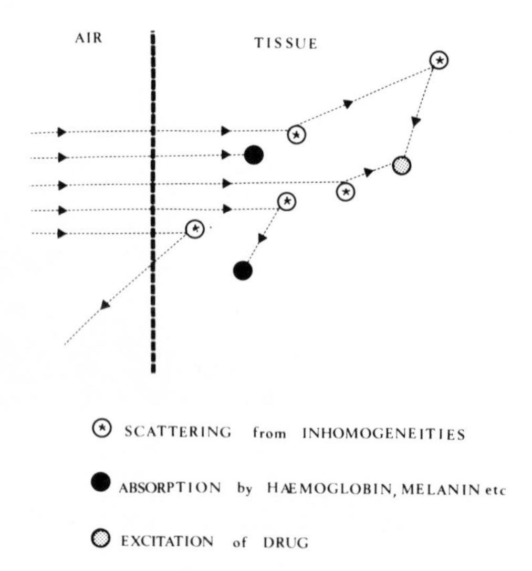

Fig. 1. Principles of light propagation in scattering and absorbing tissue.

Some of the incident photons are scattered back into the air, some are scattered to the position of an absorbing molecule and some entirely change the direction of propagation through multiple scattering.

The major part of the light reflected from most tissue is caused by scattering. The reflected light from a homogeneous non-scattering tissue such as the cornea is only about 3-5%, whereas the reflected light from a highly scattering medium such as the sclera is about 50% (Smith, Stein 1968).

The zigzag path of the photons resulting from the multiple scattering increases the probability for reaching absorbing molecules. Thus, although there is no energy loss in the scattering process itself, it will enhance the absorption of the photons.

All the photons which penetrate deeper than a few tenths of a millimeter into most tissue have been subjected to multiple scattering. These photons will thus reach any molecule from all possible directions.

The presence of scattering thus has the following three impacts upon the light distribution; an increased reflection, an enhanced absorption and a more isotropic light distribution in the region distal to the surface.

The quantity of light may be characterized by the radiance L which is defined as the flux of optical energy in a particular direction per unit solid angle and per unit area oriented normal to the direction $\vec{\ell}$ of propagation. The energy flux which reaches an infinitesimally small surface element dA oriented normal to the unit vector $\vec{\ell}$ from rays within the infinitesimally small solid angle $d\Omega$ is $LdAd\Omega$.

The next quantity to be defined is the irradiance E. This quantity is defined as the flux of energy per unit area incident onto one side of a plane area. The irradiance E onto the infinitesimally small element dA is thus

$$E = \int_{\Omega=0}^{\Omega=2\pi} L\vec{\ell}\cdot\vec{n}\, d\Omega \tag{1}$$

where $\vec{n}$ is the unit surface normal.

The quantity of light flux which is incident upon a spherical molecule may be expressed by the space irradiance

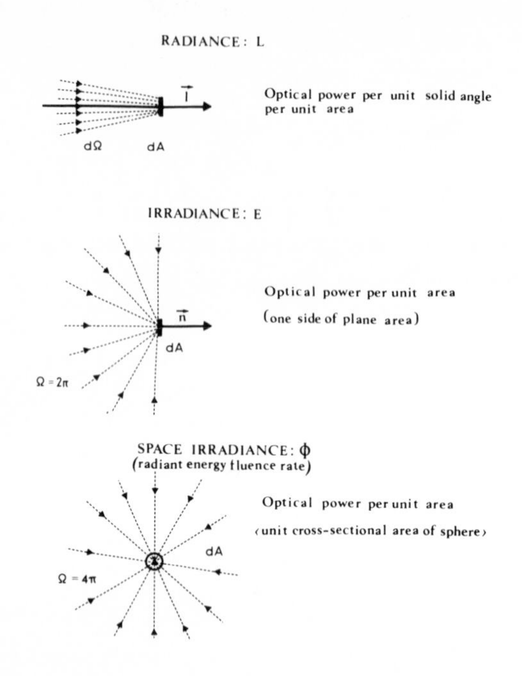

Fig. 2. Quantities of radiance.

ϕ. This quantity is defined as the flux of energy onto an infinitesimally small sphere divided by the cross-sectional area of that sphere. The space irradiance corresponds to an integral of the radiance over all solid angles.

$$\phi = \int_{\Omega=0}^{4\pi} L d\Omega \tag{2}$$

The total number N of photons which is radiated per unit time onto a spherical molecule with a cross-sectional area A is

$$N = \frac{\phi A}{h\nu} \tag{3}$$

where h is Planck's constant and ν is the optical frequency.

An exact isotropic light distribution is defined by a constant radiance in all directions. The relation between the irradiance E, the space irradiance ϕ and the radiance L may then be expressed

$$\phi = 4\pi L = 4E \tag{4}$$

The distribution of light arising from a collimated beam radiated onto a semi-infinite tissue is shown in Fig. 3.

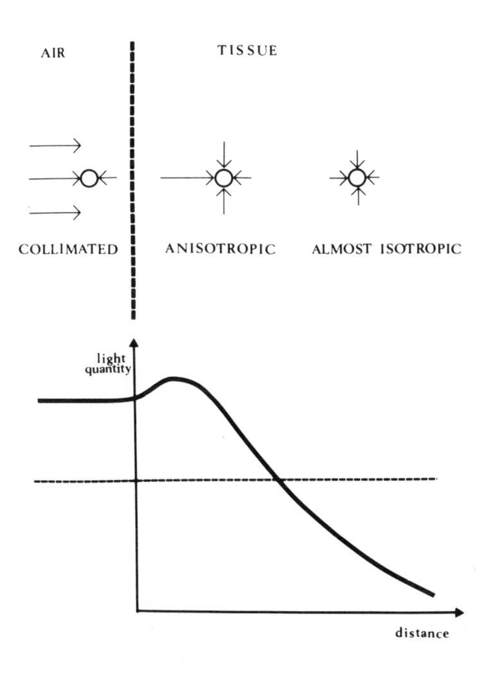

Fig. 3. Distribution of light in tissue.

The horizontal broken line corresponds to the light quantity falling onto a molecule in the absence of the tissue. The reflection of light from the tissue increases the space irradiance in the air as shown by the solid curve. The radiance in the region proximal to the surface is highly anisotropic. The space irradiance in this region might even be higher than in the air (Profio, Doiron 1981). The radiance in the region distal to the surface is almost isotropic and the space irradiance decays exponentially with distance into the tissue. The radiance in this region has a small anisotropy which arises from the net transport of optical power into the tissue.

The radiance may thus be expressed (Ishimaru 1978) (see Fig. 4)

$$L = L_m + \frac{3}{4\pi} \vec{j} \cdot \vec{\ell} \tag{5}$$

where L_m is the average radiance and $\vec{j}$ is the diffuse energy flux vector. The last term in this expression is much smaller than the average radiance; but since it represents the net flux of diffusing energy it cannot be put equal to zero. The net flux of energy flows from regions with high space irradiance and to regions with smaller space irradiance. The diffuse energy flux vector will be proportional to the decrease of space irradiance per unit length, and it will be directed in the direction of maximum decrease. The flux vector may thus be written

$$\vec{j} = -\zeta \mathrm{grad}\phi \tag{6}$$

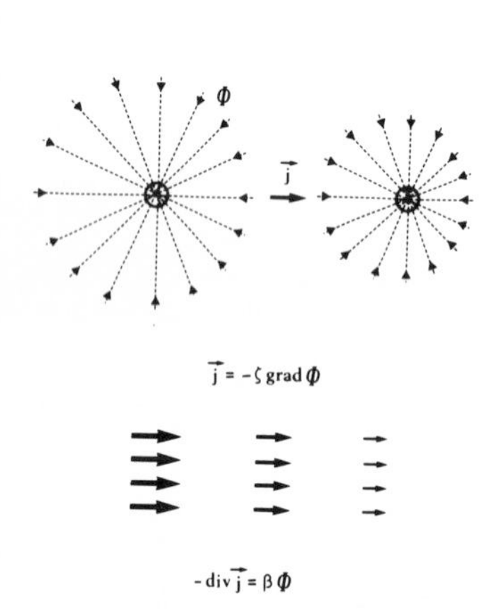

Fig. 4. Space irradiance and the diffuse energy flux vector.

where $-\mathrm{grad}\phi$ is a vector directed in the direction of maximum decrease in space irradiance. The magnitude of this vector is equal to the maximum decrease of space irradiance per unit length. The coefficient of proportionality ζ, which will be referred to as the diffusion coefficient, is a function of the scattering properties of the tissue.

The optical power absorbed per unit volume must be transported into the element of volume by the diffuse energy flux vector. The energy balance may be expressed

$$-\mathrm{div}\,\vec{j} = \beta\phi \tag{7}$$

where $-\mathrm{div}\,\vec{j}$ is the influx of energy per unit volume and $\beta\phi$ is the absorbed optical power per unit volume. The absorption coefficient β is determined by the concentration of absorbing molecules, i.e. hemoglobin, myoglobin, melanin etc.

Substitution of Eq. (6) into Eq. (7) gives

$$\mathrm{divgrad}\,\phi - \frac{\phi}{\delta^2} = 0 \tag{8}$$

where

$$\delta = \sqrt{\frac{\zeta}{\beta}} \tag{9}$$

The parameter δ which is a function of both the scattering and the absorption properties, will be referred to as the optical penetration depth. The distribution of light in the region distal to the surface may thus be expressed by a homogeneous diffusion equation (Eq. (8)). This equation which has been derived by a heuristic approach may also be expressed through a series expansion of the rigorous transport equations (Ishimaru 1978; Morse, Feshbach 1953).

COLLIMATED BEAM EXCITATION

The photoactivating light may be radiated into a subcutaneous tumor or a bronchial tumor by direct illumination of the skin or the bronchial mucosa (Kato et al 1983; Kennedy 1983). The distribution of light in tissue for these cases may be approximated by the one-dimensional solution of Eq. (8). This solution may be expressed

$$\phi = \phi_{op} e^{-\frac{x-\delta}{\delta}} \tag{10}$$

where x is the distance from the surface. The parameter ϕ_{op} is the space irradiance at a distance into the tissue equal to the penetration depth. This expression does not relate the optical distribution in the regions distal to the surface to the incident radiation. The expression in Eq. (10) might of course be extrapolated to the surface. But since the validity of the diffusion theory may be limited in the proximal region, this relation might better be expressed by a semi-empirical expression. The space irradiance may thus be written

$$\phi_{op} = k_p I \tag{11}$$

where I is the optical power density of the incident beam and k_p is a dimensionless coupling coefficient. This coupling coefficient will in principle be dependent on the tissue properties, the condition of the surface and the collimation of the incident beam. The coefficient should be determined experimentally for each particular tissue. The details of the light distribution in the region close to the surface are of course lost by this type of approach. But since the light is scattered to an almost isotropic radiation within a few tenths of a millimeter from the surface for most tissues, this approach will give an adequate description of the light distribution in the region relevant to photoradiation therapy, i.e. in the region several millimeters from the surface (see Fig. 5).

INTERSTITIAL PRT

Large localized tumors may be treated with interstitial PRT (Dougherty et al 1981). The light is here coupled into the tissue through inserted optical fibers.

These fibers usually consist of a fused silica core surrounded by a silicone cladding with a lower index of refraction. The light rays are guided in the fiber through total reflection at the interface between the core and the cladding (see Fig. 6). The numerical aperture NA of a fiber is defined by NA = sinφ where φ is the maximum acceptable angle between a ray entering the fiber from air and the fiber axis. Rays which enter the fiber at larger angles are not subjected to total reflection. These rays are transmitted into the cladding and the energy is rapidly absorbed in the cladding material or in the outer protective jacket of the fiber. A typical value for the numerical fiber is NA = 0.2 corresponding to an angle $\varphi = 11.5^{\circ}$.

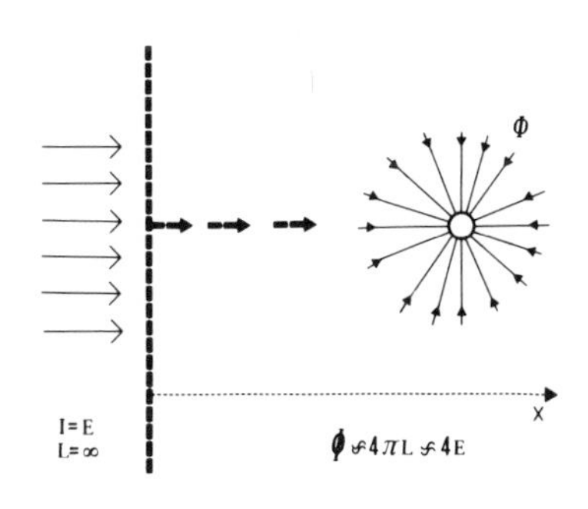

Fig. 5. Photoactivation with a collimated beam. Almost isotropic light distal to surface.

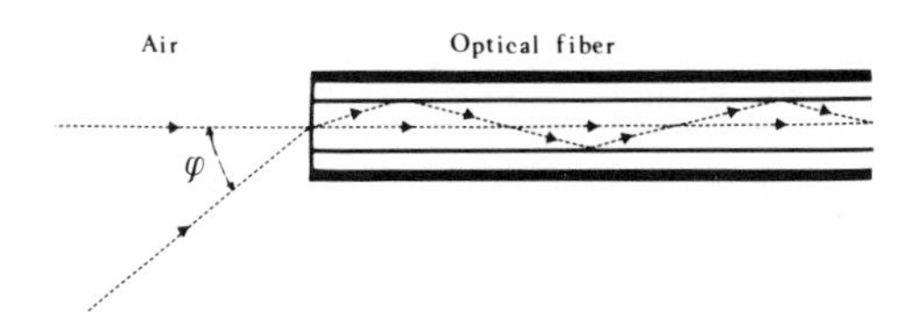

Fig. 6. Guiding of rays in an optical fiber.

The coupling of light into the tissue is shown in Fig. 7. The rays will all be coupled out within a small solid angle Ω_f given by

$$\Omega_f = \left(\frac{NA}{n}\right)^2 \pi \qquad (12)$$

where n is the index of refraction of the tissue. The radiance along the fiber axis at the interface between fiber and tissue is thus

$$L = \frac{P}{\Omega_f A_f} = \frac{P}{\left(\frac{NA}{n}\right)^2 \pi A_f} \qquad (13)$$

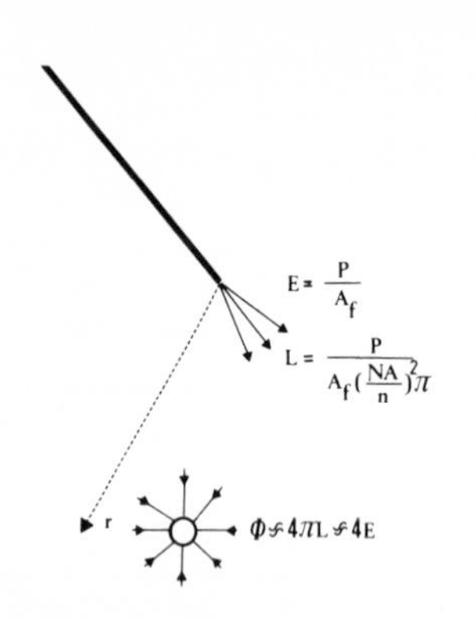

Fig. 7. Light distribution during interstitial photoradiation therapy. Single fiber.

where P is the total radiant power from the fiber and A_f is the cross-sectional area of the fiber core. This high radiance beam is rapidly scattered into an almost isotropic and spherically symmetric distribution (Doiron et al 1983; Svaasand et al, in press). The irradiance may thus be expressed as a solution of Eq. (8) for the spherically symmetric case

$$\phi = \phi_{os} \left(\frac{\delta}{r}\right) e^{-\frac{r-\delta}{\delta}} \qquad (14)$$

where r is the distance from the fiber end.

The parameter ϕ_{os} may be interpreted as the space irradiance at a distance equal to the optical penetration depth. The relation between this space irradiance and the total radiant power P may be expressed by a coupling coefficient k_s. This coefficient may be defined by

$$\phi_{os} = k_s \frac{P}{4\pi\delta^2} \qquad (15)$$

This particular form of the semi-empirical relation defines k_s as a dimensionless quantity. This coupling coefficient is dependent on both the properties of the tissue and the fiber.

The radiant power which may be coupled out of the fiber core with typical radii 200 µm or 400 µm is, for thermal reasons, rather limited. It is therefore advantageous to reduce the irradiance at the interface by increasing the size of the interface. A possible solution to this problem is to use a cylindrical applicator as shown in Fig. 8. A length ℓ of the fiber core is here covered with a nonabsorbing material which scatters the light almost isotropically. The radiance at the interface between the applicator and the tissue is now almost isotropic. Thus L becomes

$$L = \frac{P}{\pi A_c} \tag{16}$$

where A_c is the surface area of the applicator.

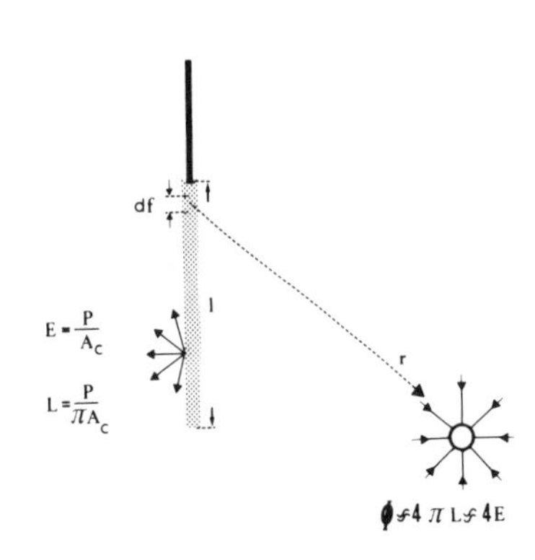

Fig. 8. Light distribution during interstitial photoradiation therapy. Cylindrical applicator.

The light distribution in the tissue may now be evaluated by considering each element of length df of the applicator as a spherically symmetrical emitter. The space irradiance may thus be expressed

$$\phi = k_{cf} \frac{P}{4\pi\delta^2 \ell} \int_0^{\ell} \left(\frac{\delta}{r}\right) e^{-\frac{r-\delta}{\delta}} df \tag{17}$$

where r is the distance from each element df and to the position in the tissue where the space irradiance is evaluated. The semi-empirical dimensionless coupling coefficient for this applicator is k_{cf}.

This coupling coefficient is thus dependent on both the properties of the tissue and the properties of the applicator. The coefficient should therefore in principle be calibrated for each applicator design and each relevant tissue.

EXPERIMENTAL RESULTS. *IN VITRO*

The light distribution within the tissue was measured with an inserted detector fiber. This fiber could be positioned at each point in the tissue from all directions. A fiber with a small numerical aperture will measure the radiance L in the direction along the fiber axes. The light energy flux P_f coupled into guided rays is then

$$P_f = L\Omega_f A_f = L\left(\frac{NA}{n}\right)^2 \pi A_f \tag{18}$$

The light coupled into the fiber in the regions with almost isotropic irradiance may thus be expressed (Fig. 9)

$$P_f = \phi \frac{\Omega_f}{4\pi} A_f \tag{19}$$

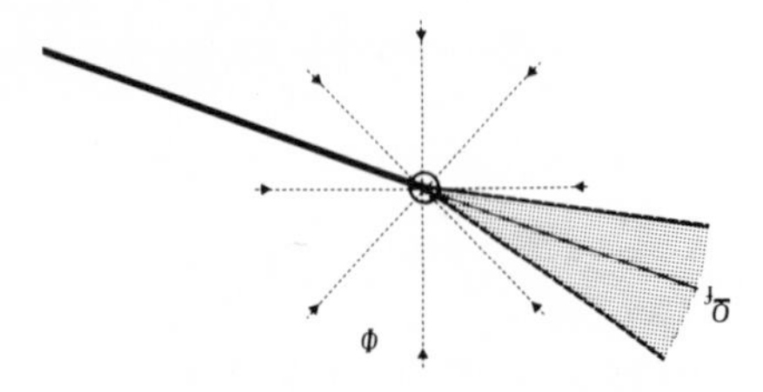

Fig. 9. Detection of light distribution in tissue.

Coupling of radiant power into the tissue and the detection of the space irradiance were both done with identical fibers* mounted into the lumen of hypodermic needles. The fiber core diameter was 200 μm and the length of the detector fiber was 1 mm. The effective numerical aperture for a fiber of this length was measured to be NA = 0.37, and the maximum light acceptance angle φ in a tissue with index of refraction n = 1.4 is thus $\varphi = \arcsin(\frac{NA}{n}) = 15.3^o$.

The space irradiance in normal neonatal human brain tissue is shown in Fig. 10. The brain was from a 14 day old fullborn child (male) and the measurements were carried out

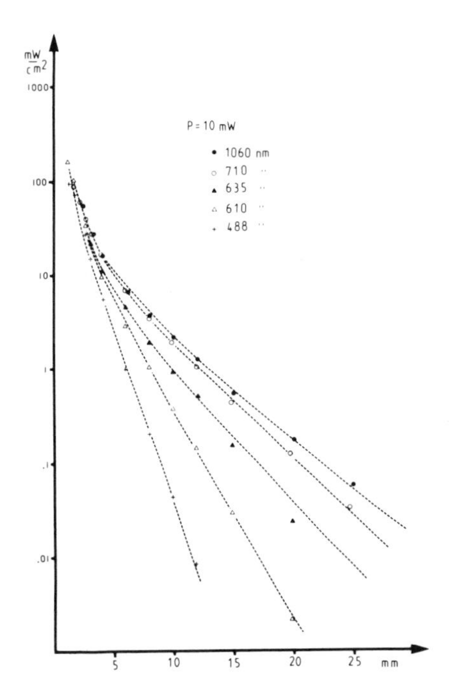

Fig. 10. Space irradiance versus distance from fiber end. Neonatal human brain (14 day old male). Radiant power P = 10 mW.

immediately after autopsy (2 days post mortem). The radiant optical power was 10 mW and the light distribution was measured at several wavelengths.

*Quartz et Silice S.A.

The broken lines in Fig. 10 correspond to the best fit between the experimental results and the functional relationship from Eq. (14). The optical penetration depth varies from δ = 1.3 mm for λ = 488 nm wavelength, δ = 4.0 mm for λ = 635 nm and to δ = 5.3 mm for λ = 1060 nm. The coupling coefficient k_s varied somewhat for different insertions. Values for wavelengths in the red part of the spectrum was in the range k_s = 2.0-3.2.

Corresponding results for normal adult human brain tissue are shown in Fig. 11. The brain was from a 68 year old male. The broken lines correspond again to the best fit

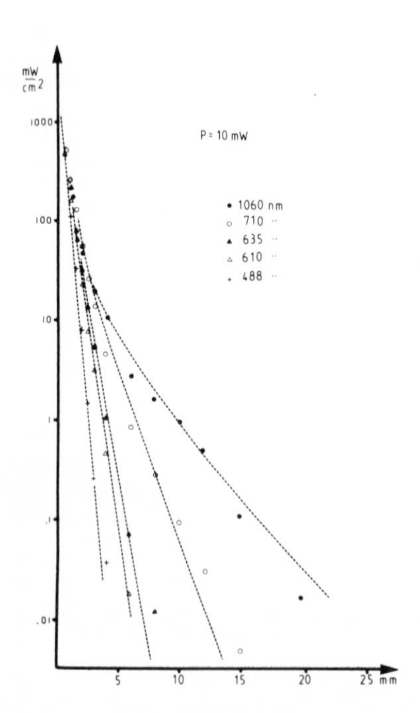

Fig. 11. Space irradiance versus distance from fiber end. Adult human brain (68 year old male). Radiant power P = 10 mW.

to the functional relationship from Eq. (14). The fit between the experimental values and the theory is here not as good as in the case of the neonatal brain. This is due to the more inhomogeneous structure of the adult brain. The optical penetration depth varies here from $\delta = 0.45$ mm for $\lambda = 488$ nm wavelength, $\delta = 0.8$ mm for $\lambda = 635$ nm and to $\delta = 3.7$ mm for $\lambda = 1060$ nm. The reduced penetration depth in adult human brain compared to that of the neonatal brain is due to the increased scattering in the fully myelinated brain (Svaasand, Ellingsen in press). The results for the adult and the neonatal brain are summarized in Table I.

Color	Wavelength nm	Penetration depth (mm)	
		68 year old male	14 day old male
Blue	488	0.45	1.3
Green	514	0.50	1.25
Red	610	0.65	2.45
Red	635	0.80	4.0
Red	710	1.5	4.7
Infrared	1060	3.7	5.3

Table I. Optical penetration depth in human brain tissue.

The coupling coefficient k_s for the wavelength in the red part of the spectrum was in the range $k_s = 3.0\text{-}3.1$. There was not found any significant difference between the coupling coefficient for the adult and the neonatal brain.

The coupling coefficient was however significantly reduced if the radiant power was sufficiently high to introduce carbonization at the fiber tip. Carbonization was introduced for radiant power levels down to $P = 50$ mW of blue or green light. But carbonization was never observed for levels of red light below $P = 100$ mW.

The coupling coefficient k_p (Eq. (11)) for collimated beam irradiation was found to be about $k_p = 0.8\text{-}1.0$ for both brains in the red part of the spectrum.

EXPERIMENTAL RESULTS. *IN VIVO*

The *in vivo* experiments were carried out on subcutaneously introduced Lewis Lung tumors in B_6D_2 mice. The tumor size became typically 5-15 mm in diameter within 7-14 days. The coupling coefficient k_s for an inserted 200 µm core diameter fiber was monitored versus time for different radiant power levels at λ = 635 nm wavelength. The results are shown in Figs. 12 and 13.

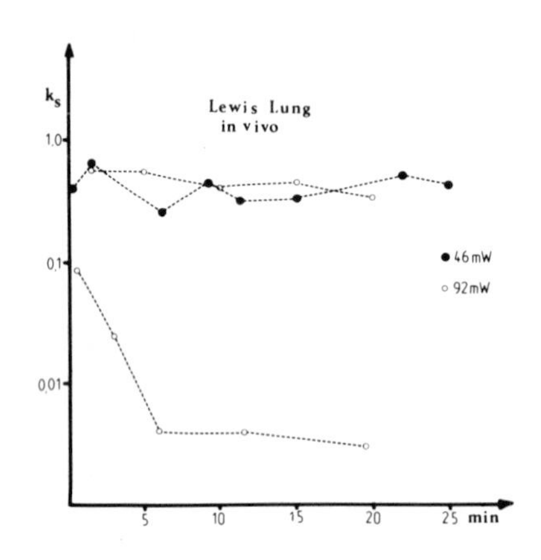

Fig. 12. Coupling coefficient k_s versus time for different levels of radiant power. Lewis Lung tumor in mice *in vivo*. Wavelength λ = 635 nm. Fiber core diameter 200 µm.

The coupling coefficient k_s for red light (λ = 635 nm) is typically in the region k_s = 0.3 - 0.5. This value is, however, only repeatedly observed for power levels less than about 20-30 mW.

The filled circles in Fig. 12 correspond to this case. The slight variation in the coefficient versus time is due to movements of the animal. If the power is increased to about 100 mW, one of two cases occurs; the coupling coefficient is either constant or the coefficient decreases with time down to a level two to three orders of magnitude below the initial a value. These cases are shown by the upper and lower open circles in Fig. 12, respectively.

The decrease in coupling coefficient is due to formation

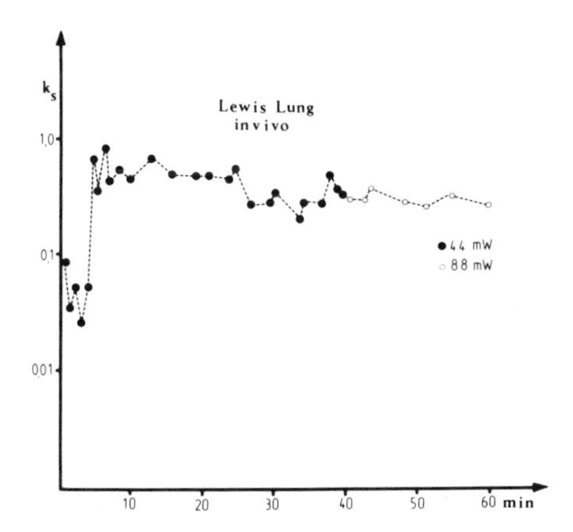

Fig. 13. Coupling coefficient k_S versus time for different levels of radiant power. Lewis Lung tumor in mice *in vivo*. Wavelength λ = 635 nm. Fiber core diameter 200 µm.

of a blackish, partly carbonized layer of clot at the fiber end. If the power was below 50 mW this layer might occasionally be cleared off the fiber end due to the movements of the animal. This case is shown in Fig. 13. The filled circles represent the coupling coefficients for 44 mW; the coefficient drops down during the first 5 minutes after the onset of power. But small displacements of the fiber due to motion of the animal result in an increase in the coefficient up to k_S = 0.3-0.5. An increase in power up to 88 mW after about 40 minutes did not change this value. This is shown by the open circles in Fig. 13.

The coupling coefficient was also measured after sacrifice of the animals. The value of k_S for λ = 635 nm wavelength was now found to be k_S = 1.8, i.e. close to the value for the brain of human cadavers.

Both the coupling coefficient and the threshold power for carbonization were found to be significantly higher for the *in vitro* case than for the *in vivo* case. The major reason for this is presumably the presence of blood flow. Seepage of blood into the lesion introduced by the hypodermic needle results in an accumulation of blood proximal to the fiber end surface. A minor temperature rise of this blood results in the formation of a blackish blood clot. The presence of this clot increases the optical absorption and this further enhances the local temperature rise.

The optical penetration depth in the mice tumor model was, however, not significantly different for the *in vivo* and the *in vitro* case. The penetration depth for λ = 635 nm wavelength was somewhat dependent on the tumor necrosis, and typically in the range δ = 1.1 - 1.4 nm.

CYLINDRICAL APPLICATORS

Measurement of the light distribution from cylindrical applicators was performed with a commercial applicator*. The emissive region of this applicator was 35 mm long and the diameter was 1 mm. The results are shown in Fig. 14.

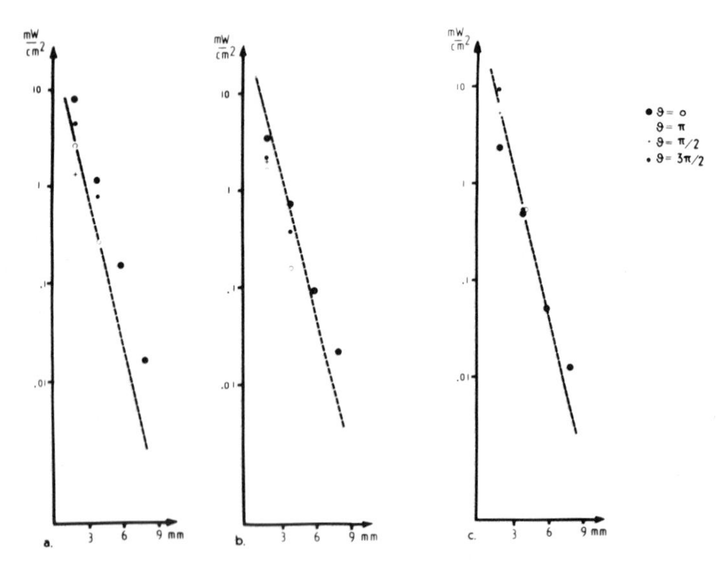

Fig. 14. Space irradiance versus distance from cylinder axis.
a) Axial coordinate z = ℓ (z = 35 mm) (applicator end)
b) Axial coordinate z = $\frac{3\ell}{4}$ (z = 27 mm)
c) Axial coordinate z = $\frac{\ell}{2}$ (z = 17 mm)
Wavelength 635 nm, total optical power P = 25 mW, human brain tissue, white matter. 68 year old male.

*Oncology Research and Development Inc.

The space irradiance is here shown as a function of distance from the applicator axis for three different positions along this axis. Fig. 14 a) shows the space irradiance in a plane normal to the axis at the end of the applicator ($z = \ell$), and Figs. 14 b) and 14 c) show the corresponding irradiance at $z = \frac{3\ell}{4}$ and $z = \frac{\ell}{2}$. The irradiance is measured in four orthogonal directions in each plane: $\theta = 0$, $\theta = \frac{\pi}{2}$, $\theta = \pi$, $\theta = \frac{3\pi}{2}$.

These values are measured in the white matter of the 68 year old adult male brain. A typical value for the penetration depth of this brain was $\delta = 0.8 - 1.0$ mm at wavelength $\lambda = 635$ nm (see Fig. 11).

The coupling coefficient k_{cf} has been determined from the experimental value $\phi = 0.5$ mW/cm^2 at position r = 4 mm, z = 17 mm, $\theta = 0$ (Fig. 14 c). The total power coupled into the fiber was determined to be P = 25 mW. The coupling coefficient is then determined from Eq. (17) upon substitution of $\delta = 0.9$ mm. The value is determined to be $k_{cf} = 2.3$.

The corresponding value for the coupling coefficient determined from the measured value $\phi = 0.04$ mW/cm^2 at position r = 6 mm, z = 17 mm, $\theta = 0$ is $k_{cf} = 2.2$ (Fig. 14 c). The optical power transmitted by the fiber was determined by cutting the fiber in front of the cylindrical applicator and measuring the transmitted power.

Since the power P thus was determined as the total power transmitted by the fiber rather than the power which is coupled out of the applicator, this value for the coupling coefficient also includes any possible optical loss in the applicator itself. This value for the coupling coefficient may thus be interpreted as the effective coupling coefficient.

The broken lines in Fig. 14 represent the space irradiance predicted from numerical solutions of Eq. (17) with the parameters: P = 25 mW, $k_{cf} = 2.2$, $\ell = 35$ mm and $\delta = 0.9$ mm.

These results show good correspondence with the measured values in Figs. 14 b) and 14 c). The observed values in Fig. 14 a) show a tendency to be above the predicted value. The reason for this is that the optical power was coupled out slightly inhomogeneously along the cylinder and there was a tendency towards a somewhat higher radiation from the region close to the end of the applicator.

TEMPERATURE RISE

The temperature rise proximal to the applicator surface is, for a given optical power, considerably smaller for the cylindrical applicator than for the single fiber applicator. The temperature rise at the surface of a cylindrical applicator will in principle be between two limits; an upper limit determined by a "worst case" approach where the total optical power P is absorbed at the surface of the applicator, and a lower limit where the radiant power is absorbed within the tissue in a region corresponding to the optical penetration depth (see Fig. 15 a)). The temperature rise ΔT_{cf} may be expressed approximately

$$\frac{P}{2\pi\kappa\ell} \ln \frac{\delta_V}{b} > \Delta T_{cf} > \frac{P}{2\pi\kappa\ell} \ln \frac{\delta_V}{b+\delta} \qquad (20)$$

where κ is the thermal conductivity of the tissue, b and ℓ are the applicator radius and length, respectively, and δ_V is the thermal penetration depth. The thermal penetration depth is defined by

$$\delta_V = \sqrt{\frac{\chi}{Q}} \qquad (21)$$

where χ is the thermal diffusivity of the tissue and Q is the volume of perfused blood per unit volume of tissue per second. The contribution to the cooling from the blood flow is insignificant if the linear dimensions of the applicator are smaller than the thermal penetration depth.

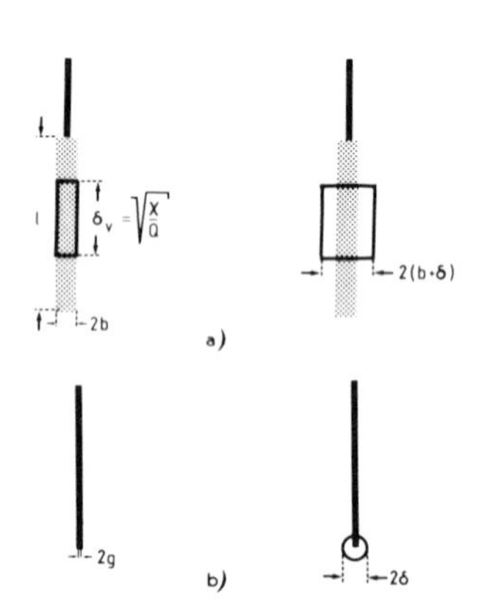

Fig. 15. Regions of optical absorption resulting in temperature rise proximal to applicator surface. a) cylindrical applicator, b) single fiber applicator.

This depth is $\delta_V = 5$ mm for a tissue with medium blood flow and $\delta_V = 10$ mm for a tissue with low blood flow (Svaasand 1982). If the length of the cylinder ℓ is smaller than the thermal penetration depth, ℓ should be substituted for δ_V in Eq. (20). The maximum temperature rise per unit power for the applicator with $\ell = 35$ mm and $b = 0.5$ mm inserted in a tissue with thermal conductivity $\kappa = 0.4$ w/mK and low blood flow ($\delta_V = 10$ mm) is thus $\Delta T_{cf}/P = 34$ K/W.

The corresponding minimum temperature rise per unit power of red light in a tissue with an optical penetration depth equal to that of the adult human brain ($\delta = 0.8$ mm) is $\Delta T_{cf}/P = 23$ K/W.

The difference between the maximum and minimum temperature rise proximal to the applicator surface is thus rather small for the cylindrical applicator. A total radiated optical power of $P = 500$ mW will result in an expected temperature rise in the region 10-15 K.

The cylindrical applicator was, because of its large physical dimensions, not calibrated in the mice tumor model. The predicted temperature rise was, however, confirmed by pressing the applicator between two fingers. The steady state temperature at the applicator surface for $P = 600$ mW radiant power at $\lambda = 514$ nm wavelength was now 46°C, corresponding to a temperature rise $\Delta T_{cf} = 10$ K.

The temperature rise ΔT_s proximal to the surface of a single fiber applicator may be expressed

$$\frac{P}{2\pi\kappa g} > \Delta T_s > \frac{P}{4\pi\kappa\delta} \tag{22}$$

where g is the radius of the fiber core. The maximum temperature rise corresponds to the case where the optical power is completely absorbed at the end-surface of the fiber core. The minimum temperature rise corresponds to the case where the power is absorbed within a sphere with radius equal to the optical penetration depth. The blood flow will have negligible influence in this case because the optical penetration depth and the fiber core radius both are much smaller than the thermal penetration depth. The maximum temperature rise per unit power for a fiber with 200 μm core diameter is thus $\Delta T_S/P = 3981$ K/W. The minimum temperature rise per unit power for red light in the adult brain is $\Delta T_S/P = 248$ K/W. There is thus a significant difference between the maximum and the minimum temperature rise for the single fiber applicator.

The minimum temperature rise for 100 mW radiant power of red light from a single fiber inserted in the Lewis Lung mice tumor model will thus be in the range $\Delta T_S = 14.2 - 18$ K ($\delta = 1.1 - 1.4$ mm).

A temperature rise of this order of magnitude easily explains the observed formation of blood clots proximal to the fiber in the *in vivo* case. This clot formation enhances the absorption of light and the temperature rise might thus be in accordance with the upper limit in Eq. (22). A radiant power of 20 mW from a 200 μm core fiber may in this limit easily bring the temperature up to carbonization levels.

CONCLUSIONS

The attenuation of light propagating in tissue is determined by scattering and absorption. The efficiency for excitation of a drug molecule in a collimated beam of red light ($\lambda = 635$ mm wavelength) will be reduced to 10% at a location between 2.6 mm to 13.2 mm from the tissue surface. The lower value is valid for the adult human brain with optical penetration depth $\delta = 0.8$ mm and the upper value is valid for the neonatal human brain ($\delta = 4$ mm). The coupling coefficient is taken to be $k_p = 1.0$.

The required power density of an incident optical beam for administering a typical therapeutic dose of $10 J/cm^2$ at these depths during 30 min. exposure is 56 mW/cm^2.

The excitation efficiency is further reduced to 1% at 4.5 mm and 22.4 mm for the adult and the neonatal brain, respectively.

The radiant optical power of red light which *in vivo* can be coupled interstitially into the tissue through a 200 μm core optical fiber is about 50 mW. A higher radiant power increases the probability for suppression of the light due to clot formation and carbonization. A radiant power of

50 mW will administer an optical dose of 10J/cm^2 during 30 min. exposure for distances from the fiber end up to 3 mm for the adult brain and up to 5.8 mm for the neonatal brain, if the coupling coefficient is $k_s = 0.5$.

The corresponding safe radiant optical power from a 35 mm long cylindrical applicator is about 500 mW. An optical dose of 10J/cm^2 during 30 min. exposure will be obtained at distances up to 3-4 mm from the cylinder axis for the adult brain if the coupling coefficient is $k_{cf} = 0.5$.

The corresponding distance for the neonatal brain is 9.5 - 14 mm if $k_{cf} = 0.5$. The optical penetration depth for most normal and neoplastic tissues has a magnitude which is in the range between the values for the adult and the neonatal human brain (Doiron et al 1983; Svaasand et al in press).

The relatively large variations in optical penetration depth for various tissues indicate that a precise dosimetry in photoradiation therapy must be based on detailed simultaneous knowledge of the drug concentration, the photosensitization and the optical properties of each particular tissue.

ACKNOWLEDGEMENTS

The author wishes to thank R. Ellingsen, D.R. Doiron, A.E. Profio and T.J. Dougherty for cooperation and stimulating discussions. The work has been supported by a NATO Research Grant no. 0342 under NATO Science Programmes. The author is thankful to B. Reitan for competent typing.

REFERENCES

Doiron DR, Svaasand LO, Profio AE (1983). Light dosimetry in tissue; application to photoradiation therapy. In Porphyrin Photosensitization: Plenum Press, pp. 63-76.

Dougherty TJ, Thoma RE, Boyle DG, Weishaupt KR (1981). Cancer Research 41, 401.

Ishimaru A (1978). In Wave Propagation and Scattering in Random Media: Academic Press, pp. 175-190.

Kato H, Konaka C, Ono J, Matsushima Y, Nishimiya K, Lay J, Sawa H, Shinohara H, Saito T, Kinoshita K, Tomono T, Aider M, Hayata Y (1983). Effectivenes of HpD radiation therapy in lung cancer. In Porphyrin Photosensitization: Plenum Press, pp. 23-40.

Kennedy J (1983). HpD photoradiation therapy for cancer at Kingston and Hamilton. In Porphyrin Photosensitization: Plenum Press, pp. 53-77.

Morse P, Feshbach H (1953). In Methods of Theoretical Physics: McGraw-Hill Book Co. Inc., pp. 171-200.

Profio AE, Doiron DR (1981). Med. Phys. 8, 190.

Smith RS, Stein MN (1968). Am. Journ. of Opthalmology, p. 21, July.

Svaasand LO, Ellingsen R (in press). Optical properties of human brain. Photochem. Photobiol.

Svaasand LO, Doiron DR, Dougherty TJ (1983). Med. Phys. 10, 17.

Svaasand LO (1982). Med. Phys. 9, 711.

Porphyrin Localization and Treatment of Tumors, pages 115–132

LIGHT PROPAGATION IN ANIMAL TISSUES IN THE WAVELENGTH RANGE 375-825 NANOMETERS

Brian C. Wilson, W. Patrick Jeeves,
Diane M. Lowe, and Gabriel Adam*
Ontario Cancer Foundation and
McMaster University
Hamilton, Ontario, Canada

INTRODUCTION

In contrast with, for example, the extensive body of knowledge on the dosimetry of ionizing radiation, the propagation of light in tissues has been little studied. An exception to this has been in the investigation of the optical properties of skin (Wan *et al* 1981a, 1981b; Anderson, Parrish 1982; Bruls, van der Leun 1982; Dubertret *et al* 1982; Parrish 1982, 1983), especially in the UV region, where a variety of therapeutic irradiation techniques have been applied. However, the current effort to implement PhotoRadiation Therapy (PRT) for the treatment of localized malignancy, both superficial and deep-seated (Cortese, Kinsey 1982; Dahlman *et al* 1983; Hayata *et al* 1983; Kessel, Dougherty 1983), provides strong motivation to improve our understanding of the optical properties of tissues throughout the visible wavelength range, in order to optimize and refine the clinical techniques.

In the most extensive published work to date (Svaasand *et al* 1981; Doiron *et al* 1983), the attenuation of light has been measured in a number of different animal and human tissues at selected wavelengths. This work, and that of other groups (Eichler *et al* 1977; Wan *et al* 1981a, 1981b; Preuss *et al* 1983) indicates that

A. the optical properties (absorption, scattering, reflectance) of mammalian tissues vary widely from tissue to tissue, and may vary within a single tissue

B. these properties are strongly wavelength-dependent in the visible range

C. considerable changes may occur between the *in-vivo* and the *post-mortem* characteristics

D. interfaces, boundaries and inhomogenities are substantial factors affecting the distribution of light in different irradiation sites.

However, these data are fragmentary, and the phenomenological and theoretical description of light propagation is, as yet, poorly developed, It has, therefore, been our aim to construct an experimental system to measure the distribution of light in different tissues *in-vivo* (or *post-mortem*), over the full visible wavelength spectrum. We will describe this apparatus here, and present preliminary experimental data to illustrate the methodology, and to provide pointers for future work. Other studies, using Monte Carlo computer modelling (Wilson, Adam 1983), and measurements in tissue-simulating phantoms, are in progress to gain an understanding of the distribution of light in media of known optical absorption and scattering.

MATERIALS AND METHODS

The use of an optical fiber coupled to a light sensitive detector has been demonstrated to provide a convenient method of measuring the distribution of light flux at selectable points within tissues (Svaasand *et al* 1981; Doiron *et al* 1983). Since an optical fiber has a limited acceptance angle (numerical aperture), the directionality of the flux may also be investigated by changing the orientation of the fiber tip with respect to the direction of the light source. The fundamental optical properties of tissues may be determined from such measurements by applying analytic or numerical models for the light propagation. Such use of single optical fibers as probes of the light distribution within irradiated tissues is the basis for the present system, a block diagram of which is shown in Figure 1.

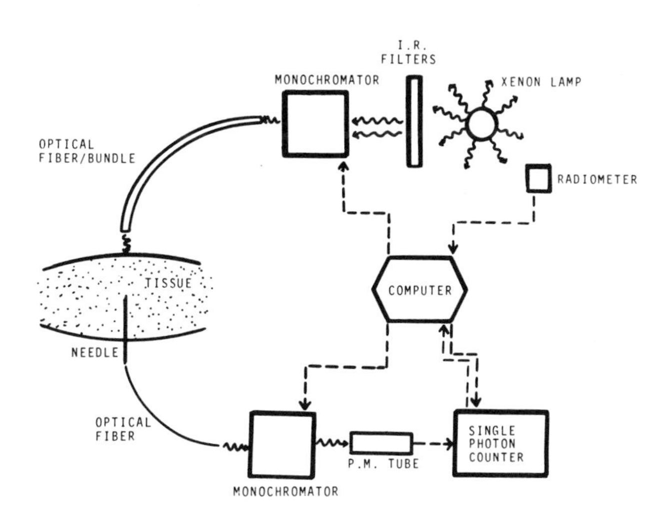

Fig. 1. Block diagram of the apparatus used to measure the photon flux at depth in tissue as a function of wavelength. (⇝ represents photon, - - → represents electrical signal)

Equipment

A 1000 Watt xenon arc lamp, filtered to remove the infra-red ($\geq$ 900 nm) is used as a broad-band source. A diffraction grating monochromator, driven by a stepping motor interfaced to a microcomputer, provides monochromatic light of selectable wavelength focussed into an optical fiber, or fiber bundle. Thus, the tissue may be irradiated, either by inserting into it the single fiber, or by using the fiber bundle to provide, via a lens collimator, an external plane light beam of variable diameter.

The detector fiber(s) (silica, 400 μm core diameter, 0.4 numerical aperture) is mounted within a rigid, 18-gauge biopsy needle fixed to a stereotaxic frame. This allows accurate placement of the fiber tip at any position and orientation with respect to the light beam. The fiber is optically coupled, via a second computer-controlled grating monochromator, to a cooled photomultiplier tube. This, in turn, is connected to a 100 MHz single photon counting system (SPC), comprising pre-amplifier, amplifier/discriminator, and counter/timer. The SPC is also bi-directionally interfaced to the computer so that its operation is under program control, and readings (counts or time) are stored on magnetic disc. Both "source" and "detector" monochromators can be driven independently, or synchronously at the same wavelength.

As presently configured, the system has a useful wavelength range of 375-825 nm and wavelength resolution (FWHM) of 6 nm (synchronous operation). Finally, the intensity of the xenon lamp is monitored by a solid-state radiometer, to allow correction for variations with time in light output.

Animals

For the muscle and liver measurements, New Zealand White rabbits of either sex, weighing between 1.5 and 3.5 kg were used. General anesthesia was induced and maintained with sodium pentobarbital (65 mg/cm^3),and lidocaine hydrochloride was used for additional local anesthesia as necessary.

For muscle, one hind leg was clamped at the foot and held horizontally with the rabbit prone. An area of skin on the upper lateral thigh was removed, exposing the muscle surface, which was then illuminated by the collimated beam from the source fiber bundle. The detector fibers were inserted at the required depths and orientations, after slitting the skin at the entry point to reduce resistance to the fiber/needle tip.

For liver, an abdominal incision with the rabbit supine allowed one lobe to be lifted from the abdominal cavity and gently clamped in the vertical plane. The underside of the lobe was the illuminated surface and, as with muscle, the exposed surfaces were kept moist throughout with frequent applications of cell culture medium.

To sacrifice the animal, an additional bolus of 1-2 cm^3 sodium pentobarbital was injected; to prevent muscle spasm, 0.5-1.0 cm^3 gallamine (20 mg/cm^3) was administered just prior to the fatal anesthetic dose.

Experiments

Most of the results below are for the geometry shown in Figure 2A, with the detector fiber in the forward orientation along the central axis of the light beam. Unless otherwise stated, the beam diameter, B, was 15 mm at the tissue surface, and the source and detector monochromators were run at the same wavelength. Thus, the detected counts, normalized at each wavelength to the relative incident light intensity (counts in air), are a measure of the attenuation of the light beam with increasing depth in tissue along the optical axis. We will also show some data for measurements at 90° to this direction for the sideways flux, and at 180° for the backward flux.

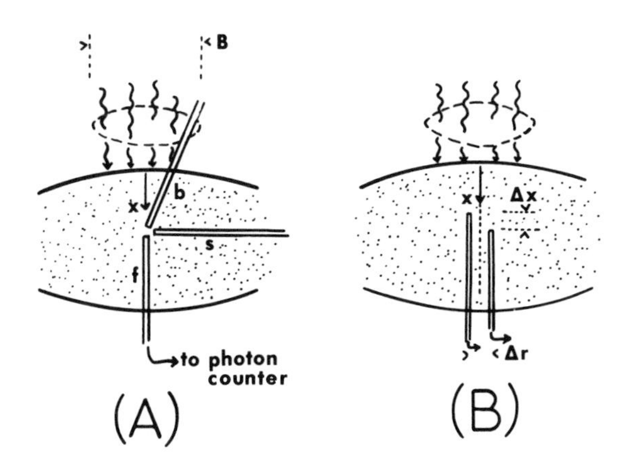

Fig. 2. Illustration of the detector-fiber geometries for light flux measurements in tissues irradiated by an external plane beam of diameter, B.
In (A), a single fiber is oriented either forwards (f), sideways (s), or backwards (b). In (B), two forward fibers, with separation (Δx,Δr), are inserted simultaneously.

To demonstrate the changes in attenuation which may occur *in-vivo*, and following death, we have also made measurements with two detector fibers placed symmetrically about the x-axis in the forward direction, as illustrated in Figure 2B. By scanning alternately with the front and rear fibers coupled to the photon counter, changes in attenuation at each wavelength, λ, were determined from the increase or decrease of the ratio,

$$R(\lambda) = \left[\frac{cps\ (x,\lambda)}{cps\ (o,\lambda)}\right]_{rear} \Big/ \left[\frac{cps\ (x,\lambda)}{cps\ (o,\lambda)}\right]_{front} \quad \ldots (1)$$

where the photon count rate at each depth, x, is normalized to the count rate in air (x=0), i.e. to the incident light beam intensity. (For fibers of identical spectral sensitivity, the denominators cancel in equation 1.) In the ideal case, for exponential attenuation, the penetration depth δ (=1/attenuation coefficient) may then be expressed as

$$\delta = -\Delta x / \ln R \quad \ldots (2)$$

where $\Delta x = x_{rear} - x_{front}$, and where R is assumed constant with depth. We have found it necessary in this double fiber technique to have $\Delta r > \Delta x$. At smaller lateral separation, the detected flux in the rear fiber is affected by the presence of the other fiber/needle.

For all experiments, measurements at different depths were always made by starting at the outer edge of the tissue and moving inwards. This avoided artefacts due to the channels left in the tissues by the biopsy needles.

RESULTS AND DISCUSSION

Spectral Attenuation Measurements

Figure 3 shows examples, for muscle and liver, of the photon flux detected by a forward-oriented fiber along the central beam axis, with the tissue irradiated by an external plane beam. These data are for measurements made *post-mortem* within three hours of death, but with the tissue *in-situ* as described above.

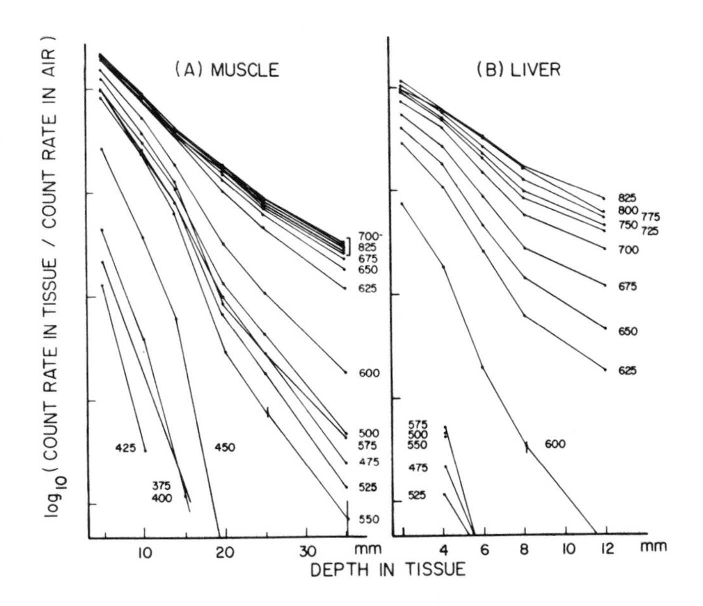

Fig. 3. Examples of the forward flux measured as a function of depth along the central beam axis. Wavelength = 375-825nm in 25 nm steps. The representative error bars are one standard deviation of photon counts.

As would be expected (Svaasand *et al* 1981, Doiron *et al* 1983), the heavily pigmented liver is much more strongly attenuating at all wavelengths than the rabbit muscle, and is also more strongly wavelength-dependent. For example, at the longest wavelength of 825 nm, the flux drops an order of magnitude for ~8 mm of liver; at the wavelength most commonly used for PRT, namely 630 nm, the thickness required is only ~3 mm. By contrast, the corresponding values in the muscle are ~14 mm and ~13 mm, respectively. We shall compare the attenuation as a function of wavelength in more detail below.

For both tissues, the curves are approximately exponential. For this particular muscle, the discontinuity at 15-20 mm was due to the fiber passing through a muscle-muscle boundary. This was confirmed by subsequent dissection of the leg. This type of interface presumably causes partial reflection of the beam, thus suddenly reducing the forward flux. Note that the effect is only slight for wavelengths above about 600 nm, but is very marked at shorter wavelengths. At all wavelengths, however, the slope of the curve is the same on both sides of the discontinuity, indicating that the fiber passed through homogeneous muscle in these regions.

The attenuation generally decreases with longer wavelength, but the presence of specific absorption bands around 425 nm and 550 nm complicate the attenuation spectra. This may be seen more clearly in Figure 4, where the normalized forward flux has been plotted as a function of wavelength for these depths.

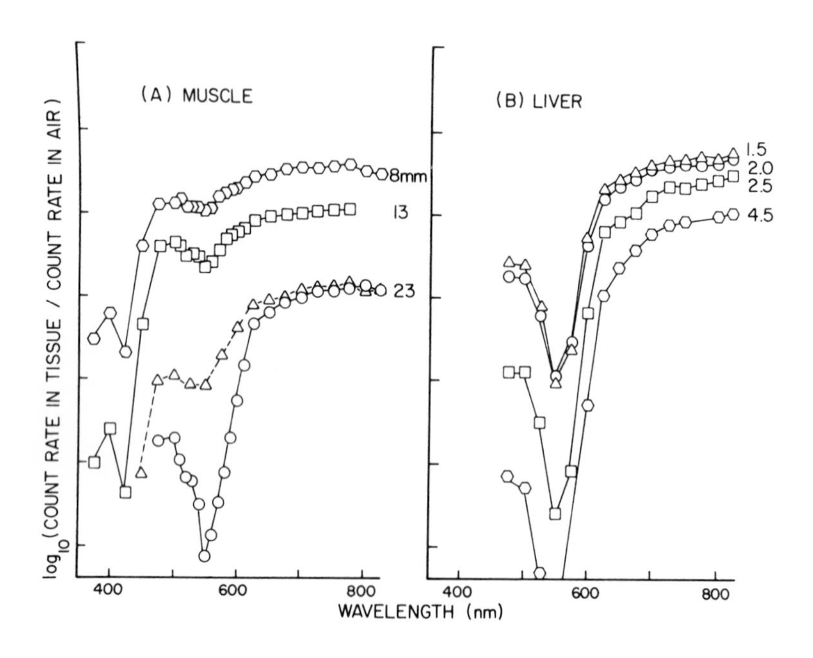

Fig. 4. Examples of the forward flux at different depths in *post-mortem* tissue, plotted as a function of wavelength. For liver, the values at 425 nm are very low and so off-scale at all depths. The dashed curve in (A) is as described in the text, normalized at 825 nm to the corresponding curve with synchronous source and detector monochromators.

We observe that

(a) above 650 nm the muscle attenuation is almost independent of wavelength, whereas the attenuation continues to decrease in liver,

(b) the "anomalous" increases in attenuation at 425 and 550 nm are substantial. As suggested by others (Gabel 1980; Anderson, Parrish 1982; Doiron *et al* 1983; Parrish 1983; Preuss *et al* 1983), and as confirmed by our own observations on the changes in the absorption bands from *in-vivo* to *post-mortem* (see below), these are due, at least in part, to the Soret (425 nm) and the α/β (550 nm) absorption bands of hemoglobin.

The overall structure of the spectral attenuation curve of muscle in Figure 4 is similar to that obtained by Preuss *et al* 1983, who measured the spectral transmission of bovine skeletal muscle using tissue sections of different thicknesses on a spectroradiometer. In our experiments, however, the absorption in the Soret band appears much stronger compared to the 550 nm absorption band. At this time, we do not know whether this is a true difference in the tissues measured, or results simply from the different geometries and instrumentation of the two methods.

We note in this regard that, when a broad-band light is used as the primary radiation source, it may be essential to have wavelength selectivity on both the source side and the detector side of the spectrum scanner. This is illustrated by the dashed curve of Figure 4, where the muscle was scanned with the detector monochromator removed. The shape of the curve is markedly distorted at shorter wavelengths. We have checked that this effect, at least in *post-mortem* tissue, is not due to auto-fluorescence. We attribute it to the greater penetration of the longer wavelength light which arises as leakage from the source monochromator, even though the leakage at any wavelength in air was measured as less than 10^{-5} of the peak intensity.

Sideways and Backward Scattering

As discussed by various authors (Profio, Doiron 1981; Wan *et al* 1981a; Svaasand *et al* 1981; Doiron *et al* 1983; Wilson, Adam 1983), the spatial distribution of photon flux, and, hence, of absorbed energy, is strongly dependent on the relative importance of scattering and absorption in the tissues. With an external light beam, the flux in the tissue is initially peaked in the forward direction. However, the distribution becomes increasingly isotropic as scattered photons contribute more to the flux at greater depth. This may be observed by changing the orientation of the fiber tip, for example, to the sideways or backwards direction.

Figure 5 (A-D) shows the detected flux in the three directions as a function of depth along the central axis in a rabbit muscle for a number of different wavelengths. These results are in qualitative agreement with published values (Svaasand *et al* 1981; Doiron *et al* 1983). The attenuation in all orientations is roughly exponential. For comparison, Figure 5 (E,F) shows results from Monte Carlo calculations in semi-infinite homogeneous media of specified absorption and isotropic scattering coefficients (Wilson, Adam 1983).

It is seen, both experimentally (A-D) and in the computer model (E,F), that the greater the scattering, the less the difference at any depth between the flux in the different directions. We expect that the ratio of sideways-to-forward and backward-to-forward flux will vary with

(a) tissue (depending on the absorption/scatter ratio),
(b) wavelength, (since the absorption and scatter coefficients vary with wavelength),
(c) depth in tissue (depending on the irradiation geometry).

This is illustrated in Figure 6, where we have plotted, as a function of wavelength and at two different depths in muscle, the ratios of sideways-to-forward and backward-to-forward photon counts in muscle.

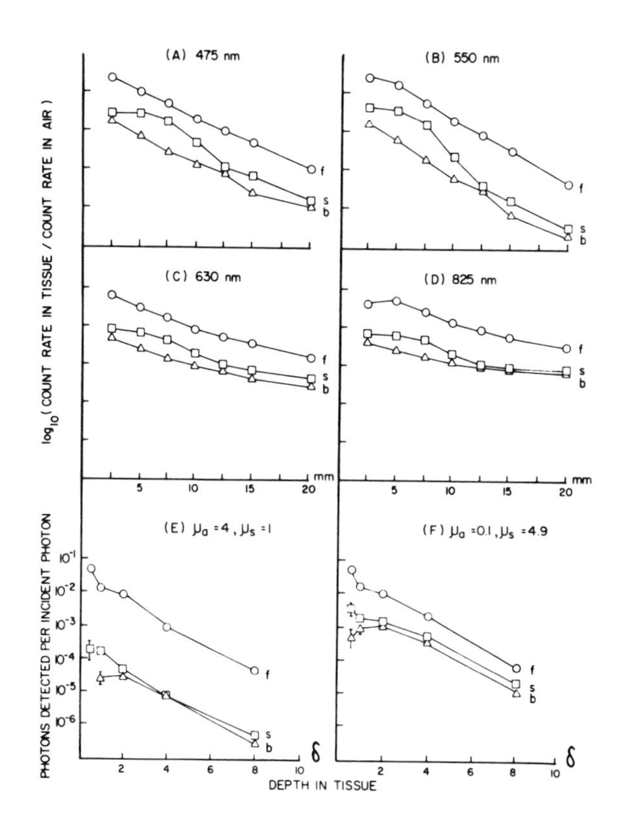

Fig. 5 (A)-(D). Examples of the forward, sideways and backward flux at selected wavelengths plotted as a function of depth in *post-mortem* muscle.

(E)-(F). Monte Carlo calculations for two "tissues". μ_a = absorption coefficient, μ_s = isotropic scattering coefficient. The values plotted are the number of photons into a detector fiber of area δ^2, per incident beam photon. The depth is expressed in units of penetration depth, $\delta=1/(\mu_a+\mu_s)$. B=5δ, detector numerical aperture = 0.4. Error bars represent 1 s.d. of photon counts.

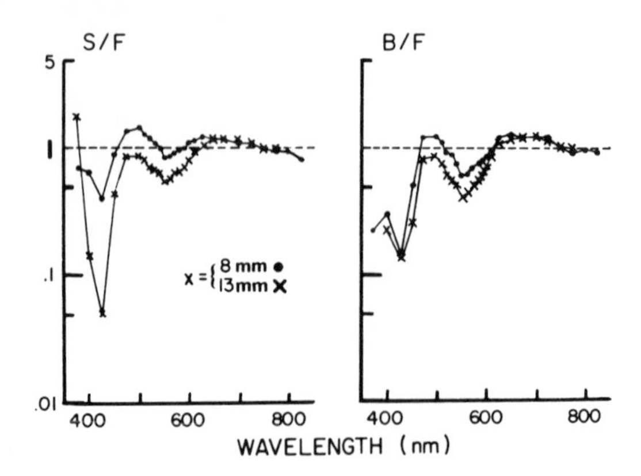

Fig. 6. Sideways-to-forward and backward-to-forward flux ratios at two different depths in muscle (*post-mortem*) plotted as a function of wavelength. Each curve is arbitrarily normalized to unity at 750 nm to show the essential wavelength-dependence.

It can be seen that the degree of non-isotropy of the photon flux reflects, in a complex way, the relative absorption/scattering ratio. Presumably, since the increased attenuation at 425 and 550 nm is due to specific absorption mechanisms, the scattering/absorption ratio is low at these wavelengths; this produces the dips in the s/f and b/f values. Further Monte Carlo studies are in progress to investigate this quantitatively for different irradiation conditions and for different detector fiber characteristics.

Penetration Depth

Forward flux measurements such as those of Figure 3, may not accurately determine the distribution of absorbed energy along the central beam axis, i.e. the exponential "depth-dose" curve, from which the true penetration depth, δ (=1/attenuation coefficient) may be determined. The slope of forward flux *vs*. depth only gives the attenuation coefficient if the ratio of detected photon flux to local absorbed dose is constant with depth. This is the case only for pure absorption, but is only an approximation in tissues where scattering is important.

In Table 1 we show Monte Carlo results for the ratio of the slope of the curve of the forward detected flux *vs.* depth, μ_{eff}(FF), to that of the absorbed dose *vs.* depth, μ_{eff}(AD), for different absorption/scatter ratios and detector numerical apertures, NA. Thus, the forward flux measurements systematically over-estimate the true attenuation coefficient. The error is highest with high scattering and with small detector numerical aperture.

NA \ μ_a/μ_s	4/1	0.1/4.9
0.2	<1.05	1.4
0.4	<1.05	1.2
0.8	<1.02	1.1

Table 1. Ratio of effective attenuation coefficients $\mu_{eff}(FF)/\mu_{eff}(AD)$. μ_a and μ_s are the absorption and isotropic scattering coefficients in the Monte Carlo program: beam diameter, $B=5/(\mu_a+\mu_s)$.

Bearing this limitation in mind, we present our results for the "penetration depth" derived from forward flux curves for rabbit muscle and liver. These are shown in Figure 7, plotted as a function of wavelength.

We note that there is considerable variation at each wavelength from experiment to experiment; this may be partly due to uncontrolled technical factors (e.g. the exact beam geometry, reproducibility of detector fiber positioning, age of tissue *post-mortem*, etc.). However it also raises the question of whether it will be possible to universally characterize the optical properties of different tissues with the degree of accuracy required for good clinical dosimetry.

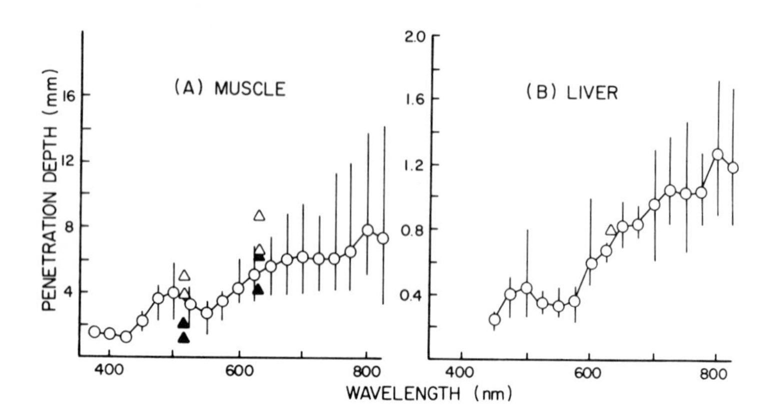

Fig. 7. "Penetration depth", *vs.* wavelength for *post-mortem* tissues (1-3 hr. *p.m.*). B = 10-15 mm. Each point is the mean of 3-6 experiments on different animals. The range of values found is indicated by the vertical bars. Also shown are the *in-vivo* (▲) and *post-mortem* (Δ) values from Doiron *et al*, 1983 for rabbit thigh muscle and pig liver.

Changes in Attenuation from *In-vivo* to *Post-mortem*

As mentioned above, the use of a double fiber technique has enabled us to observe changes in the spectral attenuation, both as a function of time *in-vivo* from placement of the detector fibers, and between *in-vivo* and *post-mortem*. The ratio, R, defined in equation (1), is shown in Figure 8 as a function of wavelength for one typical experiment in rabbit muscle.

It is not possible at this time to interpret the observed changes in detail, nor to accurately quantify them. However, several general observations can be made.

1. The changes, both during experiments *in-vivo*, but in particular immediately *post-mortem*, are considerable. It should be recalled from equation (2) that lnR is inversely proportional to the penetration depth. Thus for example, the change in R *in-vivo* to *post-mortem* at 630 nm represents a 30% increase in δ; the corresponding values at 540 nm and 825 nm are 60% and 80%, respectively.

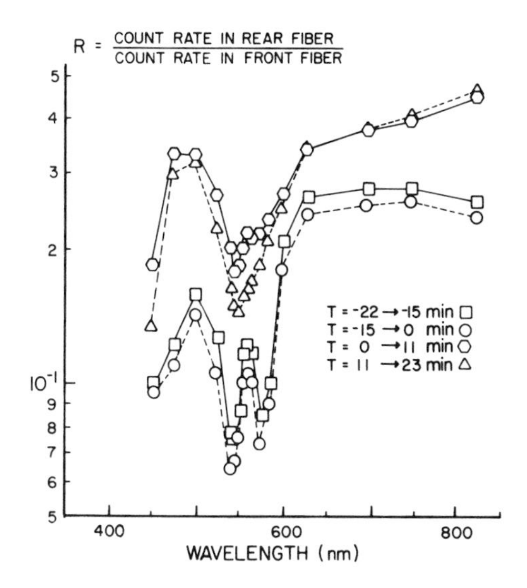

Fig. 8. Variation of R *versus* wavelength from *in-vivo* (t<o) to *post-mortem* (t>o) for rabbit thigh muscle, using the geometry of Figure 2B. Front fiber at x~2mm, Δx=3mm, Δr=4mm.

2. At least some of the changes are associated with changes in blood, probably variation in both the blood volume and the oxygen content. Specifically, the double absorption band *in-vivo* at 540/575 nm changes to a single band within a short time ($\lesssim$ 10 minutes) after death. This agrees with the changes in the attenuation spectrum of blood observed by spectophotometry (Gabel 1980; Anderson, Parrish 1982; Parrish 1983), and we have confirmed this for oxygenated and deoxygenated blood from the rabbits used here. We have not, to-date, observed the double oxyhemoglobin structure *in-vivo* in liver, possibly due to masking effect of other strongly-absorbing pigments in this wavelength region.
3. The changes which take place *in-vivo* after the placement of the detector fiber may be partly due to artefacts from movement of the tissue, or may arise from true changes resulting from the trauma of inserting the fibers. We have, for example, observed local hemorrhaging in liver at the site of insertion. Gravity

effects may also play a part in altering the local blood flow when the tissue is immobilized. Further experiments are required to resolve these issues.

4. Although not shown in Figure 8, we have found that the greatest *post-mortem* changes occur within 60 minutes of death, and that from then, up to several hours, the changes in attenuation are less marked.

CONCLUSIONS

In these preliminary experiments, our aim has been simply to delineate, as a basis for future systematic studies, the main factors which should be taken into account in measuring the optical properties of tissues. Thus, the above data should not be over-interpreted, particularly as regards the absolute values of parameters. From our results, and the work of other groups, it is clear that to achieve accurate dosimetry of light in tissues will be a complex task. Many factors appear to significantly affect the propagation of light, even at a given wavelength; irradiation geometry, inhomogeneities, interfaces/boundaries, etc.

The above results do, however, demonstrate a number of significant trends in the wavelength-dependence of the optical attenuation of muscle and liver. It is remarkable that these two tissues, which have such different attenuation coefficients at any given wavelength, nevertheless have very similar overall wavelength-dependence.

Although the measurement of light distribution is technically much simpler in excised tissues, or *in-situ* but *post-mortem*, the changes which seem to occur upon death limit the relevance of such studies to the *in-vivo* situation. Conversely, the technical complications of *in-vivo* measurements can themselves produce potential sources of error; local tissue trauma, effects of anesthesia, movement during measurement, restricted access to organs, etc. It remains to be seen whether these problems can be reduced to acceptable levels.

Finally, in carrying out future experiments of the type reported here, there is a danger of simply accumulating data on different animal tissues, which may or may not be directly applicable in the clinical situation. How then do such measurements contribute to improving clinical dosimetry? The answer lies in the use of these experiments for developing and testing generalized quantative models of the propagation of light in complex media. These will, in turn, aid in interpreting and applying light dose measurements for individual patients. As already stated, we are working on such models, and on constructing tissue-simulating phantoms with which to test the results. This is continuing in parallel with further systematic animal studies in an holistic approach to the problem of clinical light dosimetry.

ACKNOWLEDGEMENTS

This work was supported by the National Cancer Institute of Canada, and the Ontario Cancer Treatment and Research Foundation. The authors wish to thank also J. Stang and P. Thornburg for technical assistance, and A. Devries for assistance in preparing the manuscript.

*G. Adam: permanent address Department of Radiobiology, Nuclear Research Centre-Negev, Beer-Sheva, Israel.

REFERENCES

Anderson RR, Parrish JA (1982). Optical properties of human skin. In Regan JD, Parrish JA (eds): "The Science of Photomedicine", New York: Plenum Press, p 147.

Bruls WAG, van der Leun JC (1982). The use of diffusers in the measurement of transmission of human epidermal layers. Photochem Photobiol 36:709.

Cortese DA, Kinsey JH (1982). Endoscopic management of lung cancer with hematoporphyrin derivative phototherapy. Mayo Clin Proc 57:543.

Dahlman A, Wile AG, Burns RG, Mason GR, Johnson FM, Berns MW (1983). Laser photoradiation therapy of cancer. Cancer Res 43:430.

Doiron DR, Svaasand LO, Profio AE (1983). Light dosimetry in tissue: application to photoradiation therapy. In Kessel D, Dougherty TJ (eds): "Porphyrin Photosensitization", New York: Plenum Press, p 63.
Dubertret L, Santus R, Bazin M, de Sae Mela T (1982). Photochemistry in human epidermis. A quantitative approach. Photochem Photobiol 35:103.
Eichler J, Knof J, Lenz H (1977). Measurements of the depth of penetration of light (0.35-1.0μm) in tissue. Rad and Environ Biophys 14:239.
Gabel V-P (1980). Lasers in ophthalmology. In Hillenkamp F, Pratesi R, Sacchi CA (eds): "Lasers in Biology and Medicine," New York: Plenum Press, p 383.
Hayata Y, Kato H, Konaka C, Hayashi N, Tahara M, Saito T, Ono J (1983). Fiberoptic bronchoscopic photoradiation in experimentally induced canine lung cancer. Cancer Res 51:50.
Kessel D, Dougherty TJ (eds) 1983: "Porphyrin Photosensitization." New York: Plenum Press.
Parrish JA (1982). Advances in phototherapy of skin diseases. In Hélène C, Charlier M, Montenay-Garestier TH, Laustriat G (eds): "Trends in Photobiology," New York: Plenum Press, p 321.
Parrish JA (1983). Photobiologic considerations in photoradiation therapy. In Kessel D, Dougherty TJ (eds): "Porphyrin Photosensitization," New York: Plenum Press, p 91.
Preuss LE, Bolin FP, Cain BW (1983). A comment on spectral transmittance in mammalian skeletal muscle. Photochem Photobiol 37:113.
Profio AE, Doiron DR (1981). Dosimetry considerations in phototherapy. Med Phys 8:190.
Svaasand LO, Doiron DR, Profio AE (1981). Light distribution in tissue during photoradiation therapy. University of Southern California, Institute for Physics and Imaging Science, Report MISG 900-02.
Wan S, Anderson RR, Parrish JA (1981a). Analytical modeling for the optical properties of the skin with in-vitro and in-vivo application. Photochem Photobiol 34:493.
Wan S, Parrish JA, Anderson RR, Madden M (1981b). Transmittance of non-ionizing radiation in human tissues. Photochem Photobiol 34:679.
Wilson BC, Adam G (1983). A monte carlo model for the absorption and flux distributions of light in tissue. In press.

Porphyrin Localization and Treatment of Tumors, pages 133–148

PHANTOM MEASUREMENTS FOR LIGHT DOSIMETRY USING ISOTROPIC AND SMALL APERTURE DETECTORS

J.P.A. Marynissen and W.M. Star

The Dr. Daniel den Hoed Cancer Center
and Rotterdam Radio-Therapeutic Institute
Rotterdam, The Netherlands

INTRODUCTION

One of the most important practical problems of photoradiation therapy (PRT) is light dosimetry. For the proper interpretation of clinical results and results of in vivo or in vitro laboratory experiments a good knowledge of the light flux density at a given point within the medium is required. Improvement of PRT techniques in clinical use, for example simultaneous application of multiple implanted fibers, depends strongly on an optimal planning and monitoring of the light dose.

Besides the lack of relevant data on the absorption and scattering coefficients and the properties of the scattering function of biological tissues it is difficult to calculate the light distribution due to the fact that collimated beams are used for light delivery leading, e.g. to field size effects. Direct measurement of the light distribution in tissues is a possibility, but is difficult to perform and is not considered the most practical procedure to gain insight into the basics of light dosimetry.

Our approach has been the development of a liquid phantom in which the detector (plus the light delivering fibers if interstitial light delivery is being simulated) can be easily manipulated and positioned without disturbing the structure of the medium. To gather the relevant optical data in such phantoms as well as in real tissue we developed two different light detectors. One detector has an approximately isotropic response. It consists of a small resin sphere with a diameter of 1.5 mm that is mounted on a step index quartz fiber with a core diameter of 0.2 mm. Variations in the response are less than ±10% over an angle of 270 degrees (0.85 x 4 pi solid angle). The other is a small aperture detector consisting of the same type of fiber with a flat end, but a diaphragm at the photomultiplier side. This detector has a minimum aperture of 10 degrees (0.002 x 4 pi solid angle) which is much smaller than the aperture of fiber detectors that are in common use.

The phantoms are suspensions of Indian ink as absorber and polyvinylacetate (PVA) as scatterer in water. With such mixtures the absorption and scattering coefficients of the phantoms can be precisely controlled and by proper selection of the particle sizes, which determine the scattering function, in vivo situations can be simulated. All measurements reported in this paper were performed with monochromatic light (630 nm) either obtained from a dye laser or from a slide projector with an interference filter. Light beams of circular cross section, perpendicularly incident on the phantom or tissue surface, had a maximum diameter of 150 mm.

ANGULAR RESPONSE OF THE LIGHT DETECTORS

The few results reported to date on measurements of light penetration in tissue have all been obtained with a detector consisting of a quartz fiber with flat ends, directly coupled to a photomultiplier (Doiron, Svaasand, Profio 1982; Svaasand, Doiron, Dougherty 1983). The aperture of such a fiber is not large, however, (Fig.1, left side) so that the detector is not suitable to measure the flux density (space irradiance) in a tissue or phantom. On the other hand, the aperture is neither small enough to measure the radiance. Yet, such a fiber detector can be used to determine the effective attenuation coefficient at great depths where the diffusion pattern (i.e. the angular dependence of the radiance) no longer changes with depth. In that region, far from sources and boundaries, which is called the diffusion domain, the light flux density as a function of depth can be described by a simple exponential function. Near the tip of a light delivering fiber or close to boundaries such simple relationships do not hold. The angular dependence of the radiance changes with position in the medium. In this region only an isotropic detector that

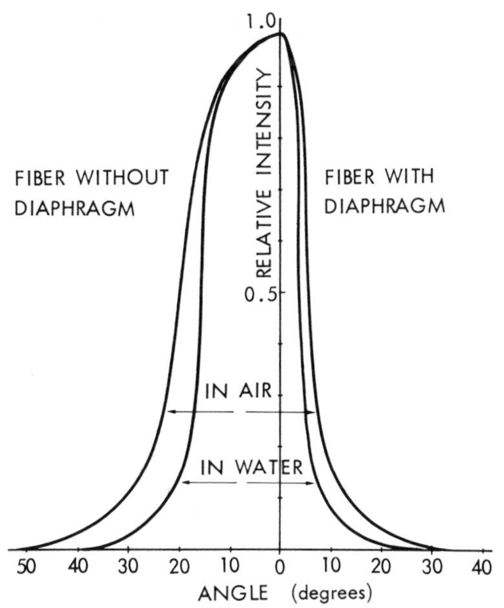

Fig. 1. Normalized response of fiber detectors as a function of the angle between the incident beam and the axis of the fiber.

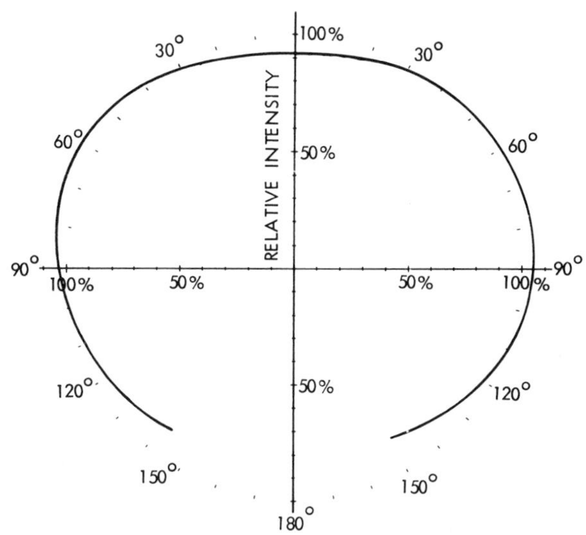

Fig. 2. Relative response of an isotropic detector as a function of the angle between the incident beam and the axis of the fiber. At zero degrees the beam faces the fiber.

covers the full solid angle can properly measure the flux density which is of interest in photoradiation therapy. Such a detector has been developed. The measured response as a function of angle for a prototype is shown in Fig.2. This isotropic detector covers about 0.85 x 4 pi solid angle in which region the response variations are not larger than ±10%. This can be improved in future models. For our present purpose, where qualitative information is more important than quantitative accuracy, the response characteristics are quite satisfactory.

In order to measure the diffusion pattern in a phantom or in tissue we also need a detector with an aperture as small as possible. The aperture can be reduced by applying a diaphragm at the photomultiplier side of a fiber with flat ends. This was done by pushing the fiber into a narrow stainless steel tube with a blackened inner wall. The response of the fiber detector with diaphragm as a function of angle is shown on the right hand side of Fig.1. It covers only 0.002 x 4 pi of solid angle, whereas a fiber detector without diaphragm covers 0.02 x 4 pi of solid angle. This still does not seem to be very much (only 2% of the full sphere), but this can be quite important when there is anisotropy, because incident angles larger than zero contribute proportionally more to the amount of light detected (see also Fig.5).

SOME RESULTS OF PHANTOM MEASUREMENTS

With the three detectors described (i.e. isotropic detector and fiber detector with and without diaphragm) some measurements in liquid phantoms were performed to get a first impression of the possible results. The phantoms consist of suspensions of Indian ink as principal absorber and polyvinylacetate as an almost perfect scatterer in water. It is thus possible to make reproducible phantoms with a desired optical mean free path and ratio of absorption to scattering. The mean free path (mfp) is defined as the inverse of the total attenuation coefficient (absorption - plus scattering - coefficient) and represents the depth at which an incident beam is reduced to 37% of its initial intensity. The albedo, c , is defined as the scattering coefficient divided by the total attenuation coefficient. In the preliminary experiments the size of the polyvinylacetate scattering particles (about 1 micrometer) was larger than the wave length of the light (630 nm) so that scattering is mainly in the forward direction. The value for g, the average cosine of the scattering angle, viz. g = 0.93 was determined from measurements that will not be discussed here.

Because in biological tissues the scattering coefficient may be large relative to the absorption coefficient (see the measurements in biological tissues) it is interesting to compare the results of measurements with the three different detectors in a phantom containing almost only scattering particles (c=1). Fig.3 shows the results of

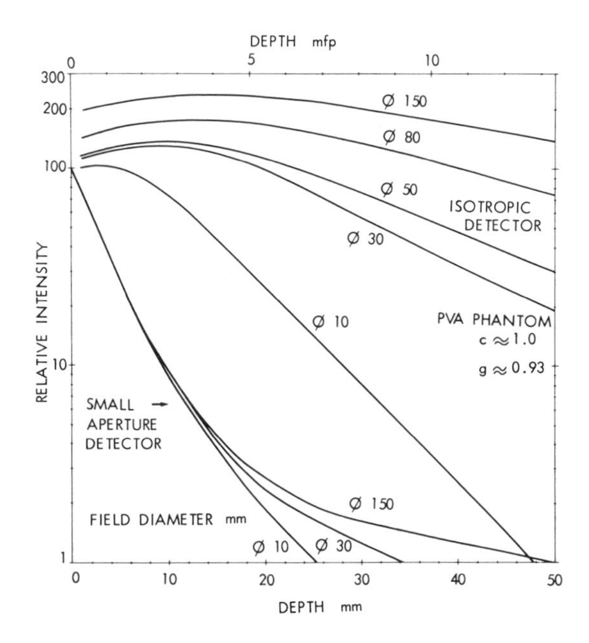

Fig. 3. Results of light intensity measurements as a function of depth and field size performed with two types of detectors in a phantom with almost only scattering particles. The data are absolute values, relative to 100 for the intensity of the incident beam. The values of c and g are discussed in the text. The photon mean free path in this phantom is 3.7 mm.

measurements with the isotropic detector and with a small aperture detector (fiber with diaphragm). The photon mean free path in this phantom is 3.7 mm. We therefore assume that disturbance of the light distribution due to the finite size of the detector (diameter 1.5 mm) can be neglected. For proper interpretation of the results it is useful to consider the mfp scale (upper horizontal) in Fig.3. If the field sizes are also expressed in units of the mean free path, the curves should be valid for all similar phantoms with the same value of c (albedo) and the same particle size, but with different mean free path. The curves tend to become straight lines beyond a depth of about 8 mfp, i.e. in the diffusion domain. Here, each measurement as a function of depth can be described by a single exponential function. The slope of each line gives the effective attenuation coefficient. It is obvious from Fig.3 that there is a large field size effect, which results in two important phenomena. Firstly, the measured effective attenuation coefficient depends strongly on field size. Note that for each field size the measurements with both detectors tend to the same slope. Secondly, at fixed depth the flux density in the medium can vary by a factor of more than two, depending on field size. In Fig.4 similar measurements are presented as in Fig.3, but now for a phantom with c=0.8, i.e. more absorption, but with about the same total attenuation coefficient (mean free path 3.2 mm). The field size effects are not as spectacular as in Fig.3. In particular at small depth the field size effect is small, but at a depth of several mean free paths it can still be considerable. Qualitatively, however, the curves in Fig.4 are similar to those in Fig.3.

In the same phantoms as those of Fig.3 and Fig.4 the diffusion pattern (radiance as a function of direction) has been measured at several depths with the small aperture detector. Results are presented in Fig.5. The curves on the left side show the diffusion pattern for c=1 at a depth of 17 mfp, well into the diffusion domain, and at a depth of 3 mfp, where the incident beam obviously dominates. The field diameter of 150 mm seems to be almost equivalent to infinity for measurements at the central axis of the field and at the depths considered here. Increasing the field diameter would only slightly increase the measured flux density at the central axis of the field. It should be mentioned that theoretically, for a pure scatterer in the diffusion domain and at infinite field size, the diffusion pattern ought to

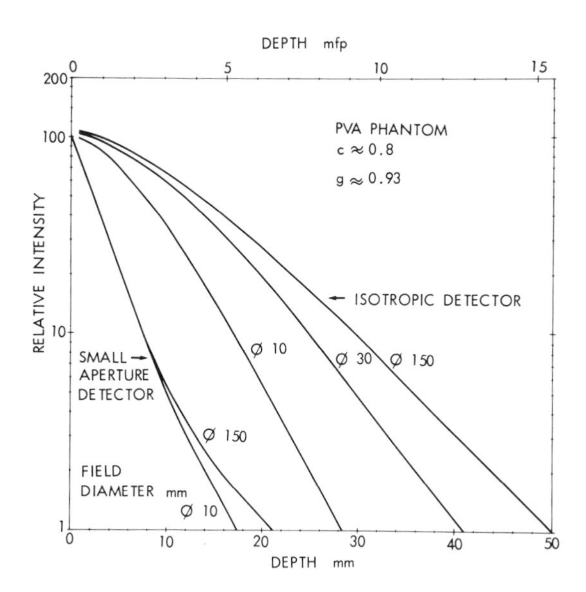

Fig. 4. Results of light intensity measurements as a function of dept and field size performed with two types of detectors in a phantom with c=0.8, i.e. the scattering coefficient is four times the absorption coefficient. The photon mean free path in this phantom is 3.2 mm, almost the same as that of Fig.3. The intensity of the incident beam is 100.

be isotropic (Van de Hulst 1980), independent of the scattering function. The fact that this is not seen in Fig.5 for c=1 at 17 mfp could either mean that the field size is still not large enough, or that the absorption of the polyvinylacetate is not zero. The latter possibility is supported by the fact that a small amount of absorption can have a great influence on the diffusion pattern when the scattering function is strongly anisotropic, as is the case with PVA (g=0.93). For more isotropic diffusion patterns see Fig.10. On the right half of Fig.5 measurements of the diffusion pattern are shown for a phantom with c=0.8 (the same as in Fig.4). Clearly, the anisotropy increases dramatically when absorption becomes more important. Again, at a depth of 3 mfp the diffusion pattern is determined by the incident beam and consequently peaked in the forward direction.

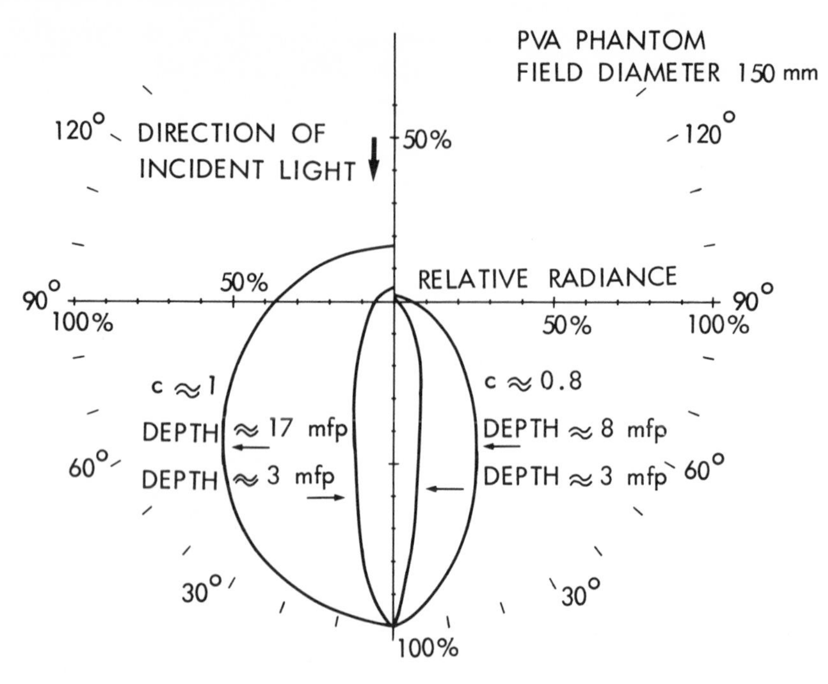

Fig. 5. Diffusion pattern (radiance as a function of direction) for the phantoms of Fig.3 and Fig.4 at two depths. Because of symmetry only one half of each curve has been drawn.

Keeping the foregoing results in mind it is instructive to compare measurements performed on the phantom with c=1 using fiber detectors with and without diaphragm as shown in Fig.6. At great depth, well into the diffusion domain, both detectors measure the same slope but near the surface of the phantom the slopes are quite different. The explanation is that close to the surface the detector with diaphragm measures only incident light, whereas the detector without diaphragm also measures scattered light. From the slope of the curves for the detector with diaphragm at small depth the total attenuation coefficient is obtained, provided the aperture is indeed small enough. In this way the absorption coefficient of Indian ink, the scattering coefficient of PVA, and the total attenuation coefficients of the mixtures have been determined. A picture qualitatively similar to

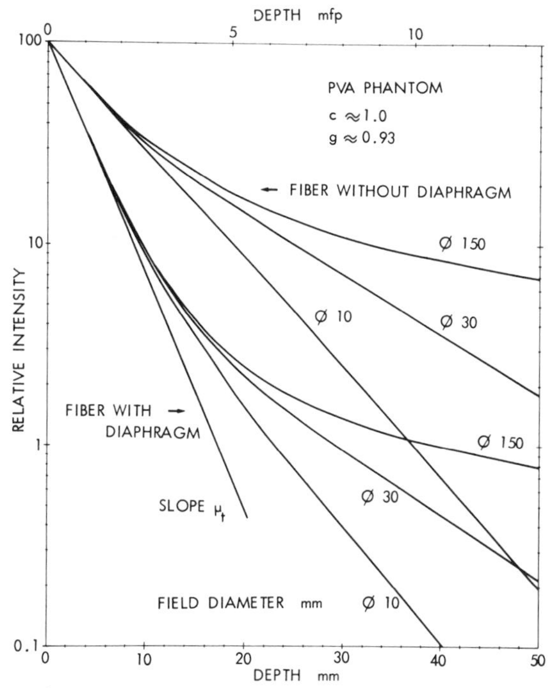

Fig. 6. Relative intensity as a function of depth and field size measured with two fiber detectors facing the incident beam in the same phantom as that of Fig.3.

Fig.6 is obtained for a phantom with c=0.8. Quantitatively, again, the differences are smaller.

RESULTS OF PRELIMINARY MEASUREMENTS IN BIOLOGICAL TISSUES

In order to see how the light flux density in real tissue compares with that in phantoms, preliminary measurements were performed in two tissues in vitro. Chicken muscle was chosen because it was expected to absorb much less than it would scatter light, and cow muscle (lean beef) because it was expected to absorb more light than chicken muscle. Both were chosen because they were readily available. Thick chunks of tissue were closely packed in a cylindrical glass beaker, as far as possible in such a way that the measurements were carried out in only one piece. The dimensions of the beaker were taken large enough to minimize unwanted boundary effects. The radius was at least 8 mean free paths and the height at least 16 mfp. Measurements with the isotropic detector were performed with the incident beam parallel to the axis of the beaker. The detector fiber was moved up and down through the bottom of the beaker and sometimes a perspex plate was used to flatten the surface of incidence. For measurements of the diffusion pattern the fiber detector was entered through the side wall of the beaker perpendicular to its axis and the fiber end was placed in the middle of the beaker. The direction of the incident beam was perpendicular to the axis of the beaker. The angular dependence of the radiance was measured by rotating the beaker about its own axis.

Results obtained with the isotropic detector in chicken muscle are plotted in Fig.7. Again, as in Fig.3 and Fig.4, the effective attenuation coefficient in the diffusion domain depends on field size. Important for clinical practice is also the increase of the flux density close to the surface by a factor of two when the field diameter is increased from 10 mm to 80 mm, keeping the incident light

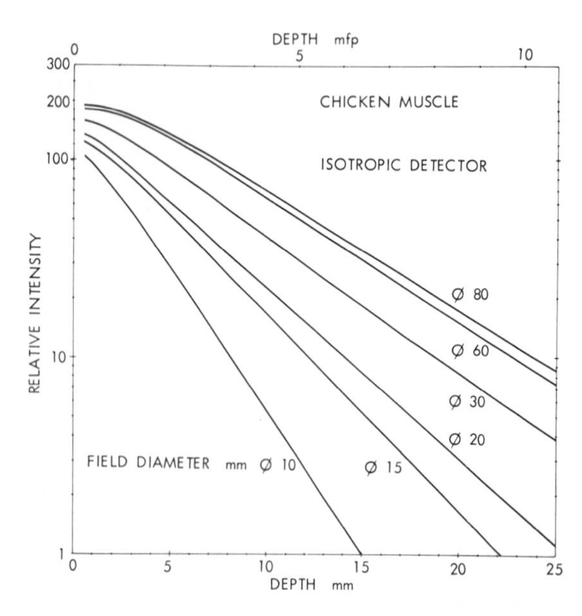

Fig. 7. Light flux density in chicken muscle in vitro, measured with an isotropic detector, as a function of depth and field size. The intensity of the incident beam is 100.

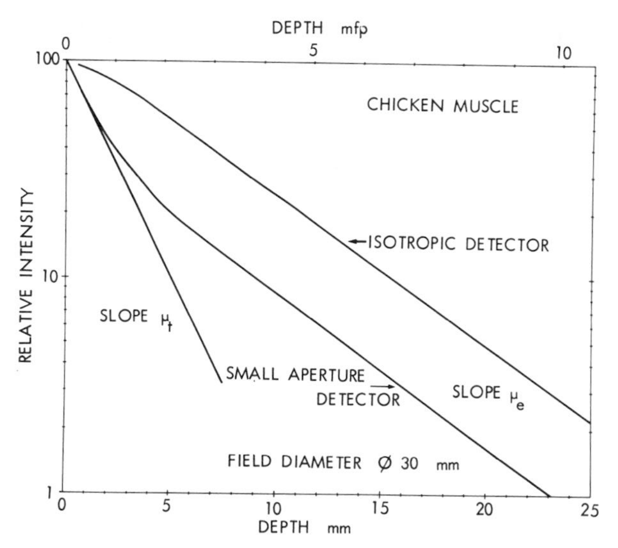

Fig. 8. Comparison of measurements in chicken muscle performed with two types of detector, for a field size of 30 mm.

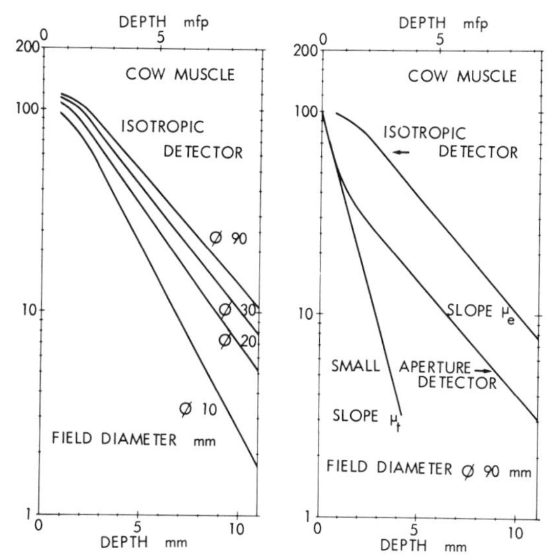

Fig. 9. Left: Light flux density in cow muscle, measured with an isotropic detector as a function of depth and field size. The intensity of the incident beam is 100.
Right: Comparison of measurements in cow muscle performed with two types of detector, for a field size of 90 mm.

intensity constant. In Fig.8 we see that the slopes of the curves measured with the small aperture and isotropic detectors at great depth are again the same, leading to the same effective attenuation coefficient. The slope measured with the small aperture detector at small depth yields a total attenuation coefficient of 0.43 mm^{-1}, i.e. a mean free path of 2.3 mm. Fig.9 shows the results for cow muscle. They are qualitatively similar to those for chicken muscle but quantitatively the field size effects are smaller. The photon mean free path in cow muscle is 1.2 mm.

An important detail of the experimental results in chicken and cow muscle is that the diffusion domain appears to begin at less than 5 mfp below the surface. This can be understood by observing the diffusion patterns measured in the diffusion domain and plotted in Fig.10. Compared with the diffusion patterns measured in the PVA phantoms (Fig.5) the patterns in the biological tissues are more isotropic. As will be explained in the next section, these results suggest that the scattering functions of the particles in the tissues are approximately isotropic. The polyvinylacetate particles on the other hand, as used in the

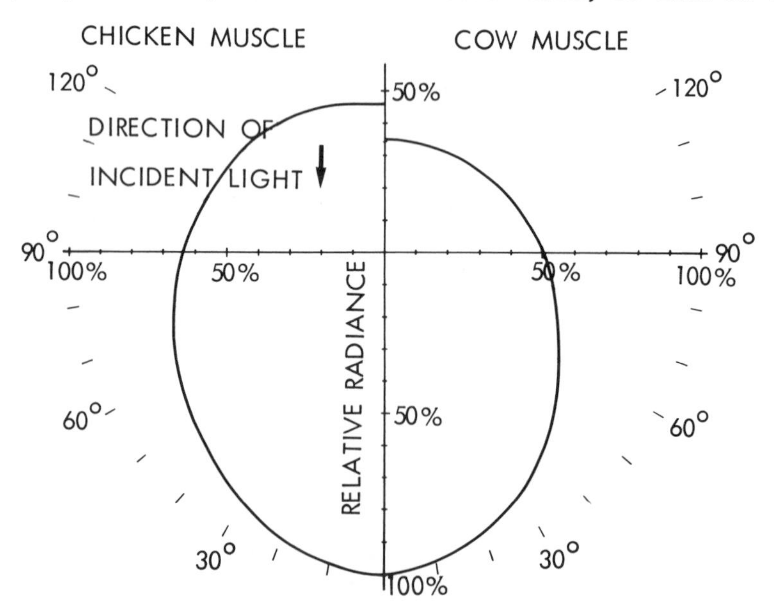

Fig. 10. Diffusion pattern of chicken muscle (left) and of cow muscle (right).

phantoms discussed before, scatter predominantly in the forward direction. The results reported here allowed us to design more tissue equivalent optical phantoms based on PVA particles of smaller size and thus a more isotropic scattering function. These phantoms are used in work that is currently in progress.

ESTIMATE OF THE OPTICAL CONSTANTS OF BIOLOGICAL TISSUES

A straightforward solution of the transport equation for photons (e.g. the Boltzmann equation) is only possible for a semi infinite medium and infinite field size. These requirements are not satisfied in the practice of photoradiation therapy, where in particular the field size can be rather small. In such a situation calculations become rather complicated and direct measurements are the method of choice. However, in laboratory studies solutions of the transport equation for a semi infinite medium and infinite field size can be quite useful for the interpretation of experimental data. We have seen in the previous sections that it is possible to experimentally determine the total attenuation coefficient, the effective attenuation coefficient and the diffusion pattern while satisfying the conditions of a semi infinite medium and infinite field size, if necessary by extrapolating the data obtained for various field sizes. In the literature (Van de Hulst 1980) solutions of the transport equation for a special type of scattering function (the Henyey-Greenstein function) have been tabulated. Values of k (= effective attenuation coefficient divided by total attenuation coefficient) are given as a function of c (albedo) and g (average cosine of the scattering angle). In addition, selected diffusion patterns as a function of c and k are tabulated. Using the experimental value of k, a limited number of paired values for c and g are obtained from the tables. Then, from the measured diffusion pattern another set of paired values for c and g is found. Combining the two sets one finds the proper combination of c and g. This procedure has been used to draw the conclusions for chicken muscle and cow muscle in the previous section.

ISOTROPIC AND SMALL APERTURE DETECTORS IN PHANTOM MEASUREMENTS

In the diffusion domain a small aperture detector can be used to measure the decrease of the flux density upon

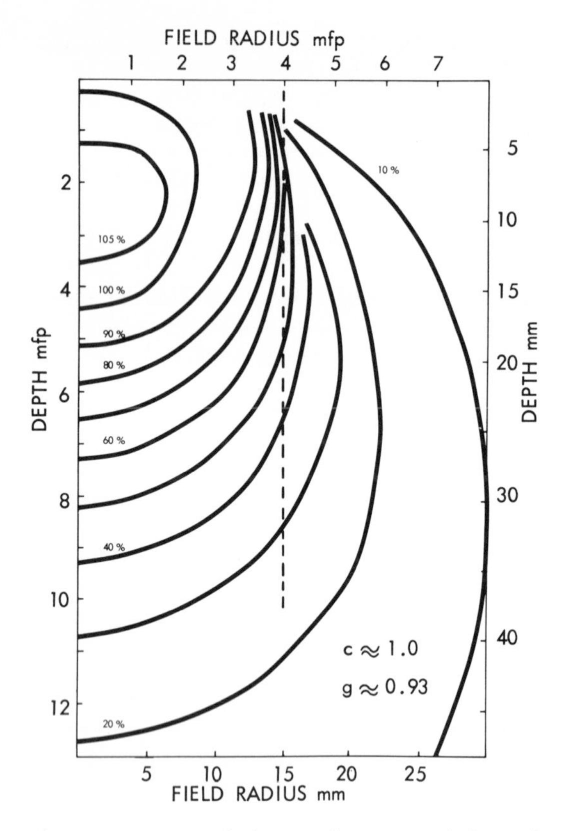

Fig. 11. Isodoses measured in a phantom with c=1. The dashed line indicates the edge of the field. The photon mean free path is 3.7 mm (total attenuation coefficient 0.27 mm^{-1}).

increasing depth, i.e. the effective attenuation coefficient. When the diffusion pattern is known, with proper calibration one can even measure the flux density. Furthermore, a small aperture detector can measure the photon mean free path. Outside the diffusion domain, an isotropic detector is always needed. An illustrative example is the measurement of isodoses, as shown in Fig.11 for a liquid phantom with c=1 and g=0.93. This phantom is not tissue equivalent, but illustrates the possibilities of and the need for phantom measurements. Taking a cross section of the beam at a certain depth in the phantom one sees that the dose falls off very slowly when moving away from the center of the beam. Within the phantom, the beam edge is hardly defined. In biological tissues these effects will be more pronounced, because the scattering is probably more isotropic. On the other hand, when absorption increases, the effects diminish. In the extreme case of absorption only the beam edge in the phantom is as well defined as the beam edge at the surface. It should be noted that the incident beam in the case of Fig.11 had a flat profile and fell off rapidly at the edge, so that the observed effects are truly attributable to scattering processes within the phantom. The light beam delivered by an optical fiber has a profile that is not flat but rather more bell shaped. This shape will influence the isodose pattern in the phantom near the edge of the beam. Measurements have shown that these effects can be considerable. It should be clear from the foregoing discussion that for proper planning of photoradiation therapy with external illumination some knowledge of the isodose pattern is essential. The same holds for the planning of interstitial photoradiation therapy with one or several implanted fibers and in both cases measurements with an isotropic detector are indispensable, which can be properly performed only in a tissue equivalent liquid phantom.

CONCLUSIONS

If scattering of light dominates absorption in biological tissues, then:

1. Field size effects are important and affect the light flux density, the penetration depth and the isodose pattern.
2. Field size and dimensions of the sample used in laboratory studies should be adapted to clinical practice.
3. Knowledge of the aperture of the detectors is necessary for proper interpretation of measured light distributions in tissues.

Using measured values of the total attenuation coefficient, the effective attenuation coefficient and the diffusion pattern, combined with tabulated solutions of the transport equation, one can estimate the scattering coefficient, the absorption coefficient and the size of the scattering particles.

In the diffusion domain a small aperture detector can

be used to measure the variation of the flux density with depth. If the diffusion pattern is known, one can even measure absolute values of the flux density. In all other situations an isotropic detector is needed.

To gain insight into the possible light distributions in tissue, phantom measurements with an isotropic detector are of great value.

REFERENCES

Doiron DR, Svaasand LO, Profio AE (1982). Light dosimetry in tissue: application to photoradiation therapy. In Kessel D, Dougherty TJ (eds): "Porphyrin photosensitization", New York: Plenum Press, p 63.

Hulst HC van de (1980). "Multiple light scattering". New York: Academic Press, p 331.

Svaasand LO, Doiron DR, Dougherty TJ (1983). Temperature rise during photoradiation therapy of malignant tumors. Med Phys 10:10.

Porphyrin Localization and Treatment of Tumors, pages 149–161

STUDIES OF LIGHT PROPAGATION THROUGH TISSUE

P.A. WILKSCH and F. JACKA
Mawson Institute for Antarctic Research
A.J. BLAKE
Department of Physics
University of Adelaide, Adelaide, South Australia.

INTRODUCTION

As part of the work of the Adelaide group in the development of cancer photoradiation therapy, studies of the physics of light propagation in tissue are being undertaken with the aim of developing guidelines and principles for dosimetry.

In order to predict the light dose delivered to each point in the exposed tissue during a treatment, we need to have a mathematical or numerical model which accurately describes the propagation of light through tissue. As a starting point, we assume that, at least on the large scale, biological tissue of a particular type behaves like a homogeneous scattering and absorbing medium. We do not focus attention on the microscopic details of its structure, but assume that these details impart to the tissue some set of large-scale average properties, which we try to determine experimentally.

It is also assumed that the behaviour of light in tissue can be treated with the standard methods of radiative transfer theory (van de Hulst 1980; Profio, Doiron 1981). In this theory, the basic parameters that need to be known are:

the absorption coefficient α_a;
the scattering coefficient α_s;
the scattering phase function p, describing the angular distribution of scattering from one particular direction into each other direction in space.

An alternative pair of parameters derived from α_a and α_s is:

the extinction coefficient $\alpha = \alpha_a + \alpha_s$;
the albedo for single scattering $\omega = \alpha_s/(\alpha_a + \alpha_s)$.

From these data one can in principle predict the light distribution for any particular configuration of the source and the medium to which they apply. Of course, these parameters will vary with the tissue type and with the wavelength of the light.

Unfortunately it is very difficult to estimate these parameters from measurements made on large thicknesses of tissue because of the complications introduced by multiple scattering. Our approach has been to try to determine them from measurements on thin tissue sections. Once their values are known, even approximately, it is possible to decide whether one or more of various approximations (e.g. isotropic scatter, diffusion theory, dominance by scattering or by absorption) might be used to simplify the model.

APPARATUS

A simple apparatus has been built which permits the measurement of the relevant optical properties of tissue sections (Fig. 1). Light from a 55 W tungsten halogen lamp

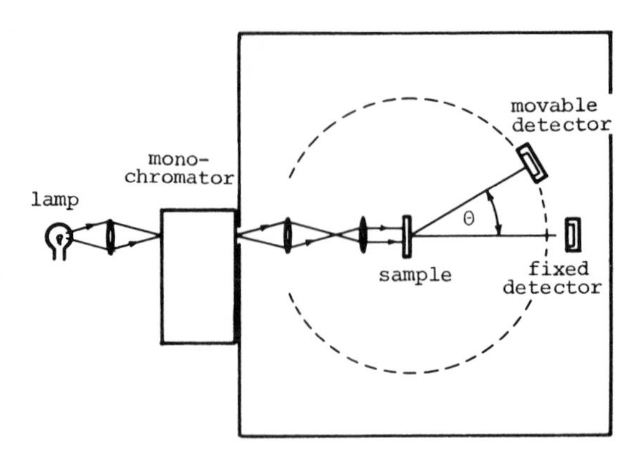

Fig. 1 Sketch of apparatus to measure optical properties of tissue sections.

is passed through a tunable grating monochromator and collimated through an area on the sample about 5 mm square. The tissue sample is held between a pair of microscope slides whose surfaces have dielectric interference coatings to minimize unwanted light reflections at the air-glass and glass-tissue interfaces. Silicon photodiodes are used as light detectors. A fixed detector is used to monitor the relative transmission of the direct beam through the sample. A movable detector can be swung in an arc around the sample in order to measure the scattered light as a function of direction in one plane. Angles up to 155° from the direct beam are accessible, and the radius of the movable detector from the sample can be adjusted. The sample and the detectors are enclosed in a light-tight box. The monochromator bandwidth is about 5 nm.

PREPARATION OF TISSUE

The satisfactory preparation and mounting of tissue slices has presented some difficulties. In order that a slice should include a representative range of microstructures, a thickness greater than about 0.1 mm is probably needed. In a thickness much smaller than this, small-scale detail and inhomogeneities are likely to lead to uncharacteristic results. On the other hand, for there to be an advantage in using the thin-slice technique for measurement of optical properties, the optical thickness ideally should be significantly less than unity (optical thickness = $\alpha \times$ thickness); that is, single scattering should predominate over multiple scattering, and the sample should remove only a small percentage of radiation from the direct beam.

It is also important that the tissue should undergo as little structural and chemical alteration as possible during its preparation, so that its optical properties are not changed. For this reason, the standard methods of fixing or embedding prior to slicing are not acceptable. Even freezing of the tissue is likely to alter it optically through the rupturing of cells, although in practice this has not been found to be a marked effect.

The samples which we have studied to date have either been cut from fresh tissue, or sliced from blocks of frozen tissue with a rotary blade. Thicknesses were measured after mounting between slides, and have ranged upwards from 0.5 mm. These samples have been adequate for the initial evaluation of the measurement technique but a good method for preparing uniform fresh sections with a smaller thickness would be valuable.

Once the tissue has been mounted, it is important to prevent loss of moisture from it, otherwise its transmission increases and its scattering decreases as it dries. This is done by injecting a sealing bead of petroleum jelly around the edges of the slides. The optical properties of the sample are then found to remain constant for at least several hours.

ANALYSIS OF MEASUREMENTS

The outcome of a measurement on a particular sample is a set of detector current readings for several different scattering angles. For the purposes of comparison of these results with each other, with model calculations and with standard tabulations for various scattering phase functions, it is convenient to express the data in terms of the parallel-slab reflection and transmission functions R and T (van de Hulst 1980). These specify as a function of direction the ratio of the light intensity scattered from the surface of the slab to that which would obtain for a perfectly diffuse (Lambertian) scatterer; they are proportional to the scattered radiance. A plot of R and T functions versus $\cos^2\Theta$ where Θ, ranging from zero to 180 degrees, is the angle between the incident and scattered directions (Fig. 1), has the convenient property that the area beneath any segment of the graph is proportional to the scattered radiant flux within that range of scattering angles, provided that symmetry about the normal to the slab is assumed. Fig. 2 illustrates this method of presentation.

The horizontal axis is linear in $\cos^2\Theta$. The corresponding scale for the angle Θ is almost linear about 45° and 135°, but is progressively compressed as one approaches 0° and 90° on the transmission side, and 180° and 90° on the reflection side. Physically, the compression near 0° and 180° compensates for the reduced solid angle encompassed by conical shells of a given angular thickness in this region, and the compression near 90° compensates for the reduced cross-section that the illuminated area of the surface of the slab presents in these directions.

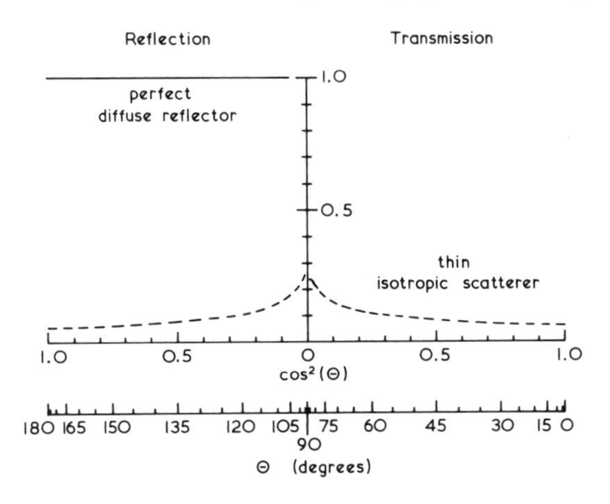

Fig. 2 Illustration of method of plotting reflection and transmission functions. The broken line is for an isotropic scatterer of optical thickness 0.25 and albedo 0.9.

As shown in Fig. 2, a perfect diffuse reflector has an R function which is 1 over the whole reflection region, and a T function which is zero. The area beneath this curve is unity, expressing the fact that all energy is reflected. An optically thin isotropic scatterer (optical thickness $\ll 1$) has R and T functions which are almost symmetrical about the centre of the graph, and approach a $1/\cos\Theta$ shape. For any optically thin slab, R and T are much less than unity over most of the range.

In the analysis of measurements, the set of detector current readings at different angles is converted into R and T functions. This requires knowledge of the total radiant flux Φ_i which falls on the plane of the sample: this quantity is measured separately with a detector replacing the sample. The light flux Φ_d remaining in the direct beam after passing through the tissue is measured with the fixed detector. The R and T functions are extrapolated into the regions of Θ near 90° and 180° which are inaccessible to measurement, and the curves are numerically integrated to obtain the total scattered flux Φ_s. The absorbed flux Φ_a is then given by $\Phi_a = \Phi_i - \Phi_d - \Phi_s$.

The situation is more complex if the scattering is not

symmetrical about the normal to the slab. Then different R and T functions are obtained for scattering in perpendicular planes through the axis, and some form of average of these has to be used in the calculation of total scattered flux.

A further complication is introduced by the fact that refraction occurs at both air-glass and glass-tissue interfaces, and total internal reflection can occur at the glass-air interface. If we assume a refractive index for the tissue of 1.3, then rays at angles of 0° to 50° from the normal inside the tissue emerge over the range of angles from 0° to 90° in air, and the rays at higher angles are internally reflected to undergo further scattering and absorption. The net effect is a significant modification in the shape of the scattered light pattern, and an increase in the amount of light absorbed. This must be taken into account before the basic parameters α, ω and p can be determined.

Our method of doing this determination has been to use an assumed set of parameters in a simple Monte Carlo numerical model which simulates the paths of photons through the tissue and its envelope and keeps account of the directions into which they are scattered upon leaving the sample. The parameters are adjusted until a good fit between the model predictions and the observations is obtained. The use of thin sections of tissue is still an advantage because the effects of variations in different parameters are still largely independent, and physical insight can still be relied upon to lead adjustments to the model in the right direction.

RESULTS

Initial measurements have been made on porcine and bovine skeletal muscle and adipose tissue. These materials were obtained fresh from the butcher, as prepared for domestic consumption, and so they had been drained of most of their blood. The aim of these measurements was firstly to explore the capabilities and the limitations of the technique, and secondly to get information on the scattering characteristics of some typical biological materials. A wavelength of 630 nm was used throughout.

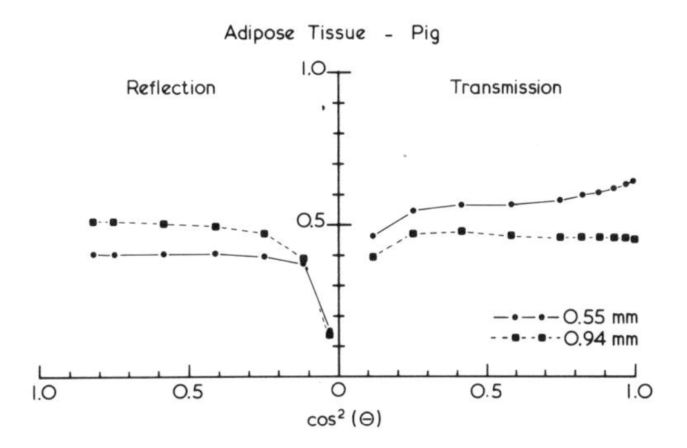

Fig. 3 Measurements of R and T functions at 630 nm for two samples of porcine adipose tissue, thicknesses 0.55 mm (circles joined by solid lines) and 0.94 mm (squares joined by broken lines).

Adipose Tissue

Fig. 3 shows typical reflection and transmission functions measured for two different thicknesses of adipose tissue of pig. The curves are more or less flat over most of their range, which means that the light leaving the surfaces is well diffused. Also the total area beneath the pair of curves for each thickness approaches unity; therefore most of the incident light is scattered out of the sample. The calculated percentages of light absorbed in the thinner and thicker sections are 7.5% and 10% respectively. There is virtually no direct beam remaining after the light has penetrated the tissue, implying that the extinction coefficient is high (greater than 10 mm^{-1}), but since the absorption is small, the albedo must be very close to unity. That is, scattering dominates by far over absorption, as might be expected from the white appearance of the tissue. The sections used here were too thick to allow precise estimates of the optical parameters, but it is apparent that the scattering must occur preferentially in the forward direction rather than isotropically, otherwise the total reflected light would be much greater than the total transmitted light. Sections of adipose tissue of cow are found to display similar characteristics.

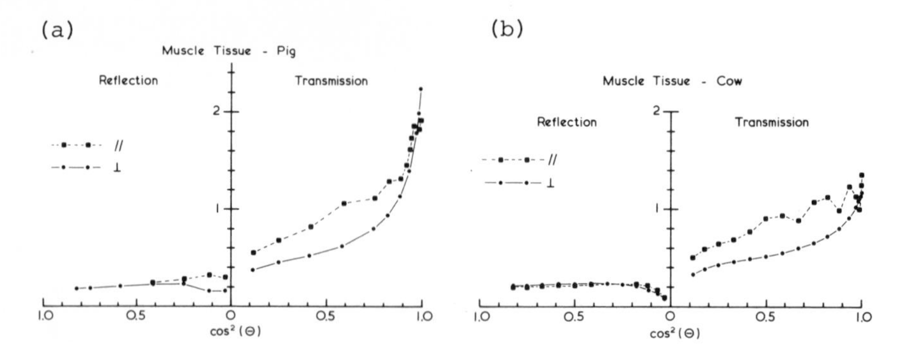

Fig. 4 Measurements of R and T functions at 630 nm for two samples of muscle tissue: (a) Pig, thickness 1.2 mm; (b) Cow, thickness 1.55 mm. Circles joined by solid lines represent measurements in a plane perpendicular to the muscle fibre direction; squares joined by broken lines represent measurements in a plane parallel to the fibres.

Muscle Tissue

It might be expected that the fibrous structure of skeletal muscle tissue would result in different scattering properties in planes at different angles to the fibre axis, and this is indeed found to be so. Fig. 4(a) shows the measured R and T functions for a 1.2 mm thick section of pig muscle, cut parallel to the fibres. For scattering in the plane perpendicular to the direction of the fibres, the reflection function is fairly flat, but the transmission function rises steeply as the direction of the direct beam is approached, implying a strong forward peak in the scattering phase function. For scattering in the plane which includes the fibre direction, the reflection function is similar but the transmission function exhibits two or three peaks superimposed upon a considerably larger background. These subsidiary peaks are interference fringes generated by the regular spacing of the sarcomeres within the myofibrils of the muscle tissue. The striated fibres behave much like a diffraction grating, giving rise to straight, parallel fringes on each side of the direct beam. The angular separation of the fringes corresponds to a sarcomere spacing near 4 μm. The prominence of these fringes is quite surprising in such a thick tissue section.

The total forward-scattered light is considerably increased by their presence.

Fig. 4(b) shows similar measurements on a 1.55 mm thick section of cow muscle. Despite the greater thickness, the interference fringes are even more marked. Their spacing is about the same as for pig muscle, and the additional contribution to the diffuse transmission is large. The prominence of the fringes increases in thinner sections.

The interference effects complicate efforts to model light propagation in muscle tissue, because they give rise to a scattering phase function which is not simply a function of the angle between incident and scattered directions, but which depends on the orientation of both these directions with respect to the axis of the muscle fibres.

MODELLING

As a first step towards estimating the extinction coefficient, albedo and scattering phase function applicable to muscle tissue, interference effects were ignored and the problem was treated as if the R and T functions were the same for all planes through the incident beam, and equal to their measured values in the plane perpendicular to the fibres. Although this is a gross simplification, it should give some insight into the character of the other scattering processes involved.

The measurements for two different thicknesses of pig muscle were chosen (Fig. 6) and a search was made for an appropriate set of parameters which, when used in a Monte Carlo simulation of the measurement as described earlier, would yield predictions which matched the observations. A model scattering phase function was used which consisted of an isotropic component plus an exponential forward peak: that is,

$$p(\cos\beta) = (1-r)/2 + (r/\gamma)\exp((\cos\beta-1)/\gamma),$$

where β is the angle between the incident and scattered direction, and $\gamma(<<1)$ is the average value of $(1-\cos\beta)$ for the exponential component. The constant r determines the relative contribution of the two terms to the phase

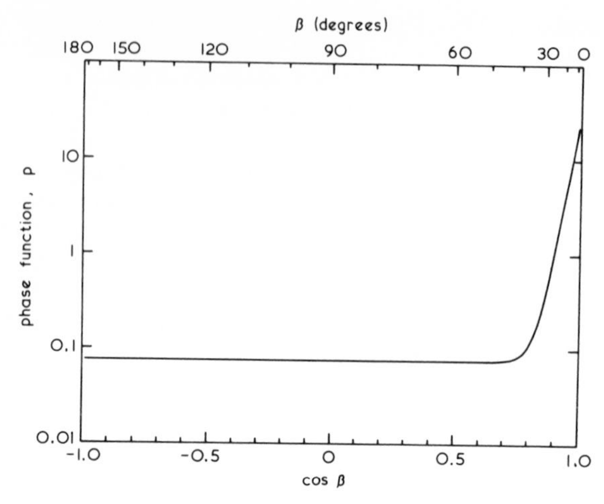

Fig. 5 Plot on a logarithmic scale of the scattering phase function p(cosβ) used in Monte Carlo modelling of light propagation in pig muscle tissue.

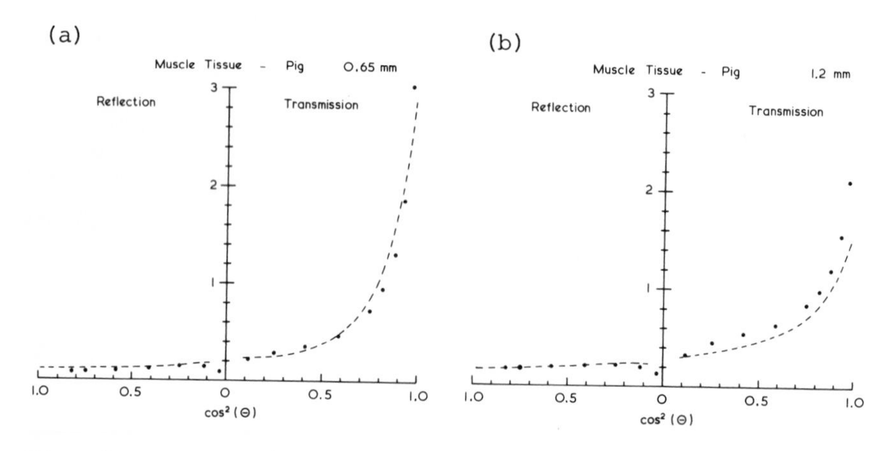

Fig. 6 A comparison between measurements of R and T functions of pig muscle tissue at 630 nm and a Monte Carlo simulation of these measurements with $\alpha = 4.1\ mm^{-1}$, $\omega = 0.98$, and the phase function shown in Fig. 5. (a) Sample thickness 0.65 mm. (b) Sample thickness 1.2 mm.

function; in the Monte Carlo model, r is the probability that exponential rather than isotropic scattering occurs. The function $p(\cos\beta)$ is normalized so that its integral over $\cos\beta$ is unity.

A reasonable fit to the observations was obtained with the parameter values $\alpha = 4.1\ mm^{-1}$, $\omega = 0.98$, $\gamma = 0.03$, $r = 0.85$. The scattering phase function corresponding to these values is shown on a logarithmic scale in Fig. 5. It is highly anisotropic, and the average scattering angle for the exponential component is about 14°. A comparison of the measured R and T functions with the model predictions is shown in Fig. 6 for the two sample thicknesses 0.65 mm and 1.2 mm. While the fit is not ideal, the model curves have the correct character, and in view of the simplifications made in the model the similarity is encouraging. Inclusion of scattering due to interference would help to increase the computed transmission function for the thicker section, in better agreement with the measurements.

Some general conclusions can be drawn on the basis of the insight gained from this exercise. An extinction coefficient α of about $4\ mm^{-1}$ is consistent with most of the measurements on pig muscle samples. The large proportion of scattered flux to absorbed flux implies an albedo very close to unity. The scattering phase function must be highly anisotropic with a strong, narrow forward peak, but scattering into the backward hemisphere is not negligible.

To illustrate the utility of the information obtained above, the same parameter values were used in a Monte Carlo simulation of the propagation and absorption of light in a thick block of tissue. Light energy was assumed to be delivered in the perpendicular direction at a point on the surface, and the relative absorption per unit volume was recorded as a function of depth below the surface and radius from the axis through the point of light delivery. Internal reflection from the surface was taken into account using a refractive index of 1.3 for the tissue. For comparison, another computation was performed using an isotropic phase function instead of the anisotropic one, but with other parameters unaltered. Results are plotted in Fig. 7 as contours of absorbed energy per unit volume, spaced logarithmically. The numbers on the contours represent energy units absorbed per cubic millimetre if the total energy in the incident beam is 7300 units. The quantity of

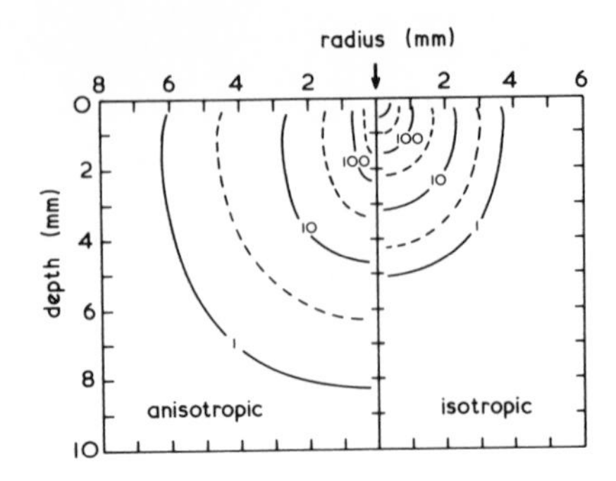

Fig. 7 Logarithmically-spaced contours of the relative energy absorbed per unit volume for light delivered at normal incidence at the position of the arrow, calculated with a Monte Carlo model using $\alpha = 4.1\ mm^{-1}$, $\omega = 0.98$ and two phase functions: anisotropic as in Fig. 5 and isotropic. See text for more details.

interest for dosimetry is the fluence Ψ (Rupert 1974) which is related to Q, the energy absorbed per unit volume, by $Q = \alpha_a \Psi$. In this case $\alpha_a = \alpha(1 - \omega) = 0.082\ mm^{-1}$, so the contours may alternatively be interpreted as representing the fluence in units per mm^2 for a total incident energy of 600 units.

A profound difference is evident for the two phase functions. The energy is dissipated throughout a much larger volume in the anisotropic case, but with a lower peak concentration. The computations also show that only 30% of the incident energy is backscattered out of the tissue, compared with 58% for isotropic scattering.

Once the fluence necessary for cell death with a given cellular concentration of photoactive drug is known, information of the type presented here can be used to establish what incident energy is required to treat a given volume.

CONCLUSIONS

A technique has been described for measurement of parameters that determine the propagation of light through tissue. For most reliable results, it is found that a range of sample thicknesses of a particular tissue type, from optically thin to optically thick, is desirable. At 630 nm wavelength, adipose tissue has a very low absorption coefficient compared to its scattering coefficient, and it will probably be reasonable to neglect absorption at least in layer thicknesses less than a few millimetres. In muscle tissues interference effects are important, and there is a high proportion of scattering at small angles, making it difficult to model the scattering phase function. Nevertheless it is important to take into account these anisotropies in the phase function because they have a strong effect on light propagation in muscle.

This work needs to be extended to other wavelengths and tissue types, and to human tissue. The effect of inclusion of blood has to be studied; it is expected that this will increase the absorption coefficient with only a minor change to the scattering properties. Predictions of light distribution at depth need to be compared with *in situ* measurements. Development of detectors for this purpose is under way. The improved understanding and information to be gained from these studies will help to provide a sound physical basis for dosimetry in cancer photoradiation therapy.

ACKNOWLEDGEMENTS

We gratefully acknowledge financial assistance from Mr. Don Schultz and from the National Health and Medical Research Council. The contributions and assistance of S.M. Jenkins are much appreciated.

REFERENCES

Profio AE, Doiron DR (1981). Dosimetry considerations in phototherapy. Med Phys 8:190.

Rupert CS (1974). Dosimetric concepts in photobiology. Photochem Photobiol 20:203.

van de Hulst HC (1980). "Multiple light scattering. Tables, formulas and applications." New York: Academic Press.

Porphyrin Localization and Treatment of Tumors, pages 163–175

FLUORESCENCE OF HpD FOR TUMOR DETECTION AND DOSIMETRY IN PHOTORADIATION THERAPY

A.E. Profio and J. Sarnaik

University of California
Santa Barbara, CA 93106

INTRODUCTION

Hematoporphyrin-derivative (HpD), after intravenous injection at a typical dosage of 2.5 mg/kg, reaches and maintains a higher concentration in malignant tumors than in most nonmalignant tissues. This differential concentration effect, together with the photodynamic action upon irradiation with visible light, is the basis of photoradiation therapy (PRT). The differential concentration, together with the fluorescence of HpD, is the basis of fluorescence diagnosis: detection, localization, sizing and perhaps staging of tumors. Fluorescence may also provide information on the concentration of HpD as a function of dosage and time after injection, hence, the optimum time for either diagnosis (maximum fluorescence contrast) or therapy (maximum therapeutic ratio). The therapeutic ratio also depends on the magnitude and spatial distribution of the light flux, as the dose rate depends on both light flux and HpD concentration.

TUMOR DETECTION

Most of our experience with detection and localization of tumors by fluorescence of HpD has been carried out with Oscar J. Balchum, M.D. and Daniel R. Doiron, Ph.D. on lung cancer patients at the Los Angeles County-University of Southern California Medical Center. Progress has been described in the literature (Profio, Doiron, and King 1979; Profio, Doiron, Balchum, and Huth, 1983; Balchum,

Doiron, Profio, and Huth 1982), and in a Ph.D. dissertation (Doiron 1982). Recent clinical results are the subject of another paper at this conference (Balchum, Profio, Doiron, and Huth, "Imaging Fluorescence Bronchoscopy in the Localization of In-Situ Bronchial Cancer.")

Thick, more advanced tumors are imaged readily by the fluorescence of HpD contained in them, after waiting about 72 hours after injection for clearance of most of the HpD from normal mucosa. Tumors thinner than 0.5 mm are harder to detect because fluorescence contrast is decreased by depth averaging (the underlying tissue has a smaller concentration of HpD than the tumor). Contrast is small anyway because of the relatively low fluorescence yield ($\sim 10^{-5}$) of HpD at the concentration expected in tumors, and the background from tissue autofluorescence and the excitation light system (minimized by the violet krypton ion laser and fused quartz fiber). Doiron in his Ph.D. dissertation calculated the contrast of a 0.1 mm thick tumor, irradiated with violet (410 nm) light from the laser. Two parameters were defined, "a" which is the ratio of the fluorescence signal from normal mucosa without HpD (thus accounting for autofluorescence and source background) to the signal from a solution of HpD, and "b" which is the assumed ratio of the concentration of HpD in normal tissue to the concentration in the tumor. The value of "a" depends on the barrier filter, among other things. The filter chosen for the fluorescence bronchoscope had a center wavelength of 695 nm and a bandwidth of 40 nm, bracketing the longer-wavelength peak of the HpD emission spectrum. The autofluorescence spectrum decreases with increasing wavelength in the red region, hence, the filter gives better contrast (smaller "a") but transmits only a fraction of the HpD spectrum, requiring a stronger violet source or greater light amplification. A typical value of "a" is 5. The value of "b" is at present only a guess, but is probably between 0.1 and 0.5. The predicted fluorescence contrast of the 0.1 mm thick tumor was between 1.2 and 1.1.

In order to test the prediction, and learn more about contrast for thin and thick tumors, an investigation was carried out by D. Rakkhit for his M.S. thesis project, using photographic densitometry. Slides (35 mm color transparencies) were selected from patient files, as we routinely photograph fluorescing sites. The optical densities were obtained by an Optronics C-4100 film scanner and digitizer, coupled to a minicomputer, in the electrical engineering department at UCSB. The ratio of the digitized brightness of a picture element within the image of the tumor (digitized on a scale of 1 to 256) was ratioed to the average brightness of 64 small picture elements immediately surrounding the tumor area. This image contrast was corrected for the gamma of the film (Ektachrome 400, push processed to effective 800 ASA speed), and for the contrast transfer function of the imaging system including the bronchoscope, image intensifier, and camera. The gamma and contrast transfer function was measured using test objects of known optical density (Kodak No. 2 transmission step tablet, calibrated on a densitometer), illuminated by a low power lamp. The object (tumor:mucosa) contrast for a thin, carcinoma _in situ_ lesion was 1.08, while the object contrast of a thicker tumor was 1.17. Although the contrast is somewhat lower than predicted by the Doiron model and data for thick tumors, it is in good agreement for a thin, carcinoma _in situ_.

Experiments are underway to measure the fluorescence yield and contrast as a function of time after injection, to be certain the optimum time for maximum fluorescence contrast with adequate tumor fluorescence yield is chosen for the examination. The method is being tested on DBA/2Ha mice with subcutaneously implanted SMT-F tumors. The mice are injected with HpD-in-saline at a dosage of 5 mg/kg, with uninjected mice as controls. At selected times after the injection, a mouse is sacrificed and a skin flap laid back for fluorescence measurements of the tumor and adjacent muscle. Fluorescence is excited by violet (405 nm) light from a 200 W mercury arc lamp equipped with a narrowband interference filter and a Schott BG-12 absorbing glass filter to minimize red background. The violet light is conducted down one leg of a bifurcated fiberoptic cable (which has randomly intermixed fibers in the common leg). The fluorescence and reflected source light are conducted up the other leg, and directed to an S-20 response photomultiplier tube after passing through a barrier filter identical to that used in the bronchoscope. The common leg of the bifurcated cable was almost in contact with the tumor, or muscle area, and the whole apparatus was calibrated against a fluorescence standard and normalized to standard violet power. Preliminary results, for large tumors, are plotted in Fig. 1.

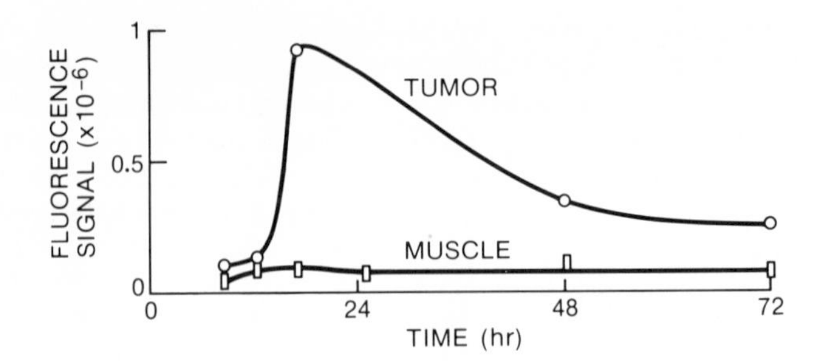

Fig. 1 Fluorescence of SMT-F tumor and muscle in DBA/2Ha mice injected with 5 mg/kg Photofrin, as function of time after injection.

In this situation, where autofluorescence and background is fairly large, maximum fluorescence contrast is achieved when the concentration (hence fluorescence signal) is highest in the tumor. Further delay only reduces contrast. The measurements were carried out by J. Sarnaik on mice prepared by Linda Wudl, Ph.D. Mice with SMT-F tumors were obtained with the kind assistance of T.J. Dougherty, Ph.D. of Roswell Park Memorial Institute. Experiments are continuing to refine the measurements, investigate small tumors, and extend the measurements to shorter times. There is evidence that there may be two narrow time intervals for maximum contrast, one shortly after injection, and another some time later. Techniques are being developed also to make *in vivo* measurements of fluorescence and study the kinetics of HpD uptake in human tumors.

DOSIMETRY

The dose in photoradiation therapy (PRT) is equal to the time integral of the dose rate, or duration of irradiation times average dose rate. The dose rate is proportional to the product of an effective space irradiance (energy flux density, W/cm^2, weighted by the probability for absorption in HpD and generation of singlet oxygen or other cytotoxic product) at the dose point, and the concentration of HpD (Profio and Doiron 1981). Table 1 lists factors expected to influence the effective space irradiance and concentration, as well as other factors that may complicate the dosimetry.

The effective space irradiance is proportional to source power, but is spectrum dependent because the absorption and scattering properties of the tissue, as well as the absorption and photodynamic yield of the photosensitizer (HpD) vary with wavelength. The diffusion of light in tissue, hence, the space irradiance at depth and the coupling between the space irradiance at the surface (or irradiator) and the source irradiance (or power), depend on the geometry. Common geometries in PRT are broad beam irradiation of a surface, and irradiation from a fiberoptic lightguide imbedded in the tumor. Assuming singlet oxygen is the predominant photodynamic product, oxygenation of tissue is significant, especially if subnormal.

Table 1. Factors Influencing the Dose Rate

Effective space irradiance:

A. Source power
B. Spectrum
C. Geometry
D. Optical properties of tissue
E. Optical properties of photosensitizer
F. Oxygenation of tissue

Concentration of photosensitizer:

A. Dosage
B. Time after injection
C. Physiological uptake and elimination rates
D. Tumor type, organ, biological variability

Other factors:

A. Heating and temperature effects
B. HpD chemical components and binding
C. Photodecomposition of photosensitizer

The HpD concentration depends on dosage, but the maximum dosage is limited by complications, in particular photosensitivity of the skin to strong light unless protected by clothing or a visible-spectrum opaque sunscreen. The new form of HpD, Photofrin II, may mitigate the problem. As shown in Fig. 1, the concentration (indicated by fluorescence) increases and then decreases in the tumor. The kinetics and maximum uptake in the nonmalignant tissue are different. Autofluorescence and source background do not matter in PRT, hence, the time delay should be chosen to maximize the concentration of HpD in tumor to the concentration in surrounding nonmalignant tissue, while allowing a reasonably high concentration in the tumor (as tumor concentration decreases the space irradiance, hence source power, or duration of irradiation, have to be increased to compensate). The actual concentration versus time profile is influenced by physiological uptake and elimination rates which may vary with tumor type and other variables.

Light, at least red light used in PRT, has no effect by itself unless the power densities and cooling by blood flow are such that there is an appreciable temperature rise (Svaasand, Doiron , and Dougherty 1983). Most photoradiation therapy procedures are designed so heating can be neglected. The chemical components and binding of HpD (Photofrin) are being studied by others. Treatment protocols should take into account the form of HpD used, and the dosage and dose for different cancers. Photodecomposition of HpD may be significant, at least for violet irradiation. The fluorescence of HpD in saline solution irradiated with violet (405 nm) at 10 mW/cm^2 was found to decrease by one-half in 18 minutes. If it turns out that photodecomposition is significant under actual PRT conditions, it may be necessary to monitor the HpD concentration and extend the duration of irradiation to compensate for the depletion.

Another consideration is that the space irradiance decreases with depth, and unless measures are taken to make the dose more uniform, it is possible that normal tissue near the source receives a larger dose than the tumor at depth, negating the differential effect of HpD concentration. Because of the shoulder in the dose-response curve, a small reduction in dose can have a profound effect. Thus, it is important to know the spatial distribution and absolute magnitude of the space irradiance.

Absolute Space Irradiance

The absolute space irradiance can be derived from measurements of the absolute radiance in a limited number of directions (even just one direction, if the angular distribution is isotropic or nearly so), by integrating the radiance over all solid angle. A nonperturbing detector is available in the form of a photomultiplier tube coupled to a single-fiber, fiberoptic lightguide imbedded in the tissue (Svaasand, Doiron, and Profio, 1981). We have calibrated such a system against a tungsten lamp fitted with a large diffuser and narrowband filter centered on 633 nm, for measurements in tissue irradiated with a helium-neon laser. The lamp was calibrated against a previously calibrated photodiode radiometer (EG&G Model 550 with flat filter and radiance lens). A correction was made for the change in numerical aperture of the fiberoptic when immersed in tissue instead of air. The calibrated fiberoptic-photomultiplier system was then used to measure the radiance in a beef brain specimen in planar geometry, irradiated with an expanded, large area, perpendicularly incident beam from the He-Ne laser. The irradiance I_o was measured as well, and the space irradiance $\phi(x)$ normalized to unit irradiance. Results are shown in Fig. 2. There is a region a few tenths of a millimeter thick, where the angular distribution is forward peaked and the reconstructed space irradiance decreases rapidly. At greater depths, after many collisions, the space irradiance decreases exponentially with an exponential decay constant $\alpha = 8.85\ cm^{-1}$, hence, 1/e-folding length of 0.113 cm, for this specimen and 633 nm wavelength. Extrapolated back to $x = 0$, we find $\phi(0) = 2.2\ I_o$.

We have found that the propagation of light in tissue, at least relatively low absorption tissue, is described rather well by diffusion theory, in regions not too close to the source or boundaries. The diffusion equation for the space irradiance is

$$D\nabla^2\phi - \mu_a\phi + s = 0 \qquad (1)$$

where D is the diffusion coefficient, ∇^2 = div grad is the Laplacian operator ($d^2\phi/dx^2$ in plane geometry), μ_a the absorption coefficient, and s the volume source (W/cm^3). In this case there is no volume source, $s = 0$, and the beam source is replaced by the boundary condition for the current

$$J^{+}(0) = \left[\frac{\phi}{4} - \frac{D}{2}\frac{d\phi}{dx}\right]_0 = I_o \tag{2}$$

The other boundary condition is that $\phi(\infty) = 0$. The solution is

$$\phi(x) = \frac{4}{1 + 2D\alpha} I_o e^{-\alpha x} \tag{3}$$

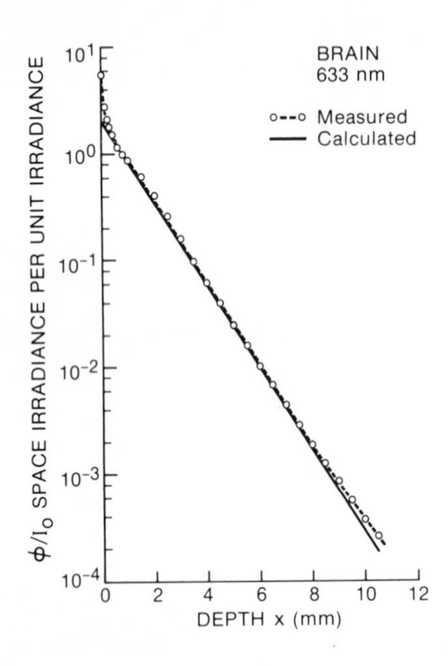

Fig. 2 . Measured space irradiance per unit source irradiance in beef brain at 633 nm wavelength, compared to diffusion theory calculation with same α, and D from the added-absorber experiment.

where
$$\alpha^2 \equiv \frac{\mu_a}{D} \tag{4}$$

Fig. 2 plots the diffusion theory result from Eq. 3, using the α from the measurements, hence, the slope in the central region is identical. In order to complete the calculation, the diffusion coefficient has to be measured independently. We have done so using a technique borrowed from neutron diffusion, where the "moderator" or scatterer has known amounts of absorber added to it.

From Eq. 4 one can see that if α^2 (derived from measurements of the spatial distribution) is plotted versus known added absorption coefficient μ_a, a straight line should be obtained with slope 1/D (as long as absorption is small enough that D is not appreciably changed) and intercept equal to the α^2 for intrinsic scattering and absorption. Figure 3 plots the results of such an experiment where the medium was the beef brain specimen, and Trypan red dye was the absorber, with its absorption coefficient at 633 nm measured on a spectrophotometer. Then from the inverse of the slope, D = 0.054 cm, and this value was used in Eq. 3, along with the α measured for no added absorber, to obtain the calculated curve in Fig. 2. Note that $\phi(0) = 2.05\ I_o$, in good agreement with the extrapolated measurements, although the transient near the surface is not reproduced.

It is realized, of course, that adding a known amount of dye and measuring spatial profiles is not very practical in patients. However, knowing a theoretical distribution and making a measurement at the surface and at perhaps one depth may be sufficient. For example, the absolute radiance (hence space irradiance) could be measured with a fiberoptic probe imbedded to the maximum depth of the tumor, and the source irradiance I_o is always known. If necessary, D might be obtained from a measurement of diffuse reflectance, which according to diffusion theory is

$$R = \frac{1 - 2D\alpha}{1 + 2D\alpha} \qquad (5)$$

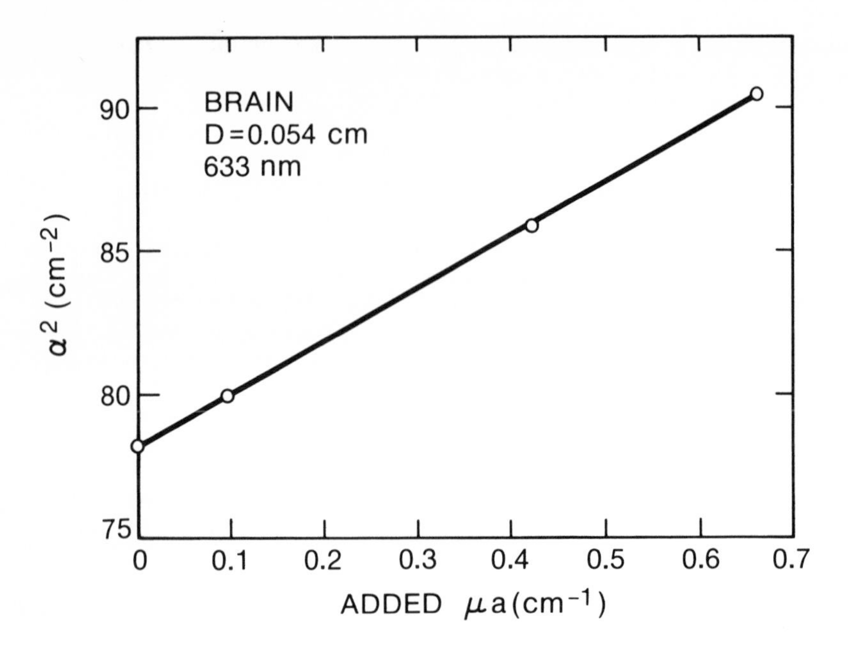

Fig. 3. Added-absorber experiment, with measured α^2 plotted against the absorption coefficient of the added dye, giving a line with slope 1/D.

Concentration of HpD

The fluorescence yield can be measured, but in order to translate it into a value for the concentration of HpD, corrections must be made for competing absorption, tissue scattering, and perhaps chemical quenching or other effects. We have developed a two-wavelength group model of fluorescence spatial distribution and emission, analogous to the model for propagation of the light from the source. If we designate the shorter-wavelength light (excitation) by subscript 1, and the longer-wavelength light (fluorescence) by subscript 2, we may write

$$D_1\nabla^2\phi_1 - \mu_{a1}\phi_1 = 0 \tag{6}$$

$$D_2\nabla^2\phi_2 - \mu_{a2}\phi_2 + y_{12}\mu_{h1}\phi_1 = 0 \tag{7}$$

where D is the diffusion coefficient and μ_a the (total) absorption coefficient averaged over the appropriate spectrum, y_{12} is the energy released as fluorescence per unit energy absorbed in HpD at the excitation wavelength, and μ_{h1} is the absorption coefficient of HpD at the excitation wavelength.

The solution of Eq. 6 is the same, for the same boundary conditions, as for Eq. 1 with s = 0. But the fluorescence is in general excited by the volume-distributed source from absorption of the excitation light. For the boundary conditions $\phi_2(0) = 0$ and $\phi_2(\infty) = 0$, we get

$$\phi_2(x) = \frac{y_{12}\mu_{h1}\phi_1(0)}{D_2(\alpha_1^2 - \alpha_2^2)} (e^{-\alpha_2 x} - e^{-\alpha_1 x}) \tag{8}$$

Very often the attenuation of the shorter-wavelength excitation light is much greater than the attenuation of the fluorescence light, $\alpha_1 >> \alpha_2$. In the limit, we have approximately

$$\phi_2(x) = \frac{y_{12}\mu_{h1}D_1}{\mu_{a1}D_2} \left(\frac{4I_o}{1 + 2D_1\alpha_1}\right) e^{-\alpha_2 x} \tag{9}$$

Furthermore, if we equate the fluorescence remittance F to the left-directed current density at the surface, we find for the fluorescence yield

$$Y = F/I_o = \frac{y_{12}\mu_{h1}D_1(1-2D_2\alpha_2)}{\mu_{a1}D_2(1-2D_1\alpha_1)} \tag{10}$$

The concentration C is related to the absorption coefficient by

$$\mu_{h1} = 2.3\ \varepsilon_1 C \tag{11}$$

where ε is the extinction coefficient. It should be noted from Eq. 10 that, aside from factors involving the diffusion coefficients and attenuation coefficients, the fluorescence yield depends on the ratio of the absorption in HpD to the total absorption (intrinsic tissue absorption plus HpD), which is reasonable. If the absorption coefficient of tissue without HpD can be measured, it should be possible to derive the concentration, and experiments are underway in our laboratory to develop the technique.

If the excitation light is the same wavelength as the light used in photoradiation therapy, and if the fluorescence efficiency has a constant relationship to the photodynamic efficiency, then the fluorescence itself is an indicator of the absorption rate and singlet-oxygen production rate. However, if the fluorescence is excited at a different wavelength, it is probably better to interpret measurements in terms of HpD concentration. For biopsy specimens, in an arrangement where background is subtracted by means of a blank, and the fluorescence signal is linearly related to concentration, the system could be calibrated by adding a known increment of HpD concentration and noting the incremental increase in fluorescence signal.

CONCLUSIONS

Carcinoma *in situ* can be detected by fluorescence of HpD, but contrast is small. The dosage and delay after injection should be chosen carefully to maximize fluorescence contrast.

The magnitude and spatial distribution of the space irradiance in tissue are described rather well by diffusion theory. Techniques have been developed to measure the diffusion coefficient and exponential attenuation coefficient in specimens. In actual practice of photoradiation therapy, monitoring the space irradiance (or radiance) at depth, with a calibrated detection system, may be necessary.

The concentration of HpD in tissue may be derived from measurements of fluorescence, corrected for absorption and scattering. It may be preferable to measure absolute concentration in biopsy specimens, given a suitable calibration procedure.

REFERENCES

Balchum OJ, Doiron DR, Profio AE, Huth GC (1982), "Fluorescence Bronchoscopy for Localizing Early Bronchial Cancer and Carcinoma in Situ." Recent Results in Cancer Research, Springer-Verlag, Berlin.

Doiron DR (1982). Fluorescence bronchoscopy for the early localization of lung cancer. Ph.D. dissertation, Univ. of California, Santa Barbara.

Profio AE, Doiron DR, King EG (1979). Laser fluorescence bronchoscope for localization of occult lung tumors. Med Phys 6: 523.

Profio AE, Doiron DR, Balchum OJ, Huth GC (1983). Fluorescence bronchoscopy for localization of carcinoma in situ. Med Phys 10:35.

Profio AE, Doiron DR (1981). Dosimetry considerations in phototherapy. Med Phys 8:190.

Svaasand LO, Doiron DR, Dougherty TJ (1983). Temperature rise during photoradiation therapy of malignant tumors. Med Phys 10:10.

Svaasand LO, Doiron DR, Profio AE (1981). Light distribution in tissue during photoradiation therapy. Report USC-MISG 900-02. Paper in preparation with recent results to be submitted to Medical Physics.

ACKNOWLEDGMENTS

The major part of this research was sponsored by the U.S. Public Health Service, National Institutes of Health, National Cancer Institute, under grant R01 CA 31865. Earlier work on fluorescence bronchoscopy was supported under NCI grant R01 CA 25582 and Department of Energy grant DE-AM03-76SF00113.

Porphyrin Localization and Treatment of Tumors, pages 177–186

PHOTOFRIN II LEVELS BY IN VIVO FLUORESCENCE PHOTOMETRY

William R. Potter and Thomas S. Mang

Division of Radiation Biology
Roswell Park Memorial Institute
Buffalo, NY 14263

To assay Photofrin II (Ph-II) levels in tissue by fluorescence, one needs to be aware of the effects of sample size and optical properties on the results. Clearly the total attenuation coefficient for light at the exciting wavelength and at the fluorescent wavelength will affect the fluorescent signal as will the size of the sample.

THEORY OF MEASUREMENT

To examine these effects, the diffusion theory of light (Svaasand, et al., 1983) was adapted to the following situation.

Consider a semi-infinite tissue with a plane wave at 630 nm incident on its surface. A fiber in contact with the surface is coupled through a lens and 690 nm interference filter to a silicon photodiode (Infrared Industries Type IRI 8016L).

The total attenuation coefficient is assumed to be the same at 690 nm and 630 nm.

A point on the surface of the tissue, beneath the face of the receiver fiber, receives a fluorescence space irradiance, F, at 690 nm determined by integration over the volume of the tissue as follows:

$$F = 2\pi K \int_0^{\pi/2} \sin\theta d\theta \int_0^{\infty} e^{-\alpha r} \frac{e}{r}^{-\alpha r\cos\theta} r^2 dr$$

Where α is the total attenuation coefficient for the diffusion dominated case

The term $e^{-\alpha r\cos\theta}$ is the exciting 630 nm irradiance at poin r.

The term $\frac{e^{-\alpha r}}{r}$ is the contribution to the 690 nm fluorescent space irradiance at r = 0 from the infinitesimal volume element dv at r. K is a constant which is the product of Φ_0, the incident irradiance at 630 nm, the porphyrin or Ph-II concentration, and the fluorescent efficiency. The absorption coefficien of Ph-II in tissues is contained in α, the total attenuation coefficient at 630 nm. The variation in total attenuation coefficient with levels of drug found in tissue is probably small. If a drug level were high enough to significantly increa α, the injected dose would have to be reduced to achieve the maximum therapeutic depth in treatment.

Integrating we find: $F = \frac{K\pi}{\alpha^2}$

Thus, it can be seen that F, the fluorescent space irradiance, varies directly as $1/\alpha^2$, the square of the attenuation length.

To investigate the effect of sample size, the integration was carried out allowing the upper limit on r to vary in multiples of $1/\alpha$. This calculation models a hemispherical fluorescent volume of radius r in the surface of a semi-infinite plane of identical optical properties. The results appear in Table 1: Where r is the radius of the fluorescent volume and $F\alpha^2/K\pi$ is the radially dependent term in F.

Table 1

$$F = \frac{K\pi(e^{-2\alpha r} - 2e^{-\alpha r} + 1)}{\alpha^2}$$

r	$F\alpha^2/K\pi$
$1/\alpha$	0.3995
$2/\alpha$	0.7476
$3/\alpha$	0.9526
∞	1.0

It is apparent that to rationally approach photoradiation therapy a knowledge of the attenuation coefficient of the tumor and neighboring tissue would be helpful. To measure the total attenuation coefficient by a non-invasive technique, a device was constructed according to the following rationale.

The end of a fiber transmitting 630 nm light chopped at 1 KHz was placed against the tissue to be measured. Chopping at 1 KHz produces a signal which when detected by a phase locked amplifier is free of interference from ambient light. The end of a receiver fiber was then placed in contact with the tissue at a distance r_1 and then at a larger distance r_2 from the transmitter fiber. The signal was detected and amplified with a photo diode and current amplifier. Alpha was determined from the measured values I_1 and I_2 at r_1 and r_2 using

$\Phi = \Phi_0/r\ e^{-\alpha r}$ twice to give,

$$\alpha = \frac{1}{r_1 - r_2} \ln \left\{\frac{\Phi_2\ r_2}{\Phi_1\ r_1}\right\}$$

It should be noted that diffusion theory applies only at distances greater than $1/\alpha$ and that several measurements at increasing r's may be necessary before consistent results for α can be obtained, i.e., one must have $1/\alpha < r_1$.

There is an unfortunate tendency to use the term "penetration depth" to denote $1/\alpha$. A literal interpretation of this phrase is extremely misleading. The term attenuation length should be used for $1/\alpha$ to avoid the impression that space irradiance goes to zero at $1/\alpha$. It should always be remembered that space irradiance is reduced by a factor of 0.37 for each $1/\alpha$ increase in distance from the illuminated surface, and is never zero.

MATERIALS AND METHODS

Instrument Description

To evaluate the usefulness of in vivo fluorescence as an assay for Ph-II it was necessary to construct a device to approximate the theory of a plane wave incident and a point detector on the surface of the tissues. This was accomplished by attaching a 630 nm transmitting fiber 1 cm from the tip of a receiving fiber. The receiver was then coupled by a lens through a 690 nm interference filter with a 10 nm band width (90% cut point) to a photo diode and current amplifier. Only the 690 nm fluorescent peak radiating from the tissue was detected (F in mv). Excitation at 630 nm was used for the same reason that treatment is done at 630 nm rather than at shorter wavelengths: penetration and sample volume is maximized. Placi the receiver fiber in contact with the tissue holds the transmitter 1 cm from the surface. To eliminate fluorescence quenching, the 630 nm excitation was switched on for two seconds only for each determination of F (Newport Research Model 845 digital shutter).

The 630 nm light was produced by a wedge tuned dye laser emitting 50 mw. Power control was achieved by feeding back to the light control circuits of the Argon laser, a signal generated by sampling the output of the dye laser (Spectra Physics 373 Dye Laser Stabilizer).

Preparation of ^{3}H Hpd and ^{3}H Ph-II

^{3}H Hpd was obtained by subjecting Hpd to tritium labeling via catalytic exchange labeling, performed by New England Nuclea Boston, Massachusetts, and as described by Gomer, et al. (1979).

^{3}H Ph-II was obtained by separating this component from Hpd using Bio-Gel P-10 (Bio-Rad, Richmond, California) as previously described by Dougherty, et al. (1983). Specific activity of thi compound was determined to be 1 mCi/mg. For in vivo experiments cold Ph-II solutions were prepared to a final concentration of 1 mg/ml, to which tritiated material was added yielding 2 to 3 x 10^{7} dpm/mg.

Tumor Systems

The EMT6 tumor line (Rockwell, et al., 1972) was propagated by intradermal inoculation of 2 to 4 x 10^{5} cells into the right axillary region of Balb/c mice. Tumors were used for measurements once they had attained a size of 6 to 8 mm in diameter.

The SMTF tumor system (Hosokawa, et al., 1973) was propagated by subcutaneous implantation of pieces of tumor, from donors, using a trochar, into the right axillary region of DBA/2 Ha-DD mice. Tumors were used for measurements once they had grown to be 6 to 8 mm in diameter.

^3H Ph-II Tissue Distribution Measurements

An i.p. injection of either 5, 10 or 20 mg/kg ^3H Ph-II was administered to the mice 24 hours prior to making the fluorescence measurements. The animals were euthanized and blood and tissue samples were removed. The "wet" weight of each tissue was then obtained. The samples were placed in individual vials to which 2 ml of Protosol (New England Nuclear, Boston, Massachusetts) was added. The samples were agitated and allowed to dissolve overnight at 37°C. Three ml of 30% H_2O_2 were then added to decolorize the sample. In preparation for liquid scintillation detection 12 ml of scintillation cocktail were added followed by addition of 1N HCl to neutralize the solution. The samples were again incubated at 37°C to reduce any chemiluminescence which may emit from the Protosol or other additives. Samples were also stored in the dark to diminish interference of phosphorescence.

RESULTS AND DISCUSSION

The results of fluorescent measurements in excised mouse tumors and tissues appears in Figs. 1-5 using tritium labeled drug to determine the tissue drug levels. Each data point represents the average of 20 mice. The results of attenuation coefficient measurements in mouse tissue appear in Table 2.

Table 2

DBA Mouse α in mm^{-1}

	α	$1/\alpha$
Liver	1.025	0.97
Tumor (SMTF)	0.289	3.4
Lung	1.68	0.59
Muscle	0.266	3.7
Brain	0.678	1.4
Kidney	1.015	0.98
Spleen	0.977	1.0

In similar experiments the EMT6 tumor from Balb/c mice as well as other tissues were also measured. Both tumor systems displayed similar sensitivities for fluorescence detection, as did muscle and lung. All of these tissues fell in the range of

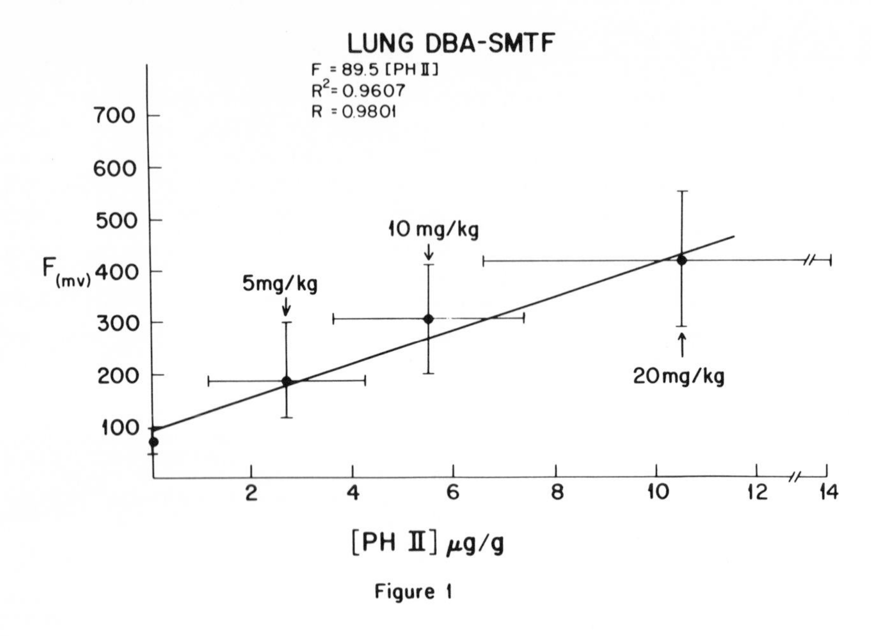

Figure 1

LEGEND: FIGURES 1-5

Abbreviations:

F = the fluorescence measured in the tissue in millivolts e.g. F = 37.5 [Ph-II]~F = 37.5 millivolts per µg/g of Ph-II

R = correlation coefficient

R^2 = coefficient of variance

The "error bars" represent the standard deviation from the mean value for a given injected drug dose (e.g. 10 mg/kg).

Each plotted point is a mean value for 20 mice. Thus, the error bars represent the variability in tissue uptake from mouse to mouse, and are very similar in magnitude for both tritium and fluorescence measurements.

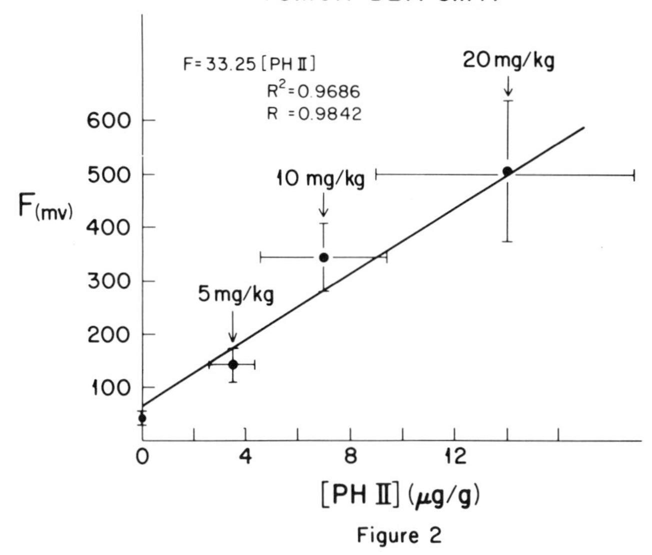

Figure 2

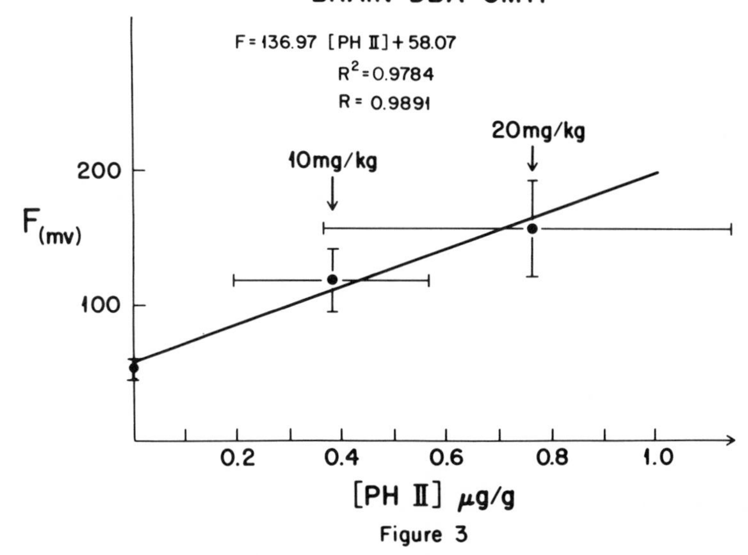

Figure 3

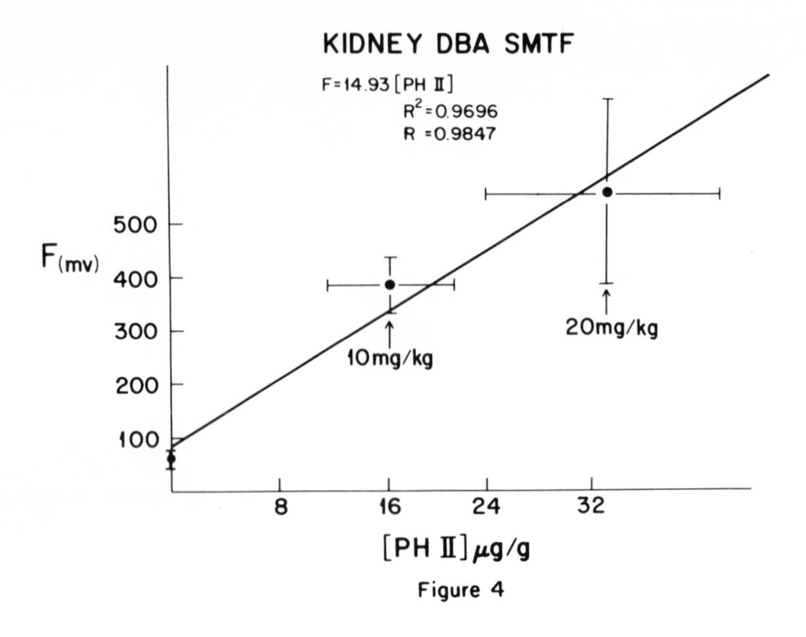

Figure 4

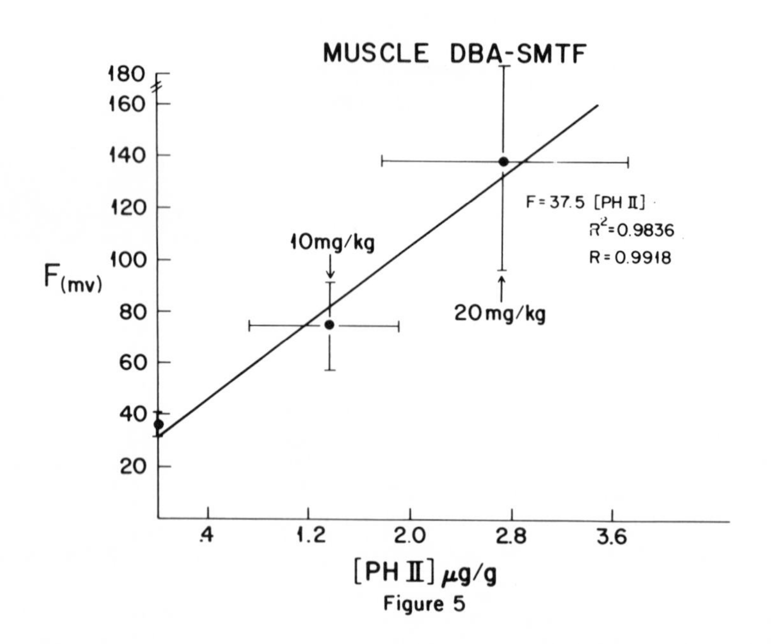

Figure 5

32 to 37 mv/μg/g. The sensitivity in the kidney was lower at 14.9 mv/μg/g as might be expected for a tissue of relatively great attenuation. The EMT6 tumor gave fluorescence values of F = 37.5 mv/μg/g [Ph-II], as compared with the SMTF tumor values of F = 38.5 mv/μg/g [Ph-II] (Fig. 2).

The brain, on the other hand, although between tumor and lung in α value, gave a value of 137.9 mv/μg/g indicating that those porphyrins retained by brain tissue (all data were at 24 hours post injection in mice) were four to five times as fluorescent as the drug in other tissues. It is possible that only the more fluorescent monomers or dimers cross the blood-brain barrier.

The time rate of serum clearance was determined in human patients for Hpd (Photofrin) and Ph-II (the isolated active component). The clearance was found to fit an exponential decay and had a half life of 10 hours for Hpd and Ph-II when levels were determined by fluorescence.

The linearity of drug fluorescence with concentration was established by adding measured amounts of drug to human serum in a 1 cm^2 plastic cuvette.

The serum was separated from the blood samples by centrifugation and placed in 1 cm^2 cuvettes. Fluorescence was measured by placing the receiver-transmitter fiber against the side of the cuvette under the same conditions as described above for tissue measurements.

Most tritium labeled compounds will deteriorate with time due to self-induced radiochemical breakdown. Indeed, this appears to be the case with Ph-II and Hpd. The deterioration manifests itself as an increase in tritium levels in the brains of mice (C. J. Gomer, personal communication) caused by a buildup of tritiated water as one of the radiochemical breakdown products.

In our work, this problem was made apparent as data were collected over time during the development of the instrument. Early data taken with fresh tritiated drug showed a strong linear correlation between fluorescence and ^3H levels. In later experiments this correlation began to deteriorate.

The use of fluorescence assay to quantify Ph-II levels is practical in the case of serum levels, skin levels, and in biopsy samples. Determination of both drug levels and attenuation coefficient is practical by non-invasive means in those situations

where a surface is accessible such as through skin, biopsy specim
surgical exposure, or possibly endoscopy.

Acknowledgment

The authors gratefully acknowledge the support and encouragement of Dr. Tom Dougherty, in whose laboratory this work was
performed. This work was supported in part by USPHS grant CA 309
02.

Dougherty TJ, Boyle DG, Weishaupt KR, Henderson BA, Potter WR, Bellnier DA, Wityk KE (1983). Photoradiation therapy - clinica and drug advances. In Kessel D, Dougherty T (eds): "Porphyrin Photosensitization," New York: Plenum Press, p 3.
Gomer CJ, Dougherty TJ (1979). Determination of ^{3}H and ^{14}C hematoporphyrin derivative distribution in malignant and normal tissue. Can Res 39:146.
Hosokawa MM, Orsini F, Mihich E (1973). Study of fast (f) and slow (s) growing transplantatable tumors derived from spontaneous tumors of the DBA/2 Ha DD mouse. Proc Am A Can Res 14:39
Rockwell SC, Kallman RF, Fajardo LF (1972). Characteristics of a serially transplanted mouse mammary tumor and its tissue cultur adapted derivative. J Nat Can Inst 49:735.
Svaasand LO, Doiron DR, Dougherty TJ (1983). Temperature rise during photoradiation therapy of malignant tumors. Med Phys 10(1):10.

Porphyrin Localization and Treatment of Tumors, pages 187–199

HpD PHOTODETECTION OF BLADDER CARCINOMA[1]

Chi-Wei Lin, David A. Bellnier, Makoto Fujime, and George R. Prout, Jr.

Urology Service, Massachusetts General Hospital and Department of Surgery, Harvard Medical School, Boston, MA 02114

INTRODUCTION

The potential application of hematoporphyrin derivative (HpD) for tumor detection is well recognized. Most malignant tumors retain higher levels of HpD than many normal tissues and subsequently emit a characteristic red fluorescence when excited with the proper wavelength of light. This has been demonstrated in a wide variety of human and animal tumors in a number of early studies (1-8). This potential application for tumor detection was greatly enhanced when more recent studies, with advanced instrumentation, demonstrated the detection of small and early stage bronchial carcinomas that were not readily detectable by convential bronchoscopy (9-13).

For bladder carcinomas, there are two important features which make the potential application of HpD-photodetection particularly appealing. The first is the accessibility of the bladder to fiberoptical instruments. The fact that the cystoscope is the tool for routine examination of bladder tumors makes the adaptation of HpD-photodetection as part of the examination relatively simple. The other is the complex nature of the disease. Carcinoma of the bladder is heterogenous, can be either single or multifocal and some with mixtures of different types of tumors. Carcinoma-in-situ (CIS) is known to associate with other types of tumors or can be present alone.

---

[1]Address communications to Dr. Chi-Wei Lin at Urology Research Laboratory, Massachusetts General Hospital, Boston, MA 02114

This type of tumor is difficult to detect and locate. A recent report by Benson et al. (14) has shown that HpD is taken up selectively by CIS, suggesting HpD-photodetection will have a good possibility for detecting CIS and therefore may have real value in clinical applications.

We now report our experience in the development of this detection. These include an attempt to use intravesical instillation as a means of HpD administration, a variation in effectiveness of HpD from different preparations and the development of HpD fluorescence in bladder tumor. A number of problems have been encountered. Continuing efforts to solve these problems will be important in the eventual success of the development of this new method of tumor detection.

INTRAVESICAL INSTILLATION OF HpD

A study was carried out to see if HpD, when instilled directly into the bladder, would be absorbed by the urothelium and concentrated in tumor tissues. This, if successful, would offer a more specific and localized distribution of HpD and therefore reduce the HpD-induced photosensitivity to other organs, including the skin.

A total of 14 cases have been studied. All were patients with transitional cell carcinoma (TCC) of the bladder and scheduled for cystectomy. HpD dosage was varied from 20 to 120 mg per patient while instillation time and volume of HpD were held constant at 90 minutes and 50 ml, respectively. The interval between HpD instillation and cystectomy was varied from 3 to 72 hours. Biopsies were taken before cystectomy from tumor and non-tumor areas of the bladder. Specimens were frozen-sectioned at 50 μ thickness and examined for HpD fluorescence under a fluorescence microscope fitted with special filters for HpD excitation and detection. Adjacent sections were cut at 5 μ thickness and stained with hemotoxylin and eosin. Both sections from the same biopsy were examined side by side so that the HpD fluorescence and the histology of the specimen could be correlated.

Results of this study are summarized in Table 1. The overall distribution of fluorescence did not correspond to the presence of tumor. Although we have observed weak to moderate levels of fluorescence in some of the tumor biopsies, the highest intensity of fluorescence in the specimens was often seen in areas of cellular degeneration, edema, and necrotic

Table 1. Summary of results on intravesical instillation of HpD and distribution of HpD-fluorescence in tumor and non-tumor biopsies of the bladder.

Case No	HpD dosage[1] (mg)	Interval[2] (hr)	Biopsies	Histology[3]	HpD-Fluorescence[4]
1	20	12	1	TCC, G II/III	+/-
			2	Chronic cystitis	+
			3	TCC, G III, invasive	+/-
2	40	12	1	TCC, G II/III	-
			2	TCC, G II/III, invasive	-
			3	Focal atypia	-
			4	Edema inflammatory, no TCC	+/-
			5	Edema inflammatory, no TCC	+
			6	TCC, G II/III, invasive	-
3	80	12	1	Atypical hyperplasia	+
			2	TCC with fibrosis	++
			3	TCC fragment	-
			4	Fibrosis, no TCC	-
			5	Focal CIS	-
4	80	12	1	Edema, degenerated epithelium	-
			2	Invasive TCC, G II, necrosis	-
			3	Invasive TCC, G II	+/-
			4	CIS	+/-
5	120	12	1	Chronic inflammation	+/-
			2	TCC, hyperplasia	+/-
			3	TCC, CIS	+/-

(Table 1 continued on next page)

Case No	HpD dosage[1] (mg)	Interval[2] (hr)	Biopsies	Histology[3]	HpD-Fluorescence[4]
6	120	12	1	Ulceration, inflammation, no TCC	+
			2	TCC, G II, non-invasive	+/-
			3	Atypia	+/-
			4	Degenerated, TCC	++ deg.,-TCC
			5	Degenerated, TCC	++ deg.,+/-TCC
			6	No abnormality	-
7	120	18	1	Squamous cell ca, invasive	+
			2	Cystitis, no tumor	+/-
			3	Cystitis, no tumor	+/-
8	120	24	1	Focal TCC, none-invasive	+/-
			2	TCC, invasive, G II	-
			3	Degenerated TCC	-
9	120	24	1	Necrosis, inflammation	-
			2	Necrosis, inflammation	+
			3	Necrosis, inlfammation	+/-
10	120	36	1	TCC	-
			2	Degenerated TCC	+
11	120	48	1	Degenerated CIS	+
			2	CIS	+/-
			3	CIS	+/-

12	120	54	1	Invasive TCC, G III	-
			2	Invasive TCC, G III	-
			3	CIS	-
13	120	72	1	Degenerated tissue	+/-
			2	Necrosis	+/-
			3	Degenerated tissue	+/-
14	120	3	1	TCC, G II	+/-
			2	Hyperplasia	-
			3	Submucosal tissue	-
			4	TCC, G I/ G II	+/-

[1] In all dosage, HpD was in saline solution with a total volume of 50 ml.

[2] Interval between the end of instillation and the time of pre-operative cytoscopy at which time biopsies were taken.

[3] Designation in histology: TCC-transitional cell carcinoma; G I to III refers to grade of tumor: G I: well differentiated, G II: moderately differentiated, G III: poorly differentiated.

[4] Intensity of fluorescence in tissue sections observed under fluorescence microscope was arbitrarily described as: strong (++), brilliant fluorescence; moderate (+), fluorescence can easily be seen; weak or trace (+/-), fluorescence can barely be seen; and negative (-), no fluorescence can be seen in the entire section.

tissue. Also, there was no fluorescence observed in a number of tumor biopsies. We conclude from these observations that intravesicular instillation most likely will not be a useful means of HpD administration due to the lack of specificity of HpD localization in tumor tissues and particularly, a significant percentage of tumor tissue did not take up HpD.

VARIATIONS IN DIFFERENT HpD PREPARATIONS

In the course of our investigation of HpD localization in bladder tumors, we have discovered that there was a marked difference among lots of HpD preparations obtained from the commercial source (Oncology Research and Development, Inc., Checktowaga, NY) in their effectiveness in localizing tumors. This observation was verified when different preparations were injected into rats bearing a transplantable bladder tumor; animals which received HpD from some preparations had intense fluorescence in their tumors while others receiving other preparations had weak or no fluorescence in the tumor.

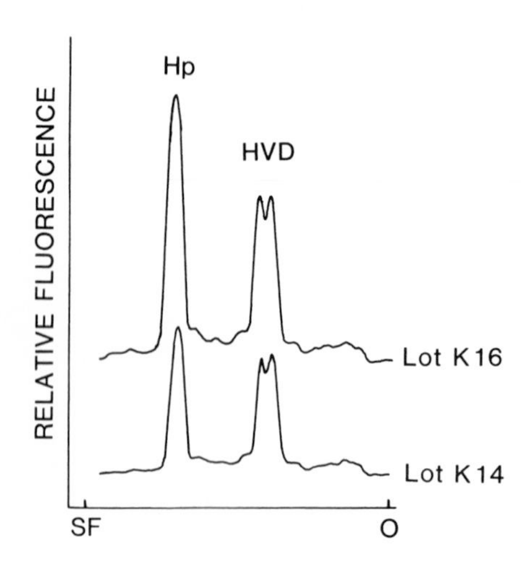

Figure 1. Reverse-phase TLC analysis of two commercial preparations (lots K16 and K14) of HpD in Whatman KC 18 TLC plates developed in solvent of methanol:3 mM tetrabutylammonium phosphate, pH 3.5 (65:35), and scanned on a Perkin-Elmer Model 650 fluorescence spectrophotometer with 400 nm excitation and 612.5 nm emission. HpD from lot K16 yielded intense fluorescence in tumors when injected into rats while lot K14 yielded no visible fluorescence.

In vitro analyses were carried out in an attempt to uncover any compositional difference between two preparations that exhibited a vast difference in their tumor localizing property. However, both TLC and HPLC analyses yielded essentially identical patterns for the two preparations (Figure 1 and Figure 2). The fluorescence yields from these two were also identical. Apparently, other factors such as physical state or molecular aggregation of the porphyrin may influence the uptake and distribution of HpD in animal and tumor tissues.

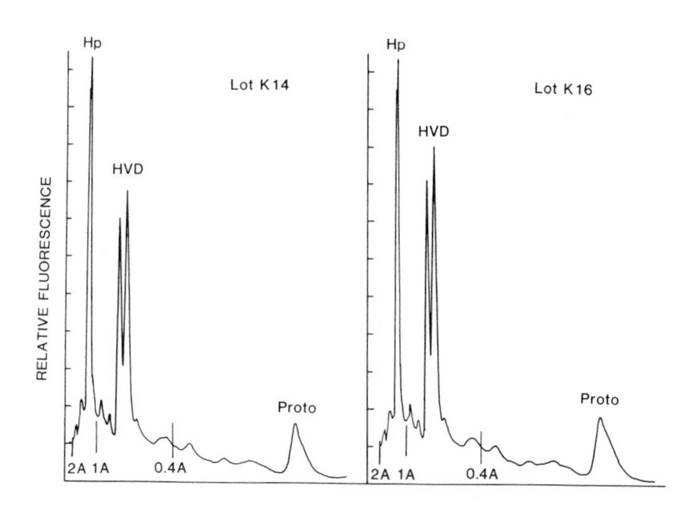

Figure 2. HPLC analysis of the same two HpD preparations described in Figure 1 using Bondapak C18 column and methanol/water/acetic acid (80:20:4) as solvent. Porphyrin components in the eluate were detected by absorption at 401 nm.

DEVELOPMENT OF AN INSTRUMENT FOR THE DETECTION OF HpD-FLUORESCENCE IN BLADDER TUMOR

An important element in the application of HpD for photodetection of tumors is the instrumentation to be used for excitation and detection of the fluorescence. In order for HpD-photodetection to have a real value in the detection of bladder tumor, where cystoscopy is the current standard method, an instrument must be able to detect tumors that are not detectable by cystoscopy. These usually are tumors that are very small, flat in shape or CIS. HpD-fluorescence generated by these types of tumors is usually very low and most likely not visible by the eye due to the thinness of the tumor and thus the small total quantity of HpD. Doiron et al. (9) have estimated that HpD-fluorescence yield by a tumor of 100 μm in thickness was only 7 X 10^{-5} red photons per violet photon incident. Therefore, strong illumimation and adequate amplification of signals are essential in detecting these types of tumors.

Modified from an instrument developed by Kinsey et al. for the detection of HpD in bronchocarcinoma (11, 12) and with the collaboration of Dyonics, Inc. (Andover, MA) we have constructed a prototype instrument for HpD-photodetection of bladder tumors. The main specifications of this instrument are listed in Table 2 and the major components are shown in Figure 3. The light source for HpD excitation is a 50-watt mercury vapor lamp, powered by an AC current of 120 Hz. A band pass filter of 405±10 nm is used to obtain the specific violet light for HpD excitation. A bifurcated light conducting fiber bundle, wrapped inside a 2.3 mm polyurethane ureteral catheter, is used to channel the exci-

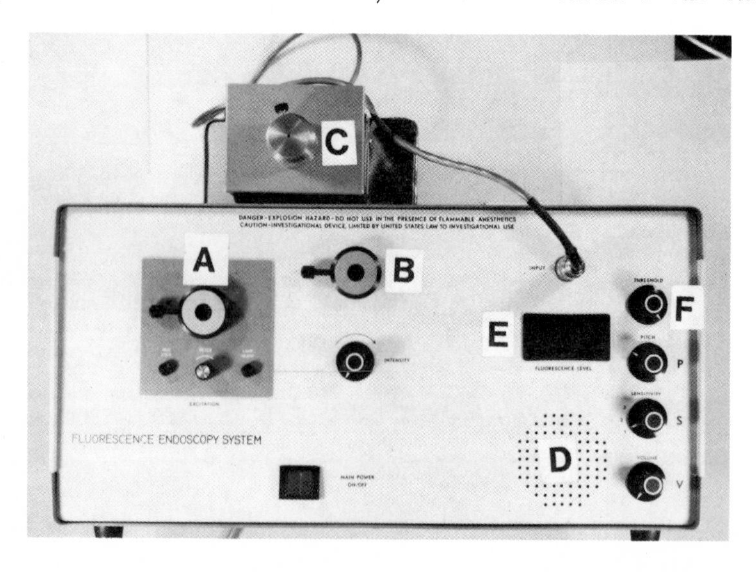

Figure 3. Major components of a prototype instrument designed for HpD-photodetection of bladder tumor. (A) Blue-violet light source for HpD excitation: (B) White light source for cystoscope illumination; (C) Photodiode light detector; (D) Audio signal output; (E) Digital voltimeter display; (F) Threshold dia

tation light into the bladder and carry the emitted fluorescence to the light detector. The latter is a 3 mm silicon photodiode and is situated behind the detector filter which has a band pass wavelength of 630±10 nm. The electronic signal is amplified and converted into an audio signal via a voltage controlled oscillator. The pitch of the audio signal is proportional to the intensity of the fluorescence detected. The catheter which contains fiber bundles for both excitation and detection can be inserted into one of the channels of a standard cystoscope. For the illumination of the cystoscope, the instrument has a separate light source which has a 150 Watts, 15 Volts incandescent lamp with 60 Hz DC current. Since the illumination light and the excitation light are in different phases, the interference of the white illuminating light to the excitation light and emitted fluorescence is eliminated by using a phase specific circuit for the light detector in phase only with the excitation light. The instrument also has a digital voltmeter in which the display is proportional to the signal output, and a threshold dial with which the background signal can be adjusted or eliminated.

This instrument is designed to allow investigators to conduct routine cystoscopy and at the same time, listen to the audio signals triggered by a HpD containing areas in the bladder. Biopsies can then be taken from these areas for pathological verification of the presence of tumors.

In vitro evaluation of the instrument using filter paper (Whatman No. 1) containing different concentrations of HpD indicated that the instrument can detect as little as 30 ng/ml of HpD, measured at a distance of 5 mm. This level of sensitivity is comparable to that of Kinsey's instrument with which detection of small, *in situ* bronchial and bladder carcinomas has been demonstrated (13, 14). The instrument also has been tested for signals generated from various background materials. More reflective surfaces such as filter paper and a mirror produced higher signals while most tissues, non-HpD containing tumors, saline solution and blood produced relatively low levels of signals. This indicates that background signals most likely will not be a significant problem to the actual operation of the instrument. The evaluation of the instrument in patients with bladder tumor is now underway.

Table 2. Summary of the specification for the prototype instrument for photodetection of bladder tumor.

Light source for HpD excitation	50W Hg arc lamp, AC, 120 Hz
Excitation filter	405±10nm, 1 in dia.
Detector optical fiber	2.3 mm (7 fr.) dia.
Excitation light intensity	0.5 mW
Detector filter	630±10 nm, 1 in dia.
Light source for cystoscope illumination	150W, 15V incandescent lamp, DC, 60 Hz
Discrimination for light sources	Phase-specific circuit
Photodetector	3 mm silicon photovoltaic photodiode
Output signals	Audio and voltage display
Minimum detectable HpD concentration	30 ng/ml at a distance of 5 mm

DISCUSSION

Not uncommon in a study at its early stage, so far we have generated more questions than answers. However, these questions do provide directions and goals for further investigations.

One of the prominent questions that remains to be answered is the mechanism of higher retention of HpD in tumor tissues. A greater understanding of this mechanism and its relation to the processes of HpD uptake and distribution in tissues and cells undoubtedly will provide a sounder basis for the development of techniques for the photodetection of tumors.

Another important and related question is the chemical and physical nature of HpD, its significance in tissue and tumor uptake and the changes resulting from interactions with tissue and cellular environments. A better understanding of these may improve the quality of HpD in tumor localization and reduce the variation in different lots of HpD preparations. In the meantime, one must devise a system for better quality evaluation so that HpD of less quality can be eliminated from studies.

Still another question remains to be answered is whether HpD localizes in all bladder tumors or only in certain types, sizes or grades of TCC. Benson et al. (14) have shown that high levels of HpD-fluorescence were generated from invasive TCC and CIS. In our small number of cases studied so far, we have observed that large, papillary tumors had very low or no detectable HpD-fluorescence. Systematic studies shall be done to examine these possible correlations.

In regards to the instrumentation for HpD-photodetection, continuous improvement must be made not only on the specificity and sensitivity of the detection, but more importantly, the reduction of dependency of distance between the object and the detector probe.

Compared to HpD-photoradiation therapy, where most of the patients are at late stages of their disease, photosensitivity is a greater concern for patients involved in HpD-photodetection studies. We consider in two groups of bladder cancer patients that HpD-photodetection will have potential clinical values. One is with CIS and the other is with tumor already resected and in a course of evaluation and treatment for residue or recurrent tumors. For these groups of patients, urologists are more concerned about maintaining their normal life activities and therefore, the possibility of photosensitivity becomes a greater problem. Reduction of photosensitivity of HpD therefore will bring about a greater degree of acceptance and possibly, a wider application of this new method of tumor detection.

ACKNOWLEDGMENTS

This study was supported by a grant (R26-CA-29529) from the National Cancer Institute.

REFERENCES

1. Figge FH, Weiland GS, Mangiello LOJ (1948). Cancer detection and therapy. Affinity of neoplastic, embryonic and traumatized tissues for porphyrins and metallophor-

phyrins. Proc Soc Exp Biol Med 68:640.

2. Rassmussen-Taxdal DS, Ward GE, Figge GHJ (1955). Fluorescence of human lymphatic and cancer tissues following high doses of intravenous hematoporphyrin. Cancer 8:78.

3. Lipson RI, Baldes EJ, Olsen AM (1961). Hematoporphyrin derivative: A new aid for endoscopic detection of malignant disease. J Thoracic & Cardiovas Sug 42:623.

4. Lipson RL, Pratt JH, Baldes EJ, Dockerty MB (1964). Hematoporphyrin derivative for detection of cervical cancer. Obst Gyn 24:78.

5. Lipson RL, Baldes EJ, Gray MJ (1967). Hematoporphyrin derivative for detection and management of cancer. Cancer 20:2255.

6. Gregorie HB, Green JF (1965). Hematoporphyrin-derivative fluorescence in malignant neoplasmas. J of S Carolina Med Ass 61:157.

7. Gregorie HB, Horger EC, Ward JL, Green JF, Rochards T, Robertson HC, Stevenson TB (1968). Hematoporphyrin derivative fluorescence in malignant neoplasms. Annals of Surg 167:820.

8. Kelly JF, Snell ME 91978). A possible aid in the diagnosis and therapy of carcinoma of the bladder. J urol 115:150.

9. Doiron DR, Profio E, Vincent RG, Dougherty, TJ (1979). Fluorescence bronchoscopy for detection of lung cancer. Chest 76:1.

10. Profio AE, Doiron DR (1977). A feasibility study of the use of fluorescence bronchoscopy for localization of small lung tumors. Phys Med Biol 22:949.

11. Kinsey JH, Cortese DA, Sanderson DR (1978). Detection of hematoporphyrin fluorescence during fiberoptic bronchoscopy to localize early bronchogenic carcinoma. Mayo Clin Proc 53:594.

12. Kinsey JH, Cortese DA (1980). Endoscopic system for

simultaneous visual examination and electronic detection of fluorescence. Rev of Sci Instr 51:1403.

13. Cortese DA, Kinsey JH, Woolner LB, Payne WS, Sanderson DR, Fontana RS (1979). Clinical application of new endoscopic technique for detection of *in situ* bronchial carcinoma. Mayo Clin Proc 54:635.

14. Denson RC, Farrow GM, Kinsey JH, Cortese DA, Zincke H, Utz DC (1982). Detection and localization of *in situ* carcinoma of the bladder with hematoporphyrin derivative. Mayo Clin Proc 57:548.

Porphyrin Localization and Treatment of Tumors, pages 201–210

PICOSECOND FLUORESCENCE OF HEMATOPORPHYRIN DERIVATIVE, ITS COMPONENTS AND RELATED PORPHYRINS

CA Hanzlik, WH Knox, TM Nordlund, R. Hilf, and
SL Gibson
University of Rochester
Rochester, N. Y. 14627

INTRODUCTION

Because of their properties of localization, (Figg et al., 1948) and photodestruction of malignant tumors (Spikes 1975), porphyrins have been investigated as photosensitizers for treatment of human cancers (Dougherty et al., 1977; Moan and Christensen 1980). In general, photoradiation therapy consists of exposing a tumor to light after injection with hematoporphyrin derivative (HPD) which is a mixture of porphyrins (Moan and Sommer 1981). The mechanisms enabling the accumulation of HPD and the subsequent photodestruction of tumors are not well understood but are thought to involve the aggregational properties of the components of HPD (Moan et al., 1982; Dougherty et al., 1982; Andreoni et al., 1982). In particular, there is evidence that the most hydrophobic, and perhaps the most aggregated porphyrin (as injected in aqueous solution) in HPD, is the most efficient photosensitizer in vivo (Moan et al., 1982; Dougherty et al., 1982). While the mixture, HPD, has been studied in regard to its absorption and fluorescence properties (Andreoni et al., 1982; Berns et al., 1982), few studies of the individual components of HPD have been done (Moan and Sommer 1981). Specifically, little information of picosecond time-resolved fluorescence has been presented (Yamashita et al., 1982).

In this paper, the study of the photophysics of HPD, its components, and related porphyrins was undertaken to obtain further evidence for the aggregation and energy transfer properties of HPD. In particular, we studied the de-excitation of the porphyrin singlet excited state through the analysis of the fast fluorescence kinetics. We demonstrated more direct evidence for the existence of aggregates through the presence of radiationless decay and and singlet-singlet annihilation terms in the fluorescent decay on a picosecond time scale. Such an approach followed the lead of Nordlund and Knox (1981) who demonstrated the existence of annihilation effects in aggregates of chlorophyll-proteins. The recent establishment of a unique laser facility enabled us to investigate fluorescence of low quantum yield samples with picosecond resolution. Fluorescence was measured using a 530 nm 30 ps pulses from a frequency doubled Nd:YAG laser with a low jitter (<2 ps) streak camera-optical multichannel analyzer detection system. Accumulation of the signal often required one hundred "shots" making low time jitter essential for aquisition of data on this fast time scale.

MATERIALS AND METHODS

HPD was prepared according to the method of Lipson et al. (1961). All porphyrins were dissolved in aqueous sodium hydroxide (0.1 N). HPD components were obtained using reverse phase chromatography (Gibson et al., 1983). The fractions were dried and resolvated in an aqueous solution of 25 mM HEPES, 10 mM $NaHCO_3$, 3 mM K_2HPO_3, 1 mM $CaCl_2 \cdot H_2O$ and 125 mM NaCl with a total osmolarity of 309 mOsM, 7.4 pH. Using this method, five fractions (1, 2, 3, 4, and 5 from most hydrophilic to most hydrophobic) were obtained. Samples of hematoporphyrin (HP), hydroxyethylvinyl deuteroporphyrin (HVD), protoporphyrin IX (PP), and mesoporphyrin (MP) were obtained from Porphyrin Products (Logan, Utah).

Absorption was performed using a Perkin-Elmer Lambda 3 Spectrophotometer. Fluorescence excitation and emission were done on a Perkin-Elmer MPF-44A Spectrophotometer. Picosecond time-resolved fluorescence was done on a source and detector as described elsewhere (Mourou et al., 1980; Stavola et al., 1980) which consist of an actively and passively mode-locked Nd:YAG laser from which one pulse is selected by a dual Pockels cell arrangement and then amplified; and a low jitter streak camera (Photochron II) with attached image intensifier (EMI) and OMA (Princeton Applied Research). A 2w crystal is used to convert the laser's fundamental 1060 nm to 530 nm. Pulse energies were typically between 5 and 20 uJ. Sample fluorescence was excited with a single 30 ps 530 nm laser pulse, repetition rate of 0.5 Hz, mounted 15 cm from a 20 cm focal length cylindrical lens. The cuvette was carefully masked to avoid scattering and secondary fluorescence detection. The line of excited fluorescence formed, in effect, the slit of the streak camera. Glass neutral density filters were used to vary the photon density incident on the sample and Schott BG 18 filters (Schott Optical Glass Inc., Duryea, PA) blocked 1060 nm light. Fluorerscence was filtered through Kodak Wratten gel neutral density filters (Eastman Kodak Co., Rochester, NY) and Schott KV 550 and OG 590 longpass before detection. The low time jitter of the streak camera allowed for simple addition of many single "shots" in the multichannel analyzer. Data consisted of the summation 100 "shots" or more with a background subtracted out in order to obtain data suitable for interpretation.

RESULTS

The absorption (1 mM solutions) and the fluorescence (10 uM solutions) spectra of the components were measured and are presented in Table 1. Figure 1 shows data obtained for fraction 5. All components showed a strong absorption peak at about 365 nm as well as a strong fluorescence excitation peak at about 396 nm. The picosecond fluorescence (1 mM solutions) of fraction 5 (Fig. 2), PP, and MP all showed a fast decay (< 20 ps) independent of intensity and a long decay (> 0.5 ns). In contrast, fraction 4 showed an intensity dependent fast decay and the slow decay (Fig. 3, A and B) whereas fractions 1, 2 (Fig. 3C), 3, HPD, HP and HVD all showed the intensity independent long decay only. In methanol, all the porphyrins showed the long decay only.

DISCUSSION

We have measured the time and excitation intensity dependence of HPD components. This, coupled with the absorption and fluorescence data, tends to confirm the existence of aggregates and monomers in equilibrium in aqueous solution (Moan and Christensen 1980; Dougherty et al., 1982; Andreoni et al., 1982). Specifically, we see evidence for aggregates in terms of the highly absorbing (365 nm) species, the quickly quenched excited state in fraction 5 and the intensity dependent decay in fraction 4. Our results also indicate a monomeric form based on a highly fluorescing species (396 nm excitation), and the long fluorescent decay present in all the components.

TABLE 1

FRACTION	ABSORPTION PEAKS ($\pm$ 1 NM)				
1	370	505	540	567	618
2	369	504	538	566	617
3	366	505	538	566	617
4	365	506	537	572	623
5	363	504	537	569	623

FLUORESCENCE PEAKS* ($\pm$ 1 NM)

FRACTION	EXCITATION					EMISSION	
1	396	502	537	560	607	614	677
2	396	501	536	559	608	614	676
3	397	503	539	562	608	617	678
4	397	504	540	563	608	617	679
5	395	593	537	563	604?	615	677

* Fluorescence excitation monitored at the shortest emission wavelength except for the longest excitation wavelength; fluorescence emission excited at 397 nm.

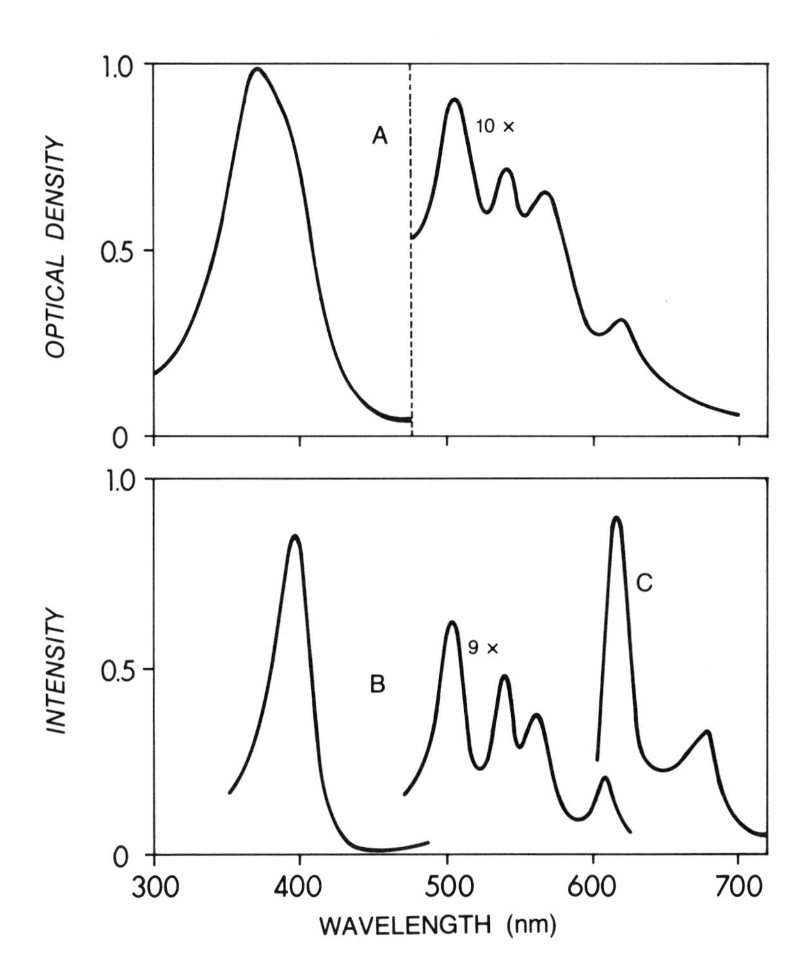

FIGURE 1

Absorption (A), fluorescence excitation (B), and emission (C) of HPD fraction 5. The long wavelength region of the absorption is shown as ten times the actual absorption; the long wavelength region of the fluorescence excitation is shown as nine times the actual fluorescence.

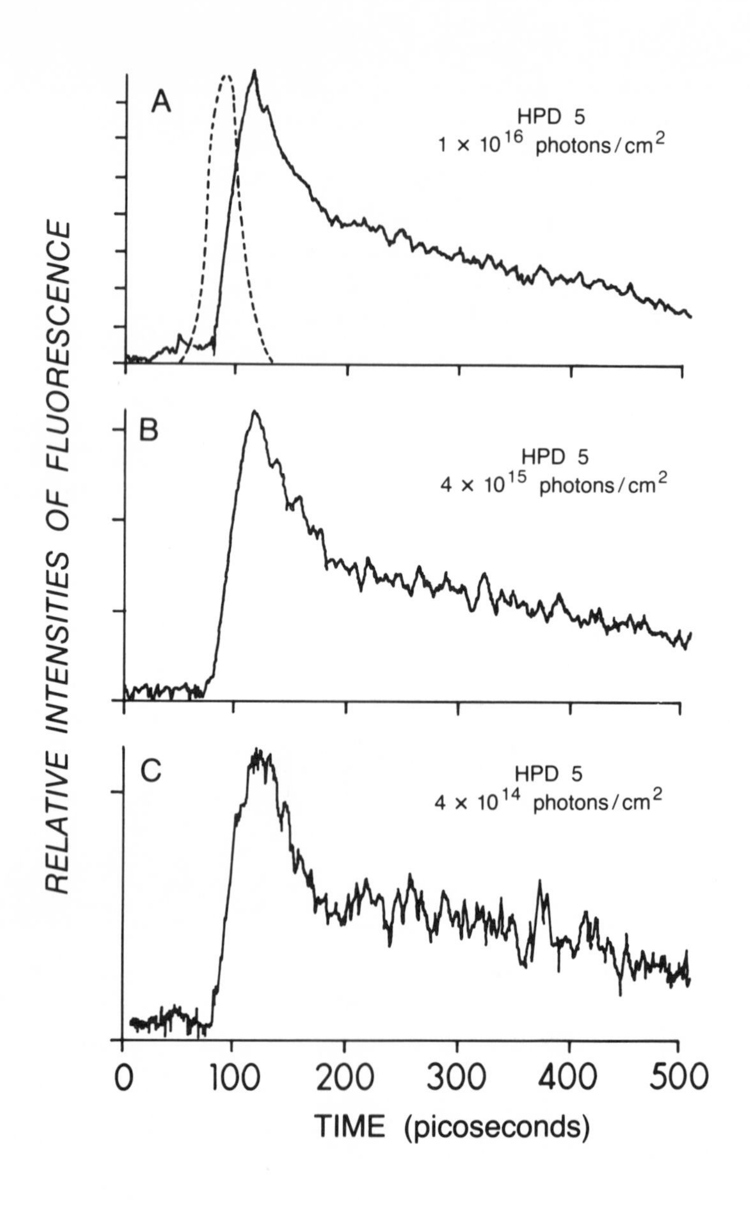

FIGURE 2

Intensity independent fast decay of HPD fraction 5. Excitation pulse from the laser is shown in dashed lines in A.

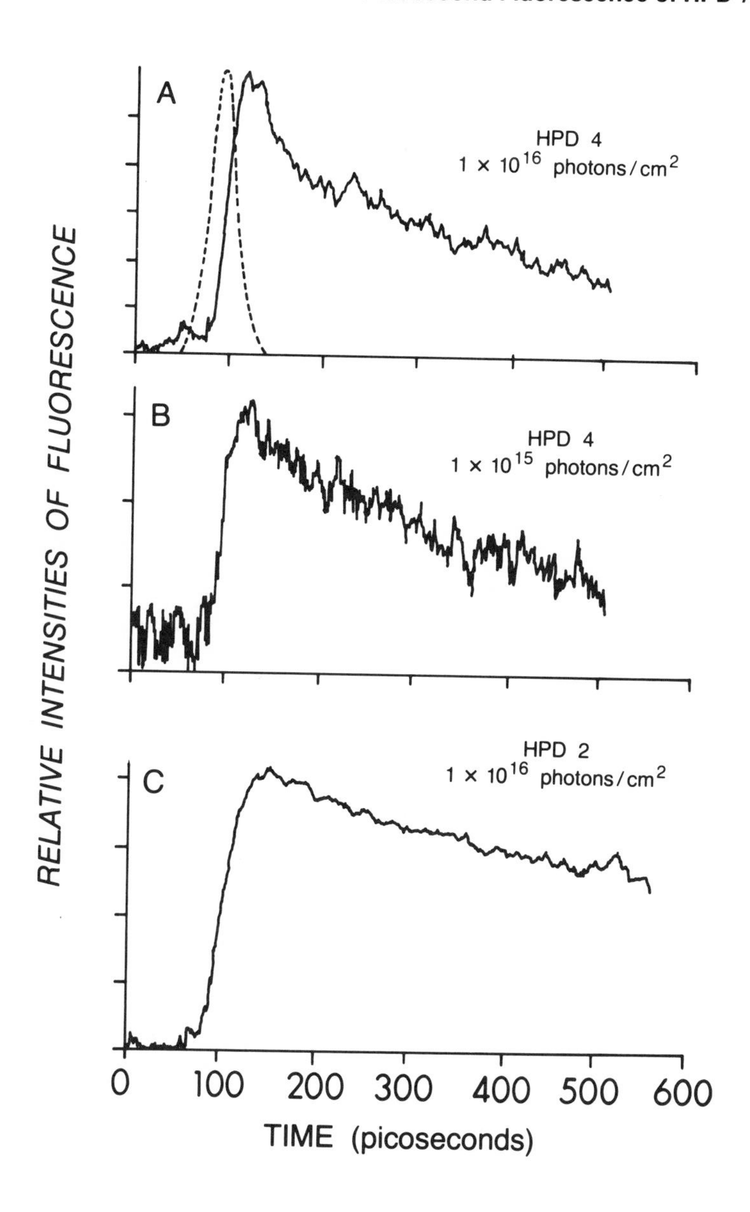

FIGURE 3

Intensity dependent fast decay of fraction 4 (A and B) and intensity independent slow decay of fraction 2 (c). Excitation pulse is in dashed lines.

In general, the fluorescence of species "i" may be evaluated in terms of a differential kinetic equation of the form:

$$dN_i/dt = [N_{io}-N_i(t)]S_i(t)I(t) - N_i(t)/T_i - G_{iss}N_i(t)^2 - K_{ij}\, N_i(t)$$

where N_{io} is the concentration of the ground state, S_i is the absorption cross section of the ground state, I(t) is the intensity of the excitation pulse, N_i is the concentration of the fluorescing excited state, T_i is the natural radiative lifetime of the species, G_{iss} is the bimolecular annihilation rate, and Kij is the radiationless energy transfer rate from state i to j. With more than one species present in solution, one must sum the effects of all "i" species to obtain the total time-resolved fluorescence of the solution.

According to our interpretation of these data, the fast intensity independent decay is evidence for the existence of a large K_{ij}; i.e., there is efficient radiationless decay of the singlet excited state to some other state j (perhaps to a triplet state via spin-orbit coupling). The intensity dependent fast decay is evidence for the existence of a G_{iss}. When the average distance between molecules is of the order of 100 angstroms or less, resonance energy transfer can exist between neighboring molecules (Geacintov and Breton 1982). Such effects are possible when the local concentration of molecules exceed 1 mM. If a sufficient density of excitation are present, bimolecular interactions with the annihilation of one or both excitations is possible (Geacintov and Breton 1982). The result of such annihilation may be the production of ground singlet or triplet states. (Breton and Geacintov 1980). The long decay in methanol attributed to the monomer species yields T_i most reflective of the natural radiative lifetime, since this is the most isolated form of the component in solution.

Thus, this preliminary study of the photophysics of HPD and its components provides a novel approach to understanding the aggregation and energy transfer properties of porphyrins important in the process of photosensitization of malignant tumors. In particular, we find evidence for two possible mechanisms for the production of the porphyrin triplet states that are thought to be involved in this process.

ACKNOWLEDGEMENTS

Supported in part by NSF (PCM-80-18488); The American Cancer Society (IN-18Wsub); the sponsors of the Laser Fusion Feasibility Project at the Laboratory for Laser Energetics; and the Standard Oil Research Center (04-129-4109-AA), Cleveland, Ohio. We acknowledge L. Forsley, R. Knox, and B. Wittmershaus for the development of a FORTH optimization routine for data analysis; and D. Williams of Xerox Corporation for the use of the fluorescence spectrophotometer.

REFERENCES

Andreoni A, Cubeddu R, De Silvestri S, Laporta P (1982). Experimental evidence for an aggregated species. Chem Phys Let 88:33.

Berns MW, Wile A, Dahlman A (1982). Cellular uptake, excretion and localization of hematoporphyrin derivative. In Kessel D, Dougherty TJ (eds): "Porphyrin Photosensitization," Plenum Publishing Corporation, p 139.

Breton J, Geacintov N (1980). Picosecond fluorescence kinetics and fast energy transfer in photosynthetic membranes. Biochim Biophys Acta 594:1.

Dougherty T, Boyle D, Weishaupt K, Gomer C, Borcicky D, Kaufman J, Goldfarb A, Grindey C (1977). Phototherapy of human tumors. In Castellani A (ed): "Research in Photobiology," Plenum Publishing Corporation, p 435.

Dougherty T, Boyle DG, Weishaupt KR, Henderson BS, Poller WR, Bellinier DA, Wityk KE (1982). Photoradiation therapy - clinical and drug advances. In Kessel D, Dougherty TJ (eds): "Porphyrin Photosensitizatioon," Plenum Publishing Corporation, p 139.

Figg FHJ, Wieland GS, Manangiello LOJ (1948). Affinity of neoplastic, embryonic and traumatized regenerating tissues for porphyrins and metalloporphyrins. Proc Soc Exp Med 68:240.

Geacintov NE, Breton J (1982). Exciton annihilation and other nonlinear high-intensity excitation effects. In Alfano RR (ed): "Biological Events Probed by Ultrafast Spectroscopy," Academic Press, Inc., p 157.

Gibson SL, Leakey PD, Crute JJ, Hilf R (1983). Photosensitization of mitochondrial cytochrome c oxidase by hematoporphyrin derivative (HpD) in vitro and in vivo. To be published in Doiron DR (ed): "Porphyrin Localization and Treatment of Tumors," New York: Alan R. Liss.

Lipson RL, Baldes EJ, Olsen AM (1961). J Natl Cancer Inst 26:1.
Moan J, Christensen T (1980). Porphyrins as tumor localizing agents and their possible use in photochemotherapy of cancer, a review. Tumor Research 15:1.
Moan J, Sommer S (1981). Fluorescence and absorption properties of the components of hematoporphyrin derivative. Photobiochem and Photobiophys 3:93.
Moan J, Christensen T, Sommer S (1982). The main photosensitizing components of hematoporphyrin derivative. Cancer Let 15:161.
Mourou G, Knox W (1980). A picosecond jitter streak camera. Appl Phys Let 36:623.
Nordlund T, Knox W (1981). Lifetime of fluorescence from light harvesting chlorophyll a/b proteins. Biophys J 36:193.
Spikes JD (1975). Porphyrins and related compounds as photodynamic sensitizers. Annals NY Acad Sci 244:496.
Stavola M, Mourou G, Knox W (1980). Picosecond time delay fluorimetry using a jitter-free streak camera. Optics Comm 34:404.
Yamashita M, Sato T, Aizawa K, Kato H (1982). Picosecond fluorescence of hematoporphyrin derivative and related porphyrins. In: "Picosecond Phenomnon III," Springer-Verlag, p 298.

Porphyrin Localization and Treatment of Tumors, pages 211–225

A COMPARISON OF SPECTRAL TRANSMITTANCE FOR SEVERAL MAMMALIAN TISSUES: EFFECTS AT PRT FREQUENCIES

F. P. Bolin, L. E. Preuss and B. W. Cain

Division of Radiation Physics
Department of Therapeutic Radiology
Henry Ford Hospital, Detroit, MI 48202

Studies were carried out in vitro to explicate the spectral transmission of visible light and the near IR by various tissues with regard to application in PRT. The experiments spanned 400 to 1100 nanometers with attention to the red, PRT region. Spectra were determined with an EG&G spectroradiometer system. Four fresh tissue samples from each of several tissue types, ranging in thickness from 1.55 mm to 16 mm were tested. A micro computer processed the raw data and plotted spectra. Interspecies studies were done. Spectra at 10, 5 and 2 nm resolution were developed. Light half value layers (hvl) derived from spectrographic data were compared with those from internal fiber optic detectors

Performing optical measurements on tissue presents an interesting but difficult challenge, since unlike the classical optical materials, with highly ordered atomic or molecular structure, this substance is best characterized in every regard as a poorly defined, inhomogeneous, turbid medium. As such it possesses a substantial light scattering capability for certain wavelengths.

Upon encountering tissue, light may undergo three types of scatter. (1) An interaction with objects of dimension one-tenth or less that of the wavelength (atoms, molecules) rarely produces a scattering event. Events of this kind (Rayleigh scattering) are weak, the scattering is isotropic, and is inversely proportional to the fourth power of the wavelength. Thus, for this interaction mode, the short light wavelengths (blue) experience the greatest scattering effect. (2) If light encounters objects whose dimensions are roughly the size of the light wavelength (for PRT red light, small cells and their larger organelles), scattering is stronger, varying inversely with wavelength, thus also favoring the more energetic light photons. It is not isotropic, but is somewhat forward directed. (3) For objects very large compared with the wavelength (macro and gross tissue structures, regular assemblies of cells, etc), scattering is highly forward directed, and is essentially wavelength independent. This is known as Mie scattering. In tissue, structures which produce all three types of scattering of PRT light abound (Ishimaru 1978). A complete description of the mix of scattering mechanisms may be somewhat complex, if polychromatic light is considered.

In general, scattering occurs when the light photon encounters tissue structures within which a change in refractive index takes place. The scattering process taking place is principally an elastic mechanism, in which the photon retains its energy, but undergoes a change of direction. The absorption of light photons is a competing process to the scatter mechanisms and is essential to PRT initiation. Absorption at a chromophore, molecule, or atom is a quantum effect with all of the photon's energy given up.

When light traverses tissue, it undergoes both absorption and scattering. Taken together, these two effects determine the mode of light transmission. If scattering is negligible, the light beam experiences absorption as it travels undeviated through the tissue and obeys the simple exponential absorption law. The attenuation rate is fixed by the object's absorption coefficient. If, on the other hand, the scattering process is appreciable, the photons are deviated and experience, on the average, a longer, maximum path length than the simple tissue thickness. They, thus, illuminate a greater volume of tissue. This greater illuminated region is important in PRT. Scattered photons have a higher probability of being absorbed before exiting the tissue, since in that longer path, possibility of an absorber encounter is enhanced. Therefore, a more complex light transport law must be used to account for both absorption and scattering (Kubelka 1954). (Both scattering and absorption coefficients are to be taken into account.)

That visible light radiation which is absorbed in tissue is in large part degraded to heat and is dissipated.

Biological effects can be produced by this heat or through photochemical processes. A few of the photons may, through non-thermal mechanisms, produce fluorescence, phosphorescence, or biochemical effects. UV is important in creating fluorescence and phosphorescence. The biochemical process may also be initiated by visible and near IR photons. Such photochemical processes are initiated by photons that experience resonance absorption with those chromophores and molecules that play a critical role in producing a specific biological effect. The PRT process is a mechanism of this type.

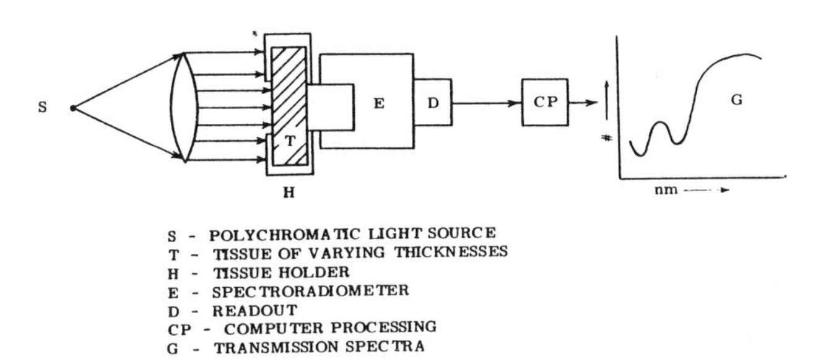

Fig. 1. Experimental system used to acquire data for tissue transmission spectra.

From the foregoing, one may anticipate that tissue light transmission is complex and varying between tissues. One tool for studying this system is by means of transmission spectra obtained with a polychromatic light source in organ sections of various thicknesses.

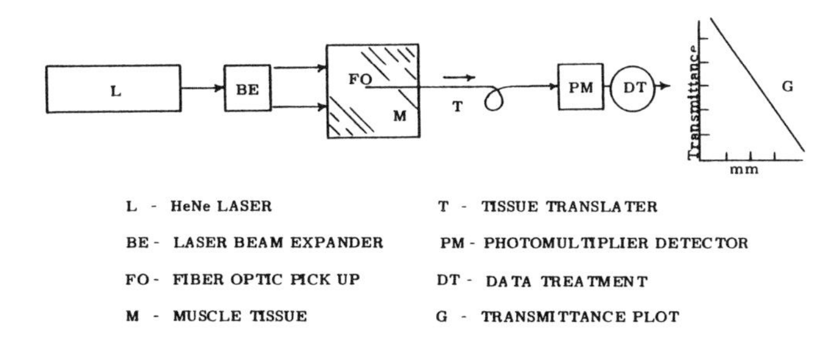

Fig. 2. Method of using a fiber optic detector for measurement of in-tissue irradiance.

In Fig. 1 is shown the experimental setup used to derive the transmission spectra for the tissue samples which are shown in the following. The polychromatic light source consisted of an incandescent projector with a maximum power output of about 100 mw/cm2 at the tissue surface. A substantial lamp output was available from 300 nm well into the near IR. A shutter system was used between readings to avoid heating the sample. The tissue holder was constructed of lucite, and allowed the sample to be rigidly held at fixed thicknesses during the measurement. The spectroradiometric system was an EG&G model with a range of 280-1100 nm. (Our experiments were limited to the 400-1100 nm range.) The readout data was manually recorded and entered into a TRS-80 microcomputer for subsequent processing and plotting. Spectra were computer generated.

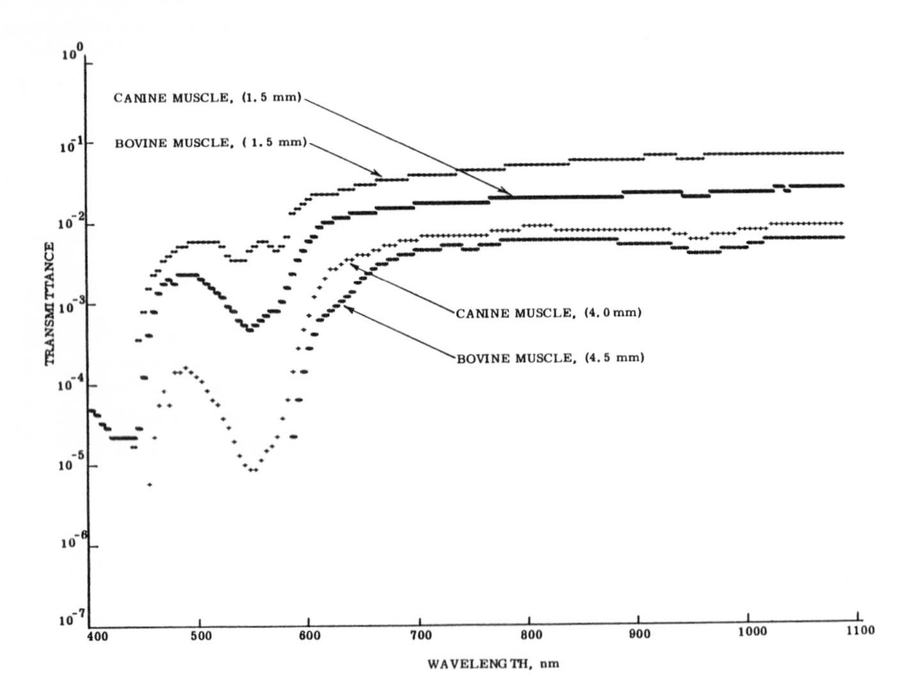

Fig. 3. Spectra from canine and bovine muscle sections compared at two thicknesses.

Fig. 2 describes the experimental arrangement used for obtaining tissue transmission data at a single wavelength.

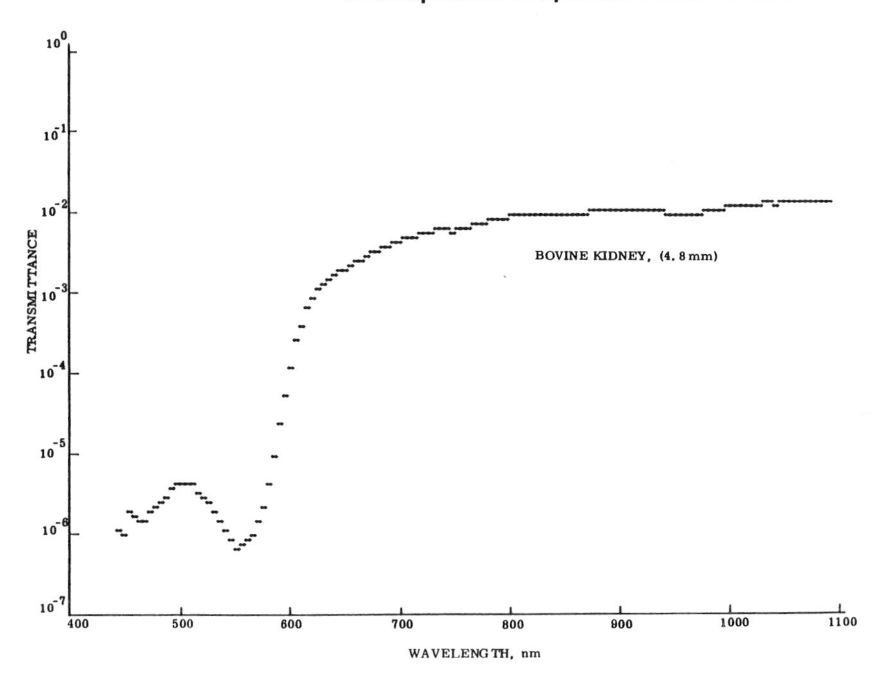

Fig. 4. Transmission spectrum of a 4.8 mm section of kidney tissue.

(The input wavelength used was 632.8 nm since this has been identified as being close to optimum for PRT applications.) (Dougherty et al. 1978). A He-Ne laser beam of approximately 10 milliwatts is expanded up to a diameter of 15-18 mm and allowed to fall on the surface of the tissue sample. An internally inserted Quartz fiber optic (200 micron diameter) connected to a photomultiplier detection system is used to sample the light power level at incremental positions within the tissue. From the resultant data a least squares line can be generated, with the light hvl derivable from the slope (Bolin et al, 1983).

Generally, transmission spectra assume characteristic, recognizable shapes, with poor transmission evident in the short wavelengths, that is, for those illustrated here, from 400-600 nm (Preuss et al. 1982). This is graphically shown in Fig. 3, where spectra for sections of both skeletal bovine and canine muscle tissue for similar thicknesses illustrate this.* Note, that for thin sections (1.5 mm) the muscle transmission spectra approach the horizontal except for the short wavelength part of the plot. This effect is somewhat lost in thicker sections, due to the encroachment into the longer wavelengths of minima manifest here below 600 nm. One may note also, in this figure, that the values between two different species are fairly well conformed.

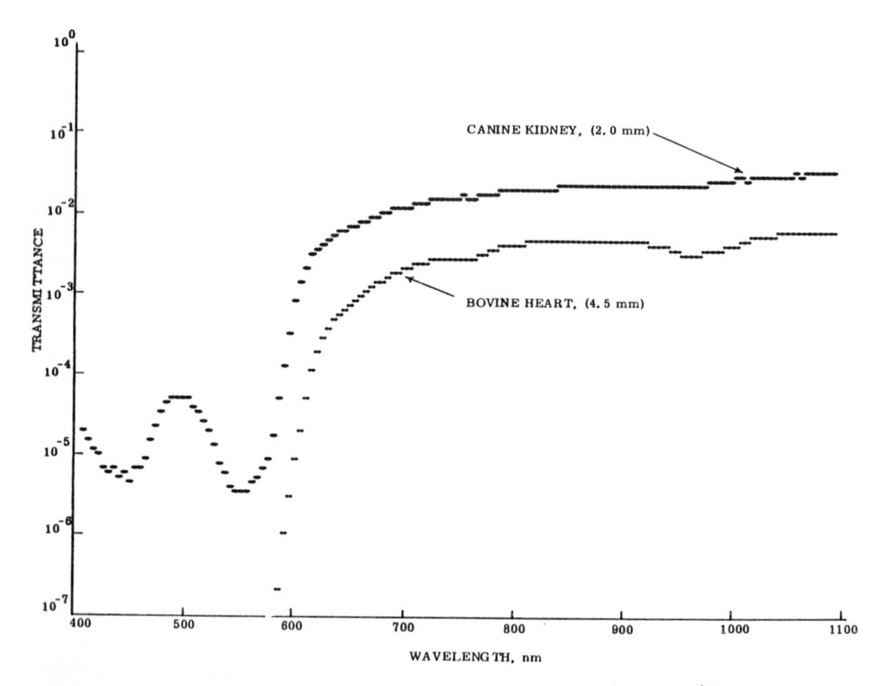

Fig. 5. Transmission spectra of canine kidney (2.0 mm upper curve) and bovine heart (4.5 mm, lower curve).

At thicknesses of about 5 mm and greater, many spectra exhibit broad and deep minima in the region from the UV to 600 nm. This is illustrated in Fig. 4 in a 4.8 mm section of bovine kidney. As a result, attenuation of blue, green, and yellow light is very heavy. The PRT red and the near IR are transmitted best. The small minima at 550 nm is apparently due to an absorption maximum of hemoglobin. Good transmission of the longer wavelengths of the red light and

* Due to remission losses and detector collimation, the transmission spectra are indicative of relative and not absolute values.

near IR contrasts with the poor penetration in the balance of the visible.

The cut-off point at about 600 nm begins to be strongly manifested in tissue at thicknesses of about 2-3 mm. It becomes quite substantial at 5 mm. This may be seen in Fig. 5 where canine kidney (2.0 mm) and bovine heart (4.5 mm) are contrasted. For the kidney section, the fall off is of three orders of magnitude, whereas for the heart it is even greater, such that the transmission values are beyond the scale of this figure. This cut-off effect is of interest because of the proximity of the PRT red region at 620-630 nm. For these curves, this region lies on a knee of the spectra. As thickness increases, this knee recedes into the longer wavelengths, thus, severely attenuating the favored red treatment frequencies. Also, it is of interest to note that this cut-off causes the curves to bear a strong resemblance to the transmission curve of a red filter commonly used in early PRT studies utilizing a continuum source and

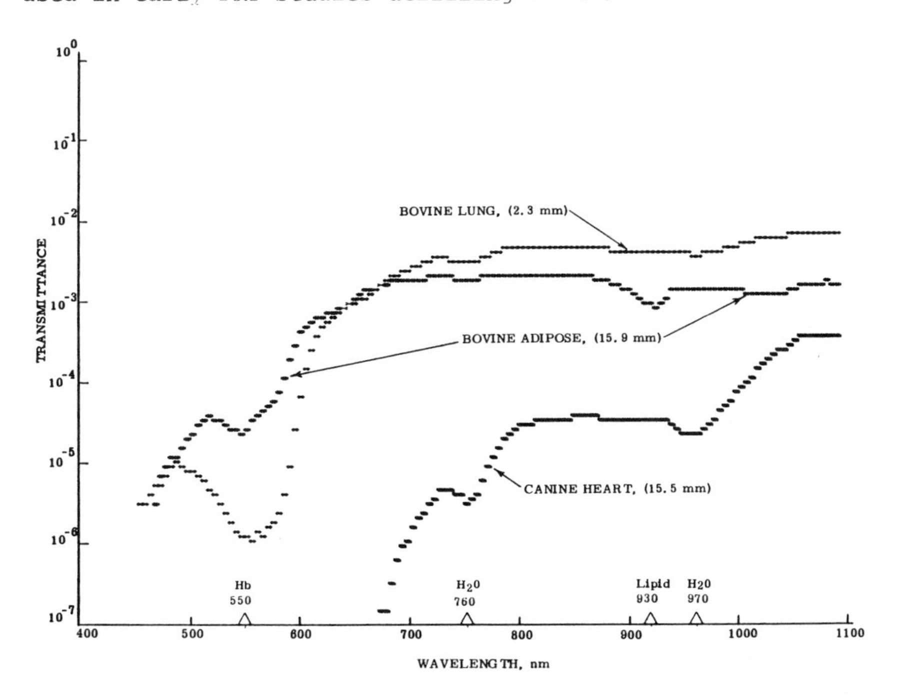

Fig. 6. Transmission of three tissue sections contrasted. In descending order: bovine lung 2.3 mm, bovine adipose 15.9 mm, and canine heart 15.5 mm.

a red glass filter (Corning #2418). This relation is especially true of the heart muscle shown here.

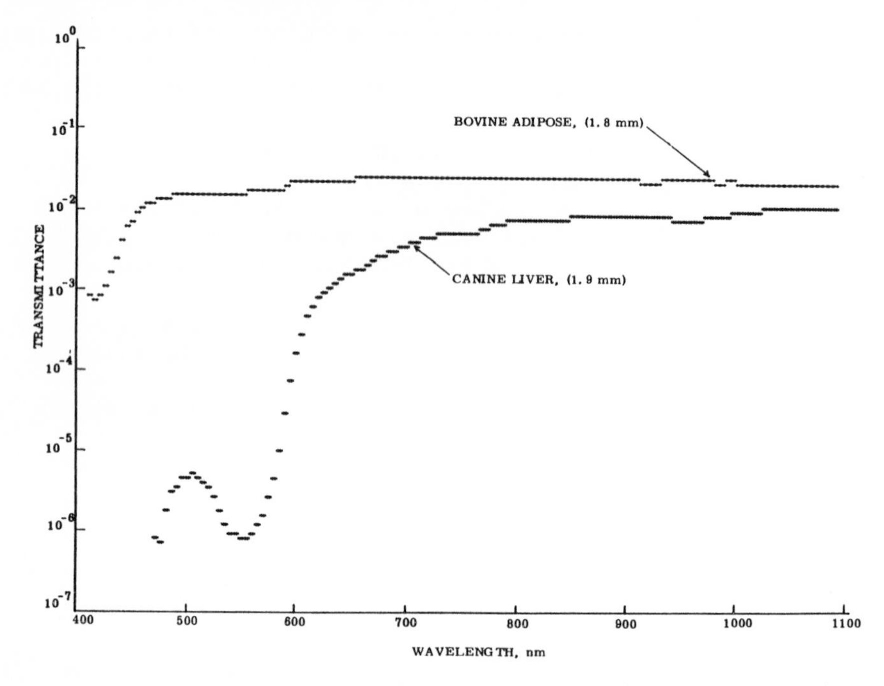

Fig. 7. Bovine adipose and canine liver light transmission compared at similar thicknesses (1.8 mm and 1.9 mm).

Absorption maxima of specific tissue components play an important role in modulating the shape of the transmission spectra and in turn may influence the intensity of the PRT band. Fig. 6 points this out with spectral transmission curves for three tissue types. Those components producing minima in the curves and which are fairly well verified are: water with strong absorption effects at 760 and 970 nm (canine heart), (Norris, Hart 1965); hemoglobin at 550 nm (bovine lung), (Edwards, Duntley 1951); and lipid at 930 nm (bovine adipose), (Massie 1976). The minima due to water and hemoglobin are pronounced, in all but very thin samples (< 2.0 mm). The lipid absorption minimum is only noted in high lipid-bearing tissues such as gross adipose (80% lipid). Hemoglobin is an important absorber in that it influences transmission near the PRT, red part of the spectrum.

Despite the fact that most spectra have a characteristic shape, some exhibit a substantially different morphology. The prime example of this is gross adipose tissue, as in Fig. 7, where a contrast is made with liver tissue in 1.8 mm and 1.9 mm sections respectively. The high transmissivity of this adipose sample gives this curve a horizontal aspect extending almost to 450 nm. In very thick sections of adipose (>15 mm), some of the features prominent in the other spectra begin to appear. This is apparent in Fig. 8, where a series of spectra, for bovine adipose sections ranging in thickness from 1.8 mm to 15.9 mm are shown. Only in the fourth specimen at almost 16 mm do some of the ubiquitous minima begin to be substantial (the hemoglobin 450 nm absorption maxima does begin to show a significant effect in the 9.8 mm sample). Note that, as expected, the water absorption effects tend to be small in adipose, in contrast to that in striated muscle, while lipid absorption occurring at 930 nm produces a prominent minimum in the two thicker adipose sections.

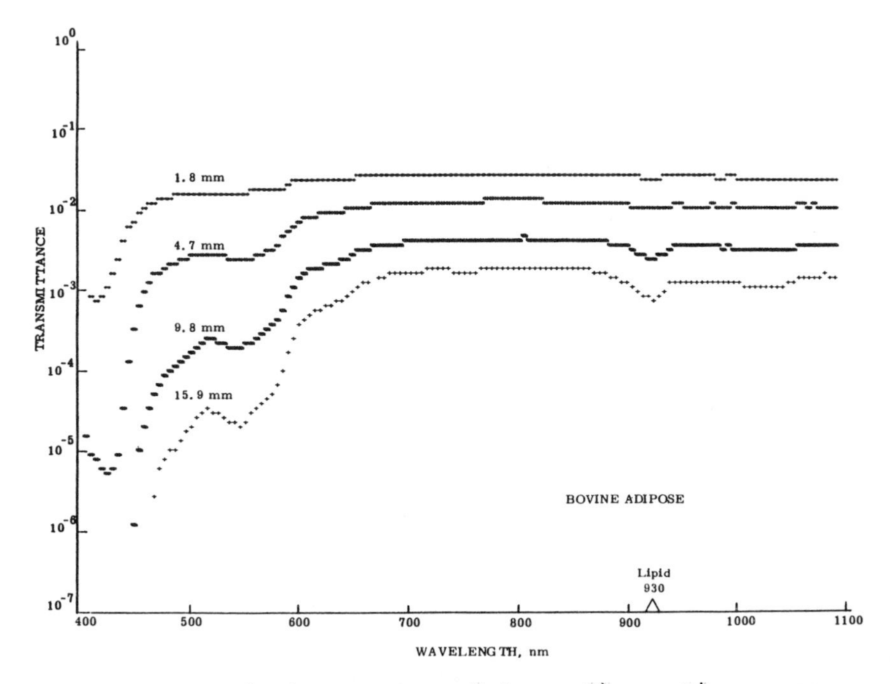

Fig. 8. Transmission spectra of four adipose tissue sections in thicknesses of 1.8, 4.7, 9.8, and 15.9 mm. Note the high transmission and the good nesting of the spectra.

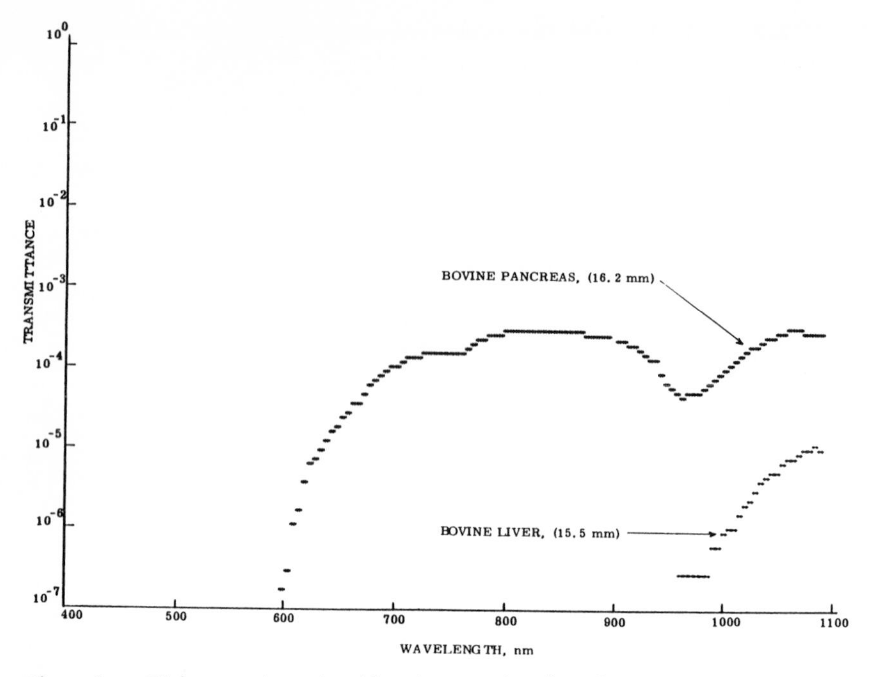

Fig. 9. This contrasts the transmission between two organs (sections of similar thickness). Upper curve is bovine pancreas (16.2 mm) and the lower curve is bovine liver (15.5 mm).

Absolute transmission values at a given wavelength may vary considerably from organ to organ. For an example of this,(see Fig. 9). Here one notes an impressive difference between similar thicknesses of pancreas and liver. The liver section, although thinner than the pancreas, has much poorer transmissivity. Especially important is the fact that the pancreatic tissue transmits a measurable amount of red light in the 630 nm region, whereas liver does not. The useful therapeutic range and tissue volume illuminated by activating light for these two tissues will differ markedly.

Spectral transmission data obtained at several thicknesses can be isolated at specific wavelengths and the data recast into a least squares line. This yields a transmission plot versus thickness at the given wavelength. Fig.

10 shows these curves for heart muscle for the 600-1100 nm range. Note that the steepest fall-off is in the 600-630 nm PRT range, with hvl's ranging from about 0.33 mm to 0.68 mm (and the maximum hvl is 1.85 mm in the IR). Fig. 11 is a similar plot, except that in this case use was made of the data from the skeletal muscle spectra, which differs measurably from heart in light transport. In addition, a separate line at 632.8 nm (obtained by the fiber optic detector method shown in Fig. 2) is included for comparison purposes. Both methods of measurement are in fairly good agreement. One difference between this set of plots and those in Fig. 9 is that the transmission here is enhanced; this is reflected in the hvl's at the PRT wave-

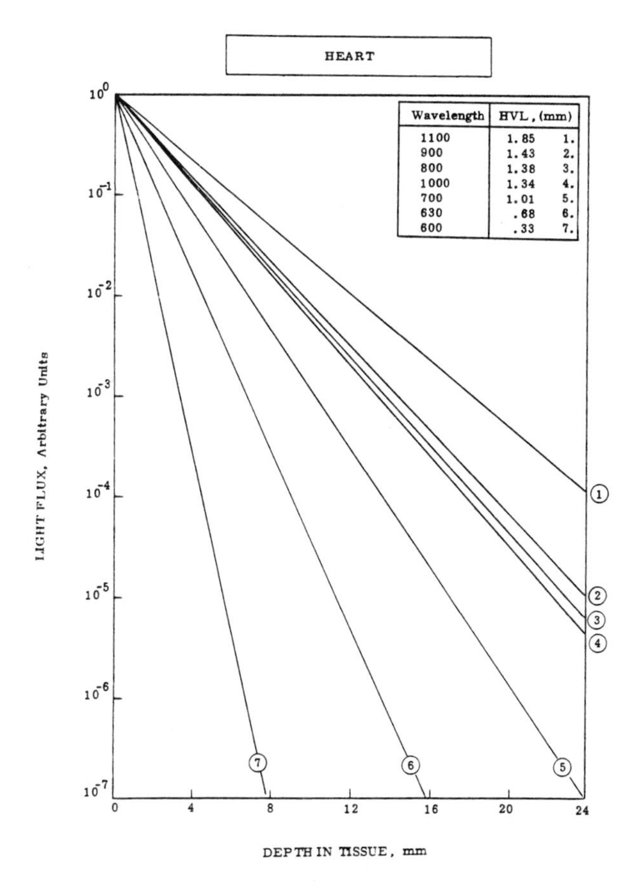

Fig. 10. Least squares lines for transmitted light power for heart muscle. Data was derived from spectral curves.

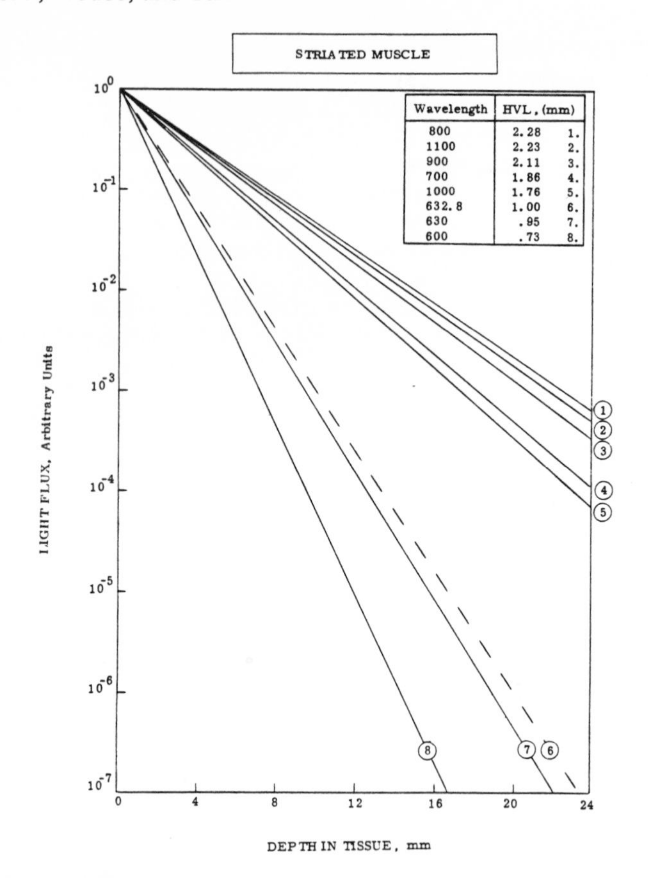

Fig. 11. Least squares curves for transmitted light through sections of skeletal muscle. Data was obtained from spectral curves. Plot at 632.8 nm was made with fiber optic method, shown in Fig. 2.

lengths which range from 0.7 mm to about 1.0 mm. One may also note that the arrangement and relative position of the least squares lines is different in these two figures. This is accounted for by the substantially different trends in transmission between striated skeletal muscle and heart. That is, from 700 nm to 1100 nm, striated skeletal muscle produces a transmission curve which tends to be almost flat, apart from minima, while heart, in the same region, develops a distinct upward trend,(see Fig. 6).

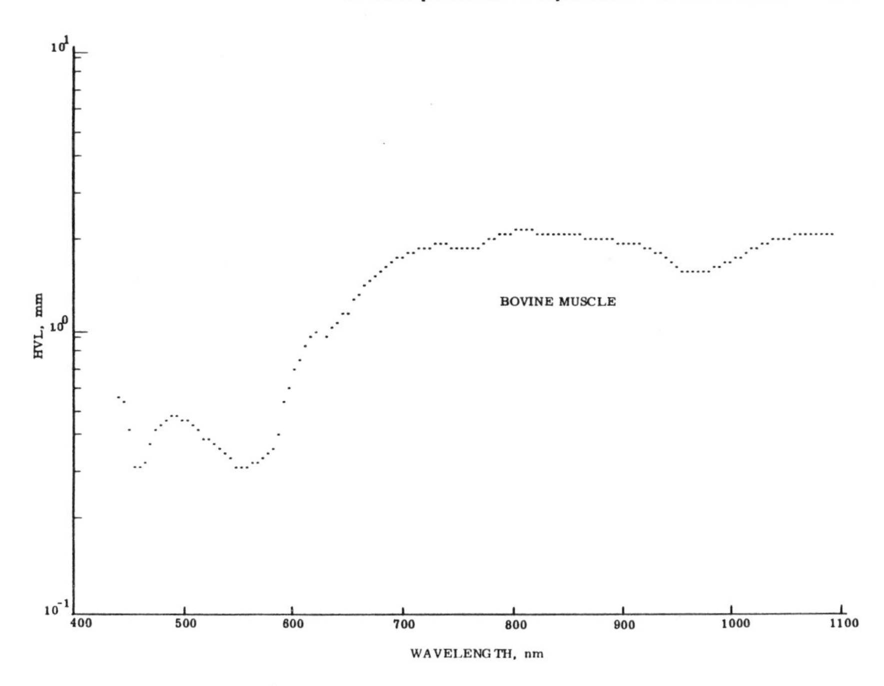

Fig. 12. Plot of half value layers, vs wavelength, for mammalian skeletal muscle.

A related quantity varying with wavelength is the half value layer. Using data derived from the spectral plots of lean, striated muscle (the lipid level was probably below 5%), a plot of the hvl vs wavelength is shown in Fig. 12. The range is seen to be from about 0.3 mm at 550 nm to 2.3 mm at 800 nm. Using data in this form may be more useful to the therapist than plots of absolute or relative transmission since the penetration depth is readily obtained. The hvl at the PRT wavelength is about 1 mm for the skeletal muscle sample.

In summary, we may state that the tissue spectra assume recognizable, characteristic shapes. At a thickness of 5 mm, the spectra exhibit broad minima in the 400-590 nm region, with a rapid rise in transmissivity from 590-630 nm and a plateauing out to 1100 nm. This spectral form is modulated by various minima due to water, hemoglobin, and other strongly absorbing components. A few tissues exhibit a generally different type of structure. Gross adipose tissue transmission spectra present a more flattened appearance without as heavy an attentuation at the short wavelengths. Despite similar, spectral morphology, absolute transmission values were found to be somewhat dependent on tissue type. In the red, the hvl's for gross adipose tissue were as much as three times larger than that of very lean, striated, skeletal muscle. Another example of such difference is liver tissue, which transmits, through a 4.5 mm section, only about 10% as much as striated muscle at 630 nm. Wavelength is a major factor in determining tissue hvl's. At 550 nm, attenuation may be more than 100 times that for 650 nm light, in a 5 mm skeletal muscle section. Tissue composition also plays a substantial role in the transport of PRT activating light. Especially important is the adipose level. This latter point was confirmed in muscle in a separate experimental sequence using a fiber optic as a detecting mechanism.

To improve PRT methodology and efficacy, it will be important to have a quantitative understanding of the interaction of light with tissues and the specific attributes of each tissue. This was shown by the fact that some tissue varieties and some tissue thicknesses adversely affect transmission in the PRT region of the spectrum. As such, further research is indicated on the effect of certain important variables (tissue type, composition, adiposity and wavelength) on light dosage, at depth.

Dougherty TJ, Kaufman JE, Goldfarb A, Weishaupt KR, Boyle D, Mittleman A (1978). Photoradiation therapy for the treatment of malignant tumors. Cancer Res 38:2628.

Edwards E, Duntley SQ (1951). Spectrophotometry of living human skin in the UV range. J Invest Derm 16:311.

Ishimaru A (1978). "Wave Propagation and Scattering in Random Media." New York: Academic Press.

Kubelka J (1954). New contributions to the optics of intensely light-scattering materials. Part II Non-homogeneous layers. J Opt Soc Am 44:330.

Massie DR (1976). Fat measurement of ground beef with a gallium arsenide infrared emitter. In ASAE publication "Quality Detection in Foods."

Norris KH, Hart JR (1965). Direct spectrophotometric determination of moisture content of grain and seeds. In "Principles and Methods of Measuring Moisture in Liquids and Solids," Vol 4, New York, NY, Reinhold, p 19.

Preuss LE, Bolin FP, Cain BW (1983). A comment on spectral transmittance in mammalian skeletal muscle. Photochem Photobiol 37:113.

Preuss LE, Bolin FP, Cain BW (1982). Tissue as a medium for laser light transport-implications for photoradiation therapy. In Goldman L (ed): "Lasers in Medicine and Surgery," Vol 357, Bellingham, WA, SPIE, p 77.

Porphyrin Localization and Treatment of Tumors, pages 227–238

A NEW DIAGNOSTIC SYSTEM FOR MALIGNANT TUMORS USING HEMATOPORPHYRIN DERIVATIVE, LASER PHOTORADIATION AND A SPECTROSCOPE

Katsuo Aizawa, M.D.*, Harubumi Kato, M.D.,
Jutaro Ono, M.D., Chimori Konaka, M.D.,
Norihiko Kawate, M.D., Kazuo Yoneyama, M.D.,
Masaki Otawa, M.D., Hideki Shinohara, M.D.,
Makoto Saito, M.D., Hidenobu Takahashi, M.D.,
Haruhiko Nakamera, M.D., Tetsushi Yamada, M.D.
Harumasa Sakai, M.D., Yoshihiro Hayata, M.D.

Departments of Physiology* and Surgery,
Tokyo Medical College, 6-7-1 Nishishinjuku,
Shinjuku-ku, Tokyo 160, Japan

Introduction

The diagnosis of central type early stage lung cancer is extremely difficult by chest X-ray examination. Such lung cancers can be detected by sputum cytology. Fiberoptic bronchoscopy facilitates the localization of the focus, but there are still cases, particularly occult cases which endoscopically show no gross abnormal findings, that present problems in localization. Recently, fluorescence bronchoscopy has been introduced for the detection of localization of early stage lung cancer using hematoporphyrin derivative (HpD) (Sanderson 1972, Profio 1977, Doiron 1979, Profio 1979, Cortese 1979, Hayata 1982). In 1960 HpD was prepared from hematoporphyrin hydrochloride treated by acetic acid and sulfuric acid (Lipson 1960) and was shown a great affinity to malignant tumors. It emits fluorescence with peaks of 630 and 690 nm wavelength. It is therefore theoretically possible to make a diagnosis of malignant tumors by detecting the fluorescence of HpD. However it is actually difficult to detect the HpD-specific fluorescence in the cases of very early stage cancers using previous image intensifier systems with a 630 nm barrier filter because the normal epithelium also has autofluorescence showing a single peak at 600 nm. Therefore the authors developed a spectroscope system compatible with fiberoptic endoscopes to analyze the shape of the fluorescence light wavelength. This permits differentiation of the two types fluorescence and greatly assists diagnosis.

Materials and methods

Equipment

The new diagnostic system consists of a light source, endoscope system and spectra-photometer (Fig. 1, 2).

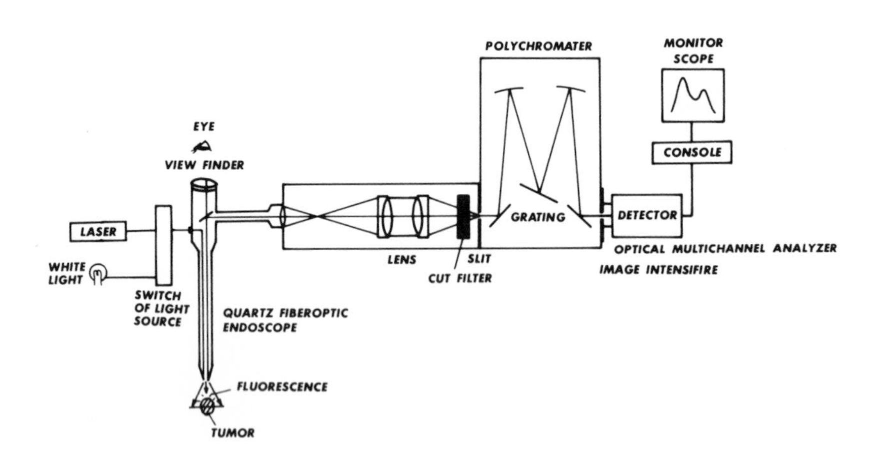

Fig. 1. Endoscope/spectroscope fluorescence detection system.

A krypton ion laser (Spectra-Physics, model 164-11, 406.7-415.4 nm wavelength, Mountain View, Ca., U.S.A.) was used as a light source for excitation of HpD and a white light (Olympus CLX-F) was used for the endoscopic observation. The laser beam is transmitted through a 400 micron quartz fiber (Fujikura Densen Co., Ltd. Japan) to a transmission appararus which is also connected to the white light source equipment (Olympus CLX-F). The

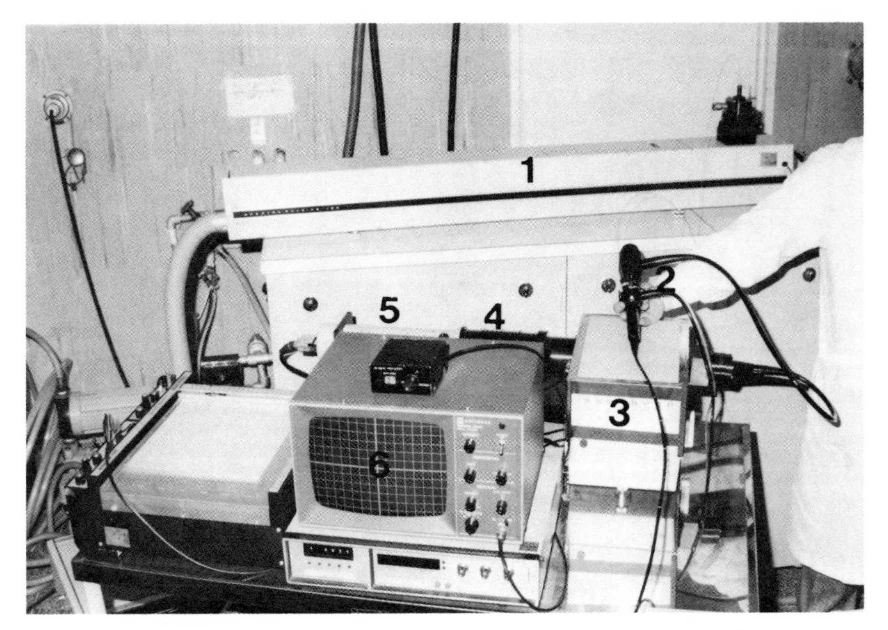

Fig. 2. The endoscopic HpD fluorescence detection system consists of a krypton ion laser (1), fiberoptic bronchoscope (2), polychrometer (3), image intensifier (4), detector (5), console (6) and monitor scope (7).

transmission apparatus was designed to alternate between the white light for endoscopic observation and the laser beam for excitation of HpD by a operation of footswitch (Fig. 3). This allows continuous fluorescence monitoring and visual observation. Both lights are guided through the quartz light guide of a fiberoptic bronchoscope which was especially designed for better laser beam transmission. HpD specific fluorescence from the cancer focus excited by the laser beam is transmitted by the endoscopic image guide, and then separated by a half mirror which yields one part for observation, the other for the fluorescence spectroscopy. Fluorescence was condensed into the spectroscope incidence slit, after excluding exciting light below 500 nm by a Y-52 cut filter (Toshiba Electric Co., Ltd). Analysis of fluorescence over a 250 nm spectrum was performed by using a 600 nm blaze holographic grating at F=4.2 (Sigma Optical Co., Ltd.). An image intensifier (Type X-1500, Philips Co.), optical multichannel analyzer (Princeton Applied Research

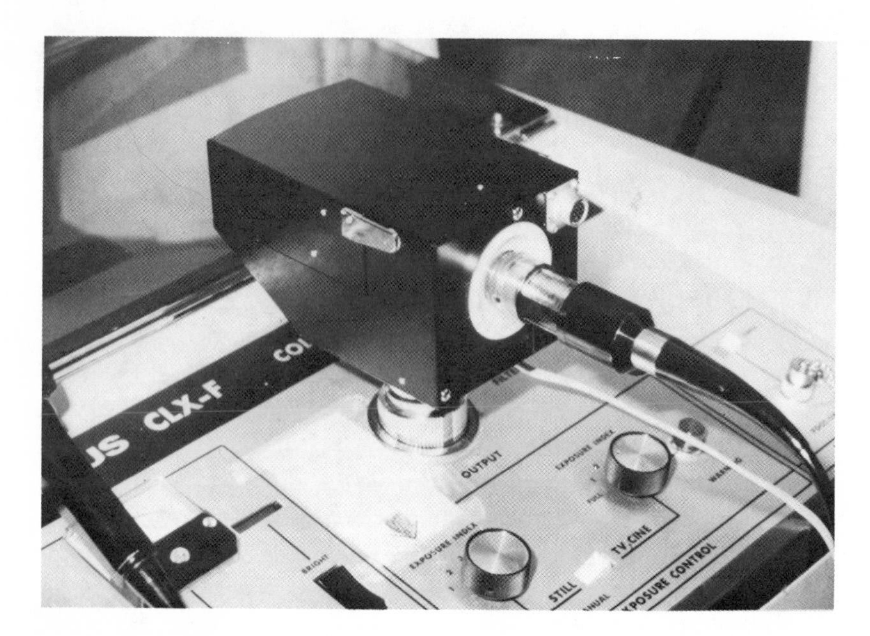

Fig. 3. Transmission of both white light and laser beam enables alternation.

Co., Inc.) and microcomputer memory were used to measure, record and integrate the curve of the fluorescence wavelength pattern. In order to obtain a clear picture of the fluorescence excited, the total signal was scanned and stored in memory A of the microcomputer and the background noise such as the autofluorescence from the normal area and excitation light was stored in memory B. Subtraction of the value in B from that in A for any given scan, therefore, yielded the differential fluorescence pattern. All fluorescence from 570-720 nm could be measured simultaneously in real time.

Hematoporphyrin derivative (HpD)

Photofrin (Oncology Research and Development Co., Ltd.) was used for this study. It was stored in darkness until used.

Basic studies

(1) Measurement of fluorescence from in vitro human

cancer cells.

Cells from a cell line (PC-10) established from a case of human lung cancer (squamous cell carcinoma) established in the Department of a Surgery, Tokyo Medical College were used. The cells were cultured on a 30x40 mm cover glass in a Petri dish with RPMI-1640 culture medium containing 10% calf serum. After 48 hours culture when the PC-10 culture cells were in the exponential proliferation phase, culture medium was substituted by pH 7.2 phosphate buffer saline solution (P.B.S.) containing 0.05 mg/ml HpD, and then culture was continued for 20 minutes. Subsequently the cover glass was removed and washed three times with RPMI-1640 culture medium. The cover glass was then placed in a flow chamber for examination, with a flow of 50 ml/min. culture medium. The fiber tip was maintained at a distance of 5 mm from the cells and excitation light spot was 2 mm in diameter. In order to evaluate the possible applicability of this method in clinical studies, the 405 nm excitation light from the krypton ion laser was transmitted via the 400 micron quartz fiber inserted through the instrumentation channel of a fiberoptic endoscope used in routine clinical practice and examination for fluorescence detection was performed by connecting the spectrophotometer apparatus to the eye piece of the instrument. Fluorescence analysis was performed by integrating 10 measurements which were recorded in memory A, while on the other hand, background fluorescence was recorded in memory B. By comparing the difference between A and B memories, the pattern and amount of HpD fluorescence retained in the cells were measured.

(2) Measurement of fluorescence from in vivo murine sarcoma.

Sarcoma-180 cells were transplanted subcutaneously to the back of 20 mice. Three weeks later, 10 mg/kg of HpD was intravenously injected into the tail vein. 48 hours after the injection, the mouse was slightly anesthetized and shaved to expose the tumor and the normal area surrounding it. Fluorescence excitation of a 2 mm diameter circle was then performed with 10 mW 405 nm krypton ion laser beam via the quartz fiber, the tip of which was maintained at a distance of 5 mm.

(3) Measurement of fluorescence from clinical cancer cases.

HpD Fluorescence of human early stage cancers of the lung, stomach, skin, urinary bladder and cervix were measured 48 hours after 2.5-3.0 mg/kg body weight HpD upon stimulation by a 10 mW 405 nm krypton ion laser beam with the fiber tip 10 mm from the target. The tumor, area of erosion and normal areas were photoradiated. Fluorometry was performed by integrating 50 measurements each.

Results

(1) Upon excitation with 405 nm light, HpD in cultured human squamous cell carcinoma of the lung (PC-10) showed fluorescence with two peaks at 627 nm and 686 nm (Fig. 4).

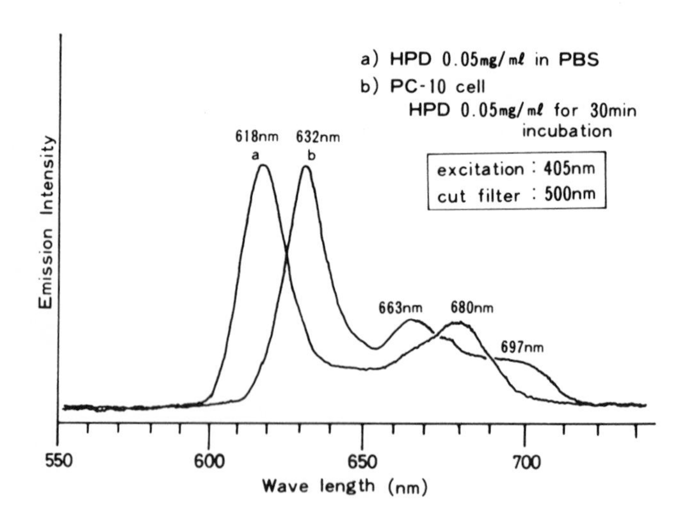

Fig. 4. Fluorescence spectra of HpD from PC-10 lung cancer cells. a) shows fluorescence from HpD in P.B.S. solution with a Hp of 7.2. b) shows fluorescence from HpD in PC-10 cells.

(2) HpD fluorescence from the mouse tumor was

clearly measured by real time (32.8 msec.) and the main peak at 627 nm and the peak shoulder at 686 nm line were identical to the HpD in the PC-10. At the same time, a peak from 660 to 670 nm was noted (Fig. 5). A significant difference of quantity of fluorescence, and therefore of HpD concentration in tissue was seen between the sarcoma and the normal areas in mouse.

(3) Fluorescence from the upper gastric corpus tumor had a distinct peak at 630 nm and at 690 nm a distinct shoulder was recognized (Fig. 6a, 7a). The peaks seen clinically coincided with those seen in cultured cells and in the transplanted sarcoma in mouse. Fluorescence in the erosion region around the tumor also had two peaks at 630 nm and 690 nm but it was quantitively recognizably less than that of the tumor (Fig. 6b, 7b). On the other hand, in normal tissue there was only a slightly recognizable elevation at 630 nm, significantly less than in the lesion.(Fig. 6c, 7c)

The fluorescence intensities obtained by integrating the area of the curve from 600 nm through 720 nm were 203,700 arbitary units in the region of tumor, 68,900 in

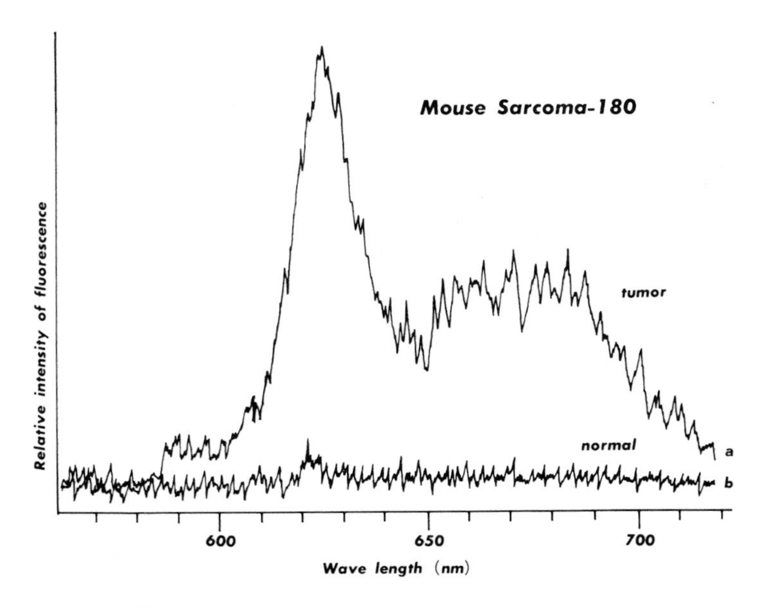

Fig. 5. HpD fluorescence from mouse sarcoma.
a) shows fluorescence from sarcoma, b) from normal area.

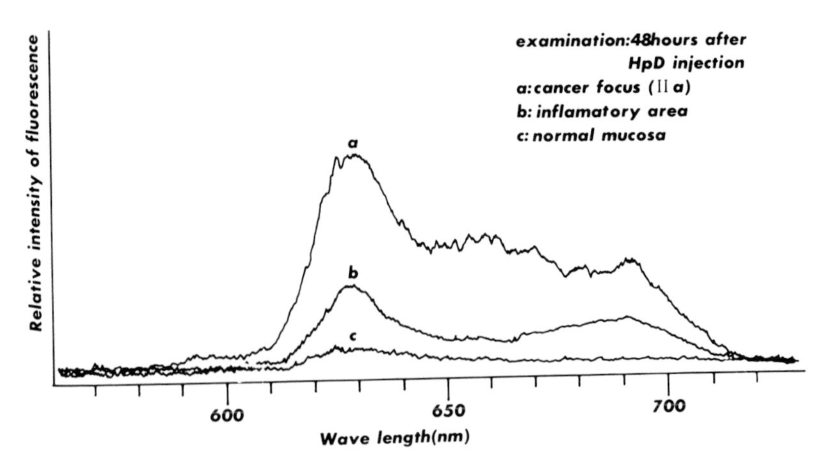

Fig. 6. HpD fluorescence from stomach.
The patient was 68 year-old male who had an early stage gastric cancer. 2.5 mg/kg body weight HpD was administered intravenously prior to the fluorescence examination. a) shows fluorescence spectra from cancer focus (IIa type), b) from erosion area, c) from normal area.

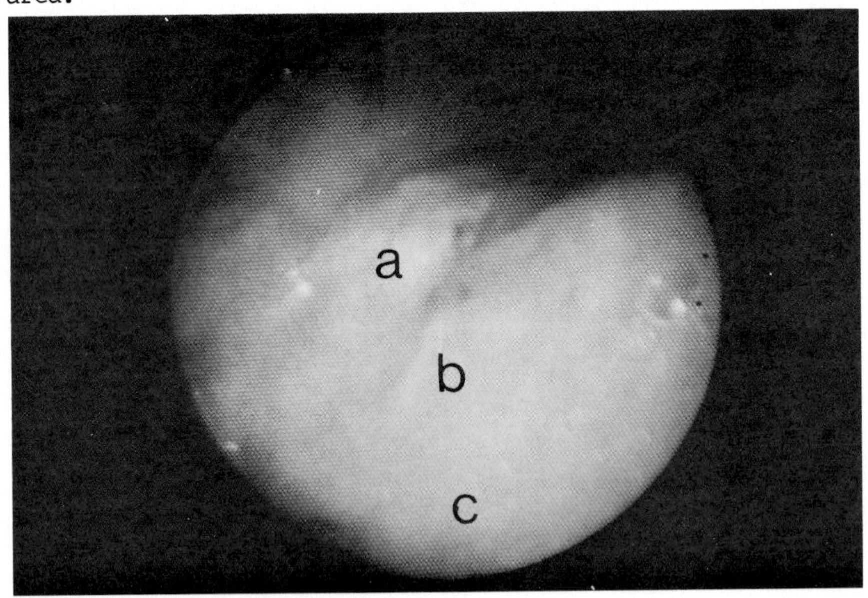

Fig. 7. Endoscopical findings of cancer focus.
a) shows IIa type early stage cancer focus, b) shows erosion area and c) reveals normal area.

the area of erosion around tumor, 16,300 in normal tissue. The values in the tumor and area of erosion were 12.5 and 4.2 times as much as in normal tissue.

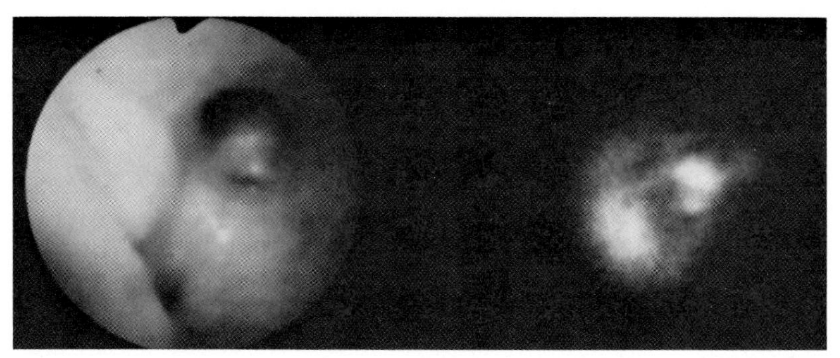

Fig. 8. Fluorescence bronchoscopy with HpD in lung cancer case.
Fluorescence was observed through the image intensifier from cancer focus.

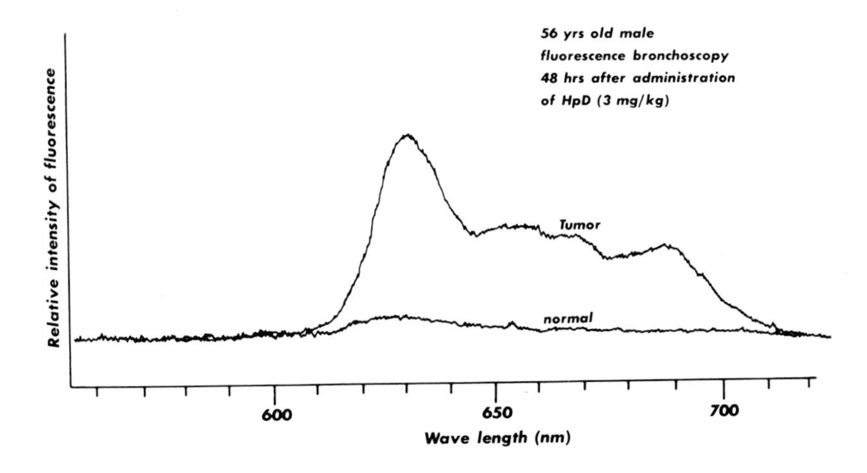

Fig. 9. HpD fluorescence spectrum from lung cancer focus as shown in Fig. 8.

Fluorescence was measured in a case of lung cancer. A patient was 56 year-old male who complained of cough and sputum. Adenocarcinoma was diagnosed by bronchoscopic biopsy. The focus was located in the left upper lobe bronchus. 3.0 mg/kg body weight HpD was administered intravenously 48 hours prior to the fluorescence measurement. Fluorescence had two peaks at 630 nm and 690nm. This fluorescence spectrum pattern was similar to that of other cancers. The value in tumor was 8.8 times as much as in normal tissue.

Discussion

Systems for the diagnosis of solid malignant diseases utilizing the tumor-specific and fluorescence properties of HpD using band pass filters have been developed (Saunderson 1972, Profio 1977, 1979, Hayata 1982, Kato 1981, 1983). In such systems autofluorescence from normal tissue has posed a significant problem. In order to avoid such confusion, the authors integrated a spectrometer with the endoscopic fluorescence detection apparatus. This displays the wavelength pattern of the fluorescence and can thereby distinguish between autofluorescence and fluorescence characteristic of HpD. The characteristic pattern with peaks at 630 and 690 nm was recognized in a cultured cell line (PC-10) and clinically in cases of gastric cancer, lung cancer and other cancers. The measurement can be made in real time. This means that the system can be used clinically. One peak was seen in the normal tissue, but with significantly less intensity. This reflects the relative affinity of HpD for malignant tissue that has been reported by many investigators (Lipson 1961, Gregorie 1968, Dougherty 1975, Carpenter 1977). The increased amount of fluorescence in malignant tissue (12.5 times) over that in normal tissue confirmed a previous report (Kinsey 1978). Their method of converting fluorescence to an audio signal still possessed the problem of distinguishing autofluorescence and HpD related fluorescence. We feel that our spectroscope system has effectively overcome this problem.

Conclusion

1) The fluorescence spectrum of hematoporphyrin derivative had two peaks at 630 nm and 690 nm in a cultured cell line (PC-10) from lung cancer after inclusion of HpD in the culture medium.

2) Measurement of fluorescence 48 hours after

intravenous injection of 10 mg/kg HpD permitted a distinction to be made between malignant and normal tissues in mice with transplanted sarcoma.

3) The ratio of fluorescence intensities in early stage adenocarcinoma of the stomach, area of erosion and normal gastric mucosa were 12.5 : 4.2 : 1.

4) The endoscopic fluorescence spectroscope is effective in demonstrating the presence of fluorescence, its wavelength pattern characteristics and distinguishing between autofluorescence of normal mucosa and that of fluorescent dyes, such as HpD. This in turn makes it feasible to calculate fluorescence intensity which reflects the relative amount of HpD absorbed in the tumor and to make a diagnosis.

References

Carpenter RJ, Ryan RJ, Sanderson DR (1977). Tumor fluorescence with hematoporphyrin derivative. Ann Otol Rhinol Laryngol 86:661.

Cortese DA, Kinsey JH, Woolner LB, Payne WS, Sanderson DR, Fontana RS (1979). Clinical application of a new endoscopic technique for detection of in situ bronchial carcinoma. Mayo Clin Proc 54: 635.

Doiron DR, Profio E, Vincent RG, Dougherty TJ (1979). Fluorescence bronchoscopy for detection of lung cancer. Chest 76:37.

Dougherty TJ, Grindey GB, Fiel R, Weishaupt KR, Boyle DG (1975). Photoradiation therapy II: cure of clinical tumor with hematoporphyrin and light. J Natl Cancer Inst 55:115.

Dougherty TJ, Lawrence G, Kaufman JH, Boyle DG, Weishaupt KR, Golgfarb A (1979). Photoradiation in the treatment of recurrent breast carcinoma. J Natl Cancer Inst 62:231.

Gregorie HB, Horger EO, Ward JL, Green JF, Richard T, Robertson HC (1968) Hematoporphyrin derivative fluorescence in malignant neoplasms. Ann Surg 167:820.

Hayata Y, Kato H, Konaka C, Ono J, Matsushima Y, Yoneyama K, Nishimiya K (1982). Fiberoptic bronchoscopic laser photoradiation for tumor localization in lung cancer. Chest 82:10.

Hayata Y, Kato H, Konaka C, Ono J, Takizawa N (1982). Hematoporphyrin derivative and laser photoradiation. Chest 81:26 .

Hayata Y, Kato H, Ono J, Matsushima Y, Hayashi N, Saito T, Kawate N (1982). Fluorescence fiberoptic bronchoscopy in the diagnosis of early stage lung cancer. Recent Results in Cancer Research Berlin-Heidelberg: Springer-Verlag, p121.

Kato H, Konaka C, Ono J, Matsushima Y, Saito T, Yoneyama K, Nishimiya K, Lay J, Tsukimura S, Otawa M, Shinohara H, Aida M, Hayata Y, Aizawa K (1981). Lung cancer diagnosis and treatment with hematoporphyrin derivative and laser photoradiation. Jap J Bronchology 3:457.

Kato H, Konaka C, Ono J, Matsushima Y, Saito T, Tahara M, Kawate N, Yoneyama K, Nishimiya K, Iimura I, Hayata Y (1981). Cancer localization by detection of fluorescence by means of HpD administration and krypton ion laser photoradiation in canine lung cancer. Lung Cancer 21:439.

Kato H, Konaka C, Ono J, Matsushima Y, Nishimiya K, Lay J, Sawa H, Shinohara H, Saito T, Kinoshita K, Tomono T, Aida M, Hayata Y (1983). Effectiveness of HpD and Radiation therapy in lung cancer. Porphyrin Photosensitization New York: Plenum, p23.

Kinsey JH, Cortese DA, Sanderson DR (1978). Detection of hematoporphyrin fluorescence during fiberoptic bronchogenic carcinoma. Mayo Clin Proc 53:594.

Lipson RL, Baldes EJ (1960). The photodynamic properties of a particular hematoporphyrin derivative. Arch Dermatol 82:508.

Lipson RL, Baldes ES, Olsen AM (1961). The use of a derivative of hematoporphyrin in tumor detection. J Natl Cancer Inst 26:1.

Profio E, Doiron ER (1977). A feasibility study of the use of fluorescence bronchoscopy for localization of small lung tumors. Phys Med Biol 2:949.

Profio E, Doiron ER, King ED (1979). Laser-fluorescence bronchoscope for localization of occult lung tumors. Med Phys 6:523.

Sanderson DR, Fontana RS, Lipson RL, Baldes EJ (1972). Hematoporphyrin as a diagnostic tool a preliminary report of new techniques. Cancer 30:1368.

Porphyrin Localization and Treatment of Tumors, pages 239–247

A TRIAL MANUFACTURE OF A MOTOR-DRIVEN LASER LIGHT SCATTERING OPTIC FOR WHOLE BLADDER WALL IRRADIATION

Haruo Hisazumi, Norio Miyoshi, and
Toshimitsu Misaki

Department of Urology, School of Medicine,
Kanazawa University, Kanazawa, 920, Japan

ABSTRACT

The detection of CIS of the bladder by fluorescence endoscopy with hematoporphyrin derivative (HpD) and argon ion laser was performed using a video monitoring system coupling with an image intensifier; however, it was difficult because of the reflecting light of the excitation light. Whole bladder wall photoradiation therapy was carried out in 3 patients with multicentric CIS of the bladder, using argon-dye laser light (630 ± 1.6 nm) of 5 to 25 J/cm^2. The anticancer effect of this therapy was followed by urinary cytology and clusters of vacuolated cancer cells were observed for 2 days after the therapy.

INTRODUCTION

Several studies have reported that at least 80% of invasive bladder tumors reveal multifocal carcinoma, carcinoma in situ (CIS) or other pre-neoplastic abnormalities in cystoscopically normal areas. These studies suggest that similar mucosal alterations are responsible for most of the subsequent tumors that develop after resection of a superficial bladder tumor (stages 0, A, and B_1).

To treat very small superficial multicentric urothelial tumors including CIS and multiple pre-neoplastic urothelial abnormalities, photoradiation therapy of a whole bladder wall lining, in which hematoporphyrin derivative (HpD) is activated by externally applied red light, may be a promising modality.

EQUIPMENT AND METHODS

I. Detection of HpD fluorescence

For endoscopic HpD fluorescence detection, an argon ion laser (Spectra Physics, model 171); operating multi-line from 457.9 to 514.5 nm was used. Light power was 40 mW. An image intensifier (Hi-Technology Trading Inc.), and a medical video system (Circon, model MV-9009s) were employed as shown in Fig. 1. The laser light was delivered through a cystoscope with a quartz fiber. The red fluorescence emission of HpD was intensified 3-5 X 10^4 times with the aid of the image intensifier and the use of a beam splitter (Circon, model MV-9864). Emitted faint HpD fluorescence was converted into a bright image through the image intensifier. The bright image was monitored on the medical video system (television: Sony, trinitron; poravideo timer: Circon, model MV-93303; color controller: Circon, model MV-9330S; videocasette recorder: Sony, model VO-5850; video camera: Circon, model MV-9290S).

For microscopic detection, the fluorescence spectra of HpD incorporated in normal and tumorous tissues of bladder biopsies 48 hrs after i.v. HpD injection (4.0 mg/kg body weight) were measured by a fluorescence spectrophotometer (Hitachi, model MPF-4). The fluorescence microphotographic studies of HpD incorporated in urinary exfoliative malignant cells 48 hours after i.v. HpD injection (4.0 mg/kg body weight) were done by using a fluorescence microphotometer (Leits, model MPV-2).

II. Whole bladder wall photoradiation

The bladder shape and size were estimated by serial sonotomography (Fig. 2; Aloka, model USI-51 and ASU-52B). The scattering optic was sonographically placed in the proper position by using a floor stand (Aloka, model PSA-3B; Fig. 3).

The motor-driven optic consists of: (a) a stainless steel tube (II); 4.1 mm in external diameter, 600 mm in length, (b) a built-in coneshaped reflecting mirror (I); 2.5 mm in base diameter, 1.3 mm in height, at the apex, (c) a small quartz tube (IV) for scattering irradiation and (d) a center adjusting piece (V) cored with a quartz fiber (III); 400 μm in diameter as shown in Fig. 4. This assembly was coupled with a driving instrument powered by a mercury battery, and was introduced into the bladder through the sheath of a cystoscope (Storz, model UR-27067) instead of a telescope (Fig. 3).

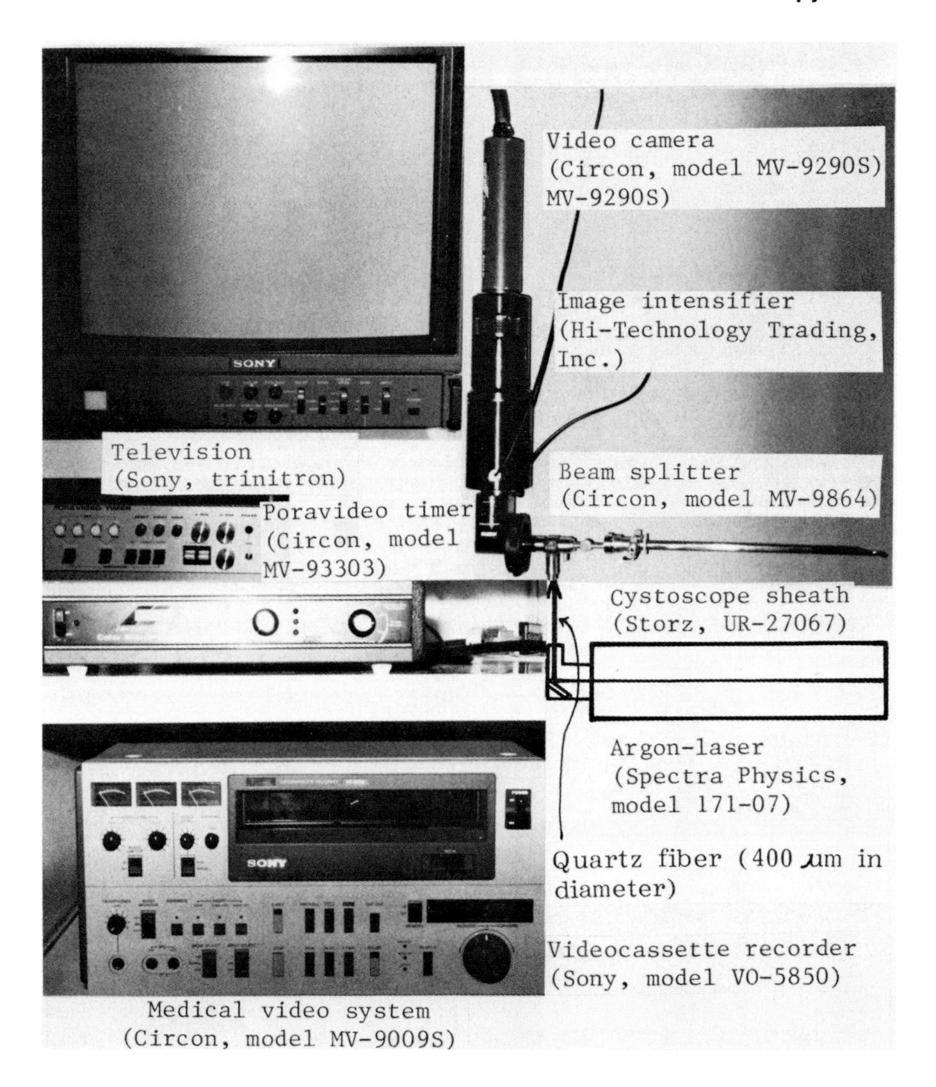

Fig. 1. Schematic diagram of the endoscopic detection of the HpD containing CIS and other urothelial abnormalities.

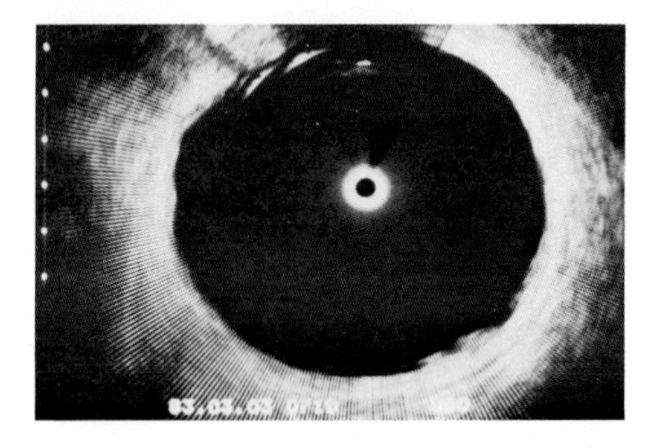

Fig. 2. A sonotomogram of a 200 ml bladder.

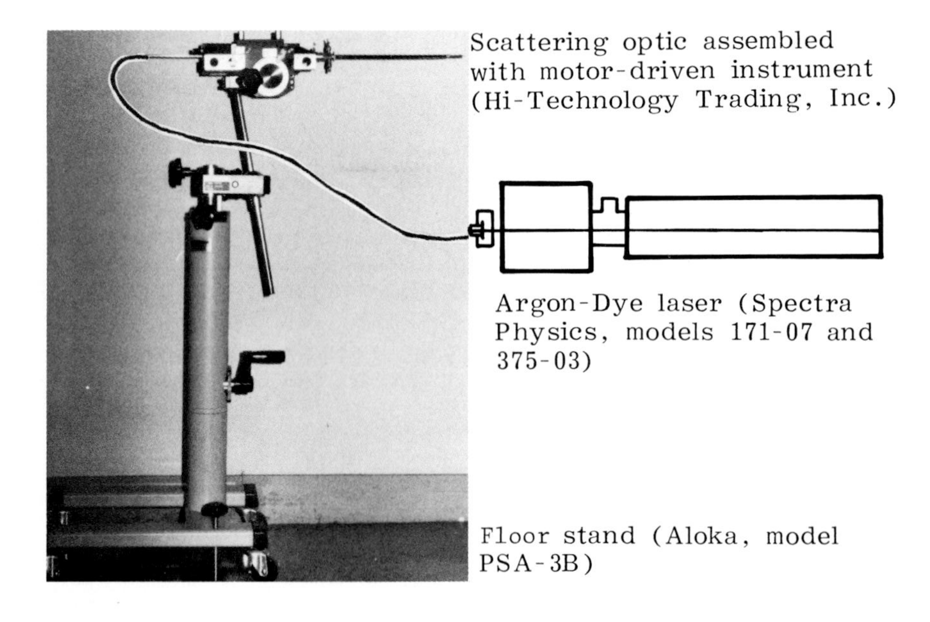

Fig. 3. Schematic diagram of the whole bladder photoradiation instrument.

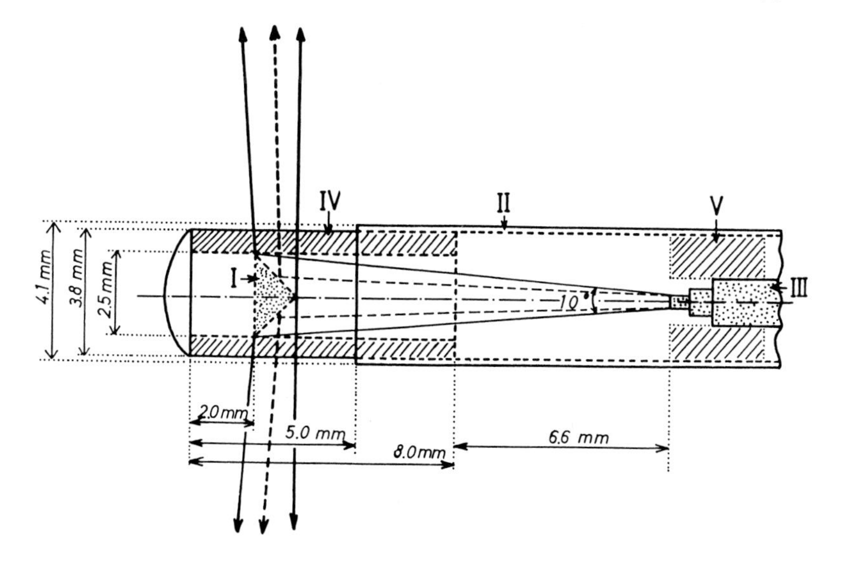

Fig. 4. Diagram of a piece of the laser light scattering optic.

Figure 5 shows the relationship between the power density of direct light or reflected light and the distance from the light source to the power detector. The light was supplied by an argon-dye laser, a Spectra Physics High Power model 375-03 dye laser pumped by a Spectra Physics model 171-07 argon laser. The wavelength of the light was 630±1.6 nm.

The optic can be variably driven at a translational speed of 40 to 160 mm/hr. When the mirror was irradiated by laser light through a quartz fiber, the scattered light was reflected in a circular band around the bladder wall. If an accumulated energy intensity of 10 J/cm^2 is necessary for the CIS therapy in patients with a 200 ml bladder, the optic must be driven for 6.3 min/10 mm to 12.5 min/10 mm for a total of 61.2 minutes with a fiber output power of 1.2 W as shown in Fig. 6.

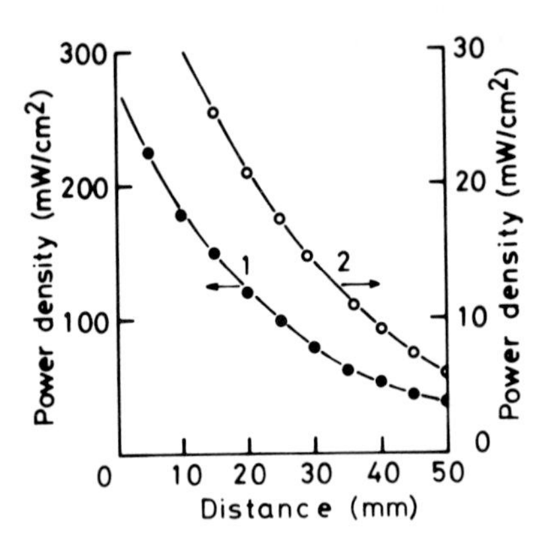

Fig. 5. Relationship between the power density of direct light or reflected light and the distance from the light source to the power detector.

1: Direct light
2: Reflected light

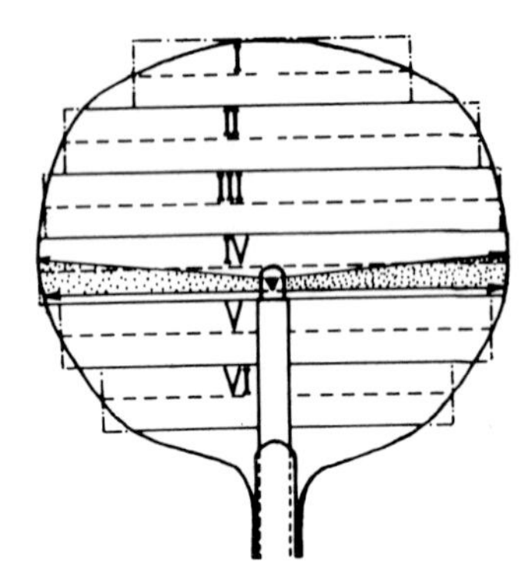

No. of horizontal sections	Radius	Power density of scattered lights	The optic speed when total light dose is 10 J/cm²
	mm	mW/cm²	min/mm
I	21	53.3	0.63
II	34	30.3	1.10
III	37	27.6	1.21
IV	38	26.8	1.25
V	35	30.2	1.10
VI	28	40.3	0.83

Fig. 6. Ultrasonographic horizontal sections of a 200 ml bladder at intervals of 10 mm.

RESULTS

I. Detection of HpD fluorescence before irradiation

Frozen sections of the biopsies were excited at 400 nm. These fluorescence spectra were corrected by using a standard phosphor, 4-dimethyl-amino-4′-nitrostilbene. HpD fluorescence spectra of bladder tumorous tissue (Curve 1) and normal tissue (Curve 2) were obtained as shown in Fig. 7. The fluorescence intensity of HpD in Curve 1 and 14 times higher than that in Curve 2. This indicates that HpD is preferentially retained by the tumor.

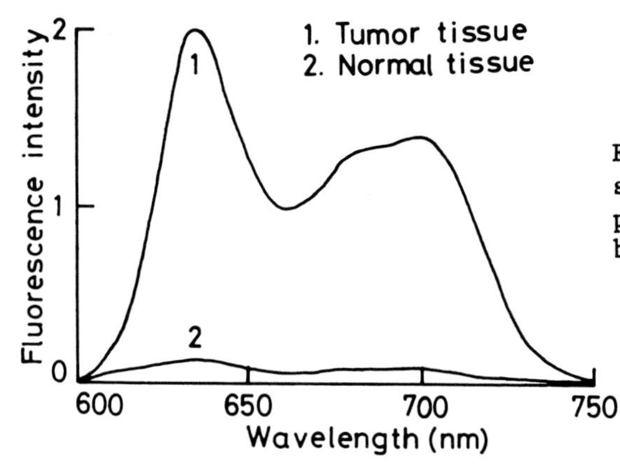

Fig. 7. Fluorescence spectra of HpD incorporated in bladder biopsies.

Urinary exfoliative cells were collected from the urine of a 30-year-old-man with CIS of the bladder 48 hrs after the i.v. injection of 4.0 mg HpD/kg body weight. Fluorescence microphotograph (Fig. 8) shows a red HpD fluorescence emission.

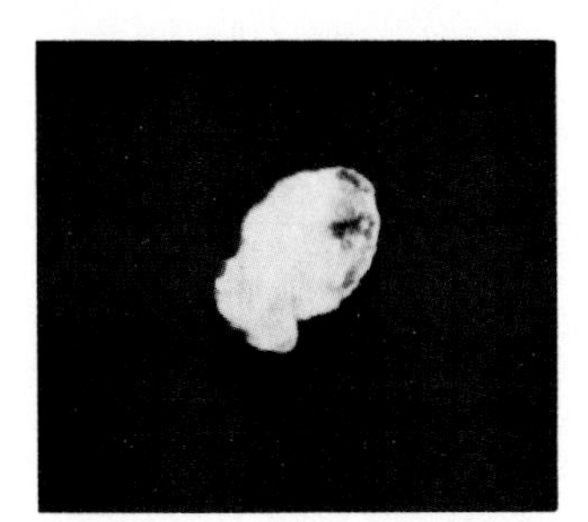

Fig. 8. Fluorescence microphotorgraph showing urinary exfoliative cells from a 30 year-old man with CIS of the bladder.

II. Whole bladder wall photoradiation for patients with CIS

The photoradiation was delivered for 6.3 min/10mm to 12.5 min/10 mm for a total of 61.2 minutes which supplies 10 J/cm^2 on the inner surface of the bladder.

III. Histological evaluation

1. Biopsy before photoradiation

Biopsies were carried out for a 30-year-old man before photoradiation. The biopsies revealed the CIS as shown in Fig. 9.

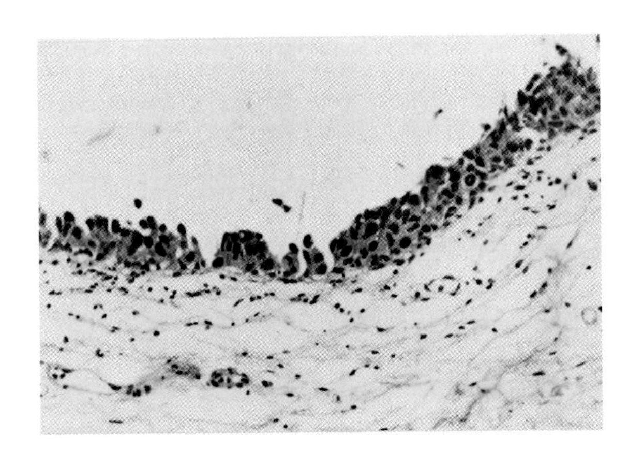

Fig. 9. Microphotograph showing CIS in a bladder biopsy from a 30-year-old man. HE stain.

2. Cytology after photoradiation

Clusters of vacuolated bladder cancer cells were found in the urine for 2 days after photoradiation (Fig. 10).

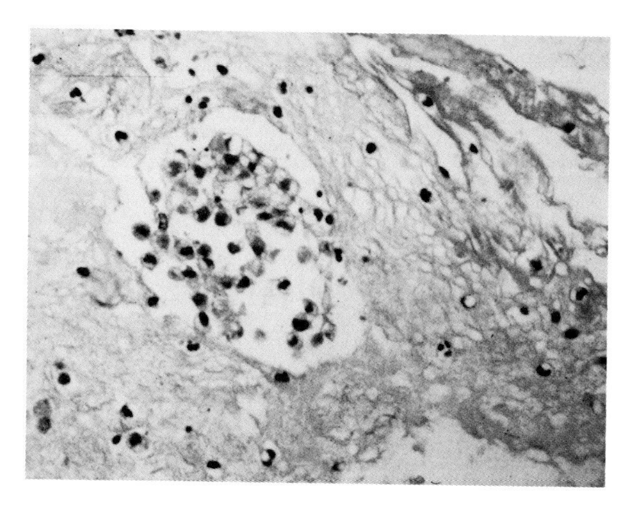

Fig. 10. Microphotograph showing a cluster of vacuolated bladder cancer cells. HE stain.

DISCUSSION

HpD has been found to accumulate in tumor tissues and was successfully used clinically as a tumor-specific compound for the localization of occult lung tumors (Profio et al. 1979). A system for imaging occult bronchogenic carcinoma was developed with HpD and a krypton ion laser as a violet excitation source by Profio et al. (1979). In our present paper, the detection of CIS of the bladder by fluorescence endoscopy with HpD and whole bladder wall photoradiation with HpD and motor-driven scattering optic are reported. The detection of the CIS by fluorescence endoscopy was difficult because of the light reflection in our video system. The accumulated energy intensity of 5-25 J/cm^2 was used for the CIS therapy in 3 cases. The anticancer effect of whole bladder wall photoradiation will hold great hope for superficial multi-centric tumors and urothelial abnormalities when the present problems are solved.

REFERENCE

Profio AE, Doiron DR, King EG (1979). Laser fluorescence bronchoscope for localization of occult lung tumors. Md Phys 6:523.

Porphyrin Localization and Treatment of Tumors, pages 249–256

Integral dye-laser irradiation of photosensitized bladder tumors with the aid of a light-scattering medium

D.Jocham , G.Staehler , Ch.Chaussy , E.Unsöld*) , C.Hammer
U.Löhrs, W.Gorisch
Urological Clinic, University of Munich, D-8000 Munich, Germany
*) Gesellschaft fuer Strahlen- und Umweltforschung mbH, D-8042 Neuherberg, Germany

Bladder carcinoma is a tumor characterized in 60% of cases by multifocal growth, and with a relapse rate of 60-80% within 20 months following conventional therapy. At present it is often not possible to spare the patient the ultimate treatment of cystectomy in order to halt the course of the disease. The rather limited success of conventional treatment methods, and the distress associated with the ultima ratio, have prompted the search for more effective methods of therapy. Since 1979 our own efforts have been aimed at developing a method for photoradiotherapy of HpD photosensitized bladder tumors.
KELLY and coworkers reported in 1976, that HpD was selectively stored by human bladder carcinoma (4). More recent investigations, for example those of BENSON and coworkers, have both confirmed these results and shown that carcinoma in situ also stores HpD selectively (1). Our concern lies with carcinoma in situ, and with the multifocal form of bladder carcinoma. These tumors present the urologist with as yet unsolved diagnostic and therapeutic problems since it is often impossible to recognise all sites of tumor growth endoscopically. Direct treatment is therefore not feasable and even local cytostatic treatment has failed to solve this therapeutic problem. Our own experiments with experimental bladder carcinomas in rabbits have shown that although photosensitized tumor tissue can be successfully destroyed by directed laser irradiation this method does not totally solve the problem of locating and reaching all tumor tissue (2). In particular it is difficult to reach tumors in the region

of the front wall of the bladder and the bladder neck. This means that full exploitation of the decisive advantage offered by photoradiotherapy - the selective destruction of tumor tissue - is limited in bladder and other hollow organs by the problems associated with directing the laser beam. Thus our aim became the development of a system whereby hollow organs could be homogeneously irradiated over the entire inner surface. Assuming that sufficient selective photosensitization of tumor areas remains after complete HpD clearance of normal tissue, homogeneous (i.e. integral) irradiation would make it possible to destroy all tumor tissue whilst maintaining intact areas of normal bladder wall.

The primary technical prerequisite for such therapy is a method of introducing the laser beam into the bladder without causing undue stress to the patient, and without offering too many technical problems to the surgeon. The use of a flexible light fiber introduced transurethrally to transport the focussed laser beam into the bladder solves this problem completely.

Initially we considered achieving homogeneous irradiation by means of a rotating light-pipe, along the same lines as the routinely-used rotating transurethral ultrasonic probe. However, this method is not only fairly complex; our experience indicated that there was a danger that certain areas of the bladder wall, in particular the front wall and the neck, might not be sufficiently irradiated. In addition it is not always possible to avoid injury to the bladder wall when using rotating devices.

Our second step was to investigate ways of scattering the laser beam homogeneously within a hollow organ. It might be possible to achieve this effect by modifying the light fiber itself. However, we tended to explore a second, and simpler possibility - scattering of the beam with a clinically harmless light scattering medium. The investigations described in the following were all aimed at developing a therapeutic method based on this principle.

In order to keep the method as simple as possible it was necessary to find a medium which would have no harmful effects if it came into contact with the mucosa or entered the vascular system. Such a medium could be used for irrigation of the bladder.

A glass model representing a hollow organ was used for preliminary in vitro tests. Using this model it was possible to test the scattering qualities of various media on a focussed beam of red light from a dye-argon laser transported via a flexible light fiber with a plane or ball-shaped tip (Fig. 1). The radiation intensity was measured by means of a photocell moved mechanically along the wall of the glass vessel and registered in relative units around the circumference of the sphere.

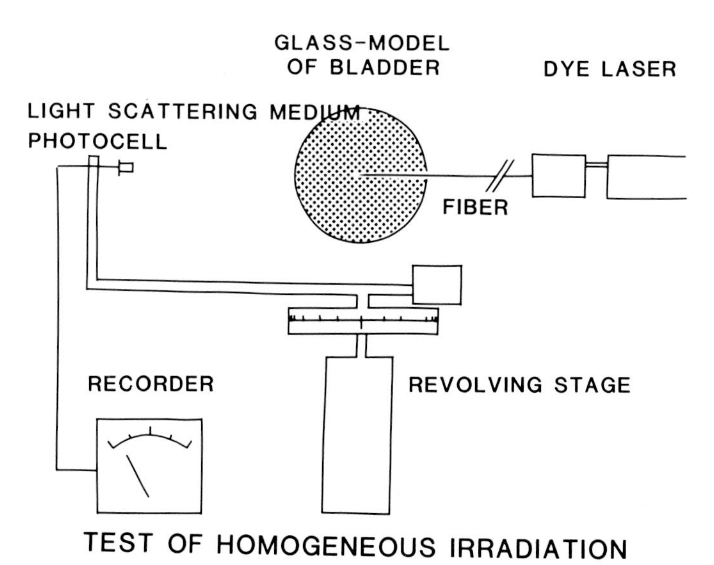

Fig. 1 : Scheme of testing homogeneous irradiation

Thus in addition to testing gold emulsions and tungsten-acid solutions, we decided to test the light scattering qualities of fat emulsions. The latter fulfil the clinical requirements completely and experience with calk suspensions encouraged us to believe that they might also create sufficient scatter.

Figure 2 shows the measured distribution of radiation around the sphere under various conditions. The upper curve shows the scatter characteristics for a flat-ended light fiber without scattering medium. The beam is only slightly broadened.

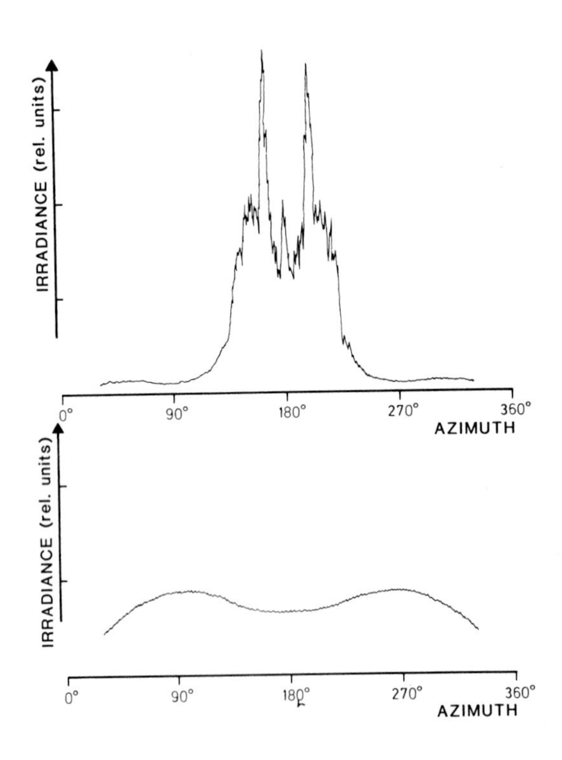

Fig. 2 : Scatter characteristics of laser light
above: flat fiber end - no medium
below: light scattering medium

The lower curve shows the light scattering characteristics of a special fat emulsion. An almost homogeneous distribution of light can be achieved, even with the flat-ended light fiber. There is a certain loss of light energy as a result of absorption which depends on the particle concentration in the solution. This loss was between 10% and 50% in those solutions with the most suitable scattering properties. The resultant warming of the solution is clinically negligible. Similarly, the theoretical possibility that the solution might alter its properties during the treatment as a result of the effect of the laser energy, or of dilution with blood or urine, can be neglected because the solution is continuously renewed throughout the period of irradiation.

The solution with the best light-scattering properties was tested in animal experiments. Tests were performed on 22 rabbits bearing Brown-Pearce-transplant bladder tumors using fat emulsion as a light scattering medium. Preliminary results of these experiments were reported last September in Seattle at the International Cancer Congress.

Animals were injected intravenously with 5 mg HpD/kg body weight. The time for integral radiation treatment was chosen such that HpD clearance from normal tissue was virtually complete, thus ensuring selective destruction of tumor tissue.

The variation in HpD concentration with time in transplant tumor and normal bladder was measured spectrofluorometrically in experiments to provide a basis for assessing the optimal period between HpD injection and treatment. The next figure shows that very little HpD is retained in normal bladder-tissue 48 h after injection, as with other normal tissues, whereas considerably higher amounts of HpD are retained in the tumor tissue (Fig. 3).

The optimal choice of radiation must also be determined. In previous experiments with chemically induced urothelial tumors in rat bladder we showed that an energy of >60 J/cm^2 is sufficient for complete destruction of tumor tissue (3, and unpublished observations). These tumors are histologically and biologically comparable with human bladder carcinoma. This radiation dose causes complete necrosis of normal mucosa when applied 6 h after injection of 10 mg/kg body weight HpD, but 48 h after injection the same dose has no effect on normal mucous lining

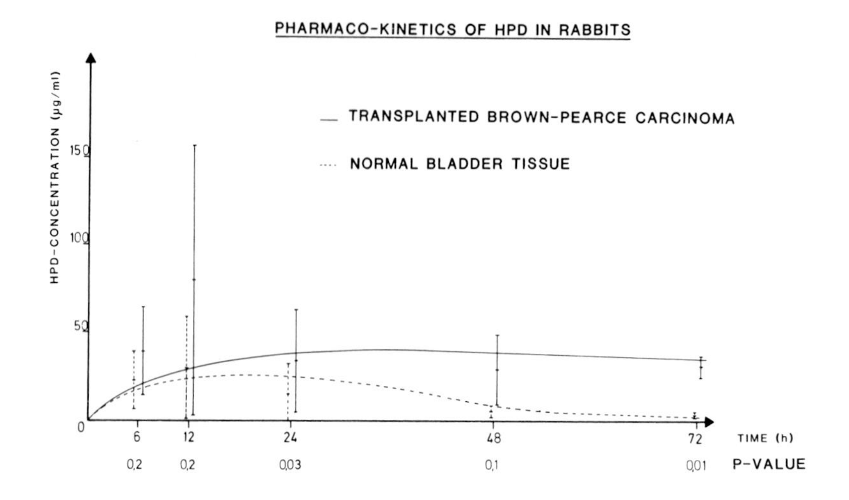

Fig. 3 : Pharmaco-Kinetics of HpD
- transplant-tumor
... normal bladder-tissue

and even at doses of up to 700 J/cm² the effects are negligible. Thus precise radiation dose employed is not critical as long as the time of treatment is suitably chosen.

Figure 4 shows the treatment scheme used in the rabbit experiments, developed on the basis of the previous observations (Fig. 4). Treatment was commenced 60 h after HpD injection. The light fiber was introduced transurethrally into the bladder and was directed past the tumor, situated at the bladder neck, into the centre of the bladder. This ensured that there was no direct irradiation of the tumor. The bladder was irrigated with light scattering medium such that the bladder volume was kept as constant as possible, and the bladder irradiated until a dose of 50 - 60 J/cm² had been applied. The average length of irradiation was 25 min.

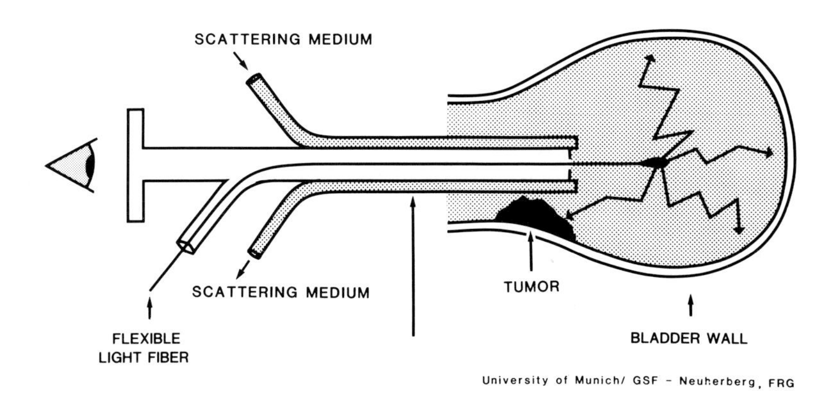

Fig. 4 : Scheme: Integral Photoradiotherapy of bladder tumors

Using this method of integral dye-laser therapy of photosensitised tumors (patent pending) we were able to completely destroy all tumor tissue in 19 of the 22 rabbits tested. In control experiments using the same treatment without injection of HpD the tumor tissue continued to grow. The normal mucous lining showed virtually no signs of damage after treatment.

Conclusion:

- Integral photoradiotherapy of hollow organs (e.g. bladder) can be successfully performed using a light scattering medium
- The light scattering medium is clinically well-tolerated
- The application is simple. There are no mechanical moving parts.

- Successful effect on tumor tissue even in areas of bladder difficult to reach.
- Normal bladder wall is not affected by laser irradiation.

At the moment we are in the process of constructing a catheter which will enable the integral irradiation to be performed without endoscopic control (patent pending) thus simplifying the method even further.

Corresponding address
Dieter Jocham M.D.
Urological Clinic and Policlinic of the University of Munich, Klinikum Großhadern
Marchioninistr. 15
8000 Munich 70
FRG

References:

1) Benson, R.C., Farrow, G.M., Kinsey, J.H., Cortese, D.A., Zincke, H., Utz, D.C.
Detection and localization of in situ carcinoma of the bladder with hematoporphyrin derivative
Mayo Clin. Proc. 57: 548-555 (1982)
2) Jocham, D., Staehler, G., Chaussy, Ch. Hammer, C., Loehrs, U.
Dye laser therapy of bladder tumors after photosensitization with hematoporphyrin derivative (HpD)
Laser-Tokyo '81, 4th Congress of the International Society of Laser surg. ed by K. Atsumi and N. Nimsakul (1981)
3) Jocham, D., Hammer, C. Loehrs, U., Staehler, G., Chaussy, Ch., Dietrich, R.
Farblasertherapie photosensibilisierter Blasentumoren
Chir. Forum 82 f. experim. u. klinische Forschung, Hrsg. S. Weller, Springer Berlin, Heidelberg, New York (1982)
4) Kelly, J.F., Snell, M.E.
Hematoporphyrin derivative: a possible aid in the diagnosis and therapy of carcinoma of the bladder
J. urol. 115: 150-151 (1976)

PORPHYRIN CHARACTERIZATION AND PHOTOCHEMISTRY

Porphyrin Localization and Treatment of Tumors, pages 259–284

DEPENDENCE OF PHOTOSENSITIZED SINGLET OXYGEN PRODUCTION ON PORPHYRIN STRUCTURE AND SOLVENT

J. G. Parker* and W. D. Stanbro

The Johns Hopkins University
Applied Physics Laboratory
Laurel, Maryland 20707

I. INTRODUCTION

Past investigations of the role of singlet molecular oxygen $^1O_2$ in physical, chemical and biological processes in the liquid phase have relied heavily on indirect methods. (Note: Throughout the text, the symbol $^1O_2$ will be used to denote molecular oxygen in the $^1\Delta_g$ state.) Foremost among these methods is the use of chemical acceptors, or traps, which react selectively with $^1O_2$ but which are essentially inert toward oxygen in the ground electronic state. Use of such indicators is hampered by three basic limitations which can be categorized roughly as follows: (1) problems of either too much or too little - too high a concentration leads to significant additional quenching and too little to serious indicator depletion, both of which cause significant problems in data interpretation (Young et al. 1973); (2) the fact that certain indicators react with both $^1O_2$ and free radicals in the same manner (Howard, Mendenhall 1975; Boyer et al. 1975); and (3) the fact that acceptors tend to be either water soluble or lipid-soluble, but not both.

The use of optical techniques is based on the fact that $^1O_2$ may undergo a collisionally induced radiative transition to the ground electronic state resulting in a relatively weak emission in a fairly narrow band (300-400Å) centered at approximately 1.27μ. Although for more than a decade detection of this emission has been used to monitor $^1O_2$ concentrations in the gas phase at relatively low pressure where quenching rates are low, application to investigation of the liquid state has only been recently accomplished (Krasnovskii 1976; Khan, Kasha 1979). Use of kinetic techniques involving pulsed laser excitation in combination with semiconductor photodiodes has occurred most recently (Salokhiddinov et al 1979; Hurst et al 1982; Parker, Stanbro 1982, Ogilby, Foote 1982; Rodgers, Snowden 1982). Notably the use of remote optical sensing of $^1O_2$ has recently led to the resolution of difficulties presented by contradictory data obtained earlier through the use of chemical acceptors (Ogilby, Foote 1981).

To date most investigations using the newer optical techniques have been concerned with the measurement of quenching rates in various solvents for conditions under which the quenching is essentially physical, i.e., no oxygen consumption. However, in most biological applications this is not the case. A good example of this is contained in a recent experimental investigation of the relative effect of different porphyrins regarding their ability to cause photo-oxidation of tryptophan on the one hand, and to inhibit respiration of intact mitochondria on the other (Sandburg, Romslo 1981). It was found for the former that the following hierarchy prevailed: uroporphyrin (UP) > coproporphyrin (CP) > protoporphyrin (PP), while in the latter case the ordering was exactly reversed. Reasons given for this behavior were essentially centered around the importance of sensitizer solubility, i.e., the ordering according to biological activity paralleled the ability of the corresponding components to partition into the lipid phase, and thus capable of membrane penetration. The importance of sensitizer solubility on the photodynamic modification of membrane transport properties has been pointed out in recent publications (Bellnier, Dougherty 1982; Kessel 1981, 1982; Moan et al 1979). The possibility of synergism between photodynamic treatment of LI2IO cells with hematoporphyrin and subsequent administration of anti-cancer drugs has been considered in recent experimental research (Creekmore, Zaharko 1983).

A comparison of the ability of several water soluble porphyrins to generate $^1O_2$ (as assessed by means of photo-oxidation of bilirubin) with their biological potency as measured by the rate of inactivation of glioma cells in culture has led to interesting results (Diamond et al 1977). The porphyrin determined to be biologically most effective was meso-tetra (4-N-methylpyridil) porphin (TMPP) [slightly more effective than hematoporphyrin (HP)], and the least effective meso-tetra (p-carboxyphenyl) porphin (TCPP), slightly less than meso-tetra (4-sulfonatophenyl) porphin (TPPS). Results of the photooxidation of bilirubin, however, indicated TMPP to be substantially less effective than the rest, all of which exhibited approximately the same efficacy. The synthetic porphyrins TMPP, TPPS and TCPP form an interesting group since they are all water soluble and essentially insoluble in lipids. Whereas TMPP remains monomeric in solution, TPPS and TCPP tend to aggregate to approximately the same degree as HP (Pasternack et al 1972; Turay 1978). Laser-excited states of both TMPP and TPPS have been investigated and described in a recent paper (Bonnett et al 1982).

II. OPTICAL MONITORING OF $^1O_2$ IN SOLUTION

A. Fundamental Energy Transfer Considerations

The generally accepted energy transfer mechanism for light-induced tumor destruction following porphyrin injection is presented schematically in Fig. 1. First the incident light is absorbed by the porphyrin in its ground singlet electronic state S_o leading to elevation to singlet S_1. In the experimental research described in this paper, an excitation wavelength of 5320 Å was used, corresponding to the frequency-doubled output of a Q-switched Nd:YAG laser; however, porphyrins absorb over a rather wide spectral range extending from 3600-6500 Å. The excited singlet S_1 ultimately returns to the ground state via three deactivation paths, the first two direct, and the third indirect. The two direct paths involve a radiationless transition with the excess energy being given off to the surrounding media as heat (very low probability) and a radiative transition resulting in emission in the spectral range 6000-7000 Å corresponding to the characteristic red porphyrin fluorescence (low probability). The third possible deactivation path (high probability) involves an intramolecular process in which a transition from S_1 to the lowest allowed triplet state T_1 occurs. Transition from T_1 to S_o occurs with extremely low probability due to poor coupling of these two states as a consequence of problems associated with spin conservation. This of course is true because the ground electronic state of the solvent molecules is also a singlet. In the unique case of oxygen, however, the ground electronic state is a triplet and the first excited state a singlet ($^1\Delta_g$), thus existing as a natural complement to the metastable dye. The ultimate result of this complementarity is that an intermolecular spin-conserving interaction takes place according to

$$T_1 + {}^3O_2 \rightarrow S_o + {}^1O_2 \qquad (I)$$

with the energy difference between T_1(1.5 - 1.6 ev) and ${}^1O_2$ (1 ev) being liberated as heat. As in the case of the porphyrin triplet state T_1, deactivation of ${}^1O_2$ to the ground ${}^3O_2$ state is extremely difficult due, again, to problems associated with spin conservation and in addition the absence of any dipole coupling. ${}^1O_2$ lifetimes therefore tend to be relatively long, ranging from microseconds for water to milliseconds for halogenated hydrocarbons.

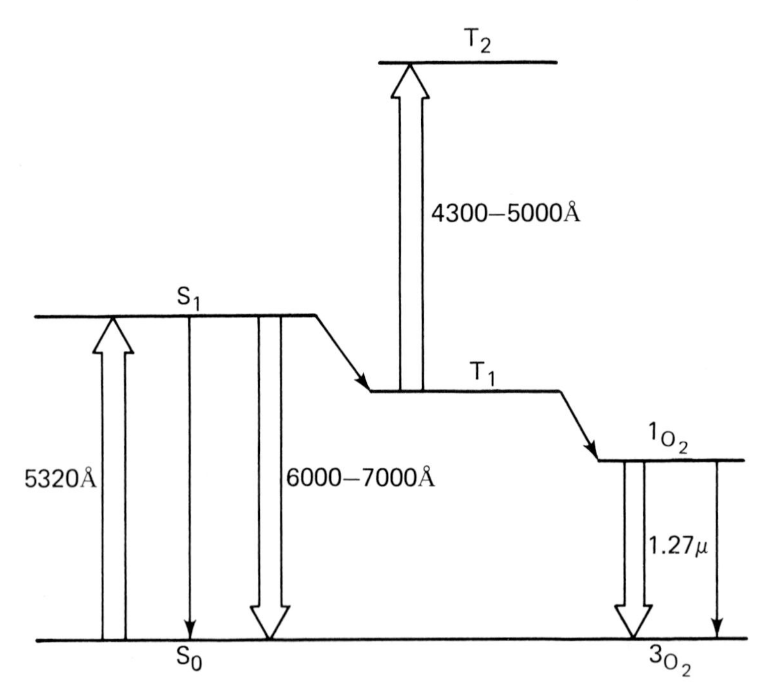

Fig. 1. Energy level diagram indicating optical transitions and energy transfer processes involved in the photosensitized generation of singlet molecular oxygen in the ${}^1O_2$ electronic state.

Three possible deactivation paths for ${}^1O_2$ exist, two

physical and one chemical. In the first, the electronic energy is converted to vibrational energy of the solvent molecules and than rapidly to heat,

$$^1O_2 + M \rightarrow {}^3O_2 + M + \text{heat}; \qquad \text{(II)}$$

in the second, a relatively weak collisionally-induced emission occurs accompanying radiative transition to the ground electronic state

$$^1O_2 + M \rightarrow {}^3O_2 + M + h\nu_E \qquad \text{(III)}$$

where the emission frequency ν_E corresponds to a wavelength of 1.27 μ; thirdly, reaction may occur

$$^1O_2 + R \rightarrow RO_2 \qquad \text{(IV)}$$

thus involving oxygen consumption, a process which under certain conditions may deplete the oxygen concentration in the vicinity of the exciting dye molecule, i.e., in the case of relatively slow oxygen diffusion rates.

Transition (III) provides the means for remote optical monitoring of $^1O_2$ in condensed media. The metastable triplet population T_1 may be independently monitored using an active optical technique based on the fact that·the absorption involved in elevating the porphyrin from the lowest triplet state T_1 to a higher state T_2 (i.e., triplet-triplet absorption) is almost as strong as that characterizing the $S_o \rightarrow S_1$ excitation and most importantly occurs in a wavelength region where the singlet absorption is weak, typically in the spectral interval from 4300-5000 Å.

Thus, using completely remote optical means, it is possible to monitor both the porphyrin triplet and resultant $^1O_2$ concentrations simultaneously. Earlier experimental measurements using hematoporphyrin in various solvents have indicated that the rate of transfer of energy from T_1 to form $^1O_2$ [Transition (I)] is essentially independent of solvent, depending only on dissolved oxygen concentration (Parker, Stanbro 1981). Thus for the air-saturated solvents methanol, ethanol and acetone, the triplet quenching times are 0.2-0.3 μsec while for H_2O and D_2O the numerical range is 2-3 μsec consistent with the fact that in the latter the ambient oxygen concentration is approximately an order of magnitude smaller than in the former.

Results of experimental data accumulated over the past decade are consistent with the energy transfer concepts described qualitatively by Reactions (I) - (IV). Testing of these concepts has mainly involved either adding or removing oxygen, thus affecting the rate of (I); substitution of deuterated for protiated solvents to slow the rate of (II); or, alternatively, addition of known $^1O_2$ physical quenchers to accelerate (II); finally, the addition of species known to react selectively with $^1O_2$ (e.g., bilirubin, β-carotene, α-tocopherol) to increase the rate of (IV). The primary means for assessing the effectiveness of $^1O_2$ has been to monitor changes in optical absorbance accompanying chemical reaction of the added acceptor or biologically to establish the effect on survival rates of various cell types.

Quantitative validation of the correctness of Reactions (I) - (IV), however, has been almost entirely lacking. Although substantial $^1O_2$ quenching rate data have been compiled (Wilkinson, Brummer 1981) for a wide variety of solvents, a complete quantitative description of the overall energy transfer process involving both the rate of transfer of dye triplet energy to form $^1O_2$ as well as the subsequent solvent quenching is not available.

If we denote the instantaneous molar triplet concentration as $[T_1]$ and that of the resultant $^1\Delta_g$ as $[^1O_2]$, reactions (I), (II) may be expressed as

$$\frac{d[T_1]}{dt} = - K_T[^3O_2]\,[T_1] \qquad (1)$$

$$\frac{d[^1O_2]}{dt} = K_T[^3O_2]\,[T_1] - k_D[^1O_2] \qquad (2)$$

in which K_T is the rate of triplet energy transfer (expressed as $M^{-1}\,sec^{-1}$), k_D the $^1O_2$ solvent quenching rate (sec^{-1}), and $[^3O_2]$ the instantaneous dissolved ground electronic state oxygen concentration. In most cases maximum values of $[^1O_2]$ are smaller than $[^3O_2]_o$, the ambient dissolved oxygen concentration and thus to a good approximation

$$K_T[^3O_2] = k_T, \qquad (3)$$

a constant.

Solution of Eqs. (1), (2) using Eq. (3) leads to

$$[{}^1O_2] = [T_1]_o \left(\frac{k_T}{k_T - k_D}\right)(e^{-k_D t} - e^{-k_T t}) \qquad (4)$$

in which $[T_1]_o$ is the porphyrin triplet concentration existing immediately after the laser pulse is applied. This of course requires the duration of the laser pulse P to be sufficiently short that $k_T P << 1$, which is usually the case.

From Eq. (4) it is clear that quantitative description of the temporal variation of $[{}^1O_2]$ involves the difference of two exponentials, one characterized by the $^3O_2$ dependent triplet quenching time $\tau_T = 1/k_T$ and the other by the $^1O_2$ decay time $\tau_D = 1/k_D$ which depends on the nature of the solvent. From an examination of the time history of the $^1O_2$ intensity, for a given set of experimental conditions it is impossible to uniquely determine the time constants τ_T and τ_D. In particular, one is only able to speak of a rise time τ_R and fall time τ_F, i.e.,

$$[{}^1O_2] = R(e^{-t/\tau_F} - e^{-t/\tau_R}) \qquad (5)$$

in which R is a constant for given solvent conditions depending on incident laser intensity and dissolved sensitizer concentration.

Given the pair of time constants τ_R, τ_F obtained from the experimental data, the problem is then to decide which of the two is τ_T and which is τ_D. This indecision can only be resolved by determining experimentally the effect of separately changing either τ_T or τ_D. From Eqs. (3), (4) it is obvious that the way to accomplish this is to change τ_T either by increasing or decreasing the ambient oxygen concentration. However, from Eq. (4) it is clear that such a modification may lead to a reversal. For example, suppose that in a run on an air-saturated sample that in actuality, (however unknown to the experimentalist), $\tau_D = \tau_R$ and $\tau_T = \tau_F$ requiring of course that $\tau_T > \tau_D$. Let us further suppose that the ambient oxygen concentration is now increased and thus τ_T shortened sufficiently that $\tau_T < \tau_D$. The consequence of this is that now $\tau_T = \tau_R$ and $\tau_D = \tau_F$ and the identity of τ_D is revealed only by the fact that of the two time constants it alone remained invariant. This sort of behavior has been encountered experimentally in the case of an ethylene glycol solvent and is discussed below.

A detailed discussion of problems associated with the determination of time constants characterizing functions of the type described by Eq. (4) from experimental data has recently appeared (Carrington 1982).

B. Experimental Test of the Theory

1. Measurements in H_2O

The two main experimental requirements to be satisfied in order that τ_D and τ_F be determined from the experimental data are: (1) fast detector response time, and (2) sufficiently high spectral resolution to maximize the discrete $^1O_2$ emission relative to that of the omnipresent continuous infrared fluorescence. The first of these conditions was satisfied through replacement of the germanium photodiode used in earlier research (Parker, Stanbro 1981, 1982) by an InGaAs photodiode designed especially for fast response (Burrus et al 1981), and the second by means of a narrow band interference filter designed for maximum transmission in the range from 1.28-1.29 μ.

One of the most important solvents to be used in testing both the predictions of theory as well as the temporal resolution capabilities of the $^1O_2$ detection system is H_2O. Previously reported values for τ_D range from 2 to 7 μs. Most recently a value of 4.2 μs has been obtained from $^1O_2$ emission measurements (Rodgers, Snowden 1982). Earlier measurements of the triplet quenching time τ_T for air-saturated samples gave values lying in the range 2-3 μs (Parker, Stanbro 1981). The closeness of these two values indicates that for the case of air-saturation, the time dependence of the $^1O_2$ emission cannot be characterized by a single time constant and that both the rise and fall times are intimately involved.

A good example of this is evidenced by the sequence of experimental traces appearing in Fig. 2, labelled A, B and C, corresponding respectively to oxygen-enriched, air-saturated and oxygen-depleted samples. For curves B and C the signal emerging from the tail end of the fast infrared fluorescence is seen to slowly rise to a peak and then decline, more rapidly for B than for C. For the oxygen-enriched sample A, the triplet quenching time has been shortened sufficiently that no peak in the $^1O_2$ emission is visible and only a long steady decay is evident, the rate of which is governed primarily by $^1O_2$ quenching. Comparison of theory and experiment in the latter case is relatively

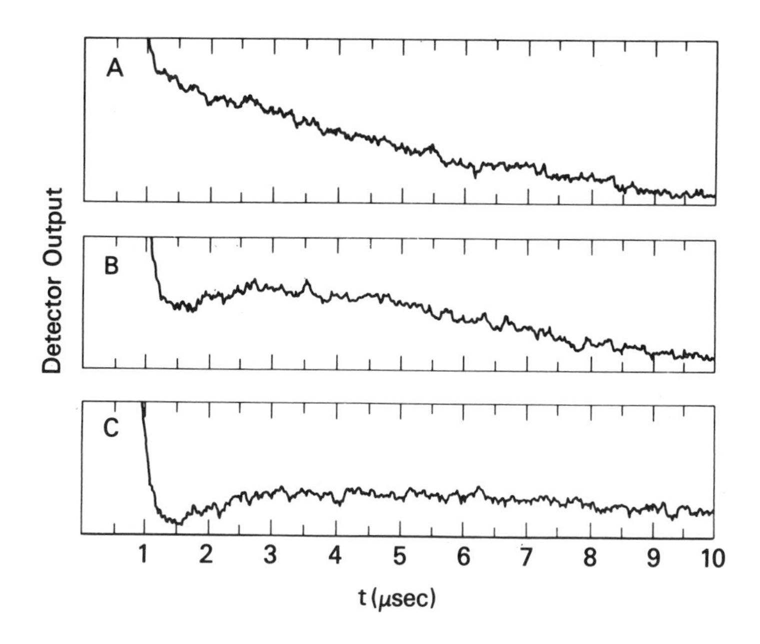

Fig. 2. Effect of changing dissolved O_2 concentration in H_2O (pH = 7.2) on $^1O_2$ emission. Curve A: sample oxygenated for 20 min; Curve B: air-saturated sample; Curve C: sample de-oxygenated by nitrogen bubbling for 60 min. Sensitizer in all cases was TPPS at a concentration of 1 x 10^{-4}M.

straightforward, yielding a decay time τ_D = 3.3 μs and τ_T = 1.0 μs. Using τ_D = 3.3 μs for cases B and C, we obtain τ_T = 2.4 μs and τ_T = 7.9 μs, respectively. A comparison of theory with the experimental data is displayed in Fig. 3.

An interesting feature of the data and theoretical calculations presented in Figs. 2, 3 is that as one progresses from the air-saturated sample B to the oxygen depleted sample C, a reversal takes place. Thus the rise time characterizing B corresponds to τ_T (2.4 μs); however, for C the rise time is τ_D (3.3 μs) and the fall time τ_F is now the lengthened triplet time τ_T = 7.9 μs. It therefore becomes apparent that values of τ_R and τ_F obtained for a sample with fixed oxygen concentration cannot yield directly the desired values of τ_D and τ_T. The only way in which this ambiguity may be resolved is from data obtained by successive determinations in which the oxygen concentration is changed.

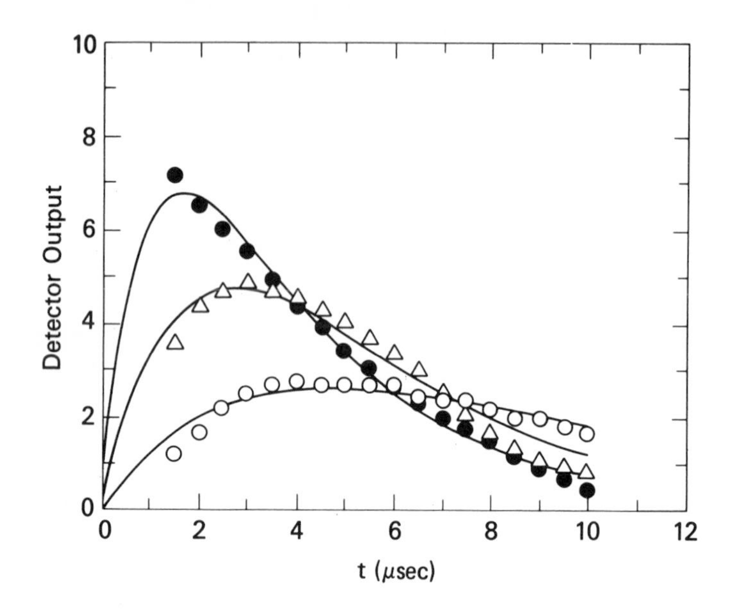

Fig. 3. Comparison of theory with experimental data displayed in Fig. 2. Fig. 2A (●); 2B (Δ); 2C (O).

The above measurements were obtained using a TPPS sensitizer at a concentration of 1×10^{-4}M with an incident laser pulse energy of 10(±10%) millijoules. To provide a basis for comparison of the $^1O_2$ generation efficiency of this porphyrin relative to other water soluble porphyrins, a series of similar runs was carried out under identical experimental conditions. Qualitatively the results may be summarized by the following ordering (acronyms appear in Section I, except for hematoporphyrin derivative HPD): TMPP > TPPS > TCPP,CP,UP > HP > HPD. As an indication of the magnitude of the degree of variation, the intensity levels recorded in the case of TMPP were approximately an order of magnitude higher than those observed in the case of HP. For HPD the signals were quite weak, being masked for the most part by instrumental noise.

2. Measurements in Ethylene Glycol

An example which highlights the problem of relating τ_R and τ_F values to τ_T and τ_D is that of ethylene glycol. Experimental data for oxygen-enriched (A) and air-saturated (B) samples are presented in Fig. 4. In both cases the

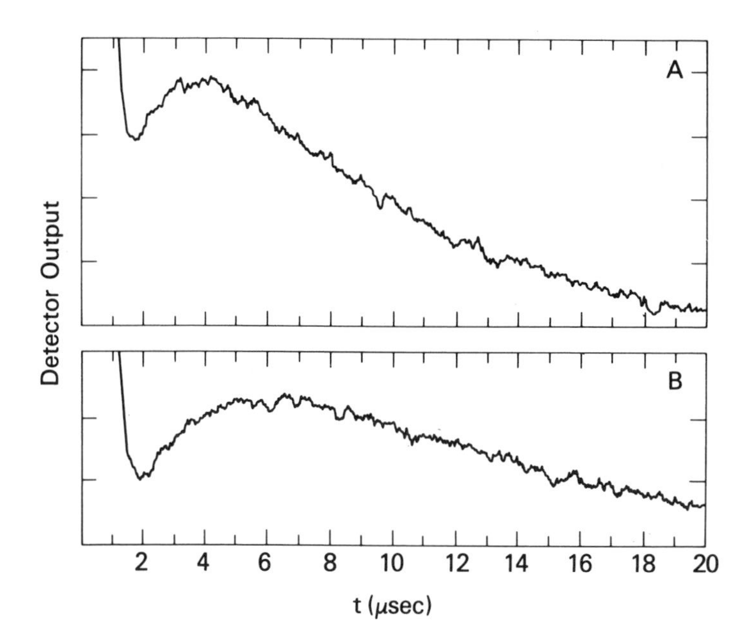

Fig. 4. Comparison of $^1O_2$ emission in an oxygenated sample of ethylene glycol (A) with that for an air-saturated sample (B).

initial behavior shows the signal to rise out of the decaying infrared fluorescence to attain a well-defined peak (earlier for A than for B) and to ultimately decay (more rapidly for A than B). The fitting of theory to the experimental data shown in Fig. 5 for the enriched sample gave τ_R = 3.3 μs and τ_F = 4.8 μs. Similar fitting for the air-saturated sample gave τ_R = 4.8 μs and τ_F = 8.6 μs. Since only τ_T can be affected by oxygen enrichment, it is thus clear that for an air-saturated sample, the rate of fall is controlled by τ_T = 8.6 μs. This substantially lengthening of the triplet quenching time relative to H_2O is probably due to a reduced O_2 diffusion rate accompanying increased solvent viscosity.

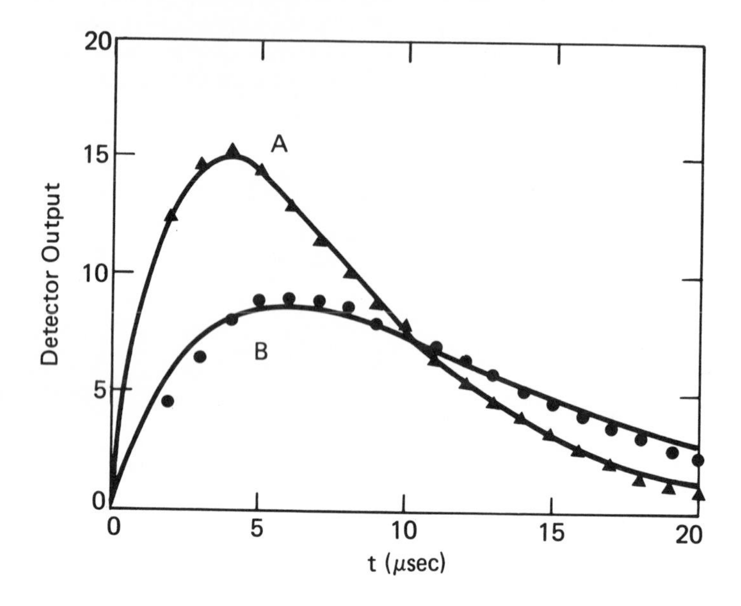

Fig. 5. Comparison of theory with experimental data displayed in Fig. 4. Fig. 4A (▲); 4B (●).

The conclusion one reaches based on the preceding comparison of theory and experiment is that the generally accepted theory appears to be quite close to the truth. However, some caution should be exercised since only cases of physical quenching have been dealt with and it has furthermore been assumed that solvent quenching of the sensitizer triplet is unimportant.

C. Effect of Substitution of D_2O for H_2O

A standard test that has been used in the past to provide indirect evidence of the participation of $^1O_2$ in various biochemical and biological reactions involves a determination of the effect of substitution of D_2O for H_2O. (Ito 1978; Sandberg, Romslo 1981). In general it has been found that this substitution brings about yield increases up to an order of magnitude which has been attributed to a longer lifetime of $^1O_2$ in D_2O than in H_2O, a value most recently determined to be 55 μs (Rodgers 1983).

The temporal behavior of $^1O_2$ in H_2O and D_2O is compared in Fig. 6. For D_2O a steady rise to an almost constant limiting value is observed. This limiting intensity is seen to be substantially in excess of the intensity observed in the case of H_2O.

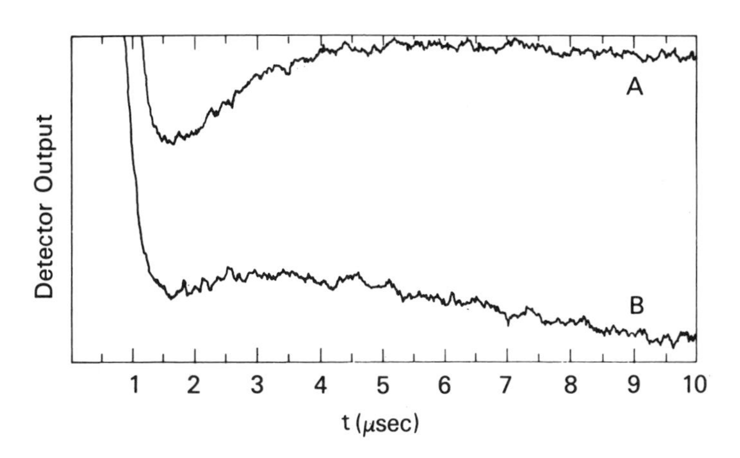

Fig. 6. Comparison of $^1O_2$ emission levels in D_2O (A) and H_2O (B). Sensitizer was TMPP at a concentration of 1 x 10^{-4}M. Both samples were air-saturated. Initial rise of (A) (following decay of initial fast infrared fluorescence component) is due almost entirely to time required to transfer energy from sensitizer triplet.

The long term decay of the $^1O_2$ signal in D_2O is displayed in Fig. 7 for a time base expanded to 50 μs. Analysis of this data yields τ_F = 68 μs which can only be equal to τ_D since τ_T in both D_2O and H_2O are known to be essentially the same (Parker, Stanbro 1981). Using the above value for τ_D, fitting of theory and experiment gives τ_T = 2.2 μs, essentially the same as for H_2O. In obtaining this data, a different sensitizer was used (TMPP) since the TPPS sensitizer used in the case of H_2O was observed to undergo significant degradation in D_2O, presumably due to reaction with $^1O_2$ which would be favored by the relatively long lifetime in this solvent No such effect was observed in the case of TMPP.

Fig. 7. $^1O_2$ emission in D_2O viewed on an expanded time-base. Solution was air-saturated and contained TMPP as a sensitizer at a concentration of 1 x 10^{-4}M. $^1O_2$ decay time was determined to be 68 μs.

D. Dependence of $^1O_2$ Yield on Solvent

An important question relates to the relative $^1O_2$ yield in different solvents for fixed sensitizer concentration and incident light intensity. In an attempt to shed light on this matter, a series of experimental runs were carried out in water, ethanol and glacial acetic acid using HP as a sensitizer at a concentration of 1 x 10^{-4} M. Data obtained for ethanol and water are presented in Fig. 8 as curves A and B, respectively. In comparing A and B, account should be taken of the reduced instrumental sensitivity in the former case. Thus to put the two on an equal basis it is necessary to multiply the A data by a factor of 2.5. $^1O_2$ levels in ethanol are thus much larger than in H_2O, even though the sensitizer concentration is the same. Some of the difference is due to the slower $^1O_2$ quenching rate in ethanol which turns out to be 15 μs. Also to be noticed from Fig. 8 is the fact that the $^1O_2$ signal in ethanol attains its maximum value rapidly, i.e., in a time less than 1 μs. [Experimental measurements indicate that for an air saturated solution τ_T = 0.2 - 0.3 μs (Parker, Stanbro 1981).] Using the appropriate experimental numbers, we determine from Eq. (4) that if all other conditions are the same, then, for example at t = 4 μs, the $^1O_2$ emission in ethanol should be twice that in H_2O. Experimentally the ratio is

30, more than an order of magnitude larger.

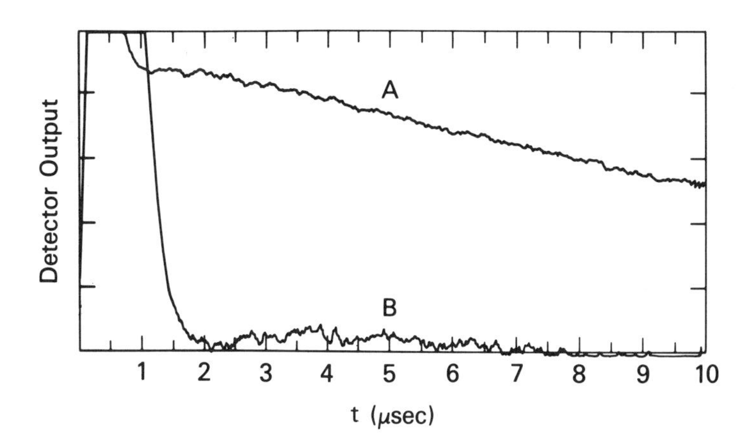

Fig. 8. Comparison of $^1O_2$ emission in ethanol (A) and water at a pH of 7.2(B). Sensitizer used was HP at a concentration of 1 x 10^{-4}M. Data indicated in curve A was obtained by reducing instrumental sensitivity by a factor of 0.40.

One possible explanation for this discrepancy is that HP in water is substantially aggregated, consistent with the known behavior of porphyrins in solution (Turay et al 1978) and that in ethanol the dye exists in basically monomer form in which it would be presumed to provide a higher triplet yield due to the absence of deactivation channels offered in the aggregated state. This would also explain why the $^1O_2$ levels for TMPP in H_2O as shown in Fig. 6 are measurably higher than the corresponding values appearing in Fig. 8, since TMPP in H_2O is known to exist as a monomer (Pasternack et al 1972). Comparison of the two sets of data indicates the reduction in the HP signal to amount to a factor of 4, which falls short of the factor of 15 required to bring the water/ethanol data into alignment.

In an attempt to elucidate this matter, experimental measurements in a 50/50 (by volume) water/ethanol mixture containing TMPP as a sensitizer at a concentration of 5 x 10^{-5} M were compared with similar data for H_2O with TMPP at the same concentration. These results are shown in Fig. 9. For the water/ethanol data (A), the peak occurs earlier in time due to the increased level of dissolved oxygen, thus shortening τ_T. For τ_D, we obtain 6-7 μs, i.e. twice that for water. Again using Eq. (4) to calculate the relative $^1O_2$ levels at t = 4 μs, we determine the ratio of ethanol/water intensities to be 1.5 which is distinctly smaller than the experimental number of 3.5.

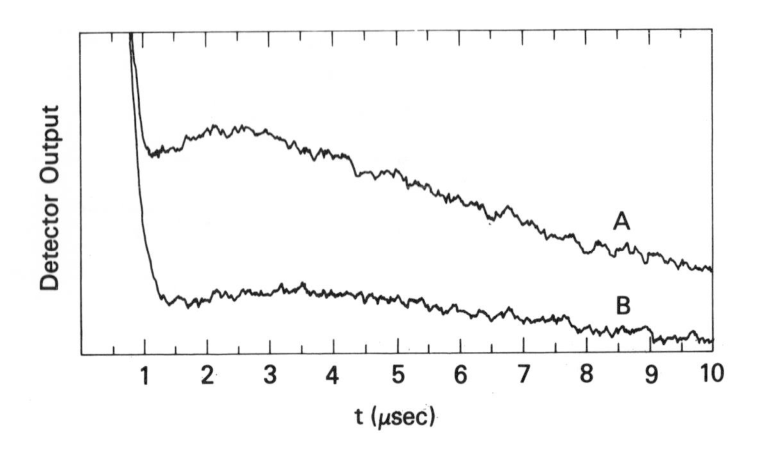

Fig. 9. Comparison of $^1O_2$ emission in H_2O (B) and a sample containing H_2O and ethanol in equal parts (by volume) (A). TMPP was used as a sensitizer at a concentration of 5×10^{-5}M

It would therefore appear that ethanol addition tends to promote the production of $^1O_2$ over and above the increase expected solely on the basis of de-aggregation.

As a third candidate in our investigation of the dependence of $^1O_2$ yield on solvent, we have chosen glacial acetic acid. This is an interesting solvent since it is involved in the first step required to convert HP into HPD. The temporal variation of the $^1O_2$ output is shown in Fig. 10. From this figure it is clear that the triplet quenching time is fairly rapid (i.e. $\leqslant$ 1 μs) whereas the τ_D is relatively long at approximately 25 μs. As in the case of ethanol, the enhancement of the $^1O_2$ emission appears too large to be accounted for simply by de-aggregation of the HP sensitizer. The most logical explanation would be in terms of chemical changes in the porphyrin side chain groups.

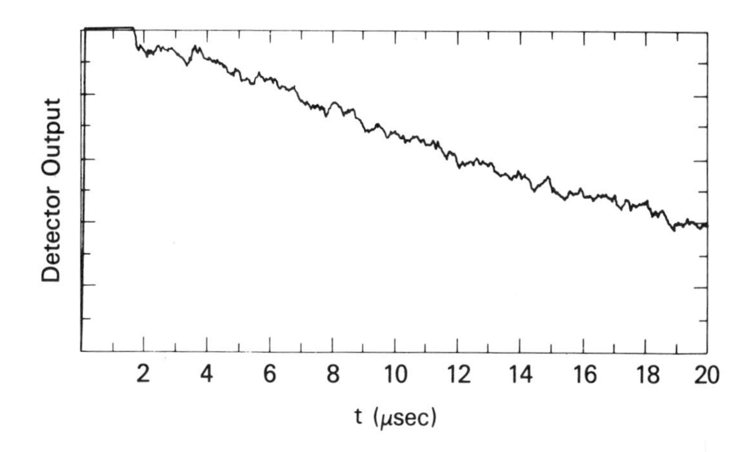

Fig. 10. $^1O_2$ emission in glacial acetic acid. Sensitizer was HP at a concentration of 1 x 10^{-4}M.

III. BIOLOGICALLY ORIENTED INVESTIGATIONS

A. Cell Growth Medium

Earlier measurements of triplet quenching in a cell growth medium containing fetal calf serum indicated problems associated with oxygen depletion (Parker, Stanbro 1981). These measurements yielded a steady increase in the triplet decay time from an initial value of several microseconds to times ultimately approaching one millisecond. Interpretation of these results led to the conclusion that, in the proximity of the sensitizer molecules, oxygen after being excited to the singlet state is consumed through chemical reaction at a rate faster than it can be resupplied through diffusion. This ultimately leads to locally anaerobic conditions and thus increased triplet lifetimes. Attempts to detect $^1O_2$ emission using an HP sensitizer were not successful.

In a renewed effort to detect $^1O_2$ emission in this biologically important medium, these experiments have been repeated using the new detector and replacing HP with TPPS as a sensitizer. First attempts using a sensitizer concentration of 1 x 10^{-4}M and air-saturation gave extremely weak signals, barely detectable above the noise level. However, oxygenation of the sample did result in measurable signal levels as indicated by the trace appearing in Fig. 11. The main characteristics of this data are: (1) small signal intensity overall; (2) relatively slow rise to the peak value; and (3) an even slower long term decay. In general the behavior of this data resembles that evidenced in Fig. 2 (C) corresponding to the case of an oxygen-depleted H_2O sample. Although the data under discussion was obtained for an oxygen enriched sample, the same effect would be observed for very slow oxygen diffusion rates corresponding to high medium viscosity. This is also consistent with the earlier observation of lengthened triplet decay rates.

It should be pointed out that in order to avoid the oxygen depletion problems experienced earlier, a circulation system was used consisting of a peristaltic pump and a flow-through cuvette.

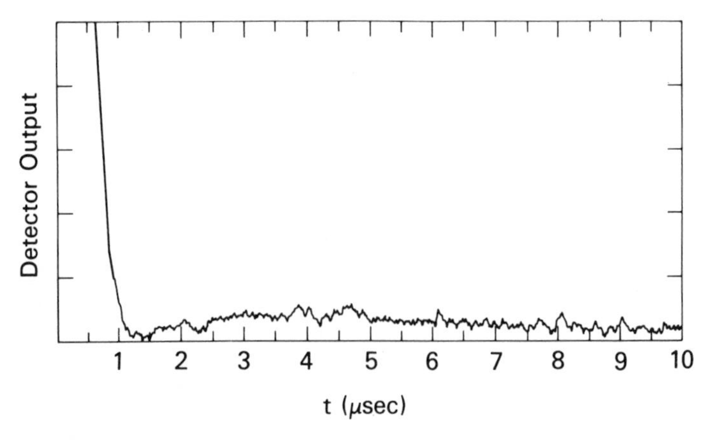

Fig. 11. $^1O_2$ emission from sample containing 1 x 10^{-4} TPPS in cell growth medium containing 15% fetal calf serum. Sample was oxygenated for 10 min.

B. H_2O-Octanol System.

Motivation for the next series of investigations was provided by recent measurements in which the biological effectiveness of a sequence of porphyrins has been classified according to their hydrophobic/hydrophilic solubility properties (Kessel 1981). As a means of quantifying the hydrophobicity of the porphyrins, the octanol:water partition ratio was used. We have carried out measurements of a similar nature; however, not with regard to biological activity, rather $^1O_2$ emission intensity.

The first step in carrying out these determinations was to fill the standard observation cuvette (1 x 1 x 4 cm) approximately half way with an aqueous solution containing the given porphyrin and then to carefully fill the upper half with neat octanol (both 1-octanol and 2-octanol were used in separate experiments). After this preparation the mixture was allowed to stand for a prescribed period of time [usually 24-48 hours) and then irradiated with the laser. The most effective means of scanning the laser from the aqueous phase through the interface to the octanol phase was to fix the relative position of the laser beam and detector and then lower the cuvette.

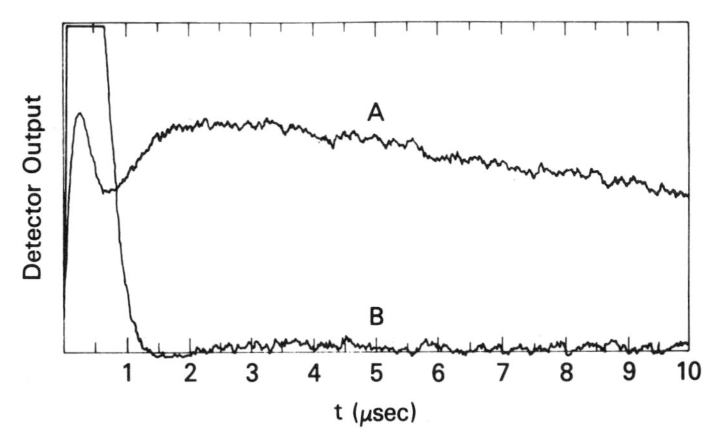

Fig. 12. Comparison of $^1O_2$ emission intensities in a heterogeneous sample composed of 1-octanol over H_2O (pH = 7.2) containing HP at a concentration of 1 x 10^{-4}M. Sample was allowed to stand for a period of 48 hours to permit HP to diffuse fully into upper octanol phase.

Typical data for the case of 1-octanol over 1 x 10^{-4} M HP in H_2O (air-saturation) is shown in Fig. 12 with the respective $^1O_2$ emission traces labelled as A and B. In the aqueous phase (B) we see the typical high initial fast infrared fluorescence followed by a rather weak $^1O_2$ emission. For the octanol phase (A), however, the infrared fluorescence is greatly diminished and the $^1O_2$ emission greatly enhanced. This reduction in the intensity of the infrared fluorescence component tends to emphasize the initial rise in the $^1O_2$ emission to well-developed peak with a subsequent gradual decay. Auxiliary measurements indicate that the triplet quenching time is approximately 1 μs and $\tau_D \simeq 20$ μs.

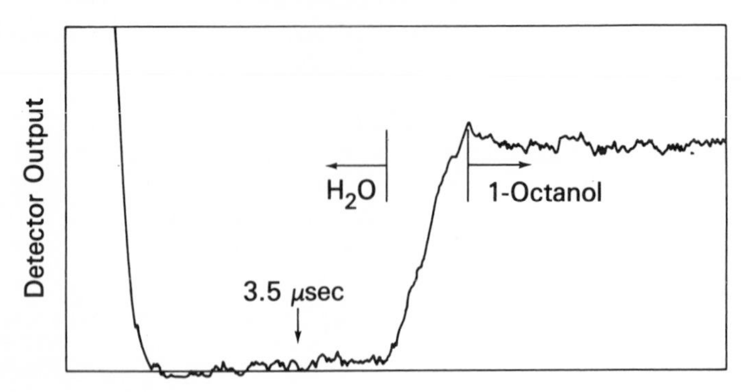

Fig. 13. Effect of simultaneously scanning laser beam and detector through H_2O/octanol interface starting in the lower H_2O phase. Boxcar integrator time delay was fixed at 3.5 μs (arrow).

From the data it is clear that a very abrupt increase in $^1O_2$ production must occur as one moves from the aqueous phase through the interface and into the octanol phase. The consequence of this strong gradient would be loss of $^1O_2$ from the octanol to the aqueous phase through diffusion, particularly in the case of a membrane of limited thickness. An example of the degree of abruptness of this transition is suggested by the data appearing in Fig. 13. In obtaining this data, the cuvette was sufficiently elevated that only the lower aqueous phase was irradiated by the laser. The PAR 160 Boxcar scan was initiated and then put on hold after 3.5 μs (see arrow in figure) into the 10 μs time base. After fixing this delay the cuvette was lowered steadily until the laser beam traversed the interface and was finally entirely in the upper octanol phase, the total distance traversed (from position of arrow to end of scan) being approximately 1.5 cm.

From Figs. 12, 13, it is clear that the ability of HP to generate $^1O_2$ depends markedly on the solvent in which it is dissolved. In this particular case, however, another factor enters since the HP that ultimately appeared in the octanol phase arrived there by diffusing through the interface from the lower aqueous phase. Thus it is possible that since HP is a multicomponent species, only the more hydrophobic components will appear in the octanol. This sort of selectivity should be even more pronounced in the case of hematoporphyrin derivative (HPD). Thus a series of experiments was carried out to determine the relative effectiveness of HP and HPD components partitioned into octanol. The results are summarized in Fig. 14 with the data for HPD indicated as the upper curve (A) and that for HP the lower (B).

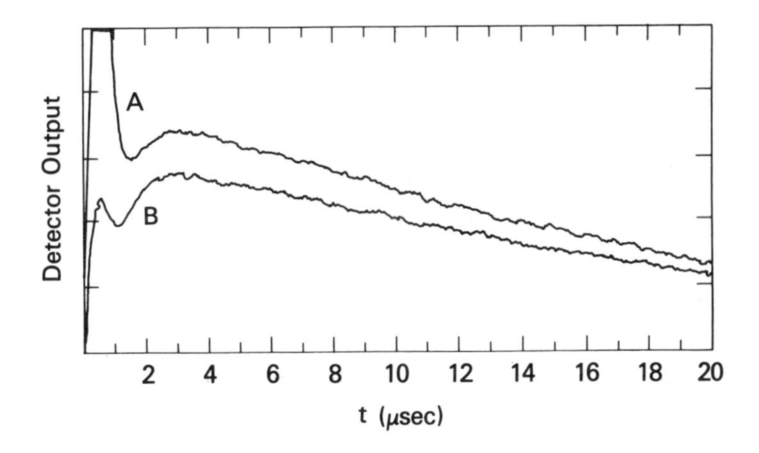

Fig. 14. Comparison of $^1O_2$ emission levels for HPD partitioned into octanol (A) and with similar data for HP (B).

The $^1O_2$ emission is seen to be slightly higher for HPD; however other features such as rise time and fall time are essentially the same. The main difference between the two cases is in the magnitude of the fast initial infrared fluorescence component which is much larger for HPD than for HP. A question obviously arises as to whether the infrared fluorescence for HPD is unusually intense and that for HP normal or is that for HPD normal and that for HP anomalously weak. Further investigation of this matter is required.

IV. DISCUSSION

Before proceeding to a discussion of the biological implications of this research, particularly in regard to photodynamic effects, we shall briefly summarize the most significant developments.

First a comment on the remote $^1O_2$ optical monitoring technique employed. Use of this method has been demonstrated to provide reliable data for a wide variety of solvents with quenching times ranging from a few microseconds to almost 100 microseconds. In addition to solvent dependent decay rate measurements, the system used has sufficient time resolution capabilities to also furnish $^1O_2$ formation rates. From the rise and fall times characterizing the total time history of the $^1O_2$ emission it is possible to extract corresponding sensitizer triplet quenching times in addition to the solvent related $^1O_2$ deactivation times. It has been pointed out, however, that caution is required in this process.

Collectively, the data provides a substantial basis for assuming the quantitative correctness of the basic energy transfer sequence taken to underlie the mechanism of $^1O_2$ production in the overall photodynamic process.

Substantial variation in the efficiency of a series of water soluble porphyrins in the generation of $^1O_2$ has been observed, with TMPP being the most efficient and HPD the least. In part this difference appears to be due to the state of aggregation of the porphyrin. TMPP exists in monomer form while HPD is known to be highly aggregated. Components of HPD partitioning into the lipid phase, however, have been shown to produce $^1O_2$ quite strongly.

In regard to the clinical relevance of the work described in this paper, it is to be noted that the optical system used consists of state-of-the-art electrooptics combined optimally with commercially available data processing instrumentation. The possibility of applying this system to *in vivo* monitoring of the effectiveness of incident visible light in the generation of $^1O_2$ is enhanced by the fortuitous coincidence that transmission through human tissue is maximal in the spectral range from 1.20 to 1.30 μ conveniently encompassing the $^1O_2$ emission band at 1.27 μ. Availability of such a monitoring system would obviously be invaluable to the clinician, particularly in the area of dosimetry.

Physiologically, the results obtained for the two-phase H_2O-octanol system appear to be of more than ordinary significance in that they bear a close relationship to processes that must be directly involved in photoradiation therapy. Thus in order that the HP or HPD sensitizers be transported throughout the body, they must be water soluble. However, to bring about cell inactivation they must be able to penetrate the cell membrane, i.e., they must be lipid soluble. The experimental data indicates that HP and HPD are ideal in this regard since in the aqueous phase (while being transported) they are poor sensitizers; however, at the point of attack (the cell membrane) they are suddenly transformed into highly effective generators of $^1O_2$ and thus presumably highly active agents of cell destruction. A consequence of this solvent-dependent behavior is that extremely steep $^1O_2$ concentration gradients will tend to develop at the interface separating the cell membranes from the surrounding medium and therefore a significant $^1O_2$ loss from the membrane will occur. This suggests in the case of these two porphyrins that the most favorable region for $^1O_2$ attack is the cell interior.

The experimentally established fact that the synthetic porphyrins TMPP and TPPS are substantially more effective in producing $^1O_2$ in an aqueous environment than HP is of considerable interest, particularly when this result is contrasted with the relative biological effectiveness of the three (Diamond et al 1977). Thus in terms of their ability to bring about photodynamic inactivation of glioma cells, TMPP proved most effective followed closely by HP, with TPPS being by far the least active. Since both TMPP and TPPS are water soluble but not lipid soluble, it might be thought from the outset that they would only be effective in causing damage to the cell membrane through external attack. However, this is not consistent with the rather large observed difference in the biological effectiveness of the two.

As pointed out in Section I, biological potency appears to be related to the ability of the porphyrin to partition into the lipid phase (Sandberg, Romslo 1981) and thus to penetrate the cell membrane. However, it is also clear that due to the relatively long $^1O_2$ collisional lifetimes, this species may migrate, by means of diffusion, over distances that substantially exceed the thickness of a typical membrane (e.g., in water this distance is roughly 0.20 μ). From this point of view, porphyrins able to generate substantial $^1O_2$ levels in the aqueous medium external to the cell should still be effective if the vulnerable cell components are located within the membrane itself. If, on the other hand, these targets reside totally within the cell, then neither TMPP nor TPPS should be effective unless selective transport of these porphyrins to the interior of the cell takes place by processes other than diffusion, i.e., binding to proteins ingested by the cell.

ACKNOWLEDGEMENTS

The authors would like to thank Dr. C. A. Burrus of Bell Laboratories, Holmdel, N. J. for his kindness in furnishing the InGaAs detector used in this research. We would also like to thank Dr. D. S. Zaharko of the National Cancer Institute, Bethesda, Md., for many enlightening discussions and valuable assistance. Appreciation is expressed Mrs. A. Landry of the Applied Physics Laboratory for her perseverance and diligence in the organization and preparation of the manuscript.

This work was supported by the U.S. Naval Sea Systems Command under Contract N00024-83-C-5301.

REFERENCES

Bellnier DA, Dougherty TJ (1982). Membrane lysis in Chinese hamster ovary cells treated with hematoporphyrin derivative plus light. Photochem Photobiol 36:43.

Bonnett R, Ridge RJ, Land EJ, Sinclair RS, Tait D, Truscott TG (1982). Pulsed irradiation of water-soluble porphyrins. J Chem Soc Far Trans I 78:127.

Boyer RF, Lindstrom CG, Darby B, Hylarides M (1975). The peracid oxidation of singlet oxygen acceptors. Tetrahedron Ltrs 47:4111.

Burrus CA, Dentai AG, Lee TP (1981). Large-area back-illuminated InGaAs/InP photodiodes for use at 1 to 1.6 μm wavelength. Opt Comm 38:124,

Carrington T (1982). Estimation of rate constants from growth and decay data. Int J Chem Kin 14:517.

Creekmore SP, Zaharko DS (1983). Modification of chemotherapeutic effects on L 1210 cells using hematoporphyrin and light. Cancer Res (to be published).

Diamond I, Granelli SG, McDonagh AF (1977). Photochemotherapy and photodynamic toxicity: Simple methods for identifying potentially active agents. Biochem Med 17:121.

Howard JA, Mendenhall GD (1975). Autooxidation and photooxidation of 1,3 diphenylisobenzofuran: A kinetic and product study. Can J Chem 53:2199.

Hurst JR, McDonald JD, Schuster GB (1982). Lifetime of singlet oxygen in solution directly determined by laser spectroscopy. J Am Chem Soc 104:2065.

Ito T (1978). Cellular and subcellular mechanisms of photodynamic action: The $^1O_2$ hypothesis as a driving force in recent research. Photochem Photobiol 28:493.

Kessel D (1981). Transport and binding of hematoporphyrin derivative and related porphyrins by murine leukemia L1210 cells. Cancer Res 41:1318.

Kessel D (1982). Determinants of hematoporphyrin-catalyzed photosensitization. Photochem Photobiol 36:99.

Khan AN, Kasha M (1979). Direct spectroscopic observation of singlet oxygen emission at 1268 μm excited by sensitizing dyes of biological interest in liquid solution. Proc Natl Acad Sci USA 76:6047.

Krasnovskii AA (1976). Photosensitized luminescence of singlet oxygen in solution. Biofizika (USSR) 21:748.

Moan J, Petterson EO, Christensen T (1979). The mechanism of photodynamic inactivation of human cells in vitro in the presence of hematoporphyrin. Brit J Cancer 39:398.

Ogilby PR, Foote CS (1981). Chemistry of singlet oxygen. 34. Unexpected solvent deuterium isotope effects on the lifetime of singlet molecular oxygen ($^1O_2$), J Am Chem Soc 103:1219.

Ogilby PR, Foote CS (1982). Chemistry of singlet oxygen. 36. Singlet molecular oxygen ($^1O_2$) luminescence in solution following pulsed laser excitation. Solvent deuterium effects on the lifetime of singlet oxygen, J Am Chem Soc 104:2009.

Parker JG, Stanbro WD (1981). Energy transfer processes accompanying laser excitation of hematoporphyrin in various solvents. Johns Hopkins APL Tech Dig 2:196.

Parker JG, Stanbro WD (1982). Optical determination of the collisional lifetime of singlet molecular oxygen $O_2(^1\Delta_g)$ in acetone and deuterated acetone. J Am Chem Soc 104:2067.

Pasternack RF, Huber PT. Boyd P, Engasser G, Francesconi L, Gibbs E, Fasella P, Venturo GC, Hinds L de C (1972). On the aggregation of meso-substituted water-soluble porphyrins. J Am Chem Soc 94:4511.

Rodgers MAJ, Snowden PT (1982). Lifetime of $O_2(^1\Delta_g)$ in liquid water as determined by time-resolved infrared luminescence measurements. J Am Chem Soc 104:5541.

Rodgers MAJ (1983). Time resolved studies of 1.27 μm luminescence from singlet oxygen generated in homogeneous and microheterogeneous fluids. Photochem Photobiol 37:99.

Salokhiddinov KI, Byteva IM, Dzhagarov BM (1979). Duration of the luminesence of singlet oxygen in solution following pulse laser excitation. Opt Spectr (USSR) 47:487.

Sandberg S, Romslo I (1981). Porphyrin-induced photodamage at the cellular and subcellular level as related to the solubility of the porphyrin. Clin Chim Acta 109:193.

Turay J, Hambright P, Datta-Gupta N (1978). Intermolecular association of natural and synthetic water soluble porphyrins. J Inorg Nucl Chem 40:1687.

Wilkinson F, Brummer JG (1981). Rate constants for the decay and reactions of the lowest electronically excited singlet state of molecular oxygen in solution. J Phys Chem Ref Data 10:809.

Young RH, Brewer D, Keller RA (1973). The determination of rate constants of reaction and lifetimes of singlet oxygen in solution by a flash photolysis technique. J Am Chem Soc 95:375.

Porphyrin Localization and Treatment of Tumors, pages 285–300

THE ANALYSIS AND SOME CHEMISTRY OF HAEMATOPORPHYRIN DERIVATIVE

A.G Swincer, V.C. Trenerry and A.D. Ward

Department of Organic Chemistry
University of Adelaide,
Adelaide, South Australia, 5000.

INTRODUCTION

The ability of porphyrins to localize in tumor tissue has been known for some time. Dougherty (1982) and others have shown that the mixture of porphyrins known as haematoporphyrin derivative (HPD) can be used for both the detection and treatment (photoradiation therapy, PRT) of a range of tumors. While investigations continue into the medical uses of PRT, basic studies on the chemistry and the physics of the process are of vital importance for there is much that is not understood in these areas. In this paper we report on further work on the analysis and separation of the components of HPD. We have divided the discussion into three main areas; namely, the further analysis of HPD by high performance liquid chromatography (HPLC), the analysis of HPD by gel filtration (gel permeation) chromatography, and the separation and purification of some of the components of HPD.

ANALYSIS OF HPD BY HPLC

In discussing the analysis of HPD by HPLC at the Washington Conference on Porphyrin Photosensitization we (Cadby, 1983) drew attention to the fact that significant amounts of unidentified material were eluted, together with protoporphyrin, by more powerful solvents than were needed to elute the major components of HPD. Further investigation of this material has shown that not all of it is eluted by aqueous methanol (1:9) at lower pH (ca. 4) and higher pH values (7.5-8.0) are necessary if all the injected HPD sample is to be eluted from the column. Aqueous tetrahydrofuran(1:9) is also an appropriate solvent (Dougherty 1983a) to elute the strongly retained fraction (ca. 30% of total material) that is not eluted by aqueous methanol at pH 4.

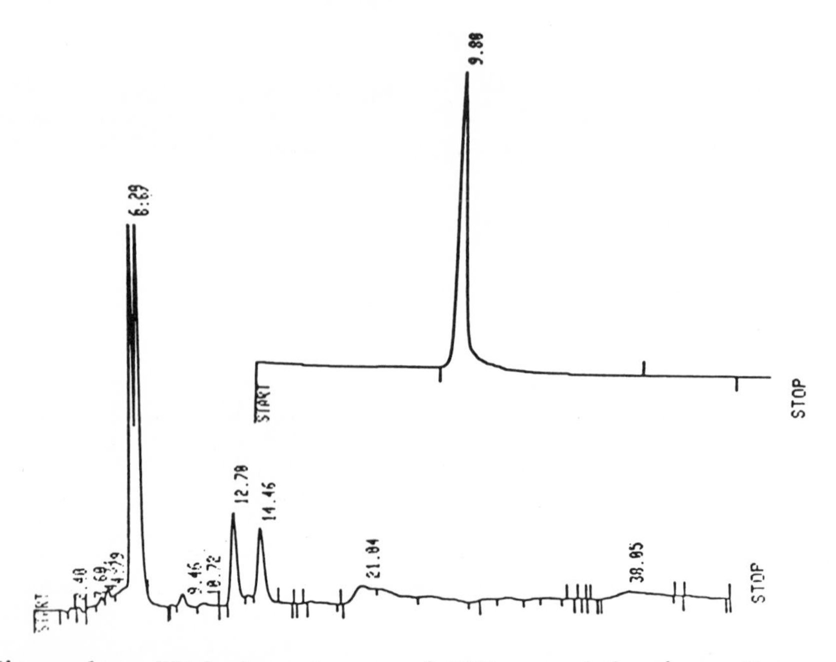

Figure 1. HPLC chromatogram of HPD recorded using a Waters 5μ RCM-100 reversed-phase silica column and 90% methanol-water (containing 2.5 mM tetrabutylammonium phosphate) adjusted to pH 4 with phosphoric acid as solvent. The upper trace refers to material subsequently eluted with 90% tetrahydrofuran-water. Figures on the chromatogram refer to time after injection.

Figure 1 shows the HPLC trace obtained from a sample of HPD using aqueous methanol (pH 4) as the eluting solvent. Haematoporphyrin (HP) appears as a well resolved doublet although under more appropriate conditions complete resolution of the two diastereomers can be achieved (Cadby 1983). The two structural isomers of α-hydroxyethyl vinyldeuteroporphyrin (HV′s) occur as a doublet (12.78, 14.46) and are well separated by this solvent system. These peaks are followed by a broad, poorly resolved set of peaks which include protoporphyrin (PP). Further elution of the column with aqueous tetrahydrofuran (1:9) brings off more porphyrin material as an apparently sharp, single peak (upper trace, Fig. 1). If the aqueous tetrahydrofuran ratio is changed to increase the amount of water then this material can be partially resolved into several broad peaks.

It thus appears that HPD contains significant amounts of porphyrin material that are considerably less polar than PP as judged by their behaviour on HPLC. Since it has generally been assumed that PP is the least polar of the known components of HPD this result was surprising and led us to consider whether other factors might be causing the retention of the material on the HPLC column. If a solution of HPD is added to a vial containing reversed-phase silica suspended in aqueous methanol (1:9) at pH 4 (i.e. the initial solvent used for the HPLC) then the porphyrin material is rapidly absorbed by the silica and "none" remains in solution. If, however, the pH is adjusted to 7.5 - 8.0 then considerable amounts remain in solution. Similarly if aqueous tetrahydrofuran is the solvent then most of the porphyrins remain in solution. This suggests that there may be a greater interaction with the column material by these porphyrins than is usual for HPLC separations. Since porphyrins tend to be associated or aggregated with each other it is not surprising to observe that "all" the porphyrin material is absorbed (or precipitated) onto the column surface, i.e. by absorbing some of the porphyrins the column surface attracts the remaining material as well. This observation could explain the poor resolution of the more slowly eluted material in Fig. 1 as, at least in part, due to the low solubility of these absorbed porphyrins in the eluting solvent. In contrast, tetrahydrofuran, which is a very good porphyrin solvent, dissolves them from the column surface very readily. Clearly, though, polarity factors are also important in the elution process.

Another explanation for the low polarity of these porphyrins would be the absence, or smaller number, of polar groups such as carboxylic acid and hydroxyl functional groups. This could occur, for instance, if an ester was formed between a carboxyl group of one porphyrin and the hydroxyl group of another leading to a covalently bound dimer (Bonnett 1983). Alternatively, ether formation between two porphyrins would remove hydroxyl functional groups from each. The ready elution of porphyrins with mildly basic solvent systems does not necessarily mean that two carboxylic acid groups need to be present in all these molecules since an alternative explanation could be that at pH values near 8 the porphyrins are largely in their monoanionic form rather than the neutral one. Hence both monocarboxylic acid as well as dicarboxylic acid containing porphyrins could elute readily under these conditions.

We had commented on the relationship between elution time and pH in our previous conference paper and it thus seemed appropriate to utilize this observation to elute the porphyrins that were strongly retained on the column at lower pH values. Accordingly, we investigated the effect of using a gradient elution system, changing only the pH of the solvent. Figure 2 shows three examples of the results we obtained when the initial solvent was aqueous methanol (1:9) at pH 4 and the final solvent was aqueous methanol (1:9) at pH 8. Under these conditions "all" the porphyrin material is eluted from the column. Although we examined a variety of gradients none of them were particularly successful in resolving the more strongly retained material which included PP. In all cases HP and HV′s were resolved and separated from the other porphyrins. For convenience and to enable an analysis to be completed in a reasonable time we eventually settled on a procedure where the eluting solvent was changed immediately after injection to the final solvent. Because of the column volume it is some time before the column conditions correspond to those of the final product. We estimate that this procedure effectively causes a uniform gradient from pH 4 to pH 8 commencing at about the time the HV peaks appear and taking approximately ten minutes to achieve the final solvent conditions. The chromatogram of HPD using these conditions is shown in Figure 3. At least ten peaks, one of which is PP (peak at 10.68) are evident in the material eluting after the HV′s. It should be appreciated that the conditions for the separation shown in Fig. 3 are a compromise between the high resolution possible for the HP diastereomers and the HV′s, and the need to elute the remaining material within a reasonable amount of time. Much better resolution could be achieved for the material at either end of the chromatogram by suitable adjustments to the conditions.

The porphyrins that are not eluted from the column using the aqueous methanol (pH 4) solvent can be eluted, as mentioned above, using aqueous tetrahydrofuran. Collection and reinjection of this material followed by elution with the gradient system produced the chromatogram shown in

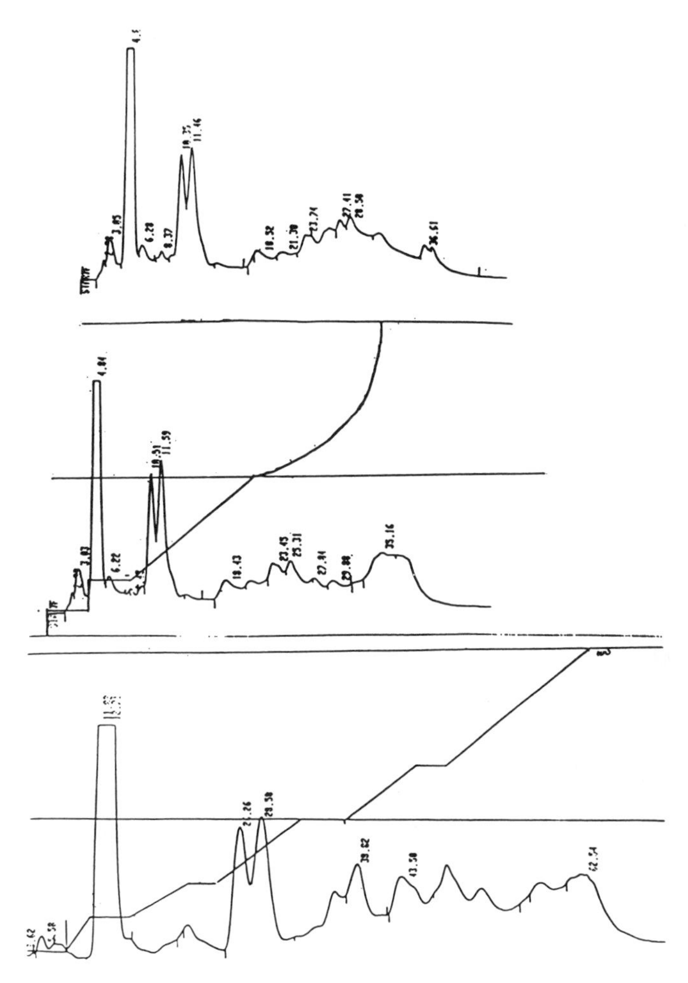

Figure 2. Examples of gradient elution of HPD using a Waters 10µ reversed-phased silica column and a Waters Z-module. Initial solvent: as described in Fig. 1. Final solvent: 90% methanol-water (containing 2.5 mM tetrabutylammonium phosphate) adjusted to pH 8 with phosphoric acid and triethylamine.

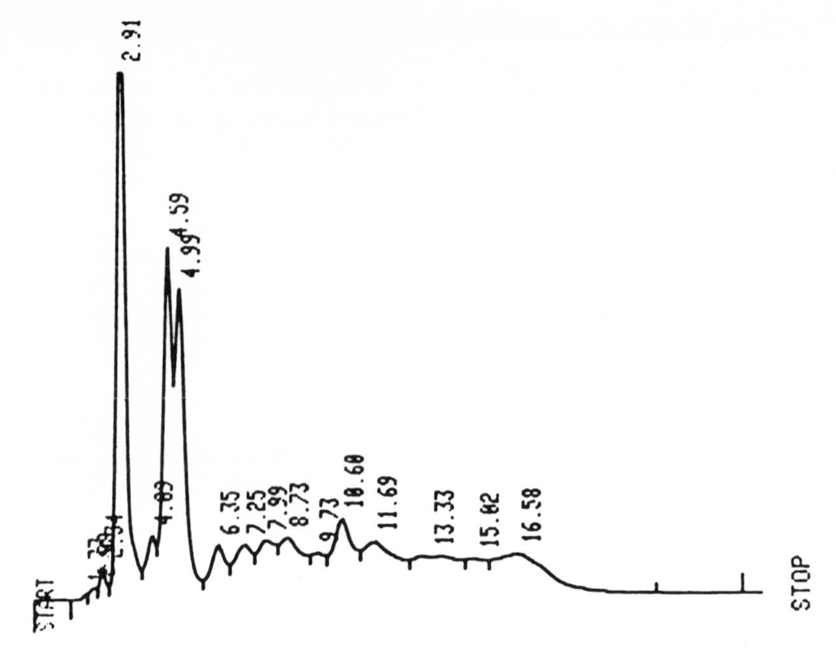

Figure 3. Gradient elution of HPD. Conditions as described in Fig. 2 with changeover to final solvent immediately after sample injection.

figure 4. Traces of HP and HV's are present in this material as well as considerable amounts of PP (peak at 11.47). The remaining material is poorly resolved but it is enriched in the most strongly retained material of Fig. 3.

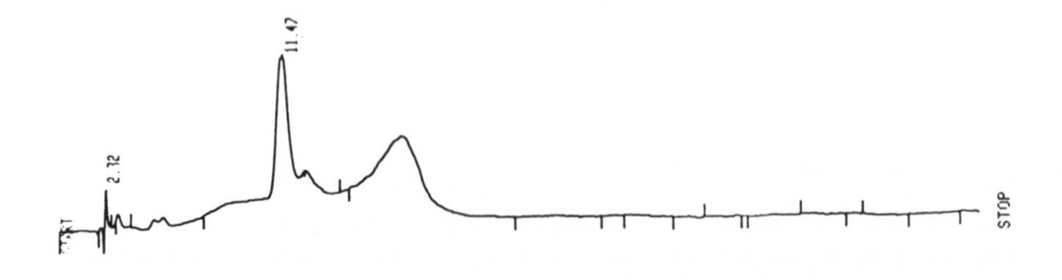

Figure 4. Aqueous tetrahydrofuran (1:9) eluted material (from Fig. 1) analysed using conditions indicated for Fig. 3.

ANALYSIS OF HPD BY GEL FILTRATION CHROMATOGRAPHY

The use of gel filtration (gel permeation) chromatography to fractionate HPD was reported recently (Dougherty, 1983b) and it was noted that only the more rapidly eluted material from a BioGel P-10 column was biologically active. We have been interested in determining the basis for the separation and the composition of the various fractions. Figure 5, upper trace, shows the result obtained when a sample of HPD is eluted from a BioGel P-10 column with water at pH 7. Two main fractions are observed and there is little porphyrin material eluting from the column between these major peaks. When the sample is eluted with water buffered to pH 7 using phosphoric acid and triethylamine then three broad fractions are observed (Fig. 5, middle trace); however, the separation between these fractions is

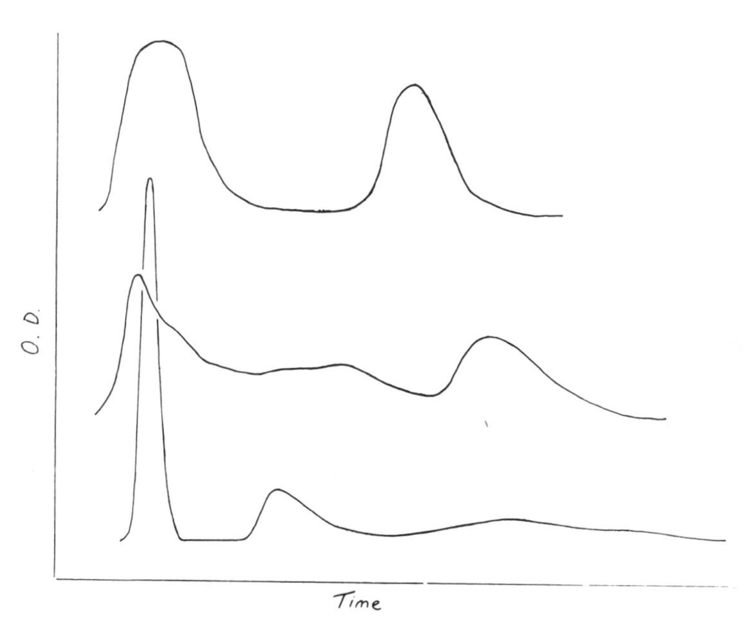

Figure 5. Gel filtration chromatograms of HPD. Upper trace using BioGel P-10 and eluting with water (pH 7). Middle trace using BioGel P-10 and eluting with water buffered to pH 7 with phosphoric acid and triethylamine. Lower trace using Sephadex G-25 and eluting with water buffered to pH 7 with phosphoric acid and triethylamine. All columns were pre-swelled with the eluting solvent.

relatively poor. In contrast, the separation on a Sephadex G-25 column (Fig. 5, lower trace) using phosphate buffered water produces a total separation between the rapidly eluted material and the remaining fractions. The final band on this column is very broad. Since the two bands on BioGel P-10 (Fig. 5, upper trace) have been referred to as "aggregate" (fast moving band) and "monomer" we have designated the middle fraction of the other traces of Fig. 5 as the "dimer" band although there is, to our knowledge, no evidence on the molecular weights of the materials in these fractions, other than their performance on these gel filtration columns.

The ratio of the aggregate to monomer material in Fig. 5, upper trace, can be determined spectrophotometrically as 59:41. It is important to measure these fractions under identical conditions since the aggregate material has a different visible spectrum to that of the monomer. We have found that the best solvent to dilute these samples with is ethanol/0.1N sodium hydroxide solution (1:1) for by using these conditions all the porphyrin fractions give a sharp Soret band, and slight variations in the conditions do not cause significant differences in spectral intensity. However, the method does assume that the porphyrin fractions have the same extinction coefficients for the Soret band and this must be a source of error in these ratios.

The ratio of the bands in Fig. 5, middle trace, is aggregate:dimer:monomer, 38:27:35. It is difficult to determine exactly the end of the aggregate band and the beginning of the dimer band since there is considerable overlap of the two bands. A comparison of the BioGel results suggests that the dimer band is largely incorporated into the aggregate band in the absence of the phosphate buffer. The ratio of the bands on the Sephadex columns is 43:27:30. These figures are very similar to those of the BioGel column using the same conditions. We have noticed that a variety of samples, ranging from low aggregate to high aggregate concentrations, all have a dimer:monomer ratio that is approximately 85:100. We tentatively suggest that this indicates that the dimer and monomer bands are in a reasonably readily established equilibrium. A corollary is that the aggregate material is not readily converted to either monomer or dimer.

An advantage of the Sephadex column is that it can be used with organic solvents. When HPD is chromatographed using ethanol:0.1N sodium hydroxide (1:1) as the solvent, the amount of aggregate is essentially unchanged from the value determined in water at pH 7. The aggregate is also unaffected by aqueous methanol (1:9), i.e. the HPLC solvent without the ion-pairing reagent. The aggregate fraction in water shows a Soret band maximum at 364 nm, the dimer has the corresponding maximum at 372 nm and the monomer at 404 nm. In ethanol:0.1N sodium hydroxide and in aqueous methanol (1:9), the aggregate has a Soret band maximum at 393 nm. The monomer band has a maximum close to 396 nm in these solvents. It has often been assumed that the position of the Soret band indicates the nature of the porphyrin material, i.e. aggregate or non-aggregate; however, the above data strongly suggest that the variation in the aggregate Soret maximum is due to solvent shifts. Hence, it would be unwise to assume for the HPD porphyrins that the nature of the material can be determined by a simple spectrophotometric measurement.

When an aqueous PP solution is chromatographed on either BioGel or Sephadex only an aggregate band is observed. This solution shows broad, poorly resolved peaks in its visible spectrum with the Soret maximum at 353 nm. This behaviour has been considered typical of an aggregated porphyrin in aqueous solution. When an aqueous PP solution is chromatographed on Sephadex using ethanol/0.1N sodium hydroxide (1:1) only an aggregate band is observed. However, the visible spectrum of PP in organic solvents has the Soret maximum at 401 nm. For this pure porphyrin also, the position of the Soret maximum does not necessarily indicate whether it is aggregated or not, if the behaviour on the gel column refers to the aggregated material. Detergent solutions have often been used to disaggregate porphyrins, but PP shows only a single aggregate band when chromatographed on BioGel in aqueous SDS solution. Clearly, more work must be done before the nature of the interaction of the porphyrins with the gel material is fully understood. However, these results suggest that the estimation of aggregate and non-aggregate material may not be clear cut in organic solvents.

We have examined the gel filtration behaviour of these porphyrins on HPLC columns. Figure 6 shows the gel filtration chromatograms of two samples of HPD separated on a

Waters I-250 column. The initial sharp peak is the aggregate which is followed by the dimer and then the monomer. The lefthand trace refers to a sample that was formed from HPD solid which was comparatively low in haematoporphyrin diacetate. This sample was also lower in biological activity than the usual HPD sample shown on the right.

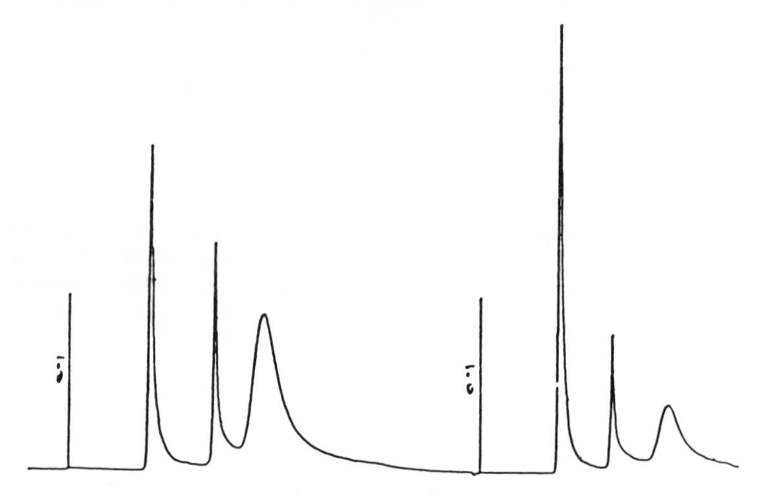

Figure 6. Gel filtration chromatograms of HpD samples using a Waters I-250 gel permeation column with water, containing 5 mM tetrabutylammonium phosphate and propan-2-ol (1:9) as solvent. Flow rate is 1 ml/minute. The initial sharp line indicates the injection point for each sample.

Figure 7 (lower trace) is typical of HPD samples that are formed from HP diacetate rather than from HPD solid. These solutions contain significantly greater amounts (> 60%) of aggregate and greater amounts (> 80%) of the total of aggregate and dimer. This trace is also typical of the aggregate fraction obtained from the BioGel column using the conditions described in Fig. 5, upper trace. The upper trace in Fig. 7 is that of a commercial sample of haematoporphyrin showing that, in these solvents at least (aqueous isopropyl alcohol), it is almost exclusively in the monomer form.

The aggregate fraction in HPD samples can most easily be obtained free from dimer and monomer by utilizing

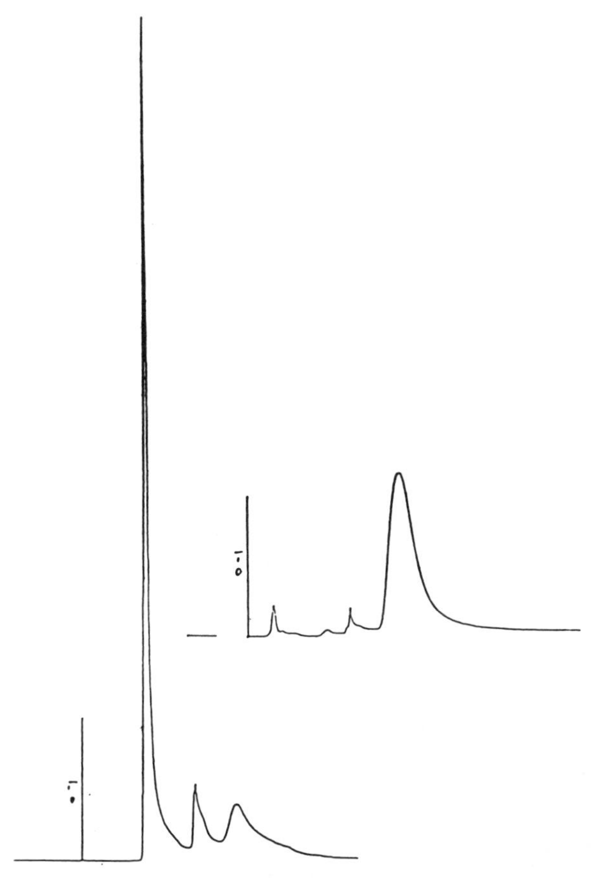

Figure 7. Gel filtration chromatograms of HPD prepared from HP diacetate (lower trace) and of commercial HP (upper trace); conditions as for Fig. 6.

Sephadex G-25 as shown in Fig. 5 (lower trace). Analysis of this fraction by HPLC produced the trace shown in Figure 8 (lower trace). Above it is displayed the monomer fraction obtained from BioGel chromatography (Fig. 5, middle trace). Both chromatograms in Fig. 8 were obtained under identical conditions. The aggregate material has only a small amount of HP, is enriched in HV's and contains the majority of the less polar material. In contrast, the monomer is composed mainly of HP and the HV's with very little of the less polar material.

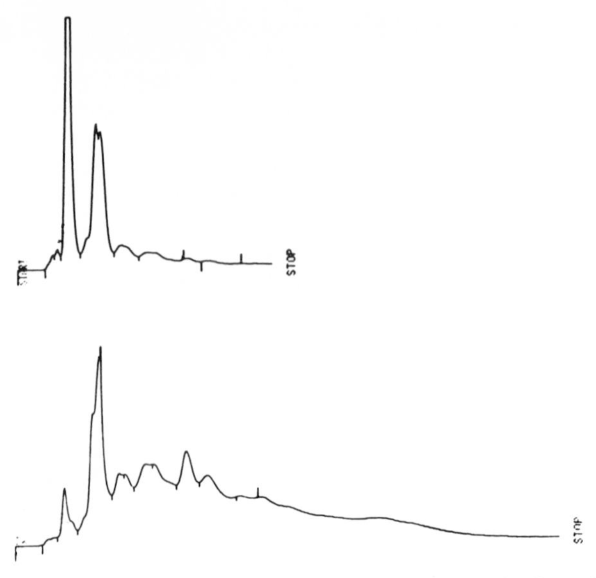

Figure 8. HPLC of aggregate band from a Sephadex G-25 column (lower trace) and the monomer band from a BioGel P-10 column (upper trace). Conditions as for Fig. 3.

SEPARATION OF THE COMPONENTS BY HPD

In order to be able to assess whether one of the components of HPD is more biologically active than the others it is necessary to obtain each of the components in a pure form and in sufficient quantity for extensive biological testing. Collection of the material from an analytical HPLC column is not, in our opinion, a valid way to achieve this goal because the amounts of material involved in each injection are so small and there is a considerable problem associated with the recovery of the porphyrin from large volumes of solvent. Unfortunately, the separation on analytical HPLC columns is not sufficiently good to make preparative scale HPLC an attractive alternative.

We have investigated the possibility of using the different base strengths of the porphyrins as a means of separating them as the acid numbers (Smith 1975) of HP, the HV's and PP are quite distinct. If a sample of HPD is added to 0.05% hydrochloric acid solution (pH ca. 1.6) in the presence of ether considerable amounts of porphyrin are extracted into the ether layer. A large portion of the porphyrin material also precipitates under these conditions. This material can be recovered by filtration and is substantially enriched in the less polar porphyrin materials. Repeated extraction of the acidic solution with ether until no further colour is extracted, followed by backwashing with 0.05% hydrochloric acid, gives an ether solution which is then extracted with 0.6% hydrochloric acid until no further colour is extracted. After backwashing this solution with ether two acidic solutions have been obtained. The first, using 0.05% hydrochloric acid contains only HP. The second (0.6% hydrochloric acid) contains only HV's. The porphyrin material can be recovered by adjusting the pH of each solution to ca. 4 at which stage all the porphyrin material precipitates. Alternatively, at this pH all the porphyrin can be extracted into ether. By this means we can obtain HP free of HV's; this is the purest sample of HP (99% by HPLC) we can obtain and it is significantly more pure than commercial samples of HP. The HPLC of this pure HP sample is shown in Figure 9, top trace. Under all conditions this HP material only shows a monomer band on a gel column.

The HV material can be dissolved or extracted from ether by dilute sodium hydroxide solution. If the basic solution is neutralized it shows only a monomer band on gel filtration. If, however, the material is precipitated by adjusting the pH to 4 and then redissolved in base and neutralized it shows both monomer and aggregate bands on the gel column. These fractions are shown in Fig. 9 where the HV samples are the middle two traces, the upper of the two being the HV aggregate material. Both samples are greater than 90% pure. PP can be prepared from haematoporphyrin by a literature procedure (Cadby 1982), the HPLC of this material is shown in Fig. 9, bottom trace.

We have found the Amicon ultrafilters are very useful for the concentration and separation of these porphyrin components. In particular, Amicon UM filters can be used to concentrate these porphyrin solutions for they pass no porphyrin at all. In contrast, Amicon YM filters pass monomer porphyrin material but not the aggregate or dimer. This allows a much cleaner separation of monomer material than is possible using gel columns. Although this behaviour is true for most filter sizes the best results are obtained using 2, 5, or 10 filters.

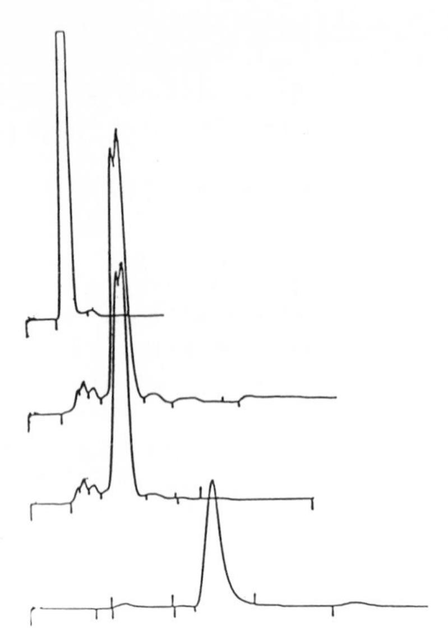

Figure 9. HPLC chromatograms of purified porphyrins. Top trace HP. Middle traces HV's (upper trace HV aggregate; lower trace HV monomer). Bottom trace PP. Conditions as for Fig. 3.

EXPERIMENTAL

HPLC was performed using a Waters Model 6000A solvent delivery system and U6K injector. The detector used was a Waters Model 440 UV absorbance type operating at 405 nm. Columns used were Waters μ Bondapak Radial Pak C_{18} (10 μm), Waters Radial-Pak C_{18} 8 mm I.D., (5 μm) and Waters I-250 Protein Column. All HPLC solvents were distilled, degassed and filtered through a 0.45 μm Millipore filter prior to use.

Haematoporphyrin was obtained in the form of its dihydrochloride from Roussel. Protoporphyrin was prepared from haematoporphyrin by brief heating dimethylformamide. Haematoporphyrin diacetate was prepared by treating haematoporphyrin with acetate anhydride and pyridine and was approximately 80% pure.

The ion-pairing reagent (IPR) was tetra-n-butylammonium phosphate (Unichrom). The reagent is prepared by diluting the Unichrom concentrate in distilled water to 1 L, providing a 5 mM solution. This solution is then diluted with distilled water (1:1) before use, providing a 2.5 mM solution whose pH was adjusted using phosphoric acid and measured with a digital pH meter.

All visible spectra were recorded on a Pye Unicam SP8-100 spectrophotometer using 10 mm quartz cells.

Generally gel separations were carried out using 7.0 x 1.5 cm glass columns packed with either Sephadex G-25 or BioGel P-10. All solvents were redistilled before use.

All optical densities (O.D.) were determined at 397 nm by diluting the sample with a 1:1 mixture of ethanol and 0.1N sodium hydroxide. Investigation has shown that all types of porphyrins, whether in aggregate form or not, give relatively sharp Soret bands close to 397 nm when diluted with this solvent.

ACKNOWLEDGEMENTS

This work was supported by grants from the National Health and Medical Research Council of Australia and a generous donation from Miss E. Laubmann to the University of Adelaide Phototherapy Fund. We wish to thank Dr. H.G. Grant for assistance with the HPLC gel permeation columns.

REFERENCES

Bonnett R., Berenbaum MC (1983). HpD - a study of its components and their properties. In Kessel D, Dougherty TJ (eds): "Porphyrin Photosensitization" New York Plenum Press p 241.

Cadby PA, Dimitriadis E, Grant HG, Ward AD (1982). Separation and analysis of haematoporphyrin derivative components by high-performance liquid chromatography. J Chromat 231:273.

Cadby PA, Dimitriadis E, Grant HG, Ward AD (1983). The analysis of haematoporphyrin derivative. In Kessel D, Dougherty TJ (eds): "Porphyrin Photosensitization" New York Plenum Press p 251.

Dougherty TJ, Weishaupt KR, Boyle DG (1982). Photosensitizers In DeVita VT, Hellman S, Rosenberg SA (eds): "Cancer Principles of Oncology" Philadelphia JB Lippincott p 1836.

Dougherty TJ (1983a) private communication.

Dougherty TJ, Boyle DG, Weishaupt KR, Henderson BA, Potter WR, Bellnier DA, Wityk KE (1983b). Photoradiation therapy - clinical and drug advances. In Kessel D, Dougherty TJ (eds): "Porphyrin Photosensitization" New York Plenum Press p 3.

Smith KM, (1975) "Porphyrins and Metalloporphyrins". Amsterdam Elsevier p 15.

Porphyrin Localization and Treatment of Tumors, pages 301–314

THE STRUCTURE OF THE ACTIVE COMPONENT OF HEMATOPORPHYRIN DERIVATIVE

T. J. Dougherty, W. R. Potter, K. R. Weishaupt*

Roswell Park Memorial Institute, Buffalo, NY
*Oncology Research & Development, Cheektowaga, NY

INTRODUCTION

Photoradiation therapy (PRT) for local treatment of malignant tumors utilizing hematoporphyrin derivative (Hpd) as photosensitizing drug is undergoing clinical trials in several centers in the U. S. and abroad. This therapy is based on the localization and retention of Hpd in most tumors, as well as its photodynamic action which likely generates singlet oxygen at least as a first step. The only drug toxicity encountered to date in over 1500 patients receiving Hpd has been a generalized photosensitivity which requires patients to avoid bright light, especially sunlight, for 30 days or more. While this toxicity is avoidable it presents a drawback to utilizing PRT for early stage disease in ambulatory and active patients.

Hpd is known to be a mixture of porphyrins first prepared by Schwartz by the mixture formed by acetic acid-sulfuric acid treatment of hematoporphyrin (Lipson, 1960). Several known porphyrins have been identified in this hydrolysis mixture, namely hematoporphyrin (Hp), hydroxyethylvinyldeuteroporphyrin (HVD), and protoporphyrin (proto). In addition, several investigators have noted an additional porphyrin of unknown structure found to be the material primarily responsible for photosensitizing activity of the Hpd mixture both in vitro and in vivo (Moan & Sommers, 1981; Berenbaum, et al., 1982; Kessel & Chow, 1983). In 1981 we described a gel filtration procedure to isolate this material in relatively pure form and showed that it was the component of Hpd solely responsible for in vivo photosensitization (Dougherty, et al., 1983). We now report the structure of this material.

MATERIALS AND METHODS

Hematoporphyrin derivative (Hpd) under the trade name, Photofrin, was obtained from Oncology Research & Development, Inc., Cheektowaga, New York. This material is supplied as 5.0 mg/ml in sterile saline at pH 7.2-7.4.

The Hpd solution was applied directly to a gel filtration column (Bio-Gel P-10, 100-200 mesh, Bio-Rad, Richmond, California) using distilled water (pH 7-8) as eluent. As previously described, this allowed the separation of a dark brown colored aggregate fraction (Fraction A) which eluted at the exclusion limit of the column well separated from the other fractions (Dougherty, et al., 1983). This same material was also purchased from Oncology Research & Development, Inc., Cheektowaga, New York under the trade name, Photofrin II, supplied in sterile saline, 2.5 mg/ml, pH 7.2-7.4. This material was identical to that separated by gel filtration by visible spectrometry, HPLC (see below) and biologic activity (see below). Its purity ranged from 80 to 90% with hydroxyethylvinyldeuteroporphyrin as the major contaminant. The methyl esters of the porphyrins were obtained by treating the free base forms (isolated from solution by adjusting the pH to 3.5) in tetrahydrofuran ethyl ether with diazomethane at ice bath temperature. Diazomethane was generated from N-Methyl-N'-nitro-N-nitrosoquanidine obtained from Aldrich Chemical Co., Inc., Milwaukee, Wisconsin.

High performance liquid chromatography (HPLC) was carried out using a reverse phase column (Altex Model 110 ultrasphere, 4.6 x 150 mm) with Constametric pump (the Anspec Co., Inc., Ann Arbor, Michigan) and visible light detection at 405 nm (AN-203 detector, the Anspec Co.). The eluent consisted of two solvent systems; A, containing equal volumes of water, tetrahydrofuran and methanol and pH adjusted to 5.7-5.8 with acetic acid and 1 N sodium hydroxide solution and B, tetrahydrofuran, water (9:1 by volume). Samples were injected into the column (20 μl) either in aqueous solutions or in solvent A. Solvent A was run at 1 ml/min for 20-30 min and then solvent B was run at the same rate until no additional peaks eluted (approximately 10 min). Under these conditions, Hp, the two HVD structural isomers, and proto were eluted at approximately 2.2 min, 4.4 min, 5.2 min and 18-20 min respectively. The final material was eluted at approximately 5 min after changing to solvent B. The corresponding dimethyl esters eluted with the same pattern, each peak being eluted slightly later than the free base forms. The fraction eluted by solvent B corresponded to Fraction A isolated from Hpd by gel filtration described above. These HPLC conditions were chosen since none of the known porphyrins were aggregated in solvent A (determined spectrophotometrically). However, Fraction A remained largely aggregated in solvent A (Soret absorption ∼365 nm) and was not eluted until solvent B was used in which it was primarily disaggregated (Soret absorption ∼400 nm).

The relative proportions of each material eluted by HPLC was determined by measuring the area under the curves and correcting for relative absorption at 405 nm; Hp, HVD and proto absorption at 405 nm was determined in solvent A and that for Photofrin II or Fraction A (gel fraction) in solvent B. Hp, HVD and proto were identified by correspondence of elution time, HPLC and thin layer chromatography (silicon plates, 80/20/1.5, benzene, methanol, water) compared to authentic samples. Hp and HVD were obtained from Porphyrin Products, Logan, Utah, and proto from Fluka, AG (Germany).

Nuclear magnetic resonance spectroscopy (^{13}C) was carried out at 25.2 MHz by the NIH facility at Syracuse University, Syracuse, New York. Proton magnetic responses were obtained at 200 MHz by the NIH facility at Carnegie-Mellon Institute in Pittsburgh, Pennsylvania. Fast atom bombardment mass spectrometry was carried out by the NIH facility at Johns Hopkins University, Baltimore, Maryland.

RESULTS

The visible spectrum of Fraction A showed a typical non-metallic porphyrin pattern (etio type) in water or organic solvents. In water absorption occurs at 620 nm $<$ 570 nm $<$ 535 nm $<$ 500 nm $<<$ 365 nm (Soret).

Hydrolysis of Fraction A

Acid: A 2 ml sample of Fraction A (or Photofrin II) (2.5 mg/ml) was mixed with 2 ml of 1 N hydrochloric acid, protected from light, and kept at room temperature. A similar sample was kept at 37°, and a third was heated in a water bath at approximately 80° under nitrogen. All samples produced a similar mixture of products although at different rates. Forty minutes at 80° C was sufficient to hydrolyze all of the porphyrin and produce a mixture of equal parts of Hp and HVD, determined by HPLC following neutralization. The same proportions of Hp and HVD were produced at 37° and room temperature although the latter required approximately 2 to 3 weeks for completion. Conversion of HVD to Hp or vice versa was insignificant in this time period for all three conditions (separate experiments with Hp and HVD were carried out as above). Similar treatment of protoporphyrin produced a small amount (10-20%) of HVD and a trace of Hp (<10%).

Base: Two ml portions of Fraction A (2.5 mg/ml) were dissolved in 2 ml 1 N sodium hydroxide solution, protected from the light and stored at room temperature. The sample was periodically analyzed by HPLC (after neutralization) over a period of four weeks. No changes were detected.

Neutral Hydrolysis: A sample of Fraction A (2.5 mg/ml) pH 7.2-7.4 was heated at 120° for 15 min in an autoclave. This sample produced small amounts of Hp and HVD but appeared not to undergo hydrolysis. Attempts to subsequently hydrolyze this sample in acid as above indicated that it was essentially insoluble in 1 N hydrochloric acid whereas prior to this procedure it had been completely soluble.

Fast Atom Bombardment (FAB) Mass Spectrometry

Figure 1 demonstrates a typical spectrum obtained from the methyl ester of Fraction A. The major peaks at 591 and 609 mass numbers correspond to the dimethyl esters of protoporphyrin and HVD respectively and are present in approximately equal amounts. A small amount of Hp dimethyl ester appears at a mass number of 627. In the higher mass range the major peaks at 1199 and 1200 mass numbers represent 7-8% of the base peaks (591, 609). Numerous other high mass peaks are also apparent. Many of the smaller mass peaks (e.g. 149, 181) correspond to the thioglycerol in which the sample was mixed for the spectral analysis. The free base form of Fraction A yielded essentially the same pattern, but all peaks were markedly less intense than those of the methyl ester. Essentially no mass peaks above HVD and proto were observed in the free base spectrum.

NMR Spectra

^{13}C: The methyl ester of Fraction A was dissolved in deuterated chloroform (50 mg/ml) for ^{13}C-NMR spectroscopy (Figure 2). The ^{13}C-NMR spectrum of the dimethyl ester of Hp

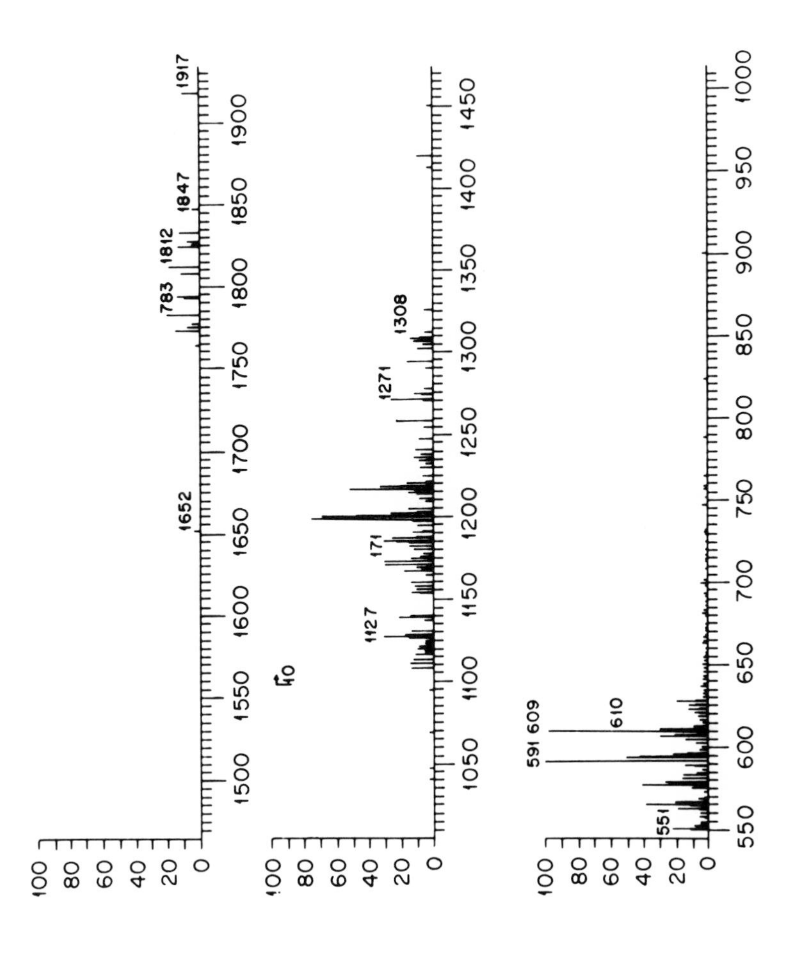

Figure 1. FAB mass spectrum of DHE

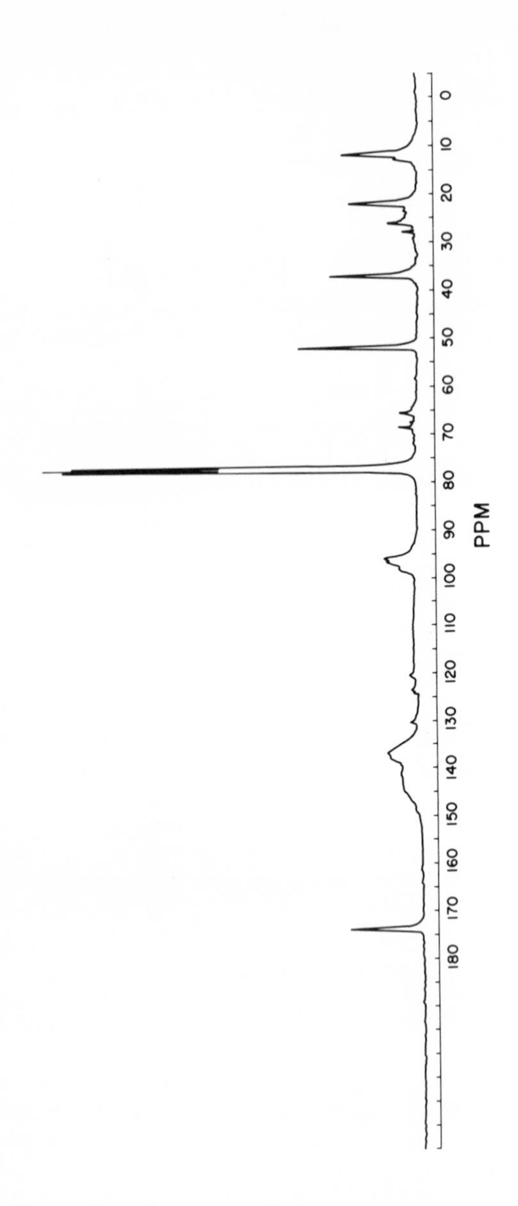

Figure 2. 13C-NMR spectrum of DHE, 50 mg/ml in $CDCl_3$

was obtained in a similar way for comparison. Peaks were identified as follows (δ, ppm, referred to tetramethylsilane); 11.48, CH_3, 21.7, - $\underline{CH_2}$ CH_2 CO_2 CH_3, 25.9, $\underline{CH_3}$ CHOH, 27.9, $\underline{CH_3}$-$\underline{C}$HO-, 36.8, - CH_2 $\underline{CH_2}$ CO_2 CH_3, 51.7, CH_2 CH_2 CO_2 $\underline{CH_3}$, 65.3, CH_3 $\underline{C}$HOH, 68.4, CH_3 $\underline{CH}$ O-, 94 to 100 methine carbons, 132 to 150 pyrrole carbons, and 173.6, -CH_2 CH_2 $\underline{CO_2}$ CH_3. Chloroform appears at δ = 77 and its inverted image near 124. The small peaks at 121 ppm and 131 ppm may be assigned to -CH = CH_2 and are likely due to HVD and proto impurity which accounted for approximately 15% of the sample. Except for the carbons at 27.9 and 68.4, the same absorbances were found in Hp dimethyl ester. The relative proportions, however, were different for the hydroxyethyl groups at 25.9 and 65.3 ppm in Fraction A compared to Hp which represent only one group per porphyrin moiety (e.g. per four methine carbons). The peaks at 27.9 and 68.4 ppm appear to have the same relative areas as those assigned to the hydroxyethyl groups. Because of an apparent wide difference in relaxation times the porportions of each type of carbon atom are difficult to establish with accuracy. In order to improve the situation, a 10 second delay was used between the accumulated scans. Under these conditions the ratio of areas was 4:2:1:1:2:2:1:1:3:9:2 in increasing ppm. Clearly the ratio of methine to pyrrole carbons should be 4:16, not 3:9 as found. Thus, even with a long delay between scans, ratios were inexact. When similar spectra were attempted with the free base form of Fraction A in DMSO-d_6, the peaks at 27.9 and 68.4 ppm were not apparent. The spectrum between 60 and 70 ppm was scanned with a 30 sec delay indicating that the absorbances at 65.3 and 68.4 ppm were of approximately equal areas.

^{1}H-NMR: Proton NMR spectra were complicated by the inability to obtain samples completely free of water which obscured the spectra from 3-4 ppm. The spectrum of the methyl ester of Fraction A in deuterated chloroform, however, revealed multiple peaks in the range of 1.85 to 2.36 ppm (major peaks appeared at 1.85 ppm, broad, 2.05, 2.2, 2.28 and 2.36 ppm, sharp). These absorbances were not discernable in the spectrum of the free base in DMSO-d_6 (30 μg/ml) which showed a single small peak at 2.16 ppm. Hematoporphyrin free base under the same conditions provided the expected spectrum with a sharp doublet at 2.13 and 2.17 ppm attributed to the methyl protons of the hydroxyethyl groups. Other absorbances apparent in the proton NMR spectra of Fraction A free base were found at approximately -4 ppm, broad, N-H, 3.8 ppm, multiplet, aromatic CH_3, 4.3 ppm, broad, $-\underline{CH_2}\ CH_2\ CO_2H$, 6.2 to 6.5 broad doublet, $CH_3\ C\underline{H}OH$ and $CH_3\ C\underline{H}O$ - and possibly $-CH = \underline{CH_2}$, 10.2 to 10.8 ppm, broad multiplet, methine protons and 12.2 ppm $CO_2\ \underline{H}$. The $CO_2\ \underline{H}$, methine, $CH_3\ C\underline{H}OH$, and $CH_3\ \underline{CH}{=}O-$ (combined) and $-\underline{CH_2}\ CH_2\ CO_2\ H$ protons were in ratio of 1:2:1:2.

Effect of Concentration on Formation of Fraction A

The usual procedure for basic hydrolysis of the acetate mixture resulting from reaction of Hp with acetic acid-sulfuric acid to produce the Hpd product mixture requires a 1 h reaction at room temperature of one part acetate mixture to 50 parts 0.1 N sodium hydroxide. This produces a mixture of porphyrins containing 40-50% Fraction A (Dougherty, et al., 1983). However, we have found that if the hydrolysis was carried out with a lower concentration of acetate in the alkaline solution, the proportion of Fraction A in the final product mixture decreased to essentially zero, when the proportion of acetate mixture to 0.1 N sodium hydroxide was 1:2500. Under these conditions the only detectable products by HPLC were Hp and HVD in ratio of approximately 2:1. Elemental analysis of Fraction A in either free base form or as the sodium salt yielded varying results because of the inability to remove all the water. We estimate that even after drying, the sample contained 5-7% water. The carbon to nitrogen ratio was found to be 7.38:1 by weight.

DISCUSSION

The structure(s) most consistent with these data are those of the ether(s) formed by linkage through the hydroxyethylvinyl groups of hematoporphyrin, Figure 5. Since there are three different ways to link the molecules, i.e., the hydroxyl groups on the two 8-, the two 3- or one 8- and one 3-hydroxyethylvinyl groups, three structural isomers are possible, each with four chiral centers. The three structural isomers are: bis-1-[3-(1-hydroxyethyl) deuteroporphyrin-8-yl] ethyl ether, bis-1-[8-(1-hydroxyethyl) deuteroporphyrin-3-yl] ethyl ether and 1-[3-(1-hydroxyethyl)-deuteroporphyrin-8-yl] -1'-[8-(1-hydroxyethyl)-deuteroporphyrin-3-yl] ethyl ether. We propose the common name dihematoporphyrin ether (DHE).

The decrease in formation of Fraction A with decreasing concentrations of acetate during hydrolysis is consistent with a bimolecular reaction involving the porphyrin acetates and/or intermediate hydrolysis products. It may indicate the necessity of a pre-formed non-covalent dimer properly aligned to allow for the formation of the ether by trapping of the intermediate carbonium ion (or other intermediate) by the hydroxyethyl group of the adjacent molecule. It has been previously reported that the porphyrin acetates can form ethers if ethyl alcohol is present during chromatography (Clezy, et al., 1980).

The acid hydrolysis to produce equal parts of Hp and HVD is consistent with the usual reaction path of ethers (Equation 1). The resistance to basic hydrolysis is also consistent with the ether structure.

$$P(CHOHCH_3)-CH(CH_3)-O-CH(CH_3)-P(CHOHCH_3) \xrightarrow[H_2O]{H^+} \underset{\text{HP}}{P(CHOHCH_3)-CH(CH_3)-OH} + \underset{\text{HVD}}{CH_2{=}CH-P(CHOHCH_3)}$$

The fast atom bombardment mass spectra analysis of the methyl esters indicates a small mass peak corresponding to DHE at 1236 (M + 1) and 1237 (M + 2). However, the major peaks in this range appear at 1199, 1200 and 1201 (7-8% of the base peaks) corresponding to the loss of two water molecules. The larger mass peaks correspond to the addition of various numbers of water molecules. It thus appears that under the conditions of the analysis, the ether undergoes dehydration followed by cleavage with H transfer as indicated.

$$P(CHOHCH_3)-CH(CH_3)-O-CH(CH_3)-P(CHOHCH_3) \xrightarrow{-2H_2O} P(CH{=}CH_2)-CH(CH_3)-O-CH(CH_3)-P(CH{=}CH_2)$$

$$\rightarrow \underset{\text{Proto}}{CH_2{=}CH-P(CH{=}CH_2)} + \underset{\text{HVD}}{HO-CH(CH_3)-P(CH{=}CH_2)}$$

Other peripheral groups deleted

Attempts to obtain mass spectra without fast atom bombardment were completely unsuccessful resulting only in a large number of small mass peaks. Initial ^{13}C-NMR spectra obtained for the free base of Fraction A in DMSO-d_6 indicated all the required structural elements of the ethers but did not reveal the necessary 'new' structure containing the ether link. Since this was considered possibly due to slow relaxation and/or aggregation, the methyl ester was prepared and ^{13}C-NMR spectra obtained in $CDCl_3$ at a slow accumulation rate, i.e., with a 10 sec delay between scans. Spectra obtained in this way revealed the ether structure with the expected absorptions occurring at 27.9 ppm and 68.4 ppm for - $C(\underline{CH_3})$-O and - $\underline{C}(CH_3)$-O respectively. A slight downfield shift for the ether structures compared to the corresponding alcohol structures (i.e. -$C(CH_3)$-OH) is predictable (Levy, 1983). Quantitative information is difficult to obtain from ^{13}C-NMR because of the wide variation in relaxation times. However, in order to verify that there were equal numbers of ether structures and alcohol structures as required for DHE, the scans were run with a 30 sec delay and expanded to observe the area in the 60-70 ppm range. Within the margin of error and noise it appeared that there was an equal number of ether carbons and alcohol carbon atoms (Figure 3).

The ^{1}H NMR spectrum demonstrated the necessary structures but was not definitive since it was not possible to remove all the water from the samples thus producing a spectrum in which the absorptions near water were at least partly obscured. However, as in the ^{13}C spectrum, the proton NMR spectrum of the free base in DMSO-d_6 did not reveal the protons of the methyl group attached to the ether group. A portion of the spectrum obtained for the methyl ester in $CDCl_3$ is shown in Figure 4, which revealed a complex set of methyl absorptions which probably result from the multiple isomeric structures possible for the ether and alcohol structures. In general, the proton spectra are much more sensitive to structural modifications than are the ^{13}C spectra which appeared to be relatively uncomplicated.

Fraction A (DHE) was not stable when heated in aqueous solution (120°, 15 min) as evidenced by its insolubility in 1 N hydrochloric acid after this procedure. This may indicate an oxidative process or possibly formation of polymeric structures. When tested in this manner, the material also lost its ability

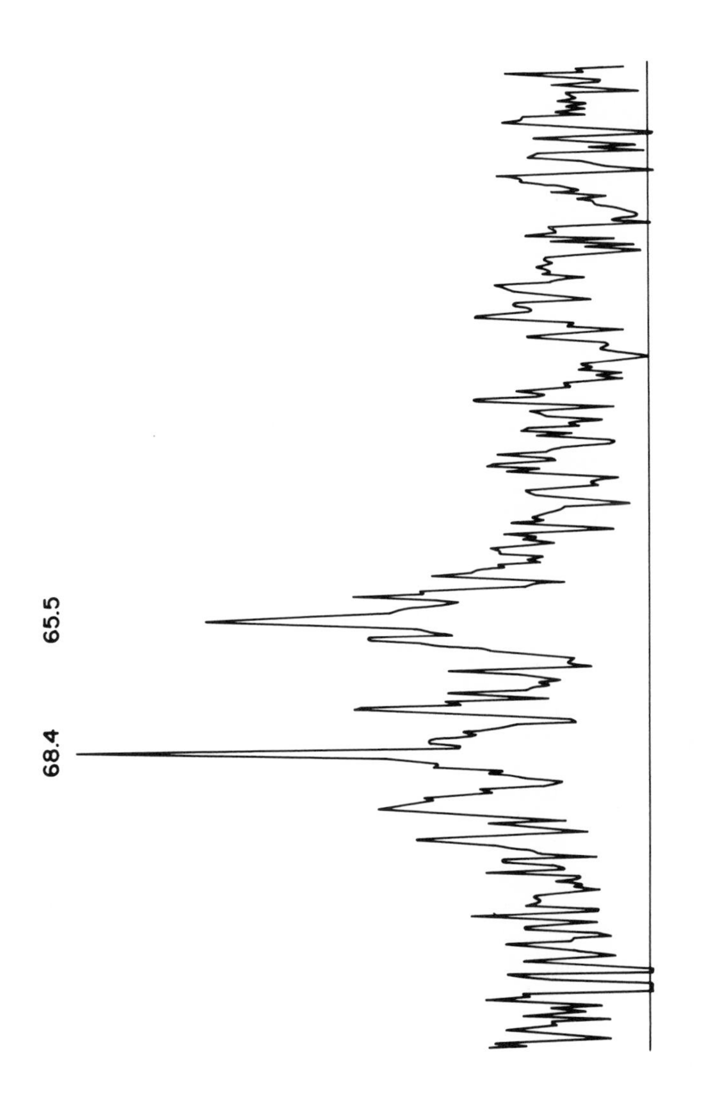

Figure 3. Expanded ^{13}C-NMR spectrum of Figure 2 with 30 sec delay between scans; 60-70 ppm region

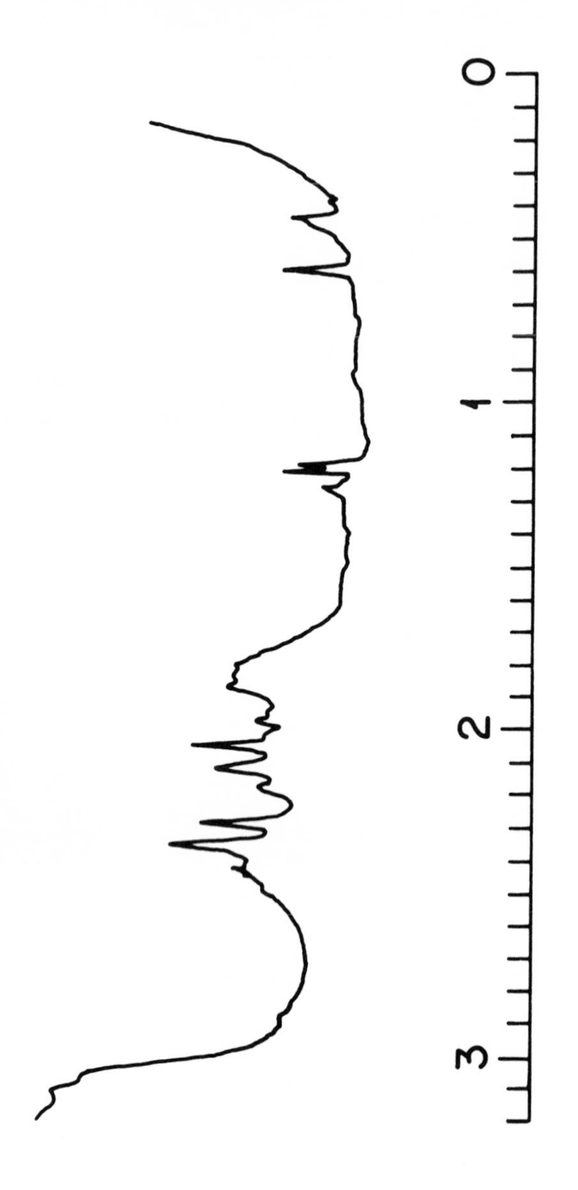

Figure 4. ^1H-NMR spectrum of DHE, 50 mg/ml in $CDCl_3$; 0-3 ppm region

Structure of Hpd-Active Component

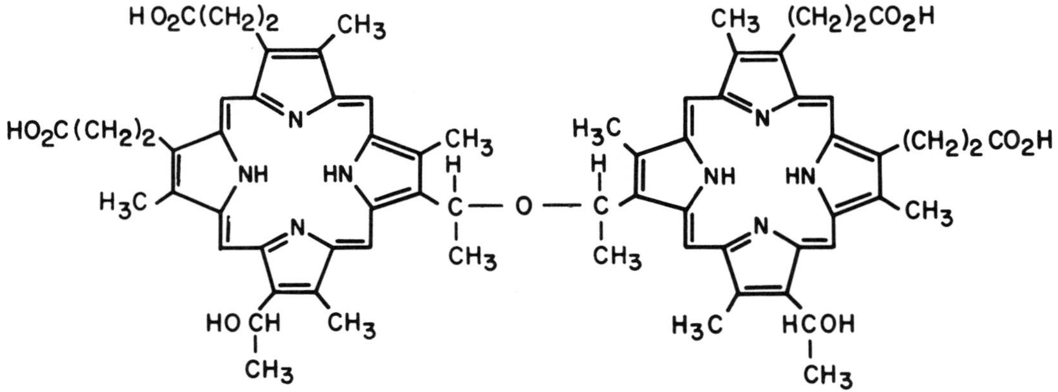

Bis-1-[8-(1-hydroxyethyl) deuteroporphyrin-3-yl] ethyl ether

Figure 5. Structure of Hpd-Active Component

to photosensitize the SMT-F tumor system used as a standard in our laboratory (Dougherty, et al., 1983).

Finally, the elemental analysis provided a carbon to nitrogen ratio of 7.38:1 compared with a calculated ratio of 7.29 to 1

ACKNOWLEDGMENTS

We would like to acknowledge the considerable assistance of Dr. George Levy and associates at Syracuse University who ran the ^{13}C-NMR spectra and assisted in interpretation, as well as Dr. R. L. Stephens at Carnegie-Mellon Institute who provided the proton NMR spectra, and to Dr. James Yergy of Johns Hopkins University, who was very helpful in determining and interpreting the FAB mass spectral data.

Berenbaum, MC, Bonnett R, Scourides PA (1982). In vivo biological activity of the components of hematoporphyrin derivative. Br J Cancer 45:571.

Clezy PS, Hai TT, Henderson RW, Thuc L (1980). The chemistry of pyrrolic components XLV. Hematoporphyrin derivative diacetate as the main product of the reaction of hematoporphyrin with a mixture of active and sulfuric acid. Aust J Chem 33: 585.

Dougherty TJ, Boyle DG, Weishaupt KR, Henderson BA, Potter WR, Bellnier DA, Wityk KE (1983). Photoradiation therapy - Clinical and drug advances. In Kessel D, Dougherty TJ (eds): "Porphyrin Photosensitization," New York: Plenum Publishing Corp, p 3.

Kessel D, Chow T (1983). Tumor-localizing components of the porphyrin preparation hematoporphyrin derivative. Can Res 43: 1994.

Levy G (1983). Private communication.

Lipson RL (1960). The photodynamic and fluorescent properties of a particular hematoporphyrin derivative and its use in tumor detection. Master's Thesis, University of Minnesota.

Moan J, Sommer SP (1981). Fluorescence and absorption properties of the components of hematoporphyrin derivative. Photobiochem Photobiophys 3:93.

Porphyrin Localization and Treatment of Tumors, pages 315–319

PHTHALOCYANINES LABELED WITH GAMMA-EMITTING RADIONUCLIDES AS POSSIBLE TUMOR SCANNING AGENTS

J.E. van Lier, H. Ali and J. Rousseau

MRC Group in the Radiation Sciences
University of Sherbrooke Medical Center
Sherbrooke, Quebec, Canada J1H 5N4

Porphin derivatives have a long and controversial history as potential tumor seeking molecules. As far back as 1924 Policard suggested their role in the fluorescence of tumors under the influence of u.v. light. Ever since, medical applications for this class of compounds have been persued resulting in their use both in phototherapy and chemotherapy of cancer (Figge *et al.* 1948; Iijima *et al.* 1961; Dougherty *et al.* 1978). With the advance of nuclear medicine, their potential use as selective carriers for γ-emitting radionuclides has also been evaluated (Winkelman *et al.* 1962; Wang *et al.* 1981). The biodistribution of ^{57}Co-labeled porphyrins was shown to depend strongly on the overall charge and solubility of the complexes, with some analogs exhibiting more favorable tumor-to-blood ratios than the well known tumor-scanning agent, ^{67}Ga-citrate (Wang *et al.* 1981).

Synthetic phthalocyanines resemble the naturally occurring porphyrins. They consist of four benzisoindole nuclei fused *via* nitrogen bridges and they form stable chelates

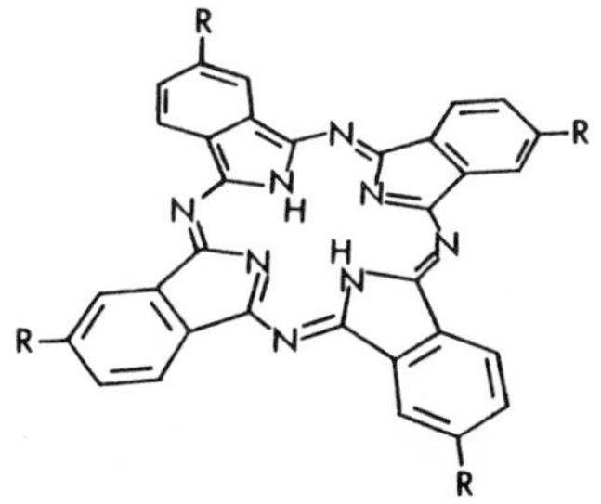

R=H	*Phthalocyanine*	*(PC)*
R=SO_3^-	*Tetrasulfo-PC*	*(TSPC)*
R=NH_2	*Tetraamino-PC*	*(TAPC)*
R=NO_2	*Tetranitro-PC*	*(TNPC)*

with metal ions and certain metal oxides. They are nontoxic and follow organ distribution patterns similar to those of comparable substituted porphin derivatives. Apart from water-soluble $^{235}UO_2$-tetrasulfophthalocyanine, which has been advanced as an agent for tumor therapy in conjunction with neutron activation (Frigerio, 1962), phthalocyanines have received little attention as to possible medical applications. As a result we initiated a program of synthesis and biological evaluation of various phthalocyanines labeled with γ-emitting radionuclides. Technetium-99m, with its convenient half-life of 6 h and a near 100% γ-emission of a single 140 keV photon, is the isotope of choice for use in diagnostic nuclear medicine. Accordingly we selected this readily available isotope for both our labeling and biodistribution studies.

CHEMISTRY

Metal derivatives of phthalocyanines can be prepared either by condensation from a monomer such as phthalic acid, phthalic anhydride, diiminoisoindoline, or phthalonitrile in the presence of a metal ion, or by exchange of the central hydrogen or metal ion from phthalocyanine or its lithium complex (Jackson 1978). The major problems encountered in direct labeling of phthalocyanines with metallo radionuclides are twofold: (a) both reactants should be soluble in the same solvent, and (b) the metal ion, or metal oxide, should have the proper dimension to allow insertion into the phthalocyanine pocket. In the case of technetium and our charged tetraamino- and tetrasulfophthalocyanines, solubility does not pose a problem. However the Tc species available for radiolabeling, including the highly oxidized pertechnetate (TcO_4^-) or reactive reduced species such as TcO_2, do not readily form stable complexes with these phthalocyanines. Accordingly we adapted the phthalic acid condensation method for the preparation of our Tc-labeled phthalocyanines (Rousseau *et al.* 1983). In this labeling procedure $^{99m}TcO_4^-$ is converted to a less stable, more reactive, reduced state of technetium *via* a reduction with hydroxylamine. In addition to sulfo-, amino- or nitrophthalic acid, hydroxylamine and pertechnetate, the reaction mixture contains urea, a trace amount of ammonium molybdate and ammonium chloride. After evaporation of the solvent the condensation occurs in the melt at 260°C. The resultant labeled phthalocyanines are purified on 0.25 mm thick silica gel TLC plates.

Three different phthalocyanines were prepared in this manner, the tetrasulfo- (TSPC), tetraamino- (TAPC), and the tetranitrophthalocyanine (TNPC). In order to permit chemical characterisation of the ^{99m}Tc-TSPC, the complex was also prepared with the longer-lived β-emitting ^{99}Tc-isotope as well as with nonradiactive Co (Rousseau *et al.* 1983). The ^{99m}Tc-TAPC and ^{99m}Tc-TNPC complexes were characterized by comparing their chromatographic properties with those of the corresponding Co-phthalocyanines, which in turn were identified by their physico-chemical properties. In the case of the Tc-TSPC complex the major condensation product was shown to contain 1 mol Tc per mol of TSPC. An out-of-plane sandwich-type structure of the complex, due to the attachment of 3 Tc-ligands to 3 out of 4 possible phthalocyanine N-atoms, was proposed. In this model, the remaining free ligands of the Tc ion are occupied by hydroxyl groups and water molecules (Rousseau *et al.* 1983). It is most likely that the ^{99m}Tc-TNPC and -TAPC complexes may be depicted as similar tridentate ligands.

BIOLOGY

Tumor uptake and organ distribution patterns were studied both in Fischer 344/CRBL female rats bearing the 13762 mammary adenocarcinoma and in healthy Dutch rabbits. The latter were selected for scintigraphic studies since their size permits total body scans with a standard γ-camera, resulting in optimal organ resolution. Activity distribution in the rats was determined by dissection and radioactivity counting of individual tissue samples after sacrificing the animals by cardiac puncture.

The tissue distribution of the phthalocyanine complexes is strongly influenced by their overall charge and water solubility. Thus, as compared to free pertechnetate, the neutral and lipophilic nitro-PC is retained in the blood, whereas the negatively charged sulfo-PC shows rapid blood clearance. The blood levels of the amino-PC remain intermediate between those of the sulfo- and nitro-PC's (Fig. 1). The amino-PC is a weak base and accordingly only partially ionized at the pH of the body fluid. The excretion pattern of these compounds through the hepatic or renal pathways is also charge-dependent, with hepatic clearance prevailing for the least water-soluble analogs. The *in vivo* stability of all three Tc-Pc complexes was evident from the absence of significant levels of radioactivity in the target organs of free pertechnetate, *e.g.* the salivary glands, thyroid and stomach.

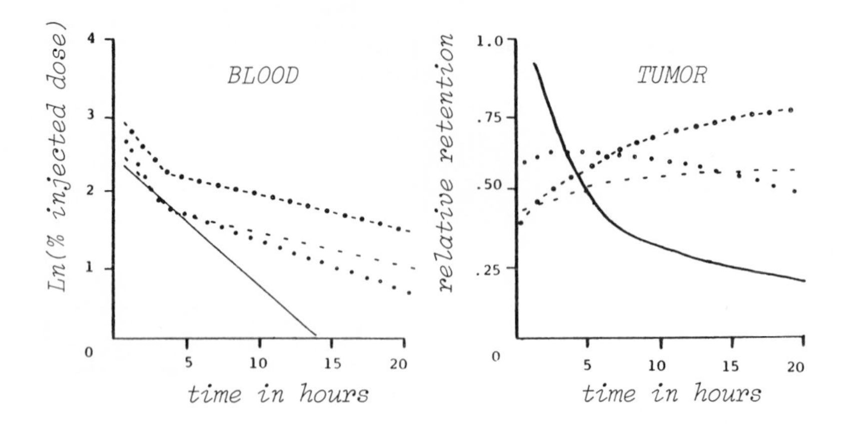

Fig. 1. Blood clearance and tumor uptake of ^{99m}Tc in Fischer 344/CRBL female rats bearing the 13762 mammary adenocarcinoma (2 g), injected *via* the caudal vein with pertechnetate (—), TSPC (···), TNPC (-·-) or TAPC (---). Radioactivity in the tumor is given in relative retention units (act per g tumor/average act per g total animal), radioactivity in total blood is expressed as Ln (% injected dose).

Overall tissue distribution patterns in experimental animals resemble those of the analogous porphin derivatives. This is particularly evident upon comparison of our organ distribution data obtained with the ^{99m}Tc-labeled TSPC and those reported by others with radiolabeled tetraphenylporphin sulfate (Wang *et al.* 1981). Both compounds exhibit exponential blood clearance rates and the bulk of the material is rapidly captured by the liver and kidneys. Tumor uptake in the outer cell layers is observed for both types of compounds. Although the total accumulation in the tumor of our ^{99m}Tc-PC complexes is rather low, the activity appears to be fixed irreversably (Fig. 1). In the case of the nitro-PC the tumor activity continues to increase with time, probably due to the prolonged high blood levels of this complex.

CONCLUSIONS

Phthalocyanines can conveniently be prepared by condensation of phthalic acids in the presence of a metal ion. They are nontoxic, and their metal chelates exhibit high *in vivo* stability. Selection of the proper monomer provides a facile route to the introduction of different functional groups into this porphin-like molecule, whereas selection of the proper central metal ion allows for labeling with useful medicinal radionuclides. The porphin-like biological behaviour of these stable metallo phthalocyanine complexes warrants a further search for derivatives with improved tumor-seeking properties.

REFERENCES

Dougherty TJ, Kaufman JE, Goldfarb A, Weishaupt KR, Boyle D, Mittelman A (1978). Photoradiation therapy for the treatment of malignant tumors. Cancer Res 38:2628.

Figge FHJ, Weiland GS, Manganiello LOJ (1948). Cancer detection and therapy: affinity of neoplastic, embryonic and traumatized tissues for porphyrins and metalloporphyrins. Proc Soc Exp Biol Med 68:640.

Frigerio NA (1962). Metal phthalocyanines. U.S. Patent 3,027,391, March 27.

Iijima NK, Matuura A, Ueno K, Fumita T, Aiba H, Ukishma F, Nohara O, Kehara T, Koshizuha Moris S (1961). Studies on merphyrin. Cancer Chemother Rep 13:47.

Jackson AH (1978). Phthalocyanines, in: The porphyrins Vol. I (Edited by Dolphin D.). Academic Press, New York 374.

Policard DA (1924). Etudes sur les aspects offerts par des tumeurs expérimentales examinées à la lumière de Woods. Compt Rend Soc Biol 91:1423.

Wang TST, Fawwaz RA, Tomashefsky P (1981). Metalloporphyrin derivatives: structure, localization, properties, in: Radiopharmaceuticals structure-activity relationships (Edited by Spencer RD) Grune & Stratton, New York 225.

Winkelman J, McAfee JG, Wagner HN, Long GR (1962). The synthesis of 57-Co-Tetraphenyl porphine sulfonate and its use in the scintillation scanning of neoplasm. J Nucl Med 3:249.

IN VITRO PORPHYRIN PHOTOBIOLOGY

Porphyrin Localization and Treatment of Tumors, pages 323–334

PHOTOSENSITIZATION OF MITOCHONDRIAL CYTOCHROME c OXIDASE BY HEMATOPORPHYRIN DERIVATIVE (HpD) IN-VITRO and IN-VIVO

SL Gibson, PB Leakey, JJ Crute and R Hilf

Biochemistry Department and UR Cancer Center
University of Rochester School of Medicine
Rochester, New York 14642

The action of hematoporphyrin derivative (HpD) as a photosensitizer of malignant tumors is well documented (Dougherty et al., 1982). This derivative of hematoporphyrin (HP), prepared by the method of Lipson et al. (1961), is composed of a mixture of porphyrin components. The mixture is treated with dilute base (0.1M NaOH) and adjusted to pH 7.3 - 7.4 with 0.1M HCl prior to its administration *in vivo* or for testing *in vitro*. The component(s) in the final preparation responsible for photoirradiation-induced cell cytotoxicity has (have) not been chemically identified, although a few reports suggest that a hydrophobic component demonstrates the greatest photosensitizing activity (Berenbaum et al., 1982; Christensen et al., 1983; Dougherty et al., 1983; Moan et al., 1983).

We have developed a method for fractionation and collection of sufficient quantitites of components from HpD to investigate the biological activity of each of these fractions. This was accomplished by use of octyl sepharose columns, which separated the HpD mixture into five selected reproducible fractions. Each fraction was assayed for its composition by analytical reverse phase high pressure liquid chromatography (HPLC). Photosensitizing activity was tested by measuring the ability to inhibit mitochondrial cytochrome c oxidase activity *in vitro*. We observed that the fraction, eluting in the most hydrophobic region (HPLC) and distinct from protoporphyrin IX (PP), demonstrated the greatest biological efficiency. This component, which we call

Fraction V, displayed similar chromatographic characteristics as Fraction 7 reported by Moan et al. (1983). Based on the mitochondrial cytochrome c oxidase assay, Fraction V was about 2 times more potent than the parent HpD mixture, and somewhat more potent than those fractions (III and IV) that contained the isomers of 2,4 hydroxyethylvinyl deuteroporphyrin (HVDP).

Materials and Methods

Animals and Tissues

The R3230AC mammary adenocarcinoma was maintained by subcutaneous transplantation into the axillary region of 60-80 gm female Fischer rats, as described earlier (Hilf et al., 1965).

Fractionation of HpD

Hpd was prepared by the method of Lipson et al. (1961). Separation of the components of HpD was carried out on octyl sepharose (CL4B), Pharmacia Fine Chemicals (Piscataway, N.J.). The octyl-sepharose beads (100ml slurry) were pre-equilibrated with 3 washes (100 ml) of 20% MeOH/H_2O, containing 10mM NaH_2PO_4 at pH 7.5. After equilibration, a glass column (2.2 cm i.d., 30 cm long) was packed to a height of 23 cm by passing 3 to 5 column volumes ($\sim$300 ml) of the equilibration buffer over the beads. HpD (10 mg) was dissolved in 0.1 ml 0.1N NaOH (1 hr at room temperature in the dark). An equal volume of 0.1N HCl was added and the volume brought to 1 ml with 10mM NaH_2PO_4, pH 7.4 (final pH = 5.5). The solvents for the mobile phase were 20% MeOH/80% H_2O, containing 10mM NaH_2PO_4, pH 7.4, and 80% MeOH/20% H_2O, with 10mM NaH_2PO_4, pH 9.4. These solvents (350 ml ea) were placed in a Gradient Elution Apparatus (Glenco Sci., Houston, Texas) and connected to the column head. One ml of HpD (10mg/ml) was carefuly pipetted onto the column, the system closed, and elution was performed at a flow rate of 1.5 to 2.0 ml/min. Fractions (3.0 to 3.5 ml) were collected, and each stored at -20° in the dark until HPLC analysis was performed. After HPLC analysis, those fractions containing the desired component were combined and taken to dryness (Savant Speed Vac Evaporator). The dried porphyrins were kept at -20° in the dark until assayed. Preparation for biological assay included resuspending the dried porphyrin fractions in 10 mM NaH_2PO_4, pH 7.5, to concentrations (based on absorbance at 397 nm) similar to that of the HpD prepared for assay.

Analytical HPLC

Analysis of the separated fractions of HpD was performed on a Beckman-Altex HPLC system (see Hilf et al., 1983).

Preparation of Mitochondria

Mammary tumors were excised, placed on ice in 0.9% NaCl, and minced with scissors. Two gm of tissue were placed in 5 ml of cold homogenization solution, containing 0.33M sucrose, 1mM dithiothreitol, 1mM EGTA, 0.3% bovine serum albumin and 100mM KCl. Homogenization was performed with a Polytron (PCU2-110, Brinkman Industries, Westbury, N.Y.) at a setting of 6 for two 15 sec intervals on ice. The homogenate was centrifuged at 500xg for 30 min at 4°C (PR-2, International Centrifuge, Needham Heights, Mass.). The supernatant was removed and centrifuged at 15,000xg for 30 min at 4°C (J-21, Beckman Instruments, Palo Alto, Calif.). The resulting pellet was resuspended in 1 ml of the homogenizing solution and sonicated for 2 min on ice (Biosonik III, Bronwill Scientific, Rochester, N.Y.), at a setting of 60. After addition of 4 ml cold homogenization solution with subsequent mixing, the mitochondrial suspension was centrifuged at 30,000xg for 30 min at 4°C. The supernatant was discarded and the pellet resuspended in 2.0 to 5.0 ml of homogenization solution to yield a final concentration of approximately 150 μg protein/ml. This suspension was separated into 0.5ml aliquots and stored at -70°C until used. There was negligible effect on enzyme activity (<2% change) after one month of storage.

Cytochrome Oxidase Assay

Cytochrome c (Sigma, St. Louis, Mo.) and dissolved in 50mM NaH_2PO_4, pH 7.5, for a final concentration of 1.2mM. This substrate was reduced by addition of a "pinch" of $Na_2S_2O_4$ followed by bubbling air through the mixture to oxidize any excess dithionite. The mitochondria, as prepared above, were sonicated on ice (2 times for 2 min each). Ten μl of the mitochondrial suspension were placed in a 1 ml (1 cm path length) quartz cuvette immediately followed by the addition of the substrate solution (0.1 ml of 1.2mM reduced cytochrome c plus 0.9 ml 50mM NaH_2PO_4) to start the reaction. Disappearance of the reduced cytochrome c was measured in a Gilford 2400S spectrophotometer (absorbance 550 nm). Rates of oxidation were linear for at least 2 min and activity was calculated from the slope of the line; mitochondrial preparations demonstrated activity of 0.5 to 1.0 μmole of cytochrome c oxidized per min per mg protein.

Photoirradiation Studies In Vitro

To the mitochondrial (sonicated) suspension was added the porphyrin fraction (in the dark) at amounts equivalent (based on absorbance, 397 nm) to 7.0 μg HpD/ml. The initial cytochrome c oxidase activity was measured. The porphyrin-mitochondrial mixture (0.5 ml) was placed in a 3 ml (1 cm pathlength) glass cuvette. The cuvette was then exposed to light generated by a 1000W Xenon arc lamp (Photochemical Research Associates, London, Ontario) filtered by an ALH1 water filter to remove the infrared emissions, a 50 ml solution of $K_2C_rO_4$ (14.3 gm/l, wavelength cutoff 505nm) placed in a 5 cm pathlength glass cuvette, appropriate bandpass and neutral density filters, and an Oriel monochrometer set at 630 nm with a 4 nm band width. The resulting incident energy on the sample was 8.0 mw/cm^2, measured by a power photoradiometer (RK5200, Laser Precision Corp.) equipped with an RD545 radiometer probe. The temperature of the area being photoirradiated was also measured and did not vary from the room temperature (21°C). Control experiments were performed with mitochondria plus light in the absence of porphyrins, and mitochondria plus porphyrins in the absence of light; no significant loss in enzyme activity (<2%) was observed during the 1 hr time used in these studies.

Primary monolayer cultures of the R3230AC mammary adenocarcinoma were grown as previously described (Gay and Hilf, 1980). After reaching confluency (3 days following plating), the monolayers were washed 3x with 1 ml 0.9% NaCl, incubated with desired concentrations of porphyrins in the dark and exposed to a 14W fluorescent light 18 cm from the monolayer surface. Determinations of cell survival were performed by the trypan blue dye exclusion method of Hilf et al. (1983).

Photoirradiation of Tumor Mitochondria from Porphyrin-Treated Animals

Fischer female rats (60-80gms) bearing the R3230AC mammary adenocarcinoma were injected i.p. with 80 mg/kg of alkalinized HpD (pH 7.4) and housed in the dark. At specific times after treatment, tumors were excised in the dark, and mitochondria were prepared as described above. Mitochondria from treated animals showed the same enzyme stability when kept frozen at -70°C as those from untreated animals. Prior to analysis, aliquots (0.5 ml) of mitochondrial suspension were thawed and subjected to sonication as described above. The suspensions were then aliquoted (30 μl) into 6x50mm glass tubes (Kimax) and placed 6 cm from a 14W fluorescent lamp. Samples (10 μl) were taken from separate tubes at various times of light exposure and assayed for cytochrome c oxidase activity.

Results

Fractionation of HpD

Chromatographic separation of the components of HpD was performed on octyl sepharose. Figure 1 is a representative tracing of the absorbance (397 nm) of the fractions collected. Five major peaks were observed and were numbered consecutively (Fig. 1). Analysis by HPLC with co-chromatography of selected porphyrin standards allowed us to identify 3 of the 5 designated fractions from the octyl sepharose column. Figure 2 depicts representative tracings of HPLC chromatograms of the HpD starting material (panel A) and Fractions I through V (panels B to F). We observed that Fraction II co-migrated with the HP standard and Fractions III and IV co-migrated with the stereoisomers of HVDP. Fraction I, the least hydrophobic, and Fraction V, the most hydrophobic were not identifiable with the porphyrin standards used. An interesting result, in agreement with Moan et al. (1983), was that Fraction V contained components eluting prior to PP (on HPLC); however, depending on the portion of Fraction V selected, small amounts of PP were identifiable. Interestingly, Fraction V was readily soluble in aqueous solution whereas PP was essentially insoluble.

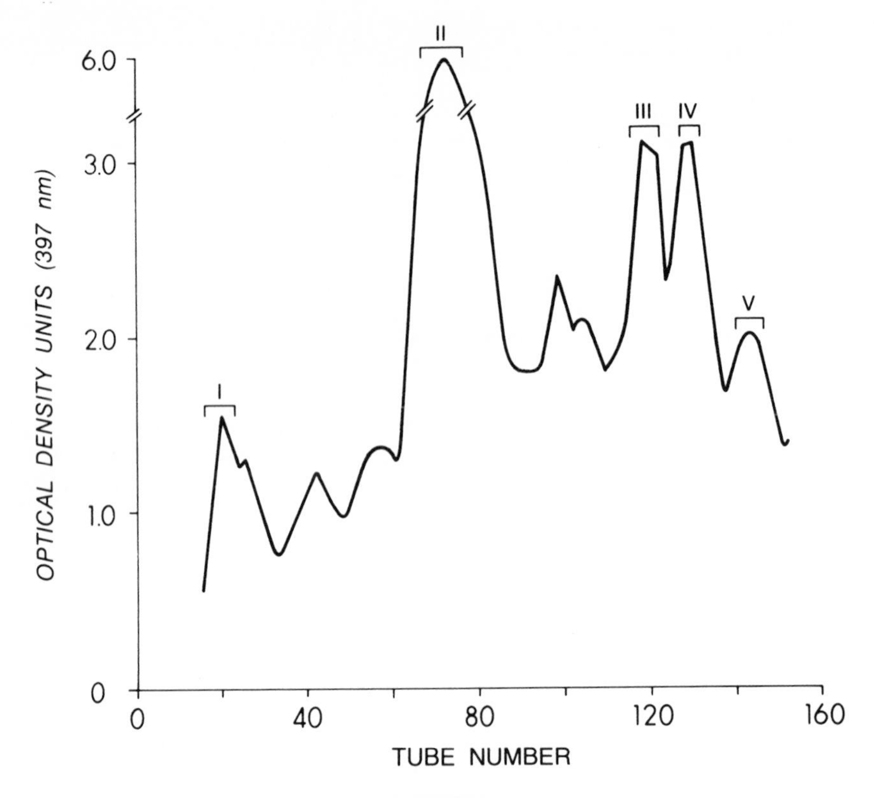

Fig. 1. Fractionation of HpD by octyl sepharose column (see text for details). Profile presented as absorbance at 397nm for each tube (3.0 to 3.5ml). Roman numerals represent the major fractions collected and pooled for HPLC analysis and photosensitization assays. Position of peaks varied < 5%, based on 20 different separations.

Effectiveness of Fractions of HpD in Photosensitizers

We examined the photosensitizing properties of HpD and the five major fractions collected by octyl sepharose chromatography. Table 1 presents results from the *in vitro* cell kill assay (Hilf et al., 1983). In this assay, Fraction V was a more potent photosensitizer in cells treated for a short period of time (5 min uptake, 1 hr light exposure). However, with longer periods of treatment (30 min uptake, 1 hr light exposure), Fractions IV, V, and the HpD preparation showed similar

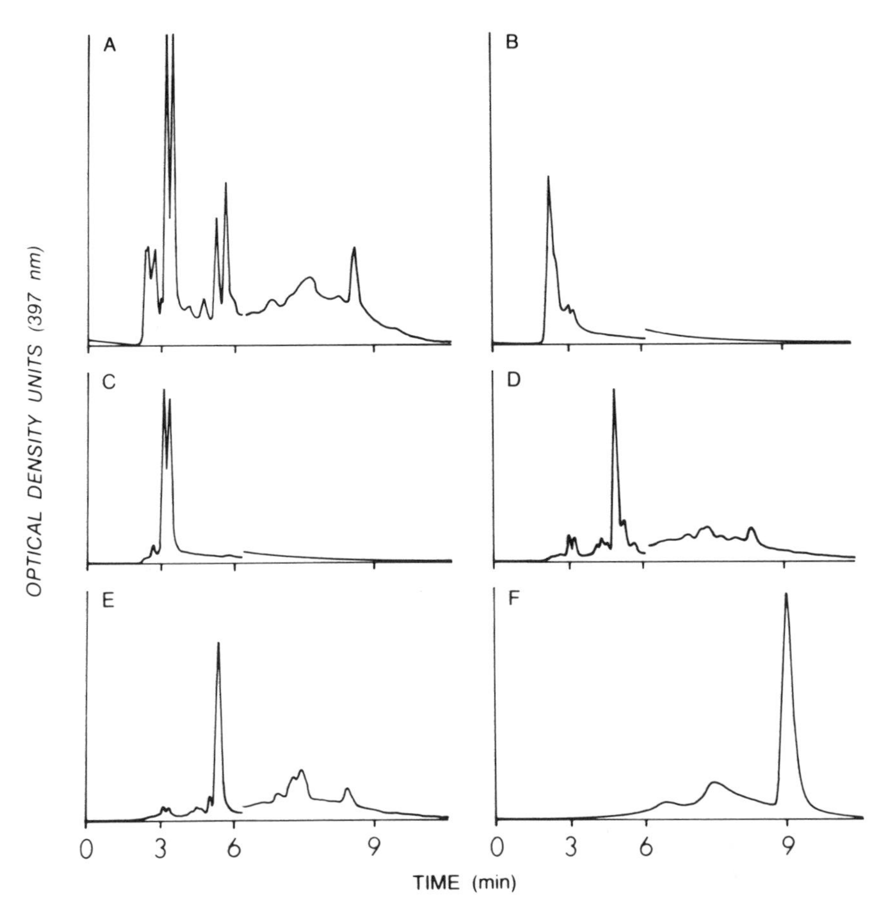

Fig. 2. HPLC chromatogram of fractions collected from octyl sepharose (Panel B, C, D, E, F represent Fractions I, II, III, IV and V in Figure 1). Panel A represents HpD prepared for column chromatography (see text).

photosensitizing activity.

We next examined the effects of these preparations on mitochondrial cytochrome c oxidase activity in vitro (Table 2). Although all of the fractions demonstrated some ability to photosensitize mitochondrial cytochrome c oxidase, i.e., inhibition of activity, Fractions III, IV and V displayed the highest potency. HpD, a mixture of these porphyrins, yielded an activity profile that was about half that seen for Fractions III, IV and V. Thus,

Table 1
Effect of Photoirradiation on Tumor Cell Survival after Treatment with HpD or its Fractions In Vitro

Fraction No.	Surviving Cell Fraction (%) Uptake 5 min hv exposure 60 min	Uptake 30 min hv exposure 30 min
I	88	82
II	82	62
III	34	16
IV	22	0
V	0	0
HpD	44	0

Cultured R3220AC mammary adenocarcinoma cells were incubated with HpD or its fractions at 7.0 μg/ml (equalized by absorbance at 397 nm). Surviving cell fraction was assayed according to Hilf et. al. (1983). Data are means of 2 experiments, each in triplicate; SEM was < 10%.

Table 2
Effect of Photoirradiation on Mitochondrial Cytochrome c Oxidase Activity After Exposure to HpD or its Fractions In Vitro

	Cytochrome C Oxidase Activity (% initial)			
	Light Exposure Time (min)			
Fraction No.	15	30	45	1 h
I	91±4	89±3	83±1	80±1
II	75±5	67±5	55±4	49±5
III	75±2	54±1	45±1	39±2
IV	67±2	49±4	42±3	39±2
V	68±1	46±3	40±2	37±2
HpD	80±5	70±2	64±6	58±1

Cytochrome c Oxidase activity after exposure to HpD or its fractions and photoirradiation (630 nm, 8.0 mw/cm^2 incident light) for various times. The HpD mixture or fraction was adjusted to equal concentrations (absorbance at 397 nm). Data are means of three experiments ± SEM (initial = 100%).

by both of these assays, the more hydrophobic components Fractions IV and V demonstrate the highest potency as photosensitizers.

Photoirradiation of Mitochondria Prepared from HpD-Treated Animals In Vivo. To address the concerns regarding studies in vitro, tumor-bearing animals were injected once with 80 mg/kg HpD and were housed in the dark until sacrificed 2, 18, 24, 72, 96 and 168 hr later. Mitochondria were prepared from the tumors and cytochrome c oxidase activity was measured after various times of exposure to light (Table 3). At 2 and 18 hr after

Table 3
Photoirradiation Effects on Mitochondrial Cytochrome c Oxidase Activity from HpD Treated Animals

Time After Injection (hr)	Cytochrome c Oxidase Activity (% initial) Photoirradiation (min)		
	30	60	120
2	96	91	77
18	96	83	80
24	69	56	38
72	79	57	35
96	90	64	39
168	95	84	60

Cytochrome c Oxidase activity was measured on tumor mitochondria prepared from animals after i.p. injection of 80 mg/kg HpD. Mitochondria were photoirradiated for various times. Data are presented as % of initial enzyme activity (initial = 100%); the values are the mean of two experiments with SEM $<10\%$.

injection of HpD, cytochrome c oxidase activity was reduced by $\sim 20\%$ after 2 hr of light exposure. Mitochondria of tumors obtained 24 to 96 hr after treatment showed a light-induced inhibition of enzyme activity that was 3 times greater than that seen at earlier times. By one week (168 hr), the ability of HpD to photosensitize mitochondrial cytochrome c oxidase returned to levels seen in tumor mitochondria obtained 2 or 18 hr after treatment. These data demonstrate that the effects of HpD treatment were optimal between 24 and 96 hr after treatment *in vivo*. After 168 hr (1 week), photosensitization had declined, indicating that less HpD (or its components and/or metabolites) was present in mitochondria of the tumor. This type of *in vivo* - *in vitro* assay provides a useful system for the detailed study of HpD and its separated components, studies that are currently underway.

Discussion and Conclusions

Data have been presented demonstrating a method to fractionate various components found in HpD. The use of column chromatography on octyl sepharose affords a method to provide reasonable separations of hydrophobic fractions in sufficient quantities for biological assays. The composition of the most active Fraction V remains to be elucidated but its behavior on HPLC confirms the findings of Moan et al. (1983); its relative cytotoxic potency against tumor cells in primary culture or as an inhibitor of mitochondrial cytochrome c oxidase activity *in vitro* indicates that considerable photosensitizing activity resides in the most hydrophobic components. These findings also agree with those of Dougherty et. al. (1983) and Berenbaum et al. (1982), using different tumor systems *in vivo*. However, our data indicate that HVDP is capable of photosensitizing cells in culture or mitochondria *in vitro*, an activity that was not considered to be significant by Dougherty et. al. (1983) and Berenbaum et al. (1982).

The finding that mitochondrial cytochrome c oxidase was inhibitable by porphyrin-sensitized photoirradiation offers a potential mechanism whereby cell cytotoxicity becomes manifest. It had been earlier noted that injected porphyrins were distributed in mitochondria and could cause uncoupling of respiration after photoirradiation (Sandberg and Romslo, 1980; Cozzani et al., 1981). Cytochrome c oxidase is a critical enzyme for oxygen consumption and its inhibition would certainly compromise cell viability (Wikstrom et al., 1981). Whether the inhibition of this enzyme and the cytotoxicity resulting from porphyrin-induced photosensitivity have a cause and effect relationship has not been elucidated. It is clear that acquisition of photosensitivity by mitochondria prepared from tumors of animals treated with HpD offers a useful assay end point to examine the pharmacokinetics of HpD and its components, an approach that has the advantage of studying the biochemical activity of porphyrin components and/or metabolites that are retained by the mitochondria.

Acknowledgements

Supported in part by CA 11193, USPHS (Animal Tumor Research Facility, University of Rochester Cancer Center) and Contract No. 04-129-4109-AA from the Standard Oil Company Research Center, Cleveland, Ohio. We wish to acknowledge Dr. Linda J. Swofford for preparation of HpD and Gary Fishbein and Steven J. Sollott for their technical assistance.

References

Berenbaum MC, Bonnett B, Scourides PA (1982). In vivo biological activity of the components of haematoporphyrin. Br. J. Cancer 45:571

Christensen T, Moan J, McGhie JB, Waksrik H, Stigum H (1983). Studies of HpD: Chemical composition on in vitro photosensitization. In Kessel D, Dougherty TJ (eds.): "Porphyrin Photosensitiation", New York: Plenum Press, p.151.

Cozzani I, Jori G, Reddi E, Fortunato A, Granati B, Felice M, Tomio L, Zorat P (1981). Distribution of endogenous and injected prophyrins at the subcellular level in rat hepatocytes and in ascites hepatoma. Chem-Biol Interactions 37:67.

Dougherty TJ, Boyle DG, Weishaupt KR, Henderson BA, Potter WR, Bellnier DA, Wityk KE (1983). Photoradiation therapy-clinical and drug advances. In Kessel D, Dougherty TJ (eds.): "Porphyrin Photosensitization", New York: Plenum Press, p.3.

Dougherty TJ, Weishaupt KR, Boyle DG (1982). Photosensitizers. In DeVita VT, Hellman AS, Rosenberg SA (eds.): "Cancer. Principles and Practice of Oncology", Philadelphia: J. B. Lippincott Co., p. 1836.

Gay RJ, Hilf R (1980). Density dependent and adaptive regulation of glucose transport in primary cultures of the R3230AC rat mammary adenocarcinoma. J. Cell Physiol. 102:155.

Hilf R, Michel I, Bell C, Freeman JJ, Borman A (1965). Biochemical and morphological properties of a new lactating mammary tumor line on the rat. Cancer Res. 25:286.

Hilf R, Leakey PB, Sollott SJ, Gibson SL (1983). Photodynamic inactivation of R3230AC mammary carcinoma in vitro with hematoporphyrin derivative: Effects of dose, time and serum on uptake and phototoxicity. Photochem. Photobiol., 37: in press.

Lipson R, Baldes E, Olsen A (1961). The use of a derivative of hematoporphyrin in tumor detection. J. Natl. Cancer Inst. 26:1.

Moan J, Sandberg S, Christensen T, Elandes S (1983). Hematoporphyrin derivative: Chemical composition, photochemical and photosensitizing properties. In Kessel D, Dougherty TJ (eds.): "Porphyrin Photosensitization", New York: Plenum Press, p. 165.

Sandberg S, Romslo I (1980). Porphyrin-sensitized photodynamic damage of isoalted rat liver mitochondria. Biochim. Biophys. Acta 593:187.

Wikstrom, M Krab K, Saraste M (1981). "Cytochrome Oxidase A Synthesis". New York: Academic Press, p. 3-13.

Porphyrin Localization and Treatment of Tumors, pages 335–350

CHEMICAL STUDIES WITH HEMATOPORPHYRIN DERIVATIVE IN BLADDER CELL LINES

M. Kreimer-Birnbaum[1,4*], J.L. Baumann[1], J.E. Klaunig[2], R. Keck[3], P.J. Goldblatt[2], and S.H. Selman[3]

[1]Research Department and Porphyrin Laboratory, St. Vincent Hospital and Medical Center, Toledo OH 43608
Departments of [2]Pathology, [3]Surgery (Div. of Urology), and [4]Biochemistry, Medical College of Ohio at Toledo

INTRODUCTION

The interaction of porphyrins (tetrapyrrolic pigments possessing the general structure depicted in Fig. 1) with light has been known since the beginning of this century to cause photosensitivity. These powerful photodynamic agents can sensitize tissues, cells, and subcellular preparations so that they are severely damaged when exposed to visible or near-ultraviolet light (Review: Spikes 1975). The photosensitizing effect seems to be related to the intense fluorescence of porphyrins, since both photosensitivity and fluorescence are evoked most effectively by long wave ultraviolet radiation (about 400 nm). When porphyrins are excited by light, several mechanisms, besides emitting fluorescence, are used for dissipating energy and to return to the ground state levels. These include heat, vibration, or formation of free radicals (Harber, Bickers 1975; Bonnett 1981). The formation of various reactive oxygen species including singlet oxygen is generally considered the main consequence of interaction between light and porphyrins, including HpD† (Review: Bonnett 1981).

*To whom correspondence should be addressed at St. Vincent Hospital and Medical Center

†See next page for abbreviations

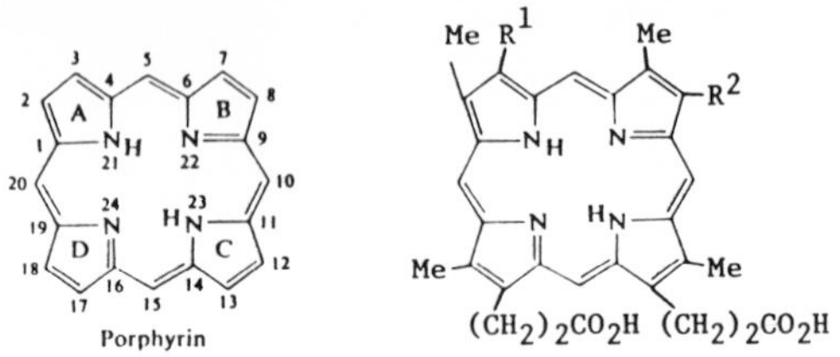

	R^1	R^2	Abbreviation in text
Hematoporphyrin	CH(OH)Me	CH(OH)Me	Hp
8(3)-(1-Acetoxyethyl)-3(8)-(1-hydroxyethyl) deuteroporphyrin isomers	CH(OH)Me	CH(OAc)Me	Monoacetyl Hp
8(3)-(1-Hydroxyethyl)-3(8)-vinyldeuteroporphyrin isomers	$CH=CH_2$	CH(OH)Me	HVD
O,O'-Diacetyl-hematoporphyrin	CH(OAc)Me	CH(OAc)Me	Diacetyl Hp
8(3)-(1-Acetoxyethyl)-3(8)-vinyldeuteroporphyrin isomers	$CH=CH_2$	CH(OAc)Me	Monoacetyl HVD
Protoporphyrin	$CH=CH_2$	$CH=CH_2$	Pp

Fig. 1: Nomenclature of some dicarboxylic porphyrins

HpD is a mixture of porphyrins with tumor localizing as well as photosensitizing properties. The active component(s) of HpD and the alkali solubilized HpD, hereafter referred to as HpD(Inj.), have been under intense study in many laboratories (Kessel, Dougherty 1983).

Furthermore, the mechanisms of porphyrin uptake by cells, their apparent preferential uptake (retention?) by tumors, and the photosensitization mechanisms remain to be elucidated. This communication summarizes our attempts to characterize the component porphyrins in HpD and various HpD(Inj.) preparations, and also describes the in vitro interactions of HpD(Inj.) with cultured rat bladder cells in the absence of light.

Abbreviations: HpD = Hematoporphyrin derivative; HpD(Inj.) = Hematoporphyrin derivative (Injectable); HVD =Hydroxyethyl vinyl deuteroporphyrin; Pp = Protoporphyrin; Cp = Coproporphyrin; HEPES = N-2-hydroxyethylpiperazine-N'-2-ethanesulfonic acid; MeOH = Methanol; DMSO = Dimethylsulfoxide; Perch = Perchloric acid; PBS = Phosphate buffered saline; TLC = Thin layer chromatography; FANFT = N-(4-(5-nitro-2-furyl)-2-thiazolyl)formamide; RP = Reverse Phase; NRBL = Normal rat bladder cells

MATERIALS AND METHODS

Porphyrins and Other Chemicals

Hp, HVD, Pp and Cp were obtained from Porphyrin Products (Logan, Utah). Cp was also obtained from Sigma (St. Louis, MO). "Photofrin"R and Hematoporphyrin diacetate were obtained from Oncology Research & Development, Inc. (Cheektowaga, NY). Diacetyl hematoporphyrin was kindly provided by Dr. Ann Smith (L.S.U. Medical Center, New Orleans). Dry porphyrin powders were kept dessicated in the dark at room temperature. HpD(Inj.) for in vitro studies with cultured cells was prepared from HpD (Porphyrin Products) as per Kessel (personal communication).

The following chemicals were used either individually or in combination: Methanol, n-Butanol and Dimethyl Sulfoxide (all three obtained from Mallinckrodt, Paris, KY), Hydrochloric Acid and Perchloric Acid (Fisher Scientific, Fair Lawn, NJ), Triton X-100 (Sigma Chemical Co., St. Louis, MO) and HEPES (Eastman Kodak Co., Rochester, NY).

Electronic and Fluorescent Spectra

Electronic spectra in the various solvents were recorded either with a UV/VIS dual beam Perkin Elmer 559 spectrophotometer, or with a Beckman Acta V-spectrophotometer. Wavelength and absorbance calibrations were performed with standards from the National Bureau of Standards, and with an Holmium Oxide filter. Apparent (εmM) millimolar absorption coefficients and (α) specific absorption coefficients (for mixtures of unknown molecular weights, at concentrations of 1 g/L) were determined (Fuhrhop, Smith 1975). Fluorescent spectra were obtained with a Farrand (MK-1) spectrofluorometer, equipped with a red sensitive photomultiplier tube. The spectra were uncorrected for the detector's response and the spectral distribution of the Xenon source. Wavelengths and relative FU at each emission maximum and minimum are presented, as well as the ratio (R) of the fluorescent intensities at each maximum. Rhodamine B in ethylene glycol (40 ng/ml) was used as a general fluorescent standard (Granick et al. 1975). Cp in 1N HCl, a relatively stable and well defined porphyrin solution, was also used as a standard.

Cells: Two types of cells were used in these studies:

I. Normal Rat Bladder Cells: A normal rat bladder cell line was developed in one of our laboratories. Bladders were excised from 100-150 g Fischer 344 male rats and placed in sterile L-15 medium. Under a laminar flow hood each bladder was cut in half. Each half was placed urothelium side upward in a 60 mm culture dish (Lux). 2.5 ml of Ham's 151 medium containing 2.5% fetal bovine serum (dialysed) and 0.1 mM calcium was added to each dish. After 2-4 days cells could be seen growing from the cut surface of the explant onto the culture dish. Medium was changed 3 times a week. After 2 weeks, the explant was removed and the attached cells (urothelial cells as evidenced by transmission electron microscopy) were removed by trypsinazation. Cells from 30 explant outgrowths were plated onto 100 mm culture dishes containing Ham's 151 medium with 2.5% serum and 0.1 mM calcium. Medium was changed 3 times a week. Cells were passed on to 30 100 mm dishes. After 2 weeks, 11 colonies of cells were observed growing on the dishes. The cells in these colonies were passaged on to 60 mm dishes. A total of 4 cell lines resulted from these efforts designated RBL-01, RBL-04, RBL-05, RBL-06. Cell line RBL-01 was used in the present study. RBL-01 has been characterized as a diploid, urothelial cell. This line has failed to grow in soft agar and syngeneic hosts through the 37th passage. It grows optimally in 2.5% serum and 0.1 mM calcium (in a variety of media). It requires serum and calcium for growth. In clonal assay it has exhibited a population doubling time of approximately 1.10 per day.

II. AY-27 Cells: The AY-27 cell line was kindly supplied by Dr. Chaplowski (U. of Mass., Worcester, MA). They are derived from a FANFT-induced transitional cell carcinoma. These cells have been further characterized in our laboratory. They can grow (in clonal assay) without serum and in very low (1 μM) calcium. They readily grow in soft agar and form tumors when injected into syngeneic hosts. They exhibit a population doubling time of approximately 1.5 per day.

Both cultured cell types showed no apparent changes in growths when maintained in serum-free media for up to 48 hrs. Cells were grown in a humidified 3% CO_2:97% air atmosphere, at 36.5°C. Stock cultures were maintained and for the individual experiments, cells were trypsinized and plated on 60 mm glass culture dishes. Prior to the HpD(Inj.)

uptake experiments, cells had the culture media removed. All subsequent processing was done in dimmed lights. Incubations with HpD(Inj.) in Dulbecco's PBS (GIBCO, Grand Island, NY), in concentrations ranging from 10 to 25 μg/ml, were run for 1 hr. After the incubations were completed, media containing porphyrins were removed, and the cells washed 3 or 4 times with PBS. Porphyrins were extracted from the cells with Triton X-100:50 mM HEPES, pH 7.0 (0.6% v/v) (abbreviated Tri:HEP) or with DMSO. Quantitation of the extracted porphyrins was done by fluorometry, using appropriate standards. Cell viability was determined by Trypan Blue exclusion, before and after incubations with HpD(Inj.). Cell counts were done on aliquots of trypsinized suspensions with the aid of a hemocytometer.

Protein was determined by the method of Lowry et al. (1951), using bovine serum albumin as the standard. From the relative fluorescent intensities and the protein levels, porphyrin uptake per mg of protein was calculated. Each value represents the average of at least triplicate plates.

Chromatographic Procedures: For identification of porphyrins, two analytical TLC systems were used.

System A. Chromatography on Silica Gel G was done, as per Ellfolk and Sievers (1966) and Henderson et al. (1980). Plates were obtained from Analtech (Newark, DE), activated at 110°C for 1 hour, and developed with a solvent composed of benzene:methanol: formic acid (84:15:1 v/v/v). Formic acid was redistilled just before use.

System B. Reverse phase chromatography on KC_{18} plates was done, as per Kessel (1982). Plates were obtained from Whatman (Clifton, NJ) or Analtech. The developing solvent was methanol: 3mM tetra-n-butylammonium phosphate (Eastman Kodak Co., Rochester, NY) (pH 3.5) (70:30 v/v), the latter functioning as the ion pair reagent.

Preparative KC_{18} plates were used for scaled up purification of some of the commercially available porphyrins. Fractions were eluted from the chromatography medium and soon after taken to dryness under a stream of N_2.

RESULTS AND DISCUSSION

Spectral studies of porphyrins: In order to study the composition of the HpD(Inj.) mixture, a systematic approach to characterize some marker porphyrins was undertaken. Cp, a tetracarboxylic porphyrin (Fig. 1, $R_1 = R_2 = CH_2\text{-}CH_2\text{-}COOH$), not known to be part of the dicarboxylic group of porphyrins found in HpD(Inj.), was nevertheless included because it is a relatively stable porphyrin. It is frequently used as a secondary standard for fluorometric determinations, especially when the porphyrins to be quantitated are unstable in solution (i.e, Pp).

Table 1: Electronic Absorption Spectra of Porphyrins: Soret Maxima and Corresponding Absorption Coefficients

	HCl		Perch:MeOH[1]		DMSO:MeOH[2]	
	λmax[3]	(εmM)	λmax	(εmM)	λmax	(εmM)
Coproporphyrin	399.5	(482)[4]	402	(472)	395	(165)
Protoporphyrin	408	(265)[5]	408	(293)	401	(92)
H.V.Deuteroporphyrin			405	(233)	397	(122)
Hematoporphyrin			402	(364)	396	(196)
HpD(Inj.)			403	(327)[6]	393	(147)[6]

[1] 1N Perchloric Acid (aq.):Methanol (1:1 v/v)
[2] DMSO:Methanol (0.2% v/v)
[3] λmax: in nm
[4] 0.1N HCl
[5] 2.7N HCl
[6] Figure in parentheses represents α = specific absorption coefficient, referring to the Absorbance of a solution whose concentration is 1g/L

Table 1 summarizes the λmax and the apparent εmM and α coefficients found for various porphyrins in different solvents.

For Cp in 0.1N HCl, the apparent εmM of 482 represents an intermediate value between 478 and 489 (Rimington 1960). In Perch:MeOH, tribanded dicationic spectra similar to the ones seen in HCl were observed (data not included in Table 1), while in DMSO:MeOH, etio type spectra (5 bands) were seen, with a blue shift of the Soret maximum and decreased apparent molar absorption coefficients.

For Pp in 2.7N HCl, an apparent εmM of 265 was calculated. Values of 275 and 262 have been reported by Grinstein and Wintrobe (1948) and Rimington (1960), respectively. An εmM of 241 (1.0N HCl) was found by Mauzerall (cited by Caughey 1972) and one of 310 (1.5N HCl) was found by Eales (1979). Some Pp preparations were found to be essentially homogeneous by two TLC methods and used without further purification.

HVD: The commercially available preparations had variable amounts of Pp and Hp besides the two main components, the HVD isomers. In DMSO:MeOH, the Soret λmax and the apparent εmM were 397 and 122, respectively. These values compare favorably with the λmax and εmM reported by Bonnett et al. (1981) for each purified HVD isomer.

Hp: The main component porphyrin was accompanied by HVD and Pp. Purification to chromatographic homogeneity was achieved by RP preparative TLC. The apparent εmM in 0.5N H_2SO_4 was 366, compared to a value of 423 (cited by Rimington 1960).

HpD(Inj.): The mixture had an apparent α of 327 in Perch:MeOH (at 403 nm) and 147 in DMSO:MeOH (at 397 nm), respectively.

Studies of Cp, Hp and Pp in aqueous solutions (PBS, pH 6.98) have been reported (Brown et al. 1976). It was shown that both the wavelength of the Soret peak (λmax) and the εmM were highly dependent on porphyrin concentration. Pp appeared to be highly aggregated under all conditions and Hp, at concentrations below 4 μM was probably dimerized. At approximately 4 μM, Hp exhibited sharp changes in spectra consistent with a "micellization" process to form large aggregates of unknown sizes. Other authors also described that in the concentration range of 0-200 μM at pH 7.0, Hp was found to be dimerized (Tipping et al. 1978). Moan and Sommer (1981) reported electronic spectra at the Soret region, of 4 HpD components: Hp, HVD isomers and Pp in neutral aqueous solution (PBS, pH 7.3). Grossweiner et al. (1982) obtained spectra of Hp in pH 7.05 phosphate buffer and in methanol:chloroform mixtures. Soret maxima and absorbances in methanol have been reported for Pp and Hp by Kessel and Rossi (1982). Most of the experimental conditions described above are different from ours, therefore, no direct comparison can be established.

Profio and Doiron (1981) and Kinsey et al. (1981) have described absorption spectra of HpD(Inj.) in saline, with and without fetal calf serum, and Grossweiner et al. (1982) reported an α of 160 for HpD in pH 7.05 phosphate buffer. Regardless of the source and mode of preparation, it is obvious that the HpD(Inj.) mixture has a lower absorption in the Soret region than the individual component porphyrins. This seems to hold true for both the Perch:MeOH and the neutral DMSO:MeOH. This point will be elaborated on in connection with fluorescence data below.

To further characterize the porphyrins under study, fluorescence spectra were determined. Some highlights of the emission spectra obtained for Pp, Hp, HVD and HpD(Inj.) are summarized in Table 2.

Table 2: Fluorescent Emission Spectra: 1N Perchloric Acid (aq.):Methanol (1:1 v/v)

	Protoporphyrin		Hematoporphyrin		HVD[1]		HpD(Inj.)	
	nm	FU	nm	FU	nm	FU	nm	FU
λ max_1	603	1.80	596	1.19	602	0.88	600	0.41
λ min	635	0.66	626	0.63	633	0.45	631	0.25
λ max_2	661	1.35	650	1.25	656	0.80	654	0.40
$\frac{FU\ \lambda\ max_1}{FU\ \lambda\ max_2}$	1.34		0.95		1.10		1.03	

[1]HVD = Hydroxyethyl Vinyl Deuteroporphyrin
Excitation spectra were maximized for the first emission peak.
All concentrations are normalized to 17 ng/ml.
FU: Fluorescent Unit = Full scale deflection of the pen covering the total height (17 cm) of the recording chart.

All the spectra showed two peaks in the 600 and 650-660 nm regions, respectively, similarly to what is observed with Cp in 1N HCl and in Perch:MeOH (data not shown). The ratio of the fluorescent peak maxima seems to be characteristic for each porphyrin for a given solvent and for the combination of excitation and emission slits as well as other specific details of a given fluorometer (Granick et al. 1975). The individual components of a mixture contribute both to the position of the λmax of emission as well as to the relative heights of the peaks. Ratios of FU λmax_1/ FU λmax_2 can thus be used as a criterion to ascertain what components might be present or predominate in a given mixture. We found that in Perch:MeOH, Pp, Hp, HVD and HpD(Inj.) had ratios of 1.34, 0.95, 1.10 and 1.03, respectively.

Fluorescence data is relative because it depends on each individual fluorometer's construction characteristics, including the exciting lamp profile, the phototube response, the set of filters that are used, if any, the slits of both the excitation and emission monochromators, etc. Moreover, some authors prefer to set the fluorometer at a fixed excitation wavelength, i.e., 400 nm, and record differences in the emission spectra only (Piomelli 1973; Granick et al. 1975), instead of scanning and determining each preparation's distinct excitation maxima. We have chosen the latter approach and thus our data can only be compared in general terms to published information. With this proviso in mind, it is interesting to note that for Pp in Perch:MeOH, a peak ratio of 1.27 has been reported by Granick et al. (1975). Fluorescent spectra in neutral solvents showed a red shifting of the λ_1 and λ_2 emission maxima, compared to the dicationic acid spectra. The DMSO:MeOH spectra present a much lower second emission peak and thus the peaks' ratios are about 2.5-3 times higher than the ones obtained in Perch:MeOH. Because the second peak is red shifted by some 30 nm, and close to the 700 nm region, their lower intensity may be reflecting a gradual loss of detecting sensitivity by the phototube on nearing this region of the spectrum. Thus, this second peak area wouldn't be a good choice for quantitative determinations.

Fluorescent spectra of four isolated fractions from HpD(Inj.) have been described recently by Moan and Sommers (1981). They have dissolved the fractions in PBS, pH 7.3, and recorded the spectra in the presence of increasing concentrations of human serum. As we mentioned in the paragraph discussing electronic spectra, the different assay conditions are critical and preclude direct comparison of peak positions, relative intensities, peak ratios, etc. Even by extrapolating the fluorescent data to zero human serum concentration, it was not possible to establish a direct comparison. Neither can we equate our data to the intracellular emission spectra of HpD reported by Berns et al. (1982), or the findings described by Kessel and Rossi (1982).

By comparing the apparent intensity of the fluorescence, it is obvious that HpD(Inj.) (Table 2) has less relative fluorescence than the other porphyrins solutions of equal concentrations. We may be dealing with (a) a characteristic low intensity both of fluorescence and electronic absorbances, (b) a solubility problem, i.e., the HpD(Inj.) was only partially solubilized in the solvents under study, or (c) that a fraction of the preparations is essentially non-fluorescent under these assay conditions. Non-fluorescing porphyrin aggregates have been reported to be present in tumor localizer solutions by Schwartz (1981) and Dougherty et al. (1983). The relative role played in tumor localizing and photosensitizing of the fraction(s) that are non-fluorescent versus the fluorescent ones remains to be elucidated with purified porphyrins. At present, extreme caution has to be exercised in calculating therapeutic doses for in vivo photoradiation therapy as well as for the in vitro photosensitizing studies.

TLC Characterization of HpD and HpD(Inj.) Preparations

Resolution of the HpD(Inj.) major components by Normal Phase TLC on Silica gel plates by System A was satisfactory. Mixtures of Pp, HVD and Hp were clearly separated and the individual components identified. However, this system worked less than satisfactorily when the HpD preparations were analyzed. Acetyl derivatives of Hp and HVD (Fig. 1) were found to have Rf coinciding with some of the tertiary alcohol markers. The HpD(Inj.) preparations, having been subjected to alkaline hydrolysis, were supposed to have lost all their acetylated derivatives (Dougherty et al. 1978; Bonnett et al. 1981). That this is indeed the case was clearly shown by reverse phase TLC (System B). HpD(Inj.) showed as its four major components Hp, HVD (isomers) and to a lesser extent Pp. HpD, on the other side, showed at least 3 main additional components that disappear during the solubilization process. One of these components runs with an Rf very close to one of the HVD isomers and it has been identified as diacetyl Hp by comparison with the compound prepared by Dr. A. Smith according to Bonnett et al. (1981). The preparation marketed as "Hp diacetate" in our TLC system showed some 9 distinctive bands and was essentially similar to preparations of HpD.

When the HpD(Inj.) prepared from various HpD powders

and Photofrin[R] were compared, their TLC profiles were found to be similar and comparable to reported profiles obtained by HPLC (Moan et al. 1982; Cadby et al. 1982; Dougherty et al. 1983). However, one HpD(Inj.) preparation showed no obvious changes in its TLC profile yet was lacking in biological activity: i.e., did not cause any significant reduction of tumor blood flows as determined by the radioactive microsphere technique in transplantable AY-27 FANFT induced bladder tumors in Fischer 344 rats (Selman et al., submitted for publication). Our observations are consistent with the remarks of Dougherty (1982) that it is not clear how the biological phenomena of localization and photosensitization are related to the porphyrin components in the mixtures, or why apparently similarly prepared injectable solutions show variable biological activity.

Table 3: Porphyrin Uptake by Cultured Rat Bladder Cells in vitro

Cell Type	Extraction Solvent	ng porphyrins /plate	μg protein /plate	μg porphyrins /mg protein
	Incubation for 1 hr			
AY-27	DMSO	927±99	1124	0.82
AY-27	Triton:HEPES	976±95	1220	0.80
NRBL	Triton:HEPES	488±64	1043	0.47
	Incubation for 2 hr			
AY-27	Triton:HEPES	1282±103	2491	0.51
NRBL	Triton:HEPES	547±29	1781	0.31

AY-27 cells derived from a transitional cell carcinoma and NRBL, normal rat bladder cells, were exposed to HpD(Inj.) in PBS (10 μg/ml) at 36.5°C. Incubation media was removed, 3 washes with PBS followed and the porphyrins were later extracted as described in Materials and Methods. Values represent averages of 3-6 plates.

Porphyrin Uptake by Cultured Rat Bladder Cells in vitro

Table 3 summarizes the apparent uptake of HpD(Inj.) components by AY-27 and NRBL cells, after incubation for 1 or 2 hr. (concentrations of HpD(Inj.) = 10 μg/ml). Two different extraction solvents have been used: Tri:HEP and DMSO. Both solvents seem to be equally efficient, inasmuch as the levels of porphyrins extracted from AY-27 cells were (mean ± S.D., n = 6) 976±95 and 927±99 ng of porphyrins per culture plate, respectively. Extracts from NRBL cells showed levels of about 50 to 60% from those found in the transformed cells. For AY-27 cells, the porphyrin uptake was 0.80 µg per mg of protein, while the comparable uptake by the NRBL was 0.47 µg per mg protein. After 2 hours of incubation, a similar pattern was observed (Table 3). Henderson et al. (1983) have reported porphyrin uptake by four different cell lines ranging from 0.06 to 0.1 (1 hr) and from 0.15 to 0.25 µg porphyrins per mg of protein, after 2 to 4 hour incubations and essentially no differences between transformed and normal cells. Our experiments were conducted with serum-free media and this may account for the apparently higher rate of uptake by both cell lines, as suggested by Chang and Dougherty (1978). The protection afforded to cells by added serum has been reported by other laboratories and confirmed by us (Klaunig et al., unpublished). Serum present in the wash media seemed to favor efflux of porphyrins from cells, while saline alone had no apparent effect (Chang and Dougherty 1978). With respect to the apparently higher levels of porphyrins found in malignant versus nontransformed cells, our results tend to confirm Mossman's et al. (1974) observations. Moan et al. (1981) also found slightly higher uptake of porphyrins per unit volume of malignant cells, but in spite of the difference in uptake, degrees of photosensitivity were comparable for both cell types.

We do not feel these short time incubations completely explain the observed higher content of porphyrins found in tumors in vivo after systemic administration of HpD(Inj.). This differential uptake or retention phenomenon constitutes the basis for photoradiation therapy. One may postulate that other mechanisms such as poor lymphatic drainage are operational in vivo as suggested by Bugelski et al. (1981). Another mechanism may be the preferential retention of porphyrins by tumors through the intracellular transformation of those porphyrins with hydrophobic side chains into more hydrophilic ones that would preclude their excretion (Kessel 1982). A very interesting and relevant observation has been described by Granick et al. (1975): when human serum albumin was added to the culture medium of hepatocytes it had an apparent direct effect by favoring the efflux of intracellular hydrophobic Pp but no effect was seen on the efflux of more hydrophilic porphyrins, such as uroporphyrins.

When we compared fluorescent emission spectra of HpD (Inj.) preparations to the spectra of the cell extracts (all in Perch:MeOH), consistent differences in both the position of the two emission peaks (600 and 653 vs. 603 and 658) and also on the peaks' height ratio (1.03 vs. 1.21) were observed (Kreimer-Birnbaum et al., unpublished observations). We have interpreted these data to mean that the porphyrin mixture extracted from the AY-27 cultured cells contained higher proportions of porphyrins with peak ratios over 1.0. According to Table 2, these findings would be consistent with having relatively more Pp and HVD associated with the cells after short term incubations in a serum-free medium. Moreover, TLC of cellular extracts show a pattern consistent with the conclusions derived from the fluorescent analysis: HVD and Pp seemed to be present in the cell extracts in higher proportions than Hp. These data would support the conclusions arrived at by other investigators (Kessel 1982; Moan et al. 1982).

Further studies are under way to identify the active component(s) of the HPD(Inj.) mixture, their fate after being internalized by the cells, and in vivo mechanisms that lead to photodestruction of tumors.

ACKNOWLEDGEMENTS

These studies were supported in part by grants from the F. M. Douglass Foundation, the American Cancer Society (Ohio Division, Inc.), and Biomedical Research Support (MCO) S0-7-RR 05700-12, and -13. Special thanks go to Ms. K. E. Schultz for her involvement in many facets of this project, including the preparation of this manuscript, to Mr. B. Barut for assisting with the cell cultures, and to Mr. M. Lust for excellent technical assistance.

REFERENCES

Barut BA, Klaunig JE (1983). In vitro effects of tumor promotors on rat urinary bladder epithelial cells. Ohio J Sci 83:100.

Berns MW, Dahlman A, Johnson FM, Burns R, Sperling D, Guiltinam M, Siemens A, Walter R, Wright W, Hammer-Wilson M, Wile, A (1982). In vitro cellular effects of hematoporphyrin derivative. Cancer Res 42:2325.

Bonnett R (1981). Oxygen activation and tetrapyrroles. Essays Biochem 17:1.

Bonnett R, Ridge RJ, Scourides PA, Berenbaum, MC (1981). On the nature of 'haematoporphyrin derivative.' J Chem Soc Perkin Trans I 37:3135.

Brown SB, Shillcock M, Jones P (1976). Equilibrium and kinetic studies of the aggregation of porphyrins in aqueous solution. Biochem J 153:279.

Bugelski PJ, Porter CW, Dougherty TJ (1981). Autoradiographic distribution of hematoporphyrin derivative in normal and tumor tissue of the mouse. Cancer Res 41:4606.

Cadby PA, Dimitriadis E, Grant HG, Ward AD, Forbes IJ (1982). Separation and analysis of haematoporphyrin derivative components by high-performance liquid chromatography. J Chromatogr 231:273.

Caughey WS (1972). "Specifications and Criteria for Biochemical Compounds." Washington: National Academy of Sciences, p 198.

Chang CT, Dougherty TJ (1978). Photoradiation therapy: kinetics and thermodynamics of porphyrin uptake and loss in normal and malignant cells in culture. Radiat Res 74: 498.

Dougherty TJ, Boyle DG, Weishaupt KR, Henderson BA, Potter WR, Bellnier DA, Wityk KE (1983). Photoradiation therapy - Clinical and drug advances. In Kessel D, Dougherty TJ (eds): "Porphyrin Photosensitization," New York: Plenum Press, p 3.

Dougherty TJ, Kaufman JE, Goldfarb A, Weishaupt KR, Boyle D, Mittelman A (1978). Photoradiation therapy for the treatment of malignant tumors. Cancer Res 38:2628.

Dougherty TJ (1982). Variability in hematoporphyrin derivative preparations. Cancer Res 42:1188.

Eales L (1979). Clinical chemistry of the porphyrins. In Dolphin D (ed): "The Porphyrins," Vol VI, New York: Academic Press, p 665.

Ellfolk N, Sievers G (1966). Thin layer chromatography of free porphyrins. J Chromatogr 25:373.

Fuhrhop JH, Smith KM (1975). Laboratory methods. In Smith KM (ed): "Porphyrins and Metalloporphyrins," Amsterdam: Elsevier, p 882.

Granick S, Sinclair P, Sassa S, Grieninger G (1975). Effects of heme, insulin, and serum albumin on heme and protein synthesis in chick embryo liver cells cultured in chemically defined medium and a spectrofluorometric assay for porphyrin composition. J Biol Chem 250:9215.

Grinstein M, Wintrobe MM (1948). Spectrophotometric micromethod for quantitative determination of free erythrocyte protoporphyrin. J Biol Chem 72:459.

Grossweiner LI, Patel AS, Grossweiner JB (1982). Type I and type II mechanisms in the photosensitized lysis of phosphatidylcholine liposomes by hematoporphyrin. Photochem Photobiol 36:159.

Harber LC, Bickers DR (1975). The porphyrias: basic science aspects, clinical diagnosis and management. Year Book of Dermatol: 9.

Henderson RW, Bellnier DA, Ziring B, Dougherty TJ (1983). Aspects of the cellular uptake and retention of hematoporphyrin derivative and their correlation with the biological response to PRT in vitro. In Kessel D, Dougherty TJ (eds): "Porphyrin Photosensitization," New York: Plenum Press, p 129.

Henderson RW, Christie GS, Clezy PS, Lineham J (1980). Haematoporphyrin diacetate: a probe to distinguish malignant from normal tissue by selective fluorescence. Brit J Exper Pathol 61:345.

Kessel D (1982). Components of hematoporphyrin derivatives and their tumor-localizing capacity. Cancer Res 42:1703.

Kessel D, Dougherty TJ (1983). "Porphyrin Photosensitization." New York: Plenum Press.

Kessel D, Rossi E (1982). Determinants of porphyrin-sensitized photooxidation characterized by fluorescence and absorption spectra. Photochem Photobiol 35:37.

Kinsey JH, Cortese DA, Moses HL, Ryan RJ, Branum EL (1981). Photodynamic effect of hematoporphyrin derivative as a function of optical spectrum and incident energy density. Cancer Res 41:5020.

Lowry OH, Rosebrough A, Farr AL, Randall RJ (1951). Protein measurement with the Folin phenol reagent. J Biol Chem 193:265.

Moan J, McGhie JB, Christensen T (1982). Hematoporphyrin derivative: photosensitizing efficiency and cellular uptake of its components. Photobiochem Photobiophys 4:337.

Moan J, Sommer S (1981). Fluorescence and absorption properties of the components of hematoporphyrin derivative. Photobiochem Photobiophys 3:93.

Moan J, Steen HB, Feren K, Christensen T (1981). Uptake of hematoporphyrin derivative and sensitized photoinactivation of C3H cells with different oncogenic potential. Cancer Lett 14:291.

Mossman BT, Gray MJ, Silberman L, Lipson L (1974). Identification of neoplastic versus normal cells in human cervical cell culture. J Obstet Gynaecol 43:635.

Piomelli S (1973). A micromethod for free erythrocyte porphyrins: the FEP test. J Lab Clin Med 81:932.

Profio AE, Doiron DR (1981). Dosimetry considerations in phototherapy. Med Phys 8:190.
Rimington C (1960). Spectral-absorption coefficients of some porphyrins in the Soret-band region. Biochem J 75:620.
Schwartz S (1981). Porphyrin photosensitization workshop. Washington DC (personal communication).
Spikes JD (1975). Porphyrins and related compounds as photodynamic sensitizers. Ann NY Acad Sci 244:496.
Tipping E, Ketterer B, Koskelo P (1978). The binding of porphyrins by ligandin. Biochem J 169:509.

From the desk of — Dr. Steve Powers

~~Draw~~

Picture of structure without heading etc.

fig 5 p 313.

---

p 422 fig 1 No headings.

p 423 Table 1 No headings.

Porphyrin Localization and Treatment of Tumors, pages 351–359

PHOTOSENSITIZER-PROTEIN CONJUGATES: POTENTIAL USE AS PHOTOIMMUNOTHERAPEUTIC AGENTS

Chi-Kit Wat, Ph.D.; Daphne Mew, B.Sc.
Julia G. Levy, Ph.D.; G.H.Neil Towers, Ph.D.
Department of Botany and Microbiology
University of British Columbia
Vancouver, B.C. Canada V6T 2B1

The photodynamic and tumor localizing properties of hematoporphyrin (Hp) and hematoporphyrin derivative (Hpd) are the rationales behind the use of these compounds in photoradiation therapy (PRT) of various malignant tumors in man and animal (Dougherty *et al.* 1978;Dougherty *et al.* 1981; Dahlman *et al.* 1983). Recently, rapid progress made in monoclonal antibody technology has added impetus to the use of specific antibodies as carriers of pharmacological agents in the medical field (Immunological Reviews vol. 62,1982). We have successfully combined these two concepts and chemically linked Hp to a monoclonal antibody directed against a tumor-specific antigen on the DBA/2J rhabdomyosarcoma M-1 in mice. We have shown that this anti-M-1-Hp conjugate is effective in the photohemolysis of human erythrocytes and, at the same time, retains its immunogenicity and photosensitizing properties towards the tumor cells *in vitro* and *in vivo*. Evidence has also been presented that the conjugate has a superior anti-cancer effect over Hp or the antibody alone and that this "magic bullet" technique allows the lowering of the minimum effective dose of Hp required for anti-cancer activity, hence minimizing adverse side effects caused by the photosensitizer (Mew *et al.* 1983). Thus, the combination of the target seeking ability of specific antibodies with the toxic effects of photosensitizers may prove to be an effective form of therapy for the treatment of tumor cells.

Damage to cell membranes (Kohn, Kessel 1979; Lamola, Doleiden 1980; Volden *et al.* 1981; Bellnier, Dougherty 1982; Moan *et al.* 1982), to DNA (Gutter *et al.* 1977; Boye, Moan

1980; Canti *et al.* 1981; Fiel *et al.* 1981) and to cell multiplication (Christensen 1981) by the photodynamic action of porphyrins has been well documented. Using red blood cells (RBC) as a membrane system, a comparative study on the hemolytic properties of Hp, Hpd and the conjugate was made.

Experimental details of the preparation of antibody-Hp conjugate, the methods for RBC isolation and conditions for hemolysis have been published (Mew *et al.* 1983). The composition of the assay system is described in the legend for each table and figure. The concentration of Hp in the conjugate was estimated by using the concentration curve of Hp (hematoporphyrin dihydrochloride, 95%, Sigma) measured at 505nm in PBS. This measurement is only an approximation because the Hp used is only 95% pure and, after the chemical manipulations during the synthesis of the conjugate, the final nature of the porphyrin is not known. Also, as demonstrated by TLC, the conjugate still has some Hp (approx. 10%) and two minor impurities adsorbed onto it (Mew *et al.* 1983).

The absorption maxima of Hp, Hpd and the conjugate used in this study are shown in Table 1. The relative optical density values in reference to the peak at 505 or 506nm of each compound are also given.

Table 1. Comparison of Absorption Maxima of Hp, Hpd and Conjugate in PBS and Their Relative Optical Densities

Hp	385	505	540	566	618nm
Rel O.D.	16	1	0.76	0.61	0.24
Hpd	370	506	539	569	620nm
Rel O.D.	14	1	0.77	0.68	0.29
Conjugate	390	506	539	572	624nm
Rel O.D.	9	1	0.68	0.58	0.26

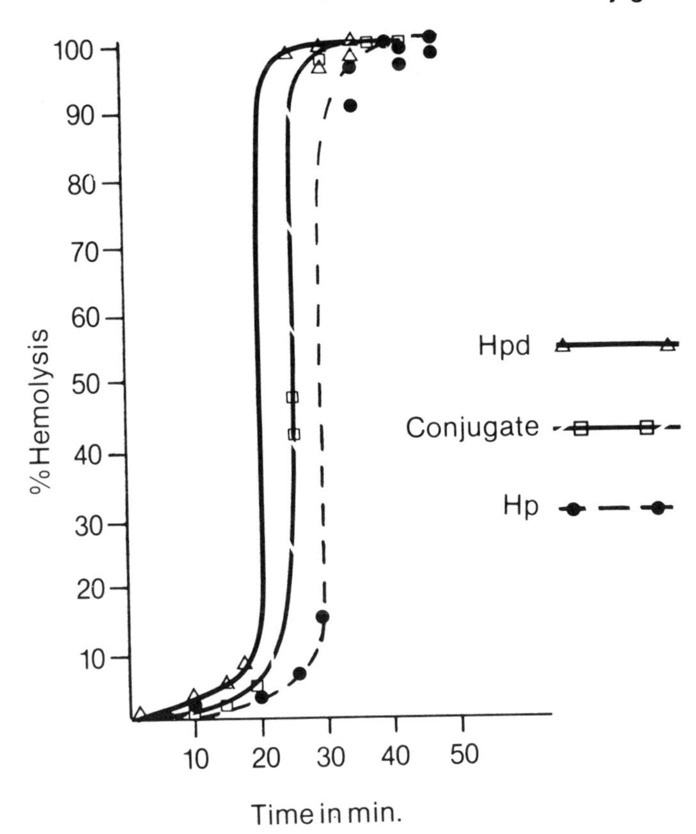

Fig. 1. Rates of Hemolysis of RBC

Assay system: RBC 200ul; photosensitizer 50ul
PBS 150ul
Stock conc. of photosensitizer: Hp 200ug/ml
Hpd 170ug/ml; Conjugate 200ug Hp/ml PBS
Light source: G.E. F20-T12-CW fluorescent lamps
intensity: 1.5mW/cm^2
Temperature: 34°
After irradiation, samples were centrifuged. 2X100ul supernatant removed. To each aliquot, 900ul Drabkin's solution added and O.D. measured at 540nm

The rates of hemolysis of RBC by these three compounds are shown in Fig. 1. The O.D. values at 505 or 506nm of the photosensitizers at concentrations used are comparable. It is obvious that the phototoxicity of these compounds decreases in this order: Hpd > conjugate > Hp. Hpd required 30min of irradiation for complete hemolysis whereas the conjugate and Hp required 35 and 40min respectively.

The efficiencies of Hp and Hpd in sensitizing human NHIK 3025 cells *in vitro* had been compared and it was found that Hpd had a sensitizing effect which was about twice that of Hp and that the least polar components of Hpd played the major role in this sensitization (Moan *et al.* 1982a). The situation becomes more complex when the various fractions and components of Hp and Hpd were purified and tested *in vitro* and *in vivo* (Kessel 1982; Kessel 1982a; Berenbaum *et al.* 1982; Moan *et al.* 1982a). In our experiment, the "Hp" in the conjugate is rendered very water soluble by being covalently bound to the antibody protein, and this conjugate is found to be more phototoxic than Hp. Also, the ratio of the O.D. at wavelengths 390:506nm for the conjugate is 9:1, whereas, the ratios for Hp and Hpd are 16:1 and 14:1 respectively (Table 1). The Soret band of "Hp" at 390nm in the conjugate seems to be suppressed and yet the hemolytic activity of the molecule as a whole is enhanced when compared with Hp. Further studies are required to explain this observation.

Singlet oxygen is most likely the major species of activated oxygen responsible for the cytotoxic action of photoactivated porphyrins (Weishaupt *et al.* 1976; Moan *et al.* 1979; Sandberg, Romslo 1981), though evidence for superoxide anion (Peterson *et al.* 1981) and hydroxyl radical production (Hariharan *et al.* 1980) has also been reported. Results from experiments carried out under nitrogen and with quenchers for various species of active oxygen (Bellus 1978; Fridovich 1978; Pryor 1980) are summarized in Table 2a and 2b. In samples kept under nitrogen and in control samples kept in dark, only negligible amounts of lysis occurred, thus demonstrating the absolute requirement of oxygen and light in the hemolysis of the RBC by the porphyrins and conjugate. Singlet oxygen quenchers, such as azide and histidine, delayed the onset of hemolysis. The presence of mannitol, a hydroxyl radical scavenger and superoxide dismutase (SOD). a scavenger for superoxide anion, were without effect (Table 2a). No D_2O effect was observed with Hp or

Table 2a. Comparison of % Hemolysis of RBC by Hp, Hpd and Conjugate

Condition	% Hemolysis		
	Hp	Hpd	Conjugate
Control	85-99	88-100	88-94
Nitrogen	0-1	2-3	2-4
SOD (50U)	95-100	98-99	90-95
Azide (10mM)	8-20	33-41	16-24
(25mM)	7-10	6-7	7-8
Histidine (10mM)	83-90	94-97	85-88
(20mM)	37-38	39-40	41-43
Mannitol (0.15mM)	93-103	88-89	87-88
(1.00mM)	94-101	84-95	98-100

Assay system: RBC 200ul;photosensitizer 50ul;quencher in PBS 50ul; PBS 100ul

Stock conc. of photosensitizer: Hp 200ug/ml; Hpd 113ug/ml
Conjugate 134ug Hp/1.1mg antibody/ml

Irradiation time: Hp 35min; Hpd 30min; Conjugate 35min

Table 2b. Comparison of % Hemolysis of RBC by Hp, Hpd and Conjugate

Condition	% Hemolysis		
	Hp	Hpd	Conjugate
Control + 0.5% Tween 80	86-90	15-17 (93-100)	86-91
BHT (0.5mM) + 0.5% Tween 80	84-87	15-32 (93-103)	78-79
DMF (10mM) + 0.5% Tween 80	0-1	1-2 (5-6)	2-3
α-Tocopherol (0.1mM) + 0.5% Tween 80	32-37	3-4 (77-81)	31-32

Assay system: same as Table 2a except quenchers were dissolved in PBS containing 0.5% Tween 80

Stock conc. of photosensitizer: same as Table 2a

Irradiation time: same as Table 2a. The values in parentheses under Hpd were determined after 40min irradiation

BHT : Butylated hydroxytoluene

DMF : 2,5-Dimethylfuran

with the conjugate either. Tween 80 was used to solubilize some of the quenchers. Its presence had no noticeable change on the rate of hemolysis sensitized by Hp or the conjugate, but with Hpd, the time for complete hemolysis was delayed by 10min (Table 2b). DMF, a hydroxyl radical and singlet oxygen scavenger, and α-tocopherol, a singlet oxygen and radical scavenger, did confer protection from lysis of the RBC. No protection was detected with BHT, a free radical quencher.

Thus, the above results showed that the conjugate prepared from hematoporphyrin and the anti-M-1 antibody retains the photosensitizing properties inherent in porphyrin molecules in general. Photohemolysis of RBC by the conjugate was shown to be mediated via singlet oxygen production, similar to the mechanism proposed for Hp and Hpd. In spite of a suppressed Soret band absorption at 390nm, its photohemolytic activity was found to be higher than that for Hp.

ACKNOWLEDGEMENTS - We thank Melinda Loo and Zyta Abramowski for excellent technical assistance and Donald MacRae for his generous donation of blood. We thank also Dr. Etsuo Yamamoto for helpful discussions on the hemolysis experiments. Hpd (Roswell Park Memorial Institute, Department of Drug Formulation and Development, Buffalo, N.Y.) was a gift from Dr. David McLean, Dermatological Oncology, Cancer Control Agency of B.C. This work was supported by grants from Cancer Research Society Inc. Montreal, B.C. Health Care Research Foundation Grant#75(82-2), NCI of Canada Grant 6048, MRC Grant 3411 and NSERC student funding.

REFERENCES

Bellnier DA, Dougherty TJ (1982). Membrane lysis in Chinese hamster ovary cells treated with hematoporphyrin derivative plus light. Photochem Photobio 36:43.

Bellus D (1978). Quenchers of singlet oxygen - a critical review. In Ranby B, Rabek JF (eds): "Singlet Oxygen: Reactions with Organic Compounds and Polymers," New York: John Wiley & Sons, p 61.

Berenbaum MC, Bonnett R, Scourides PA (1982). *In vivo* biological activity of the components of hematoporphyrin derivative. Br J Cancer 45:571.

Boye E, Moan J (1980). The photodynamic effect of hematoporphyrin on DNA. Photochem Photobiol 31:223.
Canti G, Marelli O, Ricci L, Nicolin A (1981). Haematoporphyrin-treated murine lymphocytes: *in vitro* inhibition of DNA synthesis and light-mediated inactivation of cells responsible for GVHR. Photochem Photobiol 34:589.
Christensen T (1981). Multiplication of human NHIK 3025 cells exposed to porphyrins in combination with light. Br J Cancer 44:433.
Dahlman A, Wile AG, Burns RG, Mason GR, Johnson FM, Berns MW (1983). Laser photoradiation therapy of cancer. Cancer Res 43:430.
Dougherty TJ, Kaufman JE, Goldfarb A, Weishaupt KR, Boyle D, Mittleman A (1978). Photoradiation therapy for the treatment of malignant tumors. Cancer Res 38:2628.
Dougherty TJ, Thoma RE, Boyle DG, Weishaupt KR (1981). Intersitial photoradiation therapy for primary solid tumors in pet cats and dogs. Cancer Res 41:401.
Fiel RJ, Datta-Gupta N, Mark EH, Howard JC (1981). Induction of DNA damage by porphyrin photosensitizers. Cancer Res 41:3543.
Fridovich I (1978). The biology of oxygen radicals. Science 201:875.
Gutter B, Speck WT, Rosenkranz HS (1977). The photodynamic modification of DNA by hematoporphyrin. Biochim Biophys Acta 475:307.
Hariharran PV, Courtney J, Eleczko S (1980). Production of hydroxyl radicals in cell systems exposed to haematoporphyrin and red light. Int J Radiat Biol 37:691.
Kessel D (1982). Components of hematoporphyrin derivatives and their tumor-localizing capacity. Cancer Res 42:1703.
Kessel D (1982a). Determinants of hematoporphyrin-catalyzed photosensitization. Photochem Photobiol 36:99.
Kohn K, Dessel D (1979). On the mode of cytotoxic action of photoactivated porphyrins. Biochem Pharm 28:2465.
Lamola AA, Doleiden FH (1980). Cross-linking of membrane proteins and protoporphyrin-sensitized photohemolysis. Photochem Photobiol 31:597.
Mew D, Wat CK, Towers GHN, Levy JG (1983). Photoimmunotherapy: Treatment of animal tumors with tumor-specific monoclonal antibody-hematoporphyrin conjugate. J Immunol. 130:1473.
Moan J, Christensen T, Sommer S (1982a). The main photosensitizing components of hematoporphyrin derivative. Cancer Lett 15:161.
Moan J, McGhie JB, Christensen T (1982). Hematoporphyrin

derivative: Photosensitizing efficiency and cellular uptake of its components. Photobiochem Photobiophys 4:337.
Moan J, Pettersen EO, Christensen T (1979). The mechanism of photodynamic inactivation of human cells *in vitro* in the presence of haematoporphyrin. Br J Cancer 39:398.
Peterson DA, McKelvey S, Edmondson PR (1981). A hypothesis for the molecular mechanism of tumor killing by porphyrins and light. Med Hypotheses 7:201.
Pryor WA (1980). Methods of detecting free radicals and free radical-mediated pathology in environmental toxicology. In Bhatnagar RS (ed): "Molecular Basis of Environmental Toxicity," Ann Arbor: Ann Arbor Science, p 3.
Sandberg S, Romslo I (1981). Porphyrin-induced photodamage at the cellular and the subcellular level as related to the solubility of the porphyrin. Clin Chim Acta 109:193.
Volden G, Christensen T, Moan J (1981). Photodynamic membrane damage of hematoporphyrin derivative-treated NHIK 3025 cells *in vitro*. Photobiochem Photobiophys 2:105.
Weishaupt KR, Gomer CJ, Dougherty TJ (1976). Identification of singlet oxygen as the cytotoxic agent in photo-inactivation of a murine tumor. Cancer Res 36:2326.

Porphyrin Localization and Treatment of Tumors, pages 361–371

PHOTODYNAMIC INACTIVATION OF CULTURED BLADDER TUMOR CELLS: A PRELIMINARY STUDY OF THE EFFECTS OF PORPHYRIN AGGREGATION

David A. Bellnier and Chi-Wei Lin

Urology Service and Department of Surgery
Massachusetts General Hospital, Harvard Medical School, Boston, MA 02114

INTRODUCTION

Hematoporphyrin derivative (HpD) is taken up and concentrated against an external gradient by cultured mammalian cells (Gomer, Smith 1980). Subsequent irradiation by visible light leads to cellular inactivation (Sery 1979). The molecular mechanism leading to phototoxicity, though not conclusively established, is likely the production of singlet oxygen ($^1O_2$) as an active intermediate (Weishaupt *et al.* 1976). Uncertainty also arises as to the target(s) responsible for inactivation, although damage to nuclear material (Gomer 1980, Moan *et al.* 1980) and the cell membrane (Spikes 1975, Kessel 1977, Bellnier, Dougherty 1982) have been reported.

We are currently investigating the HpD-mediated photoinactivation of a human bladder carcinoma cell line (MGH-U1), in particular the influence of intracellular porphyrin aggregation upon *in vitro* survival response. Porphyrins tend to self-associate (aggregate) as a function of solubility, pH and concentration in aqueous and organic solvents (Brown *et al.* 1980). For example, Dougherty *et al.* (1983) has reported that HpD in saline (pH $\sim$7) at a concentration of 5 mg/ml was comprised of 40-50% aggregate.

Aggregated porphyrins are inefficient photodynamic sensitizers. Moan *et al.* (1983) have shown that the relative $^1O_2$ yield for their HPLC fraction #7 of HpD, undoubtedly a porphryin aggregate, was at least an order of magnitude less than other monomer HPLC components. Further, Jori's group (1980) have reported that porphyrin dimers produce $^1O_2$ maximally. Hematoporphyrin (Hp) aggregate is a poor Type II photodynamic sensitizer ($^1O_2$-involving) in phosphatidylcholine liposomes (Grossweiner _et al_. 1982).

To examine this phenomena in our system, we have treated MGH-U1 cells with various concentrations of HpD which resulted in constant levels of intracellular porphryin, but also existed in different aggregation states, as determined through absorbance and fluorescence spectroscopy. Relative cloning efficiences, following red light (> 590 nm) exposure, were determined for these cells. Preliminary data are presented here.

MATERIALS AND METHODS

MGH-U1 was established from a human bladder carcinoma as previously described (Kato _et al_. 1978). Cells were grown in monolayer in McCoy's 5a medium supplemented with 5% fetal calf serum (FCS) at 37°C with a gas phase of 95% air/5% CO_2.

Hematoporphyrin derivative (5 mg/ml) was purchased from ORD, Inc., Cheektowaga, NY. Tritium-labeled HpD (^{3}H-HpD, 2.4 Ci/g) was prepared from the hematoporphyrin acetates by catalytic exchange (New England Nuclear, Boston, MA) followed by basic hydrolysis according to Gomer and Dougherty (1979). High and low molecular weight components of HpD (HMWF and LMWF, respectively) were isolated by polyacrylamide gel column chromatography as previously described (Dougherty _et al_. 1983). Briefly, 0.5 ml of ^{3}H-HpD was layered on a 1 x 27 cm glass column packed with P-60 gel (Bio-Rad, Richmond, CA) dissolved in 0.05 M PBS, pH 7.2. The separation of HpD into distinct bands was easily seen with the naked eye. Components were collected and diluted in PBS to yield a concentration of 2×10^{-5} M, as determined with tritium counting.

To determine HpD uptake by cells, monolayer cultures ($\sim 5 \times 10^6$/100 mm dish) were incubated with various concentrations of HpD/^{3}H-HpD (10:1 molar ratio) and FCS. At specified times (Fig. 1) individual dishes were washed three times with PBS and cells detached manually. Aliquots were removed for determination of protein by Lowry _et al_. (1951) and HpD by liquid scintillation counting of tritium activity. In one series of experiments cells were first incubated with HpD (50 µg/ml, 5% FCS) for 21 h, after which they were refed with HpD-free medium containing 5% FCS (porphyrin efflux conditions, see Henderson _et al_. 1983). Cellular HpD levels were determined as above.

Two uptake conditions and a single efflux condition were used throughout the remainder of this study, specifically: (1) 30 µg HpD/ml + 1% FCS, 2 h incubation; (2) 50 µg HpD/ml + 5% FCS, 8 h incubation; (3) 50 µg HpD/ml + 5% FCS, 21 h incubation followed by 0 µg HpD/ml + 5% FCS, 8 h incubation (efflux). These conditions resulted in equivalent intracellular porphyrin concentrations (Fig. 1, dashed lines).

Absorbance spectra for intact cells containing HpD were recorded on a Beckman UV 5270 spectrophotometer fitted with a 15 cm diameter $BaSO_4$-coated integrating sphere. Fluorescence excitation and emission spectra (not shown) were recorded on a Perkin-Elmer MPF 44E fluorescence spectrophotometer (λ_{ex}= 399 nm, λ_{em}= 612.5 nm). Relative yields of fluorescence were calculated by integration of the area under each emission peak. For comparison, absorbance and fluorescence spectra were determined for 2×10^{-5} M solutions of HpD, HMWF and LMWF.

Single, attached cells were pretreated with HpD, as above, and irradiated with graded fluences of red (> 590 nm) light. Illumination source was a bank of 8 fluorescent bulbs (GE Softwhite, 20 W/bulb) filtered with a single ruby-colored plastic sheet (Hyatt's Graphic Supply, Buffalo, NY) and three layers of small mesh wire screening (neutral density filter). The HpD-containing medium was removed prior to irradiation and cultures were refed with full growth medium immediately after. The cells were then incubated 6-8 days before staining and counting of colonies.

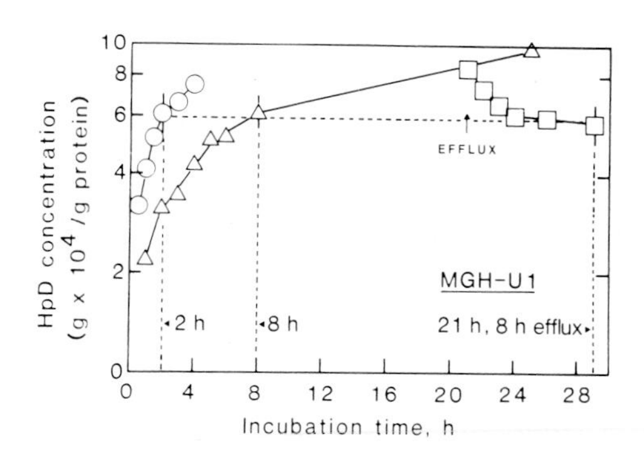

Figure 1. HpD uptake and efflux in MGH-U1 cells. Concentrations were determined with ^{3}H-HpD. (○,30 µg HpD/ml + 1% FCS; △,50 µg HpD/ml + 5% FCS; □,50 µg HpD/ml + 5% FCS, 21 h, then 0 µg/ml HpD + 5% FCS)

RESULTS AND DISCUSSION

HpD accumulation was biphasic and dependent on both the porphyrin and serum concentration in the external medium (Fig. 1). Most of the porphyrin was taken up during the initial few hours, followed by a much slower, plateau-like pattern out to 24 h. When cultures were preincubated with 50 µg HpD/ml + 5% FCS for 21 h and subsequently refed with serum-rich, HpD-free medium, there was a rapid efflux of porphyrin from the cells; by 6 h no further loss was observed. We have termed the remaining, cell-bound porphyrin the 'non-exchangeable' component. The rapid loss of HpD from the cells likely reflects the strong binding capacity of porphyrins to serum proteins (Tsutsui *et al.* 1975).

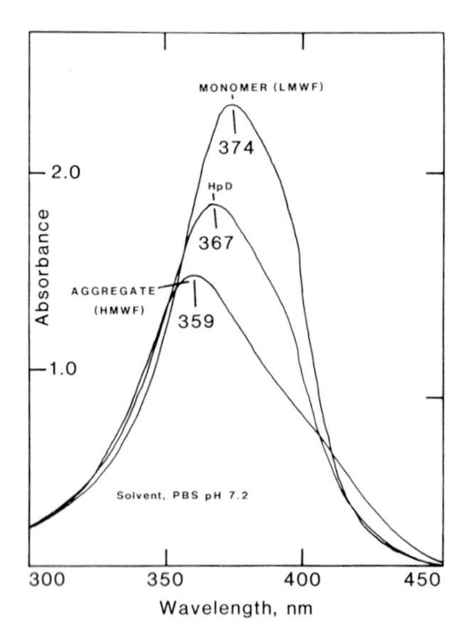

Figure 2. Absorbance spectra of polyacrylamide gel chromatography (P-60) fractions of HpD. Solvent, 0.05 M PBS, pH 7.2. Sensitizer concentration, 2×10^{-5} M.

The absorbance spectra for 2×10^{-5} HpD showed a maximum at 367 nm (Fig. 2) An equimolar concentration of HMWF had a absorbance maximum at 359 nm. For LMWF the Soret peak was 374 nm. The shift in absorbance maxima for LMWF, HpD and HMWF apparently reflects different aggregation status. If serum was added to porphyrin solutions the absorbance maxima shifted towards the red (Table 1); however, the characteristic

Table 1. Absorbance and fluorescence characteristics of porphyrin standards.

Porphyrin[a]	Solvent[b]	λ_{abs}^{max} (nm)	OD[c]	Φ_F^d	$\Phi_{^1O_2}^e$
HpD	PBS, pH 7.2	367	1.38	0.024	0.54
	PBS + BSA	385	1.40	0.026	----
	PBS + FCS	398	----	-----	----
LMWF	PBS, pH 7.2	374	2.49	0.044	1.10
	PBS + BSA	387	2.34	0.047	----
	PBS + FCS	401	----	-----	----
HMWF	PBS, pH 7.2	359	1.06	0.005	0.21
	PBS + BSA	371	1.16	0.014	----
	PBS + FCS	370	----	-----	----

[a] porphyrin concentration 2×10^{-5} M, in phosphate buffered saline (PBS)
[b] BSA, bovine serum albumin (2×10^{-5} M); FCS, fetal calf serum (50%, v/v)
[c] optical density at absorbance maximum
[d] determined from anthracene, fluorescein and Hp standards
[e] relative to 2×10^{-5} M Hp

Table 2. Spectral properties of cell-bound porphyrin.

HpD uptake conditions[a]	Φ_F[b]	λ_{abs}^{max} (nm)	OD_{Soret}[c]	$OD_{365\ nm}$	OD_{Soret}/OD_{365}
30 μg HpD/ml + 1% FCS, 2 h incubation	1.00	365,400	0.050	0.049	1.02
50 μg HpD/ml + 5% FCS, 8 h incubation	1.89	403	0.070	0.050	1.40
50 μg HpD/ml + 5% FCS, 21 h incubation, then 0 μg HpD/ml + 5% FCS, 8 h incubation	2.27	405	0.075	0.050	1.50

[a] see Fig. 1; intracellular porphyrin concentration, ~6 x 10^{-4} g HpD/g cellular protein

[b] relative to 2 h uptake (30 μg HpD/ml + 1% FCS, 2 h)

[c] optical density at approximately 400 nm

spectral differences (optical density, Soret peak) remain. These data suggest that large aggregates may remain stable even in the presence of high serum protein concentrations; e.g. the cellular environment. Quantum fluorescent yields (relative to anthracene, fluorescein and Hp standards, see Moan et al. 1983) were 0.044, 0.024 and 0.005 for LMWF, HpD and HMWF, respectively. When serum albumin was added the fluorescence yields increased slightly for LMWF and HpD and about 3-fold for HMWF, suggesting some monomerization of HpD aggregates. This is supported by the small increase in OD for HMWF after albumin addition (Table 1). Singlet oxygen yields (relative to 2 x 10^{-5} M Hp) shown in Table 1 indicate that HMWF is a poor Type II sensitizer when compared to LMWF or HpD at this concentration.

Porphyrin absorbance and fluorescence spectra were obtained for intact MGH-U1 cells (Fig. 3, Table 2). After 2 h uptake, the absorbance spectrum showed two, equivalent maxima at 365 nm and 400 nm (OD = 0.05 and 0.049, respectively). Following the 8 h uptake, λ_{abs}^{max} = 403 nm (OD = 0.07). Cellular porphyrin after 21 h uptake and 8 h efflux had a single maximum at 405 nm with an OD of 0.075. The absorbance maximum of 365 nm at 2 h probably reflects the presence of aggregated porphyrins within the cell; single Soret peaks at ∿400 nm suggest the presence of porphyrin monomers or small aggregates complexed to cellular protein. We have expressed the relative amount of cellular porphryin monomer to aggregate as the ratio $OD_{(Soret \sim 400\ nm)}/OD_{365\ nm}$ (Table 2).

Fluorescence yields for these cells, relative to 2 h uptake, were 1.0, 1.89 and 2.27 for 2 h, 8 h and 21 h + 8 h, respectively (Table 2). These data further suggest the presence of porphryin aggregates within the cells.

Fluence response curves for these cells are shown in Fig. 4. Photosensitivity corresponded quite well with both the fluorescence yields and $OD_{Soret}/OD_{365\ nm}$ ratios (compare Table 2, Fig. 4).

SUMMARY

This study was designed to determine the influence, if any, of intracellular porphyrin aggregation upon *in vitro* photosensitivity. HpD uptake conditions were manipulated in such a way as to yield three experimental groups, each containing equal levels of cellular porphyrins but having

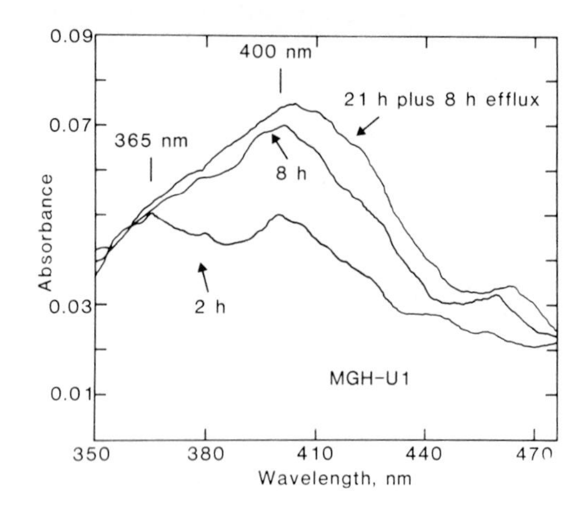

Figure 3. Absorbance spectra of intact MGH-U1 cells treated with HpD (Fig. 1). 2 h, 2 hour uptake; 8 h, 8 h uptake; 21 h plus 8 h, 21 h uptake followed by 8 h efflux conditions. Representative spectra were recorded on a spectrophotometer fitted with a $BaSO_4$-coated integrating sphere to increase resolution.

different porphryin aggregation status. Aggregation status was determined through comparison of the spectral properties of intact cells. Subsequent exposure to red light resulted in a family of survival curves which generally corresponded to cellular HpD fluorescence yields and optical densities (indicators of aggregation status). These data suggest that porphyrin aggregation within the cell may be partially responsible for the observed differential photosensitivity following various uptake conditions.

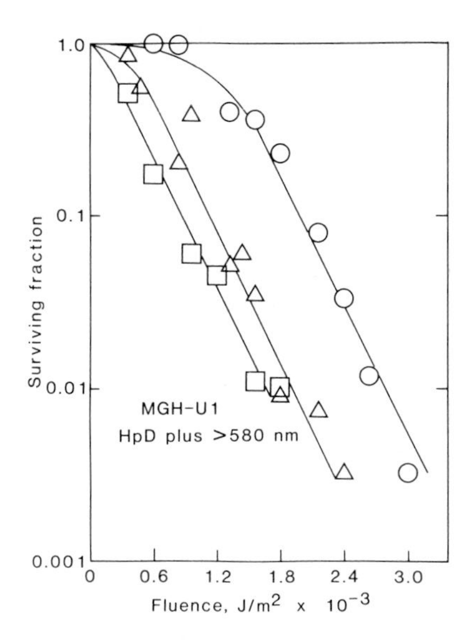

Figure 4. Relative cloning efficiency of MGH-U1 cells treated with HpD. Cells were irradiated with graded fluences of red light. Values were reproducible within 30%. Plating efficiencies of unirradiated controls were approximately 40%. (□, 50 µg HpD/ml + 5% FCS, 21 h, 0 µg HpD/ml + 5% FCS, 8 h; △, 50 µg HpD/ml + 5% FCS, 8 h; ○, 30 µg HpD/ml + 1% FCS, 2 h).; Fluence rate, 4 W/m^2.

REFERENCES

Bellnier DA, Dougherty TJ (1982). Membrane lysis in Chinese hamster ovary cells with hematoporphyrin derivative plus light. Photochem Photobiol 36:43.

Brown SB, Hatzikonstantinou H, Herries DH (1980). Int J Biochem 12:701.

Dougherty TJ, Boyle DG, Weishaupt KR, Henderson BA, Potter WR, Bellnier DA, Wityk KE (1983). In Kessel D, Dougherty TJ (ed): "Porphyrin Photosensitization", New York: Plenum Press, p 3.

Gomer CJ (1980). DNA damage and repair in CHO cells following hematoporphyrin photoradiation. Cancer Letts 11:161.

Gomer CJ, Dougherty TJ (1979). Determination of (^{3}H)- and (C-14) hematoporphyrin derivative distribution in malignant and normal tissue. Cancer Res 39:146.

Gomer CJ, Smith DM (1980). Photoinactivation of Chinese hamster cells by hematoporphyrin derivative and red light. Photochem Photobiol 32:341.

Grossweiner LI, Patel AS, Grossweiner JB (1982). Type I and Type II mechanisms in the photosensitized lysis of phosphatidylcholine liposomes by hematoporphyrin. Photochem Photobiol 36:159.

Henderson BA, Bellnier DA, Ziring B, Dougherty TJ (1983). Aspects of the cellular uptake and retention of hematoporphyrin derivative and their correlation with the biological response to PRT in vitro. In Kessel D, Dougherty TJ (ed): "Porphyrin Photosensitization", p 129.

Jori G, Reddi E, Rosani E, Cozzani I, Tomio L, Zorat PL, Pizzi GB, Calzavara F (1980). Porphyrin sensitized photoreactions and their use in cancer therapy. Med Biol Environ 8:141.

Kato T, Ishikawa K, Nemoto R, Senoo A, Amano Y (1978). Morphological characterization of two cell lines, T-24 and MGH-U1, derived from human urinary bladder carcinoma. Tohoku J Exp Med 124:339.

Kessel D (1977). Effects of photoactivated porphyrins at the cell surface of leukemia L1210 cells. Biochem 16:3443.

Lowry OH, Rosebrough NJ, Farr AL, Randall RJ (1951). Protein measurements with a Folin phenol reagent. J Biol Chem 193:265.

Moan J, Waksvik H, Christensen T (1980). DNA single strand breaks and sister chromatid exchanges induced by treatment with hematoporphyrin and light or by x-rays in human cells of the line NHIK 3025. Cancer Res 40:2915.

Moan J, Sandberg S, Christensen T, Elander S (1983). Hematoporphyrin derivative, chemical composition, photochemical and photosensitizing properties. In Kessel D, Dougherty TJ (ed): "Porphyrin Photosensitization", p 165.

Sery TW (1979). Photodynamic killing of retinoblastoma cells with hematoporphyrin and light. Cancer Res 39:96.

Spikes JD (1975). Porphyrins and related compounds as photodynamic sensitizers. Ann NY Acad Sci 244:496.

Tsutsui M, Carrano C, Tsutsui EA (1975). Tumor localizers: porphyrins and related compounds (unusual metalloporphyrins XXIII). In Adler AD (ed): "The Biological Role of Porphyrins and Related Structures", NY Acad Sci.

Weishaupt KR, Gomer CJ, Dougherty TJ (1976). Identification of singlet oxygen as the cytotoxic agent in photoinactivation of a murine tumor. Cancer Res 36:2326.

Porphyrin Localization and Treatment of Tumors, pages 373–379

PHOTOPHYSICAL AND PHOTOSENSITIZING PROPERTIES OF HEMATOPORPHYRIN BOUND WITH HUMAN SERUM ALBUMIN

E. Reddi, M.A.J. Rodgers, G. Jori

CFKR, University of Texas at Austin,
Istituto Biologia Animale,
University of Padova, Italy

Serum proteins are responsible for porphyrin transport in the bloodstream (Koskelo, Muller-Eberhard 1979) and control the endocellular concentration of hematoporphyrin (HP) and related porphyrins (Chang, Dougherty 1978; Jori et al. 1979). Conceivably, protein-bound porphyrins play a major role in photosensitized processes occuring in vivo and affect the efficiency of photoinduced cell damage. However, only few investigations have been focused on the photophysical and photosensitizing properties of porphyrins which have been complexed in the dark with macromolecular structures (Jori et al. 1983; Richard et al. 1983). In this paper, we report our findings on the HP-human serum albumin (HSA) system in neutral aqueous solution. Previous investigations (Reddi et al. 1981) indicate that HSA possesses one high-affinity site for HP with a stability constant of about 10^{6} M^{-1}; its distance from the unique tryptophyl residue of HSA is 1.7 nm, as estimated by resonance electronic energy transfer. For comparison, the photooxidation of a tryptophan derivative blocked at the α-amino and α-carboxyl groups (N-acetyl-L-tryptophanamide, NATA) sensitized by free HP has been studied under identical experimental conditions.

EXPERIMENTAL PROCEDURE

All experiments were performed in 0.05

M-phosphate-buffered aqueous solutions, pH 7.4. HP (Porphyrin Products, Logan, Utah) was either free or complexed with HSA (Sigma, fraction V). In photophysical studies, 8-17 μM HP solutions, eventually containing 20 μM HSA, were photo-excited at either 355 or 532 nm by a Quantel YG 481 Q-switched Nd:YAG laser. The transient decays were observed at 460 nm for N_2-, air- and O_2-saturated solutions using a computer-controlled kinetic spectrophotometer. The decay of singlet-oxygen, generated by energy transfer from photo-excited HP, was monitored in D_2O solutions by its luminescence emission at 1.27 μm. The quantum yields for production of triplet HP and singlet-oxygen and the rate constant for the HSA-singlet oxygen reaction were estimated as described elsewhere (Reddi et al. 1983).

In photokinetic studies, 3 ml of a 5-9 μM air-equilibrated HP solution containing 10 μM HSA or NATA were exposed to 555 nm-light from a 1,200 W halogen lamp; the optical path was 1 cm. The decrease of tryptophan concentration in both substrates was followed spectrophotofluore-metrically (Sconfienza et al. 1980) using Perkin-Elmer MPF 4 apparatus. The kinetic analysis was restricted to short irradiation times to minimize artifacts due to photodecomposition of HP.

RESULTS

Photophysical Studies

Laser irradiation (355 or 532 nm) of HSA-bound HP in N_2-saturated aqueous solution at pH 7.4 originates an HP triplet signal which decays by first-order kinetics with a rate constant of $1.2 \cdot 10^3$ s^{-1}. The triplet-triplet spectrum closely resembles those reported for monomeric porphyrins (Bonnett et al. 1980; Reddi et al. 1983) in agree-ment with the observed monomerization of HP upon binding with HSA (Reddi et al. 1981). Some pro-perties of protein-bound triplet HP are collected in Table 1 and compared with the corresponding properties of the triplet state of free HP under identical experimental conditions (Reddi et al. 1983).

Although free and HSA-bound HP appear to have essentially identical triplet quantum yields, the bound porphyrin exhibits both a remarkably longer triplet lifetime and a reduced rate constant for triplet quenching by oxygen. The photophysical properties of bound HP also appear to depend on the excitation wavelengths as previously found for

Table 1
Photophysical properties of free and albumin-bound 17 μM hematoporphyrin and neutral aqueous solution at pH 7.4.

Parameter	Hematoporphyrin (a) free	bound
$\Delta\varepsilon_{T-T}$ at 460 nm	4,200 $M^{-1}cm^{-1}$	5,000 $M^{-1}cm^{-1}$
Triplet quantum yield	0.63	0.66
Triplet decay rate constant	$1.0 \cdot 10^4\ s^{-1}$ (b)	$1.2 \cdot 10^3 s^{-1}$
	$4.5 \cdot 10^5\ s^{-1}$ (c)	$4.3 \cdot 10^4 s^{-1}$
Rate constant for O_2-quenching of the triplet	$1.8\ 10^9\ M^{-1}s^{-1}$	$1.7\ 10^8 M^{-1}s^{-1}$

(a) Data taken from Reddi et al. (1983).

(b) N_2-saturated solutions.

(c) Air-saturated solutions. For HSA-bound HP we observed also an independently decaying long-lived transient (decay rate constant = $1.7 \cdot 10^4\ s^{-1}$) upon 355 nm-excitation.

free HP (Reddi et al. 1983), 355 nm -(but not 532 nm)- irradiation of air-saturated solutions of the HSA-HP complex causes the formation of a long- lived (τ = 58 μs) transient absorbing near 700 nm which is not quenched by oxygen; probably, the long-lived species is the radical cation HP^+ arising from photoionization of HP. Changing the HP concentration or the HP/HSA molar ratio has no effect on the relative yields of the two transients.

Luminescence studies in the infrared region after photoexcitation of protein-bound HP demonstrate the formation of singlet-oxygen: the quantum yield (0.2) is almost identical with that found for the generation of singlet-oxygen by free 17 μM HP in neutral aqueous solution (Reddi et al. 1983).

Photosensitization Studies

Under our experimental conditions for irradiation, the tryptophyl moiety of both HSA and NATA undergoes photooxidative modification according to first-order kinetics at least down to 70% of the original substrate concentration (Fig. 1).

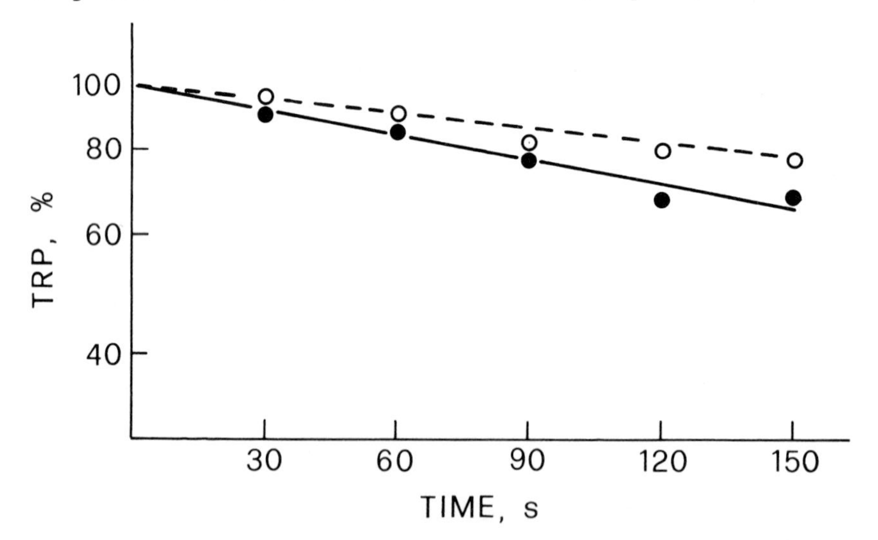

Fig.1. HP-sensitized photooxidation of tryptophan in NATA (open circles) and HSA (closed circles).

Similar observations were reported for the photo-oxidation of L-tryptophan sensitized by free or micelle-incorporated HP (Sconfienza et al. 1980). The first-order rate constants, as deduced from the slopes of the semilogarithmic plots, are $1.35 \cdot 10^{-3} s^{-1}$ and $2.88 \cdot 10^{-3} s^{-1}$, respectively, for NATA and HSA. Amino acid analysis of a sample of HSA, which had been irradiated for 10 min., indicates the photo-destruction of amino acid residues other than tryptophan, especially histidines and tyrosines.

DISCUSSION

Our data indicate that HSA-bound HP is about twofold more efficient than free HP in promoting the photooxidation of tryptophyl residues. Analogously, Richards et al. (1983) estimated that the quantum yield for the HP-sensitized photomodification of HSA is about four-fold higher than that for the HP-sensitized photoinactivation of subtilisin BPN' (the porphyrin does not bind with the latter protein). The greater photosensitizing activity of dyes complexed with macromolecules is a general phenomenon (Jori, Spikes 1981) and has been ascribed to a change of the photophysical properties of the dye, including a higher steady-state triplet population and longer triplet lifetime. In our case, the former effect is ruled out by the coincidence of the triplet quantum yields for free and HSA-bound HP. On the other hand, the longer triplet lifetime potentially endows bound HP with a greater reactivity in promoting photosensitized processes. Actually, in spite of the low rate constant for the interaction of O_2 with triplet HP, the fraction of total triplets quenched by O_2 approaches 0.97, i.e. the same value calculated for O_2-quenching of free HP triplet (Reddi et al. 1983). Hence, the maximal yield for photo-ionization of triplet HP leading to the formation of $HP^{\dagger}$ is 3%. The latter transient should play no important role in the photosensitized inactivation of HSA; it is worth emphasizing that the ratio between the rate constants for tryptophan modification in HSA and NATA is unchanged whether white light or 555 nm-light is used for irradiation. $HP^{\dot{+}}$ is not generated by 555 nm-excitation of the HSA-HP complex.

These considerations strongly suggest that singlet oxygen is the main or unique reactive intermediate involved in the photooxidation of the HSA tryptophyl residue. Similar conclusions have been drawn as regards the free HP-sensitized photooxidation of L-tryptophan (Reddi, Rodgers, Jori unpublished results). Consequently, the greater photosensitizing efficiency of HSA-bound HP should not be due to either an alteration of some photophysical properties of the complexed porphyrin or a change in the mechanism of the photoprocesses. Rather, one must take into consideration the rate constant for the reaction of singlet oxygen with the tryptophyl side chain of HSA; such a parameter is affected by the nature of the reaction medium (Reddi, Rodgers unpublished results). We determined a rate constant of 7.10^{8} M^{-1} s^{-1} for the overall reaction of singlet oxygen with the photosensitive amino acid residues of HSA as compared with a value of $7.2 \cdot 10^{7}$ $M^{-1}s^{-1}$ for the tryptophan-singlet oxygen reaction in neutral aqueous solution. We are presently extending our investigations to other HSA-dye complexes to elucidate this point in a more definite way.

Acknowlegements

The laser flash experiments and data analyses were carried out at the Center for Fast Kinetics Research at the University of Texas at Austin. The CFKR is supported jointly by NIH Grant RR00886 from the Biotechnology Branch of the Division of Research Resources and by the University of Texas at Austin. Partial support for this project came from NIH Grant GM24235 (M.A.J. Rodgers) and CNR Grant 82.01017.98 (G. Jori).

REFERENCES

Bonnett R, Charalambides AA, Land EJ (1980). Triplet states of porphyrin esteres. JCS Faraday I 76:852.

Chang CT, Dougherty TJ (1978). Photoradiation therapy. Kinetics and thermodynamics of porphyrin uptake and loss in normal and malignant cells in culture. Rad Res 74:498.

Jori G, Reddi E, Tomio L, Calzavara F (1983). Factors governing the mechanism and efficiency of porphyrin-sensitized photooxidations in homogeneous solutions and organized media. In Kessel D, Dougherty TJ (eds): "Porphyrin Photosensitization", New York: Plenum Press, p 193.

Jori G, Reddi E, Tomio L, Salvato B, Zorat PL, Calzavara F (1979). Time-course of hematoporphyrin distribution in selected tissues of normal rats and in ascites hepatoma. Tumori 65:43.

Jori G, Spikes JB (1981). Photosensitized oxidations in complex biological structures. In Rodgers MAJ, Powers EL (eds): "Oxygen and Oxy-radicals in Chemistry and Biology", New York: Academic Press, p 441.

Koskelo P, Muller-Eberhard U (1979). Interaction of porphyrins. Seminars Hematol 45:221.

Reddi E, Jori G, Rodgers MAJ, Spikes JD (1983). Flash photolysis studies of hemato- and copro-porphyrins in homogeneous and microheterogeneous aqueous dispersions. Photochem Photobiol, submitted for publication.

Reddi E, Ricchelli F, Jori G (1981). Interaction of human serum albumin with hematoporphyrin and its Zn^{2+}- and Fe^{3+}- derivatives. Int J Peptide Protein Res 18:402.

Richard P, Blum A, Grossweiner LI (1983). Hematoporphyrin photosensitization of serum albumin and subtilisin BPN'. Photochem Photobiol 37:287.

Sconfienza C, Van de Vorst A, Jori G (1980). Type I and type II mechanisms in the photooxidation of tryptophan and tryptamine sensitized by hematoporphyrin in the presence and in the absence of sodium dodecyl sulphate micelles. Photochem Photobiol 31:351.

Porphyrin Localization and Treatment of Tumors, pages 381–390

MULTICELLULAR SPHEROIDS AS AN *IN VITRO* MODEL SYSTEM FOR PHOTORADIATION THERAPY IN THE PRESENCE OF Hpd

Terje Christensen, Johan Moan, Torbjørg Sandquist and Lars Smedshammer

Departments of Biophysics and Tissue Culture, Norsk Hydro's Institute for Cancer Research, Montebello, Oslo 3, Norway

INTRODUCTION

Photoradiation therapy (PRT) is a new approach to cancer treatment, and its mode of action is therefore largely unknown. Cell cultures have been important experimental model systems in the preclinical studies performed to date. A significant amount of information about cellular effects has been obtained by exposing single cells to porphyrins and light *in vitro*. It is, however, questionable how relevant many of these studies are for PRT *in vivo*, since the physiological conditions in tumor tissue seem to be of major importance both for the specificity of Hpd retention and for the tumor destruction (Bugelski et al. 1981).

In order to get a system that resembles the physiology of a tumor to a certain degree, multicellular spheroids have been developed as a tumor model system. The use of spheroids in preclinical studies of PRT will be discussed in this presentation.

CHARACTERIZATION OF THE SPHEROIDS

The method for preparing spheroids of NHIK 3025 cells has been developed by Wibe (Wibe 1980, Wibe et al. 1981, Wibe, Oftebro 1981). Briefly, aggregates of 50 - 100 cells were formed by tilting cell suspensions for a few hours. The aggregates were inoculated in tissue culture flasks, the bottoms of which were coated with 0.4 mm of 1% agar. During

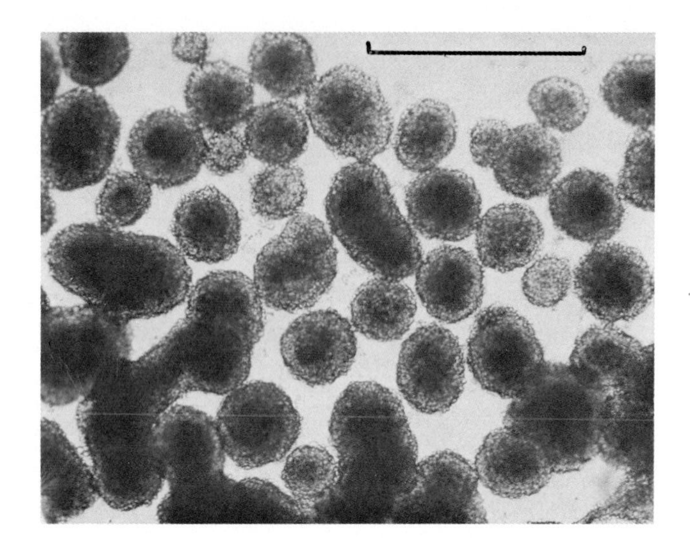

Figure 1: Spheroids in the culture flask after 18 days of cultivation. Bar 1 mm.

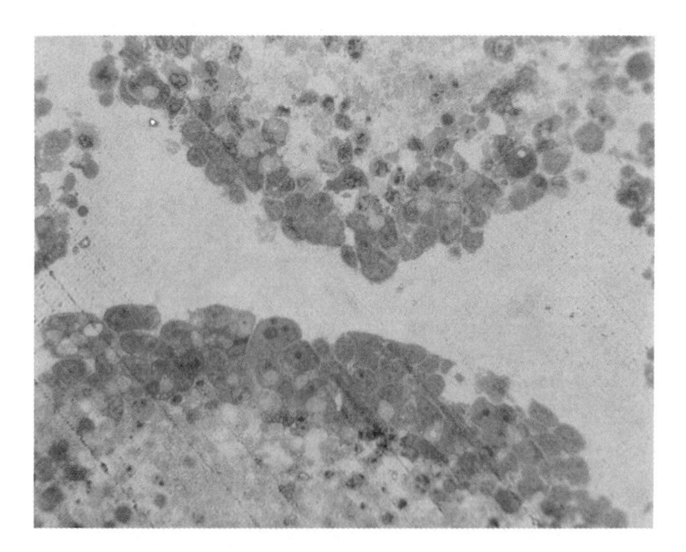

Figure 2: Central section through two adjacent spheroids. The inner regions of the spheroids are lightly stained (Toluidine blue). Bar 0.1 mm.

2 - 3 weeks of growth, the mean diameter of the spheroids reached 0.3 - 0.5 mm, and most of the cells in the center became necrotic (Figures 1 and 2).

In exponentially growing monolayer cultures of NHIK 3025 cells, practically 100% of the cells are proliferating. The mean cell cycle time is 18h. In spheroids the growth fraction is 0.6 - 0.7, and the cells are growing slower, the growth rate being somewhat dependent on the distance from the spheroid surface. Near the surface the mean cycle time is 30h while it is 41h in the central region. All phases in the cell cycle are prolonged in spheroids compared to monolayer cultures (Wibe et al. 1981). The cell loss is considerable, particularly due to the fact that the inner cells are becoming necrotic. It is not known why this happens, but it may be due to 3 factors: lack of oxygen, lack of nutrients or accumulation of metabolic products. Such an accumulation may lower the pH in the central region of the spheroid.

EXPOSURE TO Hpd AND LIGHT

The spheroids and monolayer cultures of NHIK 3025 cells were incubated for 22h in 25 μg/ml Hpd in medium E2a containing 30% serum. The cells were rinsed two times in PBS (phosphate buffered saline) and irradiated in PBS at 20°C. The light source was a bank of 4 Philips TLD/83 fluorescent light tubes. In order to obtain mainly red light, a Cinemoid 35 filter (Rand Stand Electric, Brentford, U.K.) was applied. A corrected spectrum of the filtered light is shown in figure 3. The fluence rate at the position of the cells was 14 W/m^2.

The response of the spheroids was studied by microscopy or by inoculating single cells for colony formation. In order to disperse the spheroids in single cell suspension, they were gently agitated for 8 min in trypsin-EDTA (Gibco).

The sensitivity of the spheroids was compared with that of single cells in exponential growth and in plateau phase as follows: 200 - 1000 single cells were inoculated in 25 cm^2 tissue culture flasks and incubated for 3 - 4h. The cells were then labelled with Hpd and irradiated as described above. Colonies arising from surviving cells were stained and counted after 1 week of incubation. Cultures of NHIK 3025 cells in plateau phase of growth were produced by inoculating $4 \cdot 10^5$ cells per tissue flask and incubating the cells for 1 week with a medium change after four days. The cultures were labelled with Hpd and irradiated as described above. One hour after irradiation each culture was trypsinized and the cells were inoculated for colony formation. Suspensions of Hpd labelled cells were made by removing exponentially growing cells from the substratum with a Costar cell scraper. The cells had been incubated for 22h with Hpd before being removed from the substratum. The cell suspensions were exposed to light and carefully dispersed into single cells with trypsin-ETDA and incubated 1 week for colony formation.

Quantification of Hpd uptake in spheroids and monolayer cells was performed by measuring fluorescence from the cells after homogenization in 0.2 N NaOH containing 1% Cetyltrimethylammonium bromide. The amount of proteins in each sample was determined by the Bio-Rad protein assay.

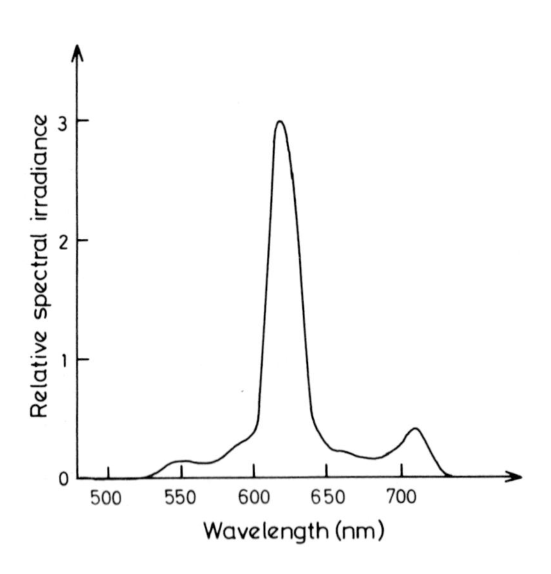

Figure 3: Corrected spectrum of the light source with filter.

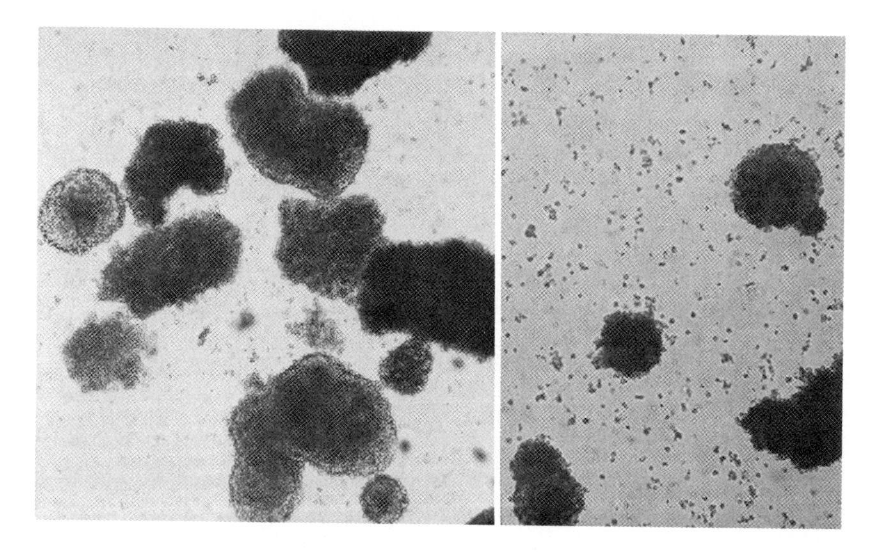

Figure 4: Spheroids in the culture flask 12h (left) and 24h (right) after exposure to $20 \cdot 10^4$ J/m^2 light.

PHOTOSENSITIZING EFFECTS OF Hpd ON SPHEROIDS

After exposure to a fluence sufficient to kill more than 95% of the cells in a spheroid, severe degeneration of the spheroid was observed (figure 4). 24h after exposure to half this fluence (75% cell kill) cells were detached from the surface and many of the remaining cells showed signs of damage (figures 5 and 6). The condensation of chromatin in damaged cells was clearly seen. This resembles the nuclear morphology of damaged cells observed previously by transmission electron microscopy (Moan et al. 1982).

Spheroids that were treated with a light fluence killing 50% of the cells or less, continued to grow after treatment. Even in spheroid populations where a majority of the cells were killed, some spheroids apparently did not break down completely (figure 4). The growth of these surviving spheroids continued after a lag period.

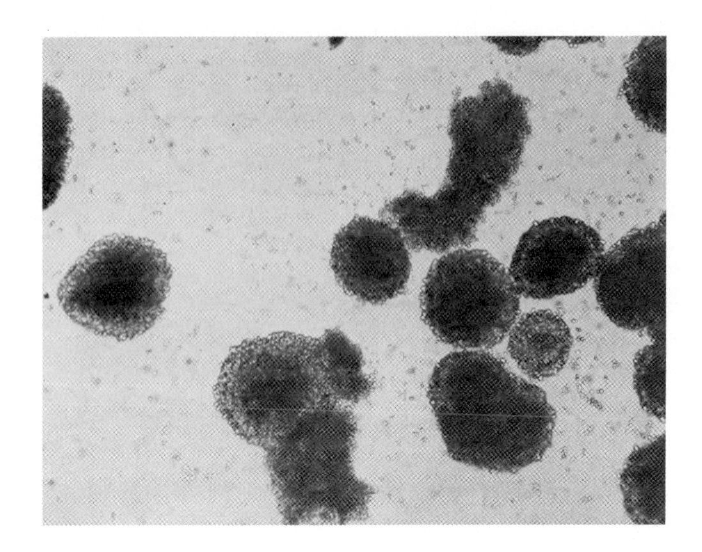

Figure 5: Spheroids in the culture flask 24h after exposure to $10 \cdot 10^4$ J/m^2 light.

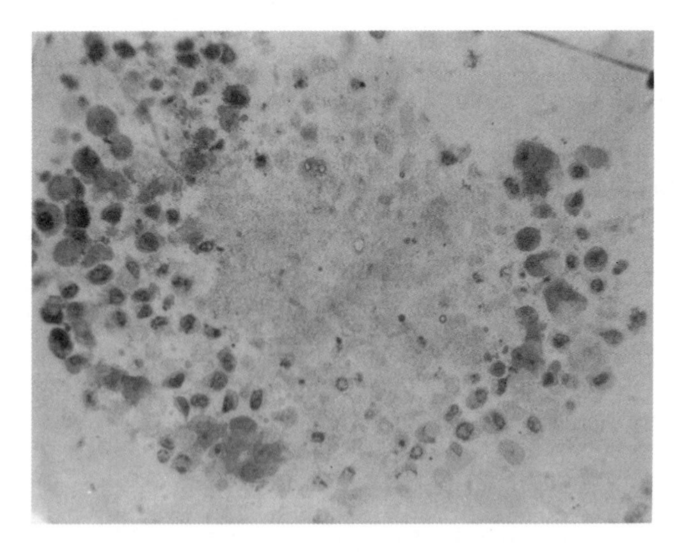

Figure 6: Central section through a spheroid 24h after exposure to $10 \cdot 10^4$ J/m^2 light (Toluidine blue).

COMPARISON OF THE SENSITIVITY OF CELLS IN SPHEROIDS WITH CELLS GROWN IN MONOLAYER

Figure 7 shows that cells in spheroids are more resistant to light than cells grown in monolayer and irradiated either in suspension or attached to the substratum. The resistance was not caused by a lower total Hpd uptake in the spheroids. On the contrary, it was found that the spheroids contained $1.02 \pm 0.13 \cdot 10^{-4}$ g Hpd/g protein compared to $0.55 \pm 0.03 \cdot 10^{-4}$ g Hpd/g protein in cells grown in monolayer. The high uptake of Hpd in spheroids may be explained by low pH in the inner region of the spheroid. It has been shown that hematoporphyrin is taken up in cells more efficiently at low pH compared to normal physiological pH (Moan et al. 1980). According to our findings a porphyrin molecule bound in spheroids is roughly 5 - 10 times less efficient in causing cell killing than when it is bound to cells grown in monolayer.

The high resistance to Hpd and light of cells in spheroids compared to cells grown in monolayer may be explained by several factors that are discussed below.

Since the spheroids contain a large amount of necrotic material, the partition of Hpd between viable and dead cells may be important for the efficiency. Spheroids labelled with Hpd were freeze sectioned and studied by fluorescence microscopy. The necrotic central parts showed more fluorescence than the viable cells in the periphery. This indicates that a large fraction of the Hpd was not localized in the viable part of the spheroid, and thus did not contribute to the photodynamic inactivation of spheroid cells.

The microenvironment in the spheroids may be another important factor. Especially the oxygen tension can probably influence the sensitivity. Lowered oxygen tension may result in a low photoinduced yield of singlet oxygen, which is probably the most important cytotoxic product generated (Moan et al. 1979).

Intercellular contact may, at least in part, explain the resistance of the spheroids. It has been shown that cell contact can modify the photosensitivity of NHIK 3025 cells labelled with hematoporphyrin (Christensen, Moan 1979). Another important factor is the cell kinetics which is different in spheroids and monolayers of exponentially growing

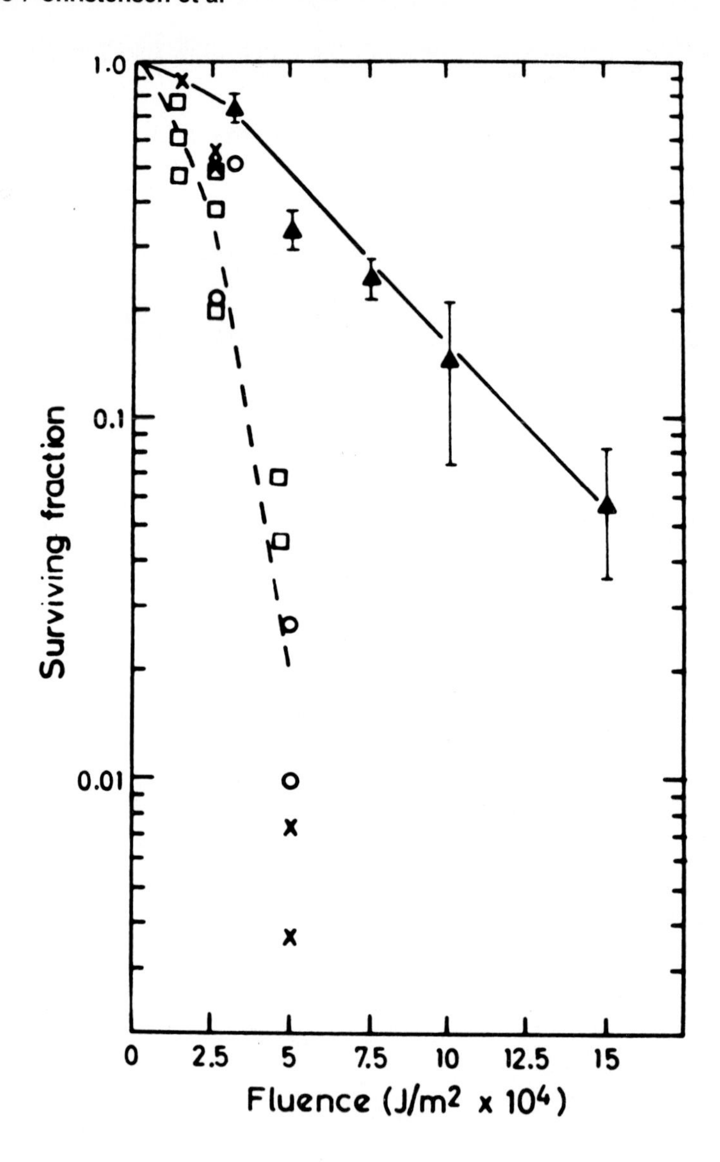

Figure 7: Single cell surviving fraction of cells exposed in spheroids (▲, mean and S.E. from 3 - 4 independent experiments), in monolayer (□, results from 3 independent experiments), in suspension (o, results from 3 independent experiments) and in monolayer in plateau phase cultures (x, results from 2 independent experiments).

cells. Resting cells and a high percentage of cells in G_1 are present in the spheroids (Wibe et al. 1981). Cells are particularly resistant in G_1 (Christensen et al. 1981). The relatively high sensitivity of cells in the plateau phase (figure 7) speaks against cell kinetic factors as the only explanation.

The lack of attachment to a firm surface, and the three-dimensional arrangement of the cells in the spheroids are probably of small importance for the resistance to Hpd and light, since single cells in suspension and in monolayer are about equally sensitive (figure 7).

Data on penetration of light through spheroids are not available. Red light penetrates efficiently through non-pigmented tissue, and published data on the penetration through different blood-free tissues (Doiron et al. 1983) leads us to the assumption that inefficient light penetration through spheroids 0.5 mm in diameter cannot explain the large difference in sensitivity of single cells and cells in spheroids.

It is well known that cells in spheroids are more resistant than single cells to several other agents like ionizing radiation (Durand, Sutherland 1972) heat (Lücke-Huhle, Dertinger 1977), Adriamycin (Durand 1967) and mitotic inhibitors (Wibe 1980, Wibe, Oftebro 1981). Lowered uptake of chemical agents in spheroids is generally not found to be the only factor responsible for the protection of the cells.

CONCLUSIONS

Multicellular spheroids may be used in studies of the basic effects of PRT. Light exposure of Hpd labelled spheroids leads to partial destruction of the spheroids. Some spheroids continue to grow despite inactivation of more than 75% of the viable cells. Spheroids take up more Hpd than cells in monolayer. Nevertheless, cells in a spheroid are more resistant to Hpd induced photoinactivation than single cells. This may be explained by preferential uptake of Hpd in the non-viable cells in the necrotic part of the spheroids. Physiological factors such as low pH may partly explain the high uptake of Hpd and the low photosensitivity of cells in spheroids compared with cells growing in the monolayer.

ACKNOWLEDGEMENTS

This study was supported by The Norwegian Research Council for Science and the Humanities. We want to thank Dr. Einar Wibe for teaching us to grow spheroids and Bjørg Nesheim and Ruth Puntervold for making sections of the spheroids.

REFERENCES

Bugelski PJ, Porter CW, Dougherty TJ (1981). Autoradiographic distribution of hematoporphyrin derivative in normal and tumor tissue of the mouse. Cancer Res 41:4606.

Christensen T, Feren K, Moan J, Pettersen EO (1981). Photodynamic effects of haematoporphyrin derivative on synchronized and asynchronous cells of different origin. Br J Cancer 44:717.

Christensen T, Moan J (1979). Photodynamic effect of hematoporphyrin (HP) on cells cultivated in vitro. Springer Series in Optical Sci 22:87.

Durand RE (1976). Adriamycin: A possible indirect radiosensitizer of hypoxic tumor cells. Radiology 119:217.

Duran RE, Sutherland RM (1972). Effects of intercellular contact on repair of radiation damage. Exp Cell Res 71:75.

Lücke-Huhle C, Dertinger H (1977). Kinetic response of an in vitro "tumor-model" (V79 spheroids) to 42°C hyperthermia. Eur J Cancer 13:23.

Moan J, Johannessen JV, Christensen T, Espevik T, McGhie JB (1982). Porphyrin-sensitized photoinactivation of human cells in vitro. Am J Pathol 109:184.

Moan J, Pettersen EO, Christensen T (1979). On the mechanism of photodynamic inactivation of human cells in vitro in the presence of haematoporphyrin. Br J Cancer 39:398.

Moan J, Smedshammer L, Christensen T (1980). Photodynamic effects on human cells exposed to light in the presence of hematoporphyrin. pH effects. Cancer Lett 9:327.

Wibe E (1980). Resistance to vincristine of human cells grown as multicellular spheroids. Br J Cancer 42:937.

Wibe E, Lidmo T, Kaalhus O (1981). Cell kinetic characteristics in different parts of multicellular spheroids of human origin. Cell Tissue Kinet 14:639.

Wibe E, Oftebro R (1981). A study of factors related to the action of 1 - propargyl - 5 - chloropyrimidin - 2 - one (NY 3170) and vincristine in human multicellular spheroids. Eur J Cancer 17:1053.

Porphyrin Localization and Treatment of Tumors, pages 391–404

MEMBRANE PHOTOSENSITIZATION BY HEMATOPORPHYRIN AND HEMATOPORPHYRIN DERIVATIVE

Leonard I. Grossweiner, Ph.D.

Biophysics Laboratory, Physics Department
Illinois Institute of Technology
Chicago, IL 60616

INTRODUCTION

The tumoricidal action in photoradiation therapy (PRT) is effected by exposing tumor tissue to strong red radiation after intravenous administration of hematoporphyrin derivative (HPD). This action has been attributed to the formation of singlet molecular oxygen ($^1O_2$) and its attack on tumor tissue membranes (Dougherty, Kaufman, Goldfarb, Weishaupt et al 1978). There is evidence for $^1O_2$ production by metal-free porphyrins (Cauzzo, Gennari, Jori, Spikes 1977) and also for erythrocyte membrane damage by $^1O_2$ (Lamola, Yamane, Trozzolo 1973), which are consistent with this mechanism. However, $^1O_2$ reacts also with DNA and proteins (Foote 1976) and there is no direct evidence that membrane damage is the only effect in PRT. HPD is believed to consist of an aggregate of hydrophobic porphyrins (Dougherty, Wityk, Bellnier, Mang 1981; Kessel 1982; Andreoni, Cubbedu, De Silvestri, LaPorta, Jori, Reddi 1982), but the detailed structure has defied analysis. The active biologic component was identified as the fastest moving fraction separated by polyacrylamide gel filtration (Dougherty, Wityk, Bellnier, Mang 1981), which is designated as HPD-A in this paper. This material is resistant to disaggregation by dilution and has the required properties of effective localization in tumor tissue and conversion to the in situ photosensitizer.

The photosensitization mechanisms of hematoporphyrin (HP), HPD and HPD-A were investigated in the present work using phosphatidylcholine (PC) liposomes as model membrane systems. The results confirm the major role of $^1O_2$ in membrane damage and provide new information about the conversion of aggregated HPD-A into the fluorescent, in situ photosensitizer. The effects of human serum albumin (HSA) on these processes provide information about the serum transport of HPD-A.

MATERIALS AND METHODS

Chemicals

Crude hematoporphyrin derivative was used as purchased from Porphyrin Products and also was prepared from hematoporphyrin dihydrochloride with the method of Dougherty et al (Dougherty, Kaufman, Goldfarb, Weishaupt et al 1978) with similar results. HPD was prepared from crude hematoporphyrin derivative by dissolving in 0.1 M NaOH, stirring for 30 min in the dark, neutralizing with 1 M HCl and adjusting to the desired pH. HPD-A was obtained by filtration of HPD on Bio-Gel P-10 columns (Bio-Rad Laboratories) using 0.01 M phosphate buffer as the eluant, at pH 7.0 or pH 7.4. The first major fraction was HPD-A, comprising 30-50% of the starting material, depending on the column loading, with the Soret band maximum at 364±1 nm and the absorbance ratio, $A_{365}/A_{395} = 1.75 \pm 0.15$. Hematoporphyrin dihydrochloride from Porphyrin Products and Sigma Chemical Co. was used with no apparent differences. The human serum albumin was from Sigma Chemical Co. (< 0.005% fatty acid) and the other chemicals were the best available grades. All solutions were made with doubly-distilled water.

Liposomes

The liposomes were prepared with L-α-phosphatidylcholine from Sigma Chemical Co. (type VII-E). The chloroform solution was evaporated to a dry film in a 10 mm test tube with a stream of nitrogen gas, dispersed in phosphate buffer by vortexing for 3 min, and the liposomes were allowed to swell for 2 hr at 3 °C. The size of the large multilamellar vesicles determined by scanning electron microscopy was 1-3 μm. The sensitizers were incorporated by swelling in a buffer solution of the porphyrin followed by removal of the external material. This was done either by three stages of centrifugation (5 min @ 15,000 g) and resuspending or by dialysis from Spectropor-1 membrane tubing, and the incorporated sensitizer was estimated by subtraction. In some experiments with HPD-A the excess was removed by elution through a Sephadex G-50 (fine) column (Pharmacia Inc.). A modified composition was used in some experiments, consisting of 78 pts PC, 16 pts dicetylphosphate (DCP), and 6 pts cholesterol (CL). The small liposomes were prepared by sonicating the large liposomes for 2 hr under nitrogen with a Heat Systems-Ultrasonics Model W-10 Sonicator. The large liposomes and probe titanium were removed by centrifugation for 2 hr at 24,000 g at 5 °C in a Sorvall RC-5B centrifuge. The sonicated liposomes eluted in a single, symmetrical band through a Sephacryl S-1000 (superfine) column (Pharmacia Inc.) preceded by a small band at the void volume, corresponding to the exclusion size of 300-400 nm. The liposomes ranged from 10-90 nm, as estimated from the published gel characteristics.

Irradiations

The large liposomes were diluted to 150-200 μg/ml PC and exposed to radiation from a 200 W mercury-xenon lamp (Hanovia 901-B1) in a 1 cm path polystyrene cuvette held at constant temperature to ±0.1 °C and bubbled with oxygen or nitrogen at 20-120 bubbles/min (1-6 ml/min). In most experiments a Corning C.S. No. 3-70 filter (> 500 nm) was used, except at low sensitizer concentrations when a C.S. No. 0-52 (> 360 nm) filter was employed with a silica cuvette. Liposome lysis was measured by the decrease of light scattering absorbance at 750 nm in a Varian model 635 spectrophotometer. The small liposomes were irradiated without dilution (≃ 2 mg/ml PC) and lipid peroxidation was determined by the increase of the absorbance ratio A_{235}/A_{215} (Klein 1970), which was measured in situ.

RESULTS

The prototype experiments were recent measurements on photosensitized lysis of large liposomes with incorporated and external hematoporphyrin (Grossweiner, Patel, Grossweiner 1982). It is convenient to summarize the results in order to delineate the similarities and differences in the new work with HPD and HPD-A. For convenience, the large liposomes are designated LMV (large multilamellar vesicles) and the small liposomes are designated SUV (small unilamellar vesicles).

Photosensitized Lysis of LMV by Hematoporphyrin

Typical lysis curves for LMV incorporating HP are shown in Fig.1. The data were obtained by dark bubbling for 60 min followed by irradiation with measurements of A_{750} at 10 min intervals. The bubbling rate was controlled at 120 bubbles/min to eliminate the dependence of lysis rate on flow rate observed in this work and in prior work with methylene blue (Grossweiner, Grossweiner 1982). Curve B shows rapid lysis

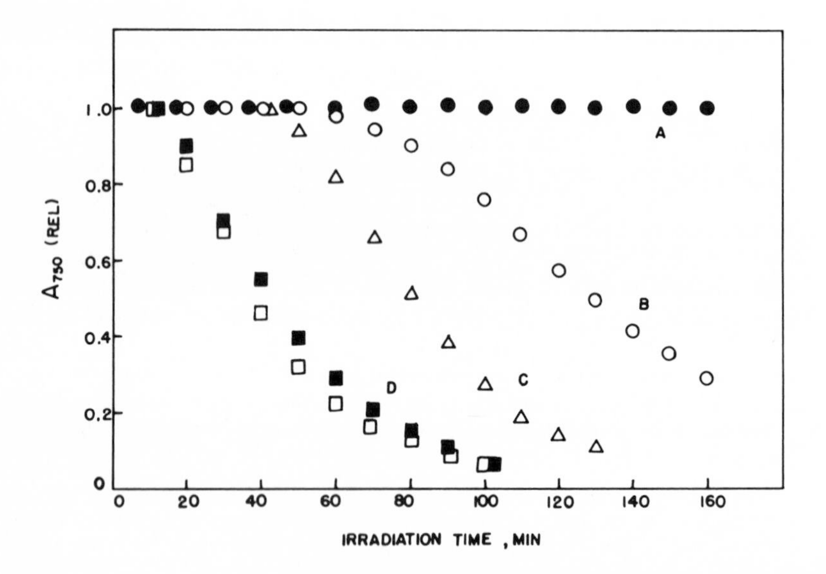

Fig. 1. Photosensitized lysis of LMV by incorporated HP: pH 7.0, 25 °C, > 500 nm, 120 bubbles/min. (A) 0.1% HP, N_2; O_2 + 0.1 M N_3^-; O_2 + 0.1 M DABCO. (B) 0.1% HP, O_2. (C) 0.1% HP + D_2O. (D) 1% HP, □ O_2; ■ N_2.

after an induction period with oxygen bubbling. Lysis was

suppressed for 0.1% HP with nitrogen bubbling and with oxygen bubbling in the presence of azide ion or 1,4-diazabicyclo[2.2.2]octane (DABCO), which are frequently employed acceptors of $^1O_2$ (Foote 1976); Curve A. Curve C shows that the lysis rate was accelerated in D_2O buffer, another standard test for $^1O_2$ based on its longer lifetime in perdeuterated solvents. However, with 1% HP lysis was not altered in the absence of oxygen (Curve D) and D_2O had no effect. Similar experiments with external HP showed that $^1O_2$ was the lytic agent with 1 μM HP and much less significant at 15 μM HP. The values of the oxygen enhancement ratio (OER) for HP, HPD and HPD-A are summarized in Table 1. In conjunction with spectral data (Brown, Shillcock, Jones 1976), The results show that

Table 1. Effect of oxygen on photosensitized lysis of LMV.

Porphyrin	Condition [a]	OER [b]
HP (0.1%)	incorporated	>> 10
HP (1.0%)	"	≃ 1
HP (1 μM)	external	1.8
HP (15 μM)	"	1.2
HPD (0.01%)	incorporated	≃ 4
HPD (5%)	"	1.6
HPD (40%)	"	1.3
HPD-A (1.8%)	incorporated	4.3
HPD-A (50 μM)	external	2.7

[a] in 0.01 M phosphate buffer, pH 7.0-7.4
[b] OER = $t_{80}(N_2)/t_{80}(O_2)$ where t_{80} is the irradiation time required for 20% lysis.

lysis was induced by $^1O_2$ with monomeric and/or dimeric HP and via an anoxic mechanism with aggregated HP.

Photosensitized Lysis of LMV by HPD

A few experiments were done with HPD incorporated in LMV, using the PC/DCP/CL preparation in pH 7.4 phosphate buffer plus 0.9% NaCl at 38 °C. The results in Table 1 show there was a higher OER than unity even at quite high HPD concentrations. The Soret band maxima and fluorescence band maxima of the different porphyrins are summarized in Table 2. The HPD absorption in buffer was similar to aggre-

Table 2. Spectral properties of HP, HPD and HPD-A in buffer and in liposomes.

Porphyrin	Condition	λ_s (nm) [a]	λ_{fl} (nm) [b]
HP (1 μM)	buffer [c]	391	615
HP (80 μM)	"	375	615
HP (80 μM)	SUV	398	620
HPD (25 μM) [d]	buffer	370	615
HPD (25 μM)	LMV	≃ 380 [f]	624
HPD (25 μM)	SUV	398 [e]	625 [h]
HPD-A (3 μM) [d]	buffer	363	615
HPD-A (20 μM)	"	364	615
HPD-A (300 μM)	"	364	-
HPD-A (3 μM)	50 μM HSA	365	625
HPD-A (25 μM)	LMV	≃ 380 [f]	625
HPD-A (25 μM)	SUV	398 [g]	630 [h]

a Soret band absorption maximum
b fluorescence maximum excited at 535 nm
c 0.01 M phosphate buffer, pH 7
d concentration estimated from $\varepsilon(max) = 7 \times 10^4 \ M^{-1} \ cm^{-1}$
e shoulder at 375 nm
f very broad band
g shoulder at 365 nm
h 2-3 fold stronger than in buffer

gated HP. In LMV there was significant broadening and evidence of a red shift, indicative of a heterogeneous environment. However, in the SUV there was a definite red shift to 398 nm with a shoulder at 375 nm. The fluorescence intensity of HPD was about 2-3 fold higher in SUV. The changes suggest that HPD is disaggregated in the membrane bilayer. Similar effects were observed with concentrated HP and HPD-A. As discussed below, the spontaneous conversion of HPD and HPD-A to a less aggregated, fluorescent state in the membrane provides a reasonable explanation for the importance of localization in generating the in situ photosensitizer.

Photosensitization of Liposome Damage by HPD-A

The stability of HPD-A against disaggregation by dilution and binding to HSA is evident in the spectral data in Table 2. Recent work has shown that the OER is 2.7 for photosensitized lysis of LMV by external HPD-A (Grossweiner, Goyal 1983). New results with 1.8% HPD-A in LMV (PC/DCP/CL preparation) are given in Fig.2 . The high OER (Table 1) and protection by azide ion and DABCO show that $^1O_2$ was the dominant lytic intermediate and indicate that it escaped from the liposomes prior to reaction. It was found also that

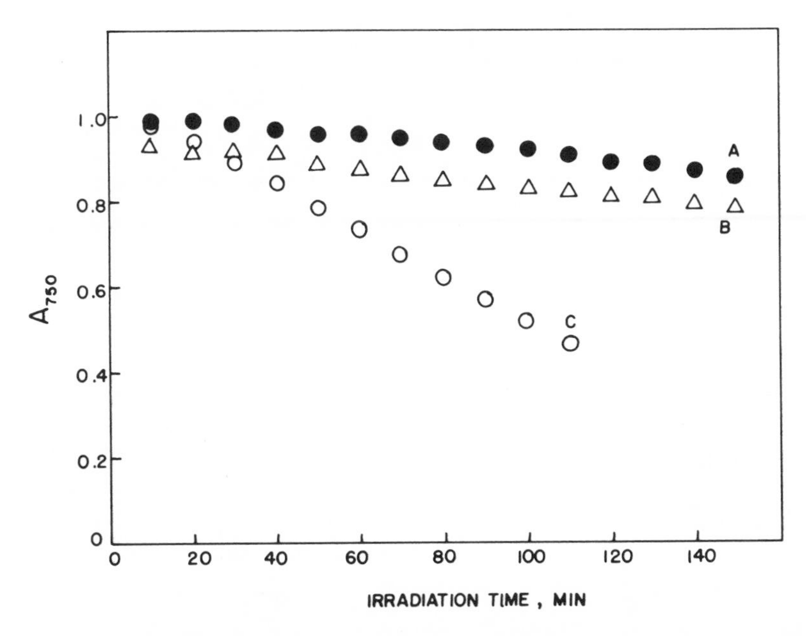

Fig. 2. Photosensitized lysis of LMV by incorporated 1.8% HPD-A: pH 7.4, 38 °C, > 360 nm, 0.9% NaCl, 120 bubbles/min. (A) N_2; O_2 + 0.1 M N_3^-. (B) O_2 + 1 M DABCO. (C) O_2.

the presence of 10 μM HSA suppressed lysis of LMV by HPD-A, which is further evidence for the escape of $^1O_2$. This protective effect is attributed to the photooxidation of HSA by $^1O_2$. The results in Fig.3 show the decrease of the tryptophan fluorescence of HSA induced by irradiation in the presence of a low HPD-A concentration. The formation of $^1O_2$ and its attack on the excess free HSA is the dominant reaction, although there was a low rate of oxidation with nitrogen saturation compared to the control (no HPD-A).

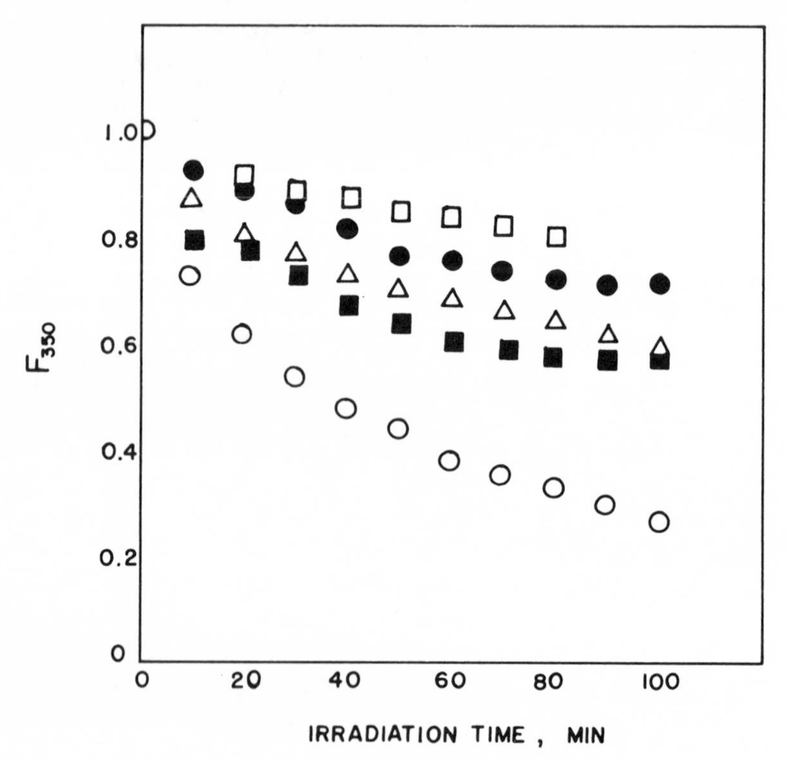

Fig. 3. Photosensitized oxidation of human serum albumin by HPD-A: pH 7.4, 30 °C, > 360 nm, 2 µM HPD-A + 50 µM HSA. □ control (O_2, no HPD-A). ● N_2. △ O_2 + 1 mM N_3^-. ○ O_2. ■ HPD-A photobleaching (O_2). F_{350} is the fluorescence at 350 nm excited at 290 nm.

Photooxidation of HSA was accompanied by photobleaching of the complexed HPD-A. The same effects were observed for HP photosensitization of HSA, in which $^1O_2$ generated by the complexed HP was responsible for the photooxidation of HSA (Richard, Blum, Grossweiner 1983). The result suggests that weak light exposure after PRT may promote HPD elimination.

When HPD-A was added to a suspension of SUV there was a rapid shift of the Soret band to 398 nm (within 1 min) and the HPD-A fluorescence was enhanced with a 5 nm red shift. The light scattering was negligible in the red region where the HPD-A is non-absorbing, and membrane damage was assayed by the increase of the lipid absorption ratio A_{235}/A_{215}. Typical results for 0.3% HPD-A in SUV are shown in Fig.4. The OER corrected for the controls is 5.3; Table 3 summarizes the data including the effects of 0.01 M azide.

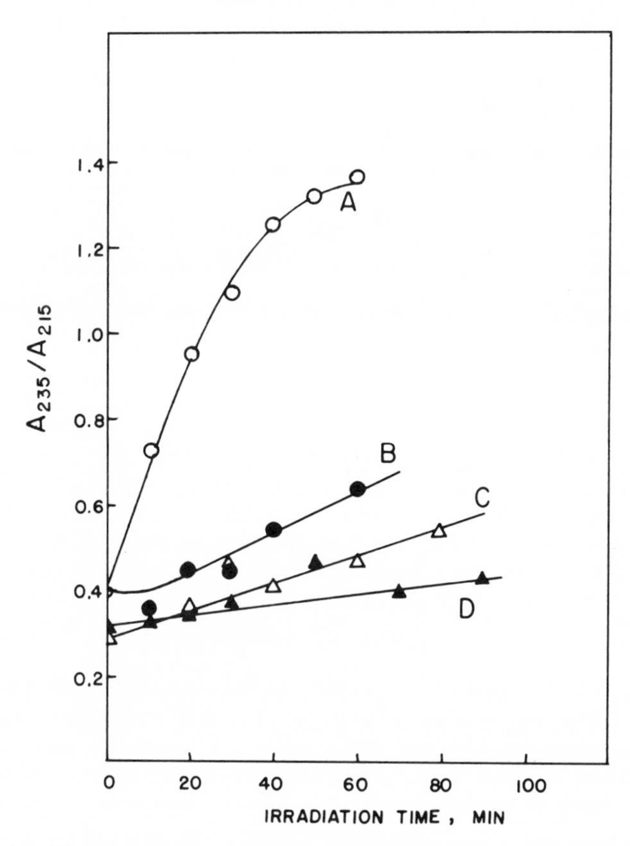

Fig. 4. Photosensitization of lipid peroxidation in SUV by 0.3% HPD-A: pH 7.0, 25 °C, > 360 nm. (A) O_2. (B) N_2. (C) O_2 control (no HPD-A). (D) N_2 control (no HPD-A). A_{235}/A_{215} was measured in situ.

Table 3. Photosensitization of lipid peroxidation in SUV by incorporated HPD-A.

Irradiation Conditions [a]	R_{ox} [b]
0.3% HPD-A/O_2	0.029
0.3% HPD-A/N_2	0.005
0.3% HPD-A/O_2/0.01 M N_3^-	0.009
0.3% HPD-A/N_2/0.01 M N_3^-	0.002
O_2 control	0.003
N_2 control	0.001

[a] pH 7.0 phosphate buffer, > 360 nm
[b] rate of increase of A_{235}/A_{215}

DISCUSSION

The present results provide unambiguous proof that HPD-A photosensitizes membrane damage in PC liposomes via the $^1O_2$ intermediary. Since HPD-A has the spectral properties of a porphyrin aggregate, this mechanism differs from concentrated HP with OER ≃ 1 (Table 1). The ease of diffusion of HPD-A into the SUV with a Soret band shift from 364 nm to 398 nm rules out the suggestion that it consists of covalently bonded porphyrin units (Berenbaum, Bonnett, Scourides 1982). We have found that HPD-A is not disaggregated by boiling, lyophilization and reconstitution in buffer, and acidification to pH 2.5 (Blum, Goyal, Grossweiner 1983). Earlier work showed that it was resistant to dilution and binding to serum protein (Dougherty, Wityk, Bellnier, Mang 1981), as shown by our results in Table 2. A recent suggestion that localization is mediated by HPD components which are hydrophobic by virtue of strong hydrogen bonding may explain these atypical properties (Kessel, Chou 1983). The formation of a fluorescent compound after localization may be explained by our observations of apparent disaggregation of HPD-A when it diffuses into the SUV. However, the high OER values were obtained also in the LMV, where the spectral properties indicate that the HPD-A was distributed between the internal aqueous and lipophilic phases. This result suggests that formation of the fluorescent material may not be required for photosensitization in vivo, and active HPD-A may be localized in both forms.

The presence of 10 μM HSA protected LMV against lysis with incorporated 1.4% HPD-A (data not shown). Similar results were reported with external 2 μg/ml HPD-A and 1.5 mg/ml HSA(Grossweiner, Goyal 1983). This concentration ratio is comparable to the ratio of HPD to serum protein in PRT. We also observed that the diffusion of 2 μg/ml HPD-A into the SUV was retarded from < 1 min to > 45 min by the presence of 15 mg/ml HSA (data not shown). These finding suggest that the diffusion of HPD-A into adventitious membranes is inhibited during serum transport prior to localization. Serum protein photooxidation should be a consequence of exposure to visible or near-ultraviolet radiation during this stage, which may be involved in the skin phototoxicity.

The different assays of membrane damage used for the LMV and SUV should not be a significant factor in connecttion with the key role of $^1O_2$. Prior work on toluidine blue photosensitization of large PC liposomes (7 PC:1 DCP:2 CL) showed that lipid peroxidation preceded lysis in the same systems (Anderson, Krinsky 1973). A similar result was observed with undyed LMV exposed to $^1O_2$ generated in the gas phase at 1 atm total pressure and bubbled into the liposome suspension (Eisenberg, Tayor, Grossweiner 1973). Recent studies on HPD photosensitization of cultured mammalian cells demonstrate the occurrence of membrane damage , although it has not been demonstrated that this process is the principal factor in cell killing (Bellnier, Dougherty 1982; Moan, McGhie, Christensen 1982). Histologic studies on implanted tumors after HPD photosensitization provide additional evidence of membrane damage (Bugelski 1980). However, $^1O_2$ is a diffusive agent and it is capable of diffusing out of SUV and reacting with substrates which are located in lipid regions of other vesicles or in the bulk medium (Rodgers, Bates 1982), and the possibility of other than membrane targets in cells cannot be ignored.

SUMMARY

HPD-A, the putative active biologic component of HPD, photosensitized lysis of large phosophatidylcholine liposomes and lipid peroxidation in small liposomes with oxygen enhancement ratios from 4 to 5. Unambiguous proof was obtained that this damage was mediated by singlet oxygen. Spectral data indicates that HPD-A localizes in a heterogeneous environment in the large liposomes and within the membrane bilayer of the small liposomes. The latter was accompanied by a shift of the Soret band from 364 nm to 398 nm and a 2-3 fold increase of the fluorescence. The presence of human serum albumin protected the large liposomes from photosensitized damage by HPD-A and inhibited the diffusion of HPD-A into the small liposomes. In connection with photoradiation therapy, these findings suggest that aggregated HPD-A and the more fluorescent, disaggregated HPD-A may act as photodynamic sensitizers after localization, and that binding of HPD-A to serum protein protects the aggregated form and inhibits photodynamic damage to membranes. However, protein photooxidation is feasible in the protein-bound state.

ACKNOWLEDGEMENTS

The author expresses his appreciation to Dr. Aleksander Blum and Dr. Greesh C. Goyal for their invaluable assistance in the preparation of this paper.

This work was supported by NIH Grant GM 20117 and Department of Energy Contract No. AC02-76EV02217. This is publication DOE/EV/02217-47.

REFERENCES

Anderson SM, Krinsky NI (1973). Protective action of carotenoid pigments against photodynamic damage to liposomes. Photochem Photobiol 18:403.

Andreoni A, Cubeddu R, De Silvestri S, LaPorta P, Jori G, Reddi E (1982). Hematoporphyrin derivative:Experimental evidence for aggregated species. Chem Phys Lett 88:33.

Bellnier DA, Dougherty TJ (1982). Membrane lysis in Chinese hamster cells treated with hematoporphyrin derivative plus light. Photochem Photobiol 36:43.

Berenbaum MC, Bonnett R, Scourides PA (1982). In vivo biological activity of the components of henatoporphyrin derivative. Br J Cancer 45:571.

Blum A, Goyal GC, Grossweiner LI (1983). Properties of HPD-A: Implications for photoradiation therapy. Photochem Photobiol

Brown SB, Shillcock M, Jones P (1976). Equilibrium and kinetic studies of the aggregation of porphyrins in aqueous

solution. Biochem J 153:279.
Bugelski PJ (1980). Morphologic aspects of the photosensitizing effects and distribution of hematoporphyrin derivative in experimental tumors and cultured cells. Thesis, State University of New York at Buffalo.
Cauzzo G, Gennari G, Jori G, Spikes JD (1977). The effect of chemical structure on the photosensitizing efficiencies of porphyrins. Photochem Photobiol 25:389.
Dougherty, TJ, Kaufman JE, Goldfarb A, Weishaupt KR, Boyle D, Mittleman A (1978). Cancer Res 38:2628.
Dougherty TJ, Wityk K, Bellnier D, Mang T (1981). Isolation and biologic activity of hematoporphyrin derivative components. Workshop of Porphyrin Photosensitization, Washington, DC, Sept. 28-29, 1981.
Eisenberg WC, Taylor K, Grossweiner LI (1983). Lysis of phosphatidylcholine liposomes by singlet oxygen generated in the gas phase. Photochem Photobiol (in press).
Foote CS (1976). Photosensitized oxidation and singlet oxygen:Consequences in biological systems. In Pryor WA (ed): "Free Radicals in Biology, Vol II" New York: Academic Press, p 85.
Grossweiner LI, Grossweiner JB (1982). Hydrodynamic effects in the photosensitized lysis of liposomes. Photochem Photobiol 35:583.
Grossweiner LI, Patel AS, Grossweiner JB (1982). Type I and type II mechanisms in the photosensitized lysis of phosphatidylcholine liposomes by hematoporphyrin. Photochem Photobiol 36:159.
Grossweiner LI, Goyal GC (1983). Photosensitized lysis of liposomes by hematoporphyrin derivative. Photochem Photobiol
Kessel D (1982). Components of hematoporphyrin derivatives and their tumor-localizing capacity. Cancer Res 42:1703.
Kessel D, Chou T-H (1983). Tumor-localizing components of the porphyrin preparation HPD. Cancer Res 43:1994.
Klein RA (1970). The detection of oxidation in liposome preparations. Biochim Biophys Acta 210:486.
Lamola AA, Yamane T, Trozzolo AM (1973). Cholesterol hydroperoxide formation in red blood cell membranes and photohemolysis associated with erythropoietic protoporphyria. Science 179:1131.
Moan J, McGhie JB, Christensen T (1982). Hematoporphyrin derivative: Photosensitizing efficiency and cellular uptake of its components. Photobiochem Photobiophys 4:337.
Richard P, Blum A, Grossweiner LI (1983). Hematoporphyrin photosensitization of serum albumin and subtilisin BPN'.

Photochem Photobiol 37:287.
Rodgers MAJ, Bates AL (1982). A laser flash kinetic spectrophotometric examination of the dynamics of singlet oxygen in unilamellar vesicles. Photochem Photobiol 35:473.

Porphyrin Localization and Treatment of Tumors, pages 405–418

CHEMICAL AND BIOCHEMICAL DETERMINANTS OF PORPHYRIN LOCALIZATION

David Kessel, Ph.D.

Departments of Medicine and Pharmacology
Wayne State University School of Medicine
and Harper-Grace Hospitals. Detroit MI 48201

The affinity of certain components of HPD for tumor loci is a requisite for diagnostic and therapeutic applications described elsewhere in this Symposium. This report will deal with phenomena associated with localization of components of HPD by Sarcoma-180 cells _in vitro_ and _in vivo_. The relevance of these observations with respect to the situation in man remains unknown. But it seems likely that in this work, as in other examples, research involving the mouse may point the way to an understanding of corresponding phenomena in man.

After administration of labeled HPD, there is a gradual accumulation of radioactivity in liver, spleen, kidney, tumor and skin (Gomer, Dougherty 1979). More detailed studies indicated localization of labeled material in tumor stroma (Bugelski, Porter, Dougherty 1981). If the distribution of radioactivity corresponds to that of the tumor-localizing components, these findings may provide clues concerning physiological determinants of tumor localization. But the chemical and biochemical mechanisms remain to be determined.

Several reports have described the composition of the products of hematoporphyrin acetylation (Berenbaum, Bonnett, Scourides 1982; Bonnett, Ridge, Scourides 1980; Clezy, Hai, Henderson, Thuc 1980) and of the subsequent alkaline hydrolysis products (Cadby, Dimitriades, Grant, Ward 1983; Bonnett, Ridge, Scourides 1981; Moan, Sandberg, Christensen, Elander 1983; Dougherty 1983). In accordance with the terminology proposed by Dougherty (1983), we term the latter material 'HPD'. Studies on accumulation of HPD components indicated preferential uptake of (and photosensitization by) what appears to be the most hydrophobic component of the product (Moan, McGhie, Christensen 1982; Moan, Christensen, Sommer, 1982; Berenbaum, Bonnett, Scourides 1982; Kessel, Chou 1983). In this report, we describe some properties of HPD components which appear to be important determinants of modes of drug uptake and retention by tumor cells.

MATERIALS AND METHODS

The murine Sarcoma S-180 tumor was maintained in suspension culture using RPMI 1640 medium supplemented with 10% horse serum, 1 μM mercaptoethanol and antibiotics.

HPD was prepared as described before (Gomer, Dougherty 1979), with minor modifications (Kessel 1981). Aqueous solutions (10 mg/ml) were stored at 4° in the dark. Other porphyrins were dissolved in DMF or water. In the latter case, compounds were dissolved in 0.1 M NaOH and neutralized to pH 7 with 1 N HCl to obtain 10 mg/ml solutions.

Porphyrins were extracted from tumor cells with 10 mM cetyl-trimethylammonium bromide (CTAB) at pH 7. The resulting homogenate was clarified by centrifugation. Levels of porphyrins were estimated in this extract by fluorescence. For analytic studies, the CTAB extract was mixed with 2 volumes of 1:1 chloroform-methanol, and the lower phase collected and evaporated. Further purification procedures are described elsewhere (Kessel, Chou 1983).

Porphyrin analysis involved use of reverse-phase TLC plates. The usual solvent was 65% methanol:35% 3 mM t-butylammonium phosphate (BAP) pH 3.5; a 75:25 system was also employed. Fluorescence patterns were examined with an Aminco fluorometer and the TLC scanning accessory. Using an excitation wavelength of 402 nm, and measuring emission at 628 nm, equal fluorescent signals were obtained from standards spotted on TLC plates, with each spot containing 0.1 μg of protoporphyrin (PP), hydroxyvinyl-deuteroporphyrin (HVD) or hematoporphyrin (HP).

HPLC separations on octadecylsilane columns were carried out using a linear gradient composed of 65% methanol: 35% BAP → 50% tetrahydrofuran:25% methanol:25% water. Optical density (500 nm) and fluorescence were monitored. No colored or fluorescent material could be detected on the columns after elution was complete. Porphyrins were concentrated from fractions by evaporation under reduced pressure. Residues were dissolved in ethyl acetate, washed with 50 mM phosphate buffer pH 3.5, and the solvent evaporated under nitrogen.

Octanol:water partitioning was carried out using equal volumes of the phases of a system containing 50% 1-octanol + 50% 50 mM phosphate buffer pH 7. Specified levels of different porphyrins initially dissolved in water or DMF were equilibrated in this system. Partition coefficients (porphyrin concentration in octanol phase expressed as % total) were measured by fluorescence, after dilution of each phase with methanol. Reverse-phase TLC analyses were carried out to determine the composition of porphyrins in each phase.

To measure accumulation of HPD components, S-180 cells were incubated with HPD (30 μg/ml) for 0.5 or 24 hr in growth medium (10% serum). The cells were then collected by centrifugation and washed for an additional 30 min in fresh medium if specified. Reverse-phase TLC was used to identify intracellular porphyrin components. In other studies, we incubated S-180 cells with different porphyrins for 10-30 min at 37°, and measured total porphyrin uptake by fluorescence assay of CTAB extracts as described above. Replicate samples of porphyrin-loaded cells were then irradiated, with viability assessed by a soft-agar clonogenic assay (Kessel 1982b) and membrane damage monitored by assessing capacity for concentrative uptake of cycloleucine (Kessel 1979).

Spectral studies were carried out with a dual-beam model 552 Perkin-Elmer spectrophotometer. Fluorescence measurements were obtained using a Perkin-Elmer 44B spectrophotometer.

RESULTS

Reverse-phase TLC analysis (65:35 methanol-BAP) of HPD indicates the presence of HP, HVD and PP, together with unidentified materials (Fig. 1A). Fluorescence scanning underestimates the amount of material at the origin of the plate; 500 nm absorbance measurements on HPLC eluates suggest that this material represents approx. 30% of HPD (Moan,

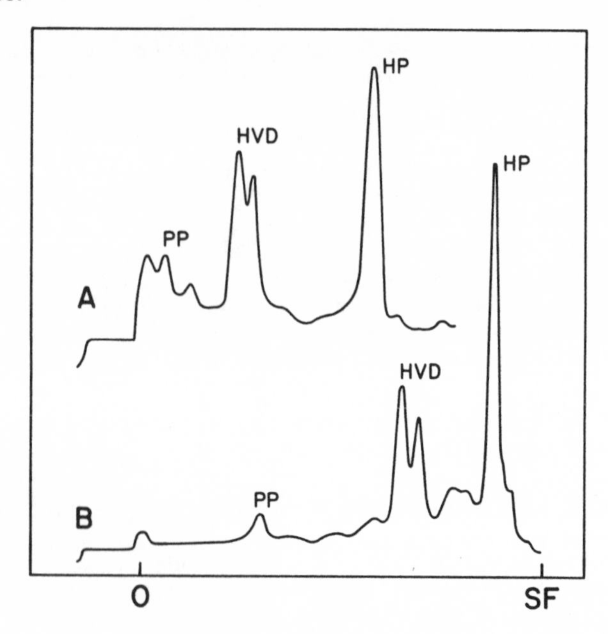

Fig. 1. Comparison of reverse-phase TLC chromatograms of HPD using 65:35 methanol-BAP (A) and with a 75:25 mixture (B). The origin (O), solvent front (SF) and Rf values of authentic samples of PP, HVD and HP are shown.

McGhie, Christensen 1982; Kessel 1983). In other studies, using the S-180 tumor _in vivo_, we found that this slowly-migrating HPD component was accumulated and retained by neoplastic tissues (Kessel, Chou 1983). The structure of this HPD component is unknown, we shall refer to it hereafter as the Π component. A different TLC pattern was obtained when a 75:25 solvent mixture was employed (Fig. 1B) This result indicates that the chromatographic pattern is highly dependent on the solvent, and that variation in results may occur as the solvent composition is altered.

We have described the use of absorbance spectra to delineate porphyrin aggregation phenomena (Kessel, Rossi 1982). Using solvents employed in reverse-phase TLC and HPLC studies, we found that all of the known components of HPD are aggregated, to varying degrees, in solvent systems commonly employed (data for HP and PP shown in Fig. 2). The tetrahydrofuran (THF):ethanol:water system disaggregates all

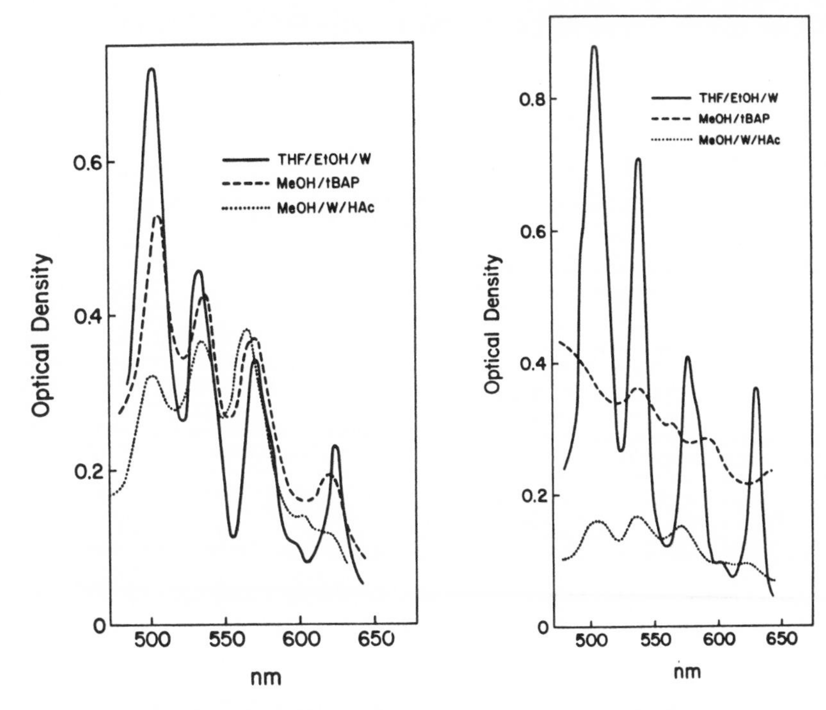

Fig. 2. Absorbance spectra of 100 μg/ml HP (left) and PP (right) dissolved in THF:ethanol:water (2:1:1), methanol:BAP (65:35) or methanol:water:acetic acid (80:20:4).

porphyrin components of HPD, judged from spectral studies.

Octanol:water partitioning studies (pH 7) indicate preferential uptake of HP, HVD and PP into the octanol phase, when solutions of these porphyrins were initially made up in DMF. At higher concentrations, the partition coefficients change, indicating corresponding alterations in water solubility with concentration-dependent drug aggregation (Table 1). For HP and HVD, these results were not altered when initial porphyrin solutions were made up in water. In the case of protoporphyrin, the solvent affected subsequent partitioning behavior. PP in DMF solution was substantially more octanol-soluble than was PP initially dissolved in water. Regardless of the initial drug solvent, spectral studies indicated that PP which partitioned into the aqueous phase was highly aggregated, while drug found in the organic phase showed spectral characteristics of monomer (Fig. 3).

Table 1. Partition Coefficients

Porphyrin	Concentration μg/ml	[Octanol/water]	
		DMF†	Water†
HP	100	17	15
	10	9.1	8.7
	1.0	4.0	3.6
HVD	100	3.3	2.9
	10	4.9	4.6
	1.0	9.3	8.7
PP	100	2	0.06
	10	37	0.85
	1.0	290	2.9

†Initial solvent for porphyrin solutions.

Studies on porphyrin uptake and photosensitization *in vitro* indicated that PP accumulation was 10-fold faster, with a corresponding enhancement of photo-toxicity when a DMF solution of drug was employed. This result indicates preferential PP uptake under conditions which lead to drug partition into the octanol phase of an octanol/water system.

With aqueous solutions of HPD, we found that the partitioning behavior was strongly dependent on concentration (Fig. 4). At a 10 μg/ml concentration, some HP was found in the aqueous phase, along with all of the Π component. As the HPD concentration was increased, progressively more of the PP, HVD and HP partitioned into the aqueous phase.

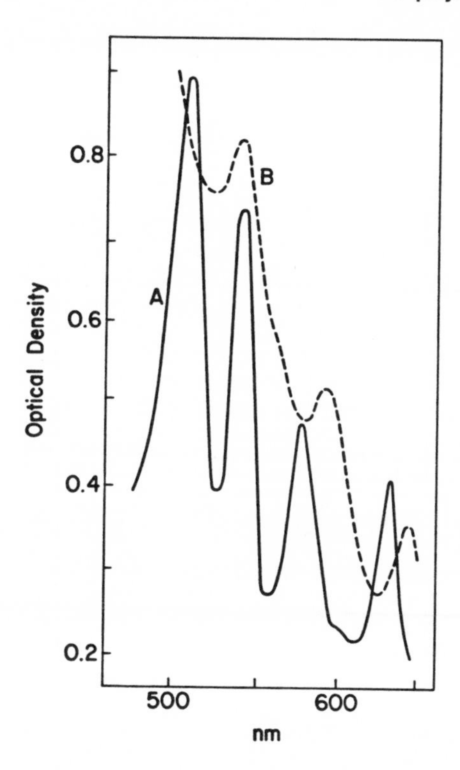

Fig. 3. Absorbance spectra of [A] octanol-rich, and [B] aqueous phases of an octanol:water system containing 100 μg/ml PP (initially dissolved in dimethylformamide).

But the Π component, which reverse-phase analysis suggests to be the most hydrophobic material present, always partitioned into the aqueous phase of the water:octanol mixture.

Uptake of HPD by S-180 cells in vitro indicated slow accumulation of the Π component with time (Fig. 5). Initially, PP was the predominant intracellular porphyrin; after 24 hr Π predominated. With either 0.5 or 24 hr loading incubations, only the Π pool was stable to washing.

DISCUSSION

We have previously described studies designed to assess modes of porphyrin-induced cytotoxicity. These have impli-

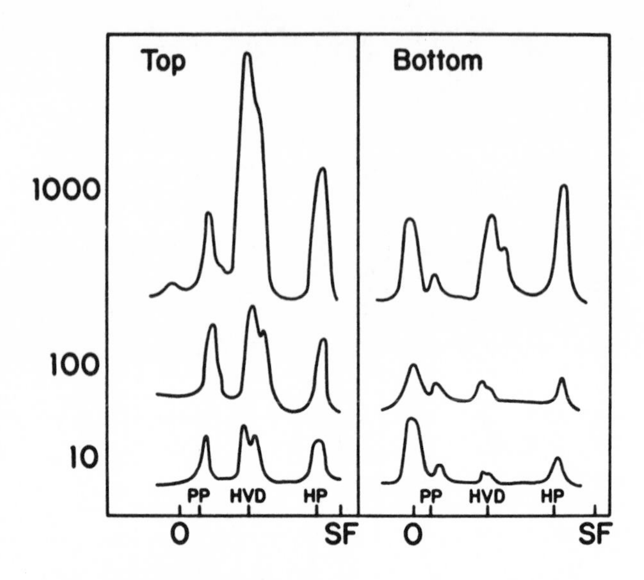

Fig. 4. Reverse-phase TLC analysis of bottom (aqueous) & top (organic) phases of an octanol:water mixture containing 1000, 100 or 10 μg/ml HPD. The origin (0), solvent front (SF) and Rf values for PP, HVD and HP are shown.

cated membrane loci (Kessel 1977; Kohn, Kessel 1979; Bellnier, Dougherty 1982) as early sites of photodamage. Binding of potentially phototoxic porphyrins at or near membrane loci was therefore postulated (Moan, Christensen 1982, Kessel 1979). There is some evidence that this damage can be repaired, if cell lysis does not occur (Moan, Petterson, Christensen 1979). Studies on membrane damage associated with photodynamic protein cross-linking have been reported from other laboratories (Dubbelman, de Goeij, van Steveninck 1978; Girotti 1976, 1979; Lamola, Doleiden 1980), as has impaired transport (Dubbelman, De Goeij, van Steveninck 1980) and membrane destruction detectable by electron microscopy (Malik, Djaldetti 1980). Other studies indicated that variations in chemical structure affect both porphyrin localization (Brun, Høvding, Romslo 1981; Moan, Sandberg, Christensen, Elander 1983; Sandberg, Romslo 1981; Kessel 1982a; Kessel, Chou 1983), and quantum yields of photo-toxic products (Jori, Reddi, Rossi, Cozzani, Tomio, Zorat, Pizzi, Calzavara 1980; Moan, McGhie, Christensen 1982; Tomio, Red-

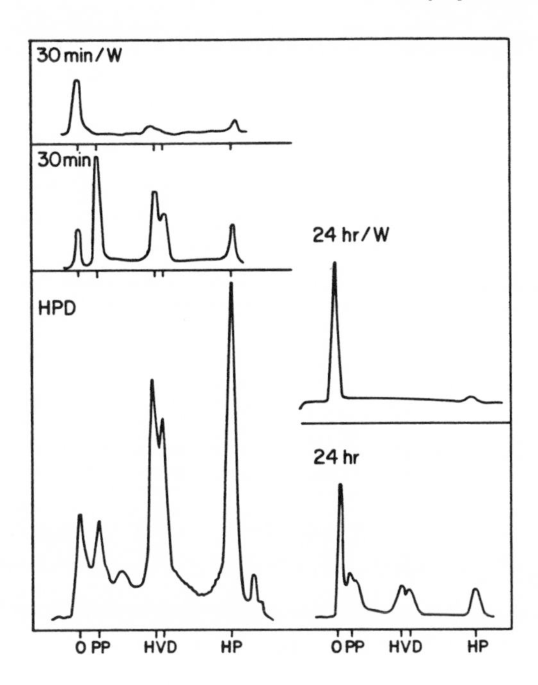

Fig. 5. Lower left: Reverse-phase TLC fluorescence scan of HPD (solvent = 65:35 methanol:BAP). Also shown are TLC separations of porphyrins accumulated by S-180 cells _in vitro_ after 30 min or 24 hr incubations (HPD level = 30 μg/ml). Effects of a subsequent 30 min wash (W) in growth medium at 37° are also shown.

di, Jori, Zorat, Pizzi, Calzavara 1980; Cauzzo, Gennari, Jori, Spikes 1977).

In studies relating to porphyrin transport and photosensitization, it is important to ascertain the composition of products employed. In the case of HP, photosensitization of intact cells was traced to presence of hydrophobic impurities mainly HVD and PP (Kessel 1982b). Some preparations of HP appear to contain substantial amounts of such impurities (Moan, McGhie, Christensen 1982).

Studies carried out using S-180 cells _in vitro_ show that the HPD component most rapidly accumulated during short incubations was protoporphyrin (Fig. 5), although this was not the most hydrophobic component. Reverse-phase analyses indicate presence of an apparently more hydrophobic component of HPD which does not migrate on reverse-phase TLC plates (65% methanol). We have termed this the Π component; it was preferentially retained when HPD-loaded cells were washed in medium containing serum, and gradually forms the predominating intracellular porphyrin pool when incubations were extended to 24 hr (Fig. 5). We interpret the data shown in Fig. 5 as an in vitro demonstration of tumor localization by HPD which occurs in vivo.

Octanol:water partitioning studies show that the Π component, in spite of its apparent hydrophobicity, has a strong affinity for the aqueous phase. Partitioning studies done with PP bear on this result and suggest that partitioning behavior is a function of the initial porphyrin solvent. PP dissolved in DMF, which spectral studies show to be present in a dissociated form, partitions into the octanol phase. In contrast, drug made up in aqueous solution, resulting in formation of a predominantly aggregated species, partitions into the aqueous phase.

An important structure-related phenomenon is the aggregation of porphyrins in different solvents (Brown, Shillcock 1976; Kessel, Rossi 1982; Moan, Sommer 1981). Dougherty initially pointed out the potential importance of aggregation phenomena in the tumor-localization process (Dougherty 1983), and reports from many other laboratories cited above have described a relationship between aggregation phenomena and properties of different porphyrins in solution and in biological systems. We have reported on effects of such aggregation on fluorescence and absorbance spectra capacity and rates of photooxidation (Kessel, Rossi 1982). Effects of structural variations on tumor-localizing phenomena have also been reported (Kessel 1982a, 1982b; Kessel, Chou 1983; Berenbaum, Bonnett, Scourides 1982; Dougherty 1983; Moan, Christensen, Sommer 1982; Moan, Sandberg, Christensen, Elander 1983; Moan, McGhie, Christensen 1982).

Results shown in Table 1 indicate that, with increasing concentration, HP forms more hydrophobic aggregates, while HVD and PP aggregates become progressively less hydrophobic. The partitioning results must derive from the forces holding these aggregates together. With hydrophobic porphyrins in aqueous solution, we conclude that aggregates are held together by hydrophobic interactions. If the carboxyl moieties tend to cluster at the outer surfaces, a high degree of affinity of such aggregates for the aqueous phase would be expected. A related phenomenon was found responsible for the solubility properties of bilirubin, wherein intermolecular hydrogen bonding transforms a relatively hydrophilic molecule into a highly water-insoluble species (Broderson 1983).

In the case of Π, apparently the most hydrophobic component of HPD, we propose that hydrophobic interactive forces would be especially strong. The preferential partition of Π from an aqueous HPD solution into the aqueous phase would thereby be explained.

Studies with PP show that material initially dissolved in DMF was better accumulated by S-180 cells *in vitro* than was drug initially dissolved in water. This result indicates that the relatively stable aggregates of a hydrophobic porphyrin, in aqueous solution, tend to be poorly accumulated by cells.

Based on studies described here, the following is postulated as the initial sequence of events in the tumor localization process.

1. While all components of HPD are more or less aggregated in aqueous solution, the Π component forms such stable aggregates that these are not readily dissociated *in vivo* or in tissue culture media.

2. Aggregates of Π are held together via strong hydrophobic forces, so that a predominantly hydrophilic surface is presented to the environment. Other porphyrin components of HPD form such aggregates, notably PP, but these are slowly dissociated upon contact with plasma proteins or lipid-rich cell surface domains. The initial rate of uptake of HPD components by cells is of the order: PP > HVD > Π >> HP. The latter compound is apparently insufficiently hydrophobic to be effectively accumulated.

3. Once accumulated by S-180 cells, PP, HVD and HP are readily removed by washing in medium containing serum. In contrast, the Π component is selectively retained, perhaps because of reaggregation to form a hydrophilic product which poorly crosses the cell membrane.

REFERENCES

Bellnier DA, Dougherty TJ (1982). Membrane lysis in Chinese hamster ovary cells treated with hematoporphyrin derivative plus light. Photochem Photobiol 36:43.
Berenbaum MC, Bonnett R, Scourides PA (1982). In vivo biological activity of the components of haematoporphyrin derivative. Br J Cancer 45:571.
Bonnett R, Ridge RJ, Scourides, P (1980). Hematoporphyrin derivative. JCS Comm 1980:1198.
Bonnett R, Ridge RJ, Scourides, PA (1981). On the nature of haematoporphyrin derivative. JCS Perkin I 3135:3140.
Broderson R (1982). Physical chemistry of bilirubin: binding macromolecules and membranes. In "Bilirubin Volume I" KPM Heirwegh SB Brown (eds) p 75. Boca Ratan CRC Press.
Brown SB, Shillcock M (1976). Equilibrium and kinetic studies of the aggregation of porphyrins in aqueous solution. Biochem J 153:279.
Brun A, Høvding G, Romslo I (1981). Protoporphyrin-induced photohemolysis: differences related to the subcellular distribution of protoporphyrin in erythropoietic protoporphyria and when added to normal red cells. Int J Biochem, 13: 225.
Bugelski PJ, Porter CW, Dougherty TJ (1981). Autoradiographic distribution of hematoporphyrin derivative in normal and tumor tissue of the mouse. Cancer Res 41:4606.
Cadby PA, Dimitriades E, Grant HG, Ward AD (1983). The analysis of derivative. In Kessel D, Dougherty TJ (eds): "Porphyrin Photosensitization", New York: Plenum Press, p 251.
Cauzzo G, Gennari G, Jori G, Spikes, JD (1977). The effect of chemical structure on the photosensitizing efficiencies of porphyrins. Photochem Photobiol 25:389.
Christensen T, Moan J, McGhie JB, Waksvik H, Stigum H (1983). Studies of HPD: chemical composition and in vitro photosensitization. In Kessel D, Dougherty TJ (eds): "Porphyrin Photosensitization", New York: Plenum Press, p 151.
Clezy P, Hai TT, Henderson RW, Thuc LV (1980). The chemistry of pyrrolic compounds XLV Hematoporphyrin derivative: haematoporphyrin diacetate as the main product of the reaction of haematoporphyrin with a mixture of acetic and sulfuric acids. Aust J Chem 33:585.
Dougherty TJ (1983). Photoradiation therapy - clinical and drug advances. In Kessel D, Dougherty TJ (eds): "Porphyrin Photosensitization", New York: Plenum Press, p 3.

Dubbelman TMAR, de Goeij AFPM, van Steveninck J (1978). Protoporphyrin-sensitized photodynamic modification of proteins in isolated human red blood cell membranes. Photochem Photobiol 28:197.

Dubbelman TMAR, De Goeij AFPM, van Steveninck J (1980). Protoporphyrin-induced photodynamic effects on transport processes across the membrane of human erythrocytes. Biochim Biophys Acta 595:133.

Girotti A (1976). Photodynamic effect of protoporphyrin IX on human erythrocytes: cross-linking of membrane proteins. Biochem Biophys Res Commun 72:1367.

Girotti A (1979). Protoporphyrin-sensitized photodamage in isolated membranes of human erythrocytes. Biochem 18:4403.

Gomer CJ, Dougherty TJ (1979). Determination of [^{3}H]- and [^{14}C] hematoporphyrin derivative distribution in malignant and normal tissue. Cancer Res 39:146.

Jori G, Reddi E, Rossi E, Cozzani I, Tomio L, Zorat PL, Pizzi GB, Calzavara F (1980). Porphyrin-sensitized photoreactions and their use in cancer therapy. Medicine, Biology, Environment 8:140.

Kessel, D (1977). Effects of photoactivated porphyrins at the cell surface of leukemia L1210 cells. Biochem 16:3443.

Kessel, D (1981). Transport and binding of hematoporphyrin derivative and related porphyrins by murine leukemia L1210 cells. Cancer Res 41:1318.

Kessel, D (1982a). Components of hematoporphyrin derivatives and their tumor-localizing capacity. Cancer Res 42:1703.

Kessel, D (1982b). Determinants of Hematoporphyrin-Catalyzed photosensitization. Photobiol Photochem 36:99.

Kessel D, Chou TH (1983). Tumor-localizing components of the porphyrin preparation HPD. Cancer Res In Press.

Kessel D, Rossi E (1982). Determinants of porphyrin-induced photo-oxidation characterized by fluorescence and absorption spectra. Photochem Photobiol 35:37.

Kohn K, Kessel D (1979). On the mode of cytotoxic action of photo-activated porphyrins. Biochem Pharmacol 28:2465.

Lamola AA, Doleiden FH (1980). Cross-linking of membrane proteins and protoporphyrin-sensitized hemolysis. Photochem Photobiol 31:597.

Malik Z, Djaldetti M (1980). Destruction of erythroleukemia, myelocytic leukemia and Burkitt lymphoma cells by photoactivated protoporphyrin. Int J Cancer 26:495.

Moan J, Christensen T (1982). Photodynamic effects on human

cells exposed to light in the presence of hematoporphyrin Localization of the active dye. Cancer Lett 11:209.

Moan J, Christensen T, Sommer S (1982). The main photosensitizing components of hematoporphyrin derivative. Cancer Lett 15:161.

Moan J, McGhie JB, Christensen T (1982). Hematoporphyrin derivative: photosensitizing efficiency and cellular uptake of its components. Photobiochem Photobiophys 4:337.

Moan J, Petterson EO, Christensen T (1979). The mechanism of photodynamic inactivation of human cells in vitro in the presence of haematoporphyrin. Br J Cancer 39:398.

Moan, J, Sandberg, S, Christensen, C, Elander, S (1983). Hematoporphyrin derivative: chemical composition, photochemical and photosensitizing properties. In Kessel D, Dougherty TJ (eds): "Porphyrin Photosensitization," New York: Plenum Press, p 165.

Moan J, Sommer S (1981). Fluorescence and absorption properties of the components of hematoporphyrin derivative. Photobiochem Photobiophys 3:93.

Sandberg S, Romslo I (1981). Porphyrin-induced photodamage at the cellular and the subcellular level as related to the solubility of the porphyrin. Clin Chim Acta 109:193.

Tomio L, Reddi E, Jori G, Zorat P, Pizzi GB, Calzavara F (1980). Hematoporphyrin as a sensitizer in tumor phototherapy: effect of medium polarity on the photosensitizing efficiency and role of the administration pathway on the distribution in normal and tumor-bearing rats. Springer Ser Opt Sci 22:76.

Porphyrin Localization and Treatment of Tumors, pages 419–442

PORPHYRIN-SENSITIZED PHOTOINACTIVATION OF CELLS IN VITRO

Johan Moan PhD, Terje Christensen MSc and
Petter Balke Jacobsen MSc

Norsk Hydro's Institute for Cancer Research,
Montebello, Oslo 3, Norway

INTRODUCTION

Selective retention of the photosensitizer Hpd in tumors is the basis for the use of this drug in diagnosis and photoradiation therapy (PRT). Its tumor localizing and photosensitizing properties have been demonstrated by a number of groups. (Lipson et al. 1961; Gomer, Dougherty 1979; Henderson et al. 1980; Benson et al. 1982; Berenbaum et al. 1982; Kelly, Snell 1976; Hayata et al. 1982; Dougherty et al. 1978; Kennedy 1983; Moan et al. 1982a). It seems that part of the Hpd content of a tumor is bound to extracellular structures (Bugelski et al. 1981). The relative contribution of the extracellular and the intracellular Hpd to the photodestruction of a tumor is unknown. Therefore, a comparison of cells of different oncogenic potential with respect to uptake of Hpd and photosensitivity is warranted.

Hpd is a mixture of many porphyrins, and one of the first aims of the applied research in this field should be to analyse each of its components further with respect to chemical identity, tumor localizing and photosensitizing properties. This may form the basis for synthesizing a porphyrin or preparing a porphyrin mixture with optimal properties.

The sensitivity of human cells in vitro to treatment with porphyrins and light varies with the experimental conditions: pH, oxygen tension, temperature and serum content of the medium surrounding the cells (Moan et al. 1980; Moan et al. 1979; Moan, Christensen 1979; Moan, Christensen 1981), incubation time (Moan, Christensen 1981) density of cells (Christensen, Moan 1980) and their stage in the cell cycle (Christensen et al. 1981). Such variations indicate that a search for improved schedules of therapy may be fruitful. We here report experiments with mammalian cells in vitro carried out to elucidate some of the subjects mentioned above.

MATERIALS AND METHODS

Chemicals

Hpd was prepared from hematoporphyrin dihydrochloride (Koch-Light) according to the methods of Lipson et al. (1961) and dissolved as described earlier (Moan et al. 1982b). Hpd prepared in our laboratory had the same HPLC chromatogram, photosensitizing efficiency and tumor-seeking ability as an injectable solution of Hpd (Photofrin) bought from Oncology Research & Development, Cheektowaga, N.Y., USA. Hematoporphyrin free base (Hp) was obtained from Sigma and protoporphyrin was bought from Porphyrin Products, Logan, Utah, USA.

Radioactively labelled thymidine (methyl-^{3}H-TdR, 5.0 Ci/mmol, 1 mCi/ml, code TRA.120, valine (L-(1-^{14}C)valine, 283 mCi/mmol, 50 μCi/ml, code CFB.75), α-aminoisobutyric acid (2-amino(1-^{14}C) isobutyric acid, 40 - 60 mCi/mmol code CFA.203) and chromate ($Na_2{}^{51}CrO_4$, 350 - 600 mCi/mg Cr, 10 - 25 mCi/ml, code CJS. 4) were obtained from the Radiochemical Centre, Amersham. Dulbeccos phosphate buffered saline (PBS) was bought from Gibco. All chemicals used were of the purest grade commercially available.

Cell Cultures

In the main part of the work we used the cell line NHIK 3025 which is derived from a human carcinoma in situ (Nordbye, Oftebro 1969; Oftebro, Nordbye 1969). The cells were cultivated in E2a medium containing 20% human serum and 10% horse serum as described elsewhere (Moan et al. 1979). In some experiments we used mouse embryo cells of the line C3H/10T1/2 - clone 8 and their 7,12-dimethylbenz(a)anthracene transformants termed type III. 10T1/2 cells do not induce tumors when inoculated in syngeneic immunosuppressed mice while 10^4 cells of type III do so (for references see Moan et al. 1981) The mouse embryo cells do not form well defined colonies. To estimate the sensitivity of such cells to a particular treatment, trypsination was carried out one or two days after the treatment and the cell number was determined by means of a Coulter counter. The NHIK 3025 cells were incubated with E2a medium containing 30% serum for one week. At that time the number of colonies was determined. Cell proliferation was followed by counting the number of cells within a well defined area on the bottom of the tissue culture flasks.

Irradiation

Cells were inoculated in 25 cm^2 tissue culture flasks. After incubation with porphyrins the cells were exposed to the light from 4 fluorescent tubes (Philips, TL 20W/09). The fluence rate at the position of the cells was 17 W/m^2 and the main part of the emission spectrum of the light overlapped the Soret band of the porphyrins. Fluence rates were measured by means of a calibrated thermopile (YSI, 65 A, Yellow Springs, OH). When the action spectrum for inactivation was estimated, we used a 1000 W high pressure Xenon lamp fitted to a Bausch & Lomb grating monochromator. The bandwidth of the light reaching the cells was 20 nm, and its fluence rate ranged from 25 W/m^2 at 360 nm to 125 W/m^2 at 580 nm.

Analytical Procedures

High pressure liquid chromatography was carried out with a reversed phase LC 18 column and a LDC liquid chromatograph with gradient elution, variable wavelength absorbance detector and fluorescence detector ($\lambda exc \sim 350$-400 nm). Mixtures of methanol and water were used as the mobile phase. Details of the chromatography may be found elsewhere (Moan et al. 1982).

Absorption-and fluorescence spectra were recorded by means of a Cary 118 spectrophotometer and a Hitachi MPF-2A fluorescence spectrometer, respectively. Spectroscopy of Hpd in cells was carried out by labelling 7.10^5 cells per tissue culture dish (28 cm^2) with 25 µg/ml Hpd in E2a medium containing 3% serum for 18 h, washing the cells four times in ice-cold PBS, removing the cells from the dishes by a rubber policeman and suspending them in 2 ml cold PBS. A similar sample of cells without Hpd was placed in the reference cuvette. The bandwidth of the fluorescence excitation light was 5 nm, and the fluorescence detection monochromator was set at 632 nm for excitation in the region 340 - 600 nm and at 680 nm for excitation in the region above 600 nm.

A premilinary action spectrum for inactivation of cells was recorded by incubating about 10^4 cells per tissue culture flask (25 cm^2) with Hpd in a similar manner as described above. Areas of 0.5 cm^2 were given different light fluences and the cell numbers in these areas were determined after two days incubation in growth medium. Boundaries between irradiated and unirradiated areas were completely sharp, showing that cellular migration caused no error. The values given for the action spectrum are proportional to D_{10}^{-1} where D_{10} is the fluence (measured in quanta per cm^2) needed to inactivate 90% of the cells.

The analytical procedures for measuring incorporation of ^{3}H-labelled thymidine in DNA, ^{14}C-valine in proteins, uptake of α-aminoisobutyric acid, and cellular loss of incorporated chromate are described elsewhere (Moan et al. in press 1983). Scanning electron microscopy was carried out as previously described (Moan et al. 1982 c).

RESULTS AND DISCUSSION

Analysis of Hpd-Components

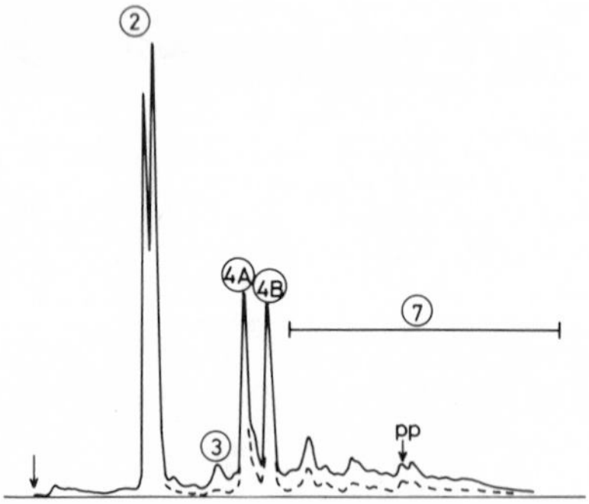

Figure 1. HPLC chromatogram of Hpd recorded as described by Moan et al. (1980). The full line represents optical absorption and the dashed line represents the fluorescence which was recorded simultaneously. The fluorescence tracing is normalized to make the peaks of component 2, 4A and 4B of the same height as the corresponding peaks in the absorption curve. The optical absorption was recorded at 393 nm.

Figure 1 shows the HPLC chromatogram of Hpd. The components marked 2, 4A, 4B and pp have been identified as hematoporphyrin [2] the isomers 2(4)-hydroxy-ethyl-4(2)-vinyldeuteroporphyrin [4A, 4B] and protoporphyrin [pp]. (See Moan et al. 1983 and references cited there). It is well known that aggregates have a low fluorescence quantum yield (Moan et al. 1983, White 1978). Figure 1, therefore, indicates that component 3 and component 7 consist of aggregated material. This has been verified by gel permeation chromatography using a P 10 column (Bio-Rad). Component 7 obviously contains a number of porphyrins. The concentration of these porphyrins was so small that it was impossible to carry out individual biologic experiments with them. Therefore, component 7 was tested as a whole. Hematoporphyrin (Hp) dissolved in water in the same way as Hpd, has a chromatogram which is similar to that of Hpd except for one feature: it contains about 60 per cent less of component 7 than Hpd does (table 1).

Table 1

	Relative amount of			Relative sensitizing efficiency
	component 2	components 4A + 4B	component 7	
Hp(NaOH-treated)	2.1	0.5	0.38	0.35
Hpd	0.9	0.6	1.0	1.0

The relative amount of different components present in Hpd and Hp (Sigma) treated for 1h with 0.1N NaOH, and the relative sensitizing efficiency of the two porphyrin mixtures. The sensitizing efficiency is defined as the inverse value of the exposure time needed to inactivate 90% of cells (t_{10}^{-1}). The cells were incubated for 30 min at 37°C with the porphyrins dissolved in PBS containing 1% human serum to a concentration of 25 μg/ml.

Since the sensitizing efficiency of Hp in a cellular system is about 65 per cent lower than that of Hpd, it seems that the sensitizing effect of both compounds is mainly due to component 7. A similar experiment was carried out with Hp dihydrochloride, Hpd and heat-treated Hpd (table 2).

Table 2

	Relative amount of			Relative Sensitizing Efficiency	
	component 2	components 4 A + B	component 7	1h incub.	18h incub.
Hp	1.7	0.2	0.3		0.1
Heat-treated Hpd	0.1	0.4	2.3	3.4	3.8
Hpd	0.9	0.6	1.0	1.0	1.0

The relative amount of different components present in Hpd, Hp-diHCl (Koch Light) and heat treated (3h, 150°C)Hpd compared with the relative sensitizing efficiencies after 1 and 18h incubation in E2a containing 12.5 μg/ml porphyrin and 3% serum. The sensitizing efficiencies are given relative to those of Hpd. For Hpd the absolute sensitizing efficiencies were t_{10}^{-1}(1h) = 0.003 s^{-1} and t_{10}^{-1}(18h) = 0.02 s^{-1} after incubation times of 1 and 18h, respectively.

The heat-treatment was initially carried out to see if Hpd could be sterilized in that way. An aqueous solution of Hpd was evaporated under reduced pressure, and the resulting solid Hpd was brought to 150°C for 3 hours. Table 2 shows that the heat-treatment gave a product that was 3 to 4 times more efficient in sensitizing cells to photoinactivation than Hpd itself. Component 7 had a different composition in Hp, Hpd and heat-treated Hpd. Notably, the heat-treated Hpd contained more of a component with a retention time equal to that of protoporphyrin than Hp and Hpd. This component may be protoporphyrin, although the heat-treated Hpd seemed to be more soluble in water than one would expect if a large fraction of it was protoporphyrin. An explanation of this could be that protoporphyrin in heat-treated Hpd is dissolved by forming mixed aggregates with more water soluble components such as Hp and hydroxyetyl-vinyl-deuteroporphyrin. The data clearly indicate that the photosensitizing ability of the three porphyrin mixtures is mainly due to their least polar fraction (i.e. to component 7). This is true both for cells incubated for 1h and 18h with the dyes, respectively. The high efficiency of component 7 is even more clearly demonstrated by the data in table 3.

Table 3

Component	Relative efficiency of sensitizing	
	inactivation	release of $[^{51}Cr]$ chromate
2	0.08	0.05
4A	0.33	0.3
4B	0.47	0.4
7	1.0	0.6

Sensitizing efficiency of the main components of Hpd, separated by HPLC. The cells were incubated for 1h at 37°C with the Hpd-components dissolved in PBS (OD_{360} = 0.2). Cells for the chromate release assay were exposed to the radioactive label in growth medium (1.7 µCi/ml) for 24h prior to porphyrin labelling and irradiation. The efficiency of inactivation is defined as in table 1 and that of sensitizing chromate release is taken as the inverse value of the fluence needed to release half of the releasable amount of chromate 5h after the irradiation.

In another contribution in this book it is shown that component 7 is the best tumor-seeker of the Hpd components (Evensen et al.).

Our findings are in correspondence with Kessel's report of the correlation between hydrophobicity and photosensitizing capacity of porphyrins (Kessel 1977). An in vivo testing of the photosensitizing properties of Hpd components was carried out by Berenbaum et al. (1982). They did not attempt to measure the tumor uptake and the photosensitizing efficiency separately. However, their data clearly showed, in correspondence with our work, that the biologic active component of Hpd was strongly retained by their reversed phase HPLC column. Furthermore, they reported that protoporphyrin, which is also unpolar and has a long retention time on the column, had no photosensitizing effect on tumors. Again, this is in correspondence with our work which shows that protoporphyrin is a less efficient photosensitizer than Hpd in a cellular system (table 8). However, protoporphyrin has a strong tendency to form large aggregates in aqueous solutions. Thus, different methods of dissolving protoporphyrin may result in different distributions of aggregate sizes. In fact, we have demonstrated that the photosensitizing efficiency of protoporphyrin is strongly dependent on the method of dissolving the compound (results not shown). Therefore,one should be careful to give general statements concerning the photosensitizing efficiency of protoporphyrin. In principle, it is also possible that the method of bringing Hpd in aqueous solution may have an influence on the nature of the aggregates, although we have not observed any such effect.

Hpd Uptake and Retention in Cells of Different Oncogenic Potential

It has been reported that malignant cells may retain more Hpd than nonmalignant ones (Mossman et al. 1974). To test if Hpd retention correlates with malignancy we performed some experiments with mouse embryo cells with widely different oncogenic potential. Our first series of experiments indicated that the malignant cells incorporated slightly more Hpd per volume unit than the nonmalignant ones (table 4).

Table 4

	Type 10T1/2			Type III		
	PBS	PBS +1% serum	PBS+10% serum	PBS	PBS+1% serum	PBS+10% serum
[Hpd]per volume unit cells (rel. units)	4.0	1.0	0.31	6.7	1.5	0.38
Sensitivity (t_{10}^{-1}, min^{-1})		0.9			0.9	
Sensitivity/ [Hpd]		0.9			0.6	

Hpd uptake and sensitivity of 10T1/2 cells and type III cells (see materials and methods). Before irradiation the cells were incubated for 30 min at 37°C with 25 µg/ml Hpd in PBS containing the given amounts of human serum. The exposure time needed to inactivate 90% of the cells is t_{10}. Results from Moan et al. (1981).

However, the two cell populations had the same sensitivity to light in the presence of Hpd. In our second series of experiments we wanted to take into account the ability of the cells to retain Hpd and exposed them to fresh medium for 4 h between Hpd labelling and irradiation (table 5).

Table 5

	Type 10 1/2	Type III
[Hpd] per g prot (rel. units)	1.7	2.5
Sensitivity (t_{10}^{-1}, min^{-1})	0.26	0.38
Sensitivity/[Hpd]	0.15	0.15

Hpd retention and sensitivity of 10T 1/2 cells and type III cells (see materials and methods). The cells were incubated for 22h at 37°C with 25 μg/ml Hpd in MEM-medium containing 10% NBC serum and then with fresh medium without Hpd for 4 hours before irradiation. (Christensen et al. Br. J. Cancer in press). The exposure time, t_{10}, is needed to inactivate 90% of the cells.

These experiments indicated that the malignant cells retained slightly more Hpd than the nonmalignant ones. In this case the malignant cells were also slightly more photosensitive than the nonmalignant ones. However, in all cases the differences were small, and we conclude that if the present results are relevant for the in vivo situation, the preferential retention of Hpd in tumors cannot be due to different porphyrin uptake in malignant and nonmalignant cells alone. This was also part of the conclusion of Bugelski et al.(1981) who proposed that the tumor uptake and retention of Hpd were due to high vascular permeability and low lymphatic drainage in tumors.

Action Spectrum for Inactivation

The absorption spectra of porphyrin aggregates, porphyrin

mono-and dimers, and protein bound porphyrins are different. Their absorption maxima are located at about 360, 394 and 405 nm, respectively. Furthermore, aggregation results in a broadening of the spectra. It is not known whether the active component of Hpd is aggregated or monomeric when it is inside or bound to the cells. Therefore, it is of interest to determine the action spectrum for inactivation. Our preliminary data shown in figure 2 indicate that the action spectrum is definitely different from the aggregate absorption spectrum in solution. However, since the absorption spectrum of cell-bound aggregates may be different from that of aggregates in solution, this

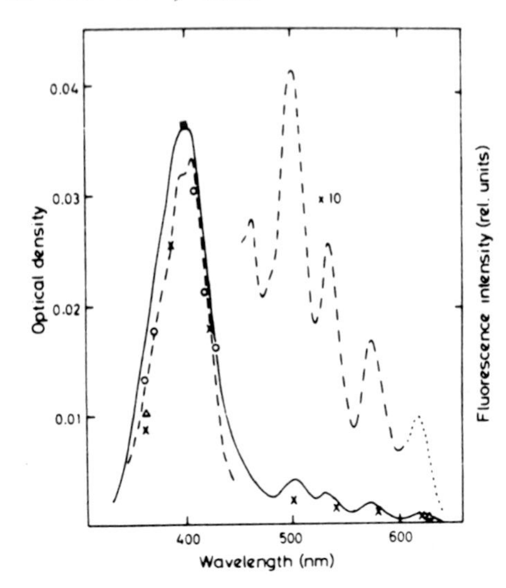

Figure 2. Absorption (———),fluorescence excitation-(-----) and action - spectrum for photoinactivation of cells containing Hpd (o, Δ, x, different symbols correspond to different experiments, all normalized to the same value at 400 nm). The fluorescence excitation spectrum was recorded with the emission monochromator set at 632 (---) and 690 nm(....); the two parts of the spectrum being normalized to the same value at 600 nm.

does not prove that the active part of Hpd in the cells is in a non-aggregated form. Light of wavelength 400 nm is 40-60 times more efficient (per incident light quantum) in killing the cells than light of wavelength 625 nm which is most frequently used in photoradiation therapy (PRT) of tumors. This is about what one would expect considering the absorption spectrum of monomeric porphyrins, and is also in correspondence with our earlier published work, where we studied photoinactivation of NHIK 3025 cells labelled with Hp (Moan, Christensen 1981). Kinsey et al. (1981), using somewhat less monochromatic light, came to a similar conclusion. Thus, it seems that one can assume that the action spectrum for cell killing is similar to the absorption spectrum of cell bound Hpd. Red light is used in PRT because of its penetration in tissue. It is obvious that a tumor-localizing porphyrin with a stronger absorbance than Hpd in the red part of the spectrum would be preferable. In fact, it is possible to synthesize such porphyrins mainly by modifying side groups. Photoprotoporphyrin, for instance, has an absorbance that is much more intense above 600 nm compared with that of Hpd. During irradiation of Hpd we have observed that another porphyrin is generated which has also a strong absorption band above 600 nm. The chromatographic properties of this porphyrin are different from those of photoprotoporphyrin. The tumor localizing and photosensitizing properties of these compounds have not been tested yet.

Figure 2 also shows that the fluorescence excitation spectrum of cell-bound Hpd is almost similar to its absorption spectrum indicating that a large fraction of the porphyrins are in a fluorescent form in the cells. The fluorescence excitation spectrum of Hpd in cells may be recorded with significantly better resolution ($\Delta\lambda < 5$ mμ in figure 2) than the corresponding absorption spectrum. Thus, two peaks are seen in the fluorescence excitation spectrum; one at about 398 nm and one at about 410 nm. The latter peak is more prominent after a long incubation time with Hpd and is less efficiently reduced when the cells are washed with fresh medium containing serum. Thus, it seems that the two peaks correspond to different porphyrin pools in the cells. This may be related to different locations in the cells and/or to different Hpd components. Experiments are being carried out in our laboratory to elucidate this.

Split Dose Experiments

It has been shown that when cells are incubated with Hp for a short time (∿30 min) they are more efficiently inactivated by a split light dose than by a similar single dose (Moan, Christensen 1979). This was shown to be due to the fact that the first part of the split dose induced a membrane damage that resulted in an increased cellular uptake of Hp. Since long incubation time with Hpd is more relevant for the clinical situation, we decided to repeat the experiment under such conditions. Figure 3 shows that the cells seem to become less sensitive if they are given a short light exposure corresponding to 0.14 t_{10} during the first part of the incubation with Hpd.

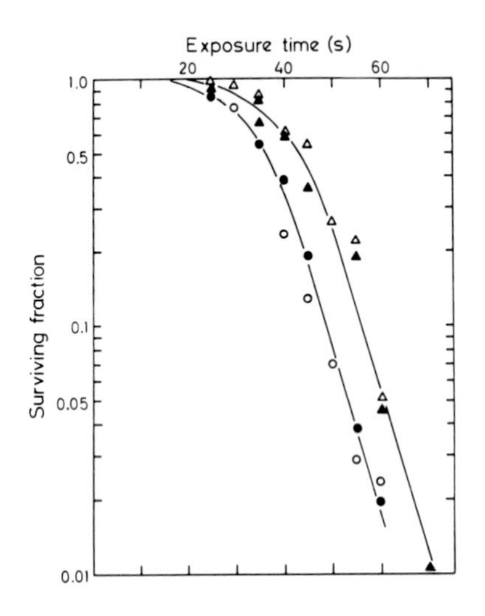

Figure 3.
Survival curves for cells incubated with 12.5 μg/ml Hpd in medium with 3% serum. Triangles correspond to cells irradiated for 50 s (~ 0.14 t_{10}) after 1 h incubation with Hpd and then incubated for another 18 h before the second irradiation. Circles correspond to cells incubated for 19 h without the first light exposure. Open and closed symbols correspond to two experiments carried out on different days.

To check if this was due to light-induced degradation of Hpd during the first light exposure, another experiment was carried out. In this case the porphyrin solution surrounding the cells was renewed immediately after the first light exposure. Furthermore, the first light dose was increased. The results are shown in table 6.

Table 6

First exposure (± Hpd 1 h)	Second exposure (+ Hpd 18 h)	Surviving fraction
3.5 min (- Hpd)	0 s	1.0
3.5 min (- Hpd)	30 s	0.12
3.5 min (- Hpd)	40 s	0.0087
3.5 min (+ Hpd)	0 s	0.5 (1.0)
3.5 min (+ Hpd)	30 s	0.064 (0.13)
3.5 min (+ Hpd)	40 s	0.0044 (0.0090)

Split-dose experiment with 20 μg/ml Hpd in medium containing 3% serum. Three hours after inoculation of the cells Hpd was added to some of the cultures. One hour later all cultures were exposed to light for 3.5 min. Then the medium on all cultures was changed to fresh Hpd-containing medium, the cultures were incubated at 37°C for 18 h and given the second light exposure. The numbers in the parenthesis are the surviving fractions calculated if the surviving fraction of cultures given only the first light exposure is normalized to 1.0.

The first light exposure in the absence of Hpd had no effect on the survival of the cells, while the same dose in the presence of Hpd inactivated 50% of the cells. Cells surviving the first light exposure with Hpd had the same sensitivity as previously unexposed cells and cells given the first light exposure in the absence of Hpd. Thus, the cells seemed to repair sublethal photodamage during 18 h at 37°C.

Degradation of Hpd in PRT

The above mentioned results indicate that the difference between the two survival curves in figure 3 is due to degradation of Hpd during the first light exposure. This is explicitly shown to be true by figure 4.

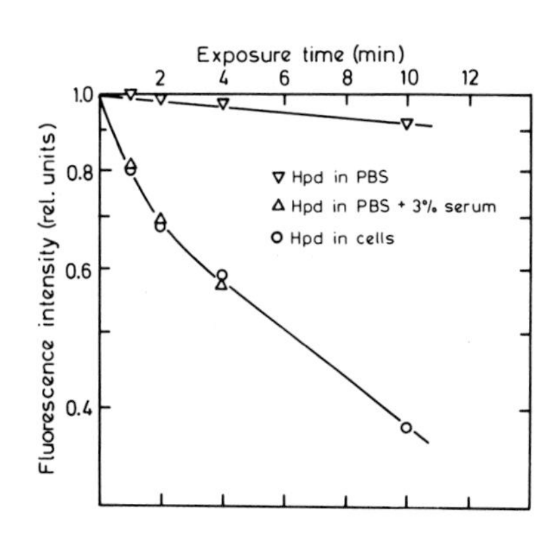

Figure 4.
Degradation of 12.5 μg/ml Hpd under conditions as used in the experiments with cells: (— Δ — in PBS with 3% human serum and — ∇ — in pure PBS). Circles correspond to an experiment where the degradation of Hpd in cells (incubated for 18 h with 12.5 μg/ml dye in medium containing 3% serum), was measured. Immediately before irradiation the cells were washed with PBS. After the irradiation they were removed from the flasks by a rubber policeman and suspended in PBS.

Hpd bound to cells as well as Hpd bound to serum proteins is rapidly degraded by light while Hpd in pure PBS is much more stable. This difference may indicate that excited porphyrin molecules react with cellular components and serum proteins. The singlet oxygen yield for bound and unbound porphyrin molecules is not much different (Moan, Sommer 1981) and it is unlikely that the generated singlet oxygen destroy bound porphyrin molecules faster than free ones.

The light induced degradation of Hpd may influence the shape of survival curves, notably for irradiation after short incubation times with Hpd, low porphyrin concentrations and high serum concentrations. Degradation will tend to reduce the slope of the survival curves, notably at high fluences.

Photodegradation of Hpd should be paid attention to in PRT. It is probable that some of the failures of PRT reported in the literature are due to Hpd bleaching. The present data indicate that PRT will be inefficient if the Hpd concentration in the tumor is too low. A low Hpd concentration in the tumor cannot be compensated for by increasing the fluence, since such an increase will result in Hpd bleaching before the tumor is inactivated. If the conditions in the tumor are comparable with the conditions used in the present work, the limiting Hpd concentration is of the order of 1 μg/g or somewhat less. (The intracellular concentration of Hpd after 1 h incubation with 12.5 μg/ml Hpd in medium with 3% serum is 10-20 μg/ml wet cells, and in that case a D_{10}-fluence degrades about 50% of the cell-bound Hpd.)

50% of the intracellular Hpd content is bleached by a fluence of about 6 kJ/m^2 from our fluorescent lamp. In terms of absorption by cellular bound Hpd this fluence corresponds to a fluence of about 100 kJ/m^2 of monochromatic light of wavelength 625-630 nm. A typical PRT fluence in this wavelength region is about 500 kJ/m^2. (Dougherty et al. 1983). Thus, bleaching is quite significant in the tumor. The quantum yield for destruction of tumor cells may probably be increased if the fluence rate is reduced and if injection of Hpd is carried out so that the drug is provided to the tumor from the blood during the irradiation.

The Shape of Dose-response Curves

Dose-response curves for photoinactivation of cells in

the presence of Hpd have different shapes depending on the experimental conditions. This is illustrated by figure 5 and table 7.

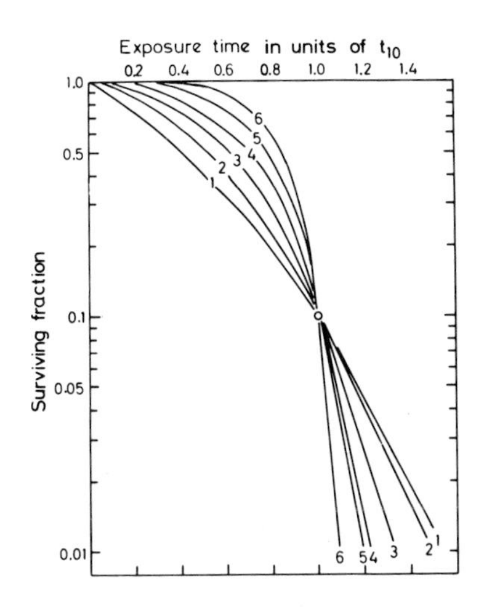

Figure 5.
Fluence response curves for cells irradiated under conditions given in table 7.

The following features should be noted: a) The shoulder of the dose-response curve is relatively more pronounced for short incubation times with Hpd than for long incubation times. 2) For a given incubation time the shoulder increases with decreasing serum concentration. 3) Washing with serum-containing medium reduces the shoulder. 4)Protoporphyrin gives a more pronounced shoulder than the more polar substances Hp and Hpd.

It is likely that the membrane is rapidly saturated with porphyrins while intracellular pools fill more slowly, notably if pinocytosis plays a significant role, as one may assume is the case for aggregated porphyrins and porphyrins strongly bound to serum proteins. Furthermore, it is likely that membrane bound porphyrins are more easily removed from the cells by washing with media containing serum than porphyrins present inside the cells. Finally, nonpolar porphyrins like protoporphyrin are probably more selectively located in the membranes than more polar porphyrins such as Hp. According to the above considerations we propose that a pronounced shoulder of the survival curve indicates that damage to the outer cell membrane plays a significant role for the photo-induced cell inactivation.

Such a model is consistent with the influence of treatment

Table 7

Curve	Porphyrin	Concentration µg/ml	Incubation time (h)	Serum concentration (%)	t_{10} (min)
1	Hp	~240	0.5	30	23
2*	Hpd	12.5	1.0	3	14
2	Hpd	250	0.5	30	7
3*	Hp	12.5	18	3	28
4	Hpd	12.5	18	3	0.8
4	Hp	~ 1.5	1.0	0	19
5	Hpd	12.5	1.0	3	6
5	Hp	12.5	18	3	11
6	Proto	12.5	18	3	1.3
6	Hpd	12.5	0.5	0	1.2

Experimental conditions for the survival curves shown in figure 5. Generally, the cells were irradiated in the presence of porphyrins. In two cases, marked with *, the porphyrin solution was removed from the cells and fresh medium with 3% serum was added 30 minutes before the irradiation. When serum was present, the cells were incubated and irradiated in E2a medium, otherwise (curves 4 and 6) in PBS.

with Hpd + light on cellular parameters discussed below.

Effects on Cellular Parameters

In order to evaluate the importance of different types of cellular damage for inactivation, we studied the effect of Hpd + light on four cellular parameters. This was done for a short (1 h) and for a long (18 h) incubation time with Hpd. In the following discussion the sensitivity of the parameters will be related to the corresponding t_{10}-value for inactivation.

Of the four cellular parameters studied, thymidine incorporation into DNA was by far the most sensitive one to treatment with Hpd + light (table 8). Thymidine incorporation was particularly sensitive after short incubation times with the dye. In this case drastic effects were observed even at light doses that had no effect on the survival of the cells. The reduction in the thymidine incorporation was mainly due to a reduction in the rate of DNA synthesis although a reduced transport rate of thymidine into the cells was also observed.

Table 8

	Efficiency in units of t_{10}^{-1}	
	1 h incubation with Hpd before irradiation	18 h incubation with Hpd before irradiation
Thymidine incorporation	17.0	3.3
Valine incorporation	1.3	1.3
Transport of AIB	2.2	0.6
Loss of chromate	0.6	$\leq$0.3

The table shows the efficiency of Hpd + light in affecting four cellular parameters: thymidine incorporation, valine incorporation, uptake of α-aminoisobutyrate (AIB), and loss of incorporated chromate. The numbers given for the three former parameters are the inverse values of the exposure times needed to reduce the parameters by 50% compared to unirradiated controls. The efficiency for loss of chromate corresponds to the inverse value of the exposure time needed to double the chromate loss compared with unirradiated controls. The chromate loss was measured 1 h after the irradiation. The values of t_{10} are the exposure times needed to inactivate 90% of the cells.

The incorporation of valine into proteins was reduced to the same degree after long and short incubation times with Hpd (table 8). In the latter case this decrease was mainly due to a reduced transport of valine into the cells. Such a reduced transport was not observed in cells irradiated with comparable fluences after long incubation time with Hpd. α-aminoisobutyric acid (AIB) is actively transported into NHIK 3025 cells since anaerobic conditions as well as low temperatures (~4°C) strongly reduce the transport. Irradiation after short incubation times with Hpd reduced the transport of AIB relatively more than irradiation after long incubation with Hpd.

Membrane damage resulting in a general increased permeability to ions is often assayed by measuring the loss of radioactivity from cells labelled with $^{51}CrO_4^{2-}$. This parameter was the least sensitive one to treatment with Hpd and light. However, in the case of irradiation after a brief incubation with Hpd a slight increase in the loss of $^{51}CrO_4^{2-}$ was observed at doses which did not inactivate more than 90% of the cells. When the chromate assay was performed one day after the irradiation (and not 1 h as discussed above) a large fraction of inactivated cells had lost their content of chromate. This is probably just a result of a general destruction of the inactivated cells.

From the observations described above, the following conclusions may be drawn: 1) Cellular parameters associated with membranes (thymidine incorporation, transport of AIB and loss of $^{51}CrO_4^{2-}$) are, relative to inactivation, more sensitive to light after short incubation times with Hpd than after long incubation times. 2) Even though thymidine incorporation is by far the most sensitive parameter, damage to this process is not the main cause of cellular inactivation under all conditions. This conclusion can be drawn since the thymidine incorporation relative to cell survival is 5 times less sensitive to irradiation after long incubation times (18 h) with Hpd than after short incubation times (1 h) (table 8). 3) The present data do not support an earlier suggestion that membrane damage and cell lysis are the general, inactivating processes for cells exposed to porphyrins and light (Moan et al. 1979). Other reports from our laboratory also indicate that lysis is of limited importance. It was shown that cell lysis was less prominent the higher the serum concentration was during irradiation (Christensen et al. 1983). Furthermore, at a given survival level a higher frequency of sister chromatid exchanges was observed for irradiation in the presence than in the absence of serum (Christensen et al. 1983). This is in correspondence with our above suggestions made on the basis of the shape of the fluence response curves.

Effect on Cell Proliferation

The division rate of cells exposed to sublethal doses of Hpd and light is reduced for some hours after the irradiation (figure 6). The lag period seems to be dose dependent, and is similar for cells irradiated after long and short incubation times with Hpd. Christensen (1981) concluded from a study on synchronized cell cultures that sublethal doses resulted in a prolongation of the first interphase and the first mitosis (mainly the metaphase) after treatment with Hpd and light.

Figure 6

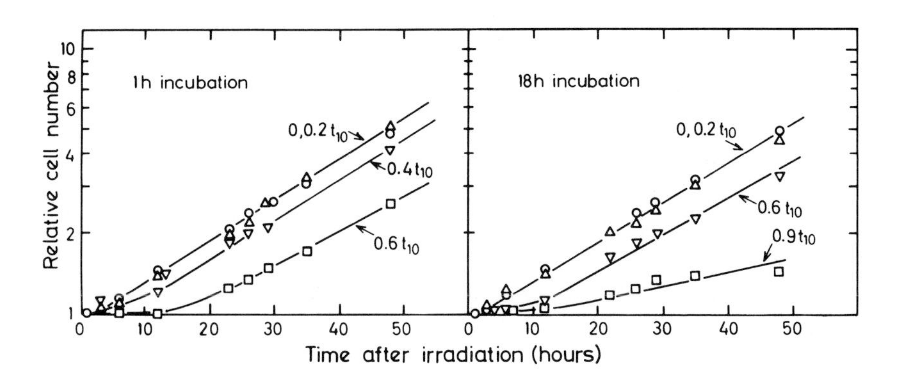

Cell proliferation after treatment with Hpd (12.5 μg/ml in medium with 3% serum) for 1 and 18 h, respectively, and fluences as given on each curve. Immediately after the irradiation the medium was changed to growth medium without Hpd.

Effect of Cell Contact

In an earlier work we studied cells exposed to Hp and light and we found that cells in microcolonies behaved differently from single cells (Christensen, Moan 1980). We decided to repeat this experiment with Hpd. The results obtained were similar to those reported of for Hp: Cells in microcolonies of 3-4 cells seem to be slightly more sensitive to the treatment than single cells (figure 7 and figure 8. As the colonies grow larger, however, the cells seem to return to their initial sensitivity. The reason for this variation in sensitivity is not known yet. Transfer of toxic products between cells in the microcolonies has been proposed as an explanation (Christensen, Moan 1980).

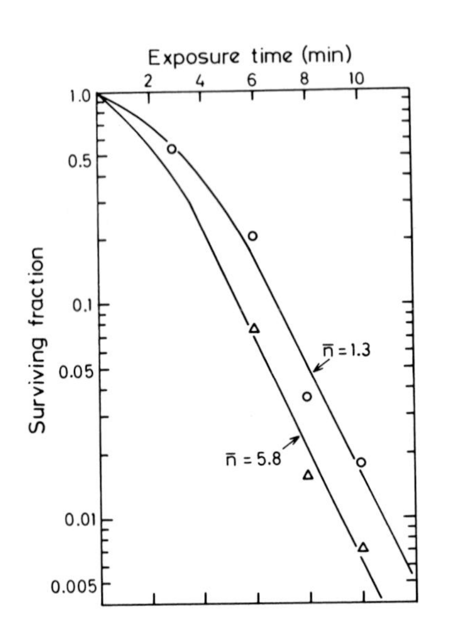

Figure 7

Survival curves for NHIK 3025 cells irradiated as microcolonies containing mean numbers of $\bar{n}$ = 1.3 and $\bar{n}$ = 5.8 cells. The values given for the survival are the calculated values for single cell survival: $s = 1 - (1 - f)^{1/\bar{n}}$, where f is the measured surviving fraction for microcolonies (Sinclair, Morton 1966). Before the irradiation, the cells were incubated for 0.5 h with 250 µg/ml Hpd in medium containing 30% serum.

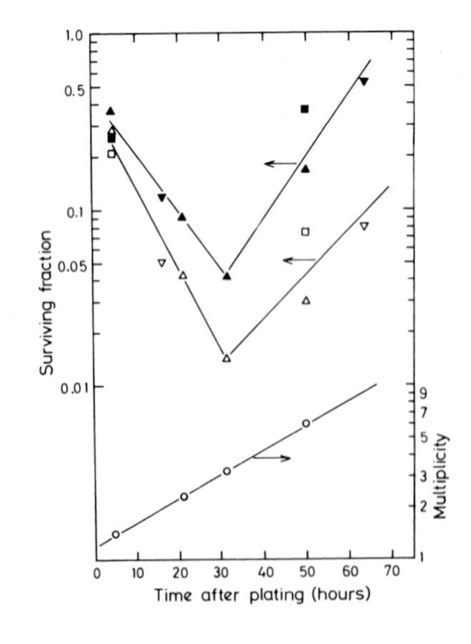

Figure 8.

Surviving fractions of microcolonies of NHIK 3025 cells incubated for 0.5 h with 250 µg/ml Hpd in medium containing 30% serum and irradiated for 6 minutes (closed symbols). The open symbols correspond to the calculated surviving fractions for single cells. The circles in the lower part of the figure show the mean multiplicity of the microcolonies at the time of irradiation as a function of time after plating. Different symbols correspond to different experiments.

Scanning electron micrographs of irradiated cells support this explanation. Figure 9 demonstrates a common feature seen in cell cultures irradiated after short incubation time with Hpd: While most microcolonies only contain cells that are indistinguishable from those of unirradiated controls, a few microcolonies are seen where all cells contain blebs that indicate membrane or cytoskeleton damage. Experiments with microbeam irradiation of cells indicated that transfer of toxic products may take place and is more efficient for malignant than for nonmalignant cells (May et al. 1971). A similar variation in sensitivity with the time between plating and irradiation has been reported for ionizing radiation (Raaphorst et al. 1979). More sophisticated experiments are being carried out to solve this problem.

Figure 9

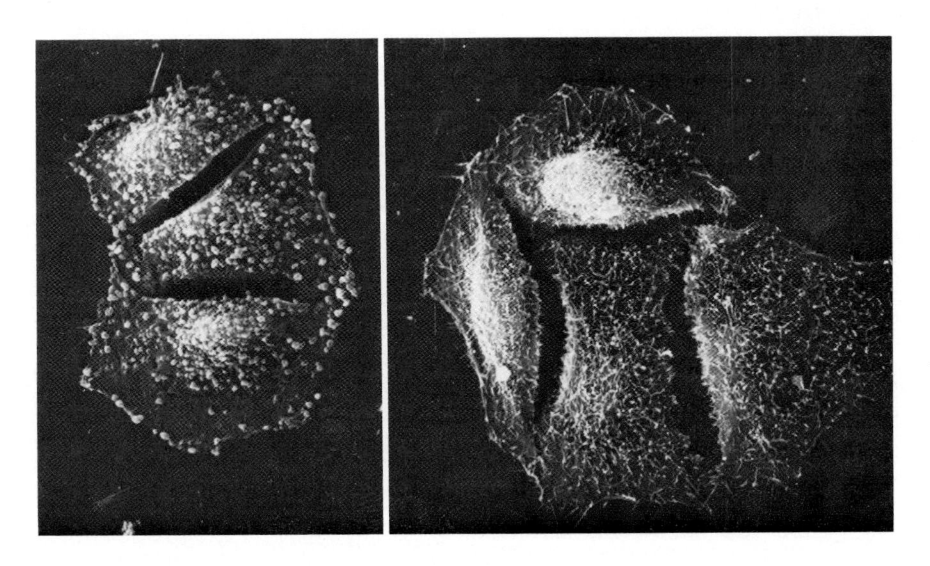

Electron micrographs of microcolonies of NHIK 3025 cells incubated for 1 h with 12.5 μg/ml Hpd in medium with 3% serum and fixed 1 h after the irradiation. The fluence corresponded to about 10% inactivation. The two colonies shown are from the same culture (X 1500).

ACKNOWLEDGEMENTS

The present work was supported by the Norwegian Cancer Society (Landsforeningen mot Kreft) and The Norwegian Research Council for Science and the Humanities.

REFERENCES

Benson RC, Farrow GM, Kinsey JH, Cortese DA, Zinke H, Utz DC (1982). Detection and localization of in situ carcinoma of the bladder with hematoporphyrin derivative. Mayo Clin Proc 57:548.

Berenbaum MC, Bonnett R, Scourides PA (1982). In vivo biological activity of the components of haematoporphyrin derivative. Br J Cancer 45:571.

Bugelski PJ, Porter CW, Dougherty TJ (1981). Autoradiographic distribution of hematoporphyrin derivative on normal and tumor tissue of the mouse. Cancer Res 41:4606.

Christensen T (1981) Multiplication of human NHIK 3025 cells exposed to porphyrins in combination with light. Br J Cancer 44:433.

Christensen T, Feren K, Moan J, Pettersen EO (1981). Photodynamic effects of haematoporphyrin derivative on synchronized and asynchronous cells of different origin. Br J Cancer 44:717.

Christensen T, Moan J (1980). Photodynamic effect of hematoporphyrin (HP) on cells cultivated in vitro. In Pratesi R, Sacci CA (eds) "Lasers in Photomedicine and Photobiology", Berlin, Heidelberg, New York: Springer, p 87.

Christensen T, Moan J,McGhie JB, Waksvik H, Stigum H (1983). Studies on HPD: Chemical Composistion and in vitro photosensitization. In Kessel D, Dougherty TJ (eds) "Porphyrin Photosensitization", New York, London, Plenum Press, p 151.

Dougherty TJ, Boyle DG, Weishaupt KR, Henderson BA, Potter WR, Bellnier DA, Wityk KE (1983). Photoradiation therapy - clinical and drug advances. In Kessel D, Dougherty TJ (eds) "Porphyrin Photosensitization", New York and London, Plenum Press, p 3.

Dougherty TJ, Kaufman JE, Goldfarb A, Weishaupt KR, Boyle D, Mittleman A (1978). Photoradiation therapy for the treatment of malignant tumors. Cancer Res 38:2628.

Gomer CJ, Dougherty TJ (1979). Determination of ^{3}H - and

^{14}C hematoporphyrin derivative distribution in malignant and normal tissue. Cancer Res 39:146.
Hayata Y, Kato H, Konaka C, Ono J, Takizawa N (1982). Hematoporphyrin derivative and laser photoradiation treatment of lung cancer. Chest 81:269.
Henderson RW, Christie GS, Clezy PS, Lineham J (1980). Haematoporphyrin diacetate: A probe to distinguish malignant from normal tissue by selective fluorescence. Br J Path 61:345.
Kelly JF, Snell ME (1976). Hematoporphyrin derivative: a possible aid in the diagnosis and therapy of carcinoma of the bladder. J Urol 115:150.
Kennedy J (1983). HPD photoradiation therapy for cancer at Kingston and Hamilton. In Kessel D, Dougherty TJ (eds): "Porphyrin photosensitization", New York and London: Plenum Press, p 53.
Kessel D (1977). Effects of photoactivated porphyrins at the cell surface of Leukemia L1210 cells. Biochemistry 16:3443.
Kinsey JH, Cortese DA, Moses HL, Ryan RJ, Branum EL (1981). Photodynamic effect of hematoporphyrin derivative as a function of optical spectrum and incident energy density. Cancer Res 41:5020.
Lipson RL, Baldes EJ, Olsen AM (1961). The use of a derivative of hematoporphyrin in tumor detection. J Natl Cancer Inst 26:1.
May JF, Rounds DE, Clarence D (1971). Intracellular transfer of toxic components after laser irradiation. J Natl Cancer Inst 46:655.
Moan J, Christensen T (1979). Photodynamic inactivation of cancer cells in vitro. Effect of irradiation temperature and dose fractionation. Cancer Lett 6:331.
Moan J, Christensen T (1981). Photodynamic effects on human cells exposed to light in the presence of hematoporphyrin. Localization of the active dye. Cancer Lett 11:209.
Moan J, Christensen T, Sommer S (1982b). The main photosensitizing components of hematoporphyrin derivative. Cancer Lett 15:161.
Moan J, Evensen JF, Christensen T, Hindar A, Sommer S, McGhie JB (1982a). Chemical composition of hematoporphyrin derivative, tumorlocalizing and photosensitizing properties of its main components. Abstr. from 10th Annual Meeting of The American Society for Photobiology. Vancouver, p 173.
Moan J, Johannessen JV, Christensen T, Espevik T, McGhie JB (1982c). Porphyrin-sensitized photoinactivation of human

cells in vitro. Am J Pathol 109:184.
Moan J, McGhie JB, Jacobsen PB (in press). Photodynamic effects on cells in vitro exposed to hematoporphyrin derivative and light. Photochem Photobiol.
Moan J, Pettersen EO, Christensen T (1979). The mechanism of photodynamic inactivation of human cells in vitro in the presence of haematoporphyrin. Br J Cancer 39:398.
Moan J, Sandberg S, Christensen T, Elander S (1983). Hematoporphyrin derivative: chemical composistion, photochemical and photosensitizing properties. In Kessel D and Dougherty TJ (eds): "Porpyrin photosensitization", New York and London: Plenum Press, p 165.
Moan J, Smedshammer L, Christensen T (1980). Photodynamic effects on human cells exposed to light in the presence of hematoporphyrin. pH effects. Cancer Lett 9:327.
Moan J, Sommer S (1981). Fluorescence and absorption properties of the components of hematoporphyrin derivative. Photobiochem Photobiophys 3:93.
Moan J, Steen HB, Feren K, Christensen T (1981). Uptake of hematoporphyrin derivative and sensitizing photoinactivation of C3H cells with different oncogeneic potential. Cancer Lett 14:291.
Mossman T, Gray MJ, Silberman L, Lipson RL (1974). Identification of neoplastic versus normal cells in cervical cell culture. J Obstet Gynecol 43:635.
Nordbye K, Oftebro R (1969). Establishments of four new cell strains from humane uterine cervix I. Exp Cell Res 58:458.
Oftebro R, Nordbye K (1969). Establishments of four new cell strains from humane uterine cervix II. Exp Cell Res 58:459.
Raaphorst GP, Sapareto SA, Follman ML, Dewey WC (1979). Changes in cellular heat and/or radiation sensitivity observed at various times after trypsinization and plating. Int J Radiat Biol 35:193.
Sinclair WK, Morton RA (1966). X-ray sensitivity during the cell generation cycle of cultured Chinese hamster cells. Rad Res 29:450.
White WI (1978). Aggregation of porphyrins and metalloporphyrins. In Dolphin D (ed): "The porphyrins", New York, San Fransisco, London: Academic Press, p 303.

Porphyrin Localization and Treatment of Tumors, pages 443–457

CELLULAR BINDING OF HEMATOPORPHYRIN DERIVATIVE (HpD) IN HUMAN BLADDER CANCER CELL LINE: KK-47

Haruo Hisazumi, Norio Miyoshi, Osamu Ueki, Kazuyoshi Nakajima

Department of Urology, School of Medicine, Kanazawa University, Kanazawa 920, Japan

ABSTRACT

In vitro cellular uptake or binding of hematoporphyrin derivative (HpD) was investigated by comparing the optical properties; absorption spectra, fluorescence spectra, fluorescence lifetime and fluorescence polarization, of HpD in KK-47 cells derived from a human bladder carcinoma, organic and micelllar solutions. In addition, the relationship between the mode of cellular HpD uptake and photodynamic cellular inactivation by argon-dye laser was studied using a clonogenic assay system. HpD molecules may weakly bind to the cell membrane within approximately 2 hrs after incubation in HpD-containing medium, and then major HpD molecules may slowly be distributed throughout the cytoplasm as a dimer form for 4 hrs after incubation in HpD-free medium. The intracellular binding loci were found to be the mitochondria and nuclear membrane by subcellular fractionation and fluorescence microscopic studies. The loci may be relatively hydrophilic on the basis of the polarity-dependence data on the fluorescence polarization. An exponential cellular photoinactivation curve was observed in relation to increasing concentrations of HpD and to intracellular strong-binding of dimeric HpD.

INTRODUCTION

As a promising new therapy, hematoporphyrin derivative (HpD) and photoradiation have been introduced for the management of malignant tumors by Dougherty and his co-workers (1977). This therapeutic effect results from the fact that HpD is retained much more by malignant tissues than by normal tissues and the cellular photoinactivation provoked by red light (630 nm) irradiation is obtained by the light-catalyzed formation of singlet oxygen and free radicals. The mechanism of the predominant retention of HpD by malignant tissues and its binding to cellular sites have not yet been fully elucidated. Kessel (1982) has studied the transport and binding of HpD by murine leukemia L1210 cells, and reported that the hydrophobic components of HpD are gradually accumulated by L1210 cells and appear to be responsible for their preferential affinity for neoplastic cells. To clarify the mode of cellular binding of HpD from the standpoint of optical properties, in the present study, we investigated HpD-containing KK-47 cells *in vitro* in terms of the fluorescence spectrum, fluorescence lifetime and fluorescence polarization, and the results obtained were compared with those of HpD dispersed into aqueous and micellar solutions as standard solvents because it has been reported that sodium dodecyl sulfate micelle exerts a monomerizing effect on HpD molecules (Anderoni et al. 1982).

Furthermore, the intracellular distribution of HpD was investigated using differential centrifugation of cell homogenates and a fluorescence microscope, and photodynamic inactivation of KK-47 cells treated with HpD and argon-dye laser light was determined using a clonogenic assay system.

MATERIALS AND METHODS

Reagents: HpD supplied by Dr. T.J. Dougherty (Roswell Park Memorial Institute, Buffalo, NY) was for clinical trials. PBS (Dulbecco´s A solution, pH 7.4) was prepared according to Dulbecco´s formula (Dulbecco and Vogt, 1954). Ethylene glycol (EG), 0.1 M cetyl trimethyl ammonium chloride (CTAC), bovine serum albumin (BSA) and EtOH were purchased from Wako Junyaku LTD, Tokyo. Ham´s F12 medium (Nissui Pharmaceutical Co. Tokyo) were used for cell cultivation. The pH value of the aqueous solutions used was adjusted with 0.1 N NaOH and 0.2 N HCl, and was determined by a pH meter (Hitachi-Horiba, model M-5). The dielectric constant of MeOH-H_2O mixtures was obtained from the data of Åkerlöb (1932). The viscosity constant of the EG-MeOH mixtures was calculated on the basis of the data of Young et al. (1971).

Cell Cultivation: KK-47 cell line was established from a human bladder carcinoma; its biological and histological properties have been reported in previous papers (Hisazumi et al. 1979, Naito et al. 1982, Misaki et al. 1980). KK-47 cells which had been grown in a log-growth phase in culture bottles were freed from the glass by gentle trypsinization and redispersed as a monocellular suspension in Ham's F12 medium at a concentration of 10^5 cell/ml. Tissue culture dishes, 6 x 1.5 mm (Falcon, Co., Ca., USA), containing 5 ml of the suspension were incubated in a humidified atmosphere of 5% CO_2 and 95% air at 37C for 48 hrs (logarithmic growth phase) and then washed with serum-free Ham's F12 medium 2 times. The doubling time of the cells was 48 hrs. HpD was added to the dishes at a final concentration of 200 μg/ml at 37C for 3 hrs in order to incorporate as much HpD as possible into the cells. The dishes were washed with serum-free Ham's medium to remove the free HpD in the culture medium and the monolayer cells in 5 ml of serum-free Ham's F12 medium were incubated in a humidified atmosphere of 5% CO_2 and 95% air at 37C for varying incubation times. Then the cells were subjected to studies of the cellular uptake of HpD, using the following optical methods. In a preliminary study of cytotoxicity *in vitro*, approximately 5% cell killing was detected at an extracellular concentration of 200 μg/ml HpD at 37C for 3 hrs. As a control, PBS containing 0.6 μg/ml HpD was used. This concentration was of an order that exhibits the same intensity of fluorescence as that of intracellular HpD. The intracellular HpD concentration was estimated to be 12 pg/ml from the calibration curve; the fluorescence intensity vs. the concentration of HpD in the PBS.

Optical Investigation Methods: The fluorescence spectrum and the degree of fluorescence polarization of HpD were determined using a fluorescence spectrophotometer (Hitachi, model MPS-4). The fluorescence spectra obtained were corrected by two standard fluorophors, 4-dimethyl-amino-4'-nitrostilbene and N, N-dimethyl-m-nitroaniline. The fluorescence lifetime of HpD was obtained using a single photon counting system (ORTEC and Northland MCA) coupling a nanosecond pulser (PRA), and were analyzed using the computer deconvolution method. HpD-fluorescence in various solutions were slightly quenched by O_2-bubbling. A large quenching effect of O_2 on the HpD fluorescence lifetime was not observed. Wavelengths for excitation were 398 nm in aqueous and PBS solutions, 403 nm in micellar solutions,

400-408 nm in KK-47 cell suspension, 408 nm in 3% BSA solution and 393 nm in EtOH. Fluorescence microphotographs of HpD-containing KK-47 cells were taken by a microphotometer (Leitz, model MPV-2).

Subcellular Fractionation: Cell homogenates in isotonic sucrose were separated by differential centrifugation into nuclear, mitochondrial, microsomal and supernatant fractions. The relative content of HpD was expressed in terms of relative fluorescence intensity: exciting wavelength, 398 nm; monitoring wavelength, 630 nm. Protein in each of the fractions was quantitatively determined using Lowry´s method (1951).

RESULTS

Fig. 1 shows the absorption spectra of HpD in various solutions. Curve 1 is an aqueous solution, curve 2 is a CTAC micellar solution and curves 3, 3´ and 3" are at 10^3, 10^4 and 10^5 cells/ml of KK-47 cell suspensions, respectively. It has been reported that absorption bands at 370 and 400 nm were assigned to dimer and monomer, respectively; these curves indicated that the dimers were partly contained in aqueous solution, and that the monomer dispensed absorbance at monomer band (Mossman et al., 1974). On the other hand, when the number of the KK-47 cells increased, cellular HpD-uptake and the proportion of HpD monomer to total HpD increased. When the micelle was formed by dissolving in an aqueous solution and, as a result, the monomerization of HpD dimers occurred, in a visible band, the 4 absorption peaks of curve 2 became sharp and the larger 2 peaks were blue-shifted. In the KK-7 cells, the absorption spectra appeared at the same positions as those in the aqueous solution.

Fig. 2 shows the fluorescence spectrum of HpD in the various samples. The emission peak in the KK-47 cells indicated by curve 3 was shifted to a longer wavelength than that of the micellar system and the peak intensity at the 630 nm band became weak compared with that at the 675-695 nm band. Therefore, it was concluded that the HpD incorporated in the KK-47 cells was appreciably dimerized.

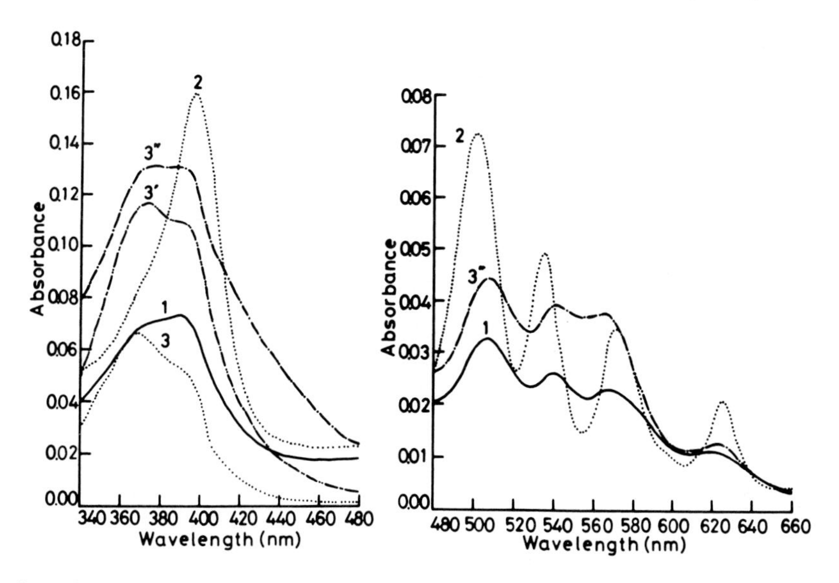

Fig. 1. Absorption spectra of HpD in aqueous and CTAC solutions and KK-47 cell suspensions.

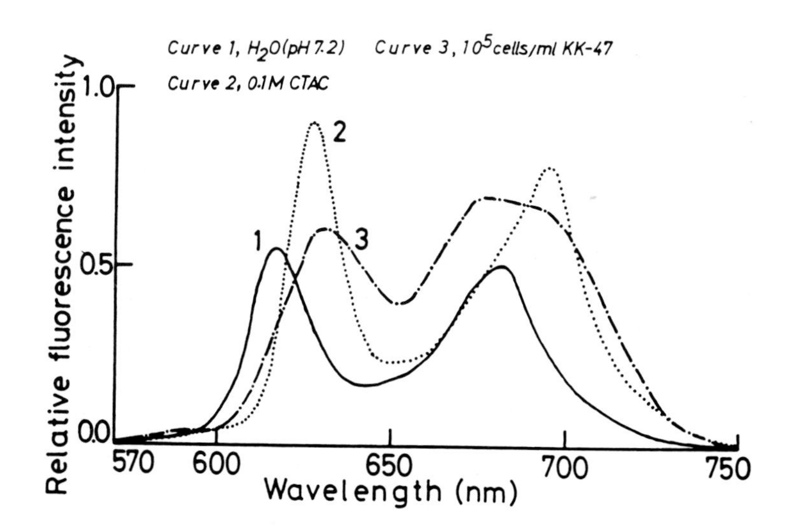

Fig. 2. Fluorescence spectra of HpD in aqueous and CTAC solutions and KK-47 cell suspensions.

Fig. 3 shows that the fluorescence decay curve of HpD in PBS and 0.1 M CTAC micellar solution exhibited a single exponential decrease. However, HpD-containing KK-47 cells showed a bi-phasic exponential decay curve as shown by curve 3. It is expected that the carboxyl group (anionic ions) of the hematoporphyrin molecule is easily bound to the cationic ammonium ions of the CTAC micelle and that the HpD dimer in aqueous solution is monomerized by the binding of HpD to the micelle in order to compare the binding mode in KK-47 cells. The slope of the decay curves was analyzed using a computer deconvolution method. The fluorescence lifetime was 16.2 ns and 17.9 ns in the PBS and micellar solutions, respectively, and 2.4 ns (shorter lifetime) and 15.2 ns (longer lifetime) in the KK-47 cells. The shorter lifetime in the KK-47 cells may result from dimeric HpD and the longer lifetime from

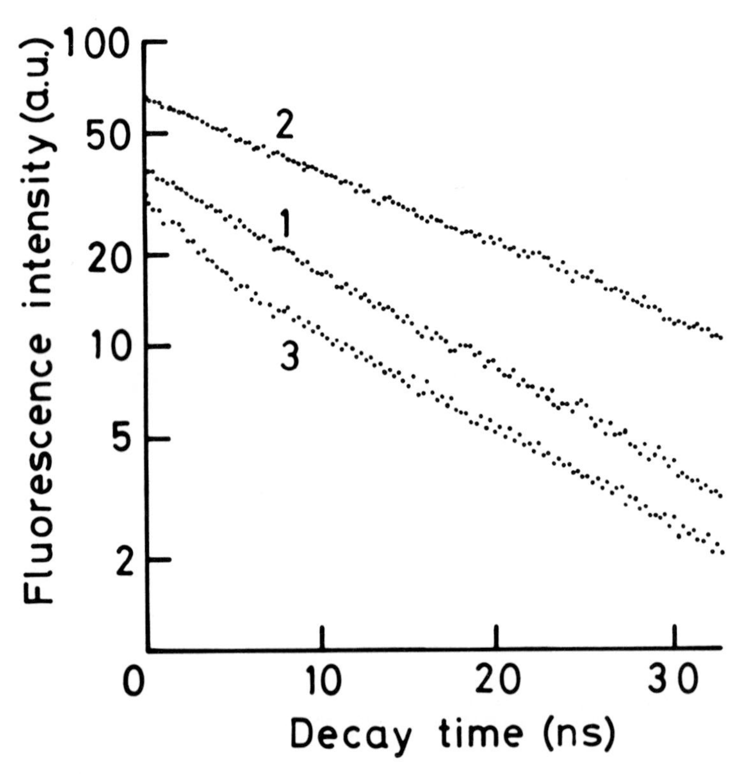

Fig. 3. Fluorescence decay curves of HpD in aqueous and CTAC solutions and KK-47 cell suspension.

monomeric HpD. Andreoni et al. (1982) determined the lifetimes of 4.1 ns and 16.5 ns in PBS containing 2 μM HpD at pH 7.4 and reported that the fast-decaying component in curves 1 and 2 may be due to the low concentration of HpD, 0.6 μg/ml. It is considered that the dimeric HpD may be emitted with low efficiency.

Fig. 4 shows the fluorescence spectra of HpD-containing PBS and KK-47 cells at varying incubation periods. This experiment was repeated 10 times. The HpD in various samples was excited by excitation maximum (398 nm, PBS; 402 nm, KK-47 cell suspensions). Immediately after washing; 0 hr-incubation (curve 1), the fluorescence bands of the cells appeared at 627 nm and 682 nm. In the case of a 2 hr-incubation (curve 2), both bands were red-shifted by 7 nm and the latter band was broadened. The 7 nm shift between curves 1 and 2 was reproducible. On the other hand, the fluorescence intensity of the cells decreased after a 24 hr-incubation time (curve 3); however, the shape and position of the bands remained unchanged as compared with those of the 2 hr-incubation. That the fluorescence spectra were changed by the periods of incubation may result from the cellular binding process of HpD. In HpD-containing KK-47 cells, the fluorescence intensity of the long wavelength band (674-698 nm) was higher than that of the short wavelength band (634 nm), while in the case Of HpD-containing PBS (curve 4), the difference between both bands was the reverse of those in curves 1, 2 and 3. This difference, the broadening of the long wavelength band and the red-shift may be optically characteristic findings of HpD-containing KK-47 cells.

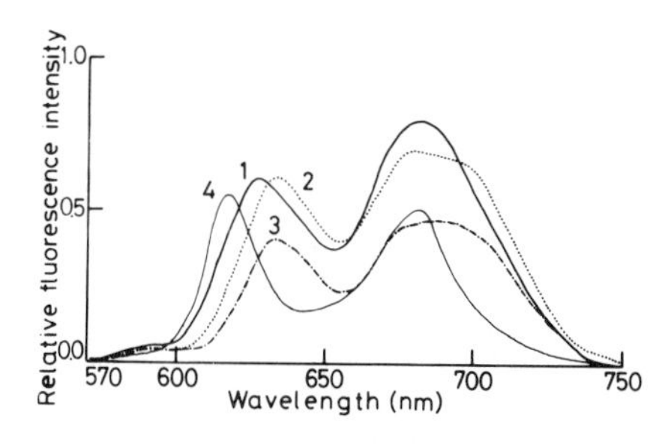

Fig. 4. Fluorescence spectra of HpD in KK-47 cell suspension ($1x10^5$ cells/ml) at different incubation times.

Fig. 5 shows changes of the fluorescence lifetime as a function of incubation time. The fluorescence lifetime of HpD was unchanged with increasing incubation time in the PBS and CTAC micellar solutions as shown by curves 1 and 2. Curves 3^f and 3^s show the lifetimes of the fast-decaying and slow-decaying components, respectively, and these lifetimes initially changed but reached a constant level 4 hrs after incubation. The initial change may result from the cellular binding process of HpD monomers and dimers during incubation. Pre-exponential factor (curve 3^r) was obtained from the ratio of the fluorescence intensity of 3^f to 3^s. Accordingly, the factor increased within 4 hrs after incubation. This increase suggested that the dimer formation progressed with increasing incubation time because the fast-decaying component (3^f) was produced by the presence of the dimer. The factor value was higher (0.70) in the cells than in the PBS (0.25).

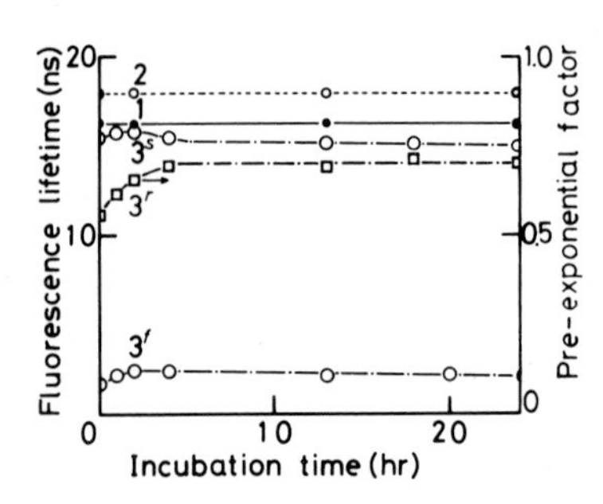

Fig. 5 Changes in fluorescence lifetime of HpD with increasing incubation time in aqueous and CTAC solutions and KK-47 cell suspension.

The degree of fluorescence polarization (p) was calculated according to the following equation (Azumi and McGlynn, 1962);

$$P = \frac{I_{//} - I_{\perp}(I_{\perp} / I_{//})}{I_{//} + I_{\perp}(I_{\perp} / I_{//})}$$

where $I_{//}$ and $I_{\perp}$ are the fluorescence intensities which were observed through a polarizer oriented parallel and perpendicular to the plane of polarization of the excitation beam. As shown in Fig. 6, the degree of polarization in the PBS, and CTAC micellar solution was 0.007 and 0.022, respectively, and was unchanged with increasing incubation time except an initial and very short incubation time in the micellar solution as indicated by curves 1 and 2. The degree of polarization in the cells rapidly increased from 0.020 to 0.11 for an incubation period of 2 hrs and then decreased gradually. Curve 3 suggests that HpD molecules may be rapidly bound to the cell membrane from the same low degree of polarization as that of CTAC micelles and then distributed throughout the cytoplasm. As indicated by curve 4, when HpD was dissolved in 3% BSA solution at 37C, a similar rapid increase was observed. This may also suggest the occurrence of HpD binding to BSA. To explain the gradual decrease in polarization in curves 3 and 4, it was considered that the HpD molecules might be stabilized with the constituents of the cytoplasm and serum ablumin.

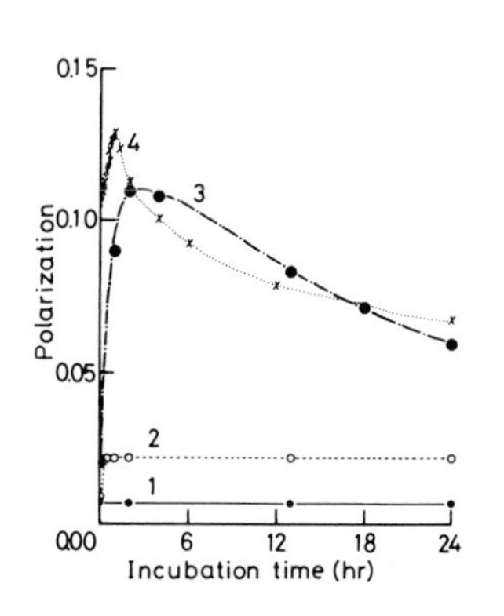

Fig. 6 Changes in the degree of polarization of HpD with increasing incubation time in aqueous and CTAC solutions and KK-47 cell suspension at 37C. Curve 1, aqueous solution, 0.6 μg/ml HpD; curve 2, 0.1M CTAC miceller solution, 0.6 μg/ml HpD; curve 3, KK-47 cell suspension, 12 pg/cell HpD; curve 4, 3% BSA solution, 0.6 μg/ml HpD. Excited at an excitation maximum; 398 nm in curve 1, 402 nm in curve 2, 403 nm in curve 3.

Relationships between the degree of fluorescence polarization and the polarity (ε) of the MeOH-H_2O mixtures or the viscosity of the EG-MeOH mixtures are shown in Fig. 7. As shown by curve 1, the degree of the polarization was demonstrated as a sigmoid curve accompanied by an increasing dielectric constant. In addition, the polarization (curve 2) gradually reached a constant level with increasing viscosity. As shown by curve 3 in Fig. 6, the degree of polarization in HpD-binding KK-47 cells after 24 hrs was 0.067 with a good reproducibility. This value corresponded to a 35 dielectric constant (ε) of curve 1 in Fig. 7. Accordingly, the polarity of cellular substances containing HpD was similar to that, ε =32.6, of MeOH. A viscosity giving a degree of polarization of 0.067 was 11 cp from curve 2. This 11 cp indicates that the viscosity of the cellular substances containing HpD 24 hrs after incubation is almost the same as that, 10.3 cp, of p-cresol. Kessel and Kohn (1980) have reported that a longer incubation resulted in an accumulation of mesoporphyrin at more hydrophobic loci, $\varepsilon=10$, due to the fact that the drug was not readily washed out. Then, it was suggested that HpD elements incorporated into the cells might be more hydrophilic than mesoporphyrin after washing.

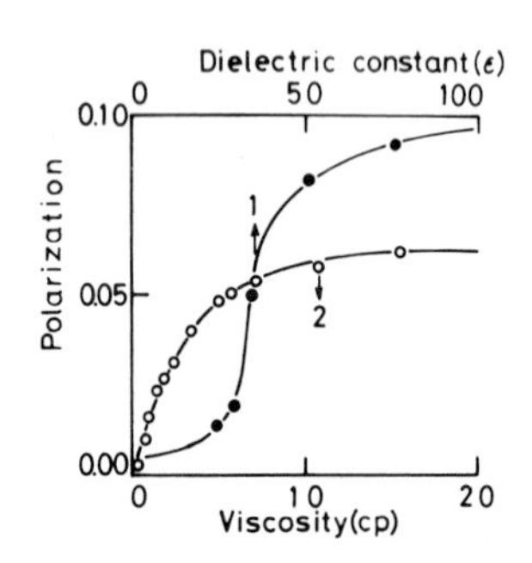

Fig. 7 Relationship between the fluorescence polarization and viscosity or dielectric constant (ε) 37C.

Fig. 8 is a fluorescence photomicrograph of HpD-containing KK-47 cells after a 6 hr incubation time resulting in the strong binding of HpD in the cells. The cells show a bright red-fluorescence of intracelluar HpD excited by blue light; wavelength, 350 nm to 410 nm. The nuclei are free of fluorescence, and are well-demarcated with HpD fluorescence of the cytoplasm.

Fig. 8 Fluorescence photomicrograph of HpD-containing KK-47 cells. X250.

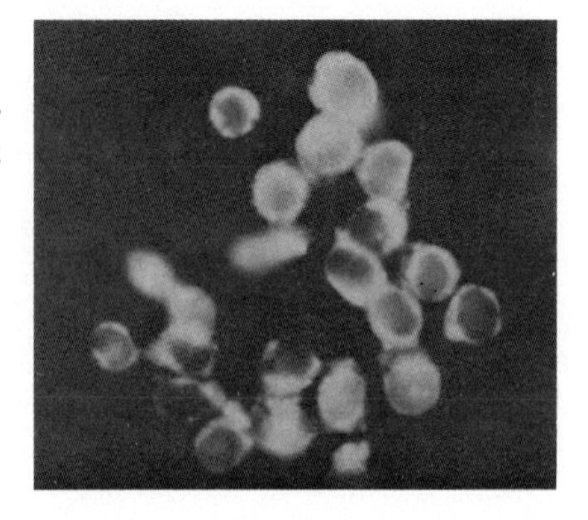

The relative content of HpD per mg protein of the 4 fractions is listed in Table 1. The HpD content of the mitochondrial and nuclear fractions was greater than that of the other fractions. However, because there was no fluorescence in the nucleus as shown in Fig. 8, the HpD content in the nucleus fraction may depend on the presence of HpD-bound nuclear membrane.

celle fraction	HpD[1]	Protein[2]	HpD / Protein
nuclear	0.541	0.049	11.0
mitochondrial	0.633	0.030	21.1
microsomal	0.126	0.026	4.85
soluble	0.007	0.031	0.23

[1] determined by relative fluorescence intesity.
[2] determined by Lowry method.

Table 1. Relative constant per unit protein.

The photodynamic inactivation of in vitro cultured KK-47 cells was studied as shown in Fig. 9.

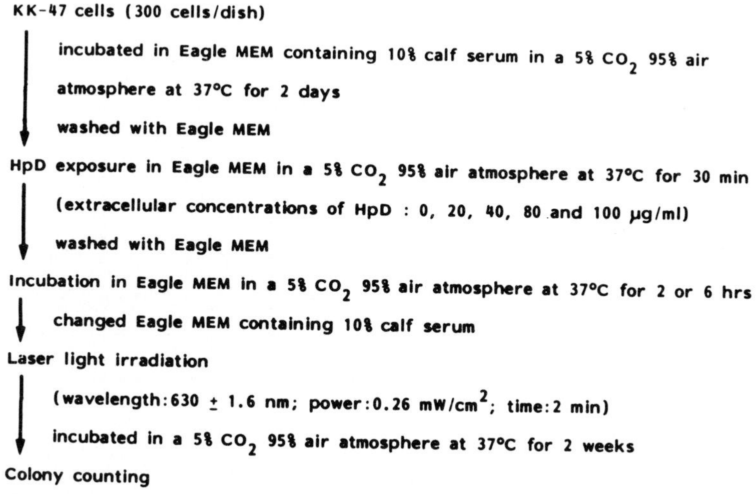

KK-47 cells (300 cells/dish)

↓ incubated in Eagle MEM containing 10% calf serum in a 5% CO_2 95% air atmosphere at 37°C for 2 days; washed with Eagle MEM

HpD exposure in Eagle MEM in a 5% CO_2 95% air atmosphere at 37°C for 30 min

↓ (extracellular concentrations of HpD : 0, 20, 40, 80 and 100 µg/ml); washed with Eagle MEM

Incubation in Eagle MEM in a 5% CO_2 95% air atmosphere at 37°C for 2 or 6 hrs

↓ changed Eagle MEM containing 10% calf serum

Laser light irradiation

↓ (wavelength: 630 ± 1.6 nm; power: 0.26 mW/cm^2; time: 2 min); incubated in a 5% CO_2 95% air atmosphere at 37°C for 2 weeks

Colony counting

Fig. 9. Experimental profile of the photodynamic inactivation of in vitro cultured KK-47 cells.

Asynchronous KK-47 cells were incubated in serum-free Eagle MEM containing various concentrations of HpD at 37C for 30 min. After washing, the cells were incubated in the serum and HpD-free medium for 2 or 6 hrs to develop the intracellular distribution of HpD. The HpD-containing cells were irradiated with laser light (0.26 mW/cm^2) for 2 min. Photodynamic cell killing was determined using a clonogenic assay system. The degree of cellular photoinactivation was found to be related to the following factors; (1) the concentration of extracellular HpD and (2) the length of the incubation time following the HpD exposure as shown in Fig. 10.

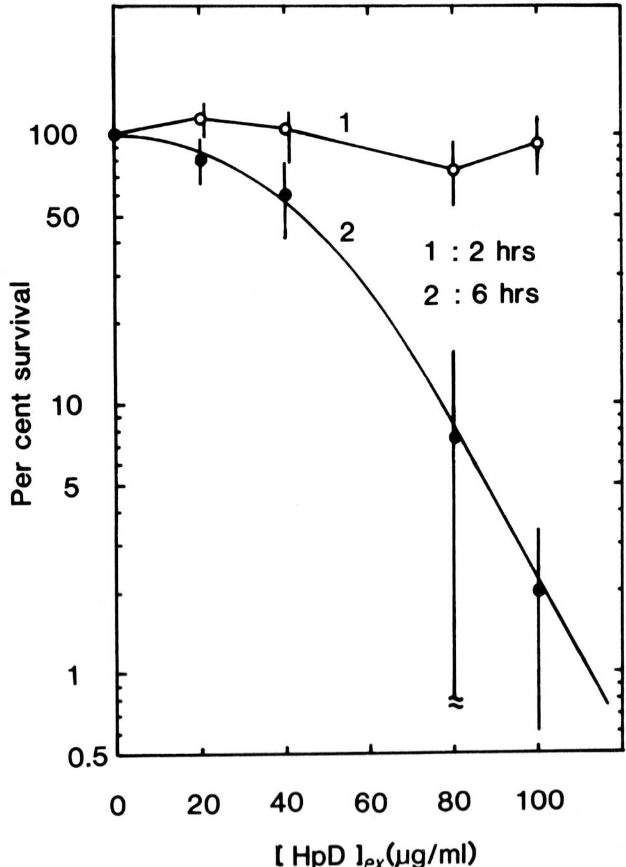

Fig. 10. Survival curves of KK-47 cells irradiated with argon-dye laser light (630 ± 1.6 nm, 0.26 mW/cm^2) at various extracellular concentration of HpD.

No cell killing was observed at 2 hr incubation resulting in the weak binding of HpD, and an exponential survival curve (curve 2) was obtained.

DISCUSSION

Dougherty et al. (1981) have reported that the cellular uptake *in vitro* of HpD was characterized by a rapid, weak binding followed by a slow phase of strong binding. A similar HpD-uptake mode was suggested by our optical investigations. It has been generally said that the fluorescence emission spectrum and degree of fluorescence polarization are changed by the environment; e.g., viscosity, polarity and so on, around the phosphor bound by the cells. The fluorescence emission peaks may be red-shifted because the excited singlet state of HpD is stabilized by binding the cells. The degree of fluorescence polarization may increase at a high viscosity environment because that degree reflects the ease of vibration of the side-chains in the HpD molecule. Accordingly, it was considered from Fig. 7 that the HpD molecules may rapidly bind to the cell membrane at the initial time and then incorporate slowly with the intracellular stable binding site after 2 hrs incubation time.

Berns et al. (1982) showed a phase-contrast micrograph and fluorescence micrograph of myocardial cells treated with HpD (25 μg/ml) for 1 hr, and concluded that the intracellular binding site of HpD was the mitochondria. However, from our fluorescence photomicrograph and subcellular fractionation studies, it was confirmed that HpD was bound by not only the mitochondria but also by the nuclear membrane.

In conclusion, the optical results obtained suggested that HpD monomers and dimers may bind weakly to the cell membrane, and then may slowly distribute into the stable binding site of KK-47 cells. In addition, from the results obtained by differential centrifugation and fluorescence microphotograph, the intracellular binding sites of HpD were considered to be the mitochondria and the nuclear membrane. Furthermore, it was found that the photodynamic cytotoxicity of the cells incorporating the stable binding HpD was larger than that of the cells to which HpD may bind weakly at the initial incubation time.

REFERENCES

Åkerlöb G (1932). Dielectric constants of some organic solvent-water mixtures at various temperatures. J Am Chem Soc 54:4125.

Andreoni A, Cubeddu R, De Silvestri S, Laporta P, Jori G, Reddi E (1982). Hematoporphyrin derivative: experimental evidence for aggregated species. Chem Phys Lett 88:33.

Azumi T, McGlynn SP (1962). Polarization of the luminescence of phananthrene. J Chem Phys 37:2413.

Berns MW, Johnson FM, Burns R, Sperling D, Guiltinan M, Siemens A, Walter R, Wright W, Hammer-Wilson M, Wile A (1982). In vitro cellular effects of hematoporphyrin derivative. Can Res 42:2325.

Dougherty TJ, Boyle D, Weishaupt K, Gomer CJ, Brocicky D, Kaufman J, Goldfarb A, Grindey G (1977). Phototherapy of human tumors. "Research in Photobiology" (Edited by Amleto Castellani), Plenum Publishing Corporation, New York, p 435-445.

Dougherty TJ, Henderson BW, Bellnier DA, Weishaupt KR, Michalakes C, Ziring B, Chang C (1981). Preliminary information pertaining to mechanisms in hematoporphyrin derivative phototherapy of malignant tissue. Abstracts of 10th Annual Meeting of American Society for Photobiology, June 14-18, Quality Inn Fort Magruder, Williamsburg, Virginia, USA, TPM-B4, p 124.

Dulbecco R, Vogt M (1954). Plaque formation and isolation of pure lines with poliomyelitis viruses. J Exp Med 99:167.

Hisazumi H, Kanokogi M, Nakajima K, Tsukahara T, Naito K, Kuroda K (1979). Established cell line of urinary bladder carcinoma (KK-47): growth, heterotransplantation. Jap J Urol 70:485.

Kessel D, Kohn KI (1980). Transport and binding of mesoporphyrin IX by leukemia L1210 cells. Cancer Res 40:303.

Kessel D (1981). Transport and binding of hematoporphyrin derivative and related porphyrins by Murine leukemia L1210 cells.

Lowry OH, Rosebrough NJ, Farr AL, Randall RJ (1951). Protein measurement with the Folin phenol reagent. J Biol Chem 193:265.

Misaki T, Hisazumi H, Kanokogi M, Kuroda K, Matsuhara F (1980). The prophylactic use of carboguone and urokinase in bladder carcinoma. Acta Urol Jap 26:683.

Mossman BT, Gray MJ, Silberman FI, Lipson RL (1974). Identification versus normal cells in human cerivical cell culture. Am College of Obstetricians and Gynecologists 43:635.
Naito K, Hisazumi H, Kanokogi M, Nakajima K, Kato T, Tsukahara T, Kobayashi T, Misaki T, Kuroka K, Matsuhara F (1982). Tissue culture cell lines established from human urogenic carcinomas. Jap J Urol 73:1019.
Young RH, Wehrly K, Martin RT (1971). Solvent effects in dye-sensitized photooxidation reactions. J Am Chem Soc 93:5774.

Porphyrin Localization and Treatment of Tumors, pages 459–469

EXAMINATION OF ACTION SPECTRUM, DOSE RATE AND MUTAGENIC PROPERTIES OF HEMATOPORPHYRIN DERIVATIVE PHOTORADIATION THERAPY

C.J. Gomer, D.R. Doiron, N. Rucker, N.J. Razum and S.W. Fountain

Clayton Center for Ocular Oncology, Childrens Hospital of Los Angeles and Departments of Pediatrics (Division of Hematology-Oncology) and Ophthalmology, USC School of Medicine, Los Angeles, CA 90027.

INTRODUCTION:

While hematoporphyrin derivative (HpD) photosensitization is exploited clinically for the treatment of solid tumors (1,2), there are still many determinants of this procedure which are not completely understood. In this manuscript we report on studies which were designed to document the action spectrum (620-640 nm), evaluate light dose rate effects and determine the mutagenic potential of HpD photosensitization. The object of this research was to obtain information which would be of benefit in the planning of current protocols for HpD photoradiation therapy (PRT).

METHODS AND MATERIALS:

Cells: Chinese hamster ovary (CHO) fibroblasts were used in all experiments (3). The cells were grown as a suspension in Hams's F-10 medium supplimented with 10% fetal calf serum (FCS) and antibiotics.

Light Sources: A rhodamine-B dye laser pumped by a 5 watt argon laser (Spectra-Physics) was the source of monochromatic light for both action spectrum and dose rate experiments. A 400 micron quartz fiberoptic cable was interfaced with the dye laser and the tip of the fiberoptic was fitted with a micro-lens. Both the light fluence and the wavelength were adjusted for specific experiments.

A parallel series of soft white 30-watt fluorescent bulbs enclosed on top with a sheet of clear plexiglass and filtered with a milar film were used as the light source for HpD mutation studies. The emission spectrum of this light source ranged from 570-650 nm and the light fluence was 0.35 mW/cm^2. Ultraviolet light at 254 nm was obtained with a 30 watt germicidal lamp and was used at a dose rate of 0.2 mW/cm^2. X-rays were obtained with a 300 kVp G.E. Maxitron and had a dose rate of 155 rads/min.

Cell Treatment Procedures: A detailed procedure for the treatment of cells and the subsequent determination of survival and/or mutation induction has been previously described (3,4). In action spectrum studies, appropriate numbers of cells were plated onto 60 mm dishes and were incubated either with 25 ug HpD/ml for 1 hour in medium supplemented with 1% serum or with 25 ug HpD/ml for 12 hours in medium supplemented with 5% serum. Following incubation, the cells were washed once with serum free F-10 medium and then were exposed to monochromatic red light over the range of 620 nm to 640 nm at a dose rate of 1 mW/cm^2. Total light doses varied from 0-1800 J/m^2. The light dose rate experiments utilized cells which were plated onto 35 mm dishes and then were incubated with 25 ug HpD/ml for 1 hour in medium supplemented with 1% serum. Individual dishes were subsequently exposed to 630 nm light delivered at light fluences which ranged from 0.5 to 60 mW/cm^2. Total light doses ranged from 0 to 5400 J/m^2. In the mutation studies, cells were plated onto 60 mm dishes and were exposed to either HpD photoradiation, x-rays or U.V. light. Incubation conditions of HpD photoradiation in these experiments were as follows: 25 ug HpD/ml for 1 hour in 1% F-10, 50 ug HpD/ml for 1 hour in 1% F-10, 25 ug HpD/ml for 12 hours in 5% F-10, 50 ug HpD/ml for 12 hours in 5% F-10, 25 ug HpD/ml for 12 hours in 10% F-10, or 50 ug HpD/ml for 12 hours in 10% F-10. Cytotoxicity and mutagenic potential were examined in cells incubated in these conditions. Mutation induction (for cells treated with HpD photoradiation, x-rays or U.V.) was measured at the HGPRT locus using resistance to 10 uM 6-thioguanine (5). A detailed description of this procedure has recently been published (4).

RESULTS:

Survival curves for CHO cells incubated with HpD for 1 hour or 12 hours and then exposed to varying wavelengths of red light are shown in Figures 1 and 2. At equal total delivered doses it was observed that 627.5 nm, 630 nm and 632.5 nm were most effective wavelengths of red light for producing cellular cytotoxicity. Conversely, light at 620 nm, 622.5 nm, 637.5 nm and 640 nm were least effective at inducing HpD photosensitization. Similar results regarding HpD action spectrum were obtained for 1 hour and 12 hour HpD incubation periods.

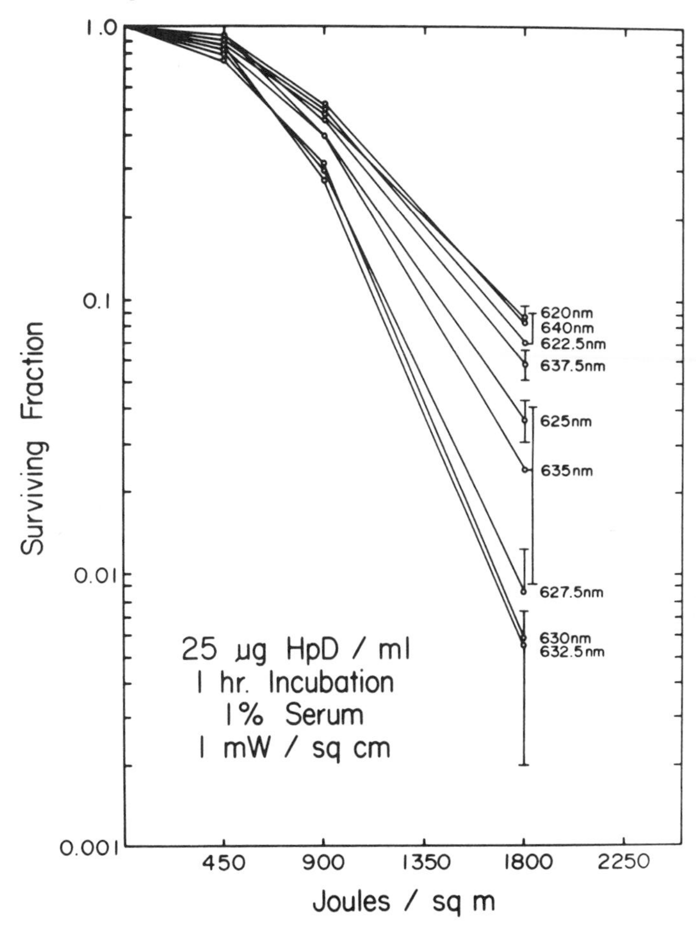

Figure 1. Surviving fraction of CHO cells as a function of total light dose. Cells were exposed to various wavelengths of red light following a 1 hr HpD incubation.

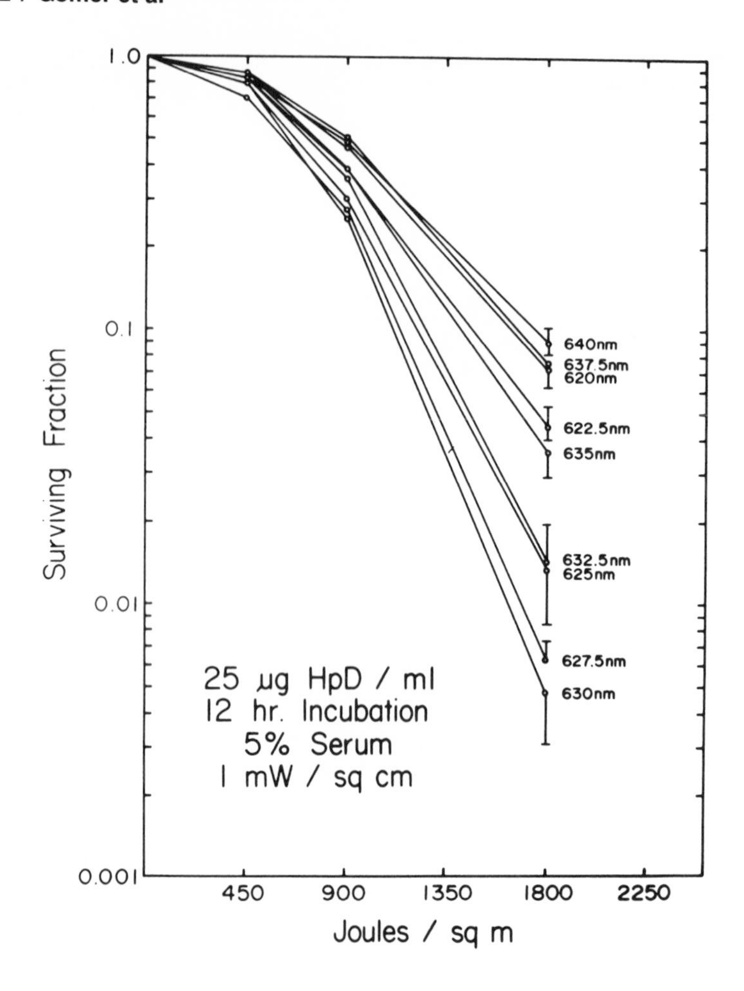

Figure 2. Surviving fraction of CHO cells as a function of total light dose. Cells were exposed to various wavelengths of red light following a 12 hr HpD incubation.

Figure 3 shows survival curves for CHO cells treated with HpD photoradiation at various dose rates. There was no significant variation in the survival levels at equal delivered light doses when the fluence varied from 0.5 to 60 mW/cm^2.

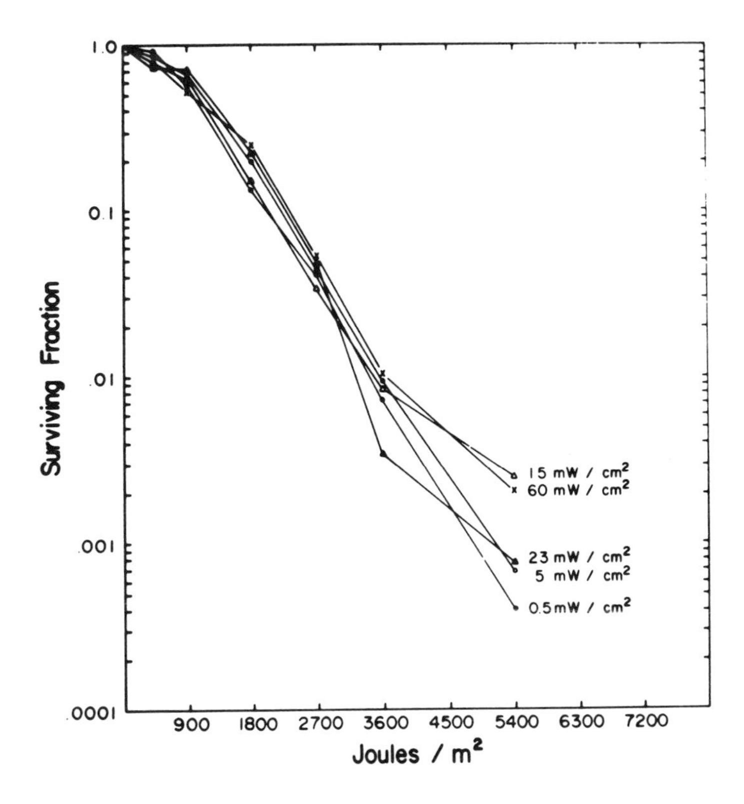

Figure 3. Surviving fraction of CHO cells as a function of total light dose. Cells were incubated with HpD (25 ug/ml) for 1 hr and then exposed to red light at various light fluences (0.5-60 mW/cm^2).

Figure 4 and 5 show survival curves for CHO cells treated with either x-rays, U.V. light, or HpD photoradiation. These survival curves are typical for mammalian cells exposed to physical agents. There were six survival curves obtained for HpD photoradiation and each represented a different HpD incubation protocol. Figure 6 shows the mutation frequency (resistance to 6-thioguanine) as a function of surviving fraction of treated cells. Treatment with x-rays or U.V. induced a dose-related increase in mutation frequency. There was no mutagenic effect (above background levels) for CHO cells exposed to HpD photoradiation with any of the HpD incubation protocols.

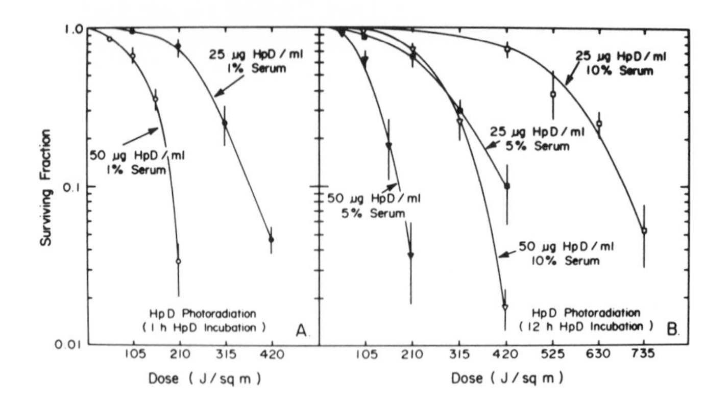

Figure 4. Surviving fraction of CHO cells as a function of total light dose. Cells were incubated with HpD for either 1 or 12 hr with either 25 or 50 ug HpD/ml in 1, 5 or 10% serum. (4)

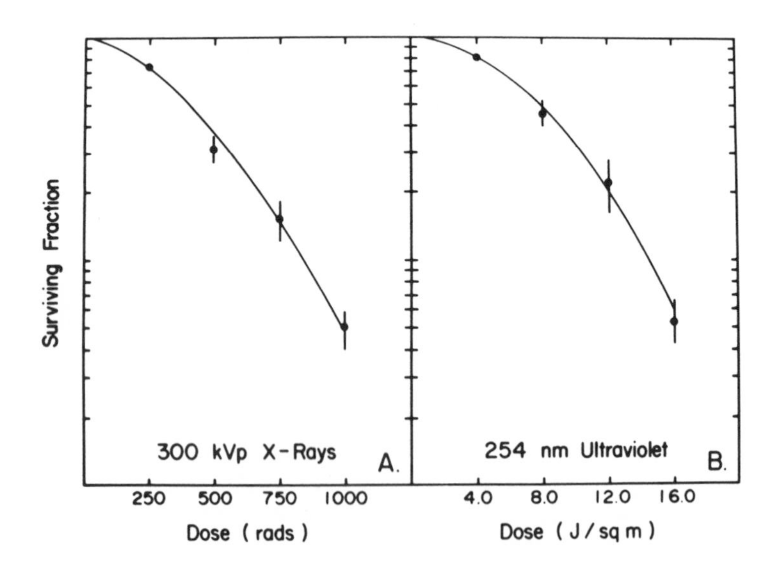

Figure 5. Surviving fraction of CHO cells as a function of treatment dose (x-rays or U.V.) (4).

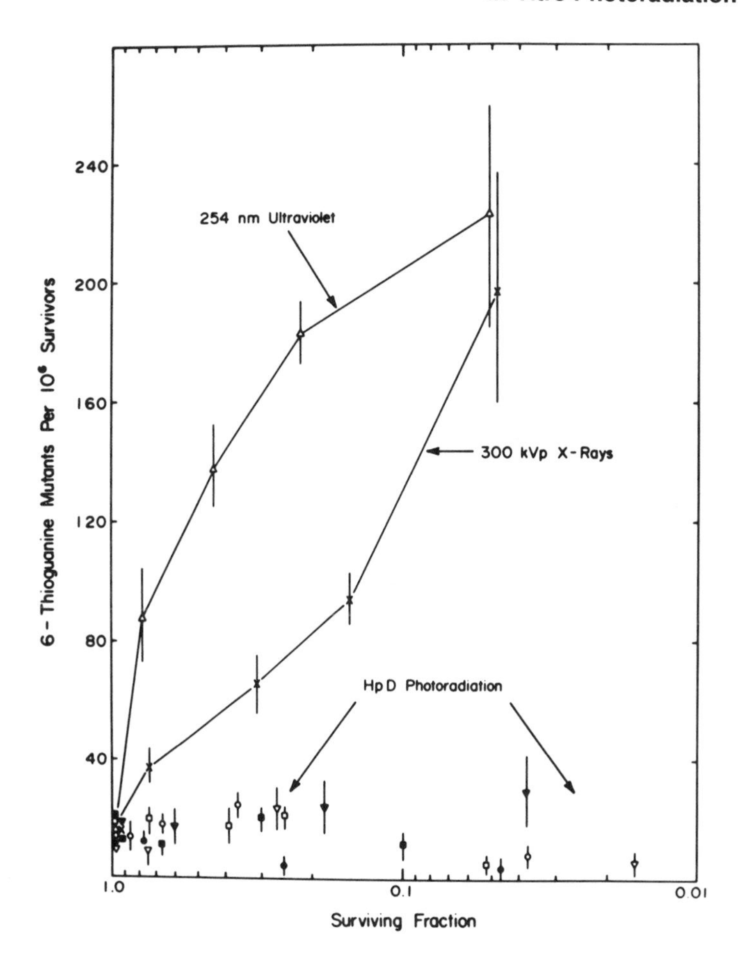

Figure 6. Mutation frequency (6-thioguanine resistant mutants per 10^6 survivors) measured as a function of surviving fraction of CHO cells treated with either U.V., x-rays or HpD photoradiation. Surviving fractions were taken from figures 4 and 5. (4).

DISCUSSION:

The clinical success of HpD PRT will depend on the continued accumulation of information related to understanding the basic mechanisms of HpD photosensitization as well as to improving the current clinical protocols. Recent studies by Henderson et al. (6) and Star et al. (7) indicate that the major in-vivo toxic response of HpD PRT is to the tumor vasculature. This information suggests that objectives and results of in-vitro studies will have to be applicable to vascular tissue as well as tumor tissue. The results from in-vitro studies described in this paper should provide helpful information related to the use of HpD PRT regardless of the actual target sites.

The increased utilization of monochromatic light in HpD PRT has necessitated the acquisition of quantitative data regarding the action spectrum of HpD photosensitization. We examined the efficiency of light from 620 nm to 640 nm since the tissue transmission properties of red light are exploited in HpD PRT (2). When HpD is dissolved in aqueous solution (including growth medium containing serum) the absorption maximum in the red region is at 620 nm. However, when cells are incubated with HpD the absorption maximum in the red region (for HpD in cells) is shifted to 630 nm. The data obtained in our action spectrum experiments correlate with HpD bound to cells. The most efficient wavelengths of red light for inducing cell killing correspond to the absorption maximum for cells incubated with HpD. This information would suggest that clinical treatment of tumors with HpD PRT should utilize red light at 630 nm.

The information obtained from the dose rate experiments indicate that reciprosity holds for fluences of 630 nm light ranging from 0.5 to 60 mW/cm^2. This observation is opposite to that observed with ionizing radiation where repair of sublethal damage at low dose rates produces a cell sparing effect (8). The lack of a dose rate effect with HpD photoradiation might be explained by two factors. First, the type of damage being repaired following ionizing radiation at low dose rates is not produced by HpD PRT or second, the lowest dose rate used in our study (0.5 mW/cm^2) is still too high for repair kinetics to play any significant role in ultimate cellular survival. We cannot draw any conclusions from our data on whether photobleaching played a role in the outcome of our results or whether photobleaching was identical at various dose rates. The results we have obtained do agree with previous anecdotal clinical results (9). In addition, we have also observed equal amounts of tumor necrosis in anterior chamber tumors treated with identical total doses of 630 nm light delivered at either 25 or 150 mW/cm^2 (unpublished results).

While HpD photosensitization was shown to be cytotoxic to CHO cells, it was not observed to be mutagenic. The resistance of mammalian cells to 6-thioguanine is an assay for point or deletion mutation at the hypoxanthine-guanine phosphoriybosyl transferase locus of the X-chromosome. HpD photosensitization does induce various types of nuclear damage (10, 11, 12) and singlet oxygen is a reported mutagenic agent (13). Therefore, the negative observation for mutagenicity with HpD photoradiation was surprising. One possible explanation for our results is that the level of nuclear damage and the resulting mutagenic induction frequency following HpD photosensitization is disproportionately low relative to the level of lethal damage induced outside the nucleus. The results of the mutagenic experiments are encouraging as they relate to potential side effects of HpD PRT. However, several photosensitizers have been shown to be carcinogenic (14) and therefore in-vivo carcinogic studies of HpD PRT should be performed.

In conclusion, in vitro experiments related to HpD photosensitization of mammalian cells have been performed. Action spectrum experiments have demonstrated that red light between 627.5 nm and 632.5 nm was the most efficient in producing HpD photosensitization. Reciprocity was observed for HpD photoradiation induced cell killing using fluences at 630 nm which ranged from 0.5 to 60 mW/cm^2. Finally, experiments indicate that HpD photosensitization was not mutagenic when assayed using resistance to 6-thioguanine.

ACKNOWLEDGMENTS:

This investigation was performed in conjunction with the Clayton Foundation for Research and was supported in part by USPHS Grants CA 31230 and CA 31885, awarded by the NCI, DHHS. We thank Albert L. Castorena for assistance in the preparation of this manuscript.

REFERENCES:

1. Dougherty, T.J., Kaufman, J.B., Goldfard, A., Weishaupt, K.R., Boyle, D.G. and Mittleman, A. Photoradiation therapy for the treatment of malignant tumors. Cancer Res., 38, 2628-2635, (1978).
2. Dougherty, T.J., Weishaupt, K.R. and Boyle, D.G. Photosensitizers, In: (eds., De Vita, V., Helman, S. and Rosenberg, S.), Cancer, Principles and Practices of Oncology, pp. 1836-1844, Philadelphia: J.B. Lippincott Co., (1982).
3. Gomer, C.J. and Smith, D.M. Photoinactivation of Chinese Hamster cells by hematoporphyrin derivative and red light. Photochem. Photobiol., 32, 341-348, (1980).
4. Gomer, C.J., Rucker, N., Banerjee, A. and Benedict, W.F. Comparison of mutagenicity and induction of sister chromatid exchange in chinese hamster cells exposed to hematoporphyrin derivative photoradiation ionizing radiation or ultraviolet radiation. Cancer Res., 43, 2622-2627, (1983).
5. O´Neill, J.P., Brimer, P.A., Machanoff, R., Hirschf, G.P. and Hsie, A.W. A quantitative assay of mutation induction at the hypoxanthine-guanine phosphoribosyl transferase locus in Chinese hamster ovary (CHO-HGPRT system): development and definition of the system. Mutat. Res., 45, 91-101, (1977).
6. Henderson, B.W., Dougherty, T.J. and Malone, P.B. Studies on the machanism of tumor destruction by photoradiation therapy. In: Porphyrin localization and Treatment of Tumors. (D.R. Doiron and C.J. Gomer, eds.), A.L. Liss Publishers, N.Y.
7. Star, W.M., Marijnissen, J.P.A., van den Berg-Blok, A.E. and Reinhold, H.S. Destruction effect of photoradiation on the microcirculation of a rat mammary tumor growing in "sandwich" observation chambers. In: Porphyrin Localization and Treatment of Tumors. (D.R. Doiron and C.J. Gomer, eds.), A.L. Liss, Publishers, N.Y.
8. Hall, E.J. Radiation dose rate: a factor of importance in radiobiology and radiotherapy. Br. J. Radiol., 45, 81-97, (1972).
9. Dahlman, A., Wile, A.G., Burns, R.G., Mason, R.G., Johnson, F.M. and Berns, M.W. Laser photoradiation therapy of cancer, Cancer Res., 43, 430-434, (1983).

10. Gomer, C.J. DNA damage and repair in CHO cells following hematoporphyrin photoradiation. Cancer Lett., Cancer Res., 11, 161-167, (1980).
11. Evensen, J.F. and Moan, J. Photodynamic action and chromosomal damage: a comparison of hematoporphyrin derivative (HpD) and light with x-irradiation. Br. J. Cancer, 45, 456-465, (1982).
12. Fiel, R.J., Datta-Gupta, N., Mark, E.H. and Howard, J.C. Induction of DNA damage by porphyrin photosensitization. Cancer Res., 41, 3543-3545, (1981).
13. Weishaupt, K.R., Gomer, C.J. and Dougherty, T.J. Identification of singlet oxygen as the cytotoxic agent in photoinactivation of a murine tumor. Cancer Res., 36, 2326-2329, (1976).
14. Santamaria,L., Bianchi, A., Arnaboldi, A. and Duffura, F. Photocarcinogenesis by methoxypsoralen, neutral red and proflavin. Possible implications in phototherapy. Med. Biol. Environ., 8, 171-181, 1980.

Porphyrin Localization and Treatment of Tumors, pages 471–482

IN VITRO AND IN VIVO STUDIES ON THE INTERACTION OF HEMATOPORPHYRIN AND ITS DIMETHYLESTER WITH NORMAL AND MALIGNANT CELLS

G. Jori, I. Cozzani, E. Reddi, E. Rossi,
L. Tomio, G. Mandoliti, G. Malvadi
Istituto Biologia Animale,
Universita di Padova and Divisione di
Radioterapia, O.C. Padova 35131, Italy

Although several porphyrins are known to localize in tumor tissues, only relatively water-soluble porphyrins are currently used for the photodiagnosis and phototherapy of tumors, mainly owing to the problems connected with their transport in the bloodstream; one, hematoporphyrin derivative (HpD), appears to be particularly efficient in photoinducing tumor regression (Dougherty 1980). In principle, also liposoluble porphyrins, such as porphyrin esters, could be utilized in the phototreatment of tumors: their photophysical properties are closely similar with those of the corresponding free acids (Bonnett et al. 1980) and the ability of porphyrin esters dissolved in organic solvents or incorporated into liposomes to photosensitize biological systems has been demonstrated (Grossweiner et al. 1982; Kessel, Chu 1983; Spikes 1983). Presumably, liposoluble porphyrins interact with cell sites other than those typical of hydrophilic porphyrins, thus originating different types of photoeffects (Sandberg, Romslo 1981). On these bases, we performed a comparative investigation on the interaction of Hp and Hp dimethylester (HpdiMe) with normal and tumor cells in vitro and in vivo. The porphyrins were transported in animals by incorporation into unilamellar vesicles of dipalmotoyl-phosphatidylcholine (DPPC). The delivery of liposome-bound drugs to specific tissues is a well-established technique (Ryman, Tyrrell 1980).

EXPERIMENTAL PROCEDURE

Hp and HpdiMe were purchased from Porphyrin Products, DPPC from Sigma. Unilamellar DPPC liposomes were prepared by dissolving the phospholipid and the porphyrin (70:1 molar ratio) in $CHCl_3$: CH_3OH 9:1 (v/v); the solvent was removed in a rotary evaporator, the film was resuspended in 0.01 M phosphate buffer pH 7.4 containing 150 mM NaCl, and the resulting suspension was sonicated at 50° C for 30 min. In this way, a fairly homogeneous dispersion of liposomes with an average outside diameter of 35 nm was obtained.

The uptake and release of $(Hp)_{aq}$, $(Hp)_{lip}$ and $(HpdiMe)_{lip}$ was tested with HeLa cells, cultivated as described by Cozzani and Spikes (1982), as well as with normal and neoplastic hepatocytes isolated from rat liver according to Jeejeeboy et al. (1975). Typically, $0.5-2 \cdot 10^6$ cells/ml were incubated either as adherent monolayers (HeLa) or suspended in PBS (hepatocytes) with 0.5-20 µg/ml porphyrin; 1% bovine serum albumin (BSA) was eventually added to the medium. At timed intervals, the cells were layered on 0.25 M sucrose, centrifuged for 3 min. at 3,500 rpm and the pellet was assayed for porphyrin content by treatment with 2% SDS and spectrophotofluoremetric analysis (Jori et al. 1979). In release experiments, the cells collected by centrifugation were further incubated with PBS eventually containing BSA and processed as above outlined.

In vivo distribution studies were performed with Wistar rats, 20±1 days old, either normal or bearing a subcutaneously grown solid Yoshida hepatoma AH-130 (Tomio et al. 1982), which had been injected i.v. with 5 mg/kg of porphyrin. At fixed times, rats were sacrificed, the tumor and selected tissues were removed, rinsed with PBS and directly assayed for the porphyrin content (Jori et al. 1979) or subjected to cell fractionation to estimate the subcellular distribution of porphyrins (Cozzani et al. 1981).

RESULTS

Uptake and release of $(Hp)_{aq}$ from normal and maligant cells.

The Uptake of Hp occurs linearly during 1-5 min. incubation at least within the 1-10 μg/ml range of Hp concentration. The addition of BSA lowers the extent of Hp uptake by normal hepatocytes (fig. 1); a similar pattern is observed with HeLa cells. Malignant hepatocytes accumulate Hp

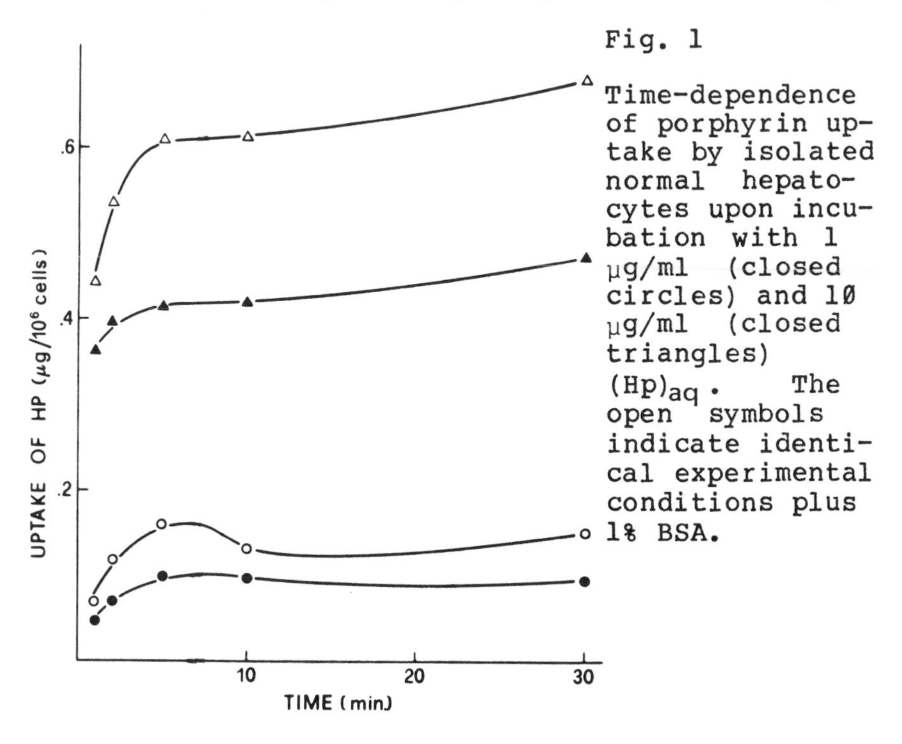

Fig. 1

Time-dependence of porphyrin uptake by isolated normal hepatocytes upon incubation with 1 μg/ml (closed circles) and 10 μg/ml (closed triangles) $(Hp)_{aq}$. The open symbols indicate identical experimental conditions plus 1% BSA.

at a higher rate and to a larger extent (fig. 2), while the presence of BSA exerts a much smaller effect especially for incubation times longer than 5 min. This behavior probably indicates a greater affinity of Hp for neoplastic cells, as it is also suggested by the observed slower release of Hp from neoplastic than from normal hepatocytes. BSA prevents only the process of Hp release from normal hepatocytes.

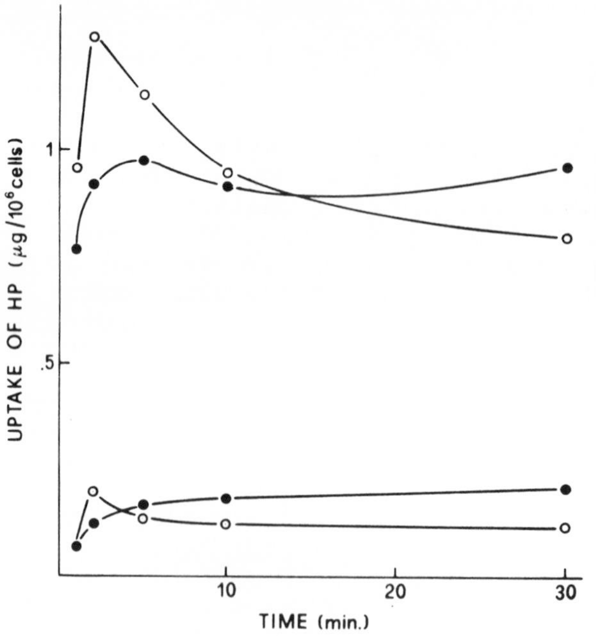

Fig. 2.

Time-dependence of porphyrin uptake by isolated malignant hepatocytes upon incubation with 1 μg/ml (lower plot) and 10 μg/ml (upper plot) $(Hp)_{aq}$. The closed circles indicate the presence of 1% BSA.

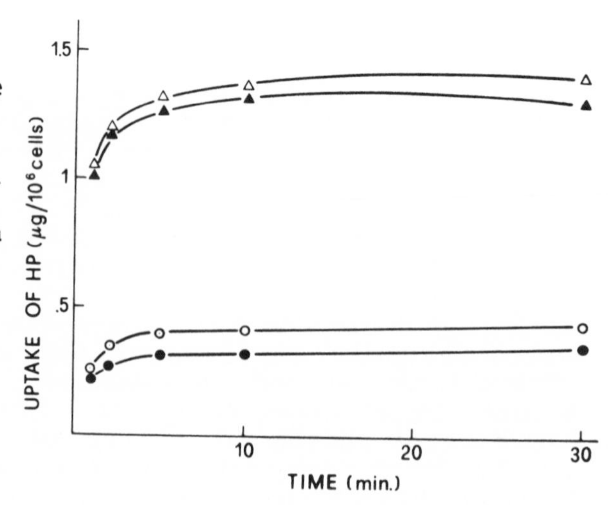

Fig. 3.

Time-dependence of porphyrin uptake by isolated normal hepatocytes upon incubation with 1 μg/ml (lower plots) and 20 μg/ml (upper plots) $(Hp)_{lip}$. The closed symbols indicate the presence of 1% BSA.

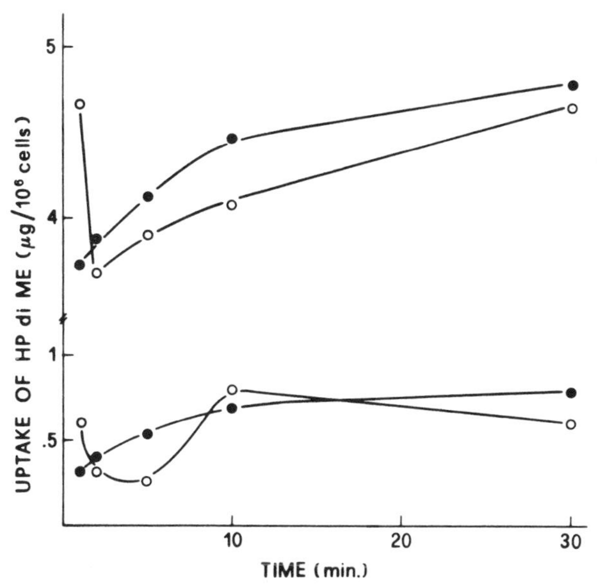

Fig. 4.

Time-dependence of porphyrin uptake by isolated normal hepatocytes upon incubation with 1 μg/ml (lower plots) and 10 μg/ml (upper plots) $(HpdiMe)_{lip}$. The open symbols indicate the presence of 1% BSA.

Uptake and release of $(Hp)_{lip}$ and $(HpdiMe)_{lip}$ from normal and malignant hepatocytes.

When normal or malignant hepatocytes are incubated with liposomal Hp, a biphasic Hp-accumulation process is observed leading to endocellular porphyrin concentrations larger than those obtained with $(Hp)_{aq}$ (fig. 3); remarkably, BSA has no effect with both cell types. Again, neoplastic cells display a higher affinity for Hp: e.g., after 5 min. incubation, malignant and normal hepatocytes uptake 4 μg and, respectively, 1.3 $\mu g/10^6$ cells. Moreover, in release experiments, the endocellular concentration of Hp gradually decreases reaching a final value of 2 μg and 0.6 $\mu g/10^6$ cells for malignant and, respectively, normal hepatocytes. If one assumes an average cellular volume of 33,500 μ^3 for normal hepatocytes and 7,250 μ^3 for malignant hepatocytes, the corresponding endocellular Hp concentrations are 28 μM and 412 μM.

Still higher levels of cell-bound porphyrin are found after incubation of hepatocytes and $(HpdiMe)_{aq}$ (fig. 4) with little or no effect of added BSA. The amount of cell-bound HpdiMe is clearly similar for normal and malignant hepatocytes; however, the latter cells release a remarkably smaller amount of porphyrin.

In vivo distribution of $(Hp)_{lip}$ and $(HpdiMe)_{lip}$.

The time-dependence of $(Hp)_{lip}$ distribution in the liver of normal rats resembles that observed for $(Hp)_{aq}$ (fig. 5); the presence of the tumor reduces the amount of Hp bound by the liver cells. However, in the case of $(Hp)_{lip}$, a substantially higher aliquot of the porphyrin migrates to the kidneys. Moreover, the tumor content of $(Hp)_{lip}$ is still increasing at 72 h after injection, whereas $(Hp)_{aq}$ begins to be cleared from tumor cells at 24 h. The phenomena are further enhanced in the case of administration of $(HpdiMe)_{lip}$. Thus, particularly favorable tumor/liver ratios of porphyrin concentration are found for a relatively long time after administration of liposome-bound Hp and HpdiMe (table 1).

Subcellular distribution of $(Hp)_{lip}$ and $(HpdiMe)_{lip}$ in vivo.

As previously shown (Cozzani et al. 1981), $(Hp)_{aq}$ injected into normal and tumor-bearing rats distributes among mitochondria, microsomes, lysosomes and the nuclear and cytoplasmic membranes; the membrane fraction contains the sites with highest affinity for Hp. The interaction of liposomal porphyrins with hepatocytes and hepatoma cells, after i.v.-administration to rats, causes a slower but tighter binding of both Hp and HpdiMe with the cytoplasmic membrane (table 2). Preliminary experiments would indicate that the largely preferential association of $(Hp)_{lip}$ and $(HpdiMe)_{lip}$ with the external cell membrane is obtained over a large range of injected porphyrin dose.

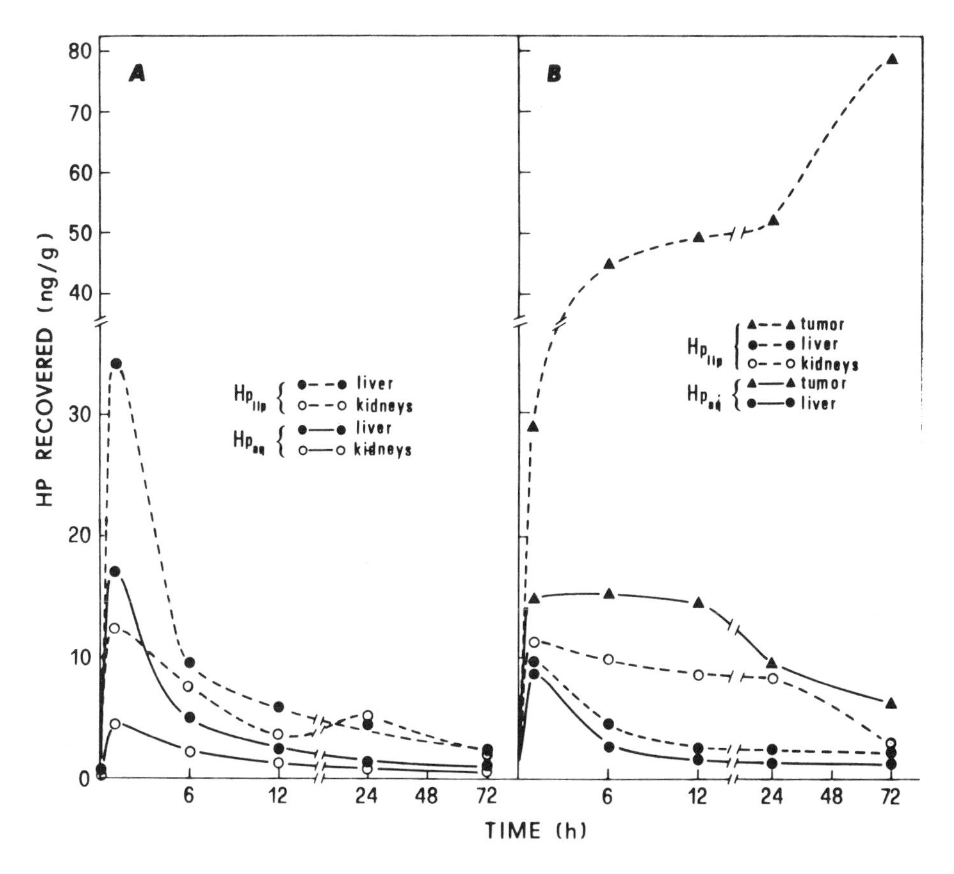

Fig. 5. Recovery of Hp from selected tissues of normal and tumor-bearing rats after i.v.-injection of 2.5 mg/kg $(Hp)_{aq}$ or $(Hp)_{lip}$.

DISCUSSION

Liposomes have been extensively studied as carrier vehicles for introducing various types of substances, including antitumor drugs, into cells either by membrane fusion or endocytosis (Gregoriadis 1976). We have used unilamellar DPPC liposomes since they are devoid of any appreciable cytotoxicity at the doses administered to rats in our experiments (Abra, Hunt 1981), readily

Table 1

Tumor-to-liver ratio of endocellular porphyrin concentration at various times after i.v.-administration to rats affected by AH-130 solid Yoshida hepatoma.

	Ratio		
Time (hours)	$(Hp)_{aq}$	$(Hp)_{lip}$	$(HpdiMe)_{lip}$
1	0.89	2.14	0.70
3	2.11	3.99	0.89
6	---	4.10	1.13
24	3.57	7.90	2.97
48	4.90	12.55	7.15
72	6.25	18.39	11.58

Table 2

Localization of Hp or HpdiMe in the cytoplasmic membrane of tumor cells at various times after i.v.-injection to rats affected by AH-130 solid Yoshida hepatoma.

	Percent of endocellular porphyrin		
Time (hours)	$(Hp)_{aq}$	$(Hp)_{lip}$	$(HpdiMe)_{lip}$
1	56.7	27.4	15.9
3	71.1	---	22.2
12	82.5	78.5	29.9
24	---	---	40.1
48	73.6	87.7	78.2
72	76.9	94.5	82.1

incorporate a large number of hydrophilic and lipophilic porphyrins (Spikes, 1983), and remain stable in the circulation below the phase transition temperature, i.e. 41° C for DPPC (Weinstein et al. 1979).

The results presented in this paper demonstrate that both Hp and HpdiMe can be delivered to cultured mammalian cells and various tissues in vivo once entrapped in the liposomal structure. In particular, the vehicular properties of liposomes lead to a greater accumulation of porphyrins in liver, as one would expect owing to the tendency of i.v.-administered colloidal suspensions to become associated with organs of the reticuloendothelial system (Ryman, Tyrrel 1980). On the other hand, appreciable amounts of liposomal porphyrins are also found in the kidneys, which suggests that an important fraction of liposome-bound Hp and HpdiMe administered to rats is not transported by serum albumin. Our hypothesis is supported by the cell studies, showing that the presence of BSA in the incubation medium affects the uptake and release of $(Hp)_{aq}$, while it has a minor influence in the case of $(Hp)_{lip}$ and $(HpdiMe)_{lip}$.

Our studies also indicate that the release of porphyrins from DPPC liposomes is not specific for any given cell type or tissue. However, neoplastic cells accumulate substantially larger amounts of liposomal Hp and HpdiMe both in vitro and in vivo. This fact must be related, at least in part, with a tighter binding of the porphyrins by the tumor cells, as shown by our competition studies between cells and BSA for Hp and HpdiMe binding, as well as by our porphyrin release experiments. In particular, the uptake of liposomal porphyrins by neoplastic cells in hepatoma-bearing rats is still increasing at 72 h after administration, while $(Hp)_{aq}$ begins to be cleared from tumor cells at 24 h after administration (fig. 5); (Jori et al. 1979). The slower migration of liposome-bound porphyrins to neoplastic tissues in vivo probably reflects the mechanisms controlling the interaction of DPPC unilamellar vesicles with tumor cells, which is still a matter of debate (Poste et al. 1982). A similar phenomenon has been observed for other liposome-carried drugs (Poste et al. 1982; Rahman et al. 1982).

We are presently investigating the detailed

process of Hp and HpdiMe transfer from liposomes to normal and tumor cells, although our subcellular fractionation experiments strongly suggest that the liposome-cell interaction involves the cytoplasmtic membrane to a major extent. In any case, it is clear that high tumor/liver ratios of Hp and MpdiMe concentrations can be obtained by our procedure especially for a relatively long period after i.v.-administration of the liposome-bound porphyrins. Thus, the possibility exists to improve the efficacy of tumor phototherapy and to minimize the onset of undesired side effects due to residual porphyrin in normal tissues. Moreover, since HpdiMe is characterized by a very high degree of purity and both Hp and HpdiMe are bound to liposomes in a monomeric form, our procedure for porphyrin delivery to tumors may overcome some problems connected with the heterogeneity and aggregation state of HpD and other porphyrins. Preliminary experiments show that rats bearing a solid AH-130 Yoshida hepatoma or mice bearing a MBL-2 lymphoma undergo a rapid regression of the tumor when they are exposed to red light at 72 h after i.v.-administration of 2.5 mg/kg liposomal Hp and HpdiMe.

Acknowledgment

This research received partial financial support from Consiglio Nazionale delle Ricerche (Italy), contract No. 82.01017.98 under the Progetto Finalizzato "Lasers in Biology and Medicine.

REFERENCES

Abra RM, Hunt CA (1981). Liposome disposition in vivo. III. Dose and vesicle size effects. Biochim Biophys Acta 666:493.

Bonnett R, Charalambides AA, Land EJ, Sinclair RS, Tait D, Truscott TG (1980). Triplet states of porphyrin esters. JCS Faraday I 76:852.

Cozzani I, Jori G, Reddi E, Tomio L, Zorat PL (1981). Distribution of endogenous and injected porphyrins at the subcellular level in rat hepatocytes and in ascites hepatoma. Chem Biol Interactions 37:67.
Cozzani I, Spikes JD (1982). Photodamage and photokilling of malignant human cells in vitro by hematoporphyrin-visible light. Med. Biol Environ 10:271.
Dougherty TJ (1980). Hematoporphyrin derivative for detection and treatment of cancer. J Surg Oncol 15:209.
Gregoriadis G (1976). The carrier potential of liposomes in biology and medicine. New Engl J Med 295:704.
Grossweiner LI, Patel AS, Grossweiner JB (1982). Type I and II mechanisms in the photosensitized lysis of phosphatidylcholine liposomes by hematopophyrin. Photochem Photobiol 36:159.
Jeeyeeboy KN, Ho J, Greenberg GR, Phillips MJ, Bruce-Robertson A, Sodtke U (19&5). Albumin, fibrinogen and transferrin synthesis in isolated rat hepatocyte suspensions. Biochem J 146:141.
Jori G, Reddi E, Tomio l, Salvato B, Zorat PL, Calzavara F (1979). Time-course of hematoporphyrin distribution in selected tissues of normal rats and in ascites hepatoma. Tumori 65:43.
Kessel D, Chou TH (1983). Porphyrin localizing phenomena. In Kessel D, Dougherty TJ (eds): "Porphyrin Photosensitization", New York: Plenum Press, p 115.
Poste G, Bucana C, Raz A, Bugelski P, Kirsh R, Fidler IJ (1982). Analysis of the fate of systemically administered liposomes and implications for their use in drug delivery. Cancer Res 42:1412.
Rohman YE, Cerny EA, Patel KR, Lau EH, Wright BJ (1982). Differential uptake of liposomes varying in size and lipid composition by parenchymal and kuppfer cells of mouse liver. Life Sci 31:2061.
Ryman BE, Tyrrel DA (1980). Liposomes-bags of potential. Essays Biochem 16:49.

Sandberg S, Romslo I (1981). Porphyrin-induced photodamage at the cellular and the subcellular level as related to the solubility of the porphyrin. Clin Chim Acta 109:193.

Spikes JD (1983). A preliminary comparison of the photosensitizing properties of porphyrins in aqueous solution and liposomal systems. In Kessel D, Dougherty TJ (eds): "Porphyrin Photosensitization", New York: Plenum Press, p 181.

Tomio L, Zorat PL, Jori G, Reddi E, Salvato B, Corti L, Calzavara F (1982). Elimination pathway of hematoporphyrin from normal and tumor-bearing rats. Tumori 68:283.

Porphyrin Localization and Treatment of Tumors, pages 483–500

PARAMETERS OF HEMATOPORPHYRIN DERIVATIVE TUMOR CELL KILLING EFFICIENCY: DECOMPOSITION OF HEMATOPORPHYRIN DERIVATIVE AT HIGH POWER DENSITIES

Robert E. Anderson, B.S.
Cerebrovascular Research

Robert E. Wharen, Jr., M.D.
Department of Neurosurgery

Charlotte A. Jones, Ph.D.
Urology Research

Edward R. Laws, Jr. M.D.
Department of Neurosurgery

Mayo Clinic
Rochester Minnesota 55905

ABSTRACT

Hematoporphyrin derivative (HpD) tumor cell killing efficiency has been evaluated as a function of power density, optical spectrum, HpD concentration, and HpD preparation using trypan blue exclusion assay. Cell survival curves for Mayo Clinic HpD and for Photofrin were equivalent at all concentrations and power densities for both violet and red light. Survival curves for red light demonstrated a two-fold increased cellular killing efficiency at higher densities (160 mW/cm^2). However, for violet light an irreversible and marked decrease in cellular killing efficiency was found as the power density was increased. This phenomena of decreased cellular killing efficiency of violet light at high power densities is consistant with a process of photo-decomposition of HpD and is most pronounced at low concentrations (< 25 μg/ml).

INTRODUCTION

The concept of photoradiation therapy of malignant tumors is based upon the fact that certain molecules can function as photosensitizing agents. The presence of such molecules within a cell makes the cell vulnerable to the application of light of appropriate wavelength and intensitity. At the turn of the century Raab (22) first observed that acridine orange was toxic to paramecia upon exposure to sunlight. In 1903, Tappenier and Jesionek (25) described the photosensitizing effect of eosin and light on superficial tumors in man. It was not until 1975, however, that photoradiation was applied to the treatment of both animal tumors and malignant tumors in man by Dougherty et al (7) and Kelly et al (14) using hematoporphyrin derivative as a photosensitizer.

The action of a photosensitizer is produced by the absorption of a photon of a wavelength sufficient to promote electrons within the sensitizer to an excited triplet state. This excited molecule may then interact either directly with substrates within the cell or indirectly with these substrates through the production of singlet oxygen. The various photochemical reactions that are elicited are subsequently capable of killing cells rather nonspecifically through multiple interactions with the cell membrane, cytoplasm, and nucleus, although currently the membrane and intracellular organelles are considered to be the primary locations of cellular damage (21).

Since 1975, investigations of photoradiation effects have almost exclusively involved the use of hematoporphyrin derivative as the photosensitizer. Hematoporphyrin derivative is a compound similar to the heme group of hemoglobin without the iron ligands. This compound has been known to concentrate in malignant tumors since the late 1800's (1,2,9). Hematoporphyrin derivative, upon irradiation by light of an appropriate wavelength, can mediate photochemical reactions that are cytotoxic to malignant cells. In addition, hematoporphyrin derivative will produce a red fluorescence (635 mm) when stimulated by violet light (405 mm). Together these properties make hematoporphyrin derivative (3) applicable both for the detection and for the treatment of tumors (9,17,18,27,29).

The application of HpD photoradiation therapy to the field of neurosurgery was initially very encouraging. A number of investigators have reported the capability of HpD to kill selectively glioma cells, both in vitro (5,6,12) and in vivo (5,12,16,19,20). In addition, more recent reports (10,16,19,20) including our own (16), have described initial attempts in the treatment of human malignant gliomas and report that HpD phototherapy is capable of tumor cell destruction in man.

Although these initial reports were encouraging, the ultimate clinical effectiveness of this technique in neurosurgery is dependent upon a more thorough understanding of the various parameters of the photodynamic process. We have investigated the parameters of HpD tumor cell killing efficiency in an effort to maximize the effectiveness of photoradiation therapy applied to malignant brain tumors. Our results thus far have yielded two interesting and previously unreported phenomena. The first is evidence suggesting the decomposition of HpD at high power densities of violet light (405 nm). The second is a comparison of the killing efficiency of different preparations of HpD, for although there have been some unsubstantiated claims regarding the relative effectiveness of different preparations of HpD, no direct comparison has yet been reported.

MATERIALS AND METHODS

Hematoporphyrin Derivative

Two preparations of HpD were investigated. Mayo Clinic HpD was prepared in accordance with the Mayo Clinic's Investigational New Drug (IND) permit (3,4) (Mayo Lung Project, Dr. R.S. Fontana), tested for sterility and used as a stock solution of 5 mg/kg in saline with the pH adjusted to 7.8 Photofrin was purchased from Oncology Research and Development. The absorption spectra of Mayo Clinic HpD and Photofrin were obtained at a concentration of 10 μg/ml HpD in 0.9% NaCl using a Tracor Northern spectrophotometer.

Cell Line and Assay Procedure

MEWO cells (a cell line derived from a human malignant melanoma were grown in 75 cm^2 flasks using Dulbecco's modified Eagle medium containing 10% fetal calf serum. Separate 75 cm^2 flasks of MEWO cells in log phase growth were exposed to HpD in medium containing 10% fetal calf serum for 6 hours. Prior to irradiation the cells were harvested with trypsin and suspended in the same medium without HpD. A 200 µl aliquot of cellular suspension was irradiated in microtiter plates in a water bath at 37°C. Survival curves were obtained 30 minutes following irradiation using a trypan blue exclusion assay. Controls were run in each case on cells not exposed to HpD but receiving identical light exposures. Each survival curve was obtained on at least five separate occasions.

Irradiation

Cell survival curves were obtained following irradiation with violet (405 nm), red (625 nm), and white light (340-680 nm) at energies of 0,5,10,20,40,80,160, and 320 joules, and at power densities of 20,40,80, and 160 mW/cm^2 for cells exposed to either Mayo Clinic HpD or Photofrin. A Hanovia 1 KW xenon lamp was used as a light source and filtered via a KG-3 (340-680 HPBW) heat absorption filter (Schott) and either a 405 ± 40 nm HPBW BG-12 violet filter or 625 ± 5 nm HPBW 5393 Oriel red filter (Figure 1). For white light only the heat absorption filter was used. The temperature of the cellular suspension during irradiation was monitored with a RM-2B infrared radiometric microscope (Barnes Engineering). Power densities were measured using a Hewlett-Packard radiant fluxmeter.

RESULTS

The absorption spectra for Mayo Clinic HpD and Photofrin at 10 µg/ml HpD in 0.9% NaCl were identical and consisted of five absorption peaks with wavelengths of 405, 508, 540, 570, and 620 nm (Figure 1).

No temperature change was observed during any of the irradiation periods with either violet, red, or white light.

Cell survival curves following violet light irradiation at a power density of 40 mW/cm^2 demonstrated a consistant pattern with an increasing slope and decreasing shoulder to the curve at increasing concentrations of HpD (Figure 2).

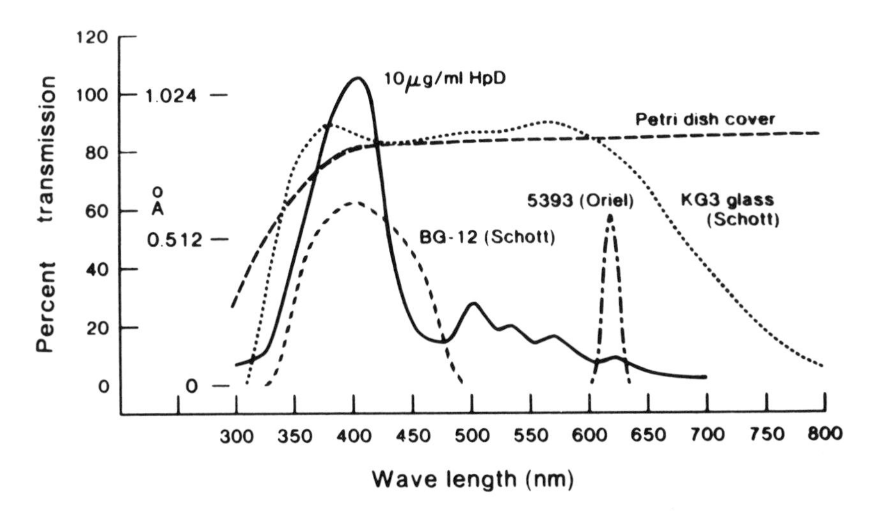

Figure 1. Absorption spectra of HpD and transmission of the optical components used in the experimental design. HpD concentration is 10 μg/ml in 0.9% NaCl. The absorption spectra demonstrated five absorption peaks. The major (405 mm) and minor (624 nm) peaks were photoilluminated separately in this study. A Schott BG-12 and an Oriel 5393 were used to filter the excitation light for the HpD peaks of 405 nm and 624 nm respectively. Schott KG3 glass and an 80 mm H_2O filter were used to filter out unwanted IR wavelengths to minimize heat radiation.

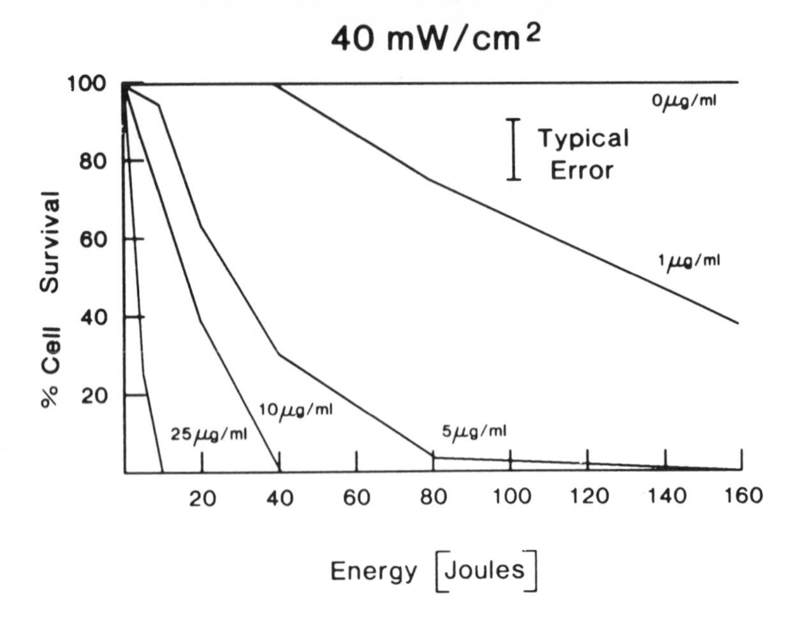

Figure 2. Survival curves of MEWO cells following irradiation with violet light (405 nm) at a power density of 40 mW/cm^2. Prior to irradiation the cells were exposed to Mayo Clinic HpD at concentrations of 0,1,5,10, and 25 μg/ml for 6 hours. Typical error represents the standard deviation of five separate determinations for each survival curve.

The cellular killing efficiency varied linearly with HpD concentration. Survival curves obtained using Photofrin (Figure 3) were identical to those obtained with Mayo Clinic HpD.

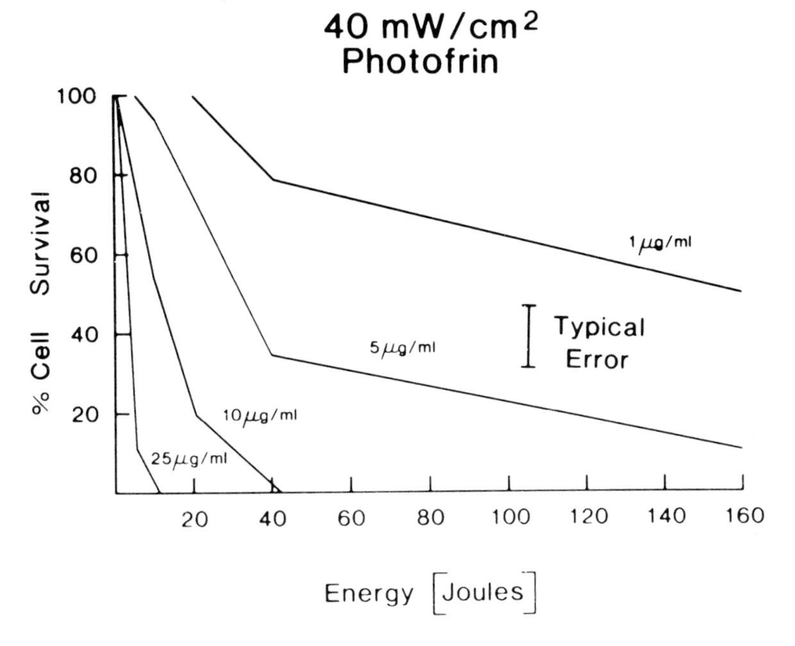

Figure 3. Survival curves of MEWO cells following irradiation with violet light (405 nm) at a power density of 40 mW/cm^2. Prior to irradiation the cells were exposed to Photofrin at concentrations of 0,1,5,10, and 25 µg/ml for 6 hours. Typical error represents the standard deviation of five separate determinations for each survival curve.

Similar survival curves were obtained following red light irradiation at a power density of 40 mW/cm^2 at various concentrations of HpD (Figure 4). The relative killing efficiency of violet light compared to red light irradiation was approximately 16:1 (Figure 5) at a power density of 40 mW/cm^2. This ratio varied somewhat depending upon the concentration with the ratio tending to decrease at higher HpD concentrations.

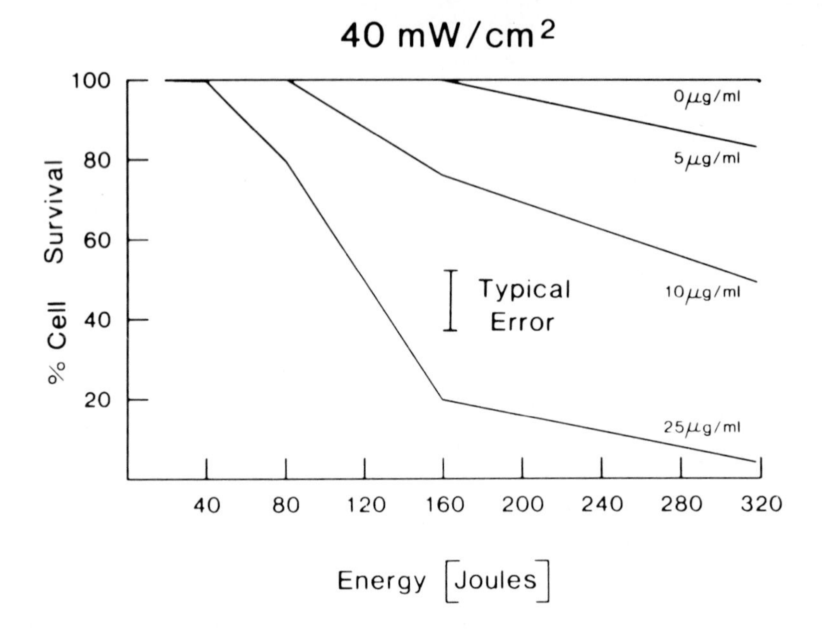

Figure 4. Survival curves for MEWO cells following irradiation with red light (625 nm) at a power density of 40 mW/cm^2. Prior to irradiation the cells were exposed to Mayo Clinic HpD at concentrations of 0,5,10, and 25 µg/ml for 6 hours. Typical error represents the standard deviation of five separate determinations for each survival curve.

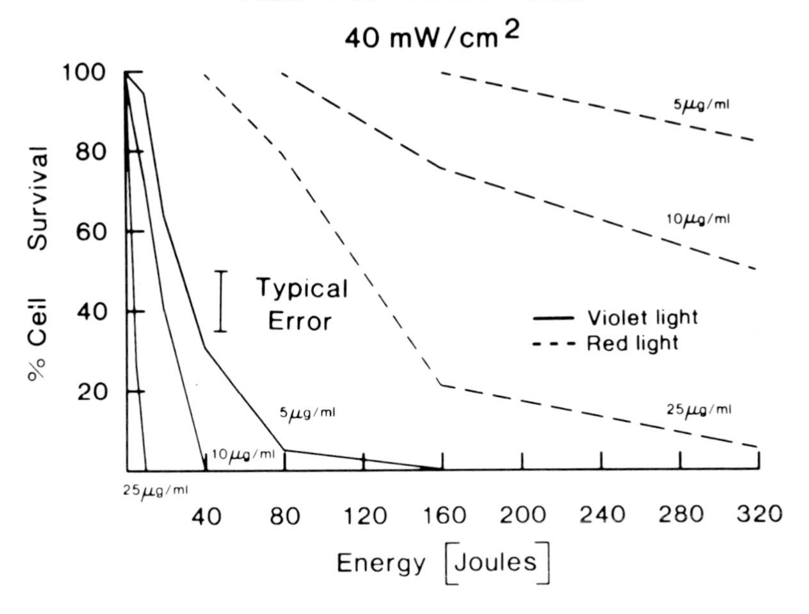

Figure 5. Comparison of HpD tumor cell killing efficiency following irradiation of MEWO cells with either violet (405 nm) or red light (625 nm) at a power density of 40 mW/cm^2. Prior to irradiation the cells were exposed to Mayo Clinic HpD at concentrations of 5,10, and 25 µg/ml for 6 hours. Typical error represents the standard deviation of five separate determinations for each survival curve.

Survival curves for red light irradiation demonstrated a two-fold increased cellular killing efficiency at higher power densities compared to lower power densities at all concentrations of HpD (Figure 6). One hundred ten joules of red light at a power density of 40 mW/cm^2 produced a 50% cell kill compared to only 55 joules required at a power density of 160 mW/cm^2 at an HpD concentration of 25 µg/ml. Identical curves were obtained for both Mayo Clinic HpD and for Photofrin.

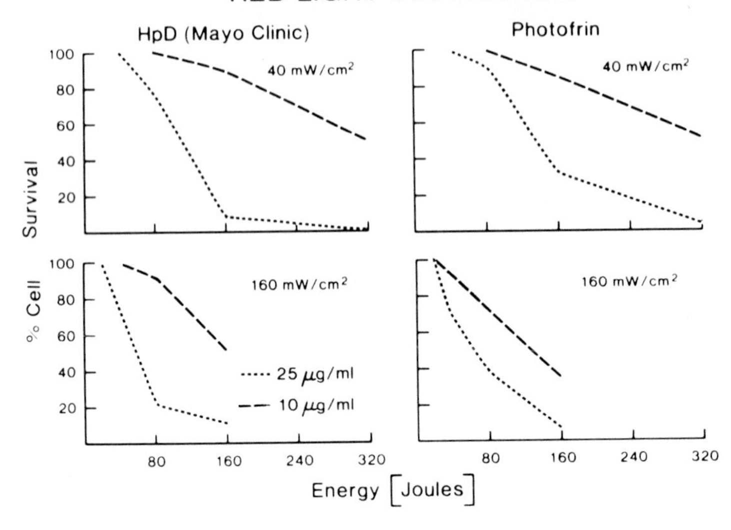

Figure 6. Survival curves of MEWO cells comparing the tumor cell killing efficiency of red light irradiation (625 nm) at power densities of 40 mW/cm^2 and 160 mW/cm^2. Prior to irradiation the cells were exposed to either Mayo Clinic HpD or Photofrin at a concentration of 10 and 25 µg/ml for 6 hours.

Survival curves following violet light irradiation, however, demonstrated a marked decrease in cellular killing efficiency at high power densities. Twenty joules of violet light at 20 mW/cm^2 produced a 50% cell kill while 105 joules of violet light were required at a power density of 160 mW/cm^2 for an equivalent cell kill at an HpD concentration of 5 µg/ml (Figure 7). This decrease in cellular killing efficiency at higher power densities of violet light varied greatly with HpD concentrations being most pronounced at low concentrations of HpD (Figure 8). Again, similar results were obtained for both Mayo Clinic HpD and for Photofrin (Figure 9).

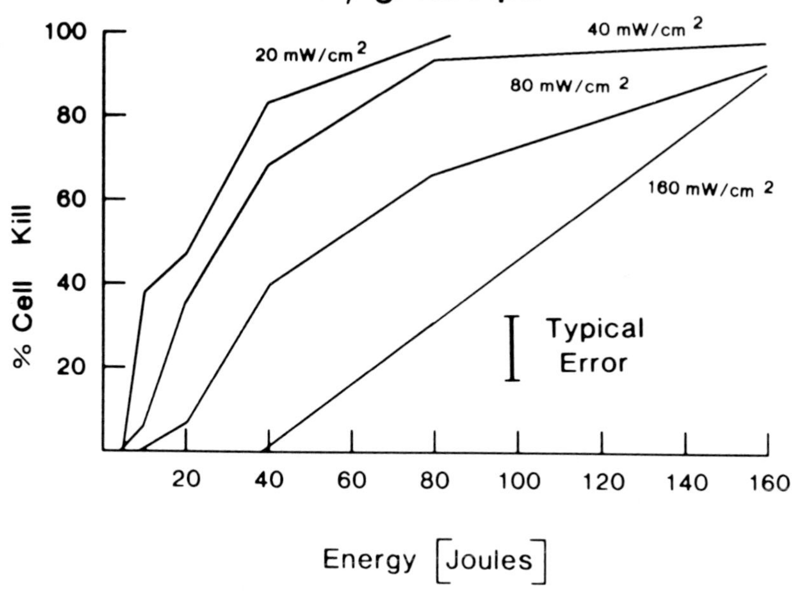

Figure 7. Cellular killing curves for MEWO cells following irradiation with violet light (405 nm) at power densities of 20, 40, 80, and 160 mW/cm^2. Prior to irradiation the cells were exposed to Mayo Clinic HpD at a concentration of 5 μg/ml for 6 hours. Typical error represents the standard deviation of 5 separate determinations for each survival curve.

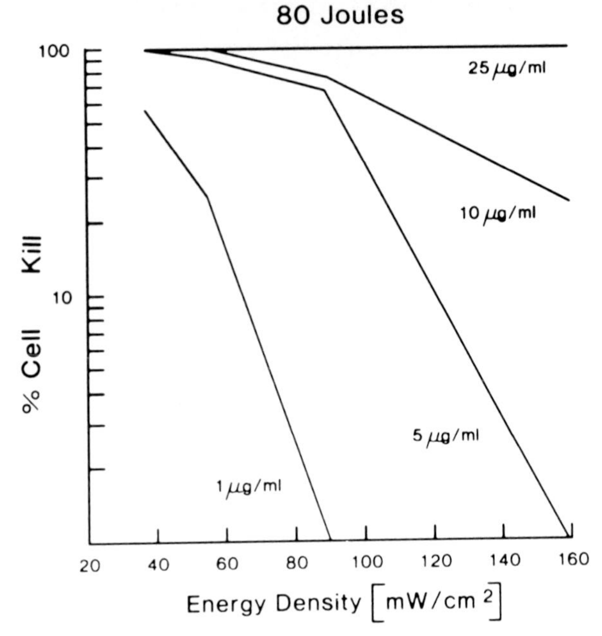

Figure 8. Cellular killing curves for MEWO cells irradiated with 80 joules of light (405 nm) at power densities ranging from 40 mW/cm^2 - 160 mW/cm^2 demonstrating a decrease in killing efficiency at high power densities. Prior to irradiation the cells were exposed to Mayo Clinic HpD at a concentration of 1,5,10, and 25 µg/ml for 6 hours.

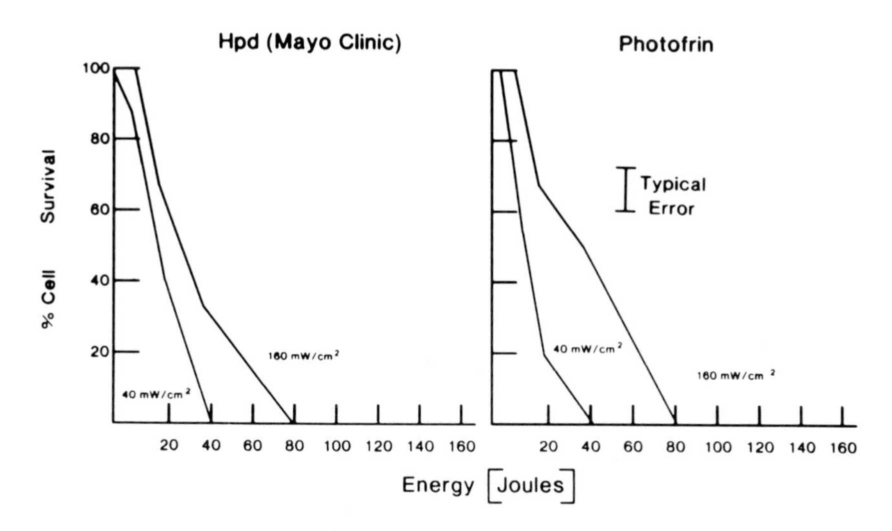

Figure 9. Survival curves for MEWO cells irradiated with violet light (405 nm) at power densities of 40 mW/cm^2 and 160 mW/cm^2 demonstrating a decrease in killing efficiency at higher power densities for both Mayo Clinic HpD and Photofrin. Prior to irradiation the cells were exposed to either Mayo Clinic HpD or Photofrin at a concentration of 10 μg/ml for 6 hours.

The decrease in cellular killing efficiency observed at high incident power densities of violet light was found to be irreversible. Following exposure to white light at a high power density of 500 mW/cm^2, cells further irradiated with violet light at 20 mW/cm^2 demonstrated survival curves with a marked decrease in cellular killing efficiency compared to cells irradiated with only violet light at 20 mW/cm^2 (Figure 10).

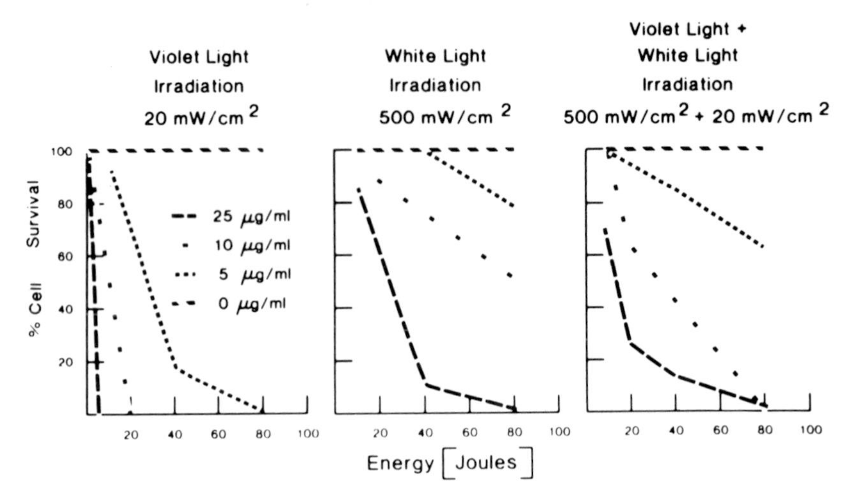

Figure 10. Cell survival curves of MEWO cells following irradiation with: a) violet light (405 nm) at a power density of 20 mW/cm^2 on the left, b) high power density (500 mW/cm^2) white light (340-680 nm) in the center, or c) high power density (500 mW/cm^2) white light (340-680 nm) followed by violet light (405 nm) at a power density of 20 mW/cm^2 on the right, demonstrating the irreversible decrease in the cellular killing efficiency of violet light by high intensity white light consistent with the photodecomposition of HpD. Prior to irradiation the cells were exposed to Mayo Clinic HpD at a concentration of 0,5,10, and 25 μg/ml for 6 hours.

DISCUSSION

The effectiveness of tumor cell destruction by HpD photoradiation therapy is dependent upon an understanding of the parameters involved in the photoradiation process. The relative HpD tumor cell killing efficiency has been examined as a function of optical spectrum, HpD concentration, HpD preparation, and power density using a reproducible trypan blue exclusion assay. Although there has been some controversy involving the adequacy of the trypan blue exclusion technique as an assay for cytotoxicity (23,24), others (15) have shown that trypan blue exclusion is an effective technique which allows quantitation of the relative effects of the various parameters of the photodynamic process.

The two preparations of HpD were found to be identical both in their absorption curves (Figure 1) and in their relative killing efficiency for both violet and red light (Figure 2,3,6). Photofrin and Mayo Clinic HpD also showed identical responses to higher power densities of both red and violet light, showing an increased cellular killing efficiency at high power densities of red light (Figure 6) and a decreased cellular killing efficiency at high power densities of violet light (Figure 9). Thus, in this experiment there is no support for the contention that one preparation of HpD is more effective than the other.

The relative decrease in killing efficiency observed at lower power densities of red light (Figure 6) has been observed previously by Dougherty et al (8) who attributed this phenomenon to a partial repair process occurring during exposure.

The phenomenon of an irreversible decrease in tumor cell killing efficiency at high power densities of violet light (Figure 7-10) is consistent with the photodecomposition of HpD. Photodecomposition of dyes at high power densities is a well established phenomenon (25) which is dependent upon the intensity of the incident light and the concentration of the dye. Decomposition occurred most prominently at low concentrations of HpD and was not observed at a concentration of 25 μg/ml (Figure 8). At a concentration of 5 μg/ml HpD the % cell kill decreased from 100% at a power density of 40 mW/cm^2 to 0% at a power density of 160 mW/cm^2.

These findings indicate that the effectiveness of HpD photoradiation therapy using high intensity violet light will be dependent upon the concentration of HpD in the tumor. Quantification of hematoporphyrin or HpD concentrations ranging from 1-40 μg/gm tissue depends upon the dose administered and the interval between drug administration and quantitative measurement. Indeed in human gliomas (28) the concentration of HpD in the tumor 24 hours following administration of 5 mg/kg HpD was 2.5 μg/gm tissue. Thus, the concentrations of HpD clinically achievable in some tumors may well be in the range where photodecomposition of HpD at high power densities must be considered in order to maximize the application of HpD-photoradiation therapy in patients.

REFERENCES

1. Anderson TM (1898). Hydroa aestivale in two brothers, complicated with the presence of hematoporphyrin in the urine. Br J Dermatol 10:1.

2. Blum HF (1941). "Photodynamic Action and Diseases Caused by Light" New York: Reinhold Publishing Corp., pp 211.

3. Carpenter RJ III, Neel HB III, Ryan RJ, Sanderson DR (1977). Tumor fluorescence with hematoporphyrin derivative. Ann Otol Rhinol Laryngol 86:661.

4. Cortese DA, Kinsey JH, Woolner LB, Payne WS, Sanderson DR, Fontana RS (1979). Clinical application of a new endoscopic technique for detection of in situ bronchial carcinoma. Mayo Clin Proc 54:635.

5. Diamond I, Granelli SA, McDonagh AF, Neilsen S, Wilson CB, Jaenicke R. (1972). Photodynamic therapy of malignant tumor. Lancet 2:1175.

6. Diamond I, Gronelli SC, McDonagh AF (1977). Photochemotherapy and photodynamic toxicity: simple methods for identifying potentially active agents. Biochem Med 17:121.

7. Dougherty TJ, Grindey GB, Fiel R, Weishaupt KR, Boyle DG (1975). Photoradiation therapy II. Cure of animal tumors with hematoporphyrin and light. J Natl Cancer Inst 55:115.

8. Dougherty TJ, Gomer CJ, Weishaupt KR (1976). Energetics and efficiency of photoinactivation of murine tumor cells containing hematoporphyrin. Cancer Res 36:2333.

9. Figge FHJ, Weiland GS, Manganiello LOJ (1948). Cancer detection and therapy: Affinity of neoplastic, embryonic, and traumatized tissues for porphyrins and metalloporphyrins. Proc Soc Exp Biol Med 68:640.

10. Forbes IJ, Cowled PA, Leong SY, Ward AP, Black RB, Blake AJ, Jacka FJ (1980). Phototherapy of human tumors using hematoporphyrin derivative. Med J Aust 2:489.

11. Gomer CJ, Dougherty TJ (1979). Determination of (^{3}H) and (^{14}H) HpD derivative distribution in malignant and normal tissue. J Cancer Res 39:146.

12. Granelli SG, Diamond I, McDonagh AF, Wilson CB, Neilson SL (1975). Photochemotherapy of glioma cells by visible light and hematoporphyrin. Cancer Res 35:2567.

13. Jori G, Pizzi G, Reddi E, Tomio L, Salvato B, Zorat P, Calzavara F (1979). Time dependence of hematoporphyrin distribution in selected tissues of normal rats and in ascistes hepatoma. Tumori 65:425.

14. Kelly JF, Snell ME, Berenbaum MC (1975). Photodynamic destruction of human bladder carcinoma. Br J Cancer 31:237.

15. Kinsey JH, Cortese DA, Moses HL, Ryan RJ, Branum EL (1981). Photodynamic effect of hematoporphyrin derivative as a function of optical spectrum and incident energy density. Cancer Res 41:5020.

16. Laws ER Jr, Cortese DA, Kinsey JH, Eagan RT, Anderson RE (1981) Photoradiation therapy in the treatment of malignant brain tumors: A phase I (feasibility) study. Neurosurgery 9:672.

17. Leonard Jr, Beek WL (1971). Hematoporphyrin fluorescence: An aid in diagnosis of malignant neoplasm. Laryngoscope 81:365.

18. Lipson RL, Baldes EJ, Olsen AM (1961). The use of a derivative of hematoporphyrin in tumor detection. JNCL 26:1.

19. Perria C, Capuzzo T, Cavagnaro G, Patti R, Franeaviglia N, Rivano C, Tercero VE (1980). First attempts at the photodynamic treatment of human gliomas. J Neurosurg Sci 24:119.

20. Perria C (1981). Photodynamic therapy of human gliomas by hematoporphyrin and He-Ne laser. IRCS Med Sci (Cancer) 9:57.

21. Pooler JP (1981). Dye-sensitized photodynamic inactivation of cells. Medical Physics 8:614.

22. Raab O (1964). Photodynamic Action and Diseases Caused by Light (ed. Blum HF). New York: Hafner.

23. Roper PR, Drewinko B (1979). Cell survival following treatment with antitumor drugs. Cancer Res 39:1428.

24. Schrek R (1979). Utility and efficiency of viable cell counts. Cancer Res 39:4288.

25. Tappenien H, Jesionek A (1903). Therapeutische versuche met fluoreszierenden stoffe. Moench Med Wochschrl 2042.

26. Undenfriend S (1971). "Fluorescence Assay in Biology and Medicine". New York: Academic Press, p 104.

27. Weishaupt KR, Gomer CJ, Dougherty TJ (1976). Identification of singlet oxygen as the cytotoxic agent in photoinactivation of a murine tumor. Cancer Res 36:2326.

28. Wharen RE Jr., Anderson RE, Laws ER Jr., (in press). Quantitation of hematoporphyrin derivative (HpD) in human gliomas, experimental central nervous system tumors and normal tissues. Neurosurgery, April.

29. Winkelman J, Rasmussen-Taxdel DS (1960). Quantitative determination of porphyrin uptake by tumor tissue following parenteral administration. Bull Johns Hopkins Hosp 107:228.

Porphyrin Localization and Treatment of Tumors, pages 501–520

UPTAKE AND LOCALIZATION OF HPD AND "ACTIVE FRACTION" IN TISSUE CULTURE AND IN SERIALLY BIOPSIED HUMAN TUMORS

M. W. Berns[1,2], Ph.D., M. Hammer-Wilson[1], R. J. Walter[2], W. Wright[1], M.-H. Chow[2], M. Nahabedian[1], A. Wile[1], M.D.
[1]Dept. of Surgery, [2]Dept. of Developmental and Cell Biology
University of California, Irvine
Irvine, California 92717

INTRODUCTION

The use of hematoporphyrin derivative (HPD) plus light as a modality for cancer detection and cancer therapy has captured the attention of many research and clinical oncologists lately. Though the concepts of using the tumor localizing capability of porphyrin for detection, and the photosensitive properties of porphyrin for a therapeutic killing effect can be traced in the literature to studies in the 1940's-1960's (Figge, Welland, 1948, 1949; Lipson, et al, 1961; Rasmussen-Taxdall et al, 1955), it wasn't until more recently that these concepts were put into substantial practice (Dougherty et al, 1978). However, as is often the case with a new complex modality, a satisfactory understanding of the processes involved may take a long time to achieve and involve many studies in numerous laboratories. In fact, a complete understanding of the processes may not occur until long after the modalities are being applied in clinical studies. This is certainly the situation with HPD-photoradiation therapy (HPD-PRT). Though HPD-PRT is being applied in clinical settings throughout the world, an adequate understanding of the mechanisms of action of HPD (and its various components) and the photophysical mechanisms involved are poorly understood. This lack of understanding exists for the complex HPD solution (as well as its individual components) and it exists at all of the critical levels of the molecular, cellular, tissue, and organ levels. In order to fully appreciate the potential of HPD-PRT, it will be necessary to characterize the processes at all of these levels bearing in mind that the cytocidal effect of HPD-PRT involves destruction of the tumor cells. This may be a direct effect on the cancer cell, or more indirect effect such as destruction of the cells in the vasculature that nourishes the tumor. Because of the complex nature of HPD-PRT (in both tumor detection and tumor-killing) we have undertaken a series of studies aimed at the cellular and tissue levels using cell cultures, animal tumor models, and patients in the clinic (Berns et al, 1982; Dahlman et al, 1983).

Materials and Methods

HPD was obtained from ORD, Inc. (Cheektowaga, N.Y.) and was diluted prior to use for cellular studies in saline and/or culture media to a dilution of 25 μg/ml or 50 μg/ml (Berns et al, 1982). A fresh dilution was made prior to each experiment. In human patient studies the HPD was injected intraveneously without dilution (Dahlman et al, 1983). A dosage of 3 mg HPD/kg b.w. was used. In animal (mouse) studies a dosage of 10.0 mg/kg b.w. was used.

Studies were also conducted with an enriched active fraction of HPD (called SPP or Photofrin II) which was graciously provided by ORD, Inc. for use in experimental non-human studies only. This sample was supplied in a solution of sterile saline in a concentration of 2.5 mg/ml. This was half the concentration of the HPD solution provided for standard use. Since it was estimated by the suppliers that SPP contained twice the concentration of the "active" ingredient, a solution with one-half the concentration of material should yield an activity equal to the standard HPD solution.

Both HPD and the SPP were analyzed using reverse phase high pressure liquid chromatography (HPLC). The samples were separated on a Beckman Ultrasphere-octyl column (4.6 mm x 15 cm) using a gradient elution. The mobile phase was composed of soln A: methanol/water (30:70) + 1% acetic acid; soln B: methanol/isopropanol (30:70) + 1% acetic acid; soln C: isopropanol + 1% acetic acid applied as follows: 80% A + 20% B for 5 min then changing to 50% A + 50% B over 45 min; 50% A + 50% B changing to 95% B over 30 min; 10 min hold; 30% B + 70% C changing to 100% C over 10 min then held in 100% C.

The effluent was analyzed by recording absorbance at 400 nm on a variable wavelength detector (Hitachi 100-10 spectrophotometer with an Altex flow cell).

Three cellular systems were used in these studies: rat kangaroo epithelial (PTK_2-ATCC #56), BALB/3T3 non-tumorigenic fibroblast clone #A31 (ATCC #163), BALB 3T12 tumorigenic fibroblast line (ATCC #164). The cells were grown in standard T25 culture flasks and in multipurpose Rose chambers for microscopy and dosimetry as described earlier (Berns et al, 1982, 1983).

The cells were seeded into the Rose chambers 24 hrs prior to exposure to HPD or SPP. At various times following exposure to the dyes, the cells were analyzed for uptake, washout, and photosensitivity.

The uptake and washout measurements were made using a laser stimulated quantitative fluorometer interfaced with a Tracor Northern #1710 multichannel analyzer and computer system (see detailed description in Berns et al, 1982; Siemens et al, 1982). Fluorescent measurements were made in individual regions within cells (extended cytoplasm and perinuclear cytoplasm) and groups of cells. Fluoresence was recorded at above 580 nm to correspond to known fluorescence emission peaks for components of the HPD solution (Berns et al, 1983).

Cell killing studies were performed in Rose culture chambers with cells being exposed to HPD at a concentration of 25-100 μg/ml for a time period of 24 hrs immediately after their return to normal culture medium. The cells were exposed to red laser light in the wavelength range of 625-630 nm at average powers ranging from 5 mw/cm^2-100 mw/cm^2. Two different laser systems were used: a Coherent INOVA 20 argon ion pumped dye laser (wavelength emission at 630 nm) and a Plasma Kinetics gold vapor laser (wavelength emission at 628 nm). The argon-dye laser emitted light as a continuous wave beam with average powers in the range of 5 mw - 4 W. The gold vapor laser emitted a pulsed beam at 8 khz with a pulse duration of 30 nanoseconds.

The average powers used in the studies for the gold laser was varied from 5 mw/cm^2 to 400 mw/cm^2. Total energy for all the lasers was in the range of .5J/cm^2 to 100 J/cm^2. Power and energies of all the lasers were continually monitored with calibrated power meters and calorimeters. Wavelength was monitored with a JY 534 monochrometer. Cell survival was determined 24 hours after laser exposure by trypan blue exclusive staining.

The effect of HPD on the cell surface was studied at the biochemical level by examining the effect of HPD on the fluorescent and mobility properties of the membrane-bound probe fluorescein conjugated soybean agglutinin (f-SBA). These measurements were made employing standard methods of fluorescence photobleaching and fluorescence recovery after photobleaching (FRAP). Cells were incubated in a solution of HPD and then exposed to the membrane-binding probe (the reverse experiments were also conducted: exposure to the membrane probe followed by HPD exposure). By monitoring the fluorescence of the membrane probe (its fluorescence emission peak is substantially shorter than that of HPD) it was possible to determine (1) the lateral mobility of the receptor to which the probe is bound in the cell membrane, (2) the % of recovery of the membrane probe in a membrane area that has been photobleached by the laser, and (3) the sensitivity of the cell surface to photobleaching. These three parameters were examined in normal and malignant cell types exposed to HPD in order to elucidate cell membrane effects of HPD. The system for making these measurements has been described elsewhere (Berns et al, 1982).

Structural studies on the cell surface were undertaken using scanning electron microscopy. As indicated in previous studies (Siemens et al, 1980), agents affecting the cell membrane often result in an alteration in the amount of microvilli that extend from the cell surface. The effect of HPD on the cell surface of 3T3 and 3T12 cells was examined using the methods described earlier (Siemens et al, 1980).

The uptake and loss of HPD from human tumors was determined by taking serial punch biopsies from a large squamous cell carcinoma growing on the outer cheek of a female patient. Biopsies were taken prior to HPD injection (3 mg/kgm b.w.) and at 24 hour intervals for five days. Biopsy was always made from an area of the tumor containing some viable tissue. The tissue was immediately placed in isotonic saline at 4°C and frozen-sectioned within 30 minutes. The frozen sections were alternately stained with H & E and left unstained, mounted on a glass slide on ice. Within one hour of sectioning, the unstained sections were examined under the laser fluorescence microscope (described earlier in this section) and individual measurements of HPD fluorescence were made at different sites within the sections. By comparison with the parallel H & E stained sections HPD fluorescence measurements were made within different cellular areas of the tumor: (1) actively proliferating cells, (2) areas of cellular necrosis, (3) areas of inflammatory tissue.

Animal studies were undertaken primarily to compare the tumoricidal effects of the gold vapor laser with the argon ion pumped dye laser. The animal tumor model employed was the RIF-1 tumor in C3H/Km mice (Twentyman et al, 1980). Mice were injected subcutaneously with an aliquot of 5×10^5 tumor cells that had been taken from previously minced and frozen tumor tissue. Between 8-12 days following injection, tumors were detected growing at the site of injection. When the tumors were .1-1 cm in diameter, the animals were injected with HPD (10.0 mg/kg b.w.). Twenty-four hours following injection the animals were anesthesized with ketamine HCl (5 mg/30 gm b.w.), shaved at the tumor site, and exposed to the laser light. The tumors were measured in diameter up to 2 months (the maximum length of survival) following laser exposure.

RESULTS AND DISCUSSION

Uptake and Excretion of HPD and SPP

As indicated in earlier studies (Berns et al, 1983), measurements on individual cells indicated that a fluorescent component(s) of HPD maximized in cells by 20 hours of exposure. The rate of uptake of the fluorescent molecules appeared similar for malignant as well as non-malignant cell types, though the final concentration per cell differed for the various cell lines tested. In the present study we have compared the uptake of the fluorescence component(s) of HPD and the "enriched" SPP compound (Fig. 1). It is evident that in the cell line tested, a fluorescent component from both solutions tested binds to the cells with similar kinetics. It should be noted that there appears to be an unequal distribution within the cell as indicated by a four-fold increase in fluorescence in the perinuclear region of the cytoplasm as compared with the more distal regions of the cytoplasm. Since the same volume of the cytoplasm is measured in all cases, the implication is that the binding sites for the fluorescent HPD component are more abundant in the perinuclear region.

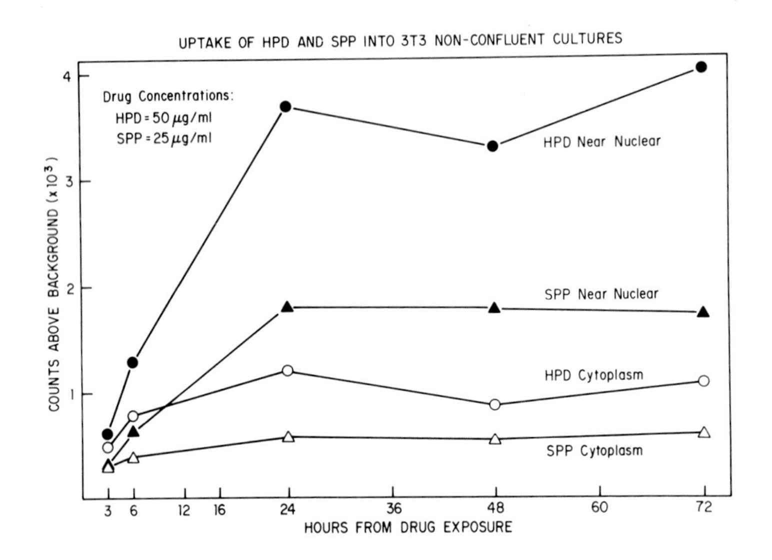

Fig. 1. Fluorescent measurement of uptake of whole HPD and enriched "active" HPD (SPP) by 3T3 cells.

These observations agree with previous studies (Berns et al, 1983; Salet, Moreno, 1981) demonstrating a selective affinity of HPD for mitochondria and lysosomes which have a heavier concentration around the nucleus.

It is of interest to note that the amount of fluorescent material localizing in the cells is about 50% as much for the SPP as the HPD. This appears logical because the concentration of HPD was 50μg/ml an SPP was 25μg/ml. However, these concentrations were chosen because the cytotoxic activity of the SPP was estimated (by the supplier) to be twice that of HPD. Our data does not address the issue of cytotoxic activity, but it does suggest that the amount of fluorescent material in both solutions (HPD and SPP) is similar (since a 50% dilution of the SPP results in 1/2 the amount of activity as the HPD). This suggests the possibility that the fluorescent fraction may be different from the light-sensitive cytotoxic fraction. The HPLC separations of both the HPD and SPP demonstrate that the SPP does contain most of the same components as the HPD, though certain of the components (particularly the large peaks eluting between 30-50 min after sample injection) are greatly reduced (Figs. 2 and 3).

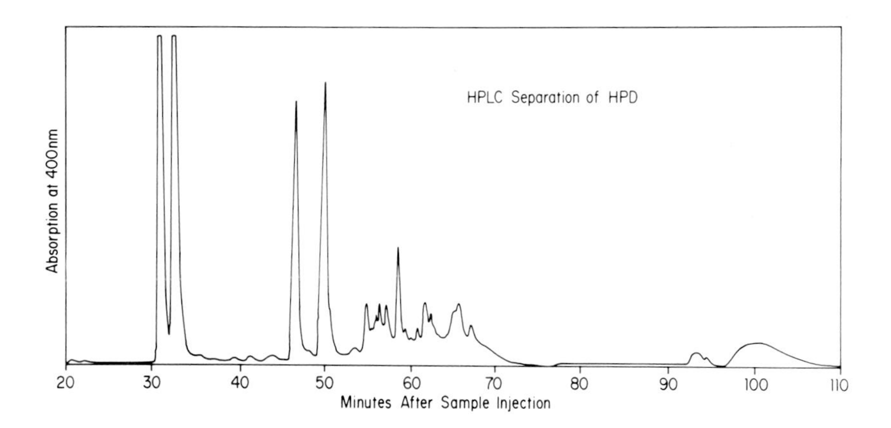

Fig. 2. Chromatogram of reverse-phase HPLC separation of whole HPD. Sample size = 25 μg; flow rate = 1.25 ml/min; O.D. = 1.0.

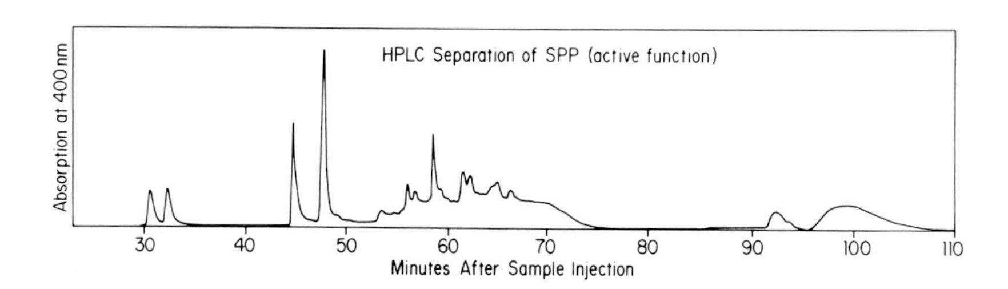

Fig. 3. Chromatogram of reverse-phase HPLC separation of enriched "active" HPD (SPP). Sample size = 12.5 μg; flow rate = 1.25 ml/min; O.D. = 1.0.

The loss of the fluorescent component from cells occurs rapidly with a 75% loss by 24 hours (Fig. 4). By 72 hours the amount of fluorescent material is less than 10% of its maximum. Previous cell killing studies (Berns, unpublished) demonstrate that even this low level of HPD (at 72 hours) is still enough to make the cells sensitive to visible red light. This corresponds to the kinetics *in vivo* as indicated by the biopsy studies (see later section).

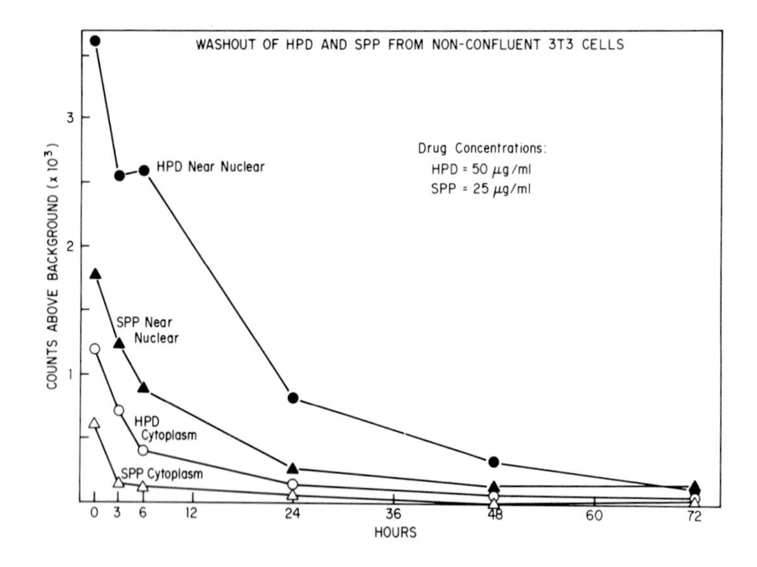

Fig. 4. Fluorescent measurement of excretion (washout) of HPD and enriched "active" HPD (SPP) from 3T3 cells following 24 hr. pretreatment of the cells with the drug.

Subcellular Localization

Our earlier studies demonstrated that mitochondria clearly bind a fluorescent component of HPD (Berns et al, 1982). The studies of Salet and Moreno (1981) also implicated the mitochondria as a binding site. However, other studies (Kessel, 1981) have strongly implicated the membrane as a binding site. In order to look more closely at the cell surface, we have examined the fluoresence of a common membrane probe, fluorescein conjugated soybean agglutinin (f-SBA), used to study the fluidity properties of the cell membrane.

The initial studies (Fig. 5) demonstrated that HPD by itself did not affect membrane fluidity in PTK_2 cells. However, if the HPD was first reacted with white light, it then had a significant effect on membrane fluidity. This suggests that some photoreactive species of HPD affects the membrane. In subsequent studies we have found that on some days the HPD alone (without light) will affect membrane fluidity and other days it will not. This result suggests that either the physiological state of the cells, or some environmental factor (such as the culture environment) may mediate the effects of HPD on the cell surface. In addition we have found in recent studies that the 3T12 cell line is more sensitive to HPD (with respect to membrane fluidity) than either the PTK_2 or 3T3 cell lines. Clearly, we need more work in this area before more definitive statements can be made.

Similarly, our preliminary results on fluorescence recovery on a photobleached spot (a measure of how much of the cell membrane is mobile with respect to the probe used) indicates that HPD (without light) can influence this parameter (Fig. 6). The result with the light-reacted HPD fraction is perplexing because it has no effect on the % mobile fraction, whereas in the previous studies this HPD component is the one that affected membrane fluidity. Clearly more work is needed in this area. In addition, studies with the other cell lines (3T3 and 3T12) do not demonstrate an HPD effect on the % mobile fraction in the cell surface.

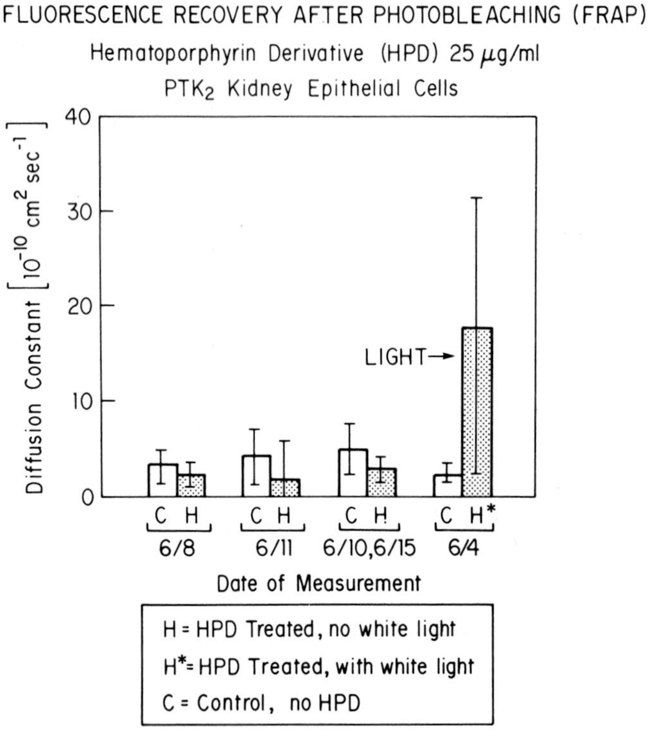

Fig. 5. Effect of exposure to light on the mobility of lectin receptors in PTK_2 cell surface membranes pretreated with HPD.

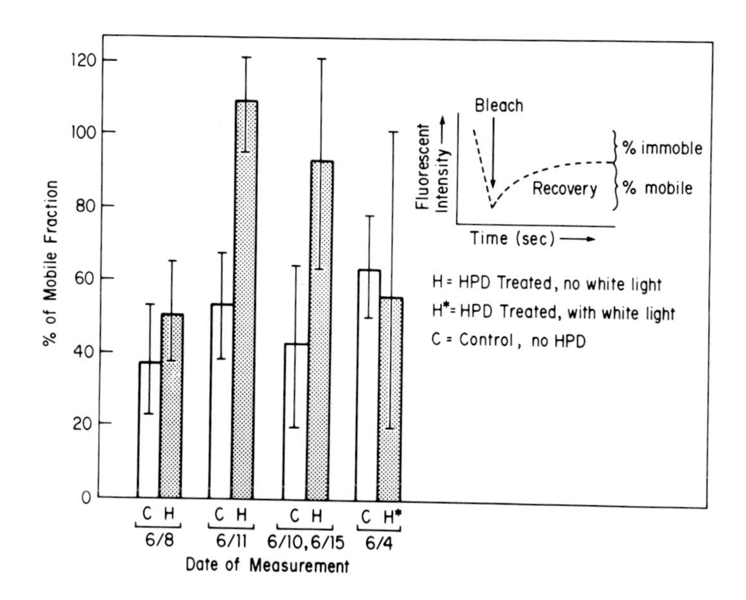

Fig. 6. Effect of exposure to light on % of mobile lectin receptors in PTK_2 cell surface membranes pretreated with HPD.

A situation where there is an HPD effect on the cell surface is termed "situation bleaching" (Fig. 7). In a normal photobleaching recovery experiment, the laser is highly attenuated and focused to a small spot on the cell surface. The fluorescence in this spot is recorded and serves as the baseline point (see Fig. 6) to which recovery in the bleached spot (when the laser power is increased) is compared. In almost all of the cases we have examined (using PTK_2, 3T3, and 3T12 cells), the cell surface is far more susceptible to photobleaching (even with the attenuated laser beam) when exposed to HPD. This is strong evidence that some component of HPD is binding directly to the cell membrane.

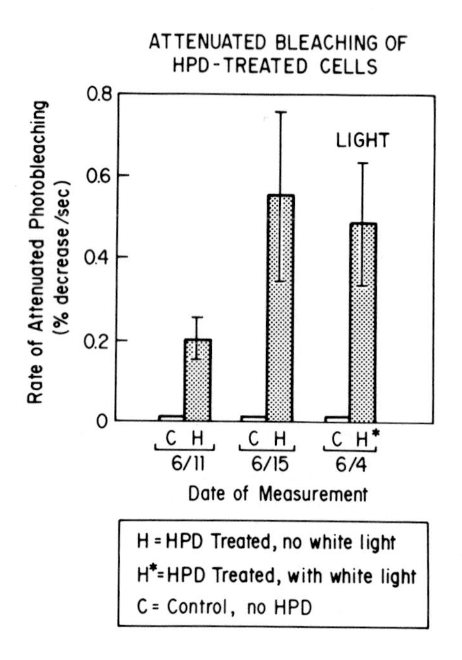

Fig. 7. Fluorescence bleaching rate of fluorescein conjugated lectin bound to surface membrane of PTK_2 cells as affected by exposure to HPD.

Structural observations on the cell surface of 3T3 and 3T12 cells with the scanning electron microscope reveal an HPD affect on the cell surface microvilli (Figs. 8 and 9). Both cell lines show a marked reduction in the occurrence of these projections in HPD-treated cells. This is consistent with results in other studies demonstrating that agents that affect the cell membrane cause a reduction in the microvilli (Siemens, et al, 1980).

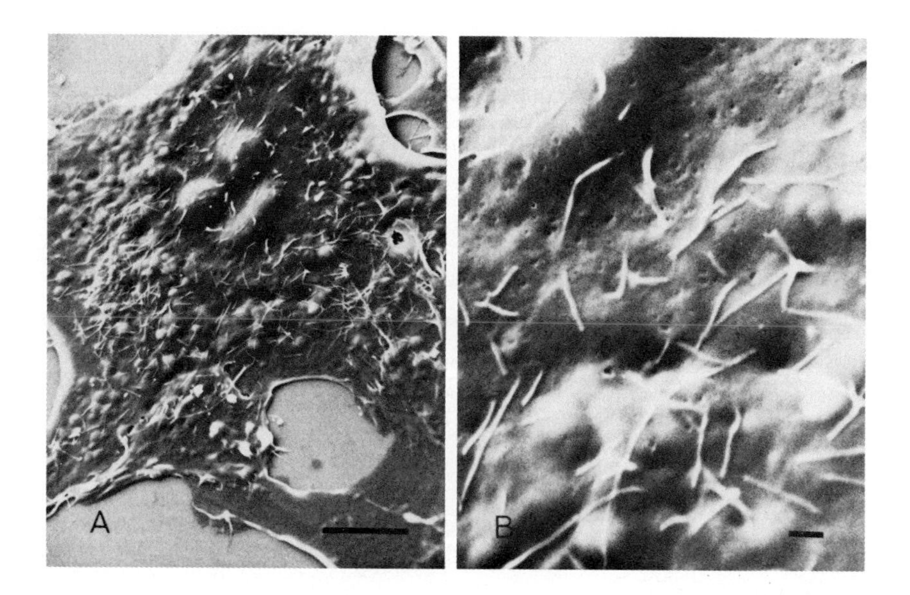

Fig. 8. Scanning electron micrograph of untreated 3T3 cells. Calibration mark for A = 5 microns and for B = 0.5 microns.

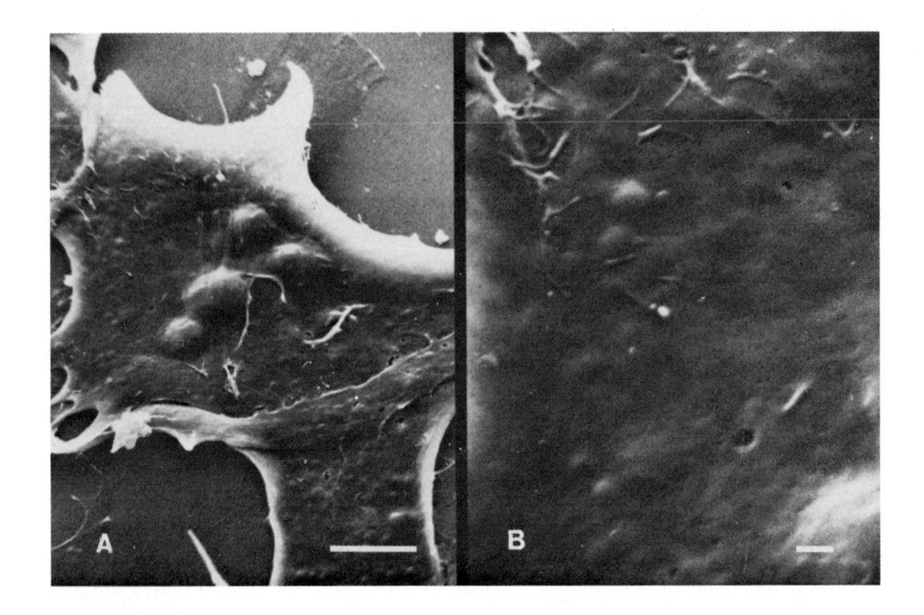

Fig. 9. Scanning electron micrograph of 3T3 cells treated with 50 μg/ml HPD for 24 hrs. Calibration bar for A = 5 microns and for B = 0.5 micron.

Tumor Biology

Fluorescent measurements on frozen sections of a human squamous cell carinoma from a patient injected with HPD demonstrate two patterns of uptake and excretion (Figs. 10 and 11).

It can be seen (Fig. 10) that the proliferative tumor tissue takes up the fluorescent component of HPD by 24 hours after injection and it rapidly comes out of this tissue over the next 24 hours. At 48 hours post injection the HPD comes out more gradually and by 96 hours the fluorescent component is between 25-35% of the 24 hour amount. It should be noted that the HPD level at 96 hours is still significantly above the pre-HPD (untreated) tissue level. These observations parallel the *in vitro* cell studies (Fig. 4) which show that at 72 hours post HPD exposure the amount of HPD per cell is 25% the level at 24 hours. Unfortunately, measurements on the biopsy tissue could not be made sooner than 24 hours post patient injection, so it is not possible to make a direct comparison between *in vivo* and *in vitro* studies.

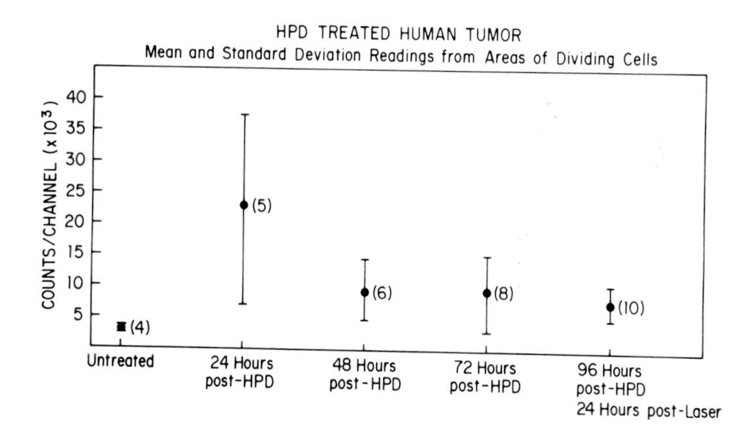

Fig. 10. Fluorescent measurement of uptake and excretion of HPD by proliferating tumor cells biopsied from a human squamous cell carcinoma following HPD injection.

The observations on the non-proliferative necrotic and inflammatory tumor-tissue reveal a different pattern than in dividing tissue (Fig. 11). It appears that significantly more fluorescent HPD component associates with these tissues (about twice as much) by 24 hours post-injection. In addition, it appears that the HPD is retained considerably longer by these components of the tumor. By 96 hours post injection the amount of fluorescent material is almost as high as the 24 hour value.

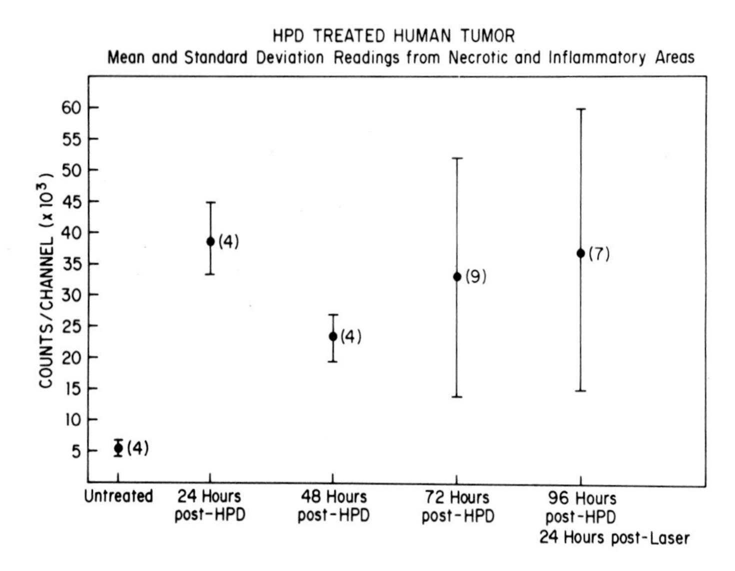

Fig. 11. Fluorescent measurement of uptake and excretion of HPD by necrotic and inflammatory tumor cells biopsied from a human squamous cell carcinoma following HPD injection (same patient as in Fig. 10).

These observations stress the importance of cautious interpretation of fluorescent measurements. Using fluorescence as an index of HPD in a tumor and for optimizing the time and amount of laser exposure, one must take into account the fact that the HPD may be distributed quite differently throughout the tumor. Furthermore, there is no clear indication that the fluorescent component(s) of HPD are the same as the photo-active components of HPD.

Gold Laser Studies

The studies with the gold vapor laser were undertaken to determine if this laser would be effective as a light source for HPD-PRT. The potential simplicity of the device (it's one laser rather than two) and the fact that the wavelength (628 nm) is within the active range for PRT suggested that this might be a useful system. However, the pulsed nature of the light (8 nsec.) and the high peak powers of the pulses raised questions with respect to comparative results with the CW argon pumped-dye lasers.

Studies on monolayer cell cultures exposed to HPD (Fig. 12) indicate that the gold laser is effective in killing cells. In fact, it appears that when compared to the CW-dye laser, the gold laser may be more effective in generating the phototoxic response in cells. At .5J/cm^2 the gold laser produced at 25% cell kill compared to almost no cell kill by the dye laser. Similarly at 5J/cm^2-10J/cm^2 the gold laser caused cell kills in the range of 70-90% compared to 20% for the CW dye laser. It wasn't until energies of 40-50J/cm^2 that both lasers gave comparable results (90% kill).

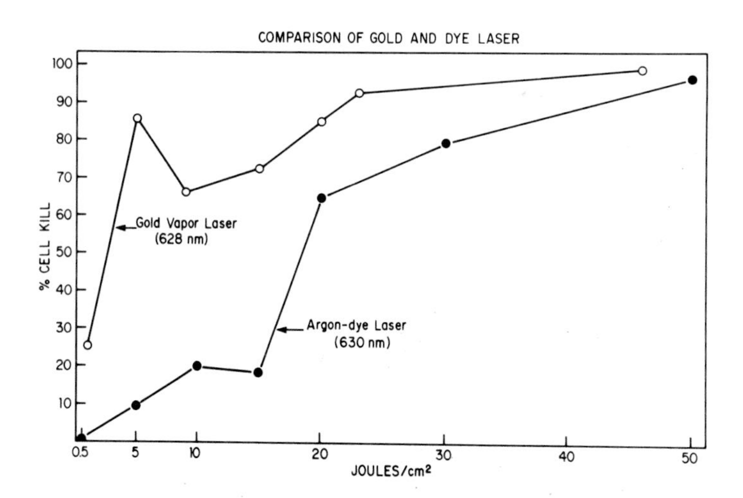

Fig. 12. Comparison of % cell killing of PTK_2 cells using a gold vapor laser and an argon dye laser system after HPD treatment.

The studies on mouse tumor irradiation demonstrated that the gold laser at the least, has a similar effect to the argon-dye laser (Figs. 13 and 14). Because of the difficulty in working with an animal model system over long distances (the mice had to be flown up to Northern California, irradiated, and returned the same day) the data is very preliminary. In Fig. 13 (gold laser), it is evident that all three HPD plus light groups exhibited a tumor growth delay. This is particularly evident at 18-22 days post tumor cell injection. Similarly, for the argon-dye laser (Fig. 14), tumor growth delay was observed at 16-24 days for all groups receiving HPD + light. In all groups studied it is apparent that at the laser energies used, cell killing was considerably less than 100%, as the tumors continued to grow eventually killing the animals.

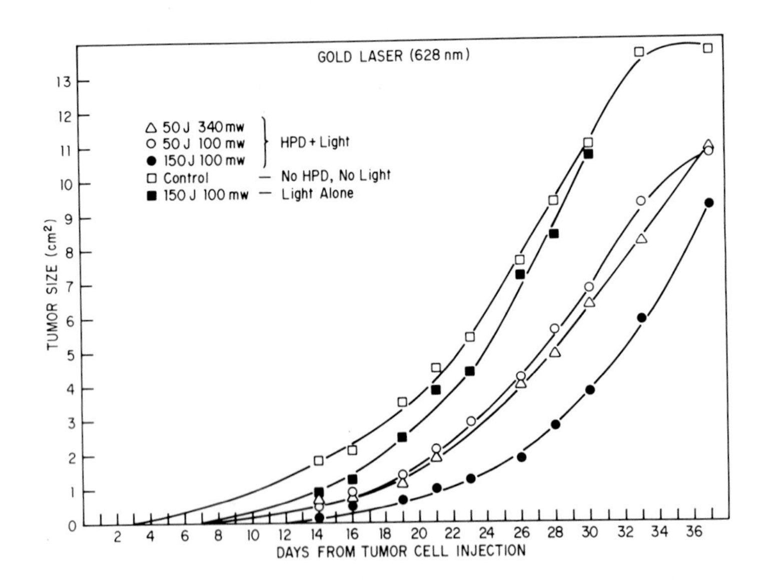

Fig. 13. Effect of gold-vapor laser irradiation on RIF-1 tumor growth in mice. Laser irradiation occurred 14 days post tumor injection and 24 hrs post HPD injection.

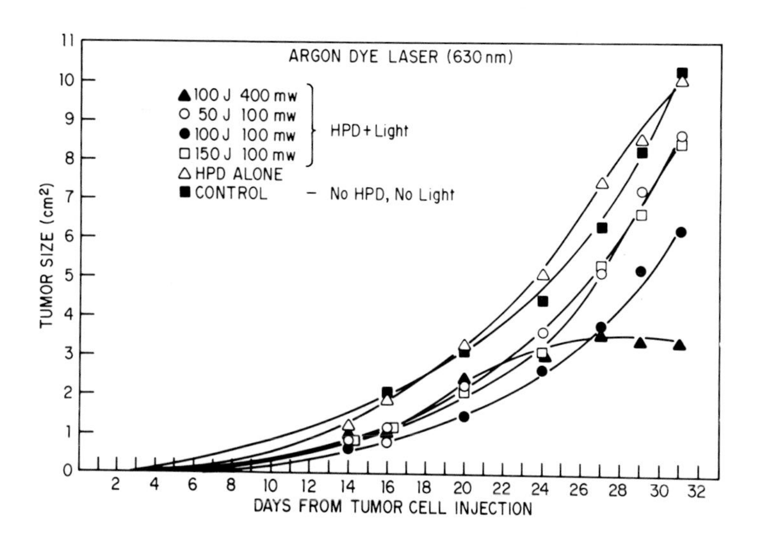

Fig. 14. Effect of argon dye laser irradiation on RIF-1 tumor growth in mice. Laser irradiation occurred 14 days post tumor injection and 24 hrs post HPD injection.

Our observations with the gold laser suggest that it is at least as effective as the argon-dye laser. It is not possible at this time to state that it is more effective.

ACKNOWLEDGMENTS

This work was supported by the following grants from the NIH: CA 32248, GM 23445, HL 15740, and RRO 1192.

REFERENCES

Berns MW, Dahlman A, Johnson FM, Burns R, Sperling D, Guiltinan M, Siemens A, Walter R, Wright W, Hammer-Wilson M, Wile A (1982). *In vitro* cellular effects of hematoporphyrin derivative. Cancer Res 42:2325-2329.

Berns MW, Wilson M, Rentzepis R, Burns R, Wile, A. Cell biology of hematoporphyrin derivative (HPD). Lasers in Surgery and Medicine 2:261-266, 1983.

Dahlman A, Wile, AG, Burns, RG, Mason GR, Johnson FM, Berns MW (1983). Laser photoradiation therapy of cancer. Cancer Res 43:430-434.

Dougherty TJ, Kaufman, JE, Goldfarb A, Weishaupt, KR, Boyle D, Mittleman A (1978). Photoradiation therapy for the treatment of malignant tumors. Cancer Res 38:2628-2635.

Figge FHJ, Welland GS (1948). The affinity of neoplastic, embryonic and traumatized tissue for porphyrins and metalloporphyrins. Anat Rec 100:669.

Figge FHJ, Welland GS (1949). Studies on cancer detection and therapy: the affinity of neoplastic embryonic and traumatized tissue for porphyrins and metalloporphyrins. Cancer Res 9:549.

Kessel D (1981). Transport and binding of hematoporphyrin derivative and related porphyrins by murine leukemia L1210 cells. Cancer Res 41:1318-1323.

Lipson RL, Baldes EJ, Olsen AM (1961). Hematoporphyrin derivative: a new aid for endoscopic detection of malignant disease. J Thoroc Cardiovasc Surg 41:623-629.

Rasmussen-Taxdall DS, Ward GE, Figge FH (1955). Fluorescence of human lymphatic and cancer tissue following high doses of intravenous hematoporphyrin. Cancer (Phila) 8:78-81.

Salet C, Moreno, G (1981). Photodynamic effects of hematoporphyrin on respiration and calcium uptake in isolated mitochondria. Int J Radiat Biol Relat Stud Phys Chem Med 39:227-230.

Siemens AE, Kitzes MC, Berns MW (1980). Hydrazine effects on vertebrate cells in vitro. Toxic Appl Pharmac 55:378-392.

Twentyman PR, Brown J, Gray JW, Franko J, Scoles MA, Kallman RF (1980). A new mouse tumor model system (RIF-1) for comparison of end-point studies. J Natl Cancer Inst 61:595-603.

Porphyrin Localization and Treatment of Tumors, pages 521–530

FLUORESCING CELLS IN SPUTUM AFTER PARENTERAL HPD

K.B. Patel, Y-N Qin, O.J. Balchum, D.R. Doiron

Pulmonary Disease Section, Dept. of Medicine, Los Angeles County-University of Southern California Medical Center, Los Angeles, CA 90033

Abstract

Cells exfoliated into sputum were examined for fluorescence after the intravenous injection of HpD. Malignant and non-malignant cells were seen to fluoresce up to 9 days post injection of HpD. Not all exfoliated squamous cell cancer cells or non-malignant cells fluoresced. Implications are discussed relative to imaging diagnostic fluorescence bronchoscopy and photoradiation therapy of obstructing endobronchial cancers and bronchial carcinoma in situ.

Introduction

Hematoporphyrin derivative (HpD) is a tumor seeker that after intravenous injection is retained in bronchial cancers in a greater concentration and far longer than in normal bronchial tissues.

HpD is therefore used for the early diagnostic localization of carcinoma in situ (CIS) of the lung (Balchum et al. 1982) and for photoradiation therapy of obstructing endobronchial cancer and CIS (Balchum, Doiron, Huth 1983). Bronchial areas of CIS are detected by imaging fluorescence bronchoscopy carried out at 72 hours after the

Supported by grant from National Cancer Institute, CA-25572-04, Request reprints from O.J. Balchum, M.D., Ph.D., USC Medical School, 2025 Zonal Avenue, Los Angeles, CA 90033.

IV injection of HpD (2.0-2.5 mg/kg). Fluorescence bronchoscopy employs violet light (406.7 nm and 314.1 nm), from a krypton ion laser, conducted via a fine quartz fiber inserted into the channel of the bronchoscope.

The HpD in mucosal cancer lesions is excited by violet light to emit red light (610 nm-740 nm) and this fluorescence is imaged and visualized on the screen of an image intensifier (40,000 luminence gain) attached to the eyepiece. This method is very sensitive and has detected small areas of CIS (1 x 2 mm) that were invisible under white light examination and where the mucosa appeared entirely normal to white-light inspection (Balchum, Profio, Huth 1983). Photoradiation employs HpD with the illumination of CIS and endobronchial cancer with red light (630 nm) from an argon pumped dye laser. Activation of HpD in tumor cells results in the formation of singlet oxygen, which oxidizes cell membranes and organelles with the potential of causing cell death.

During fluorescence bronchoscopy, it has been noted that recent biopsy sites, such as on the main carina, which were coated by white mucus, fluoresced brightly, as did the necrotic material on large endobronchial tumors, and occasionally even loose mucus in a bronchus. Although these occurrences did not interfere with diagnostic localization, it became of interest to know whether cells other than malignant cells retained HpD and fluoresced. This preliminary study was therefore aimed at understanding the types of cells in expectorated sputum and bronchial washings that retained HpD and fluoresced, whether bronchial cancer cells exfoliated into sputum also fluoresced, and if any differential in fluorescence between the various types of cells found in sputum could be detected. Such information would aid in further development and refinements in fluorescence bronchoscopy and photoradiation therapy.

Material and Methods

Incident fluorescence microscopic examination of cells, and transmission white light observations, were done using a Zeiss III microscope with a x 40 panchromatic objective. Violet light from a 50-watt high pressure mercury arc lamp (HBO) was used for fluorescence excitation. Filtering of the excitation light was by a

BG-28 red attenuation filter and a BP 405/14 interference filter, in conjunction with a FT 580 chromatic splitter. The red HpD fluorescence was observed and photographed using a LP 590 barrier filter. Phase contrast observations were done using an Optovar magnification changer, a phase contrast condenser, and a Neofluor x 40 PHZ objective. Photographs were made using 35 mm Kodak Ektachrome 400, push-developed to 800, after automatic exposure. Black and white prints were made from these transparencies. HpD was obtained from Oncology Research and Development, Inc., Cheektowaga, New York.

Collections of expectorated sputum began on day 1, the day of intravenous injection of HpD (2.0-2.5 mg/kg). Bronchial washings were also collected during fluorescence bronchoscopy 72 hours after HpD injection. All containers were opaque-shielded so that no light reached the specimens.

Three different methods were used for processing these specimens: A. Suspension in Hank's Solution with the initial examination for fluorescence, followed by phase contrast examination; B. Air-dried smears fixed in 95% alcohol with examination for fluorescence, followed by Papanicolau staining and then white-light microscopic examination; C. Air-dried smears stained with hematoxylin, prior to fluorescence and then white-light examination (Table 1).

The last method, C. Table 1, was found to be the best and will be described in more detail. The specimen was suspended in 50 ml of Saccomanno's Fixative (2% Carbo-wax in 50% alcohol), and blended, centrifuged, and then resuspended by vibration on a vortex mixer.

Smears of 2 drops of sediment between 2 slides were prepared by smoothly drawing them apart. These were air-dried, fixed in 95% alcohol, and then stained with hematoxylin (See Appendix). Slides were examined for each cell that fluoresced and photographs were taken, and then each same cell was examined under white light for cytodiagnosis and photographs taken without changing the microscope stage position. Both fluoresing and non-fluorescing cells of various types were therefore studied. All of these steps in specimen collection, processing, and microscopic observations were done in the dark.

Table 1 - Methods of Specimen Processing and Examination

Method A	Method B	Method C
Collection of Sample in Hank's Soln.	Collection of Sample in Saccomanno's Fixative	Collection of Sample in Saccomanno Fixative
↓	↓	↓
Waring Blender, 5-10 Secs, High Speed	Waring Blender, 5-10 Secs at High Speed	Waring Blender, 5-10 Secs, at High Speed
↓	↓	↓
Centrifugation, 15 min at 4000 rpm	Centrifuge for 15 min at 4000 rpm	Centrifuge for 15 min at 4000 rpm
↓	↓	↓
Resuspension Sediment with Vortex Mixer	Sediment Re-Suspended with Vortex Mixer	Sediment Re-Suspended with Vortex Mixer
↓	↓	↓
Suspension Under Cover-Slip examined by fluorescence and phase contrast methods	Two drops spread evenly between two Slides	Two drops spread evenly between two slides
	↓	↓
	Slides Air-Dried	Slides Air-Dried
	↓	↓
	Fixation in 95% alcohol for 15 min	Fixation in 95% alcohol for 30 min
	↓	↓
	Fluorescence Examination	Haematoxylin Staining
	↓	↓
	Papanicolau Staining and White-Light Examination	Fluorescence, then White-Light Examination

Intrinsic fluorescence of Saccomanno's Fixative, 95% alcohol, and haematoxylin stain was found to be absent by prior studies. Dyes such as EA 50 and Orange G, and Papanicolau's stain, either showed intrinsic fluorescence or the stained cells became fluorescent. Neither was the case with haematoxylin solution or stain. The latter did not alter the degree of fluorescence of cells by observation before compared to after staining. This was the case for both malignant and non-malignant cells. Cell and nuclear detail was good and adequate for cytodiagnosis by staining with haematoxylin alone. Fluorescence quenching was observed in Method A but not Method C, within the time required for cytodiagnosis and photography carried out entirely in the dark.

Results

Eight patients with biopsy proven endobronchial squamous cell cancer were studied by the 3 methods (Table 1). Using Method A, the following observations were made: Fluorescence of cells was observed. The outline of cells and their nuclei could be seen by the phase contrast technique, but the details of nuclear structure required for cytodiagnosis of stages of metaplasia were difficult to discern. It was not possible to reliably relocate and re-examine cells because they were in fluid suspension under the coverslip. Specimens had to be studied immediately because cells degenerated and detail was lost.

Using Method B, the fluorescence of cells could be observed. By means of coordinates on the microscope stage, most (75%) of the cells could be relocated after Papanicolau staining but this was time consuming. Although, well suited for cytodiagnosis, observations for fluorescence of cells could not be done on Papanicolau stained smears since some of the dyes in the latter fluoresced per se. Saccomanno's Fixative proved superior for preserving cells in collected specimens.

Method C proved to be the most suitable. Fluorescence of cells could be observed. The preservation of cells was excellent. Cytodiagnosis after staining with haematoxylin alone was possible, since cell and nuclear structure could be observed clearly and in detail. It was possible to first observe cells for fluorescence and photograph them, and then examine them under and carry out white-light photography. Documentation of fluorescing and non-fluorescing cells included alveolar macrophages, neutrophils, cells showing regular metaplasia, the various stages of atypical metaplasia, and malignant cells.

Results can be exemplified by the findings in the sputum of one patient, a 66-year-old man with a small mucosal squamous cell cancer on the spur between the anterior and apico-posterior segments of the right upper lobe. This was the only area found to fluoresce upon 2 fluorescence bronchoscopic examinations, 72 hours after each intravenous injection of HpD (2.0 mg/kg) on 1/26/82 and on 1/30/82. The areas of CIS that fluoresced appeared visibly to be normal on white-light examination.

Many non-cancer cells, such as alveolar macrophages, neutrophils, and regular metaplasia cells were found to fluoresce for 6 to 8 days after IV HpD. The fluorescence exhibited by alveolar macrophages was often bright. It was noted that the fluorescence of macrophages was differential, with fluorescing and non-fluorescing macrophages found beside or near each other on the same slide (Fig. 1). Cells in the various stages of atypical metaplasia fluoresced, including mild atypia (Fig. 2) and marked atypia (Fig. 3). CIS and squamous carcinoma cells (Fig. 4) fluoresced. Fig. 5 shows a non-fluorescing squamous cancer cell in this same subject. In none of the cells was fluorescence of the nucleus observed.

The sputum of 6 patients without lung cancer (sarcoid, bronchiectasis, chronic bronchitis, emphysema) and of 2 patients with squamous cell cancer of the lungs, none of whom had been injected with HpD, was also examined. In none of these 8 controls without prior HpD injection did the alveolar macrophages, neutrophils, squamous epithelial cells, bronchial epithelial cells, atypical metaplastic cells or squamous cancer cells fluoresce.

Discussion

The fluorescence of squamous carcinoma cells and alveolar macrophages, neutrophils, and cells representing the various stages of metaplastic atypia was highly likely due to the HpD they contained, since no fluorescence was found in these various cells in the sputum of patients who had not been injected with HpD. It is highly unlikely that

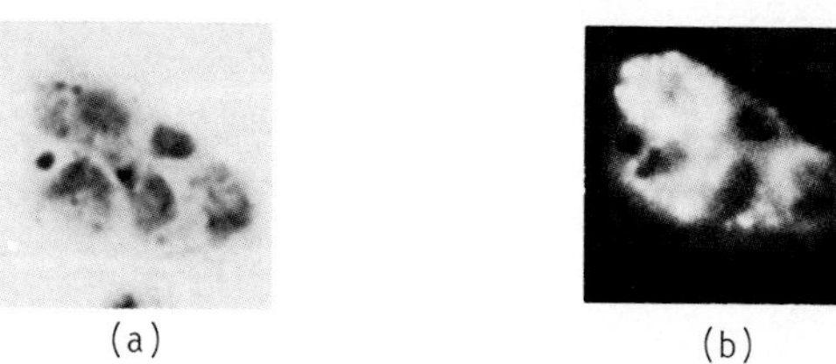

(a) (b)

Fig. 1. Alveolar Macrophages (a); fluorescing (b), with different degrees of intensity.

(a) (b)

Fig. 2. Mild Atypia (a); fluorescing cytoplasm (b). Note Non-fluorescing nucleus.

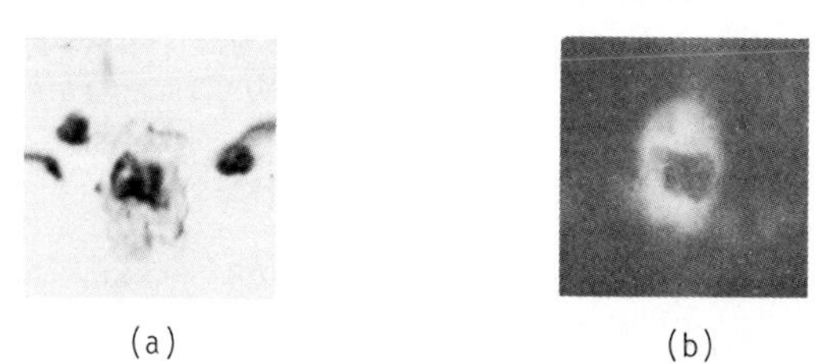

(a) (b)

Fig. 3. Marked Metaplastic Atypia (a); fluorescing brightly (b).

(a) (b)

Fig. 4. Squamous Carcinoma Cell (a) with fluorescing cytoplasm (b).

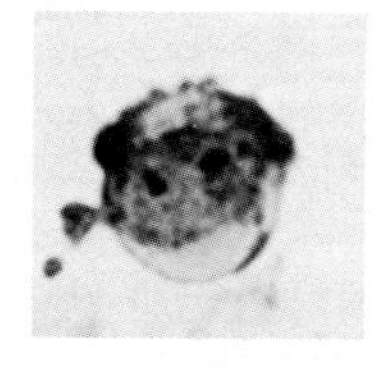

Fig. 5. Non-fluorescing Squamous Cancer Cell; (Same sputum sample as Fig. 4)

any of the fluorescence observed was due to fixatives or haematoxylin staining or other artifacts employed in method C, which we are convinced resulted in no false fluorescence or decrease in or abolition of fluorescence.

The following conclusions can be drawn from this preliminary study:

1. The fluorescence of cells in sputum persisted for at least 9 days after HpD injection. This included both malignant and non-malignant cells.

2. The same smear of sputum revealed both fluorescing and non-fluorescing malignant and non-malignant cells. This observation was made repeatedly.

3. After the initial observations were made, the slides were stored in the dark in slide boxes at room temperature. The intensity of fluorescence remained at the same level as at the initial observations for a period of at least 5 weeks, using method C.

The reason that all cancer cells in sputum did not fluoresce is unclear at present. One cause may depend upon the vagaries of cell attachment and exfoliation of cancer cells from bronchial cancer lesions into the sputum. HpD may not have reached superficial cancer cells or other cells on or near the tumor surface. The lack of fluorescence, on the other hand, does not indicate that they did not contain HpD, since fluorescence may not be directly proportional or closely correlated with the actual tissue concentration of HpD (Gomer, Dougherty 1979), although this latter study did not include bronchial tissue or cancer of the lungs in animals or man. It is likely that some components of HpD in cells fluoresce more brightly than others, when excited by violet light.

That cells other than cancer cells retain HpD and fluoresce was observed by Berns and Wile (1983), by biopsy of an oropharyngeal carcinoma of the cheek at 24, 48, 72, and 96 hours after HpD injection. Inflammatory and necrotic tissue fluoresced, as well as proliferative malignant tissue. Fluorescence was observed to decline in proliferative malignant tissue over 96 hours, but not in non-proliferative, inflammatory or necrotic tissue. That the latter fluoresced was first noted by Gregorie (1968).

Extrapolation of these results to diagnostic imaging fluorescence bronchoscopy is difficult at this time. Gross (visible) necrotic tissue on large endobronchial tumors has been observed by us (Balchum) to fluoresce as well as the mucus adherent to previous biopsy sites. These are readily recognized and have not interferred with recognition of the positive fluorescence of endobronchial tumor, since one is aware of this. The patient illustrated in Table 2 likely had carcinoma in situ. The mucosa visibly appeared normal in the area that positively fluoresced, which showed squamous cancer cells on bronchial brushings of this area, and no where else. These findings were confirmed at the time of a second imaging fluorescence bronchoscopy 2 weeks later (Balchum, Profio, Doiron, Huth 1983). The non-malignant cells which fluoresced therefore came from other than the site of this single, very small mucosal lesion. Yet no other mucosal areas fluoresced, nor was mucus seen to fluoresce. Non-malignant cells in normal bronchial tissue surrounding a small mucosal plaque of CIS could contribute to background fluorescence and impair or obscure the contrast between fluorescing CIS lesions and background, making them difficult to localize by fluorescence bronchoscopy.

These findings may be applicable to photoradiation therapy, and the estimation of dosage of red light (630 nm) used in HpD-red light photoradiation treatment of large obstructing endobronchial tumors (Balchum, Doiron, Huth 1983). The latter contain components of inflammatory and necrotic tissue, as well as proliferating tumor. Other than malignant cells may well react after the absorption of red light. It may be possible that the degree of reaction could be greater than of proliferating malignant cells per se.

These findings may also help explain the reaction of normal bronchial tissue to red light after photoradiation therapy. Higher dosages of light have been accompanied by the formation of thick mucus, exudate, and even a firm membrane or plug in the area of bronchus illuminated by red light. Even though the normal bronchial mucosal areas fluoresce less brightly than endobronchial or mucosal (CIS) tumor, the neutrophils and macrophages may be activated to excess because they contain HpD and gather in the illuminated areas, forming thick exudate and membrane in the bronchus. Low light dosage (100 Joules/cm^2) has largely resolved this problem.

References

Other pertinent references are in the bibliography of the references cited below:

Balchum OJ, Doiron DR, Profio AE, Huth GC (1982). Fluorescence bronchoscopy for localizing early bronchial cancer and carcinoma in situ. Results in Cancer Research 82:97-120.

Balchum OJ, Doiron DR, Huth GC (1983). Photoradiation therapy of obstructing endobronchial lung cancer employing the photodynamic action of hematoporphyrin derivative. Proceedings of Symposium on Porphyrin Localization and Treatment of Tumors, Santa Barbara, April 24-28, 1983 - in press.

Balchum OJ, Profio AE, Doiron DR, Huth GC (1983). Imaging fluorescence bronchoscopy for localizing early bronchial cancer and carcinoma in situ. Proceedings of Symposium on Porphyrin Localization and Treatment of Tumors, Santa Barbara, April 24-28, 1983 - in press.

Berns MW, Wile A (1983). Uptake and localization of HpD in tissue culture and in serially biopsied human tumors. Proceedings of Symposium on Porphyrin Localization and Treatment of Tumors, Santa Barbara, April 24-28, 1983 - in press.

Gomer CJ, Dougherty TJ (1979). Determination of 3H- and 14C hematoporphyrin derivative distribution in malignant and normal tissue. Cancer Research 39:146-151.

Gregorie HB, Horger EO, Ward JL, Green TF, Richards T, Robertson HC Jr, Stevenson TB (1968). Hematoporphyrin derivative fluorescence in malignant neoplasms. Ann Surg 167:820-827.

Appendix

Haematoxylin Stain (Harris)

Haematoxylin, 0.4%
Potassium Alum, 8.4%
Mercuric Oxide, Yellow, 0.2%
Acetic Acid, Glacial, 3.5%

IN VIVO PORPHYRIN PHOTOBIOLOGY

Porphyrin Localization and Treatment of Tumors, pages 533–540

HEMATOPORPHYRIN DERIVATIVE AND PULSE LASER PHOTORADIATION

David A. Bellnier, Chi-Wei Lin, John A. Parrish
and Patrick C. Mock

Urology Service and Department of Dermatology
Massachusetts General Hospital, Harvard Medical
School, Boston, MA 02114

INTRODUCTION

Photoradiation therapy (PRT) using hematoporphyrin derivative (HpD) is undergoing clinical trials for the treatment of malignant tumors. The ultimate effectiveness of this modality is a function of tumor sensitizer concentration and light flux (1). Argon-pumped continuous (CW) dye lasers tuned to emit 625-635 nm light are presently being used for the excitation of HpD *in situ*. Typical light doses are on the order of 10-100 J/cm^2 for superficial exposures and 100-1,000 J/fiber for interstitial fiberoptic implant type treatments (2). Considering total available laser output (at 630 nm), pump/dye/fiberoptic coupling efficiencies and tumor volume, treatment times may be several hours. Considerable shortening of treatment duration is theoretically possible with the use of tunable, pulse dye-lasers, since these lasers produce extremely high peak power outputs (10^7 W or greater) over short pulse widths. We have initiated a study to evaluate pulsed laser light activation of HpD *in vivo*. Preliminary results are presented here.

MATERIALS AND METHODS

Hematoporphyrin derivative (HpD) was obtained from ORD, Inc., Cheektowaga, NY. The drug was diluted with normal saline to a concentration of 1 mg/ml for injection. Solutions were kept in the dark at 4°C prior to administration

C3H/He mice (female, 4-8 weeks old) were obtained from Charles River Breeding Labs, Wilmington, MA. They were housed in plastic cages with hardwood chip bedding and were given food and water acidified with HCl (pH 2.3) *ad libitum*. Animals were maintained on a 12 hour light-dark cycle.

A transitional cell carcinoma (MBT-2) induced in the urinary bladders of syngeneic mice by the carcinogen N-(4-(5-nitro-2-furyl)-2-thiazolyl)formamide, FANFT, was used in these studies (3). Suspensions of second generation tumors were kept frozen in liquid nitrogen, and were thawed and injected into donor mice. Experimental tumors grew from minced tumor tissue implanted via 17 gauge trocar into the axilla region of the mice. When the tumors approached treatment size (∿5 x 5 mm) the mice were injected with 20 mg HpD/kg body weight and placed in a darkened room for 24 hours. Control animals (no HpD) were injected with an equal volume of normal saline.

Prior to irradiation tumors were shaved to maximize light penetration. Each mouse was positioned in an aluminum holder designed to expose the tumor while shielding the body of the animal; mice were not anesthetized. Irradiation light was provided by either of two sources: (1) a 5-watt argon laser (Model 164, Spectra-Physics, Corp., Mountain View, CA) pumping a dye laser (Model 365, Spectra-Physics) using Rhodamine B dye. The output of the dye laser was tuned with a birefringent filter to 631 nm (Fig. 1). The exit beam was collimated with a three lens laser beam expander/collimator (Oriel Corp., Stamford, CN) to give a field ∿0.8 cm^2 (1 cm dia.). The incident fluence rate was measured with a radiometer (Model 65A, YSI-Kettering, Yellow Springs, OH); (2) a pulse dye laser using Kiton 620 dye (solvent MeOH:H_2O, 50:50) pumped by a linear flashlamp (Model LFDL-2, Candela Corp., Natick, MA) tuned with a single glass prism to 632 nm (Fig. 1). The spot size was ∿0.5 cm^2 (0.8 cm dia.). Fluence and fluence rate were varied by altering energy per pulse and pulse repetition rate; pulse width was constant, τ=1.6 µsec.

Tumor temperatures were determined during photoirradiation (Fig. 2). A 29 gauge needle microprobe (Bailey Instruments, Inc., Saddle Brook, NJ) was inserted several minutes before light exposure to allow for baseline temperature determination. Temperature readings were taken every 0.5-2 minutes, including several minutes following irradiation. Two to five animals were used for each treatment protocol.

To evaluate treatment response tumor sizes were monitored daily for 12-14 days after irradiation. Values shown in Table 2 indicate the number of complete responses (no palpable tumor) per number of total animals treated at 5 days postirradiation. Some tumors were excised at 3 h, 24 h and 5 days for histological examination.

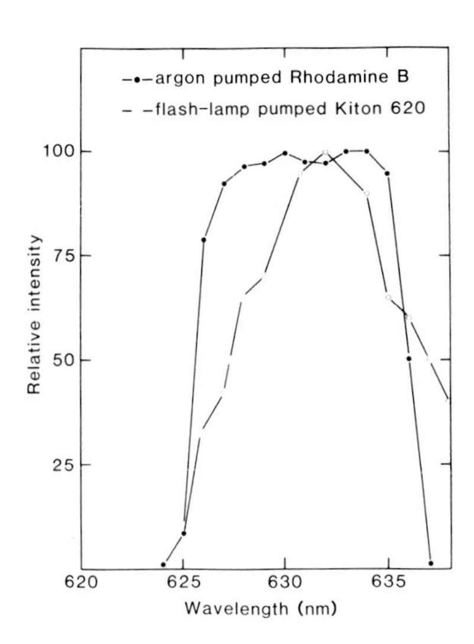

Figure 1. Spectral output of continuous (CW) and pulsed dye lasers. Units of intensity are arbitrary.

RESULTS

The effects of continuous and pulse laser PRT on our transplanted murine bladder TCC are summarized in Table 2; corresponding temperature profiles taken during light exposures are illustrated in Fig. 2.

CW Continuous light at a fluence rate of 70 mW/cm^2 over 30 minutes (126 J/cm^2) resulted in 50% (6/12) tumor control at 5 days. Some edema and slight bluish discoloration were observed immediately after irradiation. Histologic examination at 3 h revealed some tumor bleeding; tumors excised at 24 h showed extensive hemorrhage and coagulation necrosis. In most cases some viable tumor remained at day 5. Control tumors (no HpD) did not respond.

Table 1. Summary of continuous and pulse laser conditions for treatment of FANFT-induced murine transitional cell bladder carcinoma (MBT-2).

Laser	Dye¶	Wavelength (nm)	Fluence rate	Pulse width (μsec)	Rep. rate (sec^{-1})	Spot size (cm^2)
CW	R-B	631	70 mW/cm^2	---	---	0.8
Pulse(A)	R-B	620	0.25 $J/cm^2 \cdot$ pulse	1.6	4	0.5
Pulse(B)	Kiton	632	0.13 $J/cm^2 \cdot$ pulse	1.6	4	0.5
Pulse(C)	Kiton	632	0.13 $J/cm^2 \cdot$ pulse	1.6	2	0.5
Pulse(D)	Kiton	632	0.10 $J/cm^2 \cdot$ pulse	1.6	4	0.5

¶R-B is Rhodamine B dissolved in ethylene glycol; Kiton is Kiton 620 dissolved in methanol:water (50:50, v/v).

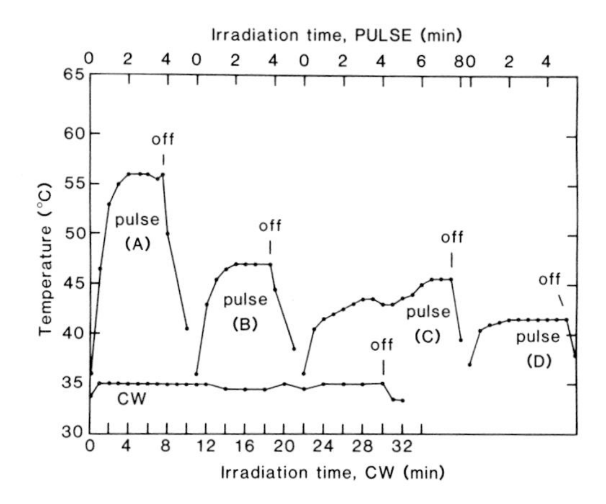

Figure 2. Tumor temperatures during continuous and pulse laser irradiation. See Table 1 for a complete description of exposure conditions.

Pulse(A) In this group, pulsed light (τ=1.6 μsec) was delivered at a rate of 0.25 J/cm^2· pulse with a repetition rate of 4 Hz; total exposure time was 3.75 minutes (225 J/cm^2 total light dose). However, we inadvertantly used the incorrect dye (Rhodamine B), which had a maximum lasing λ=620 nm with our flashlamp system. This treatment resulted in 100% response at 2 days for tumors with and without HpD. By day three, the treated area either had fallen off, or was chewed off by the animal, leaving a circular area of ulceration corresponding to the incident light field. Histologic sections taken at 24 h showed complete desquamation of the epidermis, dermal hyaline degeneration and marked bleeding and inflammatory cell infiltration of the subcutaneuos tissue;tumor tissue was partially to fully necrosed. Underlying organs (liver and bowel) showed superficial regions of necrosis.

Pulse(B) Pulsed light (632 nm) delivered at a rate of 0.13 J/cm^2· pulse (4 Hz) over 3.75 minutes (total light dose=113 J/cm^2) resulted in tumor reactions similar to the Pulse(A) group, i.e., both control (3/4) and HpD-containing (2/3) tumors showed complete responses at day 5 and histological examination revealed similar, although somewhat less extensive, damage.

Pulsed light in both (A) and (B) groups were associated with high tumor temperatures: ∿56°C and 48°C, respectively. We therefore conclude that thermal effects are responsible for

Table 2. Summary of tumor response (MBT-2 tumor) following continuous and pulse laser irradiation.

Laser[¶]	Incident fluence (J/cm^2)	Treatment time (min)	Response[#]		Comments
			with HpD	w/o HpD	
CW	126	30	6/12	0/4	Immediate edema, discoloration
Pulse(A)	225	3.75	6/6	4/4	Immediate edema, damage to underlying organs
Pulse(B)	113	3.75	2/3	3/4	Immediate edema, damage to underlying organs
Pulse(C)	113	7.5	0/2	0/2	Some immediate edema
Pulse(D)	124	5.17	0/6	0/4	No reaction

[¶] See Table 1 for laser specifications.

[#] Response indicates the number of animals with non-palpable tumors per total number of animals treated at day 5 post-irradiation.

the normal and tumor tissue damage (Fig. 2).

Pulse(C) and (D) Neither treatment resulted in tumor control although the total delivered light dose was approximately the same (125 J/cm^2) as that from the CW laser (50% response at 5 days). As noted in Table 2, there was some immediate reaction (edema) following Pulse(C); we observed no reaction whatsoever after Pulse(D) irradiation. Histologic sections were not remarkable.

DISCUSSION

The preliminary studies presented here suggest that our pulsed laser light is ineffective in activating HpD *in vivo*. Experimental tumors in the Pulse(D) group showed no reaction following a light dose sufficient to effect a 50% response (at day 5) when delivered continuously (CW).

Pulsed light delivered at high energies per pulse (Pulse (A), 0.25 J/cm^2· pulse) resulted in uniform destruction of normal and tumor tissue, regardless of the presence of HpD; these effects apparently are due to thermal damage (Fig. 2).

It has been suggested that the use of very short pulses of light might lead to an increased spatial delineation of tissue damage due to confined thermal changes following selective absorption by a chromophore (4), in this case endogenously administered HpD. Although these studies were not designed to explore this proposition, we found no differential in gross tumor heating during pulse radiation between animals given HpD or saline (control).

Although it is not clear why our pulsed light proved ineffective, we strongly suspect that the absorbing molecules have become saturated and thus we wasted most of the photons per pulse. We are also interested in the effect of varying the pulse width since extremely short pulse durations (i.e., the same order of magnitude as the HpD excited state in tissue) may contribute to sensitizer saturation. Those data will be presented at a later time.

ACKNOWLEDGEMENTS

We wish to thank Dr. Makoto Fujime and Cheryl Fay for excellent technical assistance.

REFERENCES

1. Doiron DR, Svaasand LO, Profio AE (1983). Light dosimetry in tissue: application to photoradiation therapy. In Kessel D, Dougherty TJ (ed): "Porphyrin Photosensitization", New York: Plenum Press, p 63.

2. Dougherty TJ, Boyle DG, Weishaupt KR, Henderson BA, Potter WR, Bellnier DA, Wityk KE (1983). Photoradiation therapy-clinical and drug advances. Ibid., p 3.

3. Soloway MS (1977). Intravesical and sytemic chemotherapy of murine bladder cancer. Cancer Res 37:2918.

4. Anderson RR, Parrish JA (1981). Microvasculature can be selectively damaged using dye lasers: a basic theory and experimental evidence in human skin. Lasers in Surgery and Medicine 1:263.

Porphyrin Localization and Treatment of Tumors, pages 541–562

TISSUE DISTRIBUTION OF ^{3}H-HEMATOPORPHYRIN DERIVATIVE AND ITS MAIN COMPONENTS, ^{67}Ga AND ^{131}I-ALBUMIN IN MICE BEARING LEWIS LUNG CARCINOMA

Jan Folkvard Evensen*, Johan Moan+, Atle Hindar+ and Stein Sommer+
*The Norwegian Radium Hospital
+Norsk Hydros Institute of Cancer Research
Oslo 3, Norway

INTRODUCTION

During the last 10 years, from being phenomenons observed in the laboratory, the tumor affinity and phototoxicity of porphyrins have turned out to be a promising principle in diagnostic and therapeutic oncology. Today, advanced equipment has been developed for the use of HpD in diagnosis as well as therapy of cancer. Thus, lasers, fiber optics and image intensifiers are used in searching for early/occult lung cancer (Doiron et al. 1979, Hayata et al. 1982). Furthermore, dye laser units for photoradiation therapy have been constructed and are comercially available (PRT-100, COHERENT, Palo Alto, Ca., USA). Today more than 1000 patients have been treated with photoradiation therapy.

The underlying mechanisms for tumor uptake and retention are still largely unknown. It has been shown that malignant cells _in vitro_ may take up larger amounts of porphyrin than normal cells (Mossman et al., 1974). However, this is not always so (Moan et al., 1981) and it is likely that the tumor physiology is more important for the observed tumor affinity of porphyrin (Bugelski et al., 1981). The present examination was carried out to elucidate further the uptake mechanisms of porphyrin _in vivo_ and to compare HpD with ^{67}Ga which is the most widely used tumor localizer today.

It is now well known that HpD is a mixture of a number of porphyrins (Kessel 1978, Moan et al., 1982a). Hence, in the present work, we have also studied four major components of HpD with respect to uptake in tumors and several tissues of mice bearing Lewis' lung carcinomas.

MATERIAL AND METHODS

Chemicals

Acetylated hematoporphyrin (HpA) was prepared as described by Lipson et al. (1961). HpA was labelled with tritium by New England Nuclear (NEN), Boston, Mass. Labelled HpA dissolved in chloroform/ethanol (1:1) was evaporated to dryness and cold HpA added. Then HpA was dissolved in 0.1 N NaOH to a concentration of 5 mg/ml. The solution was allowed to stand at room temperature for 1 hour. The treatment with NaOH results in hydrolysis of the acetate groups, and the product is termed hematoporphyrin derivate, HpD, in correspondence with the main trend in the literature (Moan et al., 1981). The HpD solution was neutralized with 0.1 N HCl to pH= 7.4 and diluted to a final concentration of 2.5 mg/ml with PBS. The specific activity was 0.27 MBq/mg.

^{67}Ga was supplied by NEN as carrier free ^{67}Ga citrate.

^{131}I-labelled human serum albumin was purchased from Institute of Energy Technology, Kjeller, Norway. The specific activity was about 1.5 MBq/mg.

Chromatography of HpD

HpD was analyzed and separated by means of reversed phase high pressure liquid chromatography (HPLC) with a gradient elution system (methanol/water) as described earlier (Moan et al., 1982a).

Mice and Tumors

The Lewis lung carcinoma was obtained from NIH (Bethesda, Md., USA) in 1974 and has been maintained in serial passage by i.m. injection of minced tumor tissue into the hind leg of $B_6D_2F_1$ (C57Bl/6 x DBA/2) mice from G. Bomholt gaard, Ry., Denmark.

Experimentals

In the present experiment the tumor cells were inoculated subcutanously at the back of female $B_6D_2F_1$ mice weighing 16-18 g. After 10 days the tumors reached about 10 mm in diameter and most of them were unnecrotic. The radioactive agents were injected i.v. in the tail.

Each mouse was injected with 0.07 MBq ^{3}H-HpD or with 0.15, 0.02 and 0.12 MBq of component 2, 4 and 7. Three mice respectively, were sacrificed 3, 12, 24, 48 and 72 h after injection. The mice were anaesthetized with ether and killed by cervical dislocation. Tumor and the various organs were removed from the mice and washed in physiologic saline. The tissues were then weighed and prepared for radioactive counting avoiding macroscopic necrotic regions in the tumors.

The weighed samples were dissolved in 1 ml Lumasolve (LUMAC SYSTEMS A.G., Switzerland) and kept at 60°C for 4-6 hours. The coloured samples were bleached by adding 0.3 ml isopropanol and 0.3 ml hydrogenperoxide (35%). After allowing the mixture to stand at room temperature for 30 min, 10 ml Lipoluma (LUMAC) was added. The samples were then light and temperature equilibrated for 3 hours and counted in a KONTRON MR3000 Automatic Liquid Scintillation System (Kontron, Switzerland). The cmp were converted to dpm by using the external standard ratio method. The counting efficiency was 30-40%. Finally, the dpm/mg wet tissue was calculated.

In the experiment with ^{67}Ga each mouse was injected with 0.2 MBq ^{67}Ga citrate. The mice were sacrificed at 6, 12, 24 and 48 h after injection. Immediately following washing and weighing the ^{67}Ga activity of the tissue samples were counted in a CG 4000 Intertechnique Well counter (78370 Plaisir, France) using a 90 - 300 keV window. The cpm/mg wet tissue was then calculated.

In the kidney, cortex and medulla were separated and the concentration of ^{3}H-HpD and ^{67}Ga determined by scintillation counting as described above. Finally the concentration ratio between these two parts of kidney were calculated for ^{3}H-HpD and ^{67}Ga respectively.

The experiment with ^{131}I-albumin was carried out as described for ^{67}Ga, this time using a 260-470 keV window. 50 kBq human ^{131}I-albumin was injected in each mouse.

Autoradiography

Mice were sacrificed 24h after being injected intravenously with 0.1 ml ^{3}H-HpD. The kidneys were excised and immediately afterwards fixed in 4% phosphate-buffered formalin, pH= 7.1. The samples were then washed in 0.1 M cacodylate buffer containing 0.1M sucrose, dehydrated in graded alcohols and embedded in epon. Sections, one μm thick, were cut and coated by hand dipping in Ilford Emulsion K-2. After exposure over 12 days at 5°C, the slides were developed with a Kodak D-19 developer and thereafter stained with toluidine blue. Sections were examined with a Nikon Apophot M microscope and photographed with Kodak Tri-X-pan film.

RESULTS

The chromatographic analysis of HpD showed that it contained a large number of components (Fig. 1) some of which have been identified earlier (Moan et al., in press). They are numbered 2, 4A, 4B and 7 with increasing hydrophobicity from left to right. Component 7 seems to be heavily aggregated (Evensen et al., submitted for publication).

The HPLC chromatogram of labelled and unlabelled HpD was essentially the same, this holds true for both fluorescence and absorption readings. In recording the radioactivity of the injected material recovered from the HPLC column, the number and localization of the peaks were in correspondence with those in the absorption readings, their internal height-ratio were however somewhat different.

The tissue distribution of ^{3}H-HpD and ^{67}Ga at various times following injection are shown in Table 1 and 2. The tissue/tumor concentration ratio seems to stabilize at 24h.

The concentration of ^{3}H-HpD as well as that of ^{67}Ga in liver, kidney and spleen was higher than in tumor, while the concentration in skin, heart, muscle and brain was lower. This holds true for all times following injection. Being initially higher than in tumor, the concentration of ^{3}H-HpD and ^{67}Ga in lung fell below that of tumor at about 6h and 36h, respectively.

As seen from the bottom line in Table 1 and 2, the absolute tumor uptake of ^{3}H-HpD in terms of percent of in-

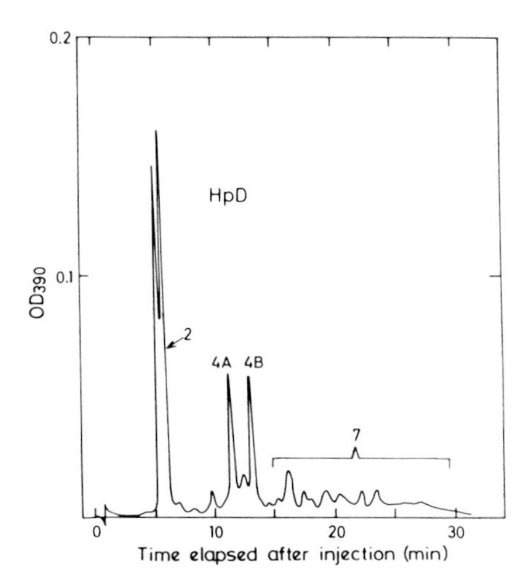

Fig. 1. High pressure liquid chromatography of HpD. Injected volume: 200 μl. Column: Supelcosil LC 18 (250 x 4.6 mm). Gradient elution with methanol and water. Conditions and chromatography described in Moan et al. 1982a. More than 95% of the injected material was eluted during a 40 min. run. The components were numbered according to an earlier publication (Moan et al. 1983).

jected dose per gram tumor tissue is 1/3 to 1/4 of that of ^{67}Ga.

The relative tissue distribution of the components at 24 h are shown in Table 3. Most striking is the accumulation of component 2 in kidney and of component 7 in the liver. Furthermore, relative to the injected amount, component 7 is the most effective tumor localizer of the components.

Fig. 2 shows the tissue distribution of ^{3}H-HpD, ^{67}Ga and ^{131}I-albumin at 24 h.

The concentration ratio between kidney cortex and medulla for ^{3}H-HpD and ^{67}Ga was 1.7 and 1.4, respectively.

Figs. 3A and 3B show the autoradiographic distribution of ^{3}H-HpD in kidney medulla (A) and cortex (B) 24h after injection. The radioactive labels are mainly localized in

Table 1: Relative tissue distribution of ^{3}H-HpD at various times following i.v. injection in mice bearing LCC. The absolute tumor uptake (in terms of percent of injected dose per gram of tumor tissue) is also given. The values are the mean of three recordings.

	3 h		12 h		24 h		48 h		72 h		24 h
	M	± SD	M	± SD	M	± SD	M	± SD	M	± SD	Gomer and Dougherty 1979
Brain	0.23	0.04	0.13	0.02	0.16	0.04	0.08	0.03	0.07	0.01	–
Muscle	0.28	0.08	0.19	0.02	0.22	0.06	0.15	0.02	0.24	0.06	0.11
Stomach	0.63	0.00	0.60	0.09	0.42	0.02	0.50	0.06	0.67	0.06	–
Heart	0.75	0.13	0.70	0.11	0.54	0.07	0.49	0.08	0.74	0.07	–
Skin	0.90	0.30	0.86	0.12	0.59	0.18	0.78	0.14	0.92	0.02	0.39
Lung	1.31	0.24	0.77	0.13	0.62	0.03	0.56	0.08	0.72	0.12	0.52
Tumor	1.00	0.00	1.00	0.00	1.00	0.00	1.00	0.00	1.00	0.00	1.00
Spleen	2.13	0.28	2.03	0.22	1.53	0.06	1.59	0.21	2.12	0.28	1.63
Kidney	4.78	0.52	4.94	0.69	3.15	0.53	3.57	0.71	4.70	0.85	2.60
Liver	6.03	0.72	6.44	0.57	4.47	0.64	4.68	0.82	6.44	0.27	4.23
% of inj. dose/g of tum.tissue	1.8		2.1		2.6		2.6		1.6		2.6*

* The weight of each mouse was supposed to be 20 gram

Table 2: Relative tissue distribution of ^{67}Ga at various times following i.v. injection in mice bearing LLC. The absolute tumor uptake (in terms of percent of injected dose per gram of tumor tissue) is also given. N is the number of mice used in each experiment.

	6 h		12 h		24 h		48 h	
	N=3		N=2		N=7		N=2	
	M	± SD	M	± SD	M	± SD	M	± SD
Brain	0.13	0.00	0.12	0.03	0.14	0.14	0.10	0.02
Muscle	0.23	0.08	0.18	0.02	0.10	0.05	0.10	0.01
Stomach	1.61	0.40	2.44	0.39	3.41	1.17	2.85	0.01
Heart	0.96	0.52	0.52	0.03	0.49	0.22	0.28	0.01
Skin	1.06	0.41	0.76	0.00	0.52	0.22	0.56	0.10
Lung	1.27	0.29	1.16	0.13	1.05	0.34	0.81	0.25
Tumor	1.00	0.00	1.00	0.00	1.00	0.00	1.00	0.00
Spleen	1.38	0.50	1.35	0.05	1.46	0.60	1.36	0.12
Kidney	1.41	0.25	1.78	0.35	2.32	0.60	1.81	0.33
Liver	1.27	0.29	1.66	0.09	2.25	0.84	2.13	0.13
% of inj. dose/g of tum.tissue	8.4		9.2		7.1		6.5	

Table 3: Relative tissue distribution of the HPLC separated components of ^{3}H-HpD in mice bearing LLC 24 h after i.v. injection. The values are the mean of three recordings.

	COMPONENTS					
	2		4A+4B		7	
Organ	M	± SD	M	± SD	M	± SD
Muscle	0.34	0.06	0.21	0.06	0.15	0.03
Skin	0.83	0.13	0.48	0.03	0.33	0.09
Lung	0.84	0.28	0.56	0.15	0.76	0.09
Tumor	1.00	0.00	1.00	0.00	1.00	0.00
Kidney	9.31	2.19	4.07	0.12	2.85	0.56
Liver	4.96	0.49	4.63	0.41	13.24	1.88
% of inj. dose/g of tum.tissue	1.2		2.3		5.7	

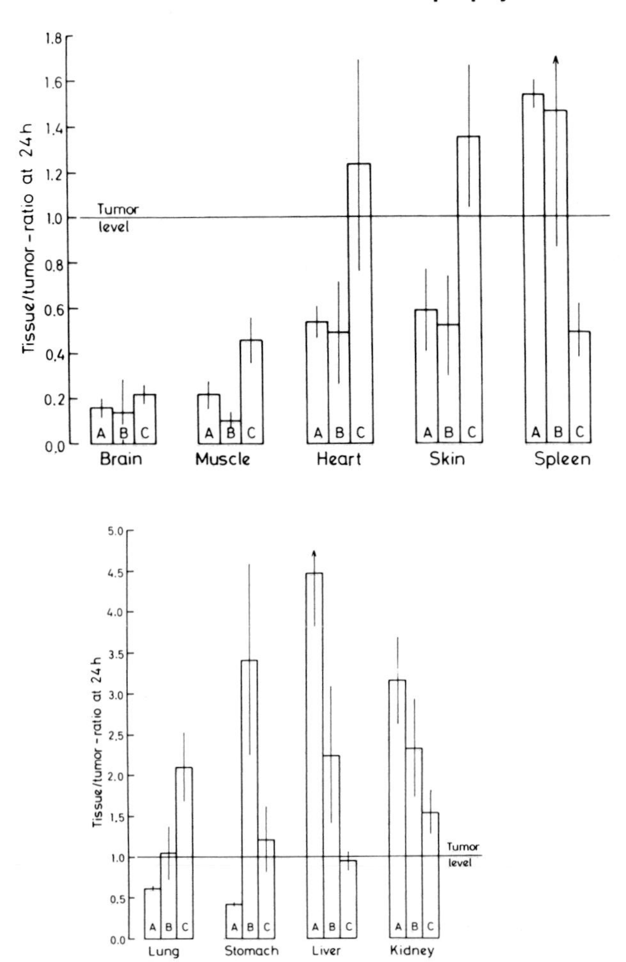

Fig. 2. Tissue/tumor-ratio of A: ^{3}H-HpD, B: ^{67}Ga, and C: ^{131}I-albumin in LLC bearing mice at 24h after i.v. injection.

the cortex, i.e. proximal tubule. The proximal/distal grain density ratio was about 5.

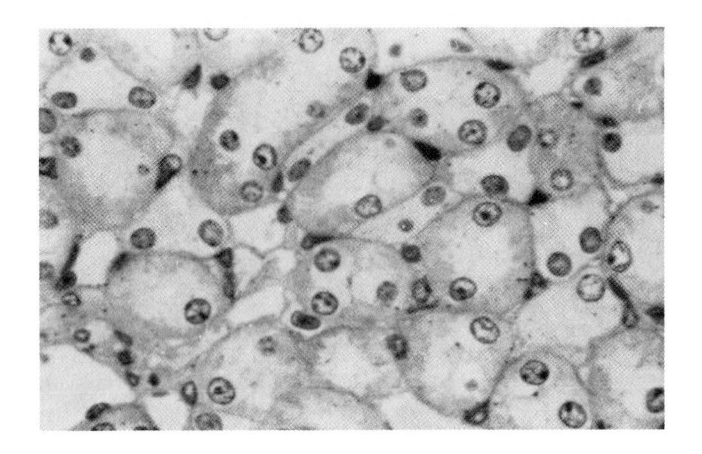

Fig. 3A. Autoradiographic distribution of ^{3}H-HpD in B_6D_2 mice kidney (medulla) x 1000.

DISCUSSION

Concentration of porphyrins in tissue are usually determined by chemical extraction and fluorescence measurements. However, when porphyrins with different polarities are present in the tissue such as after an injection of HpD, this method is unreliable due to two facts: the extractability strongly depends on the polarity of the porphyrins and so does the fluorescence quantum yield due to different degrees of aggregation. We therefore preferred to use radioactively labelled HpD. This seems to be particulary important in the present study since it has been shown that the aggregates present in HpD, some of which may be covalently linked together, are the best tumor localizers (Moan et al., 1982 a, b).

Gomer and Dougherty (1979) have tested ^{3}H-labelled HpD prepared by NEN in an in vivo system, and concluded that little if any exchange of the radioactive label took place. We therefore used ^{3}H-labelled HpD from the same source.

The organ/tumor ratios obtained in the present investigations are in accordance with those found by Gomer and

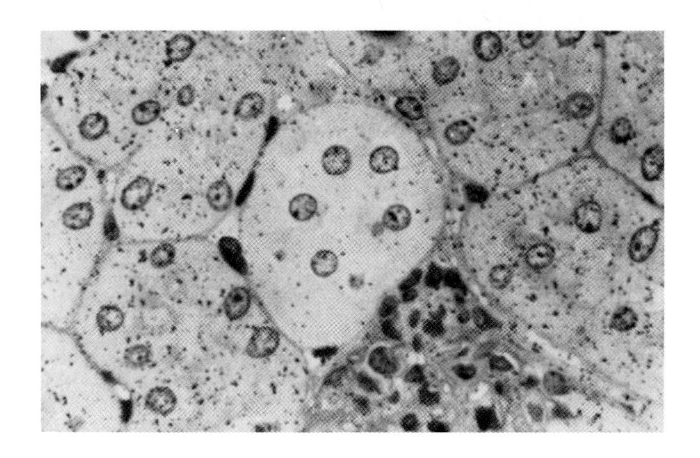

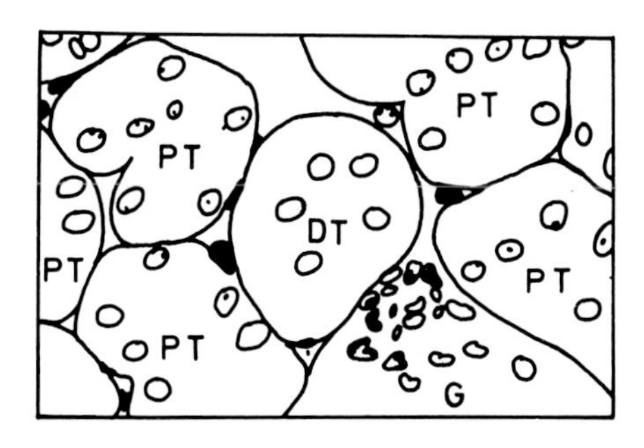

Fig. 3B. Autoradiographic distribution of ^{3}H-HpD in B_6D_2 mice kidney (cortex). PT: proximal tubules, DT: distal tubules, G: glomeruli) x 1000.

Dougherty (1979) and so are the absolute tumor uptakes, Table 1.

We found the tissue distribution of ^{67}Ga to be essentially completed 24h after injection, Table 2. The same conclusion was drawn by Hayes (1977) based on studies of humans. Furthermore, the tumor uptake in terms of percent of injected dose agrees with results published by Zanelli and Kaelin (1981). The same conclusion holds true for the inter tissue distribution (Zanelli and Kaelin 1981, Ito et al., 1971), Table 2. The organ/tumor ratios at 24h for ^{3}H-HpD and ^{67}Ga are shown in Fig. 2. to be strikingly similar.

Two exceptions (stomach and liver) may be noted and will be discussed later.

The present investigation shows that the cortex/medulla concentration ratio of ^{3}H-HpD and ^{67}Ga are 1.7 and 1.4 respectively. ^{3}H-HpD is predominantly localized in the proximal convoluted tubules (Fig. 3A, B).

Other investigators have found a similar distribution of ^{67}Ga in kidney (Swartzendruber et al., 1971). Other sites with high concentrations of ^{67}Ga were the red pulp of spleen and the Kuppfer cells in the liver. Furthermore, the macrophages contained 10-20 times more ^{67}Ga than thymocytes and lymphoma cells. A corresponding autoradiographic distribution of ^{3}H-HpD was reported by Bugelski et al. (1981).

Furthermore, porphyrins have been shown to have a high affinity for embryonic and regenerating tissues (Figge et al., 1948) as well as for growing bone (Barker et al., 1970). Similarly ^{67}Ga is accumulated in healing skin incisions, cranitomy sites, fractures and some other related neoplastic tissues (Bell et al., 1979).

^{67}Ga and HpD also behave similarly with respect to cellular uptake at different pH-values. Thus, for both compounds it has been reported that the cellular uptake increases with decreasing pH (Glickson et al., 1974, Moan and Christensen, 1981). The latter investigators studied cellular uptake of commercial hematoporphyrin, which was later shown to contain the same localizing components as HpD (Moan et al., 1982 a). Moan and Christensen have proposed that the low pH in tumors compared to non-malignant tissue (Gullino et al., 1965) may be one of the reasons for selective uptake of HpD in tumors.

The similarity of gallium and crude HpD with respect to tissue distribution may be related to their following common characteristics:

Firstly, gallium and porphyrin are both transported from the site of injection by the blood stream. Porphyrins are bound to albumin and hemopexin (Hx) (Müller-Eberhard & Morgan, 1975) and gallium is bound to transferrin (Tf) and probably haptoglobin (Clausen et al., 1974, Gunasekera 1972). To reach the interstitial space (IS) of tumors the protein-bound labels (^{67}Ga and ^{3}H-HpD) must pass the endothelium of

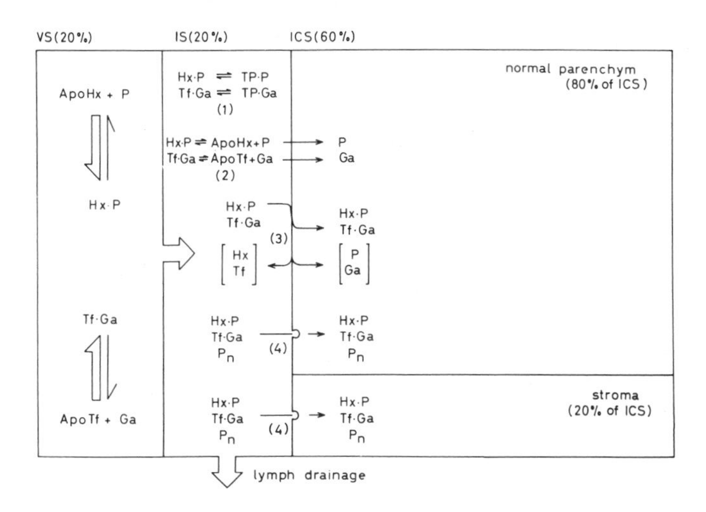

Fig. 4A. Transport and distribution of porphyrin (P) and gallium (Ga) within normal tissue.

VS: vascular space; IS: interstitial space; ICS: intracellular space; TS: tissue protein; P_n: porphyrin aggregates; H_x: Hemopexin and Tf: transferrin.

(1) Redistribution of porphyrin and gallium from transport proteins (H_x and Tf) to tissue proteins (TP).

(2) Passive diffusion (pH dependent) of porphyrin and gallium across the cellular membrane.

(3) Receptor mediated transport.

(4) Cellular uptake by phagocytosis.

The relative volumes of VS, IS and ICS refers to a hepatoma (Gullino, 1966). Furthermore, some tumors are known to contain as much as 50% of macrophages.

For further explanation, see text.

the capillary walls (Fig. 4A, B). It is known that neoplastic tissue has an increased vascular permeability to plasma

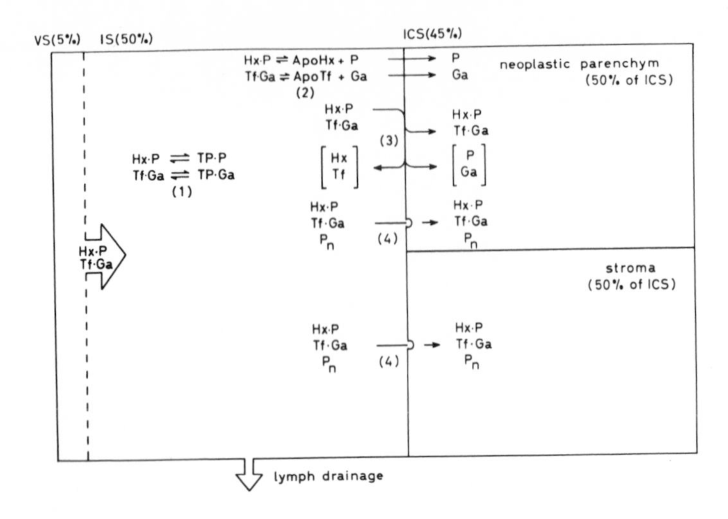

Fig. 4B. Transport and distribution of porphyrin and gallium within neoplastic tissue (abbreviations, see Fig. 4A).

proteins compared with normal tissue (Peterson & Appelgren, 1973). Furthermore, neoplastic tissue has a compromised lymphatic drainage and a larger IS than non-neoplastic tissue (Goldacre & Sylvén, 1962; Gullino, 1966). These facts may partly account for the high concentration of ^{67}Ga and ^{3}H-HpD in tumors (Fig. 4A, B). However, other factors must also be of importance since ^{131}I-albumin distributes differently from ^{3}H-HpD. Several such factors may be mentioned: The protein-bound labels (^{3}H-HpD and ^{67}Ga) may, in the IS, redistribute to other proteins. Thus it has been shown that porphyrins bind to fibrin, elastin and collagen (Musser et al., 1980, 1982; Fig. 4A, B). Furthermore, Musser et al. (1982) found evidence of a greater porphyrin binding capacity of the newly laid down collagen in neoplastic tissue than in more mature tissue. In addition to the larger IS in neoplastic than in normal tissue, this will contribute to a higher concentration of porphyrin in former than the latter tissue.

A second process related to the distribution of gallium and porphyrins in tissues is phago-/pinocytosis (Fig. 4A, B). Thus it has been shown that ^{67}Ga as well as several porphyrins are present in high concentrations in lysosomes (Swartzendruber et al., 1971; Allison et al., 1966). Cells of the reticuloendothelial system (RES) are notably prone to phagocytosis and, as mentioned, accumulation of ^{3}H-HpD and ^{67}Ga has been demonstrated in the red pulp of the spleen, Kuppfer cells in liver and macrophages in tumors (Bugelski et al., 1981; Swartzendruber et al., 1971). Recruitment of macrophages and inflammatory cells in tumors could in part account for the accumulation of HpD and ^{67}Ga in accordance with the observation that about five times more HpD is present in the tumor stroma than in the tumor cells (Bugelski et al., 1981). The fact that component 7, which is aggregated, seems to be a better tumor localizer than component 2, which is monomerized, also indicates that phagocytosis contributes to the tumor uptake of porphyrins.

Uptake by phago-/pinocytosis also seems to be the explanation of the high concentration of ^{3}H-HpD and ^{67}Ga in the kidney (i.e. proximal convoluted tubuli). Proximal tubuli are known to show a high phagocytic activity against proteins and some other molecules (Maunsbach, 1976). Filtered in the glomeruli, polar porphyrins and gallium are absorbed by the luminal face of the proximal tubule. In accordance with this is the high concentration of the most polar component (comp. 2) in the kidney.

Trapping of radioactive labels by pinocytosis in the tumor cells has been proposed by Potchen et al. (1971). However, examining 12 different tumor strains in rodent, Easty (1964) found no evidence for uptake of albumin in tumor cells. Therefore, we propose that HpD and ^{67}Ga present in tumor cells have been taken up by passive diffusion facilitated by the low tumor pH. It is well known that the passive cellular uptake of a substance is related to its hydrophobicity. Lowering pH results in a neutralization of the negative charge of the porphyrins and a displacement of the equilibrium $Ga(OH)_4^- + H^+ \leftrightharpoons Ga(OH)_3 + H_2O$ (which are the predominant forms of gallium at physiologic pH) to the right (Hayes, 1976). In both cases more hydrophobic molecules are formed, thus favouring cellular uptake (Fig. 4A, B).

Finally, a process that may govern the tissue distribution of porphyrin and gallium is cellular uptake by receptor mediated transport (Fig. 4A, B). After specific binding of the transport molecule (Hx and Tf) to the receptor, the radioactive label (^{3}H-HpD and ^{67}Ga) or label-protein complex is transferred across the cellular membrane by a carrier. Thus, according to Müller-Eberhard and Morgan (1975) and Morgan (1976) there is reason to believe that such receptors exist in hepatocytes. Furthermore, Müller-Eberhard (1970) considers hemopexin as a distributor of heme within the organism. The same conclusion may be drawn for ^{3}H-HpD, knowing that hemopexin also binds porphyrins (Morgan 1976).

The metabolism of ^{67}Ga in vivo seems to be very similar to that of iron. Other investigators have found that ^{67}Ga is stored along with iron in ferritin (Larson, 1978), binds to lactoferrin in tumors and normal tissue (Hoffer et al., 1977) and, most important, is transported in serum by transferrin (Clausen et al., 1974).

Ito et al. (1971) have found high concentration of ^{67}Ga in bone marrow. This is in correspondence with the high concentration of transferrin receptors in erythroblasts (Conrad et al., 1981 and references therein).

High concentration of transferrin receptors have also been found in malignant tissue (Larrick 1982) and the existence of transferrin receptors in tumor cells have been proposed as an explanation of the increased uptake of ^{67}Ga in tumors (Larson 1978).

Several authors have found a high uptake of porphyrins in liver (Gomer & Dougherty, 1979: Wang et al., 1981; Winkelman, 1967). In fact, the liver seems to be one of the organs most prone to accumulate porphyrins.

Furthermore, Newstead et al. (1978) report a high uptake of ^{67}Ga in human stomach. Their reported stomach/liver concentration ratio is in correspondence with the present investigation.

The dissimilar uptake of ^{67}Ga and ^{3}H-HpD in the liver and stomach is probably related to the way of degradation and/or excretion of the respective radiopharmaceuticals. While both labels are excreted partly by the kidneys, the liver is the main site of porphyrin catabolism. This is made possible by hemopexin which transports porphyrins to the liver (Müller-Eberhard, 1970).

^{67}Ga is excreted through urine and to a lesser extent through feces. This may be the explanation of the high concentration of ^{67}Ga in stomach. ^{67}Ga is probably secreted by gastric mucosa. High concentrations of ^{67}Ga in other secretory organs like the parathyroid, the breast and the gut have been reported earlier (Langhammer et al., 1978).

CONCLUSION

In reviewing the literature on this subject it turns out that there are several factors rather than one governing the uptake of ^{67}Ga and porphyrins in tumors. In different kinds of tumors the relative contribution of each factor will vary.

Concerning porphyrins, their inherent properties (polarity and thereby degree of aggregation) are also of utmost importance. This can be seen from the dissimilar tissue distribution of the HPLC separated components of ^{3}H-HpD.

The absolute tumor uptake of crude ^{3}H-HpD is less than that of ^{67}Ga at all times following intravenous injection. However, crude ^{3}H-HpD and ^{67}Ga seem to be equally effective in discriminating malignant from normal tissue,

The following processes are proposed to be the main reasons for accumulation of HpD and ^{67}Ga in tumors:

- Increased vascular permeability to plasma proteins of neoplastic tissue,

- the relatively larger IS in neoplastic tissue versus normal tissue,

- the binding of labels to tissue proteins in IS,

- the compromised lymphatic drainage in neoplastic tissue,

- the trapping of labels in tumor macrophages and

- passive diffusion into the tumor cells facilitated by the lower pH in the interstitial fluid of tumors.

The uptake of porphyrins and gallium in normal cells is

proposed to be mediated by receptors and/or phagocytosis.

The shown similarities between ^{67}Ga and ^{3}H-HpD may be helpful in evaluating patients for photoradiation therapy.

ACKNOWLEDGMENT

We would like to thank Dr. T. Christensen and Dr. M. Aas for valuable criticism and comments. We are also greatful to R. Puntervold for skilful assistance in preparing slides for autoradiography.

The present work was supported by The Norwegian Cancer Society (Landsforeningen mot Kreft).

REFERENCES

Allison AC, Magnus IA, and Young MR (1966). Role of Lysosomes and of Cell Membranes in Photosensitization. Nature 209:874.

Barker DS, Henderson RW and Storey E (1970). The In Vivo Localization of Porphyrins. Br. J. exp. Path. 51:628.

Bell G, O'Mara RE, Henry A, Subrama nian, G, McAfee JG, and Brown LC (1971). Non-Neoplastic Localization of ^{67}Ga-Citrate. J. Nucl. Med. 12:338.

Bugelski PJ, Porter, CW, and Dougherty TJ (1981). Autoradiographic Distribution of Hematoporphyrin Derivative in Normal and Tumor Tissue of the Mouse. Cancer Res. 41: 4606.

Clausen J, Edeling C-J, and Fogh J (1974). ^{67}Ga Binding tp Human Serum Proteins and Tumor Components. Cancer Res. 34:1931.

Conrad EC, Barton JC, Gams RA and Ostroy F (1981). Iron, Folic Acid, and Vitamin B_{12}. In Fairbanks VF (ed): "Current Hematology". Wiley Med, p. 123.

Doiron DR, Profio E, Vincent RG and Dougherty TJ (1979). Fluorescence Bronchoscopy for Detection of Lung Cancer. Chest 76:1.

Easty GC (1964). The Uptake of Fluorescent Labelled Proteins by Normal and Tumour Tissues in vivo. Brit J Cancer 18:368.

Figge FHJ, Weiland GS, and Manganiello LOJ (1948). Cancer detection and therapy. Affinity of neoplastic embryonic and traumatized regenerating tissues for porphyrins and metalloporphyrins. Proc. Soc. Expl. Biol. Med. 68:640.

Glickson JD, Webb J, and Gams RA (1974). Effects of buffers and pH on in vitro binding of ^{67}Ga by L 1210 leukemic cells. Cancer Res. 34:2957.

Goldacre RJ, and Sylvén V (1962). On the access of blood-borne dyes to various tumor regions. Brit. J. Cancer 16:306.

Gomer CJ, and Dougherty TJ (1979). Determination of ^{3}H - and ^{14}C Hematoporphyrin Derivative Distribution in Malignant and Normal Tissue. Cancer Res. 39:146.

Gullino PM, Grantham FH, Smith SH, and Haggerty AC (1965). Modifications of the Acid-Base Status of the Internal Milieu of Tumours. J. Nat. Cancer Inst. 34:857.

Gullino PM (1966). The Internal Milieu of Tumours. Progr. exp. Tumor Res. 8:1.

Gunasekera SW, King LJ, and Lavender PJ (1972). The behaviour of Tracer-Gallium-67 towards Serum Proteins. Clin.

Chim. Acta 39:401.
Hayata Y, Kato H, Ono J, Matsushima Y, Hayashi N, Salito T, and Kawate N (1982). Fluorescence Fiberoptic Bronchoscopy in the Diagnosis of Early Stage Lung Cancer. In Band PR (ed): "Recent Results in Cancer Research - Early Detection and Localization of Lung Tumors in High Risk Groups". Springer Verlag, Berlin, Heidelberg, New York, p 121.
Hayes RL (1976). Factors affecting uptake of radioactive agents by tumour and other tissues. In Tumor Localization with Radioactive Agents Internatl AEA, Vienna, IAEA-MG-50/14, p 29.
Hayes RL (1977). The Tissue Distribution of Gallium Radionuclides. Journal of Nuclear Medicine 18:740.
Hoffer PB, Huberty J, and Khayam-Bashi H (1977). The association of Ga-67 and Lactoferrin. J. Nucl. Med. 18:713.
Ito Y, Okuyama S, Sato K, Takahashi K, Sato T, and Kanno I (1971). ^{67}Ga Tumor Scanning and Its Mechanisms Studied in Rabbits. Radiology 100:357.
Kessel D (1978). Transport and binding of hematoporphyrin derivate and related porphyrins by murine leukemia L1210 cells. Cancer Res. 41:1318-1323.
Langhammer H, Hör G, Kempken K, Pabst HW, Heidenreich P, and Kriegel H (1976). Tumor Scintigraphy with Gallium 67. In "Tumor Localization with Radioactive Agents", Internatl. AEA, Vienna, IAEA-MG-50/14, p 69.
Larrick J (1982). Transferrin Receptors on Normal and Transformed Cells. 13th International Cancer Congress, UICC. Seattle 8-15 Sept.
Larson SM (1978). Mechanisms of Localization of Gallium-67 in Tumors. Sem, Nucl. Med. Vol. VIII, 193.
Lipson RL, Baldes EW, Olsen AM (1961). The use of a derivative of hematoporphyrin in tumor detection. J. Natl. Cancer Inst. 26:1.
Maunsbach AB (1976). Cellular Mechanisms of Tubular Protein Transport. In Thurau K (ed): "International Review of Physiology, Kidney and Urinary Tract Physiology III, Vol II, University Park Press, p 145.
Moan J, and Christensen T (1981). Cellular Uptake and Photodynamic Effect of Hematoporphyrin Photobiochemistry and Photobiophysics 2:291.
Moan J, Steen HB, Feren K and Christensen T (1981). Uptake of Hematoporphyrin Derivate and Sensitized Photoinactivation of C3H Cells with Different Oncogenic Potential. Cancer Letter, 14:291.
Moan J, Christensen T, and Sommer S (1982 a). The main photosensitizing components of hematoporphyrin. Cancer

Letters 15:161.
Moan J, Evensen JF, Christensen T, Hindar A, Sommer S, and McGhie JB (1982 b). Chemical Composition of Hematoporphyrin Deriative, Tumorlocalizing and Photosensitizing properties of its Main Components. 10th Annual Meeting, American Society for Photobiology, June 27 - July 1.
Moan J, Sandberg S, Christensen T, and Elander S (1983). Hematoporphyrin Derivate: Chemical Composition, Photochemical and Photosensitizing Properties. In Kessel D and Dougherty TJ (eds): "Porphyrin Photosensitization", Plenum Publishing Corporation, p 165.
Morgan WT (1976). The Binding and Transport of Heme by Hemopexin. Annals of Clinical Research 8:223.
Mossman BT, Gray MJ, Silberman L, and Lipson LR (1974). Identification of Neoplastic Versus Normal Cells in Human Cervical Cell Culture. Obstet, Gynecol. 43:635.
Muller-Eberhard U (1970). Hemopexin. New Eng J Med 283:1090.
Muller-Eberhard U, and Morgan WT (1975). Porphyrin-binding Proteins in Serum. NY Acad Sci 244:626.
Musser DA, Wagner JM, Weber FJ, and Datta-Gupta N (1980). The binding of tumor localizing porphyrins to a fibrin matrix and their effects following photoirradiation. Res. Commun. Chem. Pathol. Pharmacol 28:505.
Musser DA, Wagner JM, and Datta-Gupta N (1982). The interaction of tumor localizing porphyrins with collagen and elastin. Res. Commun. Chem. Pathol. Pharmacol. 36:251.
Nelson B, Hayes RL, Edwards CL, Knisely RM, and Andrews GA (1972). Distribution of gallium in human tissues after intravenous administration. J. Nucl. Med. 13:92.
Newstead G, Taylor DM, McReady, VR, and Bettelheim R (1976). Gallium-67 Deposition in the Human Gastro-Intestinal Tract. Int. J. Nucl. Med. Biol. 4:109.
Peterson H-I, and Appergren KL (1973). Experimental Studies on the Uptake and Retention of Labelled Proteins in a Rat Tumor. Eurp. J. Cancer 9:543.
Potchen EJ, Elliot AJ, Siegel BA, Studer R, and Evens RG (1971). Pathophysiologic Basis of Soft Tissue Tumor Scanning. J. Surg. Oncol. 3 (6):593.
Swartzendruber DC, Nelson B, and Hayes RL (1971). Gallium-67 Localization in Lysosomal-like Granules of Leukemic and Nonleukemic Marine Tissue, J. Nat. Cancer Inst. 46:941.
Wang TST, Fawwas RA, and Tomashefsky P (1981). Metalloporphyrin derivatives: Structure-localization properties. In Richard P. Spencer (ed): "Radiopharmaceutal: Structure-Activity Relationships", Grune & Stratton, Inc., p 225.

Winkelman J, Slater G, and Grossman J (1976). The Concentration in Tumor and Other Tissues of Parenterally administered Tritium- and ^{14}C-labelled Tetraphenylporphinesulfate. Cancer Research 27:2060.

Zanelli GD, and Kaelin AC (1981). Synthetic porphyrins as tumour-localizing agents. British Journal of Radiology, 54:403.

Porphyrin Localization and Treatment of Tumors, pages 563–569

EFFECTS OF HEMATOPORPHYRIN (HPD) AND A CHEMILUMINESCENCE SYSTEM ON THE GROWTH OF TRANSPLANTED TUMORS IN C_3H/HeJ MICE

M. J. Phillip[*1], J. D. McMahon[1], M. D. O'Hara[2], F. W. Hetzel[2], C. Amsterdamsky[3], and A. P. Schaap[*3]

[1]School of Dentistry, University of Detroit
[2]Therapeutic Radiology, Henry Ford Hospital
[3]Department of Chemistry, Wayne State University
Detroit, Michigan

ABSTRACT

Photoradiation therapy is emerging as a promising technique for combating cancer. Fundamentally, this approach consists of two steps: (1) hematoporphyrin derivative (HPD) is used to selectively sensitize cancer cells to visible light; (2) after an appropriate time interval, light is introduced into the tumor via a laser-fiber optic system to trigger the cytotoxic action of HPD. The present investigation was initiated to determine the therapeutic potential of HPD in combination with a chemiluminescent activator in treating mice which had been transplanted with tumors.

INTRODUCTION

The principles on which photoradiation therapy (PRT) is based have been known for over 80 years. In 1900 it was discovered that certain fluorescent dyes could sensitize living organisms to visible light (Raab, 1900). Some time after this discovery, several investigators reported selective retention of photosensitizing agents by malignant tumors in animals as well as in humans (Auler and Banzer, 1942; Figge et al., 1948; Lipson et al., 1961; Gregorie et al., 1968; Winkelman and Rasmussen-Taxdal, 1969).

More recently, the specific uptake and retention of hematoporphyrin derivative (HPD) by malignant tissue followed by laser irradiation has been utilized in the development of a promising modality for the treatment of cancer (Kelly et al., 1975; Granelli et al., 1975; Dougherty et al., 1975; Dougherty et al., 1978). Although the mechanism of tumor destruction by PRT has not yet been fully delineated, there is substantial evidence that singlet oxygen is involved as the active agent in the oxidation of biomolecules (Weishaupt et al., 1976; Moan et al., 1979).

Our present investigation was initiated to determine the therapeutic potential of HPD in combination with a chemiluminescent activator. This new approach is targeted at improving the delivery of light to the HPD by replacing the laser with an efficient chemiluminescent system (CLS) that can be injected directly into the tumor. Preliminary results have been obtained in treatment of transplanted tumors in C_3H/HeJ mice.

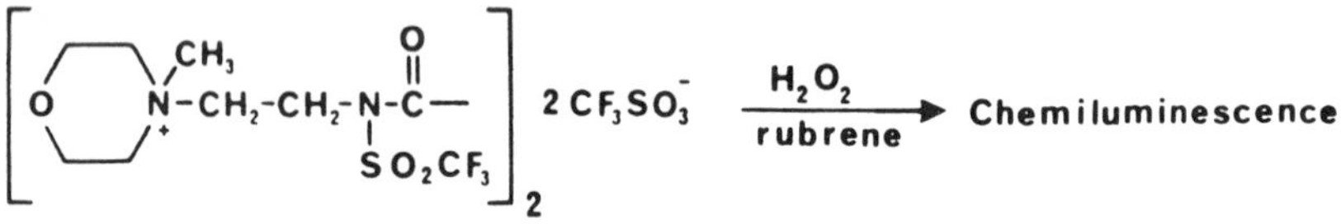

1

sulfonated rubrene =

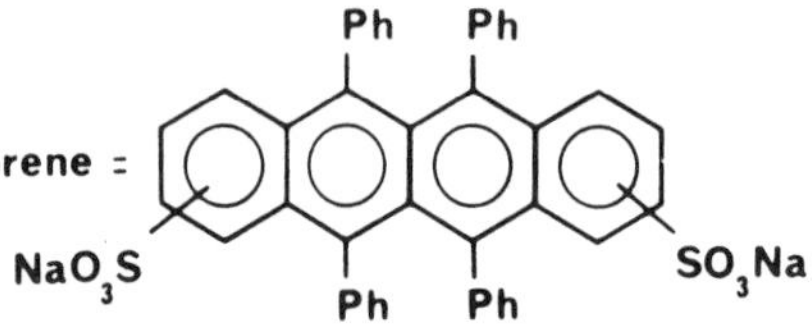

2

As the chemical light source for this study, we have utilized a system related to the peroxyoxalate chemiluminescence developed at American Cyanamid (Rauhut, 1966; Rauhut et al., 1975; Tseng et al., 1979). The luminescence is produced in aqueous solution by treatment of the substitutued oxamide 1 with 1% hydrogen peroxide in the presence of sulfonated rubrene 2 as fluorescer. The reaction is initiated by the addition of the hydrogen peroxide and the surfactant Deceresol N1 to 1 and 2. The intense yellow-red light from this reaction lasts for 10-20 min. The mechanism for this chemiluminescent reaction is thought to involve the formation of the high-energy cyclic peroxide, 1,2-dioxetanedione (3). Subsequent decomposition of peroxide 3 in the presence of rubrene 2 gives singlet excited 2, fluorescence from which provides the observed light.

OXAMIDE 1 + H_2O_2 ⟶ 3 (1,2-dioxetanedione: O—O ring with C—C, each C=O)

3 + RUBRENE ⟶ $^{1}\text{RUBRENE}^{*}$ + 2 CO_2

$^{1}\text{RUBRENE}^{*}$ ↓ HV

MATERIALS AND METHODS

A transplanted mammary adenocarcinoma from a female C_3H/HeJ mouse obtained from Henry Ford Hospital was excised after sacrifice of the animal. The tumor was carefully dissected and transplanted into the left axillary fold of 70 C_3H/HeJ male and female mice using the technique described by Phillip et al. (1971, 1973). When the transplanted tumors became palpable, the animals were sensitized with 0.2 mL hematoporphyrin derivative (HPD) obtained from Henry Ford Hospital. The concentration of the HPD was 10 mg/kg body weight. Twenty-four h after sensitization the animals were treated with 0.2 mL of the chemiluminescence system (CLS) described below. The CLS was injected subcutaneously in the area of tumor localization.

Initial samples of oxamide 1, sulfonated rubrene 2, and Deceresol N1 were generously provided by American Cyanamid. We also thank Dr. A. G. Mohan for a description of the procedures for the sytheses of 1 and 2. The chemiluminescence system was prepared by adding 180 mg of oxamide 1, 25 mg of sulfonated rubrene 2, and 0.1 mL of Deceresol N1 to 5 mL of 1% hydrogen peroxide.

RESULTS AND DISCUSSION

This investigation was conducted to evaluate the therapeutic potential of HPD in combination with chemiluminescence. A standardized suspension of viable adenocarcinoma cells excised from a tumor-bearing C_3H/HeJ mouse was injected into the axillary fold of young healthy mice. A group of animals were given HPD/CLS therapy as soon as the transplanted tumors were palpable. The treatment was administered on two successive days. On the first day the animals were sensitized with HPD and 24 h later the animals were treated with the CLS, injecting directly into the tumor to activate the HPD.

There were four different groups of animals in the investigation. Group I was treated with HPD + CLS; group II was treated with CLS only; group III was transplanted and untreated; and group IV was neither transplanted nor treated and served as a control against the development of spontaneous tumors. The animals in each group were carefully examined at weekly intervals and tumor development was monitored by computing tumor volume.

Average tumor volumes for each group eight weeks after transplantation are shown in Table I. Significantly, group I which was treated with HPD + CLS exhibited approximately four times smaller tumors than the tumors in group III (transplanted and untreated controls). Photograph I shows a typical mouse with a large tumor that was not treated with the chemiluminescence system. In contrast, photograph II demonstrates the typical reduction in tumor volume that is effected by the combined HPD/CLS treatment.

Table I. Average Tumor Volumes Eight Weeks After Transplantation

Group	Volume (cm^3)
I (HPD + CLS)	.6
II (PCS only)	1.4
III (transplanted and untreated)	2.4
IV (normal, not transplanted)	no tumors developed

These results are presently interpreted in terms of a mechanism for tumor destruction involving: (1) localization of HPD in malignant mouse tissue; (2) absorption by HPD of the CLS-produced luminescence; (3) energy transfer from excited HPD to oxygen to generate singlet oxygen; and (4) oxidation of biomolecules in the tumor by singlet oxygen.

Although the results described above are preliminary, we have demonstrated that a chemiluminescent system in combination with HPD may provide an alternate approach to treatment of malignant tumors. This investigation is continuing with studies of various chemiluminescent reactions with other photosensitizing dyes.

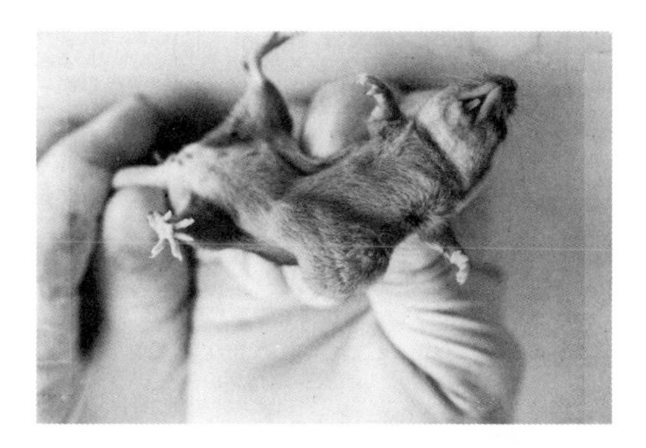

Photograph I. Typical mouse with tumor not treated with chemiluminescence system.

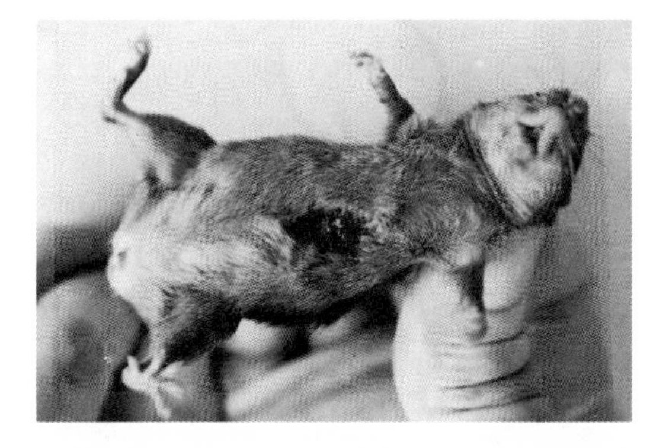

Photograph II. Typical reduction in tumor volume upon treatment with HPD and chemiluminescence system.

ACKNOWLEDGEMENTS

The authors wish to express their appreciation to Dr. Arthur G. Mohan, American Cyanamid Company, for his assistance with the oxamide chemiluminescence system. A. Paul Schaap also gratefully acknowledges support from the U.S. Office of Naval Research (N00014-82-K-0696).

REFERENCES

Auler, H. and Banzer, G. (1942) Untersuchungen uber die Rolle der Porphyrine bei Geschwulstkranken Menschen und Tieren. Z. Krebsforsch. 53:65.

Dougherty, T., Grindey, G., Fiel, R., Weishaupt, K., and Boyle, D. (1975) Photoradiation therapy II. Cure of animal tumors with hematoporphyrin and light. J. Natl. Cancer Inst. 55:115.

Dougherty, T., Kaufman, J., Goldfarb, A., Weisshaupt, K., Boyle, D., and Mittleman, A. (1978) Photoradiation therapy for the treatment of malignant tumors. Cancer Res. 38:2628.

Figge, F. H. J., Weiland, G. S., and Manganiello, L. O. (1948) Cancer detection and therapy: affinity of neoplastic, embryonic and traumatized tissues for porphyrins and metalloporphyrins. Proc. Soc. Exp. Biol. Med. 68:640.

Gregorie, H. B., Jr, Horger, E. U., Ward, J., Green, J., Richards, T., Robertson, H., and Stevenson, T. (1968) Hematoporphyrin-derivative fluorescence in malignant neoplasms. Ann. Surg. 167:820.

Granelli, S., Diamond, I., McDonough, A., Wilson, C., and Nielsen, S. (1975) Photochemotherapy of glioma cells by visible light and hematoporphyrin. Cancer Res. 35:2567.

Kelly, J. F., Snell, M. E., and Berenbaum, M. C. (1975) Photodynamic destruction of human bladder carcinoma. Br. J. Cancer 31:237.

Lipson, R. L., Baldes, E. J., and Olsen, A. M. (1961) The use of a derivative of hematoporphyrin in tumor detection. J. Natl. Cancer Inst. 26:1.

Moan, J., Pettersen, E., and Christensen, T. (1979) The mechanism of photodynamic inactivation of human cells _in vitro_ in the presence of hematoporphyrin. Br. J. Cancer 39:398.

Phillip, M. J., Lewis, A. J., and Daily, N. H. (1971) Effects of a coupled tumor protein antigen (Lewis CTPA) on the growth of transplanted tumors in inbred mice. Oncology 25:528.

Phillip, M. J., Cultrona, S. M., and Porcelli, D. L. (1973) An immunological study on the effects of a coupled tumor protein antigen on the growth of transplanted tumors in inbred mice. Oncology 28:306.

Raab, O. (1900) Uber die Wirkung Fluorescirender Staffe auf Infusoriea. Z. Biol. 39:524.

Rauhut, M. M., Roberts, B. G., and Semsel, A. M. (1966) A study of chemiluminescence from reaction of oxalyl chloride, hydrogen peroxide, and fluorescent compounds. J. Am. Chem. Soc. 88:3604.

Rauhut, M. M., Roberts, B. G., Maulding, D. R., Bergmark, W., and Coleman, R. (1975) Infrared liquid-phase chemiluminescence from reactions of bis(2,4,6-trichlorophenyl)-oxalate, hydrogen peroxide, and infrared fluorescent compounds. J. Org. Chem. 40:330.

Tseng, S. S., Mohan, A. G., Haines, L. G., Vizcarra, L. S., and Rauhut, M. M. (1979) Efficient peroxyoxalate chemiluminescence from reaction of N-(trifluoromethylsulfonyl)-oxamides with hydrogen peroxide and fluorescers. J. Org. Chem. 44:4113.

Weisshaupt, K., Gomer, C., and Dougherty, T. (1976) Identification of singlet oxygen as the cytotoxic agent in photoinactivation of a murine tumor. Cancer Res. 36:2326.

Winkelman, J. and Rasmussen-Taxdal, D. S. (1969) Quantitative determination of porphyrin uptake by tumor tissue following paranterol administration. Bull. Johns Hopkins Hosp. 107:228.

Porphyrin Localization and Treatment of Tumors, pages 571–581

TISSULAR DISTRIBUTION AND PHARMACOKINETICS OF HEMATOPORPHYRIN DIACETATE ON RATS

Gervais P., Band P., Moisan R., Mailhot S.,
Besner J.G.
Faculty of Pharmacy, University of Montreal
C.P. 6128, Succ. A, Montreal
Quebec, Canada H3C 3J7

Hematoporphyrin derivative (HpD) from Dougherty (HpDD) (Dougherty et al 1978) is a mixture of porphyrins obtained after alcali treatment of HpD from Lipson (HpDL). HpDL is prepared by treating commercial hematoporphyrin dihydrochloride (Hp) with sulfuric acid in glacial acetic acid as illustrated in fig. 1. HpDD is known for its rapid growing tissue and tumor accumulation properties (Barker et al 1970), for the emission of fluorescence from a tumor irradiated with ultra-violet light (Profio et al 1977) and for the cytotoxic photochemical action induced by singlet oxygen (Weishaupt et al 1976) when irradiated with red light at 630 nm (Gomer 1980, Kessel 1977). The energy source is an argon dye laser and the light is usually carried by an endoscope (Kinsey et al 1978). All these properties are used for treating patients with lung tumors at different stages of differentiation (Dougherty et al 1978, Hayata et al 1982).

HpDD is a complex mixture of porphyrins in which the hypothetical active portion is a dimer or oligomer (Berenbaum 1982), and little is known about its fundamental properties and pharmacokinetics. Our work has been done with hematoporphyrin diacetate (HpDA) which has been reported to be the main product of the reaction producing HpDD (Clezy et al 1980); it has also been reported to be capable of selective accumulation in malignant tissue (Henderson et al 1980, Berenbaum et al 1982) and the advantages of working with a pure substance are obvious. This paper reports the plasma and bile pharmacokinetics following single intravenous injection, and the specific tumor localizing property of an HpDA *in vivo* metabolite in a transplantable rat tumor.

MATERIALS AND METHOD

HpDA was obtained by 4 consecutive reactions carried on Hp; the process and yield are illustrated in figure 1. Following the forth reaction, the mixture was constituted of 95% HpDA and 5% hematoporphyrin monoacetate (HpMA). All constituants can be eluted and separated on an HPLC system. The intravenous solution was prepared by dissolving 40 mg of HpDA in 2 ml of ethanol, and 2 ml of 0.5 M acetate buffer pH adjusted at 6.8 was added slowly. This solution was stable for 4 hours. The use of sodium dydroxide to solubilized HpDA was avoided to prevent rapid hydrolysis of acetyl groups.

Two groups of 15 Holtzman rats weighing 300 to 450 g were used for the pharmacokinetics studies. In the first group the bile duct was canulated with PE-10 tubing and the carotid artery with PE-50 tubing. In the second group only the carotid artery was canulated. A single dose of 10 mg per kg of HpDA had been administered via jugular vein. Blood samples were obtained from the carotid artery before administration every 2 min. for 15 min., then every 5 min. for 15 min. and finally every 10 min. for 60 min. Bile was collected in weighed test tubes every 10 min. for the first hour and every 20 min. for the next 2 hours.

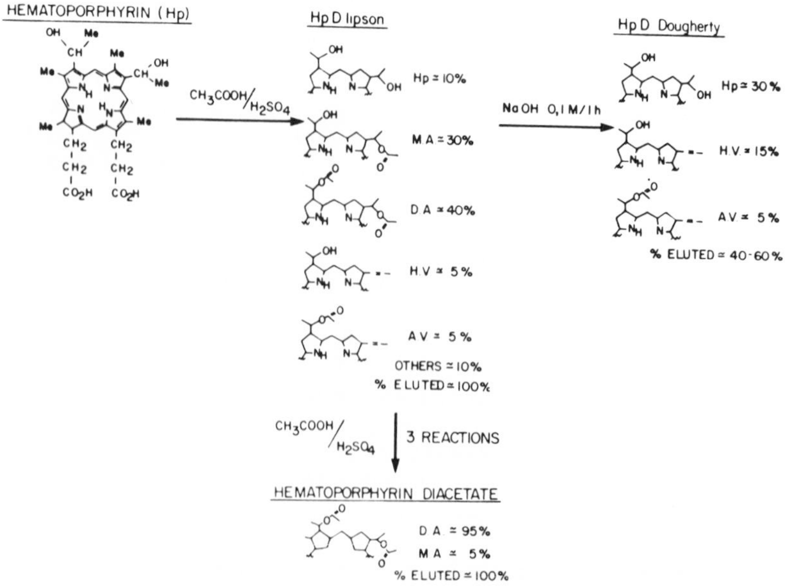

HEMATOPORPHYRIN (Hp)

CH_3COOH/H_2SO_4

HpD lipson

Hp = 10%

M.A = 30%

D.A = 40%

H.V = 5%

A.V = 5%

OTHERS = 10%

% ELUTED = 100%

NaOH 0,1 M/1h

HpD Dougherty

Hp = 30%

H.V = 15%

A.V = 5%

% ELUTED = 40-60%

CH_3COOH/H_2SO_4 3 REACTIONS

HEMATOPORPHYRIN DIACETATE

D.A = 95%

M.A = 5%

% ELUTED = 100%

FIGURE 1: SCHEMATIC SYNTHESIS OF D.A. AND 2 DIFFERENTS HpD

For the porphyrin tissue distribution, we used forty Fisher 344 female rats (90 to 110 g) with a subcutaneously implanted AC 3230 experimental tumor obtained from Mason Research Institute (Worchester, Mass.). Once tumors had attained the optimum size of 1 x 2 cm (approximately 14 days after implantation), 8 groups of 5 animals were sacrificed at 3, 24, 36, 48, 60, 72, 84 and 96 hours after administration of a single intravenous dose (40 mg of HpDA per kg). Analyses were performed on liver, kidneys, heart, lungs, tumor and underlying muscle. Tissue samples were weighed, homogenized in bidistilled water and the concentration was adjusted at 100 mg of fresh tissue per ml with bidistilled water. A post-experimental group (7 cells of 3 rats) was sacrificed at 54, 60, 66, 72, 78, 84 and 96 hours to confirm the special tumor distribution pattern.

For the pharmacokinetics study, 10 μl of plasma, obtained after centrifugation at 2000 G for 10 min. and 1 μl of bile were diluted in 2 ml of HCL 1.7 N and analyzed by spectrofluorometry. Excitation and emission wavelengths were 395 and 630 nm respectively. The pharmacokinetics initial estimates from plasma and bile data were obtained from AUTOAN subroutine and final estimates were calculated with the nonlinear regression program NON-LIN (Wagner 1975).

For tissue distribution studies, 1 ml of tissue homogenate was placed in a glass test tube with bakelite screw-caps (for HPLA analysis, 10 μl of coproporphyrin were used as internal standard); 1 ml of methanol, 2 ml of acetate buffer 0.5 M, pH = 4.0, and 2 ml of a mixture of ethyl acetate/methanol (in proportion of 9:1) respectively were added. Tubes were shaken for 5 min. then centrifuged at 800 G for 5 min., the organic phases were then pipetted into glass test tubes. The aqueous phases were reextracted with 2 ml of fresh mixture of ethyl acetate/methanol (9:1). After agitation (5 min.) and centrifugation (5 min.) the organic phases were combined and 2 ml of HCl 1.7 were added. Tubes were shaken for one minute, centrifuged and the organic phases were discarded. In most cases, fluorescence was measured with a spectrofluorometer, excitation and emission wavelengths being respectively fixed at 395 and 630 nm. When samples were analysed by HPLC, 3 ml of ammonium acetate 10 M were added within 2 min. from HCl addition and after adding 2 ml of ether, tubes were shaken for 5 min., rapidly centrifuged and the aqueous phases were discarded. Ether was evaporated under nitrogen flow in darkness.

Samples were solubilized with 100 µl of methanol and 25 µl were used for the HPLC assay. Our HPLC system was equipped with a constant flow pump (2 ml/min.), a rheodyne injector and a variable wavelength spectrophotometer set at 395 nm. The mobile phase used was a mixture of 63% of acetonitrile and 37% of ammonium acetate 0.1 M and methano- sulfonic acid 0.01 M, pH adjusted at 3.1. The column used was a 250 x 4 mn i.d. stainless column packed with ODS-2 5 µm and temperature was kept constant at 50°.

RESULTS AND DISCUSSION

A distinction has to be made between the two HpD compounds depending on whether the derivative was hydrolysed or not. Figure 1 shows the difference in composition and in percentage of eluted material between HpDL and HpDD. The two main compounds of HpDL are excreted in bile and involved in enterohepatic reabsorption. All the components are eluted and separated by HPLC. Components of HpDD are nonpolar, so enterohepatic reabsorption should not occur; furthermore 40 to 60% of the mixture is not eluted by HPLC and remains strongly absorbed on the column packing.

After intravenous administration, the plasma concentration curves presented 2 exponential phases described by an open 2 compartment model with i.v. injection; only 2 sets of data presented one exponential phase. Figure 2 repre-

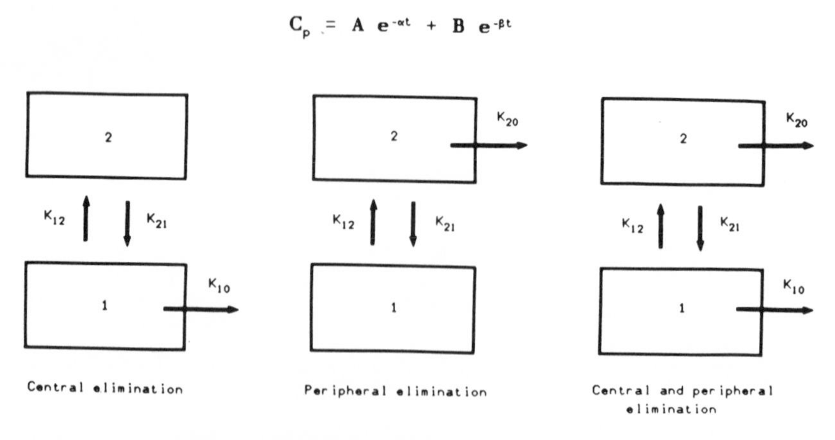

FIGURE 2: THEORICAL PHARMACOKINETICS 2 COMPARTMENT MODELS

sents 3 different pharmacokinetic models for which the distribution and elimination after i.v. injection are described by a bicompartment model. HpDA pharmacokinetic is described by the model with central elimination because the liver is included in the central compartment and elimination is mainly hepatic.

Logarithms of plasma concentration curves versus time are presented in figure 3. In both cases (with and without bile duct cannulation) distribution (α) and elimination (β) phases are rapid. Pharmacokinetic parameters describing

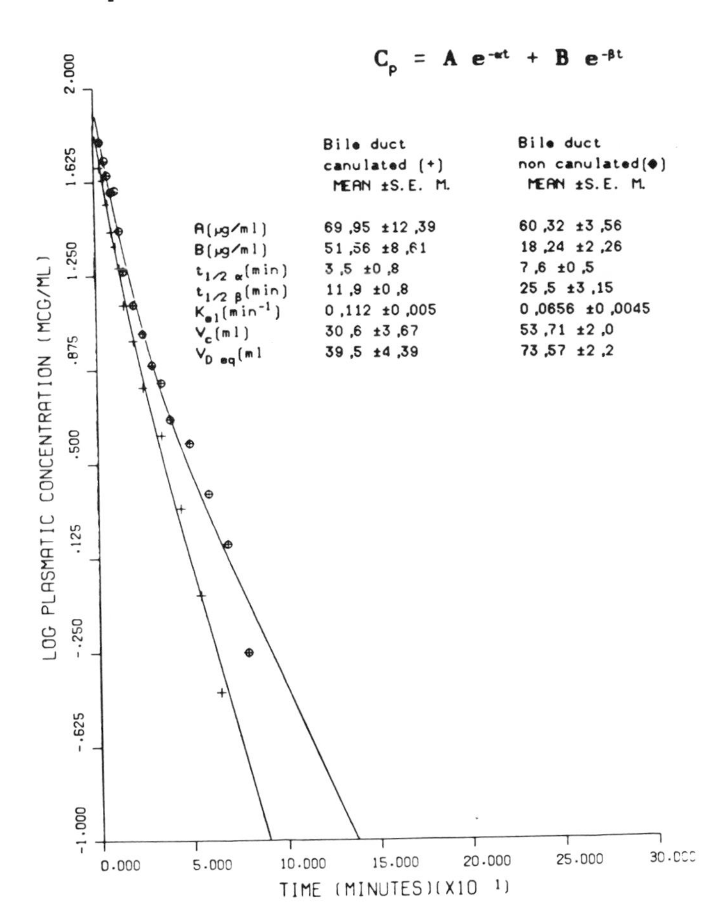

FIGURE 3: LOG OF PLASMA CONCENTRATION OF PORPHYRINS WITH (+) AND WITHOUT (◆) BILE DUCT CANULATION AFTER SINGLE I.V. DOSE (10 mg of D.A./kg)

those curves are all statistically different except the α intercept (A). These differences stress the importance of biliary excretion and reabsorption of HpDA on plasma kinetics. Molecular weight of 682 and lipophilic nature of HpDA are probably responsible for the enterohepatic reabsorption in the first hours following administration. Figure 4 illustrates the logarithm of biliary excretion rate of porphyrins versus time. For animals the mathematical model is described with a lag time of 2.6 min. before the onset of biliary elimination, and for the others, the curves have been forced through zero when stripped and no lag time was needed. The biliary excretion rate curve is described by 3

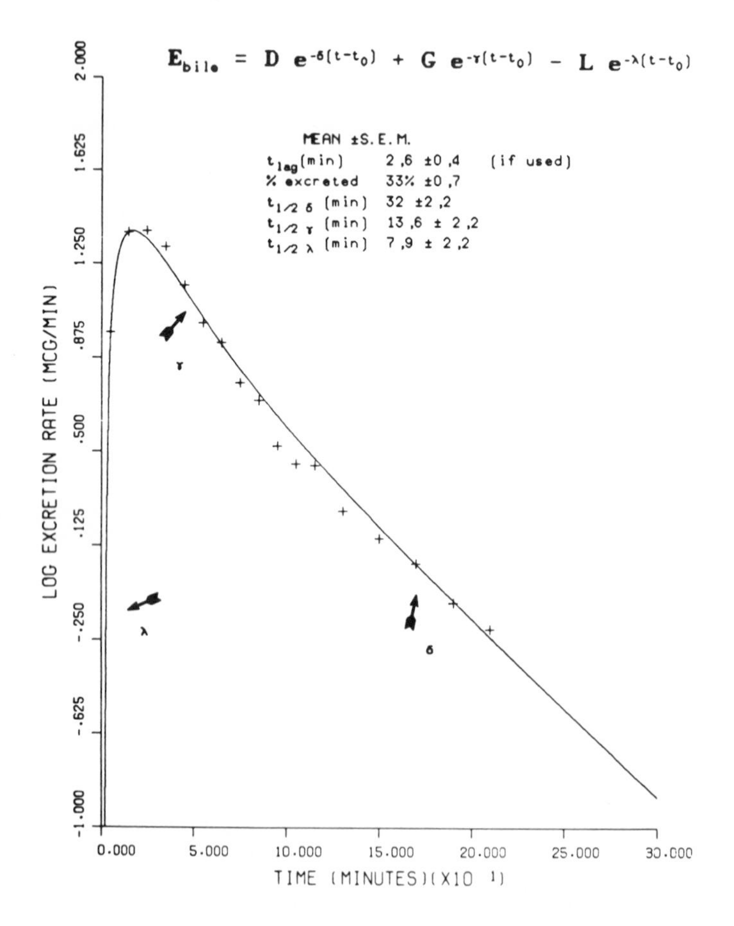

FIGURE 4: LOG OF BILIARY EXCRETION RATE VS TIME OF PORPHYRINS AFTER A SINGLE I.V. DOSE (10 mg D.A./kg)

exponentials: the "λ" appearance phase has a half-life of 7.9 min. and is followed by a "γ" distribution phase of 13.6 min. half-life and by a "δ" elimination phase of 32 min. half-life. The maximum excretion rate (t_{max}) occurs at 25 min. The pattern of excretion rate is described by the same pharmacokinetic model used for plasma kinetics. After 3 hours the bile concentration of porphyrins is barely detectable and approximately 30% of the dose is excreted.

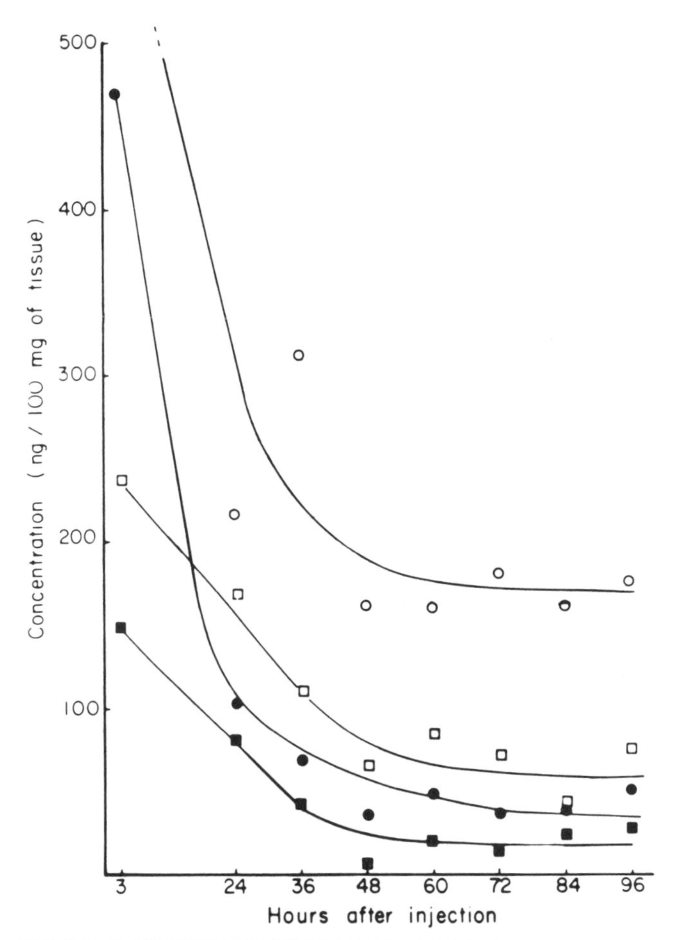

FIGURE 5: TISSUE DISTRIBUTION IN LIVER (o—o), KIDNEYS (□—□), HEART (■—■) AND LUNGS (●—●) OF D.A. (40 mg/kg)

Almost 70% of the dose is distributed and strongly bound in tissues and only a very sensitive analytical method will permit the determination of the plasma concentrations after 3 hours.

Figure 5 represents the porphyrin distribution versus time in four tissues. After injection, porphyrins are slowly cleared and concentrations reached a plateau after 48 hours, levels were then 1.75, 0.6, 0.4 and 0.25 μg per g of liver, kidney, lung and heart. The elimination pattern in neoplastic tissue and muscle (fig. 6) is almost the same, but at 60 hours the neoplastic porphyrin levels rise to reach a maximal concentration of 1.25 μg per g of tumor at 72 hours for group 1. The same experiment performed on a second group gave similar results: the maximum value is slightly less (1 μg/g) and occurs at 78 HpDA hours after

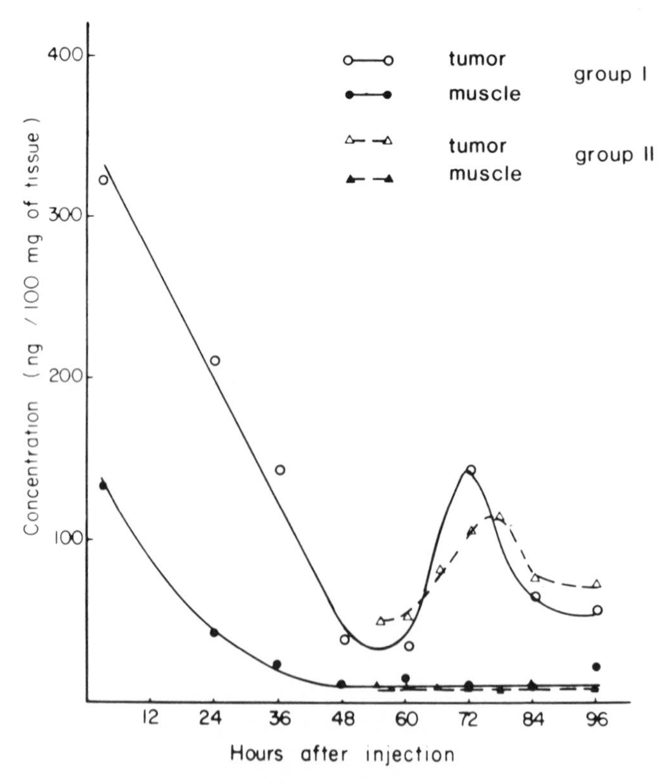

FIGURE 6: TUMOR AND MUSCLE DISTRIBUTION OF D.A. (40mg/kg)

HpDA administration. This difference could be explained by a tumor degeneration due to several transplantations.

All tissue was analysed by spectrofluorometry and tumor content at 3, 24, 48 and 72 hours after administration was also analysed by HPLC. The last technique is specific to each known porphyrin. Hp, HpMA and HpDA levels decreased gradually and 72 hours after administration, no porphyrins were eluted by the system but fluorescent material was still detectable by spectrofluorometry. Chromatogram 1 (fig. 7) represents a tumor extract at 72 hours and only the coproporphyrin used as internal standard was eluted; chromatogram 2 represents an extract of a spiked tumor (125 ng of HpDA/ 100 mg) and coproporphyrin. The peak is sharp with no interference and clearly demonstrates that the absence of

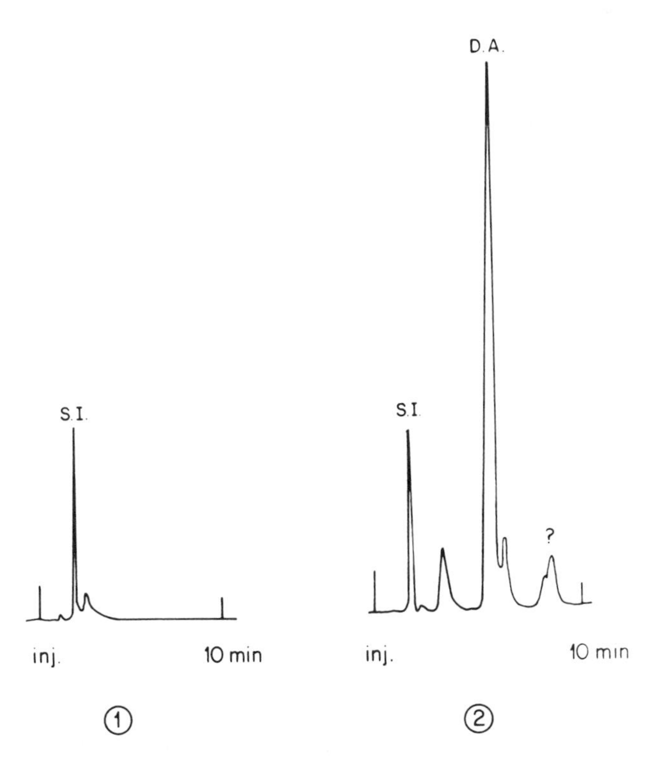

FIGURE 7: CHROMATOGRAM OF TUMORAL CONTENT AT 72 HOURS (1) AND EXTRACT OF TUMOR SPIKE (125 ng of D.A./100mg) (2)

known porphyrins in chromatogram 1 is not an artefact. One must conclude that the intense fluorescence of the tumor is due to a transformed product adsorbed on the column.

CONCLUSION

The plasma elimination half-life is very short (less than 30 min.) but tissue distribution indicated a strong fixation for at least 4 days. Such a contradiction may be explained only by another elimination phase (γ) with a low tissue plasma constant and high plasma-tissue constant. Only a very sensitive analytical method performed on large samples (1 ml for instance) would permit calculation of such a phase. If an HPLC method is to be used, it should overcome the strong adsorption on the column packing encountered with the porphyrins present in the tumor and in HpDD.

ACKNOWLEDGEMENTS

This work was supported by a Grant of National Cancer Institute of Canada. The authors wish to express their gratitude to the Faculty of Pharmacy, University of Montreal and particularly Dean J. Gagne for moral and financial support.

REFERENCES

Barker DS, Henderson RW, Storey E (1970). The in vivo localization of porphyrins. Br J Exp Path 51:628.

Berenbaum MC, Bonnet R, Scourides PA (1982). In vivo biological activity of haematoporphyrin derivative. Br J Cancer 45:571.

Bonnett R, Ridge RJ, Scourides PA (1980). Haematoporphyrin Derivative, J.C.S. Chem Comm 1198.

Clezy PS, That Hai T, Henderson RW, Thuc L (1980). The chemistry of pyrrolic compounds XLV. Haematoporphyrin derivative: haematoporphyrin diacetate as the main product of the reaction of haematoporphyrin with a mixture of acetic and sulfuric acids. Aust J Chem 33:585-597.

Dougherty TJ, Kaufman JE, Goldfarb A, Weishaupt KR, Boyle D, Mittleman A (1978). Photoradiation therapy for the treatment of malignant tumors. Can Res 38:2628.

Gomer CJ (1980). DNA damage and repair in CHO cells following hematoporphyrin photoradiation. Cancer Letters 11:161-167.

Hayata Y, Kato H, Konaka C, Ono J, Takizawa N (1982). Hematoporphyrin derivative and laser photoradiation in the treatment of lung cancer. Chest 81:269-277.

Henderson RW, Christie GS, Clezy PS, Lineham J (1980). Haematoporphyrin diacetate: a probe to distinguish malignant from normal tissue by selective fluorescence. Br J Exp Path 61:345.

Kessel D (1977). Effects of photoactivated porphyrins at the cell surface of Leukemia L 1210 cells. Biochem Vol 16(15):3443-3449.

Kinsey JH, Cortese DA, Sanderson DR (1978). Detection of hematoporphyrin fluorescence during fiberoptic bronchoscopy to localize early bronchogenic carcinoma. Mayo Clin Proc 53:594-600.

Lipson RL, Baldes EJ (1960). The photodynamic properties of a particular hematoporphyrin derivative. Arch Dermatol 82:508.

Profio AE, Doiron DR (1977). Fluorescence bronchoscopy for localization of small lung tumors. Phys Med Biol Vol 22 No 5:949-957.

Ritschel WA (1980). Single dose pharmacokinetics. In Handbood of Basic Pharmacokinetics, Second Edition. Drug Intelligence Publications, Hamilton Ill.

Rowland M, Tozer TN (1980). Hepatic clearance and elimination, in Clinical Pharmacokinetics, Concepts and Applications, Lea and Febiger, Philadelphia.

Wagner JG (1975). Fundamentals of Clinical Pharmacokinetics. Drug Intelligence Publications, Hamilton, Ill.

Weishaupt KR, Gomer CJ, Dougherty TJ (1976). Identification of singlet oxygen as the cytotoxic agent in the photo-inactivation of a murine tumor. Can Res 36:2326-2329.

Porphyrin Localization and Treatment of Tumors, pages 583–590

DOSE EFFECT RELATIONSHIPS IN A MOUSE MAMMARY TUMOR

Fred W. Hetzel, Ph.D., and Harvey Farmer, B.A.

Therapeutic Radiology
Henry Ford Hospital
Detroit, MI 48202

INTRODUCTION

For a number of years various researchers have studied the effects of HpD on murine tumor systems (for example, Dougherty, et al. 1983; Dougherty, et al. 1975). While the object of many of these studies has been to assess tumor regression, only a few have reported long term "cures" (Dougherty, et al. 1983). The object of this study was to employ a well defined murine tumor system in a standard series of radiobiological studies to determine tumor regrowth parameters, tumor cure (TCD_{50}) and concomitant normal tissue responses as a function of drug and/or light dose. Also, in parallel experiments, direct microphysiological measurements of intralesional oxygen tension were obtained under controlled parameter conditions in an attempt to elucidate at least one of the possible mechanisms accounting for tumor response.

MATERIALS AND METHODS

Animals and Tumor Systems

A C_3H (SED/BH) mouse system with an implanted mammary carcinoma has been employed in all experiments. The radiobiological properties of this system have been well defined by Suit and Maeda (Suit, Maeda 1967). The routine method for handling the above system in this laboratory has been reported (Bicher, et al. 1980) and hence will be discussed only briefly. In the mouse system the tumor is transplanted in the lower section of the right hind limb which allows us to use a specially designed holder accomodating three or more mice for treatment simultaneously. At the time of treatment all tumors were 6 and 7 mm in diameter since data obtained here and that of Sedlacek (personal communication) indicates that when the tumor diameter reaches 8mm, animal death may result from metastatic disease not related to the treatment of the primary tumor.

HpD Preparation and Administration

HpD is obtained from ORD Inc. Buffalo, N.Y. Final dilution of the HpD is in sterile, isotonic saline. In any given experimental series an adjustment is always made with saline to ensure that the total bolus volume is constant as is the dose (mg/kg). The drug is injected either intraperitoneally or intravenously (tail vein) depending upon the protocol.

Light Sources and Exposures

An arc lamp filtered with special filters is used as the source of red light for in vivo work. An Oriel 2500 watt arc lamp with a mercury lamp and a 4-inch f/0.7 lens system is employed because of its large treatment field size. The arc lamp output is filtered through an IR filter. The filters were obtained from PTR optics and Fred S. Hickey Corp. The power (mW/cm^2) is checked prior to each experiment and unless otherwise noted was 150 mW/cm^2.

Oxygen Measurements

The O_2 ultramicroelectrodes employed were the "gold in glass" type, as described by Cater and colleagues (Cater, et al. 1960) and used extensively by this laboratory in previous work (Bicher, et al. 1980; Kaufman, et al. 1981, Hetzel, et al. 1981). They are made by pulling a glass tube (KG-33,ID 1.5mm, OD 2.0mm, Garner Glass Co., Claremont, California), encasing a 20μ gold wire (Sigmund Cohn Corp., Mt. Vernon, New York) in a David Kopf Model 700C vertical pipette puller. The exposed gold tip is about 10μ in diameter, and is coated with Rhoplex (Rhom Haas,

Philadelphia, Pennsylvania) membrane as previously described (Hetzel, et al. 1981). The electronic circuitry to measure the polarographic current is provided by a Model 1200 Chemical Microsensor System (Transidyne General Corporation, Ann Arbor, Michigan) and the results are recorded on a Model T Grass polygraph.

Electrodes were calibrated as described (Kaufman, et al. 1981) in buffered saline solutions of known PO_2 values. The electrodes are then 'conditioned' by placing them in buffered saline and applying 0.8 V potential for 2 hrs. After this treatment they are usually very stable. The current reading at zero oxygen tension is very low (residual current) and the response of the microelectrode to changes of oxygen tension is very rapid (95 percent response time of the order of 0.5 sec.)

RESULTS

In the majority of experiments our desired end point was the determination of the TCD_{50} dose (i.e., the dose which

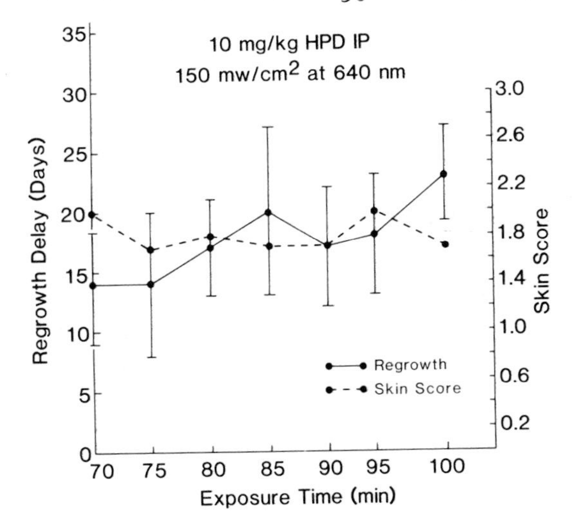

Figure 1. Tumor regrowth (solid line, left abscissa) and skin score (broken line, right abscissa) as a function of exposure time to 150 mW/cm^2 light following a 10 mg/kg IP injection. Each data point was obtained from a minimum of 6 mice.

will result in 50% cures after 120 days). Throughout all of the experiments, concomitant measurements of normal tissue response (skin score) were also made on a scale from 0 (no reaction) to 3.5 (loss of leg). While the ultimate establishment of the TCD_{50} values has yet to be obtained, a second parameter-regrowth delay has been obtained. Regrowth delay is, for our purposes, defined as the time required for the treated tumor to regrow to twice its original treatment size. For the 6 to 7mm tumors treated this regrowth size has been arbitrarily established at 15 mm. Figure 1 shows the results obtained when the animals were injected IP with 10 mg/kg HpD and treated 24 hours later with graded exposure times to red light (150 mW/cm^2). Regrowth delay for 70 minutes of treatment was 14 days (not significantly different than untreated controls) and increased up to 23 days following a 100 minute treatment. It is important to note that no significant increase in normal tissue reaction was seen over this dose range. Another interesting observation that was made in the experiment was the time pattern of response. In all cases there was an almost immediate (within 24 hours) erythema over the entire treated region. The magnitude of the erythema was dose dependent but the

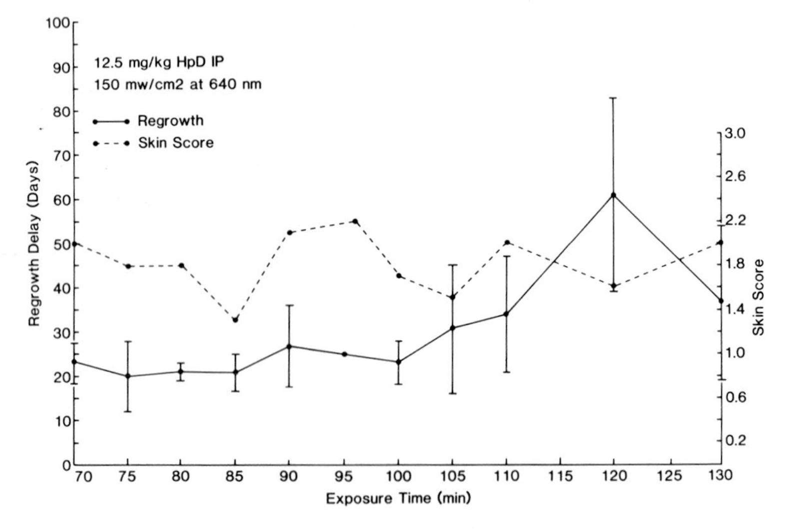

Figure 2. Tumor regrowth (solid line, left abscissa) and skin score (broken line, right abscissa) as a function of exposure time to 150 mW/cm^2 light following a 12.5 mg/kg IP injection. Each data point was obtained from a minimum of 6 mice.

appearance time and regression time (7-10 days) seemed independent of dose.

Figure 2 shows the results of a more rigorous course of treatment. In this study the animals received 12.5 mg/kg and were again treated 24 hours after injection. Although there is some scatter in the data, it is clear that a significant improvement in tumor response is obtained in the series although very few "cures" were obtained. Again, as was seen in previous studies, no change is observed in the severity of the normal tissue response which ultimately will control the therapeutic ratio.

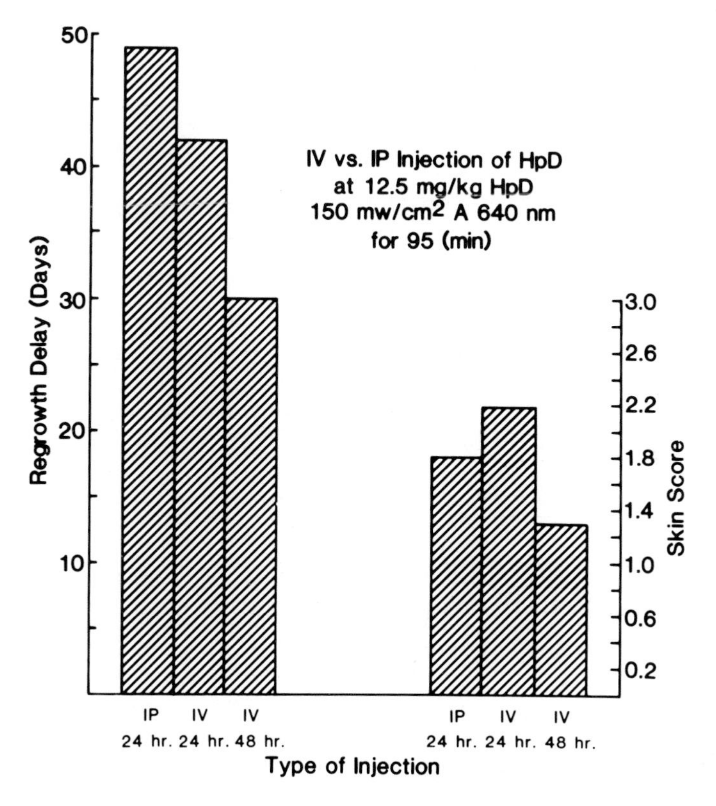

Figure 3. Comparison of tumor response (left) and concomitant normal tissue reaction (right) for either IV or IP injections and various time intervals between injection and treatment.

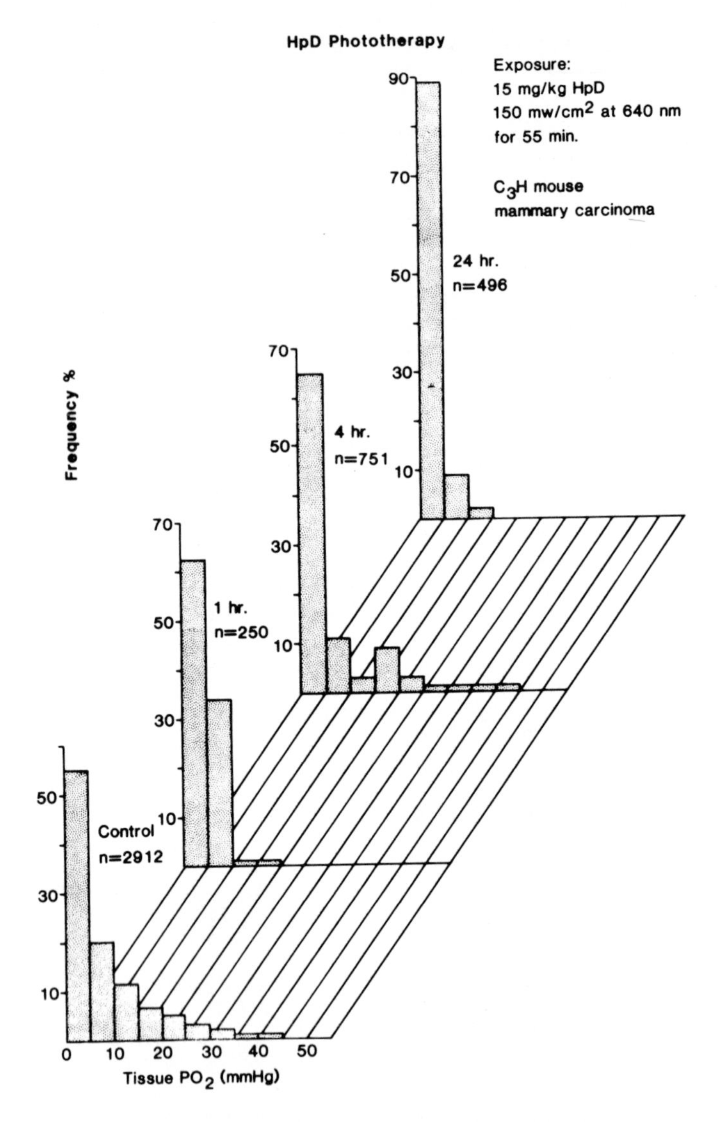

Figure 4. Frequency distribution of intratumor oxygenation before and 1, 4 and 24 hours post phototherapy. Values below 10 mm Hg O_2 are considered radiobiologically hypoxic.

A third experimental series was developed to investigate the variation in response in both normal and tumor tissue as the route of injection or the treatment time post injection were varied. The results are shown in Figure 3. While it would appear that both the IV and IP administrations gave similar, good tumor responses (treatment 24 hours post injection), the magnitude of the normal tissue damage also has to be considered. This is most evident when the 72 hour post IV results are examined. In this case a good tumor response was obtained (with 120 minute exposure) while the normal tissue effects were greatly reduced. Experiments are now in progress in which the tumor dose will be adjusted to achieve an isoeffect response in the skin. Only then can the true therapeutic gain be obtained.

Finally, in an attempt to elucidate one of the possible mechanisms of action of phototherapy on the physiological characteristics of the tumor we determined intratumor oxygen levels prior to and at various times following a single treatment. The results are shown in Figure 4. The results are very dramatic since almost all areas within the tumor become extremely hypoxic after treatment. Equally important, there seems to be no reoxygenation of the tumor for up to 24 hours post treatment.

DISCUSSION

The results presented in Figures 1 and 2 show clearly that a true dose-effect relationship exists for HpD phototherapy with a curve shape similar to that seen for ionizing radiation (Hall 1978). It should, therefore, be possible to establish true TCD_{50} values for this treatment dependent upon total drug and/or light dose. It is also clear from the data presented that conditions can be selected such that normal tissue damage can be reduced while still achieving good tumor response. This possibility is being pursued under more controlled conditions.

The intratumor oxygen profile determinations are extremely interesting. Earlier reports indicate that intratumor coagulation necrosis is observed following treatment (Bugelski, Porter, Dougherty 1981). The results presented here indirectly confirm these observations since a cessation in blood flow will coincide with the development of the observed hypoxia. The persistence of the hypoxia also suggests that revascularization and hence, reoxygenation is a slow phenomenon. This may have great significance when thoughts are directed to combining HpD phototherapy with other modalities in particular radiation therapy or chemotherapy where effectiveness would likely be reduced or with hyperthermia where enhancement could be expected. Experiments are in progress to examine these possibilities.

REFERENCES

Bicher HI, Hetzel FW, Sandhu TS, Frinak S, Vaupel P, O'Hara O'Brien T (1980). Effects of hyperthermia on normal and tumor microenvironment. Radiol 137:523.

Bugelski PJ, Porter CW, Dougherty TJ (1981). Autoradiographic distribution of hematoporphyrin derivative in normal and tumor tissue of the mouse. Cancer Research 41:4606.

Cater DB, Silver IA, Wilson GM (1959, 1960). Apparatus and techniques for the quantitative measurement of oxygen tension in living tissues. Proc Royal Soc London, Series B, 151:256.

Dougherty TJ, Grindey GB, Fiel R, Weishaupt KR, Boyle DG (1975). Photoradiation therapy. II Cure of animal tumors with hematoporphyrin and light. J Natl Cancer Inst 55:1 p 115.

Dougherty TJ, Boyle DG, Weishaupt KR, Henderson BA, Potter WR, Bellnier DA, Wityk KE (1983). Photoradiation therapy - clinical and drug advances. In Kessel D, Dougherty TJ (eds): "Porphyrin Photosensitization", New York: Plenum Press, p 3.

Hall EJ, (1978). "Radiobiology for the Radiobiologist 2nd ed." Hagerstown: Harper and Row, p 222.

Hetzel FW, Brown M, Kaufman N, Bicher HI (1981). Radiation sensitivity modification by chemotherapeutic agents. Cancer Clin Trials 4:177.

Kaufman N, Bicher HI, Hetzel FW Brown M (1981). A system for determining the pharmacology of indirect radiation sensitizer drugs on multicellular spheroids. Cancer Clin Trials 4:199.

Suit HD, Maeda M (1967). Hyperbaric oxygen and radiobiology of a C_3H mouse mammary carcinoma. J Natl Cancer Inst 39:639.

Porphyrin Localization and Treatment of Tumors, pages 591–600

DOCUMENTATION OF RADIOACTIVE CONTAMINANTS IN TRITIUM LABELED HEMATOPORPHYRIN DERIVATIVE

C.J. Gomer[1,2,3] and F.M. Little[4]

Clayton Center for Ocular Oncology (1), Childrens Hospital of Los Angeles and Departments of Pediatrics (2), Ophthalmology (3) and Neurosurgery (4), USC School of Medicine, Los Angeles, CA 90027

INTRODUCTION:

The continued use of hematoporphyrin derivative (HpD) for the photosensitization of solid tumors has lead to an increased need to obtain quantitative measurements of this drug in tissues and cells (1). Radioactively labeled HpD (^{14}C or ^{3}H) has been used in the past for this purpose (2,3). Initial studies comparing the tissue localization of ^{3}H-HpD with that of ^{14}C-HpD suggested that there was minimal tritium exchange occuring with ^{3}H-HpD (2). However, recent experiments have indicated that there can be considerable radioactive degradation of ^{3}H-HpD. During preliminary studies (designed to compare the radioactivity and fluorescence of ^{3}H-HpD in normal and malignant tissue) we observed that qualitative fluorescence observations of HpD localization did not correlate with quantitative ^{3}H-HpD tissue uptake data. This paper describes tissue distribution, fluorescence photography, and gel filtration studies which indicate that ^{3}H-HpD can degrade and produce radioactive contaminants.

METHODS:

<u>Drugs:</u> Sterile HpD and Photofrin II (the putative active component of HpD) were obtained from Oncology Research and Development, Cheektowaga, N.Y. Tritium labeled HpD (^{3}H-HpD) was obtained from New England Nuclear as a custom tritium catalytic exchange procedure (2). A

working solution of ^{3}H-HpD (3.0 mg/ml, 1.6 x 10^4 DPM/ug) was obtained by diluting a stock solution of ^{3}H-HpD with cold HpD. Tritium labeled photofrin II (^{3}H-photofrin II) was obtained by gel filtration of ^{3}H-HpD (as described below). The ^{3}H-photofrin II fraction was lyophilized and then diluted with cold photofrin II. A working solution of ^{3}H-photofrin II had a concentration of 2.5 mg/ml and a specific activity of 1.45 x 10^3 DPM/ug.

Gel Chromatography: Gel filtration of porphyrins was performed using Bio-Gel P-6 columns (Bio-Rad Laboratory, Richmond, CA). The column dimensions were 50 x 1 cm and distilled water was used as the eluent. A flow rate of 10 ml per hour was used and the volume of each fraction was 0.65 ml. Two major porphyrin fractions were obtained from gel filtration with the Bio Gel P-6 column. The fastest moving fraction is dark brown in color and elutes at the void volume of the column. This material has previously been reported to be photofrin II (4). The soret absorption maximum for photofrin II was 368-372 nm (indicating an aggregated material). The soret absorption maximum for the slow eluting component was 395 nm (indicating a monomer or dimer).

Animal and Tumor Systems: Male Fisher 344 rats bearing intracerebral 9L brain tumors were used in all in-vivo experiments (5,6). Porphyrin localization studies were performed 10-13 days after tumor implantation. The weight of individaul tumors ranged from 40 to 160 mg at the time of sacrifice.

Porphyrin Tissue Distribution: Animals received a single I.P. injection (7.5 mg/kg) of either ^{3}H-HpD or ^{3}H-photofrin II and were sacrificed 24 hours following injection. Photograhic documentation of the brain (described below) was obtained and then tissue samples (tumor, normal brain, liver, kidney, spleen, lung and muscle) were collected. These specimens were rinsed with PBS, weighed and then dissolved in protosol (New England Nuclear, Boston, MA). Following solubilization, the solutions were neutralized with 0.5 N HCl and the radioactivity in each sample was determined by standard liquid scintillation counting. Control tissue samples were also counted and served as background measurements. The resulting CPM´s were converted to DPM´s using the method of external standard quenching. The concentration of porphyrins per weight of tissue were then calculated from the specific activity of ^{3}H-HpD or ^{3}H-photofrin II. Three rats were injected with ^{3}H-photofrin II and two rats were injected with ^{3}H-HpD. One or two samples of each test tissue were analyzed from each animal.

<u>Photographic Documentation of Porphyrin Localization:</u> White light and fluorescence photographs were taken of brains (tumor tissue and surrounding normal brain) immediately following sacrifice. A 35 mm Olympus OM 2 camera (fitted with a 50 mm macro lens) was used for all photography. A 590 nm cut off filter was inserted in the lens during all fluorescent photographs. Tissue fluorescence (depicting the areas of porphyrin localization) was obtained by exciting the drug with violet light (407 and 413 nm) generated from a krypton laser (Coherent Radiation). The light fluence at the surface of the tissue was approximately 25 mW/cm^2.

RESULTS

The tissue distribution of ^{3}H-HpD and ^{3}H-photofrin II are shown in Fig. 1. Quantitative levels of porphyrins were obtained in various tissues at 24 hours following a 7.5 mg/kg dose of tritium labeled HpD or photofrin II. The concentration of ^{3}H-HpD in tumor tissue was twice that observed in normal brain tissue (3.75 mg/gm versus 1.85 ug/gm). In contrast, there was a 20 fold increase in the ^{3}H-photofrin concentration in tumor tissue compared to normal brain tissue (5.38 ug/gm versus 0.26 ug/gm). The concentration of ^{3}H-photofrin II in tumor tissue was approximately 1.5 times greater than the concentration of ^{3}H-HpD in tumor tissue.

White light and fluorescence photographs of the tumor and brain tissue which were examined for ^{3}H-HpD or ^{3}H-photofrin II localization are shown in Figures 2 and 3. It is possible to distinguish normal brain from tumor tissue using white light photography. The fluorescent photographs (following administration of either ^{3}H-HpD or ^{3}H-photofrin II) show significant porphyrin induced fluorescence in the tumor tissue and no observable fluorescence in the areas of normal brain tissue.

Figure 4 shows the absorption and radioactivity profiles for each fraction of ^{3}H-HpD separated by gel filtration. The absorption maximum for the first eluted material peak was 370 nm and the absorption maximum for the slower eluting component was 395 nm. There were significant levels of radioactivity which did not correspond to porphyrin absorption peaks.

Figure 5 shows the absorption and radioactivity profiles for each fraction of 3H-photofrin II separated by gel filtration. The radioactivity profile was identical to the absorption profile for the complete separation of ^{3}H-photofrin II.

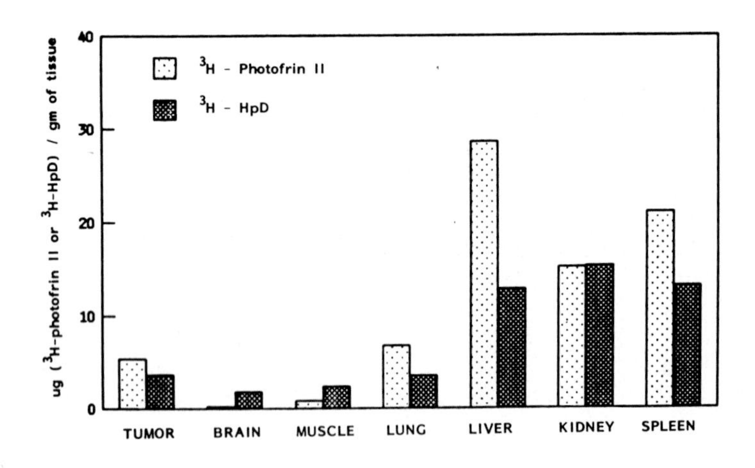

Figure 1. Tissue levels of ^{3}H-HpD and ^{3}H-Photofrin II in Fischer 344 rats 24 hours following an I.P. injection of the drug (7.5 mg/kg).

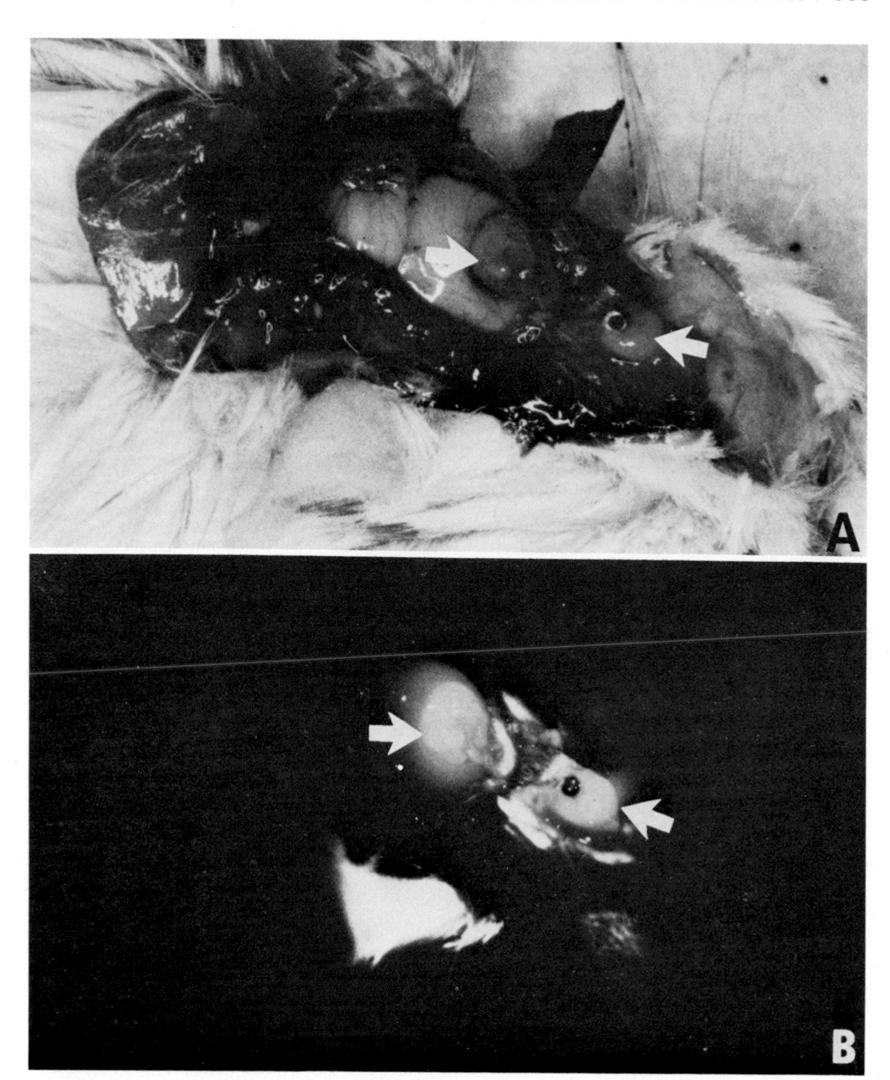

Figure 2. White light (A) and fluorescence (B) photographs of the brain of a Fischer 344 rat with a 9L-brain tumor. Photographs were taken 24 hours following the I.P. injection of ^{3}H-HpD (7.5 mg/kg). Arrow indicates the areas of tumor tissue.

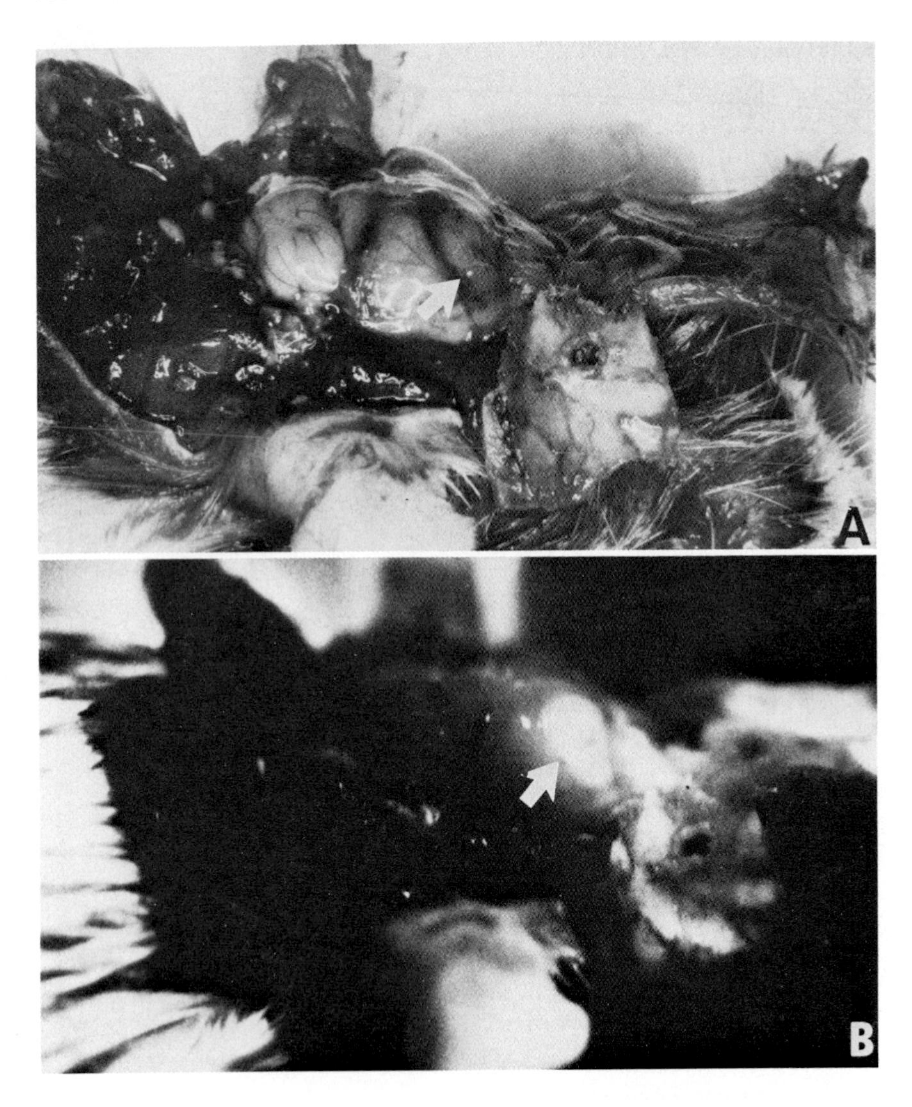

Figure 3. White light (A) and fluorescence (B) photographs of the brain of a Fischer 344 rat with a 9L-brain tumor. Photographs were taken 24 hours following the I.P. injection of ^{3}H-photofrin II (7.5 mg/kg). Arrow indicates the area of tumor tissue.

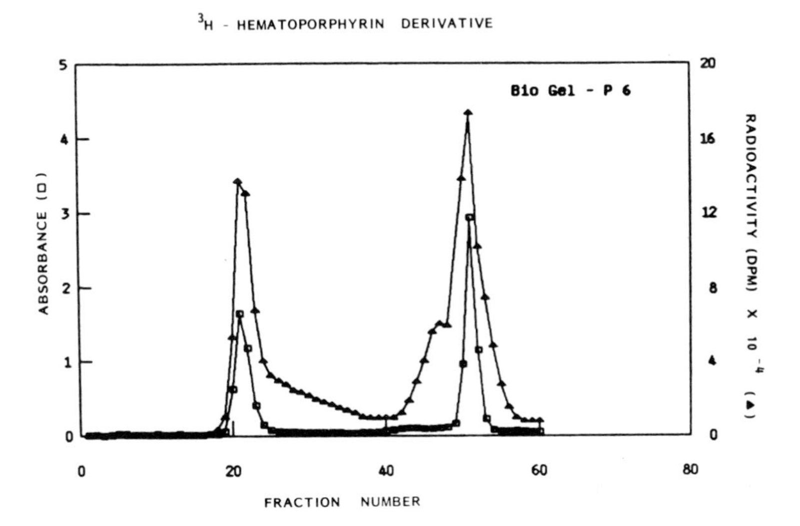

Figure 4. Separation of ^{3}H-HpD by gel filtration. Profiles represent absorbance (□) and radioactivity (▲).

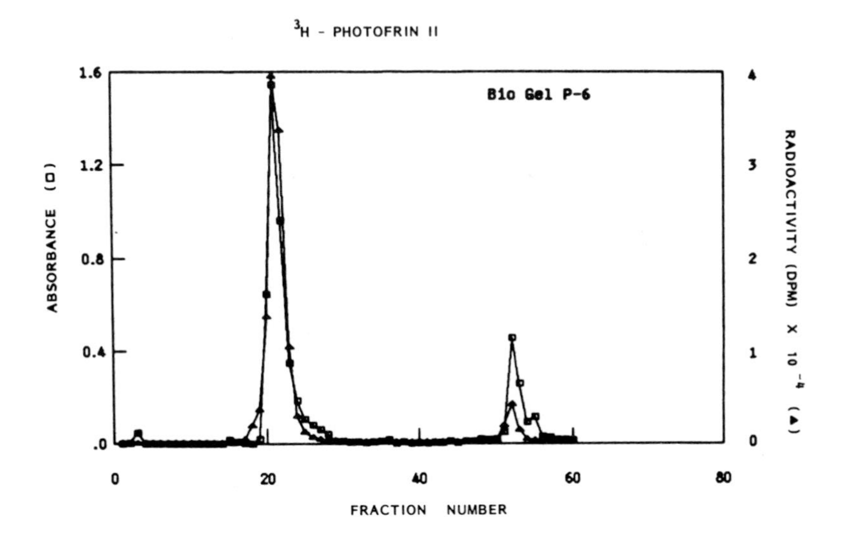

Figure 5. Separation of ^{3}H-photofrin II by gel filtration. Profiles represent absorbance (□) and radioactivity (▲).

DISCUSSION

Quantitative information on drug pharmacology can be obtained accurately and rapidly with radioactively labelled compounds. This procedure requires that the radiochemical purity of the drug is known and that the purity of the drug is reanalyzed on a routine basis. The labeling of HpD with tritium has been performed using the technique of custom catalytic exchange (2). This procedure will undoubtedly result in varying levels of radiation induced decomposion of porphyrin and radiochemical impurities. When ^{3}H-HpD was first tested against ^{14}C-HpD there were no observable differences in tissue distribution which suggested that tritium exchange had not occurred. (2). However, the current results of tissue distribution and chromatographic separation using gel filtration indicate that impurities can be found in ^{3}H-HpD.

The observation of differential HpD fluorescence for tumor and normal brain tissue is in agreement with previously published reports (7, 8). HpD and photofrin II demonstrate a high degree of preferential localization in tumor tissue of the brain when fluorescence is used to document drug distribution. Normal brain tissue did not accumulate large enough quantities of drug to be detected using fluorescence. Previous studies observed this finding and concluded that HpD did not cross the blood brain barrier (9).

The quantitative results of porphyrin uptake in normal and malignant tissue of the brain (obtained using ^{3}H-HpD) did not correlate with the fluorescence data. When ^{3}H-HpD was administered to tumor bearing rats there was no observable porphyrin fluorescence in normal brain tissue but there were significant levels of tritium detected in the same normal brain tissue. However, the quantitative documentation of ^{3}H-photofrin II did correlate with porphyrin fluorescence. The data from gel filtration of both tritium labeled materials provides an explanation for these findings. The separation of ^{3}H-HpD by gel filtration demonstrated significant levels of radioactive contaminants whereas the separation profiles (absorption and radioactivity) for ^{3}H-photofrin II demonstrated a high degree of radiochemical purity.

It is concluded that tritium labeled porphyrins can degrade (probably as a function of time in aqueous solution) and produce radioactive contaminants. The purity of tritium labeled HpD and all porphyrins should be tested repeatedly during the course of all experiments.

ACKNOWLEDGEMENTS:

This investigation was performed in conjunction with the Clayton Foundation for Research and was supported in part by USPHS Grants CA 31230 and CA 31885, awarded by the NCI, DHHS. We thank N.J. Razum for technical assistance, B.C. Szirth for photographic assistance and A.L. Castorena for assistance in the preparation of this manuscript.

REFERENCES

1. Dougherty, T.J., Weishaupt, K.R. and Boyle, D.G., Photosensitizers. In: Cancer Principles and Practices of Oncology. (V. DeVita, S. Hellman and S. Rosenberg, eds.), pp 1836-1844, J.B. Lippincott, Co., Philadelphia, 1982.
2. Gomer C.J. and Dougherty, T.J., Determination of ^{3}H and ^{14}C Hematoporphyrin Derivative Distribution in Malignant and Normal Tissue, Cancer Res., 39:146-151, 1979.
3. Gomer, C.J., Rucker, N., Mark, C., Benedict, W.F. and Murphree, A.L., Tissue Distribution of ^{3}H-Hematoporphyrin Derivative in Athymic "Nude" Mice Heterotransplanted with Human Retinoblastoma. Invest. Ophthalmol. Vis. Sci., 22:118-120, 1982.
4. Dougherty, T.J., Boyle, D.G., Weishaupt, K.R., Henderson, B.A., Potter, W.R., Bellnier, D.A. and Wityk, K.E. Photoradiation Therapy - Clinical and Drug Advances. In: Porphyrin Photosensitization (D. Kessel and T.J. Dougherty, eds.) pp 3-14, Plenum Press, New York, N.Y., 1983.
5. Wierowski, J.V., Thomas, R.R., Ritter, P. and Wheeler, K.T. Critical periods during the in-situ repair of radiation-induced DNA damage in rat cerebellar neurons and 9L brain tumor cells. Radiation Res., 90:479-488, 1982.
6. Barker, M., Hoshino, T., Gurcay, O., Wilson, C.B., Nielsen, S.L., Downie, R. and Eliason, J. Development of an animal brain tumor model and its response to therapy with 1,3-bis(2-chloroethyl)-1-nitrosourea.

Cancer Res., 33:976-986, 1973.

7. Wharen, R.E., Anderson, R.E. and Laws, E.R. Quantitation of Hematoporphyrin Derivative in Human Gliomas, Experimental Control Nervous System Tumors, and Normal Tissues. Neurosurgery, 12:446-450, 1983.
8. Perria, C., Delitala, G., Fraucaviglia, N. and Altomonte, M. The Uptake of Hematoporphyrin Derivative by Cells of Human Gliomas: Determination by Fluorescence Microscopy IRCS Medical Science, 11:46-47, 1983.
9. Wise, B.L. and Taxdol, D.R. Studies of the Blood-Brain Barrier Utilizing Hematoporphyrin. Brain Res., 4:387-389, 1967.

Porphyrin Localization and Treatment of Tumors, pages 601–612

STUDIES ON THE MECHANISM OF TUMOR DESTRUCTION BY PHOTORADIATION THERAPY

Barbara W. Henderson, Thomas J. Dougherty
and Patrick B. Malone
Division of Radiation Biology
Roswell Park Memorial Institute
Buffalo, New York 14263

Although photoradiation therapy (PRT) is being employed in a growing number of institutions for the localized treatment of a variety of malignant tumors, the mechanisms of its cytocidal action and of tumor destruction are still largely unknown. *In vitro* experiments have clearly demonstrated that cells which have been exposed to hematoporphyrin derivative (Hpd), the photosensitizer currently used in PRT, as well as to other porphyrin compounds, are very effectively killed upon subsequent exposure to light (Christensen, et al., 1981; Berns, et al., 1982; Salvatore, et al., 1982; Henderson, et al., 1983). However, attempts to prove *in vitro* preferential binding of photosensitizer by tumor cells and/or increased sensitivity of tumor cells versus normal cells towards photodynamic damage have been inconclusive at best (Christensen, et al., 1981; Berns, et al., 1982; Salvatore, et al., 1982; Henderson, et al., 1983).

In vivo, Hpd seems to accumulate to a higher degree in tumor tissue than in surrounding normal tissue as shown by fluorescence and autoradiographic studies (Cortese, et al., 1979; Bugelski, et al., 1981). However, it has not been proven that the death of tumor cells caused by PRT *in vivo* is the result of direct photodynamic action upon them. On the contrary, macroscopic and microscopic observation of tumors during and after PRT suggests that the initial effect of the photodynamic process may be exerted upon the vasculature. Bugelski, et al. (1981) noted in autoradiographic studies that tritiated Hpd was distributed at a ratio of 5:1 in favor of the vascular stroma in experimental animal tumors, and that vascular damage with escape of red blood cells from the vessels throughout the tumor mass occurred within 15 minutes after PRT. Star, et al. (1982) have reported that the predominant immediate effect of PRT on tumors grown in "sandwich" observation chambers in rats is a reduction or complete stop of blood circulation with subsequent necrosis of tumor tissue. Bicher, et al. (1981) using oxygen electrode measurements found experimental mouse tumors anoxic within 1 hour after PRT apparently due to vascular collapse.

It was the purpose of this study to investigate the relationship between tumor eradication and tumor cell survival following PRT *in vivo*.

MATERIALS AND METHODS

Tumor Systems

The EMT6 tumor line (Rockwell, et al., 1972) was propagated by intradermal inoculation of 2 to 4 x 10^5 cells into the right flank of Balb/c mice. Tumors were used for treatment and/or explantation 8 to 10 days after inoculation when they had attained a size of 5 to 8 mm in diameter. For the purpose of comparison, the SMT mammary carcinoma, carried in DBA/2Ha mice and transplanted by trocar, was used in some experiments.

Preparation of Cell Suspensions and Colony Formation Assay

EMT6 tumors, either untreated or treated, were excised under sterile conditions from mice killed by cervical dislocation. Tumors from two mice were pooled, weighed and minced with scissors. Single cell suspensions were then prepared by agitation of the minced tissue for 30 minutes at 37°C in Hank's balanced salt solution (w/o Ca,Mg) containing the following enzyme mixture: 0.05% pronase (Calbiochem, B Grade, 45,000 PUK/g), 0.01% DNAse I (Sigma, 2650 Kunitz U/mg) and 0.02% collagenase (Millipore Corp., 146 U/mg). Following cell dispersion, the cell suspension was strained through a wire mesh screen (Cellector, E-C Apparatus, St. Petersburg, FL) to eliminate all remaining cell clumps and aggregates, and spun down at 1500 rpm for 10 minutes. Cells were then washed once with Hank's BSS and resuspended in growth medium. The number of cells was assessed by using a Coulter Counter, and the cell yield per g tissue calculated. Cell viability was determined by trypan blue exclusion. Appropriate numbers of cells were then plated for colony formation in the following growth medium: BME with Hank's salts and L-glutamine (Gibco, Grand Island, NY), to which was added 1.5 g per L of sodium bicarbonate. This was further supplemented with 15% fetal bovine serum (also from Gibco) and additional L-glutamine and antibiotics.

Cultures were incubated in the dark in a humidified atmosphere of 5% CO_2 in air for 7 to 8 days, at which time they were fixed, stained, and macroscopic colonies were counted.

In those experiments where cells had been exposed to photosensitizer *in vivo*, cell isolation and plating were carried out in subdued light filtered through a sharp cut filter (Corning, Corning, NY, 1% transmittance at 630 nm, no transmittance below that wavelength).

Photosensitizers

Hpd (Photofrin) was obtained from Oncology Research & Development, Inc. (Cheektowaga, NY). The active component which is responsible for the entire *in vivo* photosensitizing activity of Hpd was obtained by gel exclusion chromatography on P-10 columns (Dougherty, et al., 1983).

In Vitro PRT

Cultured cells in exponential growth, derived from untreated tumors, were exposed to 10 μg/ml of photosensitizer ("active component") in growth medium containing 1% fetal bovine serum for 24 hours in the dark. Following the removal of the porphyrin-containing medium they were exposed to graded doses of light (GTE lamp, 4 mW/cm^2, 590-640 nm). After light exposure the cultures were supplied with complete, porphyrin-free medium and incubated for colony formation as described before.

In Vivo PRT

Tumor-bearing mice were injected i.p. with 7.5 mg/kg

photosensitizer ("active component" or Hpd). Twenty-four hours later tumors were given localized, external light treatment using an Argon dye laser system (75 mW/cm^2, 630 nm, 45 min). Light was delivered to two tumors simultaneously through 200 μM quartz fibers. Following light treatment mice were either killed immediately or at varying times later, and tumors were explanted for cell survival assays as described. For each experiment two additional mice were treated as above, following which they were observed for 30 days to determine tumor response to treatment.

Uptake of Photosensitizers by EMT6 Tumor Tissue

^3H Hpd prepared by catalytic exchange labeling was obtained from New England Nuclear, Boston, MA and prepared for use according to Gomer, et al. (1979). ^3H "active component" was obtained from ^3H Hpd by gel exclusion chromatography as previously described (Dougherty, et al., 1983). Specific activity for both compounds was determined to be 1 mCi/mg. For injection, cold photosensitizer solutions (Hpd or active component) were prepared to give a concentration of 1 mg/ml to which was added tritiated material to give 2 to 3 x 10^7 dpm/mg. Of these stock solutions varying amounts were injected i.p. into tumor-bearing mice (5 mg/kg or 10 mg/kg). Twenty-four hours later the mice were sacrificed by cervical dislocation and tumor tissue was removed and weighed. The radioactivity was determined either by tissue combustion using a Packard Tri-Carb Model 306 sample oxidizer or by dissolving the tissue in Protosol, both methods yielding the same results. Tritium content of the samples was determined by liquid scintillation counting, and results were calculated as mg photosensitizer per g tissue.

RESULTS

Cell Survival Following *In Vitro* PRT

Figure 1 shows the kinetics of cell death of exponentially growing EMT6 cells following exposure to photosensitizer ("active component") and light treatment *in vitro*. The resulting survival curve has a broad shoulder with a D_q of 1 J/cm^2 and a steep exponential portion with a D_o of 0.2 J/cm^2.

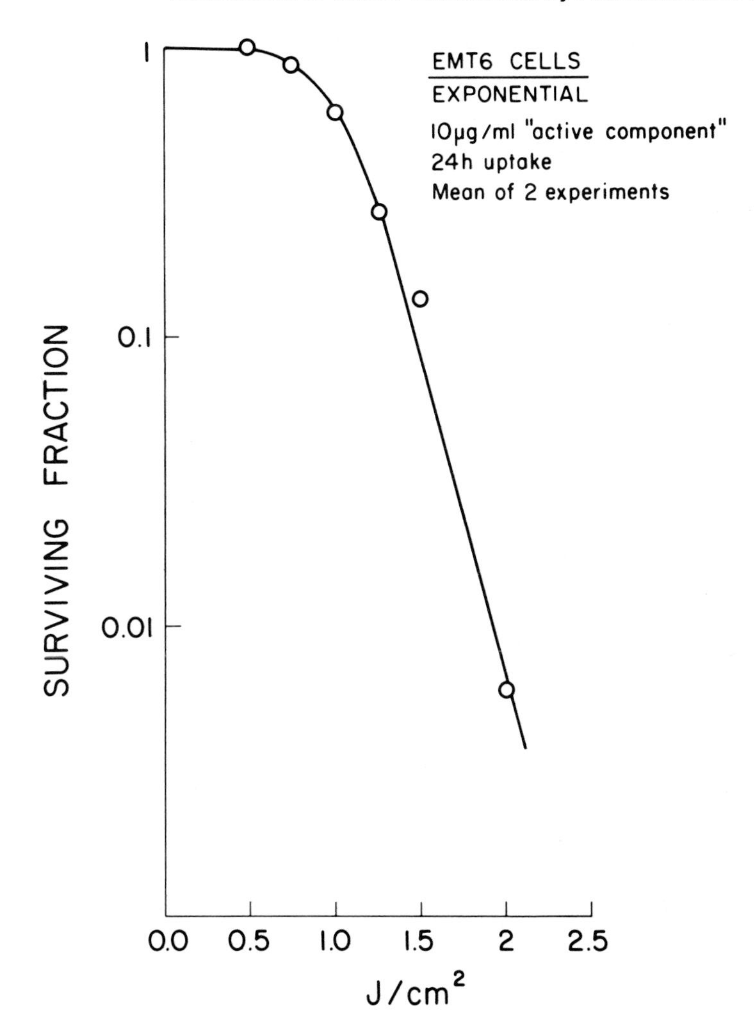

Figure 1. Effect of in vitro PRT on the survival of EMT6 cells in vitro as a function of light dose, 3 replicate cultures per point per experiment. Plating efficiency of controls exposed to photosensitizer w/o light was 65 ± 5%.

Uptake of Photosensitizers by Tumor Tissue

Tumor levels of Hpd and "active component" following uptake *in vivo* were determined in EMT6 and SMT tumors and were found to be nearly identical. Twenty-four hours after administration of 10 mg/kg Hpd, EMT6 tumors contained 3.7 $\pm$ 0.41 μg/g of tissue, which compared well with data reported by Gomer (1979) for the SMT tumor.

With administration of "active component" the same tumor levels were obtained with only half the dose. 5 mg/kg "active component" resulted in 3.2 $\pm$ 1.6 μg/g of tissue in EMT6 and 3.5 $\pm$ 1.2 μg/g of tissue in SMT tumor.

EMT6-tumor Response to PRT *In Vivo*

PRT using "active component" as photosensitizer as described led to massive necrosis of the tumor mass within 24 hours in all animals (n = 50) which resulted in 52% tumor cures (i.e. 90 days). Tumor regrowth, if it occurred, started at about the fifth day post treatment. Response to PRT was clearly restricted to the tumor area without involving the surrounding normal tissue or eliciting any general toxicity.

Gross hemorrhage was observed in some tumors by the end of light treatment, and by one hour after completion of treatment all animals demonstrated hemorrhage in and around the tumor.

Tumor Cell Survival Following PRT *In Vivo*

For each cell survival experiment three parameters were recorded: Total cell yield per gram tumor tissue, plating efficiency (PE) of recovered cells, and total number of clonogenic cells per gram tumor tissue (= cell yield/g x PE/100). Table 1 shows these data for control tumors and tumors explanted immediately following completion of PRT-light treatment. It is apparent that exposure of tumor cells to the photosensitizer alone *in vivo* does not impair their clonogenicity. But, most importantly, it shows that despite a tendency towards lower PE values, the clonogenicity of tumor cells is also not significantly impaired following PRT, if the tumor tissue is excised immediately following treatment.

Table 1

CELL SURVIVAL PARAMETERS OF EMT6 TUMORS

	No. of Experiments	Cell yield/g $\times 10^7$	PE (%)	Clonogenic cells/g* $\times 10^6$
Untreated Controls	15	3.51 ± 1.08	28 ± 6	9.86 ± 3.77
Controls 7.5 mg/kg "active component" no light	9	3.73 ± 1.02	26 ± 8	10.01 ± 4.55
PRT 7.5 mg/kg "active component" 45' light+	8	3.78 ± 0.96	24 ± 7	9.54 ± 4.35

Cell survival data determined by *in vivo* to *in vitro* colony formation assay

*Clonogenic cells/g = cell yield/g tissue x PE/100

+Excision of tumor tissue immediately following completion of PRT

Figure 2 shows the changes with time in cell survival in tumors left _in situ_ following completion of PRT. Within 1 hour after treatment both cell yield and PE values had dropped which resulted in a decrease of clonogenicity to about 33% of controls. This decrease in cell survival parameters continued with increasing time intervals between treatment and tumor excision. Tumor clonogenicity was reduced by a factor of 100 when tumor excision was delayed by 10 hours.

For comparison, some experiments were carried out using 7.5 mg/kg Hpd as photosensitizer which yielded the same results as the "active component".

To determine changes in tumor cell survival following circulatory deprivation, clonogenicity was assayed in tumors which simply had remained in a dead host for varying periods of time before being explanted. These results are compared with the changes in clonogenicity after PRT in Figure 3. Although it appears that during the first 2 hours tumor cells following PRT are dying at a slightly faster rate than those deprived of a functioning blood circulation, the decrease in clonogenicity at 3 hours of delayed tumor excision was identical.

DISCUSSION

The study presented here examined the relationship between the destruction of tumor mass and/or tumor cure caused by PRT and tumor cell survival. The EMT6 tumor system was found to be suitable for this purpose because (a) it behaved similarly to other previously evaluated cell and tumor systems in several respects, i.e. cellular response to _in vitro_ PRT (Henderson, et al., 1983), uptake of photosensitizer by tumor tissue and tumor response to PRT (Dougherty, et al., 1983), and (b) it allowed an _in vivo_ to _in vitro_ colony formation assay for the assessment of cell survival following PRT _in vivo_.

From the observation that tumor cell clonogenicity was not reduced when tumors were explanted immediately after PRT treatment, one must assume that tumor destruction did not reflect a direct lethal photodynamic effect exerted upon the tumor cells. Other factors must be involved to produce the observed delayed tumor cell death which was responsible for the entire tumor response. The rate of tumor cell death fol-

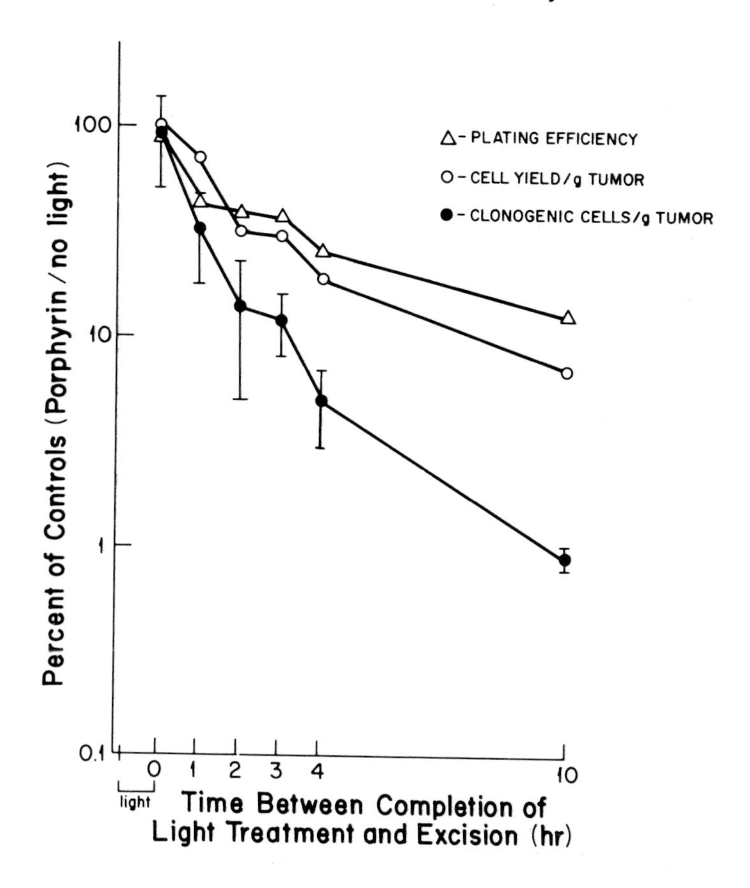

Figure 2. Cell survival kinetics following PRT *in vivo* (7.5 mg/kg "active component" followed 24 h later by 45 min of light) of EMT6 tumor cells as a function of time between treatment and tumor excision, assessed by *in vitro* colony formation. Two tumors were pooled for each experiment. Data represent the average plus S. E. of 3 to 8 experiments, 3 replicate cultures per point per experiment.

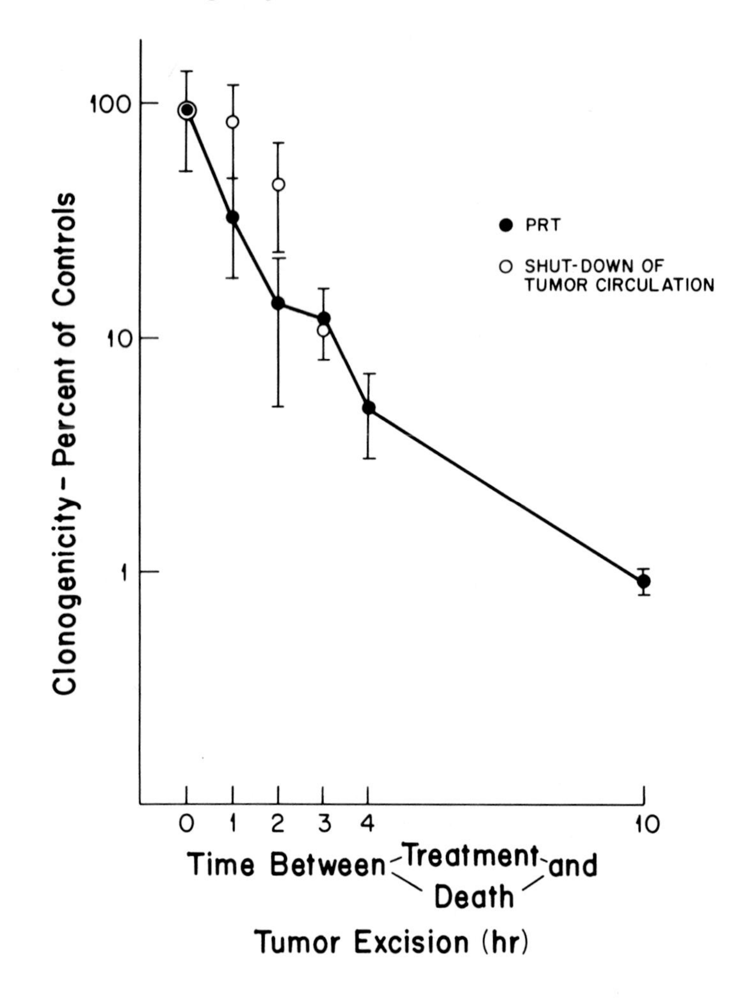

Figure 3. Comparison of the cell survival kinetics following PRT *in vivo* (see also Figure 2) and shutdown of tumor blood circulation following death of the host. Survival was determined by *in vitro* colony formation. Two tumors were pooled for each experiment. Data represent the average plus S. E. of 3 to 8 experiments, 3 replicate cultures per point per experiment.

lowing PRT resembled that following shutdown of the tumor blood circulation, which implies that vascular damage may be of primary importance.

The mechanism of tumor destruction observed in this investigation shows striking similarities as well as certain differences to the mechanism of tumor destruction following hyperthermia. It has to be emphasized, however, that PRT as administered in these experiments does not cause significant hyperthermia, i.e., light treatment raises tumor temperature only slightly to a maximum of 36.8°C from a base temperature of 35°C (Waldow, Dougherty, 1983). The phenomenon of delayed cell death following hyperthermic treatment, the kinetics of which are almost identical to those of delayed cell death following PRT, has been described by several authors (Marmor, et al., 1979; Kang, et al., 1980). It is believed to be caused by vascular occlusion due to heat. Additionally, however, during hyperthermia at temperatures over 42°C a large portion of tumor cells is killed through the direct effect of heat upon them which is reflected in a reduction of the number of clonogenic cells in the tumor immediately after completion of heat treatment (Marmor, et al., 1979; Kang, et al., 1980). No such reduction could be found following PRT in the present study. It is therefore concluded that tumor destruction by PRT in the experimental animal tumor system investigated here is entirely due to delayed cell death probably caused by vascular damage. Further studies are in progress expanding our observations to other tumor systems as well as combined treatments of PRT and heat.

REFERENCES

Berns MW, Dahlman A, Johnson FM, Burns R, Sperling D, Guiltinan M, Siemens A, Walter R, Wright W, Hammer-Wilson M, Wile A (1982). In vitro cellular effects of hematoporphyrin derivative. Can Res 42:2325.

Bicher HI, Hetzel FW, Vaupel P, Sandhu TS (1981). Microcirculation modifications by localized microwave hyperthermia and hematoporphyrin phototherapy. Biblthca Anat 20:628.

Bugelski PJ, Porter CW, Dougherty TJ (1981). Autoradiographic distribution of hematoporphyrin in normal and tumor tissue of the mouse. Can Res 41:4606.

Christensen T, Feren K, Moan J, Pettersen E (1981). Photodynamic effects of hematoporphyrin derivative on synchronized and asynchronous cells of different origin. Brit J Can 44:717.

Cortese DA, Kinsey JH, Woolner LB, Payne WS, Sanderson DR, Fontana RS (1979). Clinical application of a new endoscopic technique for detection of in situ bronchial carcinoma. Mayo Clin Proc 54:635.

Dougherty TJ, Boyle DG, Weishaupt KR, Henderson BA, Potter WR, Bellnier DA, Wityk KE (1983). Photoradiation therapy - clinical and drug advances. In Kessel D, Dougherty T (eds): "Porphyrin Photosensitization," New York: Plenum Press, p 3.

Gomer CJ, Dougherty TJ (1979). Determination of ^{3}H and ^{14}C hematoporphyrin derivative distribution in malignant and normal tissue. Can Res 39:146.

Henderson BW, Bellnier DA, Ziring B, Dougherty TJ (1983). Aspects of the cellular uptake and retention of hematoporphyrin derivative and their correlation with the biological response to PRT *in vitro*. In Kessel D, Dougherty T (eds): "Porphyrin Photosensitization," New York: Plenum Press, p 129.

Kang MS, Song CW, Levitt SH (1980). Role of vascular function in response of tumors *in vivo* to hyperthermia. Can Res 40:1130.

Marmor JB, Hilerio FJ, Hahn GM (1979). Tumor irradication and cell survival after localized hyperthermia induced by ultrasound. Can Res 39:2166.

Rockwell SC, Kallman RF, Fajardo LF (1972). Characteristics of a serially transplanted mouse mammary tumor and its tissue culture-adapted derivative. J Nat Can Inst 49:735.

Salvatore G, Ambesi-Impiombato SS, Tramontano D, Esposito M, Mastrocinque M, Andreoni A, Cubeddu R, DeSilvestri S, Laporta P (1982). Effects of laser irradiation on hematoporphyrin-treated normal and transformed thyroid cells in culture. Proc 13 Int Can Cong, p 237.

Star WM, van Rijsoord A, Reinhold HS (1982). Destructive effect of photoradiation on the microcirculation of a rat mammary tumor growing in a "sandwich" observation chamber - preliminary results. (Unpublished communication)

Waldow SM, Dougherty TJ (1983). Interaction of hyperthermia and photoradiation therapy. Rad Res (submitted).

Porphyrin Localization and Treatment of Tumors, pages 613–628

PHOTOTOXICITY OF BRAIN TISSUE IN HEMATOPORPHYRIN DERIVATIVE TREATED MICE

D.E. Rounds, D.R. Doiron*, D.B. Jacques,
C.H. Shelden and R.S. Olson
Huntington Medical Research Institutes,
Pasadena, California and *Children's
Hospital, Los Angeles, California

SUMMARY

Photoradiation therapy conditions which have been used to treat subcutaneous and breast tumors are lethal when applied to the heads of mice. Treatment of control mice with laser light at 631 nm over an energy density range of 0-90j/cm^2 had no measurable effect but mice photosensitized with 5 mg HPD/kg 72 hrs prior to laser treatment showed a threshold for brain damage at 56j/cm^2, above which the mice developed cerebral edema and died. Laser treatment caused the same rate and magnitude of temperature rise in both control and HPD-photosensitized mice. Moreover, studies using mice whose brain temperature was kept below 37^{0}C during laser treatment showed a greater phototoxicity than mice without temperature regulation. Therefore, temperature rise in cerebral tissue was not associated with phototoxicity in the brain. In contrast the oxygen consumption rate in a brain cell suspension from an HPD-treated mouse was only 54% of that from a control mouse following treatment with laser light. This observation, when taken with supporting data from other investigations, suggests that one mechanism for the phototoxic response in brain tissue is oxygen deprivation resulting from mitochondrial damage.

INTRODUCTION

Hematoporphyrin derivative (HPD) has long been considered to show preferential retention in tumor tissue, based on its fluorescence intensity distribution (Lipson et al, 1961, 1967, Mossman et al, 1974, Dougherty et al, 1975, Granelli et al, 1975, Tsutsui et al, Kelly and Snell, 1976, Carpenter et al, 1977 and Benson et al, 1982). Using tritium or ^{14}C-labeled HPD, Gomer and Dougherty (1974) and Bugelski, Porter and Dougherty (1981) have made quantitative studies of HPD distribution in tumor-bearing mice and have noted that significant levels of HPD are retained in non-malignant tissues, as well. In early clinical applications of photoradiation therapy (PRT), Dougherty et al (1978, 1979) reported that skin response in the target areas ranged from slight erythema to marked edema, extensive discoloration and heavy scab formation. These responses were considered of minor importance if the tumor tissue was destroyed.

The application of PRT to gliomas presents special considerations, however. Wise and Taxdal (1967) were unable to detect HPD fluorescence in intact brain tissue, but observed strong fluorescence in gliomas and in sites of injury. They concluded that HPD did not pass an intact blood-brain barrier, therefore, it could be used as a specific marker for delineating the margins of the tumor. In contrast, we have concluded that non-fluorescing intact non-tumor bearing brain tissue from HPD-treated mice contains enough HPD to cause profound phototoxicity following exposure to light (Rounds et al, 1982). Although edema formation in skin overlying a treated tumor may be inconsequential, a similar response in the brains of mice is often fatal.

Considering the fact that clinical PRT trials with glioma patients are currently in progress (Perria et al, 1980, Laws et al, 1981 and McCulloch et al, 1983), it is important to evaluate the magnitude of this phototoxic response in brain tissue and attempt to identify the mechanism by which it occurs. We have addressed these problems by quantitating the degree of phototoxicity in the brains of mice induced by a set of PRT conditions commonly used in clinical practice, and we have evaluated two potential mechanisms which hypothetically could lead to the observed phototoxic response.

MATERIALS AND METHODS

Phototoxicity was evaluated in male Swiss mice, ranging from 25 to 35 gms, whose scalp hair was removed with a commercial depilatory to minimize scatter and reflection of the laser light. Control mice were compared with mice photosensitized with an i.p. injection of 5 mg HPD (Oncology Research and Development, Inc., Cheektowaga, N.Y.) per kilogram body weight. In order to accentuate the significance of the phototoxic reaction, HPD-treated mice were kept in subdued illumination for 72 hrs before exposure to laser light. The light source was an argon-pumped dye laser (Coherent, model PRT-92, Palo Alto, CA). The wavelength used for all experiments was 631 nm, usually at an exit power of 400 mW. Light exposure was accomplished by placing each unanesthetized mouse in a molded styrofoam restrainer with an open window over its cerebral hemispheres. Laser light at a power density of 319 mW/cm^2 was delivered through a step index quartz fiber (Quartz Products, Plainfield, N.J.) onto the surface of the scalp. The exposure times (0-15 minutes) were regulated with a model 845 automatic timer (Newport Research, Fountain Valley, CA) which controlled a shutter in the path of the beam. An estimate of intracranial power was made by measuring the power of laser light with a model 210 thermopile (Coherent, Palo Alto, CA) before and after transmittance through a freshly excised mouse cranium.

After light exposure, brain edema was allowed to maximize over a 48 hr period. At that time, all mice were given a marker for brain edema (Rounds et al, 1982) in the form of an i.p. injection of 0.05 μ Ci of ^{3}H-tetracycline per gram of body weight, using a stock solution containing 10 μ Ci/ml saline (New England Nuclear, Boston, MA). Four hours later, the mice were sacrificed by cervical dislocation, then 0.1-0.2 gm of tissue from their cerebral hemispheres was collected and weighed wet. The brain tissue was transferred to glass counting vials and macerated in 1.0 ml of protosol (New England Nuclear). When the tissue was dissolved completely, the protosol was diluted with 10.0 ml of aquasol (New England Nuclear) and was buffered with 2.0 ml of 1 M tris-HCl (Sigma Biochemical Co., Saint Louis, MO). The radioactivity of the ^{3}H-tetracycline was counted in an LS 9000 liquid scintillation counter (Beckman Instruments, Anaheim, CA). The disintegrations per minute (DPM) were normalized as DPM/g (wet weight) of brain tissue.

Mice used in the temperature studies also had their scalps denuded with a commercial depilatory then were anesthetized with an i.p. injection of 50 mg sodium nembutal (Abbott Laboratories, North Chicago, IL) per kilogram body weight. A type T microthermocouple was inserted in the rectum for monitoring body temperature and a second type T microthermocouple was embedded in a 30-gauge needle which was placed in the center of the brain through the temporal lobe. The two thermocouples were connected to a meter (model 4128, Omega Engineering, Stamford, CN) with a digital display of the temperature and an accuracy of $\pm 0.1^0$C. Both control mice and HPD-treated mice (5 mg HPD/kg given 72 hrs earlier) were exposed over an area of 1.33 cm^2 to 100, 200, 300 and 400 mW for 5-10 minutes each, until the intracranial temperature stabilized. This sequence of measurements was repeated once or twice for each mouse.

Brain cell suspensions were prepared from control and HPD-treated mice for respiration measurements, as previously reported (Rounds and Olson, 1967, 1968). The mice were sacrificed by cervical dislocation then the cerebral hemispheres were removed and placed in 2 ml Dulbecco's minimal essential culture medium supplemented with 5% fetal calf serum (Gibco, Grand Island, N.Y.). The brain tissue was aspirated in and out of a 13-gauge needle until the tissue had dissociated. Then the larger fragments were allowed to settle for 1 minute and the finely dispersed cells in the supernatant fluid were drawn off for use.

Aliquots of the cell suspensions were placed in a glass conical centrifuge tube positioned in a water bath. When the temperature had equilibrated at 37^0C $\pm$ 0.1^0C, the suspension was agitated to admit oxygen into the fluid, then the O-ring at the tip of a Clark oxygen electrode was seated into the tapered end of the centrifuge tube taking care to exclude all bubbles from the enclosed cell suspension. The rate of change in the dissolved oxygen content due to cellular respiration, was monitored with a Beckman Model 160 Physiological Gas Analyzer (Beckman Instruments, Inc., Spinco Division, Palo Alto, CA) and recorded on a Beckman strip chart recorder.

After repeated recordings had been made, the cell suspension was exposed to laser light at 631 nm, with an exit energy from the optical fiber of 400 mW, distributed over the surface of the suspension, for an exposure period of 15 minutes. The cell suspensions were mixed by agitation at 5 minute intervals throughout this treatment. Immediately after laser treatment, new recordings of oxygen consumption rates were made in replicate. Control and experimental cell suspensions having identical pretreatment O_2 consumption rates were compared for laser effects on cellular respiration.

RESULTS

Male Swiss mice were exposed to an intracranial energy density range of 0-90j/cm^2 at 631 nm, 72 hrs after HPD photosensitization. The power density transmitted through

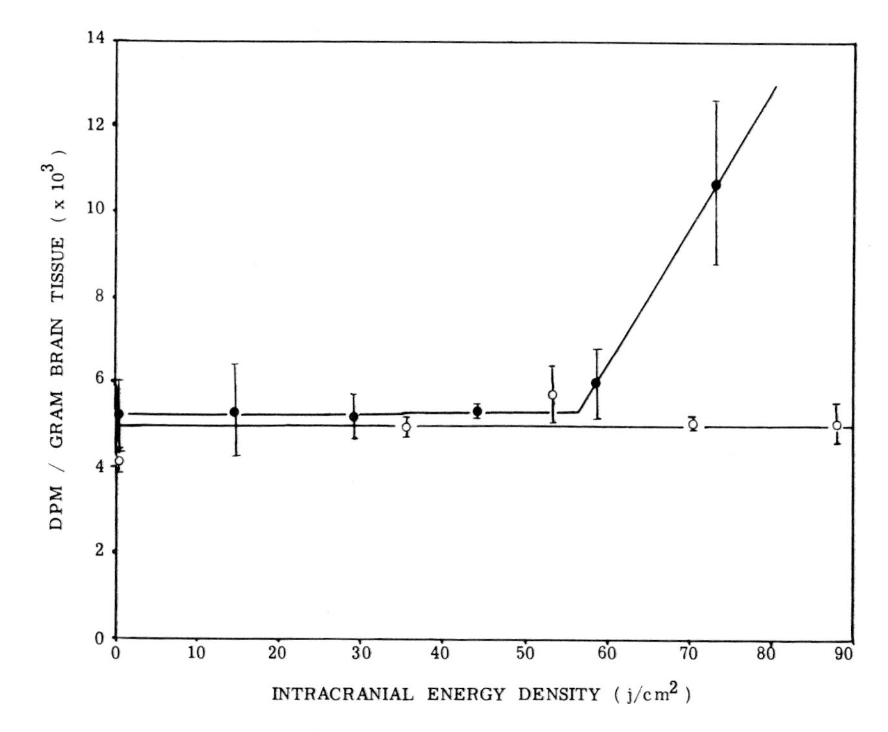

Fig. 1. Phototoxicity of brain tissue from control mice (open circles) and mice photosensitized with 5 mg HPD/kg 72 hrs before laser treatment (closed circles). Abscissa: imposed laser light at 631 nm. Ordinate: radioactivity of ^{3}H-tetracycline content of brain specimens. Four mice per data point (mean ± SE).

the scalp and skull was 30.6% of the imposed energy density, or 98 mW/cm^2. Mice which were exposed to energy densities below 56j/cm^2 showed control levels of ^{3}H-tetracycline incorporation, which indicated no phototoxicity had occurred. Mice exposed to energy densities above 56j/cm^2 showed an energy-dependent increase in ^{3}H-tetracycline incorporation (Fig. 1) and an increase in cerebral damage (Fig. 2).

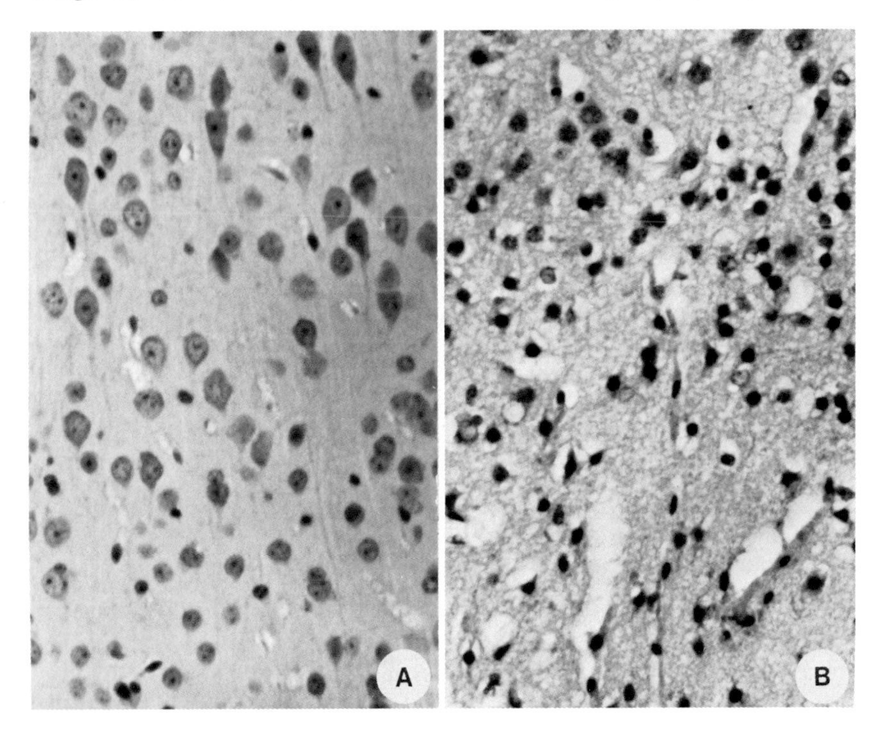

Fig. 2. The phototoxic response of cells in the cerebral cortex of mice treated with an intracranial energy density of 78j/cm^2 at 631 nm, 72 hrs after (A) no HPD and (B) 5 mg HPD/kg.

These mice also showed marked edema of the soft tissues surrounding the head and neck and appeared to be somewhat moribund. Non-HPD-treated control mice which were also exposed to an energy density range of laser light from 0-90j/cm^2 showed no skin erythema or edema nor an altered behavior pattern. Levels of ^{3}H-tetracycline incorporation throughout this range were not significantly different from that in non-exposed brain tissue (Fig. 1). These data demonstrate that HPD concentrations and light operating in the 600-700 nm range being used to treat tumors (Dougherty et al, 1978, 1979) are phototoxic to intact brain tissue.

Most interactions of laser light in biological tissues are accompanied by elevated temperatures in the target site (Rounds, 1982). The possibility that elevated temperatures could cause the phototoxic response was tested in a series of HPD-treated and control mice anesthetized with nembutal. It was observed that sequential exposures to the crania of mice with 100, 200, 300 and 400 mW of laser power caused

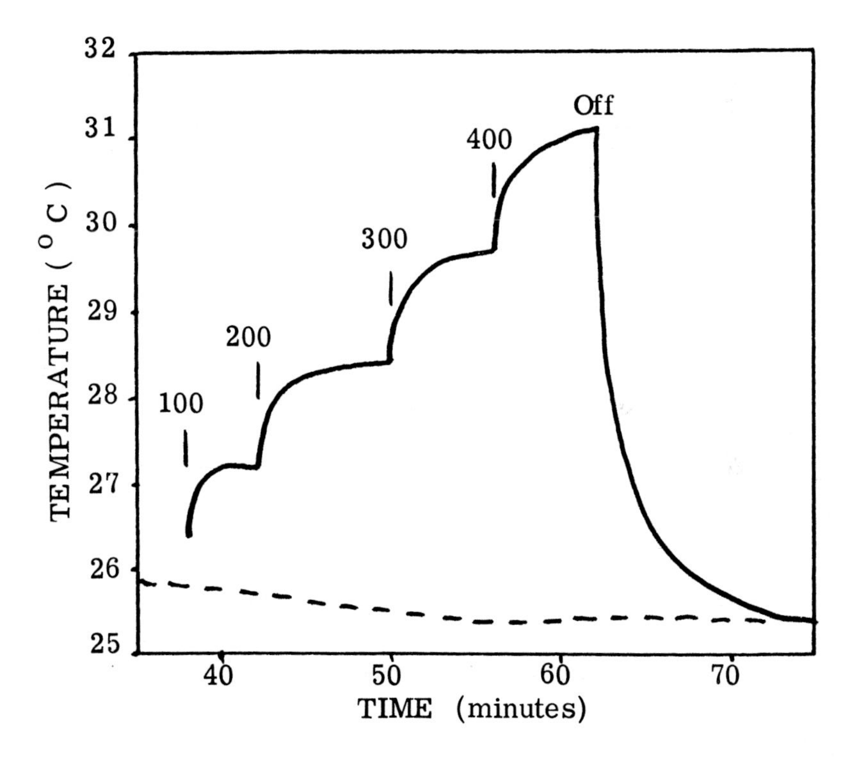

Fig. 3. Representative changes in core temperature (broken line) versus brain temperature (solid line) of a nembutal anesthetized mouse during successive cranial exposures to 100 mW, 200 mW, 300 mW and 400 mW laser power (631 nm) at the times indicated.

brain temperature elevations of 1.3 - 1.5 degrees C above core (rectal) temperature per 100 mW power imposed on the scalp (Fig. 3). These rate changes were not significantly different for control and HPD-treated mice (Fig. 4). Therefore, it became evident that temperature change was not correlated with the selective phototoxicity observed only in HPD-treated mice (Fig. 1).

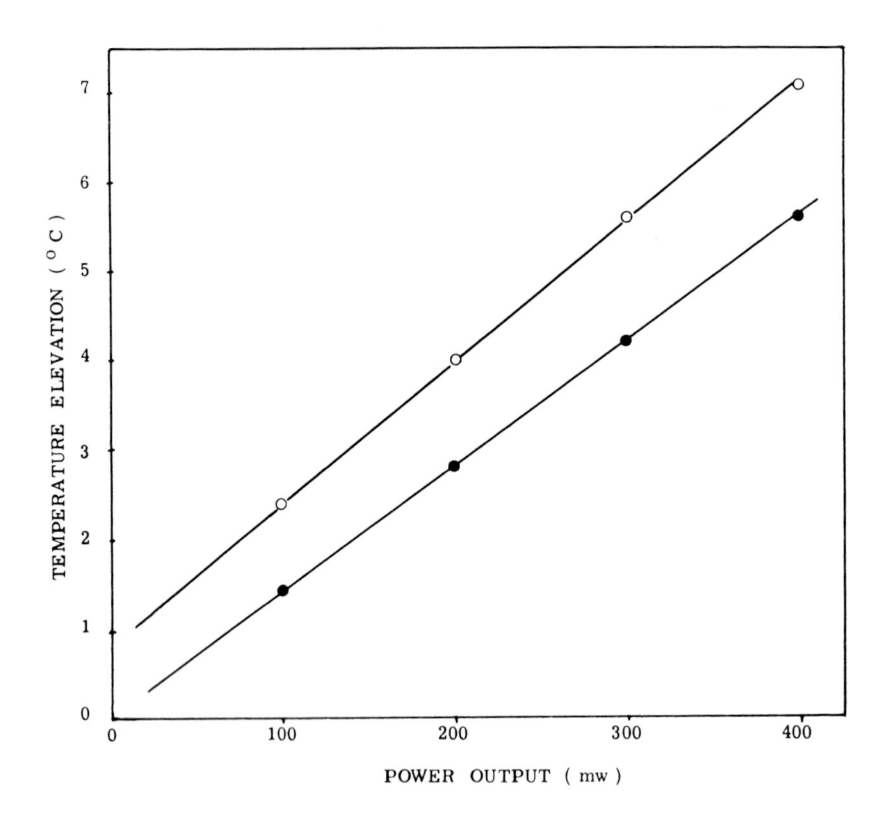

Fig. 4. Representative brain temperatures (above core temperature) in a control mouse (open circles) versus an HPD-photosensitized mouse (closed circles) during exposures to 631 nm laser light over a 0-400 mW power range.

To further substantiate this conclusion, a phototoxicity study was conducted on HPD-treated mice whose body temperature was lowered by nembutal anesthesia to 31^{0}C or below, as monitored by rectal temperature measurements. As shown in Fig. 3, brain temperatures in these mice would not be expected to rise above normal body temperature of 37^{0}C during laser exposures.

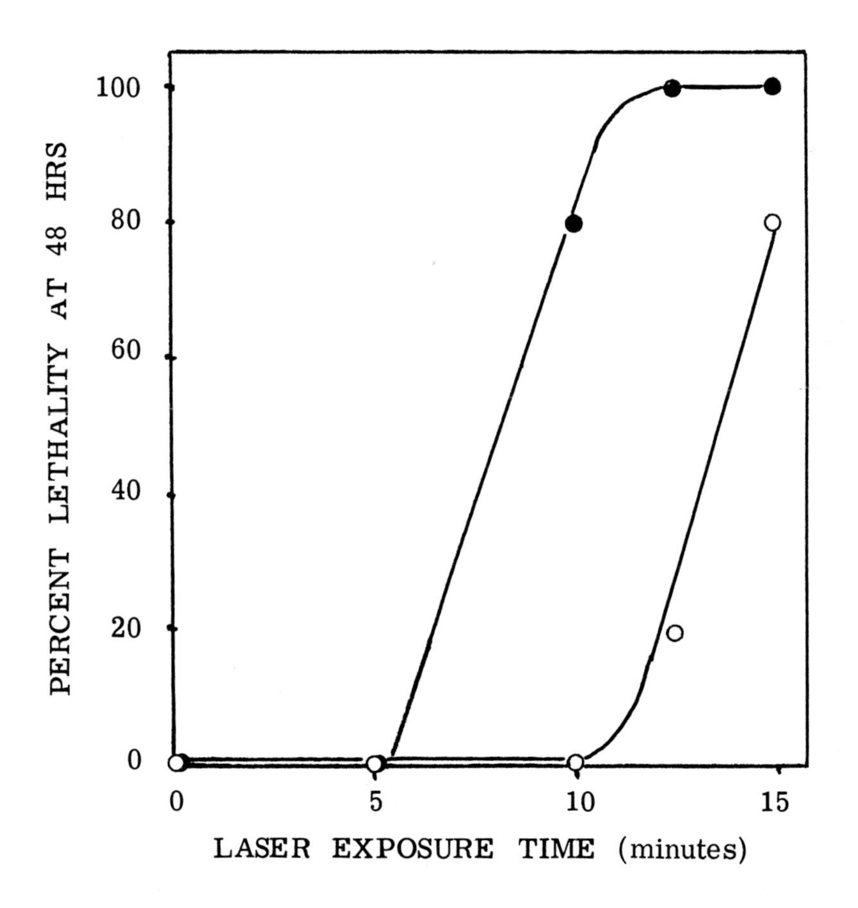

Fig. 5. Morbidity rate in sets of 5 HPD-photosensitized mice (each point) 48 hrs after exposure to 319 mW/cm^2 at 631 nm. Open circles represent mice with normal body temperatures. Closed circles represent nembutal-anesthetized mice with reduced body temperatures during laser treatment.

Instead of protection from elevated temperatures, anesthetized animals showed a significant increase in phototoxicity. Morbidity in this series was substantially increased at 48 hrs after laser exposure, as compared with unanesthetized mice (Fig. 5). Further, the threshold for brain damage in the low temperature mice was $47j/cm^2$, in contrast to a threshold of $56j/cm^2$ for mice treated at normal body temperature (Fig. 6).

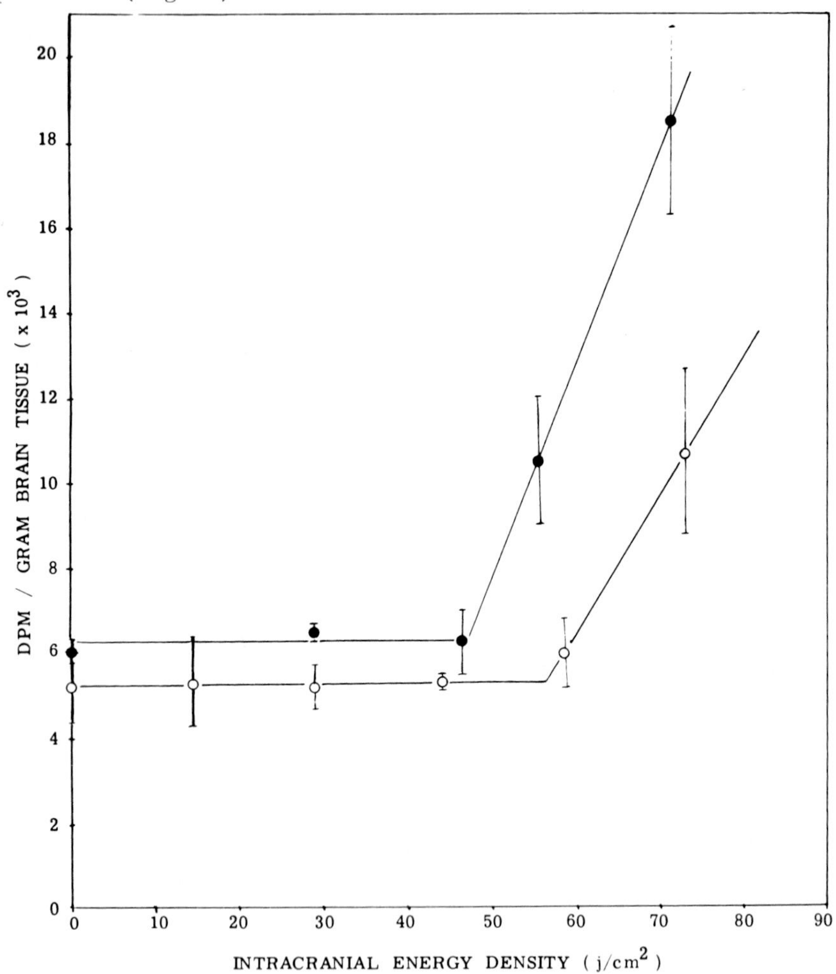

Fig. 6. Phototoxicity in brain tissue from unanesthetized mice (open circles) and in brains from anesthetized mice with lowered body temperature following HPD and light treatment.

A preliminary study was conducted to test the hypothesis that phototoxicity in HPD-treated mouse brain was associated with an inhibition in oxygen consumption. Brain cell suspensions were prepared from a control mouse and from a mouse treated with 5 mg HPD/kg 72 hrs before being sacrificed. Oxygen consumption rates were measured for aliquots of these suspensions before and immediately after 15 minutes exposure over a constant area to 400 mW of laser light. Control and HPD-containing cell suspensions showing the same pre-treatment O_2 consumption rates were compared after exposure to laser light. During the time that the brain cell suspensions from the control mouse consumed 100% of available oxygen, laser-treated cells from the HPD-containing mouse brain consumed only 54% of their available oxygen (Fig. 7). These data suggest that one mechanism for the phototoxic response in brain tissue is oxygen deprivation resulting from mitochondrial damage.

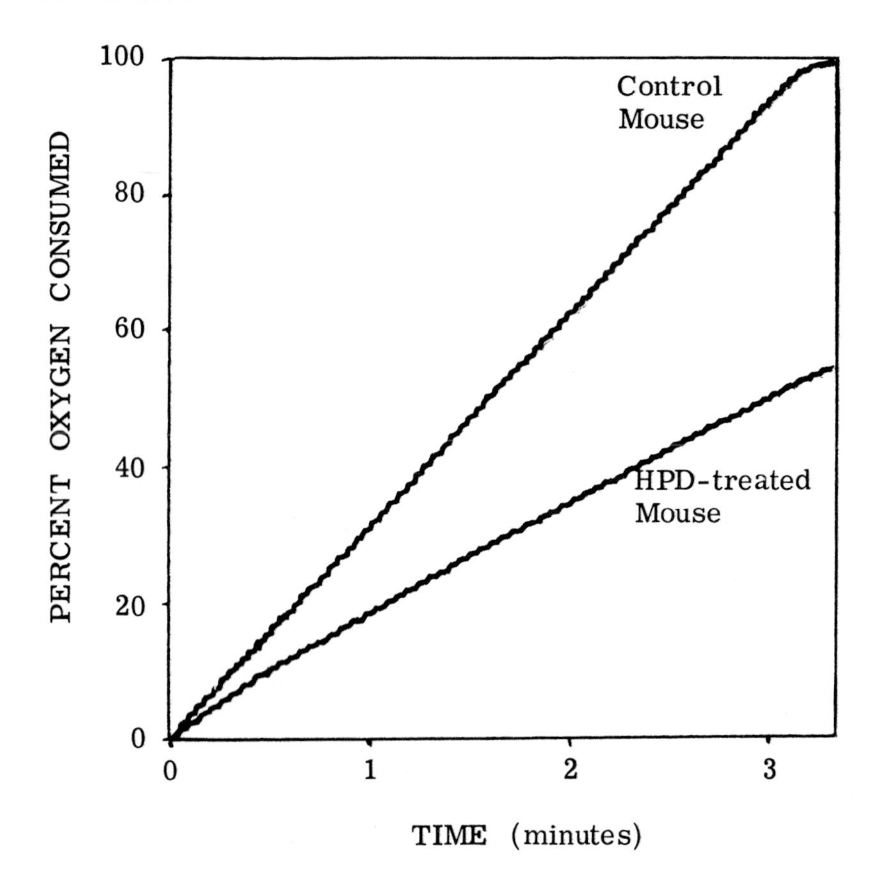

Fig. 7. Relative respiration rates of brain cell suspensions from a control mouse (upper recording) and from an HPD-treated mouse after laser treatment (lower recording). See text for details.

DISCUSSION

Based on the selective phototoxicity in HPD-treated mice, we had predicted that treated brain tissue would show a greater heat rise than control brains during laser treatment. That assumption further suggested that phototoxicity could be controlled if the temperature rise was controlled. The data in Figs. 4, 5 and 6 showed both assumptions were incorrect, so another mechanism was called for.

Cozzani et al (1981) reported that hematoporphyrin accumulates in the mitochondria of rat hepatocytes *in vivo*. Berns et al (1982) were able to demonstrate by fluorescence microscopy that HPD selectively stain the specialized mitochondria (sarcosomes) of rat myocardial cells *in vitro*. Copolla et al (1980) showed that ultrastructural changes were first detected in the form of mitochondrial damage in HPD-photosensitized lymphoma cells when exposed to light. At this conference, Gibson et al (1983) have shown that light treatment of HPD-photosensitized mitochondrial suspensions inhibited cytochrome c oxidase activity. Finally, we observed that brain cell suspensions from an HPD-treated mouse showed significant reduction in their O_2 consumption rate following treatment with laser light (Fig. 7). These data, suggesting compartmentalization of HPD in mitochondria, a vital cellular organelle, is consistent with the observations that lethal phototoxicity can occur in oxygen-sensitive brain tissue with HPD concentrations that are difficult to detect by direct visualization of fluorescence (Wise and Taxdal, 1967).

Most clinical trials of photoradiation therapy of tumors use either 2.5 or 5.0 mg HPD/kg, followed 24 - 48 hrs later with exposure to 120 - 150j/cm^2 using red light (600 - 700 nm). These conditions, even 72 hrs after HPD treatment, are lethal in Swiss mice. It must be recognized, however, that destruction of 1 cc of brain tissue in a human could be tolerated but that same volume would encompass the entire brain of the mouse. Therefore, instead of seeking conditions in which the glioma can be destroyed at an energy density of 56j/cm^2 or less, the phototoxicity data in Fig. 1 can be used to estimate the amount of brain damage associated with PRT. For example, if PRT was considered only as adjuvant therapy for a surgically removed glioma and if laser light was transmitted via fiber optics into the center of the cavity in a manner which would deliver 90j/cm^2 to all surfaces of the tumor bed, then the residual tumor cells and surrounding brain tissue would be destroyed to an estimated depth of about 1.9 mm (using an extrapolation of the 633 nm light penetration data reported by Doiron et al, 1983) before the energy density would reach the subthreshold level for brain tissue. If these theoretical considerations can be verified experimentally, this would provide a reasonable basis for adjuvant therapy of gliomas. This parameter, as well as other contributing factors are under continuing investigation in an effort to optimize conditions for applying photoradiation therapy to gliomas.

This study was supported by the Ralph M. Parsons Foundation.

REFERENCES

Benson RC Jr, Farrow GM, Kinsey JH, Cortese DA, Zincke H, Utz DC (1982). Detection and localization of in situ carcinoma of the bladder with hematoporphyrin derivative. Mayo Clin Proc 57:548.

Berns MW, Dahlman A, Johnson FM, Burns R, Sperling D, Guiltinan M, Siemens A, Walter R, Wright W, Hammer-Wilson M, Wile A (1982). In vitro cellular effects of hematoporphyrin derivative. Cancer Res 42:2325.

Bugelski PJ, Porter CW, Dougherty TJ (1981). Autoradiographic distribution of hematoporphyrin derivative in normal and tumor tissue of the mouse. Cancer Res 41:4606.

Carpenter RJ, Ryan RJ, Neel HB, Sanderson DR (1977). Tumor fluorescence with hematoporphyrin derivative. Ann Otol Rinol Larynagol 85:661.

Coppola A, Viggiani E, Salzarulo L, Rasile G (1980). Ultrastructural changes in lymphoma cells treated with hematoporphyrin and light. Am J Path 99:175.

Cozzani I, Jori G, Reddi E, Fortunato A, Granati B, Felice M, Tomio L, Zorat P (1981). Distribution of endogenous and injected porphyrins at the subcellular level in rat hepatocytes and in ascites hepatoma. Chem Biol Interactions 37:67.

Doiron DR, Svaasand LO, Profio AE (1983). Light dosimetry in tissue: application to photoradiation therapy. In Kessel D, Dougherty TJ (eds): "Porphyrin Photosensitization", New York, Plenum Press, p 63.

Dougherty TJ, Grindey GB, Fiel KR, Weishaupt KR, Boyle DG (1975). Photoradiation therapy II cures of animal tumors with hematoporphyrin and light. J Natl Cancer Inst 55:115.

Dougherty TJ, Kaufman J, Goldfarb A, Weishaupt KR, Boyle D, Mittleman A (1978). Photoradiation therapy for the treatment of malignant tumors. Cancer Res 38:2628.

Dougherty TJ, Lawrence G, Kaufman J, Boyle D, Weishaupt KR, Goldfarb A (1979). Photoradiation in the treatment of recurrent breast carcinoma. J Natl Cancer Inst 62:231.

Gibson SL, Leakey PB, Sollott SJ, Crute J, Fishbein G, Hilf R (1983). Photosensitization of mitochondrial cytochrome C oxidase by hematoporphyrin derivative (HPD) in vitro and in vivo. In Doiron DR (ed): "Porphyrin Localization and Treatment of Tumors", New York: Alan R. Liss, p

Gomer CJ, Dougherty TJ (1979). Determination of ^{3}H-and ^{14}C- hematoporphyrin derivative distribution in malignant and normal tissue. Cancer Res 39:146.

Granelli SG, Diamond I, McDonagh AF, Wilson CB, Nielsen SL (1975). Photochemotherapy of glioma cells by visible light and hematoporphyrin. Cancer Res 35:2567.

Kelly JF, Snell ME (1976). Hematoporphyrin derivative: a possible aid in the diagnosis and therapy of carcinoma of the bladder. J Urol 115:150.

Laws ER Jr, Cortese DA, Kinsey JH, Eagan RT, Anderson RE (1981). Photoradiation therapy in the treatment of malignant brain tumors: a phase I (feasibility) study. Neurosurgery 9:672.

Lipson RL, Baldes EJ, Olsen AM (1961). The use of a derivative of hematoporphyrin in tumor detection. J Natl Cancer Inst 26:1.

Lipson RL, Baldes EJ, Gray MJ (1967). Hematoporphyrin derivative for detection and management of cancer. Cancer 20: 2255.

McCulloch GAJ, Forbes IJ, Lee KS, Cowled PA, Jacka FJ, Ward AD (1983). Phototherapy in malignant brain tumors. In Doiron DR (ed): "Porphyrin Localization and Treatment of Tumors", New York: Alan R. Liss, p

Mossman BT, Gray MJ, Silberman L, Lipson RL (1974). Identification of neoplastic versus normal cells in human cervical cell culture. Obstet Gynecol 43:635.

Perria C, Capuzzo T, Cavagnaro G, Datti R, Francaviglia N, Rivano C, Tercero VE (1980). First attempts at the photodynamic treatment of human gliomas. J Neurosurg Sci 24: 119.

Rounds DE, Olson RS (1967). The effect of intense visible light on cellular respiration. Life Sci 6:359.

Rounds DE, Olson RS (1968). The effect of the laser on cellular respiration. Zeit F. Zellforsch 87:193.

Rounds DE (1982). Laser applications in medicine. In Regan JD, Parrish JA (eds): "The Science of Photomedicine", New York, Plenum Press, p 533.

Rounds DE, Jacques S, Shelden CH, Shaller BS, Olson RS (1982). Development of a protocol for photoradiation therapy of malignant brain tumors. Part I Photosensitization of normal brain tissue with hematoporphyrin derivative. Neurosurgery 11:500.

Tsutsui M, Carrano C, Tsutsui EA (1975). Tumor Localizers: porphyrins and related compounds. Ann NY Acad Sci 244:674.

Wise BL, Taxdal DR (1967). Studies of the blood-brain barrier utilizing hematoporphyrin. Brain Res 4:387.

Porphyrin Localization and Treatment of Tumors, pages 629–636

^{64}Cu LABELLING OF HEMATOPORPHYRIN DERIVATIVE FOR NON-INVASIVE IN-VIVO MEASUREMENTS OF TUMOUR UPTAKE

G. Firnau, D. Maass, B.C. Wilson,
and W.P. Jeeves
Department of Nuclear Medicine
McMaster University, and
Ontario Cancer Treatment and
Research Foundation
Hamilton, Ontario Canada

INTRODUCTION

In Photoradiation Therapy (PRT), successful tumour sterilization depends on achieving both adequate concentration of photosensitizer (hematoporphyrin derivative, HpD), and delivery of appropriate light dose. Currently, there is no satisfactory, non-invasive technique to measure the concentration of HpD in solid tumours, although some success has been obtained using fluorescence endoscopy (Doiron *et al* 1979; Profio *et al* 1979).

The aim of the present work is to label HpD with radio-isotopes suitable for quantitative nuclear scanning. These may be either gamma-emitting isotopes, for which single photon emission computed tomography (SPECT) is the scanning technique of choice, or positron emitters, for which positron emission tomography (PET) may be used. These imaging techniques allow both for tumour localization, and for determination of regional HpD concentration, through measurement of the distribution of specific activity.

In this paper, we shall present preliminary studies made using the isotope copper-64; this is a positron emitter, with a physical half-life of 12.8 hours. Copper has been shown to complex strongly with the pyrrol nitrogens of porphyrins (Falk 1964; Cohen, Schwartz 1966).

For ^{64}Cu-HpD to be clinically useful, it is necessary that the labelling procedure, and the presence of the label do not significantly alter the biodistribution properties of HpD *in-vivo*. It is also desirable that the phototoxicity of the HpD be unchanged, although, for measuring the tumour uptake, it would be adequate to have only a small fraction of the therapeutic HpD radiolabelled.

In the following sections, we shall briefly describe the labelling procedure, and the assessment of the ^{64}Cu-HpD by gel and thin-layer chromatography. Preliminary data will be presented on

a. the distribution of ^{64}Cu-HpD *in-vivo*
b. the phototoxicity of (^{64}Cu-HpD + light) compared with that of (HpD + light), as assessed by *in-vitro* studies using multicell spheroids,
c. the tumour visualization of ^{64}Cu-HpD in a rat model using Positron Emission Tomography.

LABELLING PROCEDURE AND CHROMATOGRAPHIC ANALYSIS

^{64}Cu was produced in the McMaster University Nuclear Reactor by neutron bombardment of natural, pure copper. The irradiated copper was made into an aqueous solution of [^{64}Cu] $CuCl_2$ (15.6 µmol). This was then added to an isotonic solution of HpD (Photofrin-I; 46.6 µmol). The pH of this mixture was adjusted to 11.0 with NaOH, and then neutralized with hydrochloric acid. The resulting ^{64}Cu-HpD had a specific activity of 30 MBq/mg HpD.

To determine if the ^{64}Cu was indeed associated with the HpD, the labelled solution was passed through a gel chromatographic column (BIO-RAD P-10, 20 x 1 cm), eluted with saline. The eluate was monitored, both for ^{64}Cu radioactivity and for light absorption. The resultant elution curves for ^{64}Cu-HpD were identical to those of unlabelled HpD.

Radioactivity measurements on reverse-phase thin-layer chromatographs (Whatman KC18; 20 cm; THF:H_2O:MeOH=6:34:60) also demonstrated that no free copper remained in solution after labelling. These chromatographs show the same five fluorescent components as unlabelled HpD. With ^{64}Cu-HpD, two additional non-fluorescent radioactive components are observed; these are thought to be ^{64}Cu-HpD monomers.

IN-VIVO DISTRIBUTION OF ^{64}Cu-HpD

In order to investigate the tissue distribution of the ^{64}Cu-HpD, a 290 g rat bearing a 40 g subcutaneous tumour in the shoulder (induced by injection of Adenovirus Type-12 transformed baby rat kidney cells) was injected intra-peritoneally with ^{64}Cu-HpD (18 mg/kg body weight). The animal was sacrificed at 72 hours and 2-5 samples of each tissue were counted to determine their specific activity. The results are shown in Table 1, where the data from similar studies using ^{3}H- and ^{14}C- labelled HpD in tumour-bearing mice (Gomer, Dougherty 1979) are included for comparison.

μg labelled HPD/g tissue								
Isotope	Blood	Liver	Kidney	Spleen	Lung	Skin	Muscle	Tumour
^{64}Cu	1.15 ±0.76	21.0 ±3.24	20.0 ±3.16	14.5 ±2.70	2.90 ±1.78	1.12 ±0.75	0.48 ±0.48	2.50 ±1.60
^{3}H	0.38 ±0.07	18.56 ±2.60	9.25 ±3.56	8.36 ±1.59	2.17 ±0.21	1.42 ±0.66	0.93 ±0.31	2.20 ±0.01
^{14}C	0.96 ±0.37	23.51 ±4.12	6.34 ±2.08	8.70 ±3.21	1.53 ±0.55	1.88 ±0.26	0.00	2.39 ±0.47

Table 1. Tissue and blood distribution of radiolabelled HpD 72 hours following i.p. injection

^{64}Cu-HpD: *This study*: tumour-bearing rat; 18 mg HpD/kg body weight; standard deviations are between samples from same tissue.

^{3}H-HpD, ^{14}C-HpD: *Gomer, Dougherty 1979*: tumour-bearing mice; 10 mg HpD/kg body weight; standard deviations are between animals.

It is seen that the overall distribution of ^{64}Cu-HpD in the major organs is consistent with the more extensive studies done with ^{3}H-HpD, and ^{14}C-HpD. In addition, the tumour shows uptake and retention of ^{64}Cu-HpD, at a concentration similar to the corresponding results in the tumour-bearing mice.

We have also, at this time, carried out some preliminary studies of the tissue distribution of ^{64}Cu-HpD and ^{3}H-HpD in healthy rats; the early results indicate that the biodistributions of the two labelled compounds are closely similar.

IN-VITRO PHOTOCYTOTOXICITY

In order to test the phototoxicity of ^{64}Cu-HpD compared with unlabelled HpD, we have used a multicell spheroid *in-vitro* tumour model, measuring the effect on cell clonogenicity of the combined action of drug and visible light. The spheroids (V79 Chinese Hamster cells) were grown in spinner culture according to published methods (Durand 1975; Sutherland, Durand 1976). After growth to 150-650 µm diameter, the spheroids were plated into tissue culture dishes, and incubated at 37°C for 16 hours in the presence of growth medium containing variable concentrations of either ^{64}Cu-HpD or HpD. These spheroids were thoroughly washed prior to their subsequent irradiation, in the culture dishes, with light from a tungsten-filament lamp, filtered to remove wavelengths below 590 nm. The spheroids were then re-incubated for four days in growth medium to permit colony development. Cultures were subsequently fixed, stained, and the colonies were counted to determine survival fractions.

Figure 1 shows the results of survival fractions for different concentrations of drug, with and without light exposure. Control experiments were also performed using drug but no light, or ^{64}Cu in basic aqueous solution alone. These data indicate that

a. the light alone has no cytotoxic effect
b. the drug alone, either HpD or ^{64}Cu-HpD at 10 or 20 µg/ml produced no significant cytotoxicity
c. ^{64}Cu-HpD and HpD produced essentially the same marked reduction in cell colony-forming ability at the light dose used.

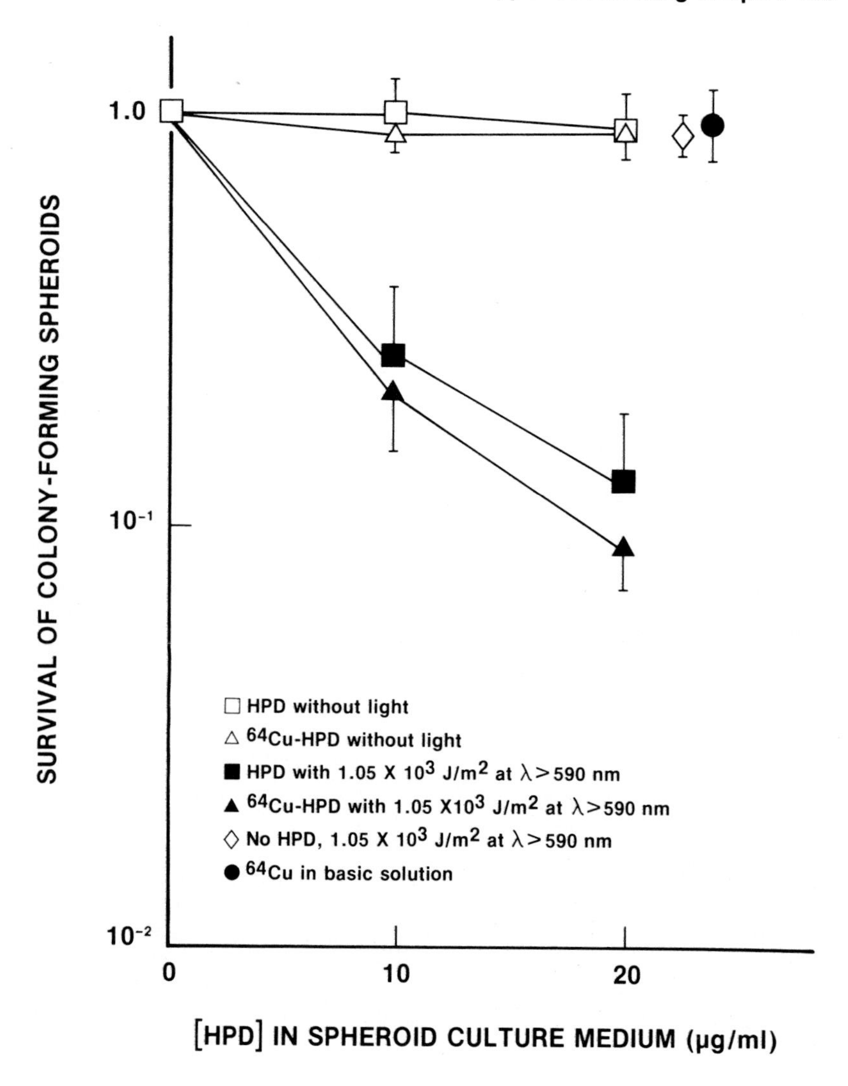

Fig. 1. Survival curves for V79 spheroids exposed to light and to various concentrations of HpD or ^{64}Cu-HpD.

TUMOUR VISUALIZATION

Prior to sacrifice of the tumour-bearing rat used for the biodistribution studies, the animal was anesthetized, and imaged in the McMaster University PET scanner (spatial resolution $\lesssim$ 8mm). The animal was positioned such that the spine was perpendicular to the plane of the detector ring. This produced a series of transverse cross-sectional slices 1 cm thick. Figure 2 shows the image obtained at the level of the tumour in the shoulder. The tumour is clearly visualized.

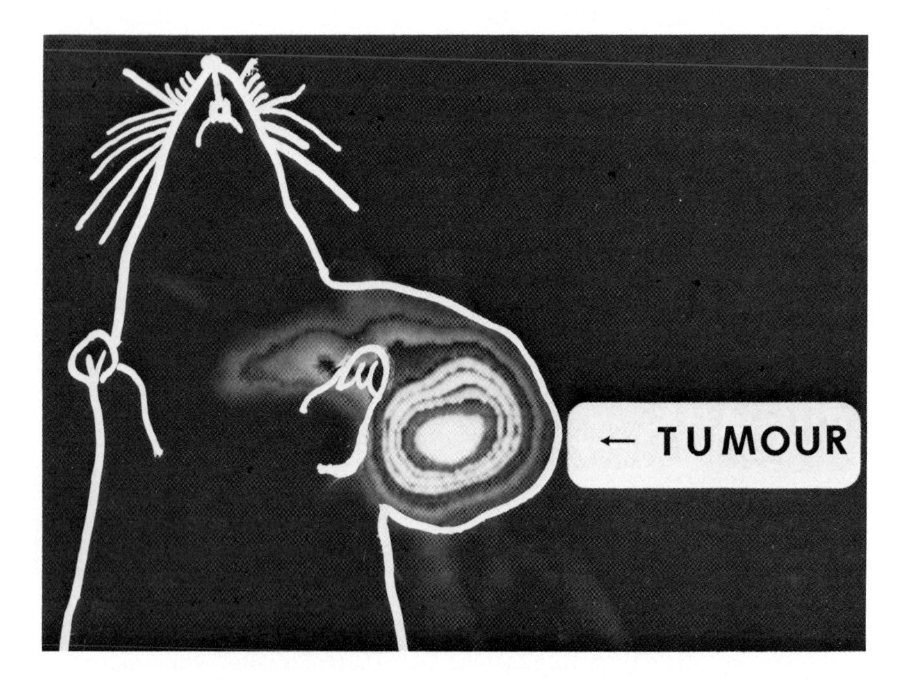

Fig. 2. Tomogram of tumour-bearing rat. The ^{64}Cu activity shows bright in the tumour. This PET scan was taken at 72 hours post-injection of ^{64}Cu-HpD.

CONCLUSIONS

These experiments show, in a preliminary manner, that it is possible to label HpD with a radionuclide suitable for (quantitative) tumour imaging. The early studies of chromatographic properties, *in-vivo* biodistribution, and photocytotoxicity suggest that the labelling procedure does not significantly affect the main characteristics of the HpD, which are important for the potential use of the labelled material for *in-vivo* quantitation during PRT.

Further studies are in progress:

a. to confirm, by experiments using ^{3}HpD and/or ^{14}C-HpD as controls, the detailed biodistribution properties of ^{64}Cu-HpD,
b. to extend the range of parameter values for the toxicity studies using the spheroid tumour model,
c. to correlate quantitative SPECT and/or PET measurements with the local concentration of HpD in tissues, for a range of tumour-bearing animals.

Finally, this work has given grounds for optimism that gamma- or positron-labelled HpD can be prepared. Ultimately this will find wide application in improving and controlling the delivery of PRT in the clinical situation, by enabling the HpD concentration *in-vivo* to be measured at the time of light irradiation.

ACKNOWLEDGEMENTS

This work was supported by the National Cancer Institute of Canada, and the Ontario Cancer Treatment and Research Foundation. The authors wish to thank also the staff of the PET scanner unit at McMaster University for their cooperation, and A. Devries for assistance in preparing the manuscript.

REFERENCES

Cohen L, Schwartz S (1966). Modification of radiosensitivity by porphyrins II transplanted rhabdomyosarcoma in mice. Cancer Res 26:1969.

Doiron DR, Profio E, Vincent RG, Dougherty TJ (1979). Fluorescence bronchoscopy for detection of lung cancer. Chest 76:27.

Durand RE (1975). Cure, regression and cell survival: a comparison of common radiobiological endpoints using an in vitro tumour model. Br J Radiol 48:556.

Falk JE (1964). "Porphyrins and Metalloporphyrins." New York: Elsevier Press.

Gomer CJ, Dougherty TJ (1979). Determination of ^{3}H- and ^{14}C- hematoporphyrin derivative distribution in malignant and normal tissue. Cancer Res 39:146.

Profio AE, Doiron DR, King FG (1979). Laser fluorescence bronchoscope for localization of occult lung tumours. Med Phys 6:523.

Sutherland RM, Durand RE (1976). Radiation response of multicell spheroids- an in vitro tumour model. Curr Top Radiat Res Quart 11:87.

Porphyrin Localization and Treatment of Tumors, pages 637–645

DESTRUCTIVE EFFECT OF PHOTORADIATION ON THE MICROCIRCULATION OF A RAT MAMMARY TUMOR GROWING IN "SANDWICH" OBSERVATION CHAMBERS

W.M. Star[1], J.P.A. Marijnissen[1],
A.E. van den Berg-Blok[2] and H.S. Reinhold[2,3].

[1]The Dr. Daniel den Hoed Cancer Centre and Rotterdam Radio-Therapeutic Institute, Rotterdam, The Netherlands.
[2]Department of Experimental Radiotherapy, Erasmus University, Rotterdam, The Netherlands.
[3]Radiobiological Institute TNO, Rijswijk, The Netherlands.

INTRODUCTION

Soon after a patient with a superficial tumour has been treated with photoradiation a purple discoloration appears in the target area. This may be observed one day after light exposure, but the earliest signs may appear already several hours after light exposure. The effect can be seen by comparing Fig. 6 and 7 of the paper by Dougherty et al. (Dougherty, Kaufman, Goldfarb, Weishaupt, Boyle, Mittleman 1978). The picture is reminiscent of a hematoma and suggests that effects on the blood circulation may play an important role in photoradiation therapy.

A model system being available (Reinhold, Blachiewicz, Berg-Blok 1979) that was specifically designed for the study of the microcirculation in tumours, it was decided to use this system to obtain information on the mechanism of photoradiation therapy.

MATERIALS AND METHODS

Female WAG/Rij rats bearing transparent observation chambers (diameter 9 mm) were prepared as reported previously (Reinhold, Blachiewicz, Berg-Blok 1979) and isologous RMA mammary tumours were implanted. When the diameter of a tumour was about 3 mm the animal was injected i.p. with 15 mg/kg Hpd (obtained from Oncology Research and Development Inc.). One day later the chamber with tumour was exposed to red light (632 $\pm$ 2 nm) from a dye laser at 60 mW/cm^2. Exposure times were 0, 6½, 10, 15, 22½, 35 and 50 minutes, 5 animals at each exposure value. As an additional normal tissue reference, to exclude possible chamber artifacts, the left ear was exposed during 15 minutes (standard) following the treatment of the chamber. Subsequently the right ear received the same exposure as the chamber, except for the 50 min. exposure, in which case both ears were exposed sequentially for 15 minutes. This was done because initial experience indicated that after 35 minutes exposure ears were damaged to such an extent that they became completely lost and that a longer exposure would not make much sense. Additional "controls" included chambers with tumour but without Hpd, exposed for 50 minutes. Possible hyperthermic effects were excluded by monitoring the temperature of the illuminated sites.

The bloodcirculation of the tumor and normal tissue in the chambers was observed with a microscope and photographed before, during and after light exposure. In addition, the effects on the chambers and the ears were described and scored. Follow-up was done every day for one week and subsequently every other day for as long as was considered useful, in view of the limited "life span" of the chambers. Usually this follow-up time lasted two to three weeks altogether.

In a separate series of experiments the viability of the tumours in the chambers was tested by removing the tumour from the chamber immediately after light exposure and re-implanting it into the flank of the same animal. This was done after 50 and 110 minutes exposure at 60 mW/cm^2.

RESULTS

After about 10-15 minutes exposure blood vessels in the tumour begin to "disappear". They seem to empty and tumour appears bleached. This is shown in Fig. 1 and Fig. 2 for a tumour before and after 35 minutes of light exposure. If illumination is stopped after about 20 minutes the circulation in the tumour (if affected) usually recovers partly or fully after a few minutes. If illumination is continued, after about 30-50 minutes the tumour vessels "reappear" but now the

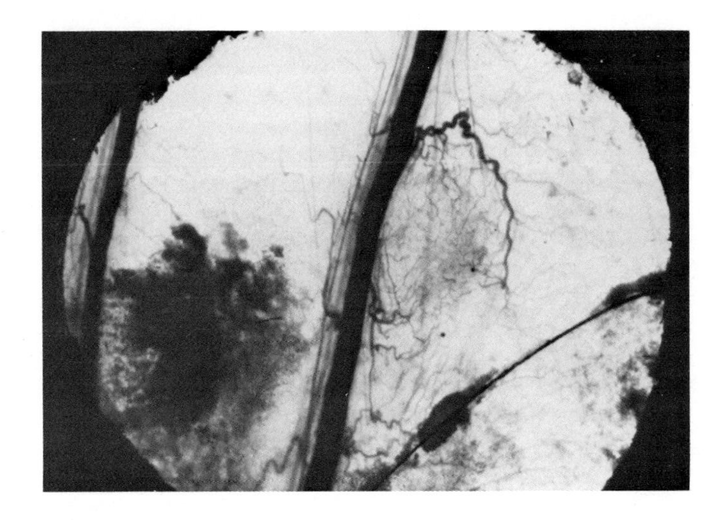

Fig. 1. Tumour in chamber close to a large normal tissue vessel(centre). The small tumour vessels are clearly visible. The circulation is normal. In the chamber a blood stain (below left) and a crack in the glass cover (below right).

Fig. 2. The same tumour as in Fig. 1 after 35 minutes of light exposure at 60 mW/cm^2. Many tumour vessels have "disappeared", some still circulate. Recovery of circulation after a few minutes. No effect on normal tissue visible.

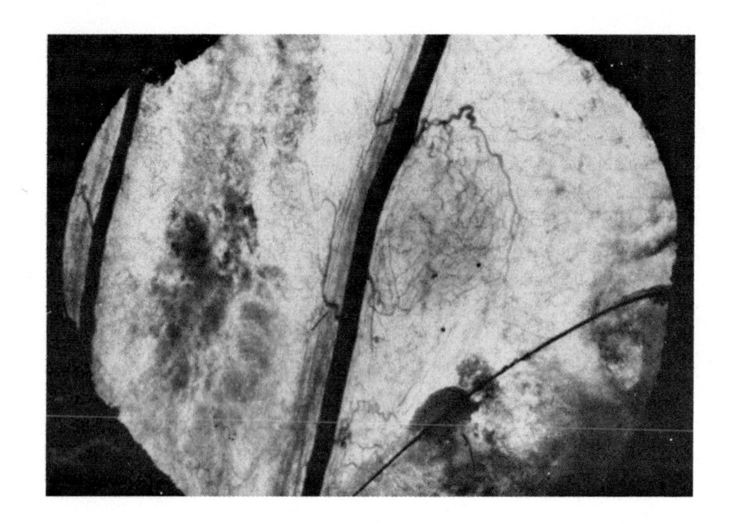

Fig. 3. The same tumour as in Fig. 2, two hours after light exposure. Blood vessels in the tumour appear filled again, but there is no circulation. Circulation stop in normal tissue as well, except large vessels and some small ones in their vicinity.

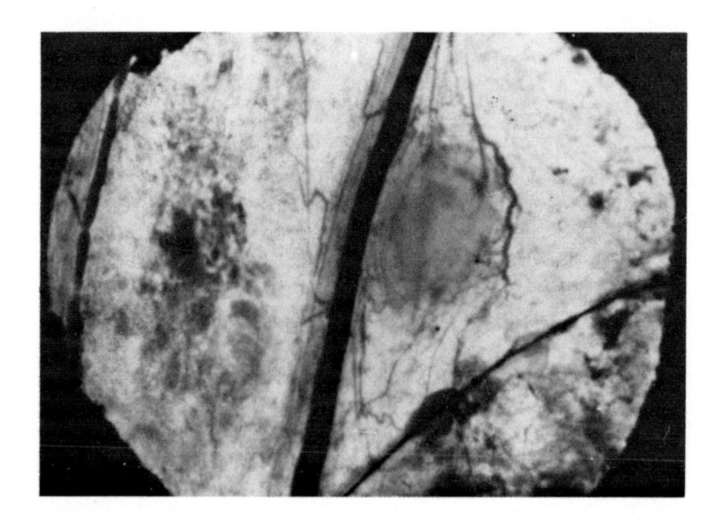

Fig. 4. The same tumour as in fig. 3, one day after light exposure. Necrosis in tumour. Circulation only in the two large normal tissue vessels.

circulation is slow and gradually decreases toward a complete stop. This may be followed by (diffuse) hemorrhages. The "re-appearance" of blood vessels is also observed when illumination is stopped in the "bleached" phase and may take about one hour to occur. This is shown in Fig. 3 for a tumour two hours after a 35 minute light exposure.

If any effect on the tumour circulation is observed during or immediately after light exposure, a complete stop of the circulation is usually seen after one day, with subsequent necrosis, as shown in Fig. 4. In this particular case there was no "cure". After about 10 days tumour regrowth was apparent, starting from the large normal tissue vessel in which the circulation had remained.

The blood vessels of the normal tissue in the chamber show effects similar to the tumour, but this seems to require a more prolonged exposure. Bleaching of the normal tissue microcirculation occurs after 30-40 minutes of light exposure. The larger normal tissue vessels appear to narrow. This is illustrated in Fig. 5 and Fig. 6. After 50 minutes exposure the tumour vessels are more visible than before, but there is practically no circulation. Vasodilatation is apparent. The larger normal tissue vessels, on the other hand, show vasoconstriction and are partly empty (Fig. 6, right). There is still circulation in about half of the normal tissue vessels. The empty normal tissue vessels in this case were filled with blood again 3 hours later, but then only the two largest vessels showed intact circulation. One day later there was a complete stop of circulation, with necrosis of the tumour tissue and no recovery occurred in the tumour nor in the normal tissue.

Occasionally white clots are seen, adhering to a normal tissue vessel wall. If there is still circulation, such a clot may be released and transported out of the chamber. These are probably platelet aggregates of the type reported by Bugelski (Bugelski, Porter, Dougherty 1981). When the circulation has stopped white clots occur more frequently but it is difficult to decide from "in vivo" observation of the chamber whether all these are composed of solid material or small pools of plasma without erythrocytes.

In Fig. 7 the percentage of tumours with a complete stop of circulation after at least one day is plotted versus exposure time. An open circle below a black circle means

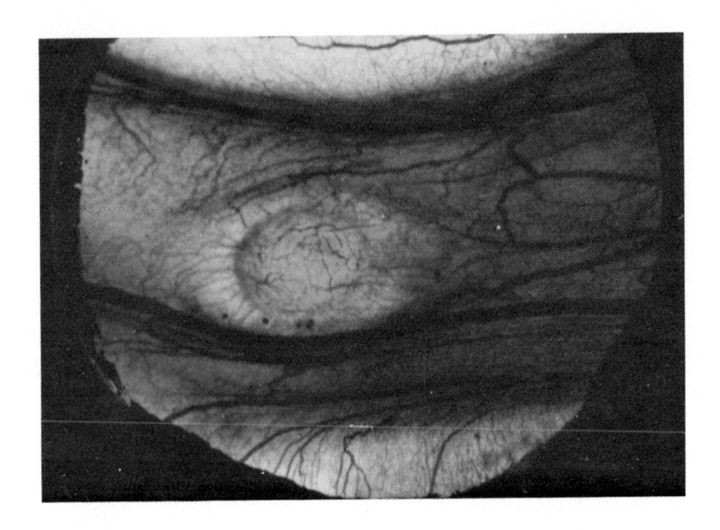

Fig. 5. Chamber with tumour before light exposure.

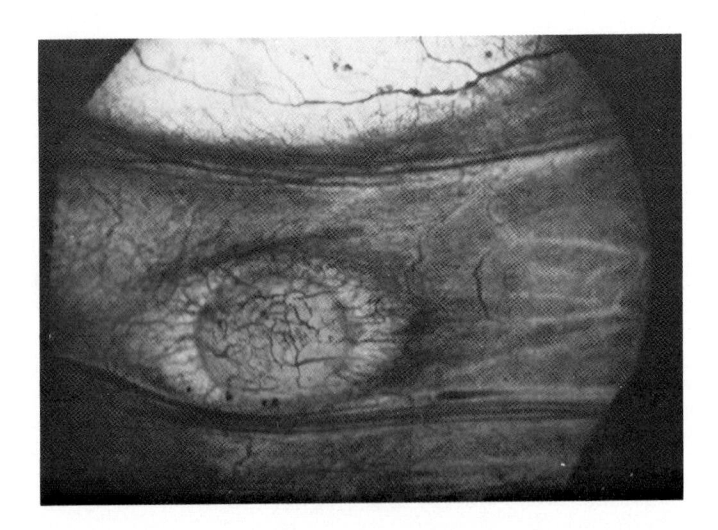

Fig. 6. The same tumour as in Fig. 5, after 50 minutes of light exposure at 60 mW/cm^2. Vasodilatation, but practically no circulation in tumour. Vasoconstriction, empty blood vessels and 50% circulation in normal tissue.

that one or more tumours did regrow during the period of observation following day one. Probit curves were computed for the two series of data represented by the black and the open circles.

A complete stop of normal tissue circulation was seen in all (5 of 5) chambers one day after 50 minutes exposure and in two (2 of 5) chambers after 35 minutes exposure. In all other cases there was only a partial stop of normal tissue circulation. A complete stop of normal tissue circulation one day after light exposure was never followed by recovery within the chamber.

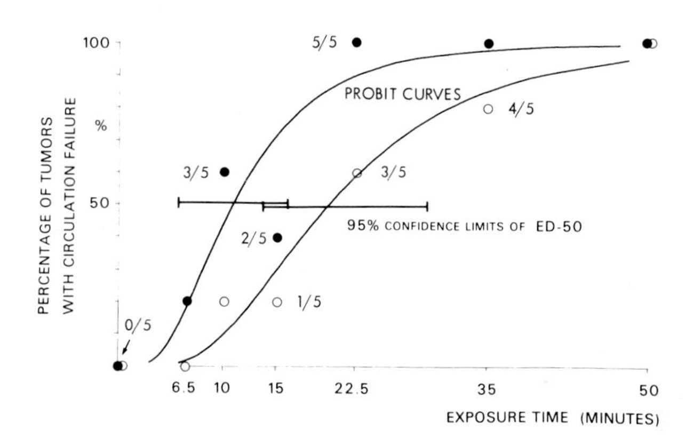

Fig. 7 Black circles: percentage of tumours with a complete stop of circulation one day after light exposure. Open circles: percentage of tumours with a complete stop of circulation after one day, without recurrence during the full period of follow-up. For each set of datum points a probit curve was computed.

The effects of photoradiation on the ears as a normal tissue reference will be reported only briefly here. The thickness of a normal rat ear is 0.35-0.40 mm. Immediately after light exposure the ear is pale and slightly purple, depending on exposure. Up till several hours after exposure no increase of thickness larger than 0.1 mm has been measured. One day after 6½ minutes exposure at 60 mW/cm^2, edema leads to a thickness of 1 mm and some erythema is seen, with complete recovery after 4-5 days. Upon 35 minutes exposure the ear eventually is completely or almost completely (more than 50%) lost. In between, after 15 minutes exposure, the average end result is loss of hair and loss of some tissue, viz. 1-2 mm from the edge of the ear but less than ¼ of the ear.

DISCUSSION

The effects reported in the previous section are not the result of induced artefacts. The temperature of the chambers and the ears was always a few degrees below normal body temperature during light exposure.

Tissue thickness in the chambers was 0.3-0.5 mm. Since the ears were 0.4 mm. thick the light dose values in both tissues should be the same. Due to backscatter this would have been different if, e.g. the skin of the flank had been used as a normal tissue reference. If we assume that the Hpd concentrations in the ears and in the normal tissue of the chamber are about the same, effects should be the same for the same light exposure. Indeed we have seen that after 35 minutes exposure an ear is damaged to such an extent that it will become almost completely necrotic. The normal tissue circulation in the chamber is almost completely destroyed by this dose as well.

Edema did probably not contribute to any of the effects on the tumours and normal tissues observed in the chambers because these effects occurred during a period after exposure when no increase of ear thickness (edema) could be detected yet.

From the exposure times after which "bleaching" of the smallest blood vessels in the tumour and the normal tissue is observed, one can roughly estimate a ratio of 2 between the Hpd concentration in the two tissues. This is about the same as the ratio measured by Gomer (Gomer, Dougherty 1979) for a mammary tumour in mice compared to skin 24 hours after Hpd injection.

The chamber and tumour of Fig. 1-4 showed a complete stop of circulation one day after light exposure, except for the two largest normal tissue vessels, one of which was close to the tumour. The tumour did recur. Considering the tumour data in Fig. 7 and the normal tissue effects mentioned in the previous section, one is tempted to think that a tumour nodule can only be cured when the normal tissue circulation close to a tumour is destroyed as well. Therefore a series of experiments was started whereby the tumour was removed from the chamber immediately after light exposure and re-implanted into the flank of the same animal. To date this has been done with 3 animals after 50 minutes of exposure, with 2 animals after 110 minutes exposure and for 5 controls without light exposure (110 minutes fits in our series of exposure times, which increase by factors 1.5). All tumours did regrow. There was no striking difference between the times it took for the various tumours to attain dimensions between 5 and 10 mm, viz., 2-3 weeks.

Our conclusion is therefore that tumour cure following photoradiation is not the result of direct tumour cell kill, but is probably secondary to destruction of the tumour microcirculation.

REFERENCES

Bugelski PJ, Porter, CW, Dougherty TJ (1981). Autoradiographic distribution of hematoporphyrin derivative in normal and tumour tissue of the mouse. Cancer Res 41:4606.

Dougherty TJ, Kaufman JE, Goldfarb A, Weishaupt KR, Boyle D, Mittleman A (1978). Photoradiation therapy for the treatment of malignant tumours. Cancer Res 38:2628.

Gomer CJ, Dougherty TJ (1979). Determination of (^{3}H)-and (^{14}C) hematoporphyrin derivative distribution in malignant and normal tissue. Cancer Res 39:146.

Reinhold HS, Blachiewicz B, Berg-Blok A (1979). Reoxygenation of tumours in "Sandwich" chambers. Eur.J.Cancer 15:481.

Porphyrin Localization and Treatment of Tumors, pages 647–655

PHOTORADIATION THERAPY OF LEWIS LUNG TUMORS IN MICE WITH LOW OPTICAL DOSE RATE AND HIGH HpD-DOSE

R. Ellingsen, L.O. Svaasand and A. Ødegaard*
Division of Physical Electronics
University of Trondheim
N-7034 Trondheim - NTH, Norway

*Department of Pathology
University of Trondheim
N-7000 Trondheim, Norway

SUMMARY

Lewis Lung carcinomas in B_6D_2 mice have been treated by photoradiation therapy (PRT). All irradiations were performed at 630 nm. The optical power levels were typically varied from 45-100 mW with a radiation duration of 60 min. This gave an optical energy of 160-360 J delivered through a centrally inserted optical fiber into the tumor during treatment. The optical power density and the temperature were monitored during exposure.

After treatment the animals were observed for at least two days before sacrifice. Biopsies have been taken for the histological follow-up. The experimental series were composed of non-treated controls and treated cases with variations in the chemical and optical dose scheme.

Only the high (50 mg/kg) Hematoporphyrin derivative (HpD)-dose gave curative response from the tumors. The low dose (5 mg/kg) gave no significantly different response from the control case by visual observation; however, histologically there was some difference.

INTRODUCTION

This work discusses PRT of Lewis Lung carcinoma in mice performed at low optical power levels combined with rather high doses of photosensitizer. Hematoporphyrin derivative (HpD) has been used.

We consider light and temperature dosimetry to be of great importance in PRT. For this reason those parameters have been monitored during the irradiation period.

The tumor response to this treatment is assumed to be dependent on the administered drug concentration (Kinsey et al 1981). This parameter was varied by one order of magnitude in two steps.

The response to treatment was evaluated by visual inspection for 2-6 days. The animals were then sacrificed and pathological sections were examined histologically.

MATERIALS AND METHODS

The tumors were rapidly growing Lewis Lung carcinomas. Tumor cells were injected subcutaneously and grew up in the shape of spheroids. 5 experimental series are reported. Each series was composed of one control and two HpD-injected animals, 5 mg/kg and 50 mg/kg, respectively. The tumor bearing animals were 6-8 week old B_6D_2 female mice. Some preliminary trials showed that a 6-9 day time lapse from tumor injection to start of treatment gave the best compromise between tumor dimension and the naturally developed necrosis of the center region. This necrosis is difficult to distinguish from necrotic regions induced by PRT in the succeeding histological examination.

HpD was manufactured by Oncology Research & Development, Inc. (Cheektowaga, NY, USA). The drug was injected into the subcutaneous tumor and the irradiation was given 24 hours later. At the time of treatment, the typical tumor dimension was 6-10 mm in diameter.

In each experimental series there were non-treated controls whose purpose was to represent a typical basis of spontaneously developing necrosis for the following examination by histology.

The light source was an argon-ion laser pumped dye-laser. One wavelength was used, 630 nm. Optical power delivered into the tumors varied from 45-100 mW with a 60 min. duration of the irradiation. This corresponds to an optical energy of 160-360 J. The fiber was threaded through a 21-gauge (0.8 mm outer diameter) needle which had been inserted into the tumor (see Fig. 1). No multiple excitation fiber insertions were used.

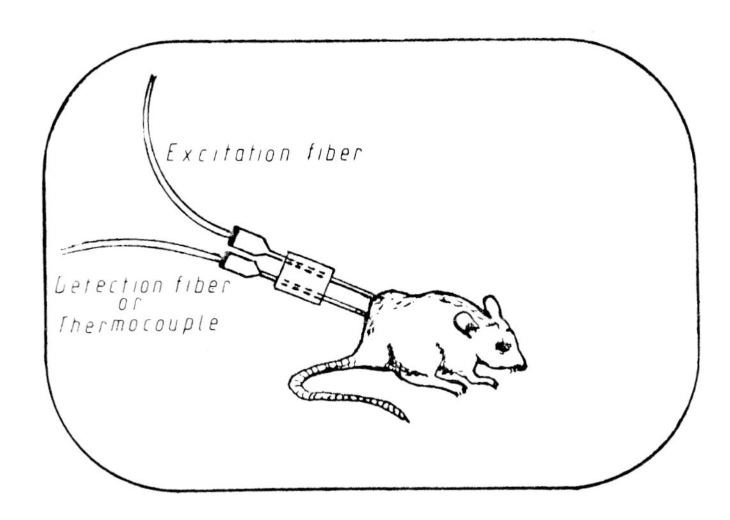

Fig. 1. Irradiation with simultaneous measurement of light intensity or temperature.

Excitation and detection fibers were silicone clad fused silica core fibers, core diameter 200 µm, from Quartz & Silice (Pithiviers, France). The fiber ends were cleaved optically flat. The thermal sensor was a sheathed chromel-alumel thermocouple wire of outer diameter 250 µm from Sodern, Thermocoax (Suresnes, France).

Detected light was transmitted through approximately 1 m of fiber and measured by a silicon PIN detector; further analyses of signals were performed by standard lock-in amplifier technique with light chopping.

An observation period succeeded treatment; in the cases referenced in this report the animals were sacrificed 2-6 days later.

The tumor response was described by visual inspection and measurement of tumor diameter. In order to attain a description of tumor cell development, the post-treatment histological evaluation was performed. A slice of 3-4 mm thickness of tumor tissue was dissected and fixed for 12 hours in a solution of 80% EtOH, 15% formaldehyde and 5% acetic acid. Subsequently the specimens were kept in 80% EtOH in H_2O until the final preparations (sectioning, staining) for light microscopy took place.

RESULTS

Preliminary experiments showed that 14 day old tumors (time measured from subcutaneous tumor cell injection) developed a necrotic center volume of approximately 1/2 - 3/4 of the total tumor volume; in those cases the approximate tumor diameter was about 15 mm. In order to reduce this problem of natural necrosis, the animals were entered into treatment 6-9 days after tumor cell injection. This scheme improved the model to an acceptable level.

Figure 2 shows measured temperatures at 4 mm distance in 3 cases; two animals were irradiated with 100 mW optical power and the third with 45 mW optical power at 630 nm for 60 min. A 4 mm distance away from the light source corresponds to approximately 2.5 optical penetration depths at this excitation wavelength. (See Svaasand, Ellingsen in press, for definition of penetration depth). The optical penetration depth for those tumors has been measured to be between 1.2 - 1.4 mm (*in vivo* measurements) at 630 nm (Ellingsen et al to be published).

The temperature measurements were carried out with the thermocouple situated at the body side of the tumor.

Both cases irradiated by 100 mW optical power (# 1 and # 2) show maximum temperature rises of approximately 5.0°C above body temperature at 4 mm distance. Whether this situation will induce hyperthermia or not depends on the body temperature. Hyperthermal effects are assumed to appear at 43°C for this irradiation duration (Short, Turner 1981).

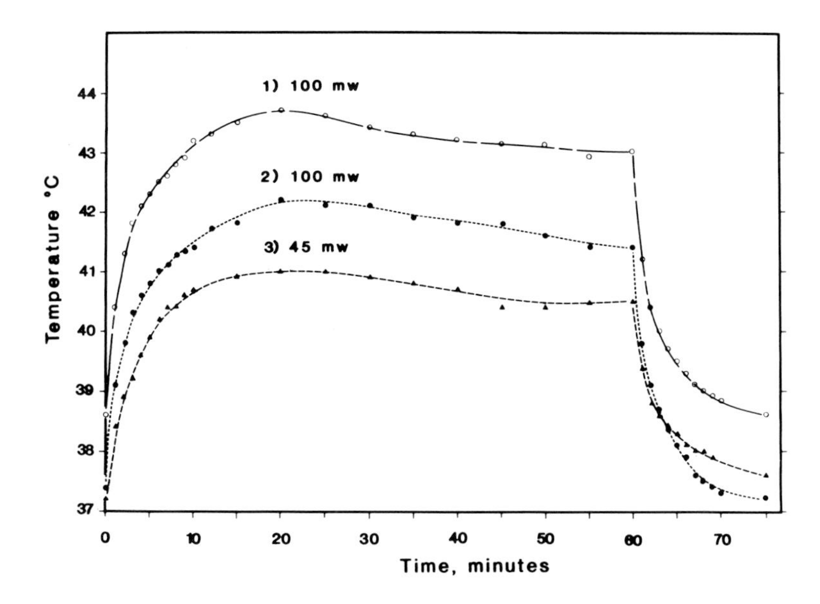

Fig. 2. Tumor temperature at 4 mm distance from fiber end versus time. Irradiation period of 60 min. Two cases irradiated by 100 mW transmitted power: # 1) initial body temperature of 38.6°C (o); # 2) initial body temperature of 37.3°C (●). One case irradiated by 45 mW; # 3) initial body temperature of 37.2°C (▲). Wavelength λ = 630 nm.

Case # 3 (45 mW) shows a temperature rise of only 3.8°C above body temperature at 4 mm distance.

From the diagrams in Fig. 2 one can also observe a typical tendency towards a slow temperature reduction following the maximum temperature rise that is reached after 20-25 min. from energy onset. This phenomenon is probably due to an increased heat loss as a result of an overall vasodilatation induced by the energy supplied to the animal.

The light intensity at 3 mm distance from the light source expressed by space irradiance in arbitrary units is shown in Fig. 3 for two cases of transmitted optical power; a) 100 mW and b) 45 mW.

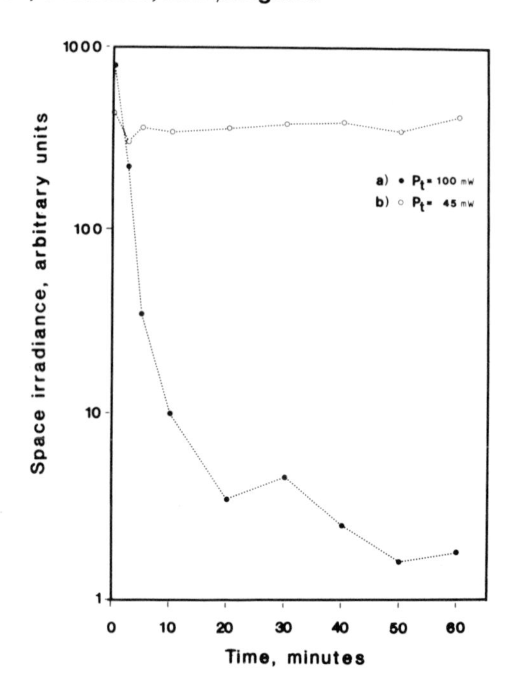

Fig. 3. Space irradiance at 3 mm distance from fiber end monitored during 60 min. irradiation of Lewis Lung carcinoma; B_6D_2 mice. Two cases; irradiated by a) 100 mW (●) and b) 45 mW (o). Both animals were injected with 5 mg/kg HpD 24 hrs. prior to irradiation. Wavelength λ = 630 nm.

Repeated experiments gave typically the same time course for the 100 mW case. This gives evidence that 100 mW transmitted optical power at λ = 630 nm through a single, simply cut optical fiber produces carbonization and increase of optical absorption around the fiber end. This affects the space irradiance at 3 mm distance to be reduced by 2-3 orders of magnitude during the 60 min. period; the main drop takes place within 10 min. from the start.

Case b) shows the time course at 45 mW optical power. The space irradiance is kept almost constant throughout the irradiation. It has been found that 50 mW represents a maximum limit for transmitted optical power at the 630 nm wavelength.

Figure 4 presents micrographs of: A) non-treated control, and 3 treated cases; B) -5mg/kg HpD; C) -50 mg/kg

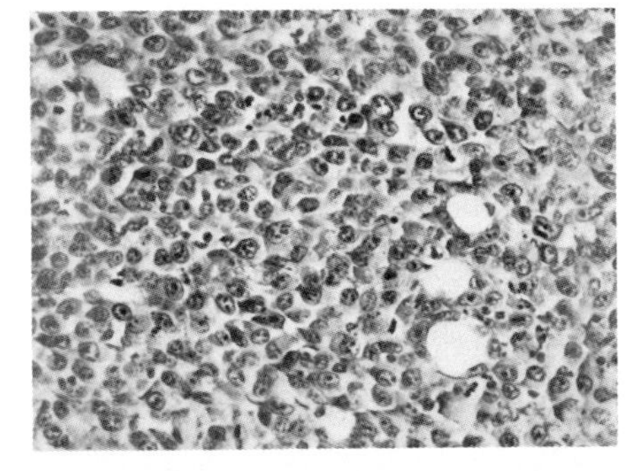

Fig. 4. A) Non-treated control. Lewis Lung carcinoma infiltrating fat tissue. × 110

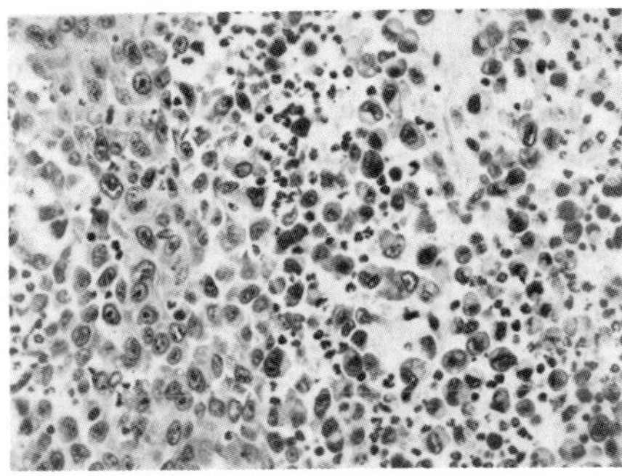

Fig. 4. B) Treated, 5 mg/kg HpD + 2 days. Necrotic tumor tissue with acute inflammatory reaction (infiltration of neutrophilic granulocytes). In the left field unaltered tumor tissue. × 110

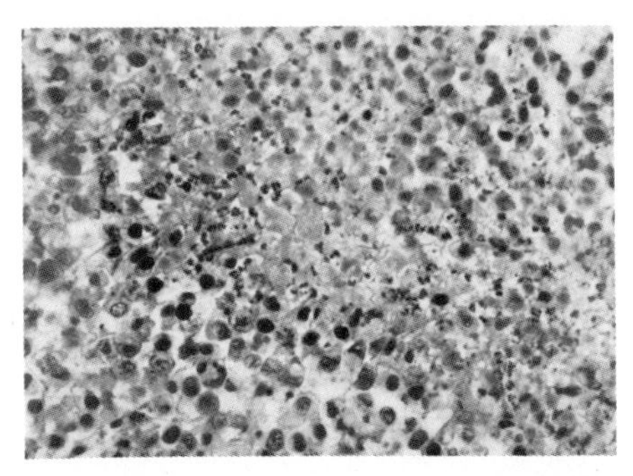

Fig. 4. C) Treated, 50 mg/kg HpD + 2 days. Extensive tumor necrosis with intense inflammatory reaction. To the left large numbers of degenerated tumor cells with pycnotic nuclei. × 110

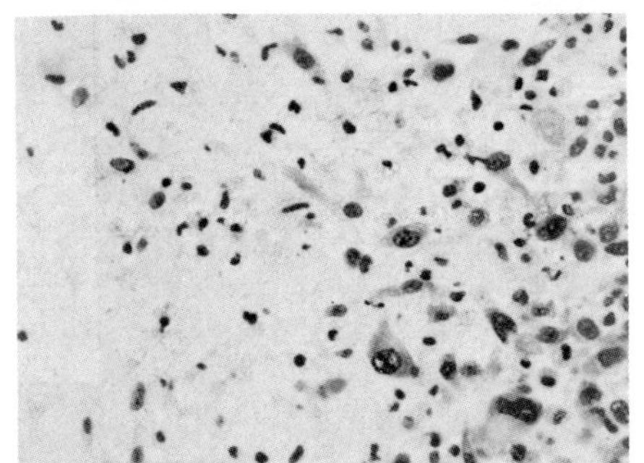

Fig. 4. D) Treated, 50 mg/kg HpD + 6 days. Acellular, post-necrotic part of the tumor with distinct fibroblastic activity. In the right field small focus of probably vital profilerating tumor tissue. × 110

HpD; D) -50 mg/kg HpD. B), C) and D) were all irradiated for 60 min. with 45 mW transmitted optical power at 630 nm wavelength. A), B) and C) were sacrificed 2 days post treatment. D) was sacrificed 6 days post treatment.

The visual observations (summarized in Table 1) are mainly confirmed by these micrographs. The extensive tumor necrosis and inflammatory reaction of C) is specific. There are, however, differences between A) (control) and B) (5 mg/kg HpD). While little or no necrosis can be observed in A) there are both necrotic areas and an acute inflammatory reaction observed in B). This state is probably induced by the phototherapy.

D) shows a situation where similar conditions as in C) are allowed to develop for 6 days. The area is dominated by an acellular, post-necrotic part of the original tumor with distinct fibroblastic activity. But still a focus of proliferating tumor tissue may be seen. Whether this is a viable tumor rest or not is an unanswered question, since no long term follow-ups have been performed.

Table 1 summarizes the visual observations of 5 series, each consisting of the three cases listed in column 1.

Injected HpD-dose [mg/kg]	Number of treated cases	Tumor volume relative to control at sacrifice	Time from irradiation to sacrifice [days]
50	5	3 cases - 0.7 1 case - 0.5 1 case - 0.2	2 3 6
5	5	1	2-6
Control	5	1	2-6

Table 1. Summary of visually observed results of PRT of Lewis Lung tumors in mice.

DISCUSSION

This *in vivo* experiment concerning the relation of tumor response to optical and chemical doses, indicates that for this specific tumor and the light delivery system consisting of one single fiber, the optical power has to be limited to maximum 50 mW. An initially higher optical power induces clotting and carbonization at the fiber tip. This increases light absorption to a non-acceptable level for phototherapy of the tumor.

In order to compensate for this rather low optical dose rate the HpD-dose is increased by a factor of 10 relative to the one usually used, 5 mg/kg. This treatment scheme is shown to be effective both by visual inspection and by histology. Further, this scheme should be regarded as dominated by photodynamic mechanisms.

REFERENCES

Ellingsen R, Svaasand LO, Ødegaard A (1983). To be published at International Symposium on Porphyrins in Tumor Phototherapy, Milan, Italy, May. Plenum Press.

Kinsey JH, Cortese DA, Moses HL, Ryan RJ, Branum EL (1981). Cancer Res. 41, 5020-26, December.

Short JG, Turner PF (1981). Proc. IEEE, 68, No. 1, 133-142, January.

Svaasand LO, Ellingsen R (1983). Optical properties of human brain. Photochem. Photobiol., in press.

Porphyrin Localization and Treatment of Tumors, pages 657–660

PORPHYRIN:PROTOPORPHYRIN:PLASMA PROTEIN INTERACTION - THE METABOLIC BASIS FOR THE TUMOR LOCALIZATION OF HEMATOPORPHYRIN DERIVATIVE - A PRELIMINARY REPORT

Mohamed El-Far, M.Sc., Ph.D.; FRSC
Visiting Senior Fulbright Scholar (Egypt)
Division of Gastroenterology/Nutrition
Department of Medicine
University of California, Davis
Davis, California 95616

Neville Pimstone, M.D., F.C.P. (S.A.)
Professor of Medicine and Chief,
Division of Gastroenterology/Nutrition
Department of Medicine
University of California, Davis
Davis, California 95616

INTRODUCTION

Many researchers have attempted to define the mechanism whereby hematoporphyrin derivative localizes in epithelial and mesenchymal tumors. The anatomic, vascular and lymphatic structure of tumors may differ and the retention of porphyrins in tumor may relate to its binding to tumor protein constituents (Bugelski et al., 1981). Moan et al. (1982) reported that crude hematoporphyrin (HP) and HPD were similar complex mixtures of porphyrins but that HPD had relatively more protoporphyrin and less pure HP. In vitro, HPD proved superior to HP in his tumor cell model with regard to porphyrin uptake. Berenbaum et al. (1982) reported that HP, hydroxylethylvinyldeuteroporphyrin, and protoporphyrin are individually inactive whereas the mono- and diacetates of HP as well as the acetoxy-ethylvinyldeuteroporphyrin were active. However, the activation required alkaline treatment which destroys the acetoxy function and the products of alkaki treatment are HP, hydroxylethylvinyldeuteroporphyrin and protoporphyrin. When the products were tested in their in vivo tumor model, there was effective tumor necrosis.

We have reported data which suggests strongly that the ability of injected porphyrin to enter tumor tissue relates to the relative binding affinity between the porphyrin binding proteins in plasma and tumor constituents (El-Far, Pimstone 1983). Thus, non-tumor localizing porphyrins have higher binding affinity to human serum albumin (HSA) and mouse serum proteins (MSP) (El-Far, Pimstone 1983). Of the group of non-tumor localizing porphyrins studied, protoporphyrin had the highest affinity to both HSA and MSP. This led to the hypothesis that HPD works because it is a complex mixture of porphyrins which contain protoporphyrin, and that the high binding affinity of the latter to plasma proteins allows the other porphyrin components to circulate "unbound" thereby allowing their tumor uptake. To test this hypothesis, we conducted two series of experiments. The first was to compare the HPD porphyrin components in the native solution injected with that retained by tumor and the second to pretreat animals with protoporphyrin, following the latter with injections of a series of non-tumor localizing porphyrins to see whether by prior saturation of porphyrin binding sites in circulating plasma proteins, the porphyrins under study would enter tumor tissues.

MATERIALS AND METHODS

The tumor model was the transplantable KHJJ mammary carcinoma in the mouse (Balb C) as described in an accompanying paper (El-Far, Pimstone 1983). The porphyrins under study were uroporphyrin isomers I and III, 7-,6-,5-, 4-COOH porphyrins of the isomer I series, and the deuteroporphyrin 2,4 disulfonic acid (DP). These were administered as previously reported (El-Far, Pimstone 1983). Protoporphyrin (PP) was dissolved in warm dimethyl sulfoxide.

Animals received PP intravenously at a dose of 40mg/kg. One hr later, each of the porphyrins under study was given separately. The animals were sacrificed 10 hours later, the tumor excised, screened for porphyrin fluorescence and the porphyrins were quantitated as described previously (El-Far, Pimstone 1983).

In another series of experiments, HPD (Photofrin) and Photofrin II were given intravenously and 10 hours post injection, the tumors were excised and the porphyrins characterized as their methyl esters. The native solutions, i.e., both HPD and Photofrin II also were esterified and

and their porphyrins characterized by thin layer chromatography (TLC).

Biologic photodynamic activity was assessed following exposure of the tumor to 200 joules of red light derived from a tunable dye laser as reported elsewhere (El-Far, Pimstone 1983).

RESULTS

Tumor Uro-PI content in protoporphyrin treated animals (200μg/g dry wt) was 10 times greater than in those animals receiving Uro-PI alone. Non-tumor localizing porphyrins studied all were present in the tumor at levels greater than 120μg/g dry wt. When characterized, the porphyrin methyl esters in the tumor corresponded to the test porphyrin given and in no case was protopoprhyrin detected. In one specific experiment where uroporphyrin was mixed with 7-COOH porphyrin at a ratio of 9:1, these porphyrins were recovered in the tumor in their same relative concentrations.

The profile of porphyrin components in the tumors of animals receiving HPD (Photofrin) and Photofrin II showed multiple components identical to those seen in the native solution injected with the exception that in every case, a band of fluorescence corresponding in mobility with protoporphyrin was not identified in tumor tissues.

Photodynamic necrosis of the 1 cm tumor was total, i.e., 1 cm with all test porphyrins studied.

DISCUSSION

We have demonstrated that whereas the native HPD solution injected has five main bands corresponding to different porphyrins, the tumor has only four - the missing band having the same mobility as protoporphyrin. This is consistent with protoporphyrin remaining in the circulation bound to proteins such as albumin, but does not rule out the possibility that protoporphyrin enters the tissue and rapidly is metabolized to heme and bilirubin. However, our finding that protoporphyrin pretreatment promotes entry of non-tumor localizing porphyrins into tumor and augments 10-fold Uro-PI uptake, strongly supports the hypothesis that high binding affinity between plasma proteins and protoporphyrin allow porphyrins injected to circulate unbound, permitting greater affinity to tumor constituents than to blood. We believe this to be the basis for the tumor localization of HPD. Protoporphyrin pretreatment provides a novel method to modulate tumor porphyrin content. This finding has major clinical significance in that the increased fluorescence achieved should make porphyrins more effective markers of early malignancy. Moreover, the amplified tumor uptake of porphyrins of an order of magnitude greater than with HPD and in the same experimental conditions, should considerably enhance the efficacy of photoradiation therapy.

REFERENCES

Berenbaum MC, Bonnett R, and Scourides PA (1982). In Vivo Biological activity of the components of Haematoporphyrin Derivative. Br J Cancer 45:571-581.

Bugelski PJ, Porter CW, and Dougherty TJ (1981). Autoradiographic distribution of Haematoporphyrin derivative in normal and tumor tissue of the mouse. Cancer Res 41:4606-4612.

El-Far MA, Pimstone NR (1983). Tumor localization of Uroporphyrin Isomers I and III and their correlation to Albumin binding. 1st International Meeting on Cell Biochemistry and Function, March 1983, U.K.

El-Far MA, Pimstone NR (1983). Superiority of Uroporphyrin Isomer I Over other Porphyrins in Selective Tumor Localization in Mouse Carcinoma. Cancer Res. (Submitted).

Moan J, Christensen T, and Sommer S (1982). The Main Photosensitizing Components of Hematoporphyrin Derivative. Cancer Letters 15:161-166.

ACKNOWLEDGEMENTS

This work has been supported by generous grants in part from the Elsa U. Pardee Foundation, University of California Cancer Research Coordinating Committee and the Fulbright Commission. Please direct requests for reprints to Dr. Neville Pimstone, Professor of Medicine and Chief, Division of Gastroenterology, UCD Professional Building, 4301 X Street, Sacramento, CA 95817, U.S.A.

Porphyrin Localization and Treatment of Tumors, pages 661–672

A COMPARATIVE STUDY OF 28 PORPHYRINS AND THEIR ABILITIES TO LOCALIZE IN MAMMARY MOUSE CARCINOMA: UROPORPHYRIN I SUPERIOR TO HEMATOPORPHYRIN DERIVATIVE.

Mohamed El-Far, M.Sc., Ph.D.; FRSC
Visiting Senior Fulbright Scholar (Egypt)
Division of Gastroenterology/Nutrition
Department of Medicine
University of California, Davis
Davis, California 95616

Neville Pimstone, M.D., F.C.P.(S.A.)
Professor of Medicine and Chief,
Division of Gastroenterology/Nutrition
Department of Medicine
University of California, Davis
Davis, California 95616

ABSTRACT

Hematoporphyrin derivative (HPD), a complex mixture of porphyrins has been used clinically as a tumor localizer both for diagnostic and therapeutic purposes. Relative lack of tumor specific uptake limits widespread clinical application. In an attempt to circumvent this problem, we studied 28 porphyrins with widely differing properties using a transplantable KHJJ mammary carcinoma in the mouse (Balb C) as a tumor model. Twenty hours after porphyrin (P) administration, the tumor, skin and gastrointestinal tract were excised and the latter lavaged with physiologic saline. Tissue porphyrin content was assessed visually by red U.V. fluorescence, and by quantitative fluorometric extraction, and photodynamic activity was evaluated in vivo using a tuneable dye laser emitting red light (615-640 nm). (1) Of the five porphyrins which were taken up by tumor tissue, i.e. HPD (photofrin), photofrin II, meso-tetra (4-

carboxyphenyl) porphine (TCPP), tetra sodium-meso-tetra (4 sulfonatophenyl) porphine (TPSS), and uroporphyrin I (UROP I), skin and intestinal fluorescence also was marked with the notable exception of those mice receiving UROP I. (2) UROP I clearly was superior to HPD in this study in that the tumor to skin porphyrin content ratio was more than seven times greater than the ratio observed with both HPD and photofrin II, (3) As no measureable UROP was present in the gut, the tumor:intestinal P ratio under conditions of assay was for practical purposes infinity. (4) Photodynamic necrosis in the tumors of UROP I treated animals was similar to that seen following treatment with HPD. We conclude that UROP I appears the most specific tumor localizing porphyrin yet studied. This specificity suggests major potential clinical application both as a diagnostic marker for early mucosal cancer, and in photoradiation therapy. Moreover, the prolonged photocutaneous side effects as seen with HPD are unlikely.

INTRODUCTION

Broad application of hematoporphyrin derivative (HPD) in clinical oncology both for photoradiation therapy (Dahlman, Wile et al 1983; Dougherty 1981) and for the diagnosis of pre- and early malignancy (Benson, Farow et al 1982; Gregorie, Horger et al 1968; Kelly, Snell, 1976; Lipson, Baldes et al 1970) has been hampered by lack of understanding of what active component is the tumor localizer. Like its parent compound crude hematoporphyrin (HP), HPD is a complex mixture of vinyl porphyrins, HP monoacetate, HP diacetate, protoporphyrin and several additional derivatives of deuteroporphyrin (Berenbaum, Bonnett et al 1982; Clezy, Hai et al 1980; Moan, Christensen 1980). Even though there is an increased tumor:neighboring tissue porphyrin content ratio following HPD administration, the amount retained by normal tissue such as skin is significant. This is sufficient to result in nonspecific skin necrosis following photoradiation therapy of cutaneous malignancies (Dahlman, Wile et al 1983), and photocutaneous side effects which persist 4 to 6 weeks after HPD administration (Berns 1983; Dahlman, Wile et al 1983). Moreover, autofluorescence in non-malignant mucosa limits fluorescent descrimination of premalignant lesions in the bronchus (Balchum, Doiron 1982). Thus, over the past decade, much effort has been devoted to define which of the components of HPD are responsible for photosensitizing properties in cancer (Schwartz, Absolon 1955; Berenbaum, Bonnett et al 1982). A major drawback to contemporary studies has been the quantitation of tumor and tissue HP content where methods used have been described as tedious, where complete porphyrin extraction is difficult and accuracy of measurements hindered by the quenching of fluorescence by tissue components (Gomer, Dougherty 1979).

In this manuscript, we report a rapid, simple and reproducible assay for tissue porphyrin content without the above disadvantages. We also, on the basis of a study on the tumor localizing behavior of 28 different porphyrins, were the first to identify that uroporphyrin I (UROP I), a single pure porphyrin exhibits properties superior to other known tumor localizing porphyrins such as HPD when tested in animals (El-Far, Pimstone 1983; El-Far, Pimstone 1983).

MATERIALS AND METHODS

Porphyrins Under Study

The following 28 porphyrins studied: uroporphyrin I; uroporphyrin I and III octamethyl ester; heptacarboxylic porphyrin I; pentacarboxylic porphyrin I; coproporphyrin I, II* and III; deuteroporphyrin IX, deuteroporphyrin 2,4 disulfonic acid; deuteroporphyrin 2,4 bisglycol; deuteroporphyrin 2,4 bisacetal; deuteroporphyrin 2 acetyl*; diacetyl deuteroporphyrin*; deuteroporphyrin dimethyl ester; diacetyl deuteroporphyrin dimethyl ester*; deuteroporphyrin 2 acetyl dimethyl ester*; proto- and Zn protoporphyrin; mesoporphyrin IX; meso-tetraphenylporphine; meso-tetra (4-pyridyl) porphine; meso-tetra (4-carboxyphenyl) porphine (TCPP); tetra sodium-meso-tetra (4-sulfonatophenyl) porphine (TPPS); porphine; phthalocyanine; hematoporphyrin derivate (HPD); and Photofrin II. The latter two products were purchased from Oncology Research & Development, Inc., Cheektowaga, N.Y. and those porphyrins annotated by asterisks were synthesized by Dr. Kevin Smith, Professor of Chemistry, UC Davis, who kindly supplied these for the study. Other porphyrin compounds were purchased from Porphyrin Products, Logan, Utah.

Four mg of each porphyrin was dissolved in 100 μl of dimethyl sulfoxide and diluted to a final concentration of 4 mg/ml with 0.5% sodium bicarbonate in Dulbecco's phosphate-buffered physiological saline according to the method of Berenbaum, Bonnett (1982).

The pH of the final solution was adjusted to 7.3 and injections were made a minimum of 3 hr following preparation of the porphyrin solution.

Characterization of "Native" Porphyrins Injected

The porphyrin solution prepared for injection, as above, was spotted on Silica Gel G (Merck) plates and characterized by using thin layer chromatography (TLC). The plates were developed in the solvent system benzene, methanol, formic acid (98-100%), 84:15:1.2 (by volume) respectively according to Henderson and Christie, 1980.

Experimental Protocol

The tumor model used was a transplantable KHJJ mammary carcinoma (Rockwell, Kallman et al 1972). Balb C mice weighing approximately 25 gm were the recipients of subcutaneously transplanted tumor particles which 7-14 days post transplantation grew to a tumor nodule 0.25-1.0 cm in diameter. At this size, the small tumor was homogenously white, and spontaneous necrosis minimal or absent.

Porphyrins tested were injected intravenously (I.V.) in a final volume of 0.25 ml and at a dose of 40 mg/kg body weight. A minimum of four animals bearing tumor were used for testing each porphyrin. Twenty hours after porphyrin administration, the tumor was excised in toto, the skin dissected free of hair and the gastrointestinal tract excised following intraluminal lavage with physiologic saline. In all studies, tissue porphyrin content was assessed visually by red U.V. fluorescence and graded 1+ (trace) to 3+ (marked). In selected models, tissues were freeze dried, porphyrins extracted quantitatively and measured fluorometrically against a standard pure porphyrin.

Tissue Porphyrin Quantitation

Porphyrin content in tumor, skin and gut were quantitated as we recently have reported (El-Far, Pimstone 1983). Tissues were freeze dried in the dark for 4 hours using Virtis research equipment, Gardiner, New York. The freeze dried tumor was ground into a fine powder and a known weight subjected to porphyrin extraction by homogenization with 1 N aqueous perchloric acid and methanol (1:1, V/V) in the dark using a Potter-Elvenhjem homogenizer with a Teflon pestle. The homogenate was then centrifuged (4000 rpm) using a Sorval bench top centrifuge for 10 minutes. The tissue precipitate was re-extracted until no fluorescence was observed under U.V. light, and the combined supernatant filtered through Whatman #1 filter paper and its volume recorded. The supernatant content was quantitated fluorometrically using a Farrand spectrofluorimeter Model MK1. The concentration was determined using the porphyrin under study as the standard under optimal conditions for photoexcitation and emission fluorescence.

Biological Photodynamic Activity

The photodynamic activity of UROP I, HPD and Photofrin II treated mouse mammary carcinoma was tested 18-20 hours post porphyrin injection. An animal receiving a non-tumor localizing porphyrin, monoacetal mono-hydroxy deuteroporphyrin served as control. The skin overlying the tumor was shaved and the tumor exposed to 300 joules (0.5 watt/sq.cm.) of red light (615 or 625 nm) emitted from a circulating dye laser (Spectra Physics, Mountain View, CA, Model #375, DCM (Exciton Chemical Co., Inc., Dayton, OH) being the lasing medium. The dye was excited by an Argon laser (Spectra Physics Model #171). Wavelength was callibrated using a spectroscope (Instruments SA, Inc.). 615 nm was the exciting waveband with UROP I which corresponded with its maximum absorption peak in the red region. Twenty hours post photoradiation treatment, the animal was sacrificed, the tumor excised and examined histologically by routine microscopy following hematoxylin/eosin staining.

RESULTS

Obvious tumor fluorescence was noted in only 5 of the 28 porphyrins tested namely HPD, Photofrin II, TCPP, TPPS and UROP I. The fluorescence in 30% of the animals studied was not homogeneous but was more intense in the periphery than in the center of the tumor. With the exception of UROP I, at the time the animal was sacrificed, both skin and intestinal fluorescence was marked (3+).

TABLE 1

Tumor:Non-Tumor Porphyrin Content Ratios

Porphyrins Studied (40mg/Kg/I.V.)	Tumor:Skin	Tumor:GI* Mucosa
UROP-I	20.4	∞
HPD (Photofrin)	2.5	1
Photofrin II (SPP)	2.9	1

*GI (gastrointestinal) mucosa data identical for stomach, small and large intestine.

The tumor, skin and gut porphyrins were quantitated in those animals receiving UROP I, HPD (Photofrin) and a derivative thereof Photofrin II. Tumor porphyrin content for UROP I was 16.93 ± 3.58 (n = 15), or HPD 25.0 ± 2.55 (n = 5), and for Photofrin II 20.6 ± 3.36 (n = 5). Corresponding levels for skin for UROP I was 0.83 ± .19 (n = 15), for HPD 10.0 ± 2.45 (n = 5), and for Photofrin II 7.0 ± 0.8 (n = 5). Whereas no detectable UROP I was present in the gastrointestinal tract (stomach, colon and small intestine) for HPD, the levels ranged between 17.5-28.5 (n = 5) and for Photofrin II 19-26 (n = 5). Thus, the absence of detectable porphyrins in the gut in animals treated with UROP I indicated tumor:gut porphyrin ratio which for practical purposes is infinity (Table 1). This is in striking contrast to the 1:1 ratio observed using HPD and Photofrin II. In the skin, despite the absence of visual fluorescence, minor amounts of porphyrins were detectable by quantitative analysis following UROP I injection. The tumor:skin porphyrin content ratio with UROP I of 18 was more than seven times greater than that of the 2.5:1 ratio observed with HPD and Photofrin II (Table 1). This is consistent with a rapid clearance of UROP I

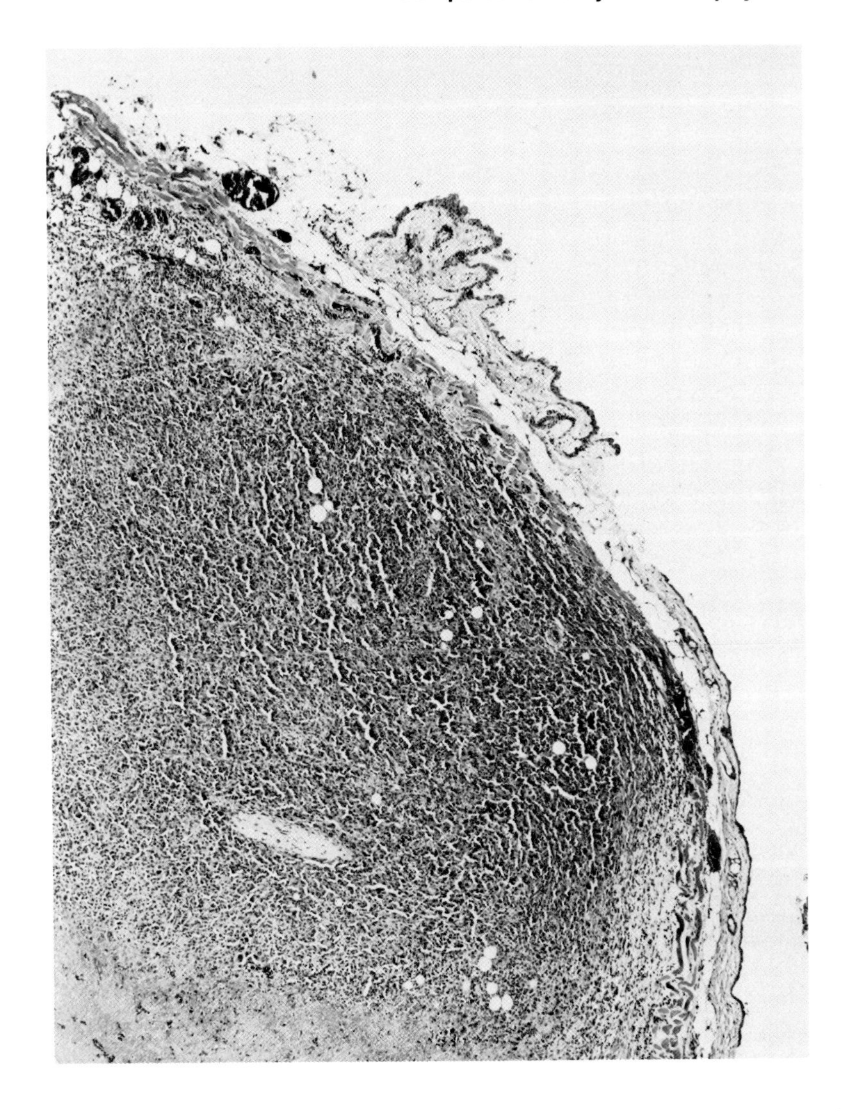

Figure 1. Photodynamic action of uroporphyrin I in Transplanted mammary carcinoma (anaplastic squamous) (x152).

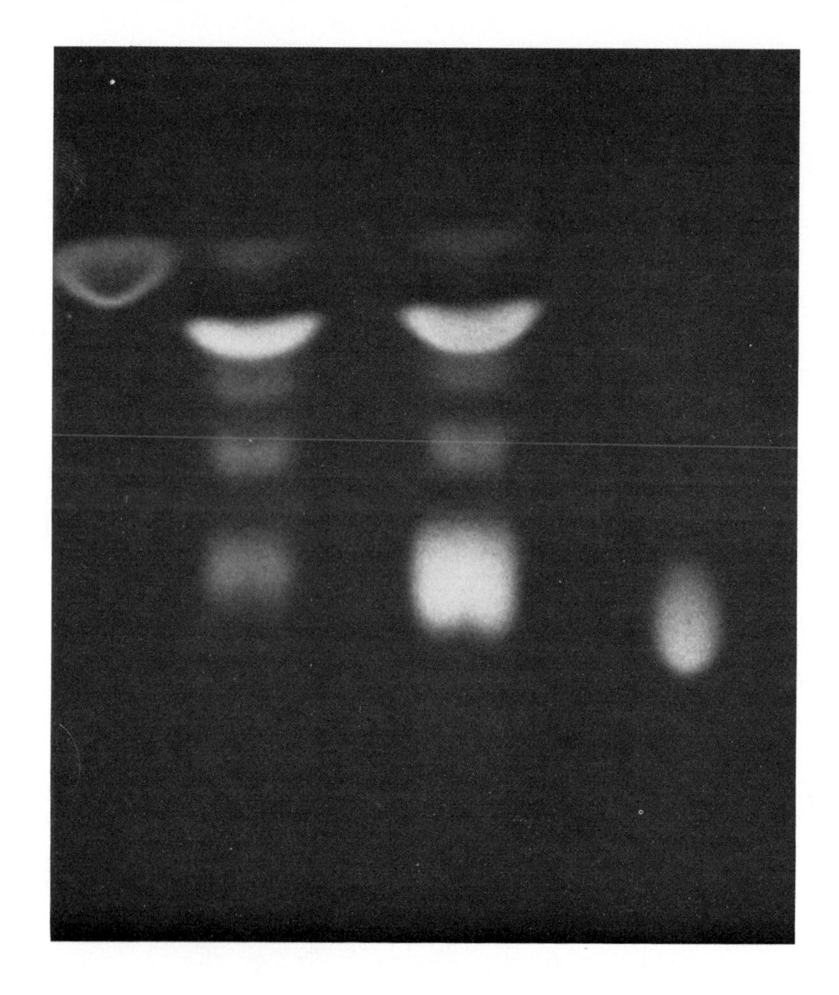

Figure 2. Characterization by thin layer chromatography of Protoporphyrin (PP), Photofrin II (SPP) Hematoporphyrin derivative (HPD) and Coproporphyrin (CP).

from the body. We also observed that the intense porphyrin fluorescence of the ears, tail and urine of the treated mice rapidly decreased over 2 to 3 hr following I.V. injection of UROP I and virtually had disappeared beyond 6 hours.

As can be seen in Figure 1, UROP I has significant photodynamic activity exhibited by necrosis in tumor receiving 300 joules of red light (wavelength 615 nm). The necrosis extended to a depth of 1/2 cm with islands of necrotic cells adjacent to areas of anaplastic squamous carcinoma. The depth and distribution of necrosis was similar to that observed with HPD or Photofrin II treated animals. In contrast, tumors illuminated with laser light of equivalent intensity, wavelength and total energy exhibited no necrosis in those animals receiving an injection of a non-tumor localizing porphyrin, monoacetal mono-hydroxy deuteroporphyrin.

The purity of the native porphyrin solutions injected was assessed by TLC analysis. UROP I prior to injection manifested as a single band whereas HPD (Photofrin) and Photofrin II exhibited five main bands of fluorescence, differing only in the relative proportions (Figure 2). The mobility of protoporphyrin (PP) and coproporphyrin (CP) also are shown.

DISCUSSION

Of the five porphyrins which were retained by tumor tissue, in two, TCPP and TPPS, there were no carboxyl groups linked to the porphyrin ring. HPD and Photofrin II are dicarboxylic porphyrin complexes (Berenbaum, Bonnett et al 1982; Clezy, Hai et al 1980; Moan, Christensen et al 1982) and octacarboxyl UROP I a single pure porphyrin. These, porphyrins had widely differing mobilities on TLC, which suggests that tumor uptake of porphyrins is not dependent on lipophilicity or the number of carboxyl side chains.

Of the neoplasms that fluoresced, more than 30% exhibited patchy fluorescence more intense in the periphery of the tumor. This is not a novel observation, and has been reported in rodent tumor following HPD diacetate (Gomer, Dougherty 1979) and HPD (Cortese, Kinsey et al 1979) as well as in human patients with bronchial carcinoma receiving HPD (Balchum, Doiron 1982). Thus, UROP I appears to behave similarly in this regard to other porphyrins studied.

We have confirmed the limited tumor selective uptake of HPD in vivo in mice as described by Gomer and Dougherty (Gomer, Dougherty 1979) who, using radiolabeled HPD reported a tumor:skin porphyrin content ratio of between 1 and 2 up to 72 hr following injection. The fluorometric method we report in this paper indicates a ratio which for HPD was 2.5 and Photofrin II, 2.9. However unlike Gomer et al (Gomer, Dougherty 1979), we observed striking (3+) visible red U.V. porphyrin fluorescence of the normal skin which correlated to measured skin porphyrin content.

The most striking immediate difference between UROP I and HPD was the virtual disappearance of UROP I fluorescence in the ears, skin, feet and urine over the initial 6 hr following injection indicating rapid elimination of the porphyrin from the body of the intact animal. This was in contrast to HPD and Photofrin II where skin fluorescence diminished slowly over 20 hr. Thus, we observed that 18-20 hr, a tumor:skin uroporphyrin ratio of approximately 20:1, more than 7 times that observed for HPD (Table 1). The absence of measureable porphyrin in the gastrointestinal tract with UROP I was in striking contradistinction to both HPD and Photofrin II where gut and tumor porphyrin levels were of similar magnitude.

The diagnostic implications of these observations are major. Carcinoma of the bronchus, bladder, cervix and gastrointestinal tract comprise approximately 50% of all malignant neoplasms occurring in man (Cancer Journal for Clinicians 1983), and colonic malignancy ranks second in mortality from cancer in both men and women (Silverberg 1981). As UROP I is virtually absent in the gut, if extrapolated to man, this should lead to selective tumor uptake, the specificity of which would permit clearly defined demarcation depicting normal and neoplastic tissue. This could readily amplify the current abilities to detect microscopic cancer by endoscopy, the porphyrin fluorescence being elicited by the blue light emitted from a Krypton laser. In addition, the photodynamic activity in vivo of UROP I still to be optimized, suggests that its diagnostic potential for tumors may be extended to photoradiation therapy. We predict from these observations that UROP I will have major advantages over HPD both as a marker of early malignancy and photoradiation therapy, but without the prolonged photocutaneous side effects.

Acknowledgements

This work has been supported by generous grants in part from the Elsa U. Pardee Foundation, University of California Cancer Research Coordinating Committee and the Fulbright Commission. We wish also to express our thanks to Dr. Kevin Smith, Professor, Department of Chemistry, University of California, Davis Medical School for some of the porphyrins provided and for his keen interest at all times. Direct requests for reprints to N. Pimstone, M.D.

References

Balchum OJ, Doiron DR (1982). Photoradiation therapy of obstructing endobronchial lung cancer. SPIE Vol. 357 Lasers in Medicine and Surgery, pp. 53-55.

Benson RC Jr., Farow GM, Kinsey JH, Cortese DA, Zincke H and Utz DC (1982). Detection and Localization of In Situ Carcinoma of the Bladder with Hematoporphyrin Derivative. Mayo Clin. Proc. 57:548-555.

Berenbaum MC, Bonnett R, Scourides PA (1982). In Vivo Biological Activity of the Components of Haematoporphyrin Derivative. Br. J. Cancer 45:571-581.

Berns MW (1983). Personal communication.

Cancer Journal for Clinicians (1983) 33:16. (Based on incidence estimates from NCI SEER Program.)

Clezy PS, Hai TT, Henderson RW, Thuc LV (1980). The Chemistry of Pyrrolic Compounds. XLV Haematoporphyrin Derivative: Haematoporphyrin Diacetate as the Main Product of the Reaction of Haematoporphyrinwith a Mixture of Acetate and Sulfuric Acids. Aust. J. Chem. 33:585-597.

Cortese DA, Kinsey JH, Woolner LB, Payne WS, Sanderson DR, Fontana RS (1979). Clinical application of a new endoscopic technique for detection of in situ bronchial carcinoma. Mayo Clin. Proc. 54:635-641.

Dahlman A, Wile AG, Burns RG, Mason GR, Johnson FM, Berns MW (1983). Laser Photoradiation Therapy of Cancer. Cancer Res. 43:430-434.

Dougherty TJ (1981). Photoradiation therapy for cutaneous and subcutaneous malignancies. J. Invest. Dermatol. 77:122-124.

Dougherty TJ, Lawrence G, Kaufman JH, Boyle D, Weishaupt KR, Goldfarb A (1979). Photoradiation in the treatment of recurrent breast carcinoma. J. Natl. Cancer Inst. 62:231-237.

El-Far MA, Pimstone NR (1983). Uroporphyrin I is a Better Tumor Localizer Than Hematoporphyrin Derivative in Mouse Mammary Carcinoma. Clin. Res. Vol. 31, No. 2.

El-Far MA, Pimstone NR (1983). Superiority of Uroporphyrin Isomer I Over Other Porphyrins in Selective Tumor Localization in Mouse Carcinoma. Cancer Res. (submitted).

Gomer CJ, Dougherty TJ (1979). Determination of [^{3}H]- and [^{14}C] Hematoporphyrin Derivative Distribution in Malignant and Normal Tissue. Cancer Res. 39:146-151.

Gregorie HB Jr., Horger EO, Ward JL, Green JF, Richards T, Robertson HC Jr., Stevenson TB (1968). Hematoporphyrin-derivative fluorescence in malignant neoplasms. Ann. Surg. 167:820-828.

Kelly JF, Snell ME (1976). Hematoporphyrin derivative: a possible aid in the diagnosis and therapy of carcinoma of the bladder. J. Urol 115:150-151.

Moan J, Christensen T (1980). Porphyrins as Tumor Localizing Agents and Their Possible Use in Photochemotherapy of Cancer. Tumor Res. 15:1-10.

Moan J, Christensen T., Sommer S (1982). The Main Photosensitizing Components of Hematoporphyrin Derivative. Cancer Letters 15:161-166.

Rockwell SC, Kallman RF, Fajardo LF (1972). Characteristics of a Serially Transplanted Mouse Mammary Tumor and Its Tissue-Culture-Adapted Derivative. J. Natl. Cancer Inst. 49:735-749.

Schwartz S, Absolon D, Vermund H (1955). Some relationships of porphyrins, X-rays and tumors. University of Minnesota, Med. Bull. 27:5.

Silverberg E (1981). Cancer Statistics 31-13.

Porphyrin Localization and Treatment of Tumors, pages 673–678

OPTIMAL PHOTODYNAMIC BAND OF RED LIGHT IN HEMATOPORPHYRIN DERIVATIVE (HPD) PHOTORADIATION THERAPY OF CANCER

Neville R. Pimstone, M.D., F.C.P.(S.A.)

Professor of Medicine and Chief
Division of Gastroentrology
Department of Internal Medicine
University of California, Davis, CA 95616

Shobha N. Gandhi, M.D.

Research Fellow
Division of Gastroenterology
Department of Internal Medicine
University of California, Davis, CA 95616

INTRODUCTION

Hematoporphyrin derivative (HPD) localizes in tumor tissue, a property exploited both as a fluorescent diagnostic marker of early malignancy, (Benson, Farrow et al 1982; Profio, Doiron et al 1982) or in photoradiation therapy (Balchum and Doiron 1982; Dahlman, Wile et al 1983; Dougherty, 1981). Krypton laser light (blue/violet, 406nm) is used to photoexcite HPD to fluoresce whereas photodynamic necrosis is effected by red light which penetrates tissue more effectively than short wavelength blue (Pimstone, Horner et al 1982). In clinical and animal studies, a tunable dye laser has been the source of red light (620-640nm) (Balchum, Doiron, 1983; Dahlman, Wile et al 1983; Dougherty, 1981), the wavelength of which corresponds to the mini-peak of HPD light absorption (Figure 1).

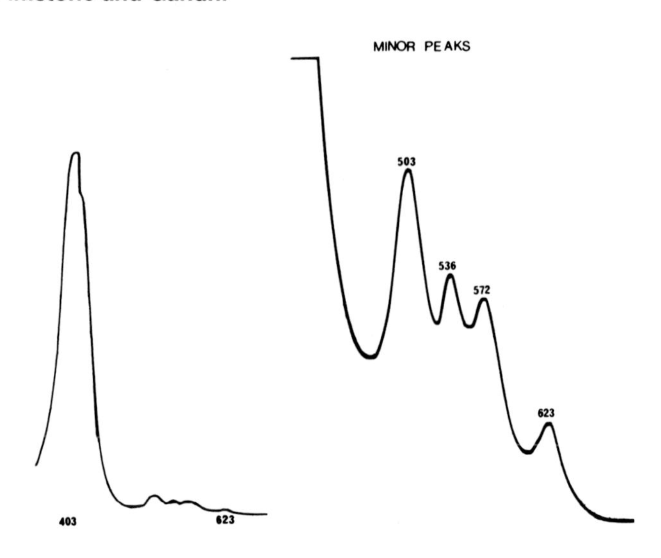

Fig.1. Absorption spectrum of hematoporphyrin derivative. Note the "mini peak" at 623nm.

As biologic tissues have increasing lucency to light of longer wavelength (Pimstone, Horner et al 1982) and HPD absorbs energy beyond its mini-peak, we examined the photodynamic effect of light and HPD between 612-690nm to define the optimal photodynamic band of red light. As a tumor model, we used the liver as described previously (Pimstone NR, Horner IJ et al 1982). The purpose of the study to be reported was to define the optimal photodynamic band of red light to be used in HPD photoradiation therapy.

MATERIALS AND METHODS

Twenty-four hours prior to the study, rats (Sprague Dawley, male, 400 gm wt) received HPD (Photofrin, Oncology Research and Development, Inc.) I.V. at a dose of 20mg/kg.

They were then anesthetized with pentabarbitol (40mg/kg I.P.). The liver was exposed by laparotomy and light of differing wavelengths was administered, but energy was kept constant at 200 joules. A screen was placed over the liver so that only a circular area approximately 1 cm in diameter was exposed to the light beam. The light source was a tunable dye laser (Spectra Physics, Mountain View, CA, Model 375), DCM (Exciton Chemical Co Inc, Dayton, OH) being the lasing medium. The dye was excited by an Argon laser (Spectra Physics, Model 171). Wavelength was calibrated using a general-purpose monochromator (Model H-20, Instruments SA, Inc). After receiving the light, some animals were sacrificed immediately thereafter; in others, the abdomens were surgically closed and the animals were sacrificed 24 hours later and the depth of necrosis observed both macroscopically and histologically. For the krypton laser experiments, the liver received 200 joules of light (647nm) using a krypton laser (Spectra Physics, Model 171). The liver necrosis was evaluated as above.

RESULTS

The immediate and delayed effects of light on the liver were easily evident both to the naked eye and on histology (Figure 2) with clear demarcation between necrotic and viable liver.

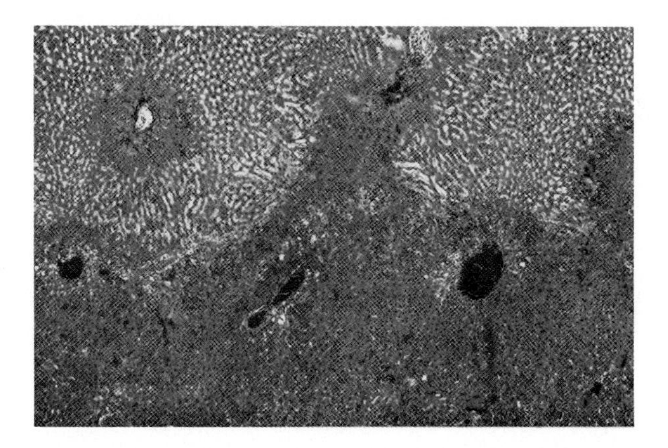

Fig. 2. Photodynamic necrosis of receiving 200 joules of light at 647nm seen histologically (40x). Note the clear line of demarcation between necrotic liver (above) and viable liver (below).

The depth of necrosis which could be measured accurately to a fraction of a millimeter was assessed at varying wavelengths and the results are presented in Table 1. Note that necrosis of the liver was essentially the same from wavelengths 620nm through 650nm.

Table 1

Wavelength	Necrosis
611 nm	1.8mm $\pm$ .1
620 nm	2.5mm $\pm$ .2
625 nm	2.5mm $\pm$ .2
630 nm	2.5mm $\pm$ .2
640 nm	3.0mm $\pm$ .2
647 nm	3.0mm $\pm$ .2
650 nm	2.5mm $\pm$ .2
660 nm	1.8mm $\pm$ .1
670 nm	1.2mm $\pm$ .1
676 nm	0.5mm $\pm$ .05

This finding suggested to us that a krypton laser which also can emit light at 647nm might be as effective as a tunable dye laser, and in a separate experiment using a krypton laser, the necrosis produced by 200 joules of light (647nm) was 2.5 mm.

Thermal necrosis of the liver did not occur based on the following observations. First, at all wavebands studied, the depth of necrosis in HpD treated animals receiving 200 joules light energy was identical at an intensity range of 50 to 1000 mW/cm^2. Second, in control animals receiving light but no HpD, no necrosis was observed.

DISCUSSION

We have demonstrated that the photodynamic activity of HPD is not limited to red light corresponding to the mini-peak of HPD absorption between 620-630nm), but was present at longer wavelengths where there was more effective penetration of light energy by the tissue. However, despite increasing tissue lucency, there was no enhanced necrosis at wavelengths longer than 650 nm, presumably due to diminished absorption of light energy by HPD in this region. The finding that photodynamic activity was equal in magnitude when light of the same energy was delivered at wavelengths ranging from 620-650nm suggested that a krypton laser which emits light at 647nm could be equally effectively used as the source of light energy in photoradiation therapy. This was confirmed with the krypton laser experiment described above.

We believe this observation is of major potential clinical importance in that the krypton laser, used traditionally only for diagnosis, can also be used for photoradiation therapy. This should prove both cost effective, as one laser can do the work of two, and more efficient, as the energy emitted by the krypton laser is more stable and predictable than that derived from a tunable dye laser.

ACKNOWLEDGMENTS: We wish to acknowledge that this project was funded in part from grants from the Elsa U. Pardee Foundation and from the University of California Cancer Coordinating Research Committee. Reprint requests should be directed to NR Pimstone, Department of Internal Medicine, School of Medicine, University of California, Davis, CA 95616.

REFERENCES

Balchum OJ, Doiron DR (1982). Photoradiation therapy of obstructing endobronchial lung cancer. Proc of SPIE 357:53.

Benson RC, Farrow GM, Kinsey JH, Cortese DA, Zincke H, Utz DC (1982). Detection and localization of in situ carcinoma of the bladder with hematoporphyrin derivative. Mayo Clinic Proc 57:548.

Dahlman A, Wile AG, Burns RG, Mason, Johnson FM, Berns, MW (1982). Laser photoradiation therapy of cancer. Cancer Research 43:430.

Dougherty TJ (1981). Photoradiation therapy for cutaneous and subcutaneous malignancies. J Invest Derm 77:122.

Pimstone NR, Horner IJ, Shaylor-Billings J, Gandhi SN (1982). Hematoporphyrin-augmented phototherapy: dosimetric studies in experimental liver cancer in the rat. Proc SPIE 357:60.

Profio AE, Doiron DR, Balchum OJ, Huth GC (1982). Fluorescence bronchoscopy for localization of carcinoma in situ. Med Physics 10:35.

CLINICAL

Porphyrin Localization and Treatment of Tumors, pages 681–691

PHOTORADIATION THERAPY OF HEAD AND NECK CANCER

Alan G. Wile, MD, Jean Novotny, G. Robert Mason, MD
Victor Passy, MD, Michael W. Berns PhD.

University of California, Irvine Medical Center
Department of Surgery, 101 City Drive South
Orange, CA 92668

ABSTRACT

We have embarked upon a pilot study of photoradiation therapy (PRT) in the treatment of persistent or recurrent cancer of the head and neck utilizing the photosensitizing agent hematoporphyrin derivative (HPD). This treatment is based upon selective concentration of HPD within malignant tissue with resultant necrosis upon illumination with light of the appropriate wave length (630 nm). Patients entered in this trial have failed all forms of conventional therapy. Twenty-one patients with local recurrence were treated. Sites of recurrence were: tongue (9), nasopharynx (3), floor of mouth (2), soft palate (2), oropharynx (1), buccal mucosa (1), maxilla (1), larynx (1), basal cell nevus (1). There were six complete responses and twelve partial responses (greater than 50% reduction). These responses are clinically significant with some complete responses lasting over one year after a single course of therapy. Ten patients with cutaneous metastases from head and neck primary tumors were also treated. There were two complete responses and three partial responses. However, these patients rapidly developed new tumors in areas adjacent to those previously treated. Less than complete responses could be augmented by repeated applications of this technique. The success of this pilot study combined with the accessibility of head and neck primaries suggest that HPD-PRT should be given a clinical trial in early mucosal cancer of the head and neck region.

INTRODUCTION

Photoradiation therapy (PRT) is a new therapeutic modality which relies upon the selective concentration of porphyrin compounds within malignant tissue for its tumoricidal effect (Berns 1982). It was noted that a mixture of porphyrin compounds, termed hematoporphyrin derivative (HPD) would localize within malignant tissue and fluoresce when illuminated with a blue light (Gregorie 1968; Lipson 1961). Numerous investigators have utilized this phenomenon in the detection of occult malignancies, particularly in the lung (Cortese 1982; Doiron 1979; Profio 1979). Dougherty has demonstrated that a red light of the appropriate wave length (630 nm) illuminating a cancer containing HPD will cause selective necrosis (Dougherty 1978). This process has been termed hematoporphyrin derivative-photoradiation therapy (HPD-PRT).

We were intrigued by the therapeutic potential of photoradiation therapy and have developed a program of HPD-PRT. The initial good results reported by Dougherty have been confirmed by us even in patients heavily pretreated with other modes of cancer therapy (Dahlman 1982; Wile 1982; Wile 1983). The treatment parameters have been established and refined. The remaining task is to elucidate those clinical situations in which HPD-PRT will be beneficial. It is with this last goal in mind that we describe our initial experience utilizing HPD-PRT in the treatment of refractory cancer of the head and neck.

MATERIALS AND METHODS

Patients

Since photoradiation therapy is an experimental modality, it is currently reserved for those patients who have failed all forms of conventional therapy. Patients with cancers of the head and neck refractory to conventional therapy with accessible lesions were entered into the study. An interval of at least three months was required between the last course of ionizing radiation therapy and HPD-PRT for those patients with demonstrated progression of disease. Patients were accepted for HPD-PRT who were receiving chemotherapy, but had progression of disease while under therapy. No new modes of therapy were begun during treatment with HPD-PRT in the one month period of observation.

Although cure of small lesions is feasible, it was not considered likely. Hence, all patients were informed that the intent of the treatment was palliative rather than curative. A consent, approved by the UCI Human Subjects Review Committee was then signed by the patient authorizing therapy.

Thirty one patients were entered into this study between May of 1981 and June of 1982. Twenty-one patients had persistent or recurrent cancer in the primary site of origin without evidence of regional or distant spread. Ten additional patients had recurrence in the skin and soft tissues of the head and neck with the primary lesions controlled. All patients except two had squamous cell carcinoma. One patient had an adenoid cystic carcinoma involving the nasopharynx and another had basal cell nevus syndrome.

Technique of HPD-PRT

Patients selected for this study received injections of HPD (Oncology Research and Development, Cheektowaga, N.Y.) at a dose of 3 mg/kg of body weight. This was administered as an intravenous bolus. They were then cautioned to avoid sun light and bright artificial light for a one month period of time in order to minimize phototoxicity. Seventy two hours later the patients returned in order to have the lesion illuminated with red light. They returned one day later for evaluation of response. Reexamination occurred at one, two, and four weeks following initial illumination. At the four week examination a final determination of tumor response was made. The response was graded as: complete response, complete resolution of tumor; partial response, greater than 50% reduction in the product of the cross-sectional diameters; stable disease, no change in the size of the lesion during the one month observation period; and progression, a 20% or greater increase in the size of the tumor during the observation. When possible, patients were then followed at bimonthly intervals.

An Argon-ion pumped dye laser (Spectra-Physics models #164, 171, 375) was used to generate a continuous red beam at 625-635 nm. The dye laser employed Rhodamine B dye and the argon excitation wavelength employed was 514 nm at 6-8 watts. Output from the dye laser was varied from 500-1500 mw. This wavelength range corresponded to the longer wavelength absorption peak of HPD. The output of the dye laser was focused onto an optical fiber (400-600 microns, Math

Associates, Port Washington, N. Y.). Coupling of the laser light and the optical fiber was facilitated by means of an alignment device manipulated in two dimensions by micrometer controls (Spectra-Physics). This system generated a maximum power of 1.5 watts of red light measured with a calorimeter (Model #404 Spectra-Physics). Wavelength was monitored using a JY #5-354 monochronometer.

The tumor bearing surfaces were illuminated in one of three different ways. Those lesions easily accessible were illuminated by superficial illumination. The fiberoptic was brought close to the surface to be treated and clamped in a mechanical holder. By knowing the size of the area to be treated and the intensity of the light as it emanated from the fiber, both the intensity of light at the surface (mw/cm^2) and the total light dose ($joules/cm^2$) could be calculated. The intensity used was in the range of 25-500 mw/cm^2 and the total dose was in the range of 17-91 $joules/cm^2$.

Those tumors persisting in the recesses of the head and neck region that were not directly accessible to light were illuminated by means of an endoscope (Pentax bronchoscope #FB-15A, light source LH-150). This group of patients consisted of three patients with cancer of the nasopharynx and one patient with a second primary cancer involving the true vocal cord. A 400 μm fiberoptic was passed through the biopsy port of the endoscope to achieve illumination. Neither the area to be treated nor the distance from the fiber to the surface could be accurately measured. In these four patients estimates of the illuminated area were made to permit a computation of treatment time.

Two patients had large (>8 cm) exophytic tumors metastatic to the skin of the head and neck region. These patients were treated by implanting the fiber directly into the tumor mass. No measurement of output was possible during this mode of therapy. What were considered reasonable periods of illuminations were used (15-30 minutes per treatment). These were periods of time that were comparable to those utilizted with the higher light doses in the superficial mode of illumination.

RESULTS

Of the twenty-one patients with cancer of the head and

neck recurrent in the primary site there were six complete responses and 11 partial responses as determined by the examination made four weeks following illumination.

Table 1.

RESPONSE OF HEAD AND NECK CANCER

RECURRENT IN THE PRIMARY SITE

	No	CR	PR	SD	NR
Tongue	9	2	6	0	1
Nasopharynx	3	1	1	0	1
Floor of Mouth	2	1	1	0	0
Soft Palate	2	1	1	0	0
Oropharynx	1	0	1	0	0
Buccal Mucosa	1	1	0	0	0
Maxilla	1	0	1	0	0
Vocal cord	1		improved		
B.C. Nevus Syndrome	1		improved		
TOTAL	21	6	11	0	2 + 2 improved

Many of these responses were transient in that regrowth of tumor was observed in subsequent followup examinations. However, a less than complete response could be augmented by repeated cycles of injection and illumination. Eight patients in this group of 21 were retreated on subsequent occasions. The largest number of retreatment cycles received by an individual was three.

Recurrent cancer of the tongue generally responded favorably. The one patient failing to respond had a tumor at the base of the tongue. An endoscope was not used to visualize this tumor. The fiber was passed through the nasopharynx and positioned above the tumor. The most likely explanation for the lack of response was inadequate illumination of the tumor.

The two patients with complete resolution of tongue cancer enjoyed sustained responses. The first was a man with an extensive cancer of the anterior two-thirds of the tongue who had progressed despite external beam radiation therapy and interstitial radiation. When first seen an exophytic tumor protruded from the dorsal surface of the tongue. There was marked weight loss secondary to dysphagia. The patient was treated only one time with a total light dose of 20 joules/cm^2 and demonstrated marked necrosis at 24 hours. Response was complete at four weeks. The patient was free of disease at 18 months following therapy when he succumbed to the sequela of a mesenteric artery thrombosis. At autopsy no residual tumor was found after histologic examination of the tongue.

The second complete response occurred in a woman with a similar history of an anterior tongue cancer who developed an unresectable recurrence after radiation therapy. She was treated a single time and remains alive and well at the time of this report without recurrence twelve months following treatment. The remaining six patients had responses which lasted for intermediate periods of time, but eventually resulted in tumor regrowth.

The responses to therapy in those patients with cancer of the nasopharynx were based on estimates of the change in size of the visible tumor. Although the one patient who enjoyed a complete response is doing well one year following HPD-PRT, a CAT scan of the nasopharynx demonstrates changes in the bony confines of the region suggesting the presence of residual disease.

The remaining nine patients were a mixture containing various sites of residual cancer. Generally, these sites were easily accessible to light and responded to HPD-PRT. One patient developed a third primary cancer at the junction of the soft and hard palate. This was a small (8 mm) invasive squamous cell carcinoma arising in a field which had previously been irradiated. Surgery would have been extensive, increasing the risk of osteoradionecrosis. The patient was treated with a single cycle of HPD-PRT, responded completely at 30 days, and was without recurrence at the 8 month examination.

Another patient received 5600 rads for a T_1 lesion of the right true vocal cord. Seven years later he developed a second primary T_1 cancer of the left vocal cord. He received two cycles of HPD-PRT over a three month time period receiving three courses of illumination (27, 30, and 55 joules/cm^2). While there was sloughing of tissue in the area of the tumor, subsequent biopsies revealed residual cancer and surgical extirpation was undertaken.

A young woman with basal cell nevus syndrome involving the face was treated with HPD-PRT. This woman had been plagued with hundreds of small (1-2 mm) lesions which had in the past become aggressive requiring multiple surgical resections of basal cancers. Several 3 cm patches were treated with complete resolution of the lesions in the treated areas.

The remaining ten patients with head and neck cancer with regional recurrence had less favorable courses than those with only recurrence in the primary site. While there were two complete and three partial responses (Table 2), the natural history of the disease in these patients did not appear to be substantially altered. Tumor could be successfully eradicated in the treated areas. However, at subsequent examination new areas of tumor involvement were observed immediately adjacent to treated areas. The two patients underoging fiberoptic implantation into large exophytic tumors failed to respond. The primary sites of origin in these two patients were tonsil and skin.

Table 2.

RESPONSE OF HEAD AND NECK CANCER

METASTATIC TO THE SOFT TISSUES OF THE HEAD AND NECK

	No	CR	PR	SD	NR
Tonsil	3	1	1	0	1
Larynx	1	0	1	0	0
Tongue	2	0	0	0	1
Parotid	1	0	0	1	0
Gingiva	2	1	0	0	0
Skin	1	0	0	0	1
Floor of Mouth	1	0	1	0	0
Unknown 1°	1	0	0	0	1
TOTAL	10	2	3	1	4

DISCUSSION

Depth of penetration in tissue by visible light is dependent upon the wavelength. Longer wavelengths of visible light penetrate to greater depths. The amount of tissue penetrated by light at wavelength 630 nm in sufficient amount to initiate a photodynamic response is reported to be 2 cm (Doiron 1983). However, a number of factors influence this penetration including density of tissue, presence of melanin, and tissue scattering. It has been noted by us that when treating dermal recurrence of cancer, dark complected patients respond less well than fair complected ones.

It would then be reasonable to expect better results in treating recurrent cancer of the head and neck involving the mucosa than in treating areas of the body with intact skin as melanin is either absent completely or present in minimal quantities. Furthermore, the keratinized layer of the epidermis is not present in the mucosa of the head and neck region, eliminating another obstacle to HPD-PRT.

These facts were borne out in the results of treatments in those patients who had recurrent cancer in the primary site of origin. Those patients with tumors that were biologically favorable, that is, where there was no evidence of regional or distant disease at the time of local recurrence, seemed to benefit substantially. The two patients with recurrent tongue cancer experienced long term remissions.

The patients with cancer recurrent in soft tissues in the head and neck region generally had much more aggressive tumors or were in advanced stages of disease as compared to those with only local recurrence. It is unreasonable to expect a local mode of therapy such as HPD-PRT to alter the natural history of a disease that has become regional or systemic.

The two patients undergoing implantation of the fiber directly into the tumor failed to respond. Other investigators have described success with this implantation method. However, they stress that multiple fibers must be implanted simultaneously into large masses in order to achieve sufficient light exposure to initiate a photodynamic response (Dougherty 1981). There is some controversy as to whether the destruction of tumor in these cases is due to HPD-PRT or is the result of mild hyperthermia induced by optical heating of tissue (Svaasand 1983).

We have demonstrated that the histologic types of malignancies involving the head and neck region respond to HPD-PRT in the majority of cases. We conclude that HPD-PRT is well suited for the treatment of mucosal malignancies. Factors which interfere with successful therapy in other parts of the body are absent in these types of malignancies. The regions are easily examined and illuminated with light. If therapy is unsuccessful the results are known as early as 30 days following examination at which time the treatment cycle may be repeated or alternative therapy may be administered. Lesions which are currently considered responsive to HPD-PRT are local recurrences of cancer of the head and neck in the absence of evidence of disseminated disease and second primary cancers arising in a field of previous radiation therapy for which a surgical approach would be extensive. We believe that at this time it is now appropriate to begin trials of HPD-PRT in the management of previously untreated cancer of the head and neck region. It is anticipated that HPD-PRT could become the initial treatment in selected early cancer of the head and neck region.

Berns MW, Wile A, Dahlman A, Johnson FM, Burns R, Sperling D, Guiltiman M, Siemans A, Walter R, Wright W, Hammer-Wilson M (1982). In vitro effects of hematoporphyrin derivative (HPD). Cancer Res 42:2325-2329.

Cortese DA, Kinsey JH (1982). Endoscopic management of lung cancer with hematoporphyrin phototherapy. Mayo Clin Proc 57:543-547.

Dahlman A, Wile AG, Burns RG, Mason GR, Johnson FM, Berns MW (1982). Laser photoradiation therapy of cancer. Cancer Res 43:430-434.

Doiron DR, Profio E, Vincent RG, Dougherty TJ (1979). Laser fluorescence bronchoscopy for detection of lung cancer. Chest 76:27.

Doiron DR, Svaasand LO, Profio AE (1983). Light Dosimetry in tissue: application to photoradiation therapy. In Kessel D and Dougherty TJ (eds): "Porphyrin photosentization," New York: Plenum Press, p 160.

Dougherty TJ, Kaufman JE, Goldfarb A, Weishaupt KR, Boyle D, and Mittleman A (1978). Photoradiation therapy for the treatemnt of malignant tumors. Cancer Res 38:2628-2635.

Dougherty TH, Thoma RE, Boyle DB, Weishaupt KR (1981). Interstitial photoradiation therapy for primary solid tumors in pet cats and dogs. Cancer Res 41:401-404.

Gregorie HB Jr, Horger EO, Ward JL, Green JF, Richards T, Robertson HF Jr, Stevenson, TB (1968). Hematoporphyrin derivative fluorescence in malignant neoplasms. Ann Surg 167:820-828.

Lipson RL, Baldes EJ and Olsen AM (1961). Hematoporphyrin derivative: a new aid for endoscopic detection of malignant disease. J Thorac Cardiovasc Surg 42:623-629.

Profio AE, Doiron DR and King EG (1979). Laser fluorescence bronchoscope for localiziation of occult lung tumors. Med Phys 6:523.

Svaasand LO and Doiron DR (1983) Thermal distribution during photoradiation therapy. In Kessel D and Dougherty TJ (eds): "Porphyrin photosensitization" New York: Plenum Press, Vol 160, p 77-90.

Wile A, Dahlman A, Burns B, and Berns MW (1982). Laser photoradiation therapy following hematoporphyrin sensitization.

Lasers in surgery and medicine 2:163-168.
Wile AG, Dahlman A, Burns RG, Mason GR, Johnson FM, and Berns MW. Laser Photoradiation therapy of recurrent human breast cancer and cancer of the head and neck. In Kessel D and Dougherty TJ (eds): "Porphyrin photosensitication," New York: Plenum Press.

Porphyrin Localization and Treatment of Tumors, pages 693–708

MULTIDISCIPLINARY APPROACH TO PHOTOTHERAPY OF HUMAN CANCERS.

I.J. Forbes[1] A.D. Ward[2] F.J. Jacka[3] A.J. Blake[4]
A.G. Swincer[2] P.A. Wilksch[3] P.A. Cowled[1] and K. Lee See[1]

Departments of Medicine[1], Organic Chemistry[2], Physics[4], and the Mawson Institute[3], University of Adelaide.

That tumours can be selectively ablated by photochemotherapy using haematoporphyrin derivative (HPD) has now been amply demonstrated. However, before this technique takes its place with surgery, radiotherapy and chemotherapy as standard treatment of malignant tumours a number of problems must be solved. These include provision of equipment and the definitive preparation of HPD. Dosimetry presents unsolved problems. Then acceptable clinical approaches must be defined, and proof of efficacy of treatments must be obtained in a form acceptable to clinicians. In addition, an understanding of the basic mechanisms of cytotoxicity, and of the pharmacology of HPD are essential to a full exploitation of the potential of this therapeutic modality.

The aim of this paper is to describe approaches by our group to some of these problems.

DEVELOPMENT OF SUITABLE LIGHT SOURCES

a) Filtered Incandescent Phototherapy Lamp

The prototype lamp illustrated in Figure 1 was built in the workshops of the Department of Physics, The University of Adelaide using components from several sources including A.G. Thompson & Co. (SA) Pty. Limited of Adelaide who are undertaking commercial production. It incorporates an incandescent filament source of 1000 Watts, a non-imaging reflector system which is gold plated and filters.

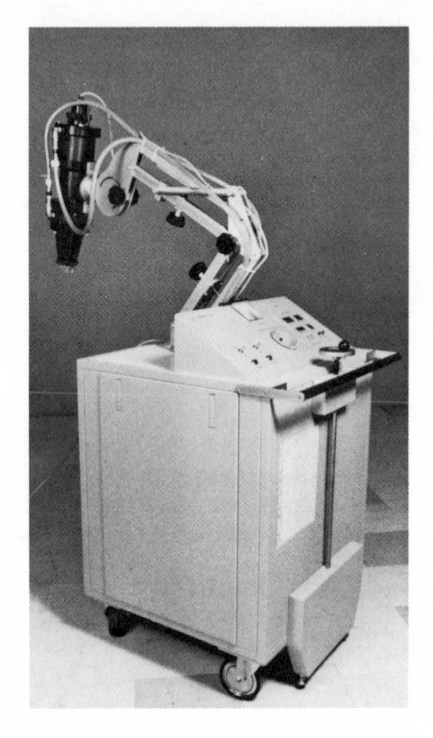

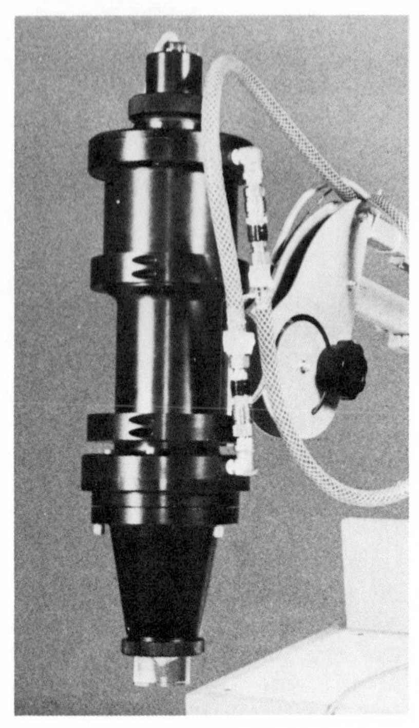

Figure la. Prototype incandescent lamp.

Figure lb. Head of the lamp, with refracting "lens"

The desired wave length range is selected by separate long and short-wave pass filters which may readily be changed to vary the lower and upper cut-off length independently. The long wave pass filter has a 50% of peak transmission at 620nm. The short wave pass dielectric filter has 50% of peak transmission of 730nm. A further transmission beyond 950nm is absorbed in the 200mm water column. The calculated total output power with the filters described is 24W, and the effective flux at the 630nm peak for HPD is calculated to be 9W.

The output from the lamp may be further modified by the addition of a refracting component. One such "lens" makes use of the total internal reflection to concentrate the output through a smaller aperture by increasing the beam divergence of 60° in human tissue when the output face is in contact with the skin, and 90° when the output face is in the air. Another lens in the form of a cylindrical rod 10mm diameter and 50mm long selects the central bright spot, scrambles it through several internal reflections and delivers 2.5W uniformly distributed over the 10mm aperture. This is very suitable for treatment of tumours in laboratory animals.

Because of the high power output, care must be taken to avoid burning skin, by spraying with water by nebulizer and by switching the lamp off periodically, using an electronic timer.

2. Metal Vapour Laser

A new metal vapour laser has been designed by Quentron Optics Pty. Limited, Adelaide, in association with Assoc. Prof. J. Piper (Macquarie University) with the aid of an Industrial Research and Development Grant from the Commonwealth of Australia, and a prototype is in use. The prototype instrument illustrated in Figure 2 delivers pulsed monochromatic light at 627.8nm wavelength. Mean operating power of the prototype is approximately 3W and the subsequent models are planned to have a 5W output. The pulse width at base is 50ns and the pulse repetition frequency used so far has been 10-15kHz.

This instrument has considerable advantages over an argon ion-dye laser system in being mobile and having lower requirements for electric power (less than 4kW). The instrument is coupled to a beam splitting device of new design allowing the selection of numbers of illuminated fibres from 1 through 9 with the exception of 5, 7 and 8.

3. Comparison of Light Sources

A transplantable Lewis lung carcinoma was used in C57 Black mice. One tenth ml of tumour suspension (approximately 10^{6} cells) was injected subcutaneously into the back of each mouse. After 7 to 10 days when tumours were 5-7mm in diameter groups of 10 mice were given HPD 25-50mg/kg intraperitoneally. After 24 hours mice were anaesthetized with Sagatal, the fur over the tumour was shaved and a spot 1cm diameter over the tumour was irradiated. Light doses for the incandescent beam lamp were 470-785 Joules/cm^2 over 620-730nm, delivered in 50 second pulses with 10 second pauses. For continuous wave (argon ion-dye laser, Spectra Physics) and pulsed wave (metal vapour laser, Quentron) source light doses of 100-300 Joules/cm^2 were given. Twenty-four hours after treatment using any of these light and HPD doses 9-10 of the tumours in each group of mice were impalpable. The number of days after treatment required for 5 of the 10 tumours to become palpable was designated the TC_{50}, i.e. the time for which 50% of tumours were controlled. This end point was proportional both to the HPD dosage and to the light dosage. Thus for a dose of HPD of 25mg/kg, using a light dose of 240 Joules/cm^2 the TC_{50} was 2 days, and for 60mg/kg it was 6 days, with a linear response for intermediate doses. Similarly, a linear response was obtained with light doses from 100 Joules/cm^2 (TC_{50} 2 days) to 250 Joules/cm^2 (TC_{50} 6 days).

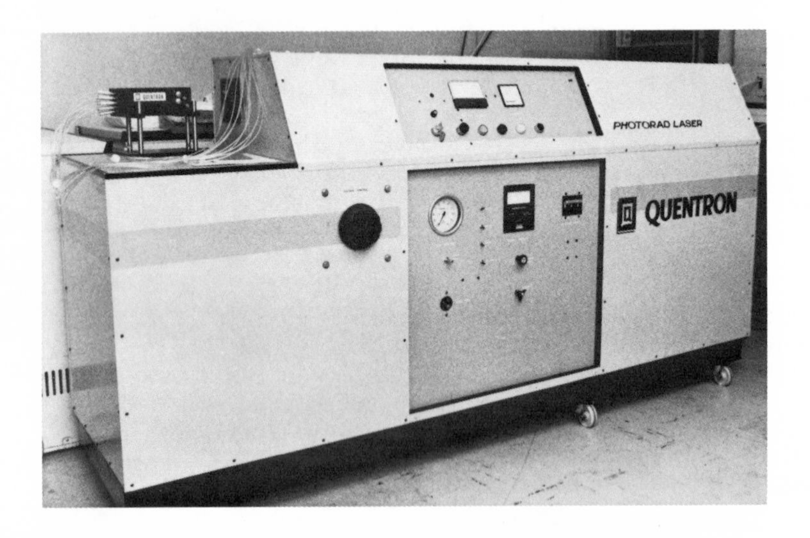

Figure 2. Prototype of metal vapour laser and beam splitting device.

Having established the characteristics of the Lewis lung transplantable tumour system the efficiency of light from different sources was compared.

The efficiency of the incandescent beam source as a stimulus to HPD-induced phototoxicity is considerably less per Joule of irradiated light (9W at 630nm of total output 24W). The high output of this instrument reduces treatment times greatly, but entails considerable risk of skin damage because of the high flux density.

The TC_{50} for the incandescent beam source using an HPD dose of 50mg/kg was 5 days with a light dose of 640 Joules/cm^2 (corresponding to 240 Joules/cm^2 at 630nm). Using the same light dose (i.e. 200 Joules/cm^2) the TC_{50} with continuous wave laser light at 630nm was 4.5 days. The corresponding TC_{50} for pulsed laser light at 627.8nm (200 Joules/cm^2) was 5 days. At other doses of HPD or light the pulsed laser was equal to or marginally more effective than continuous wave laser light.

Preliminary studies of phototoxicity in vitro, using ^{51}Cr release from Raji cells sensitized with HPD, using techniques described in the next section, also showed no significant difference in the efficiency of continuous wave and pulsed laser light, with a linear relationship between cell death and light dose.

NATURE OF THE PHOTOTOXIC SPECIES

Weishaupt et al (1976), Stenstrøm (1980), Moan (1981) and others have reported results indicating that singlet oxygen ($^1O_2$) is the phototoxic agent in photochemotherapy with HPD. Recent interest has been shown in the role of oxygen derived free radicals (ODFR) particularly the hydroxyl radical (OH·). Since the capacity to manipulate the phototoxic process may add greatly to the effectiveness of therapy, its mode of action must be understood.

Raji cells were maintained at 37°C in 5% CO_2 in RPMI 1640 buffered with 25mM Hepes and 2.3mM sodium bicarbonate and supplemented with 10% foetal calf serum (Flow Labs), and subcultured twice weekly when cell density was approximately 10^6/ml. They were labelled with ^{51}Cr (Amersham Australia) and resuspended at 10^7/ml in RPMI 1640. One ml suspensions containing specified reagents were irradiated at 25°C in polycarbonate tubes on a horizontal platform revolving around a 500W quartz iodide lamp shielded with red perspex. The flux density at the point of irradiation was 8mW/sq.cm. between 600 and 650nm. Percentage release of ^{51}Cr was calculated, taking into account background release and maximal release by 3% acetic acid. Each point was the mean of triplicate determinations. Graphs of HPD concentration against percentage release of ^{51}Cr were plotted and the effectiveness of various scavengers and quenchers were measured by comparing slopes with and without inhibitor.

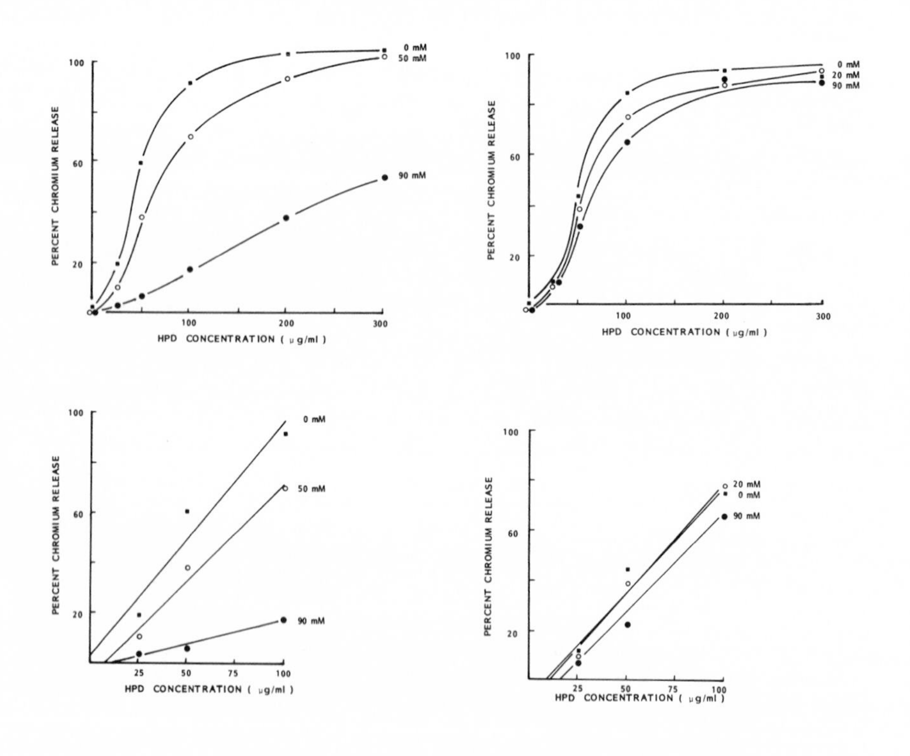

Figure 3a. Strong inhibition of phototoxicity by azide showing dose response curve (above) and regression curve (below) for the linear portion of the dose response curve.

Figure 3b. Corresponding results with a weak inhibitor of phototoxicity sodium benzoate.

Although scavengers of $^1O_2$ potently inhibit phototoxicity inhibitors of other ODFR exert an effect. Most striking is the complete inhibition of phototoxicity in the presence of sodium dithionite, which causes reduction of oxygen tension to zero.

TABLE 1

INHIBITION OF PHOTOTOXIC ACTIVITY OF HPD IN VITRO

Inhibitor	Inhibitor minus Reference Slope	Relative Inhibitory Activity	Putative ODFR Inhibited
Catalase 1000mM	0.0143	0	H_2O_2
Azide	0.7343	+++	$^1O_2$
Histidine 90mM	0.06372	+++	$^1O_2$
Benzoate 90mM	0.0914	0	OH·
Mannitol	0.4829	++	OH·
Sodium dithionite 20mM	0.8120	+++	O_2 depletion

* Difference < 0.1 + 0
0.1-0.3 = +
0.3-0.5 = ++
> 0.5 = +++

These results have implications (a) for enhancing and diminishing phototoxity e.g. in reduction of skin damage and (b) for necrotic tumours with anaerobic parts. They suggest that anoxic parts of tumours cannot be eradicated by photochemotherapy with HPD.

NATURE OF ACTIVE COMPONENTS OF HPD

Using the ^{51}Cr release assay we have investigated the phototoxicity of aggregated and non-aggregated components of

HPD, and of aggregates and non-aggregated compounds of the individual components of HPD. Methods of preparation of components will be described in an accompanying paper (Ward et al).

The activity of the monomeric fraction of HPD depends on the method of preparation. That prepared by gel chromatography is moderately active, and more active than the product obtained by membrane filtration. Pure haematoporphyrin monomer is inactive. An aggregate of haematoporphrin cannot be prepared in pure form. The impure aggregate is about as active as an equivalent concentration of HPD. Protoporphyrin is always aggregated and is active. The aggregated fraction of HPD is always very active, although it is not always shown to be more active than an equivalent concentration of standard HPD. Hydroxyethyl vinyl deuteroporphyrin preparations are relatively unstable and aggregates of this porphyrin are more active than the monomer. Observation of cells after incubation in different porphyrin solutions by fluorescence microscopy suggests that the phototoxicity is related to incorporation as shown by cytoplasmic fluorescence.

Table 2

RELATIVE ACTIVITIES OF COMPONENTS OF HPD IN VITRO

COMPONENT	ACTIVITY RELATIVE TO EQUIVALENT CONCENTRATION OF HPD*	
Haematoporphyrin	Monomer	-
	Aggregate (impure)	++
Hydroxyethyl vinyl deuteroporphyrin	Monomer	±
	Aggregate	+
Protoporphyrin	Aggregate	++
HPD	Monomer	±
	Aggregate	++

* Arbitrary scale - = inactive
+ = less active than HPD
++ = same order of activity as HPD

These studies take no account of differences of the pharmacokinetics of the components of HPD in vivo. They suggest (1) that since haematoporphyrin monomer lacks phototoxic activity it would be better if it were removed from HPD and (2) any advantages of individual aggregated species may be related to capacity to accumulate in the tumour. Studies in vivo await preparation of sufficient amounts of individual compounds.

CLINICAL APPLICATION OF PHOTOTHERAPY

The total experience in 3½ years is summarized in Table 3 and includes early results previously published (Forbes et al 1980). Not included are cases of brain tumours to be described separately by Dr.G.McCulloch. A total of 48 patients have received 99 infusions of haematoporphyrin derivative (HPD) made according to the standard procedure (described in Forbes et al 1980). Dosages have ranged from 2.5-7.5mg/kg.

Photosensitivity is the only major side-effect of HPD seen by us. This constitutes a serious deterrent to the use of phototherapy in sunny places like South Australia. Four patients had severe facial oedema, proceeding to vesicle formation with crusting in two, with subsequent complete healing. These resulted from failure to observe instructions. No constitutional or systemic side-effects have been observed. Two patients who received phototherapy to lesions of the forehead had severe oedema of the eyelids taking two or three days to subside.

1. Cutaneous Metastases of Breast Carcinoma

Thirty-five infusions of HPD in doses from 2.5 to 3.5mg/kg have been given to 16 patients. The effect on the tumour was assessed visually in all cases, with the aid of photographs. Tumour necrosis was observed in all cases except on a second treatment in one case. Light was delivered as a beam to cutaneous metastases from either incandescent lamp or laser in doses from 40 to 250 Joules/sq.cm. on days 3 and 4 after infusion. Four treatments were given by interstitial placement of optical fibre into cervical nodes. Total light doses were 400 to 600 Joules measured at the fibre tip.

Table 3

ACCUMULATED EXPERIENCE OF TREATMENT OF MALIGNANT TUMOURS WITH HPD 1979-82

	Patients	Treatments	Necrosis	Benefit
Carcinoma breast metastases	16	35	16	7
Melanoma	8	20	6	?
Carcinoma of bronchus	7	8	1	1
Vaginal tumours	5	5	5	2
Miscellaneous	13	30	10	5
	49	98	38	15

The aim of treatment of cutaneous metastases is to ablate metastases without destruction of skin. In no case of multiple tumours was this aim perfectly achieved in all tumours, but the result was near to perfection in 3. Many tumours were eradicated perfectly in 5 patients, marginal tumour was left in many, and slowly healing cutaneous necrosis occurred in others. In 3 cases in our early experience the treatment caused discomfort and was attended by ulceration which failed to heal, so that the treatment was definitely worse than the disease. These were patients with advanced massive confluent chest wall metastases. Such are not suitable for treatment at this stage of development of the modality.

Assessment of benefit is difficult and is a value judgement. No weight is given to the possible satisfaction to the patient in achieving temporary cosmetic benefit, and reassurance that treatment is being given when other modalities are not available. In the case of cutaneous metastases of breast carcinoma factors favouring an assessment that treatment has been beneficial are :-

1. If ulceration of advancing chest wall metastases has been prevented.
2. The need for other treatments has been delayed significantly.

3. Tumour ablation has been achieved, without painful chest wall necrosis, in patients in whom conventional treatments cannot be given.

A judgement that there has been no benefit would be made if :-

1. Treatment caused distress greater than evident before treatment.
2. Advancing metastases in skin or elsewhere negated benefit.

Treatment is of no benefit if new metastases become evident in high number within a short time of treatment, or if metastases elsewhere become manifest and cause serious symptoms.

By these criteria, in our judgement, treatment was beneficial in 7 patients.

Reasons for treatment failure include massive tumour and extensive confluent chest wall tumour. An area of skin necrosis within an extensive area of infiltrated epidermis will not re-epithelialize. Rapidly advancing metastases should not be treated, unless systemic treatment can also be introduced to control the disease.

Skin necrosis may be caused by heat due to infra-red irradiation from incandescent light sources or too high flux density. This is avoided by monitoring skin temperature by thermocouple and by use of aerosol water sprays.

Non involved skin is masked.

Combined treatment with either hormone or chemotherapy has not yet been evaluated, although the results in two cases suggest that this may be effective. In these cases hormonal treatment (androgen and tamoxifen respectively) was given within a month of phototherapy with excellent results.

Phototherapy has not yet been applied to primary breast cancer. This may be realistic when equipment improves and combined treatment, e.g. after limited resection, can be assessed. Its place in breast cancer at present is for cutaneous,subcutaneous and possibly lymph node metastases, of limited extent, with circumscribed tumours, not advancing rapidly. If it is offered only when other treatment modalities have been exhausted it is unlikely to be of major benefit. Its use should be explored in conjunction with other treatments, e.g. to spare the use of other agents until they can be administered with maximum effect, as in indolent metastatic disease involving only skin.

2. Vaginal Tumours

Five with vaginal tumours have been treated, 2 with excellent results. Necrosis was seen in all cases, but (by judgement criteria stated for breast cancer) was not beneficial in 3. The 2 successful cases had recurrent malignant melanoma and a nodule of ovarian carcinoma respectively. Tumours were eradicated completely in both of these cases. The patient with melanoma died subsequently of metastases elsewhere without recurrence in the treated area, and in the other patient there was no local recurrence but disease advanced elsewhere after more than 12 months. These cases have been reported (Ward et al 1982).

3. Malignant Melanoma

Eight cases received 20 infusions. Tumour necrosis was seen in 6 (all had multiple cutaneous metastases). These studies show that melanoma, if not too deeply pigmented, will respond to phototherapy, but it is doubtful if any patient benefited substantially. All died from advancing disease. Much effort was expended, always appreciated by the patient (the benefit of the caring doctor with a new approach) who was fearful yet ignorant of the impending manifestations.

4. Carcinoma of Bronchus

Eight treatments were given to 7 patients. In only one was benefit obtained. Despite this poor result, carcinoma of bronchus is considered by us to be one of the most important diseases for the application of phototherapy. We consider our first 6 failures to be due to technical inadequacies. The seventh case, a 72-year-old man who had right lower lobectomy for adenocarcinoma of bronchus in 1978 had haemoptysis leading to diagnosis of adenocarcinoma in the bronchial stump, estimated volume 1ml. He received 7.5mg HPD/kg. Forty-eight hours later light was delivered by continuous wave laser by beam and intratumoural insertion to a total dose of 900 Joules. Tumour necrosis was observed and three months after treatment biopsy specimens of the healed tumour bed showed no tumour. Recurrence was detected at 10 months and a second treatment (7.5mg HPD/kg, light delivery at 48 hours) was given, on this occasion using the pulsed metal vapour laser. Extensive necrosis was observed with erythema and sloughing of normal mucosa. The latter did not result in symptoms. Biopsies three and five months after the second treatment showed no tumour and the mucosa was healed. Fourteen months after the first treatment a para-oesophageal tumour was detected.

5. Miscellaneous Cases

No necrosis was evident in treated metastases of renal cell carcinoma, chondrosarcoma and very little in a case of carcinoma of prostate. Two patients with slowly advancing cutaneous metastases of mucoepidermoid carcinoma of parotid were treated. In both there was temporary control when radiotherapy had failed. One patient with advanced squamous cell carcinoma of head and neck primary pyriform fossa, having escaped from control by all attempted chemotherapy regimes and radiotherapy was treated and his case illustrates important points. Ulcerating tumour below the left ear was treated on five occasions with diminishing response. Regrowth of tumour was progressively rapid. Tumour also advanced on other non-treated sites (pyriform fossa, left mandible) during this time. This patient should not have been treated, for a number of reasons. Firstly, if all detectable tumour could have been necrosed the result would have been devastating with a defect from the pharynx to the outside. Secondly, the treatment plan was not rational. Limited necrosis avoiding the abovementioned disaster, would achieve nothing and would lead to an increase in the area of slough. Thirdly, phototherapy was acting like a cautery in this situation and it apparently stimulated tumour growth.

A case of extensive basal cell carcinoma of scalp was treated. Tumour necrosis was confirmed histologically, but the patient was subsequently treated by partial excision of the skull because of invasion of bone.

A 67-year-old woman had previously had plastic surgery for basal cell carcinoma of the nose with excision and elevation of a flap from the cheek, with excellent results. Nevertheless she declined surgery and requested phototherapy. The lesion occupied a circle of 7mm diameter, with rolled edge and atrophic central epithelium. It was treated with the greatest caution to avoid necrosis with four consecutive infusions. At one year there is a slight depression only with no appearance of recurrence. The patients regarded the treatment as much less disturbing and arduous than surgery. Having the desire to avoid surgery she found the restrictions on sun exposure no problem.

Another patient had squamous cell carcinoma of leg recurrent after two excisions and radiotherapy. The lesion measured 14x5mm approximately. It was treated by a harsh dosage of HPD and light, 72 Joules/sq.cm. with the intention of necrosing the tumour and surrounding skin. The black eschar gradually sloughed and the defect was completely epithelialized in 3 months. It has not recurred in 2.5 years. While the lesion was healing it was painful and unpleasant. The patient treated it by daily sea baths.

6. Fibreoptic Delivery Tips

Much needs to be done to maximise light delivery to the appropriate points. Unsolved problems are high tip temperature with tissue charring and lack of knowledge about light intensity in relation to tumour. The use of a bevel point usually reduces the extent of charring to a small spot. The Dougherty-Potter end has considerable advantages in the illumination of a bronchus or intra-tumoural insertion provided that in the latter case the tip can be made to penetrate the tumour. A light detector on the skin can monitor the fall off in intensity resulting from charring or change in translucency of the tumour or surrounding tissue. Interstitial treatment of subcutaneous tumours and lymph node, frequently results in partial necrosis of tumour in our hands.

CONCLUSIONS

After this exploratory phase of phototherapy for human cancers some of the problems have been defined. This treatment cannot be introduced widely in medicine until the machines and accessories are adequate, until dosimetry is based on measurement and established principles and until situations have been defined for proper clinical trials by established methods.

Many situations can now be defined in which phototherapy should not be used, at least at this stage of the development of the modality. Almost all cases of advanced malignancy should be avoided. Cases of head and neck malignancy where all else has failed should not be treated by phototherapy at this stage. Treatment cannot be used in any situation where total destruction of a trans-mural lesion will cause a fistula.

There are several clinical situations in which phototherapy seems reasonable, but which have not been exploited or explored.

1. Where orthodox surgery at a later date would not be prejudiced if treatment fails. For example in an attempt to save a nose grossly involved by basal cell carcinoma.
2. Where the patient refuses surgery.
3. As an adjunct to other orthodox procedures in situations where present day results are poor. This is our current approach to brain tumours which will be described by Dr. McCulloch. I believe that this approach should be strongly considered both in the case of brain tumours and in other situations, for example in primary treatments in head and neck disease.

The times have changed greatly since radiotherapy was introduced, when attitude to experimentation in treatment of humans were liberal and relatively uncritical. Some acceptance of poor results with radiotherapy has survived to this day. Such poor results would not be tolerated in the case of phototherapy. There are now many and serious inhibitions to the introduction of phototherapy. It must now be shown that the patient suffers no harm or undue harm. In this context photosensitivity is a major hindrance to the use of phototherapy for the treatment of cutaneous and other lesions which are currently dealt with by surgery or radiotherapy. If this side effect could be overcome I believe phototherapy would become the primary modality for the treatment of most cutaneous malignancies.

REFERENCES

Forbes IJ, Cowled PA, Leong AS-Y, Ward AD, Black RB, Blake AJ, Jacka FJ (1980), Phototherapy of human tumours using haematoporphyrin derivative. Med. J. Aust. 2 : 489-493.

Moan J, Boye E (1981), Photodynamic effect of DNA and cell survival of human cells sensitized by hematoporphyrin. Photobiochem. and Photobiophys. 2 : 301.

Stenstrøm AGK, Moan J, Brunborg G, Eklund T (1980), Photodynamic inactivation of yeast cells sensitized by haematoporphyrin. Photochem. Photobiol. 32 : 301.

Ward BG, Forbes IJ, Cowled P, McEvoy MM, Cox LW (1982), The treatment of vaginal recurrences of gynecologic malignancy with phototherapy following hematoporphyrin derivative pretreatment. Am. J. Obstet. Gynecol. 112 : 356-357.

Weishaupt KR, Gomer CH, Dougherty TJ (1976), Identification of singlet oxygen as the cytotoxic agent in photoinactivation of a murine tumor. Cancer Research. 36 : 2326.

ACKNOWLEDGEMENTS

We acknowledge with gratitude the support of benefactors to The University of Adelaide Cancer Phototherapy Research Fund, particularly Miss E. Laubman and Mr. D. Schultz. This work was also supported by grants from The National Health and Medical Research Council of Australia.

Porphyrin Localization and Treatment of Tumors, pages 709–717

PHOTOTHERAPY IN MALIGNANT BRAIN TUMORS

G.A.J.McCulloch, I.J.Forbes, K.Lee See,
P.A.Cowled, F.J.Jacka, A.D.Ward.
Departments of Medicine and Organic Chemistry
and Mawson Institute of the University of
Adelaide, South Australia.

INTRODUCTION

Since 1979 we have been treating patients with malignant brain tumors and metastatic deposits in the brain with H-P.D. and phototherapy. The rationale for using this experimental form of treatment was that these conditions are invariably fatal. Furthermore, in regard to malignant gliomas the survival time by the currently used means of treatment is very poor. Salcman, in a retrospective analysis found that in patients with astrocytoma Grade IV, only 7.5% lived longer than 2 years, even with radical surgery and radiotherapy.

Earlier reports by Perria et al used H-P.D. and phototherapy treatment in 9 patients with various types of malignant brain tumors using a dose of H-P.D. between 2.5 and 10 mg/kgm body weight and a light dosage of 9 joules/ sq.cm. in most cases. Their patients were treated at the time of open operation using a helium neon laser and fiberoptic system. No post-operative radiotherapy was used. They were unable to show any real benefit from the treatment.

Laws et al reported 5 recurrent glioblastomata treated by H-P.D. and phototherapy using an H-P.D. dose of 5 mg/kgm body weight and a total light dose between 500 and 1260 joules. They used a rhodamine 610 dye laser pumped by an argon laser. The light was delivered by a fiberoptic system introduced into the middle of the tumor by a needle placed in a burrhole. They showed a decrease in size of the tumor in some cases and no evidence of worsening of the clinical state. All of their patients had had previous radiotherapy.

Subsequently Laws' group has changed to treatment of the tumor bed at open operation (rather than via a needle) with removal of bulk of tumor.

METHOD

All patients have undergone a preliminary burrhole biopsy with confirmation of the malignant nature of the tumor by histological examination. All patients have been told of the experimental nature of the treatment and signed a suitably worded consent form. 48 hours prior to the planned surgery an infusion of hematoporphyrin derivative in the dose of 5 mg/kgm body weight was given. The method of preparation has been previously described (Forbes et al 1980).

The patients have then undergone conventional neurosurgical treatment with craniotomy and radical excision of the tumor. Frozen sections were examined for fluorescence at 400-490 nM and was found in 4 of 9 glioblastomata examined, 3 of 3 metastases examined, and not in the single astrocytoma Grade II examined. At the completion of the operative procedure the tumor bed which contained either visible or microscopic amounts of tumor was then treated with light according to 3 different programmes which have evolved during the course of this study :

1. The first light source was a rhodamine dye laser driven by an argon ion laser. This was delivered via a fiberoptic system to the surface of the brain and held in place either physically by the surgeon or by means of a specially designed clamp. The total time of the treatment was often in the vicinity of 90 to 120 minutes because of the relatively low output of the system (varying between 280 and 460 mW).
2. In the middle part of the series the patients were treated with a special lamp constructed in our laboratories. This has been described previously in these proceedings by Dr. I.J. Forbes. Essentially it consists of a 1000 watt incandescent light source with a series of long and short wave pass filters resulting in a total output of 24 watts in the 620 to 720 nM band. A series of plastic lenses have been constructed to give a variety of output patterns of light intensity for use in

varying clinical situations. The main purpose for using this lamp system is that a higher rate of light delivery has been possible, thus reducing the total exposure time.

3. More recently, a gold laser constructed by Quentron Electronics has been utilized with a wave length of 627.8 nM with a total output of 1.5 watts. This has retained the flexibility of a fiberoptic delivery system and has also ensured that the light delivered to the tumor bed is entirely within the absorption spectrum of the H-P.D. Exposure times have not been excessive with this method (about 20 minutes).

Post-operatively the patients have been treated in the routine manner but special attention has been directed towards the treatment of cerebral edema which appears to be more common in these patients. Those with a diagnosis of astrocytoma grades III or IV have also undergone whole brain irradiation of 5000 rads.

In summary, the patients have all undergone routine neurosurgical and medical treatment for their condition but with the added feature of H-P.D. injection and light treatment. The exceptions to these are those patients who have had metastatic tumors, that we have noted in other studies to be rather sensitive to H-P.D. treatment. These patients have not undergone post-operative radiotherapy and constitute an important group in demonstrating the effectiveness of this treatment.

RESULTS

Table 1 shows the details of patients with metastases. B.C. was a somewhat desperate effort in that he had a very large left parietal deposit of a malignant melanoma . Although a good clearance of tumor was obtained surgically, the tumor recurred early, possibly due to the poor light absorption by the black tumor tissue.

E.P. had had a previous craniotomy and removal of the right frontal lobe deposit. Post-operative radiotherapy did not prevent a recurrence at the same site 10 months later. There was no clinical evidence of local recurrence at death some 11 months later. An autopsy was not obtained.

The other 2 cases have shown no clinical or radiological evidence of tumor recurrence despite H-P.D. - phototherapy

Table 1. METASTASES: H-P.D. TREATMENT

ORIGIN	LIGHT DOSE	RESULT	COMMENTS
E.P. Ca Kidney (Fl.+)	1680J Laser	Died 11 months post treatment. No clinical cerebral recurrence.	Second recurrence at same site. Previous treatment craniotomy & R.T.
B.K. Ca Breast (Fl.+)	1260J Lamp	Alive & well. No C.T. scan evidence of local recurrence at 16 months.	No post-op R.T.
H.M. Ca Lung (Assumed) (Fl.+)	1890J Lamp	Alive & well 10 months. No clinical evidence of recurrence.	No post-op R.T.
B.C. Malignant Melanoma (Fl. ND)	1785J Lamp	Died 2 months post-op with large local recurrence.	No post-op R.T. ? poor light penetration.

Table 2. OTHER GLIOMATA: H-P.D. TREATMENT

ORIGIN	LIGHT DOSE	RESULT	COMMENTS
W.H. Oligo-Dendroglioma (Fl. ND)	2100J Laser	Died 2 months post-op 1 month improved.	Re-operation previous R.T.
M.F. Oligo-Dendroglioma (Fl.+)	2400J Laser	Alive. Has occasional fit. 10 months post-op - no C.T. scan evidence of tumor.	Post-op R.T.
D.S. Astrocytoma Grade II ?III (Fl.-)	1925J Lamp	Alive & well - 13 months.	Thermal burn

Fl.- = Fluorescence negative.
Fl.+ = Fluorescence positive.
Fl. ND = Fluorescence not done.

Table 3. GLIOBLASTOMATA : H-P.D. TREATMENT

CASE	LIGHT DOSE	RESULT	COMMENTS
M.G. 1) 2.5mg/kgm operation 120 hours later. (Fl.+) 2) 5.0mg/kgm operation 48 hours later. (Fl-)	1860J Laser 2100J Laser	Died 7 months after second treatment from recurrence.	2 operations & 2 H-P.D. treatments. No R.T. after first treatment. R.T. after second operation.
J.A. (Fl.+)	2288J Laser	Alive 42 months post-op. (Occasional fit) No clinical or C.T. evidence of recurrence.	Part Australian Aboriginal. Post-op R.T.
T.G. (Fl.+)	2484J Laser	Died 5 days post-op.	Second operation (Pre-op R.T.) P.M. evidence of tumor necrosis.
J.P. (Fl.-)	2100J Laser	Alive 35 months post-op. (Occasional fit) No clinical or C.T. evidence of recurrence.	Post-op R.T.
M.L. (Fl. ND)	2100J Lamp	C.T.scan - tumor recurrence at 15 months. Alive at 17 months.	? bone edge necrosis 10 days post-op. Post-op R.T.
E.W. (Fl.+)	2300J Lamp	Died 11 months post-op with recurrence.	Post-op R.T.
B.L. (Fl.-)	2520J Lamp	Died 11 months post-op with recurrence.	Post-op R.T.
B.S. (Fl.-)	2205J Lamp	Died 9 months post-op with recurrence.	Post-op R.T.
B.C. (Fl.-)	1620J Gold Laser	Alive 2 months	Post-op R.T.

being the only treatment given to the local tumor.

Table 2 shows the result of the rather heterogenous group of "other gliomata". Their responses to treatment under conventional therapy is varied and we do not feel that any conclusions can be drawn from these cases. In the case of D.S. we learnt that our lamp was capable of burning the brain tissue. Subsequent monitoring of the temperature of the brain tissue showed that a rise from 37°C to 45°C would occur in the first 10 seconds of the lamp being activated. We have subsequently irrigated the surface with cool Ringer's solution during light treatment with no rise in brain temperature.

Table 3 shows the results in the 9 patients with glioblastomata. After the first 4 cases we changed from the fiberoptic laser system to the lamp as the treatment times could be markedly reduced (from 2 hours to 10 minutes). However, after 2 disappointing results (M.G. and B.L.) we considered and subsequently changed to a more powerful laser system (the gold laser). These 2 cases had frontal lobectomies with excellent clearance of tumor but reappeared at a disappointingly early stage. Cases J.A. and J.P. are sources of encouragement in that they have survived for 42 and 35 months with no known tumor recurrence. In both cases a good tumor clearance was obtained by radical surgery. Thus, 2 of the first 4 glioblastomata cases done more than 2 years ago are still alive. Salcman's study gives a comparable figure of 7.5% at 2 years. Our numbers are not of statistical significance but do give us some encouragement.

COMPLICATIONS

Cerebral Edema : This has been more prominent in our view with all of these patients. In the early cases when the treatment times were long it was often noted by the surgeon that the treated brain surface started to swell during the treatment. On 2 occasions the remainder of tumor swelled more so than the surrounding brain surface such that although it was initially flush with the surface of the brain it became proud of the surface. The greatest potential for treatment appears to be in those patients in which a large resection of tumor is possible (e.g. in patients in whom a frontal lobectomy is performed) which thus leaves a large cavity within the skull for the swollen brain to expand into.

Those patients in whom a limited resection of tumor only has been possible have generally not done well post-operatively and have in 2 cases (B.S. and E.W. Table 3) developed increased neurological deficit - this we think is due in part to the surgical trauma and in part to the edema caused by the H-P.D. light interaction. The edema is vigorously treated with steroids and hypertonic mannitol (i.e. conventional means of combating cerebral edema). B.S. had an immediate post-operative worsening of her neurological deficit but this may have been due to the surgical trauma as the tumor involved in this case was very close to the motor area of the brain. Prior to operation she had a mild degree of weakness in the right arm and leg which was considerably more pronounced post-operatively. E.W. had a slight increase in her dysphasia post-operatively but in a few days it had reverted to the pre-operative state.

Early Death : We have had one early post-operative death that is worthy of comment (T.G. Table 3). In this patient his pre-operative condition was poor in that he was semi-conscious and had a severe degree of cerebral edema already. A limited resection of tumor was possible with a significant degree of cerebral edema still being present at the conclusion of the procedure. Post-operatively his condition worsened and he died one week post-operatively. Histological evidence at post mortem showed death of tumor tissue which the pathologists felt corresponded to fields of light penetration rather than vascular damage.

Damage to Other Tissues : With the use of the new lamp we encountered an unusual illness in the first patient who was treated with this (M.L. Table 3). This woman had a very good resection of a right frontal lobe tumor and initially had no post-operative problems. However, some 10 days after discharge from hospital, she was readmitted with head ache, local tenderness over the bone incision, a low grade fever and an elevated E.S.R. There was no evidence of bacterial infection and a repeat C.T. scan did not show any abnormality other than the cavity caused by the resection of tumor. Her symptoms gradually settled over about a month. One proposed mechanism for this illness was that the unshielded area of bone may have been affected by the spread of light from the lens. In subsequent cases care has been taken to shield the bone edge from the incident light and a similar illness has not recurred.

Skin Photosensitivity : Photosensitization of the skin has of course occurred, but with adequate protective measures, this has not been a significant problem.

Radiotherapy Sensitivity : No effect has been noted on the efficacy or side effects of the subsequent radiotherapy.

THE FUTURE

We intend continuing this means of treatment in suitable cases, concentrating on patients in whom a large tumor resection is possible, leaving an adequate cavity for expansion of the swollen brain. We do not at the moment consider performing a randomized study but intend comparing our end point of long term survival (i.e. at 2 years) with other large published series. We do not see this as a means of killing off large volumes of tumor tissue. We believe that it may be possible, if suitable cases arise, to use this method of treatment for deeply placed relatively small brain lesions by a stereotactic means. We would also like to use this method of treatment on patients with small but malignant tumors of the brain stem which would otherwise be quite inoperable. The risks of this would be greater than the risks of tumors elsewhere in the brain in that swelling of the brain stem would probably result in death of the patient. A lower dose may thus be required.

In the early part of this series, the dosage of light involved has been rather empirically chosen. In the latter part of the series however, we have been aiming at the concentration of 100 to 150 joules/sq.cm. Delivery of light may be adjusted when studies by the physicists have established principles of light dosimetry.

We believe this approach to gliomas is reasonable, as the patient is not deprived of conventional therapy and is offered the chance of longer survival with little added risk or inconvenience.

Forbes IJ, Cowled PA, Leong AS-Y, Ward AD, Black RB, Blake AJ, Facka FJ (1980). Phototherapy of human tumors using haematoporphyrin derivative. Med J Aust 2: 489-493.
Laws ER Jr, Cortese DA, Kinsey JH, Eagan RT, Anderson RE (1981). Photoradiation therapy in the treatment of

malignant brain tumors: A Phase I (feasibility) study. Neurosurgery 9:672-678.

Perria C, Capuzzo T, Cavagnaro G, Datti R, Francaviglia N, Rivano C, Tercero VE (1980). First attempts at the photodynamic treatment of human gliomas. J Neurosurg Sci 24: 119-129.

Salcman M (1980). Survival in glioblastoma; Historical perspective. Neurosurgery 7:435-439.

Acknowledgement : We wish to acknowledge the financial support of Mr. D. Schultz and Miss E. Laubman in this research.

Porphyrin Localization and Treatment of Tumors, pages 719–726

PHOTORADIATION AND LASERS IN THE TREATMENT OF CENTRAL NERVOUS SYSTEM NEOPLASMS: ADAPTATION TO THE SHELDEN/JACQUES CT-BASED COMPUTERIZED MICRONEUROSURGICAL STEREOTACTIC SYSTEM

Skip Jacques, M.D.

Director, Neurosurgical Research
Huntington Medical Research Institutes
734 Fairmount, Pasadena, California 91105

With the advent of computerized tomographic (CT) scanning, and more recently nuclear magnetic resonance (NMR) scanning, neurosurgery has now experienced the ability to diagnose intracranial tumors based on direct computerized information. This is a remarkable transformation in that neurosurgeons can now find central nervous system pathology at such an early stage that conventional freehand techniques may not be successful in biopsy or obliteration of the lesion. Because of this problem the authors, in conjunction with scientists at the California Institute of Technology, have recently developed a new stereotactic system originally based on the General Electric 8800 CT scanner, and more recently updated to the NMR scanner. It consists of a unique ring system with a series of graduated pins or holes which allows computer-assisted localization in the X, Y and Z axes. Preliminary reports of its development and use have been reported elsewhere (Shelden, McCann, Jacques, et al, 1980, Jacques, Shelden, Freshwater, Rand 1980, Jacques, Shelden, McCann, Linn 1980, Shelden, Jacques, McCann 1981). Finding central nervous system pathology, and particularly gliomas, at an early stage is probably essential when dealing with tumors. Reduction of the immunological "tumor burden" is the crucial step to any successful adjuvant therapy program. To this end we have recently added to our stereotactic system a CO_2 laser and a tuneable dye laser, fully mated to a binocular optical system. We are now concentrating on newer methods of locally applied adjuvant therapy.

MATERIALS AND METHODS.

Computer Programs.

Multiple computer programs have been developed in conjunction with the Engineering and Applied Science Divisions at the California Institute of Technology for reformatting routine CT scan digital data. Research is now in progress to do the same for NMR tomographic scanning. These include algorithms to three-dimensionally reconstruct, magnify, digitally subtract, color code, and exactly determine volumetrically, central nervous system pathologies of various kinds. When these algorithms are applied to CT or NMR scan data prior to surgery, they afford the surgeon a more accurate preoperative delineation of the pathology. The surgeon can now preoperatively assess the size of the tumor in terms of its volume, three-dimensional configuration, density and contour boundary analysis. Three-dimension reconfiguration allows a very accurate determination of the volume and proximity and relationships of contiguous structures. In the process of being developed is a digitized, computerized brain map based on routine stereotactic atlases which will, in turn, relate various lesions to very precise anatomical features. This allows the surgeon not only to find a lesion but also allows him to approach the pathology at the most innocuous angle. Future sophistication, particularly with the NMR scanner, will allow pathological correlations to permit one to make exact predictions as to pre- and postoperative symptomatology and cognitive rehabilitation techniques.

Surgical Technique.

The Shelden/Jacques stereotactic system is based on a newly designed ring system which is placed on the patient's head by means of four pins which pass through the plane of the ring and attach to the patient's skull (Fig. 1). The ring consists of an exact duplicate of a ring which is on the phantom system, and made of an aluminum alloy which only minimally affects the CT scanner. The dimensions of the ring are entered into the computer. A similar plastic type ring system is now in development for use with the NMR scanner. A series of 24 pins, each varying in height by 1 mm. from its nearest neighbor, are aligned vertically from the superior surface of the ring with quadrant pins at 0, 90,

180, and 270°. These quadrant pins align the ring in the scanner prior to surgery.

Fig. 1. Shelden-Jacques stereotactic system with tulip resectoscope.

Prior to the actual operation, a single scan is obtained with the ring in place and the patient and ring mounted to a newly designed base mounting system for attachment to the CT scan table. This allows the computer to accurately define all areas of the intracranial space in X, Y and Z axes relative to the ring. The coordinates so obtained by our newly designed computer algorithms are set on the phantom device at surgery and transferred to the patient at the actual operative procedure by means of a typical stereotactic arc system and micromanipulator which is, in turn, coupled to our "tulip" resectoscope (Fig. 2). The laser in turn is mounted to the binocular optical system which is joined to the tulip.

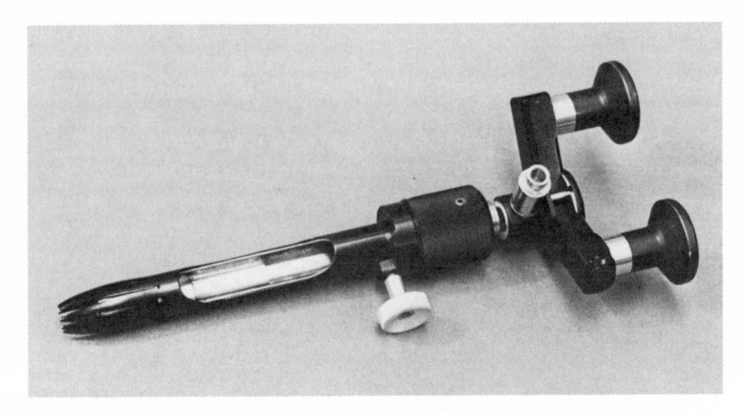

Fig. 2. Tulip resectoscope with binocular optical system.

After the CT scan the patient is transferred to surgery where the patient and head ring are attached to the operating room with the same mounting system that attached the patient to the CT scan table. Therefore, all angles have been preserved. After selection of the appropriate angle of entry and transfer of the arc system between the phantom and patient's skull, a small skin and craniotomy incision is made and the lesion approached by means of the newly designed "tulip" resectoscope with its inherent stereotactic binocular system and CO_2 laser. Multiple instruments have been developed for use down through the tulip including the CO_2 laser, the tuneable dye pump argon laser system for photoradiation therapy, small rotodissectors, microdissecting probes, hypothermic devices, and new biopsy devices.

SURGICAL EXAMPLES.

We have approached many types of central nervous system pathology using the technique as previously described (Shelden, McCann, Jacques et al, 1980, Jacques, Shelden, McCann, Freshwater, Rand, 1980). These have consisted of intracranial gliomas, metastatic tumors of the brain, intracranial hemorrhages and small arteriovenous malformations. The technique would also be extremely useful in the removal of foreign bodies and abscesses. The technique has proved very innocuous in our hands with postoperative hospital care decreased by one-half to one-third. The technique has been very accurate as well. We have been able to approach and remove as many as four separate lesions at one surgical setting.

We believe that eventually all tumors will be much smaller when first visualized by CT or NMR methodology. The next revolution will occur when a brain tumor blood-borne antigenic marker is routinely detected by routine blood tests, as many early central nervous system lesions may be asymptomatic. These smaller pathologies will make the stereotactic method of localization and removal mandatory. We believe that with the advent of modern CT scanners and NMR on the horizon, all tumors, large or small, should be approached first stereotactically with resort to freehand techniques along conventional lines if necessary. This eliminates blind transcortical probing to locate a subcortical tumor and obviously decreases the risk of the procedure substantially. Blind probing, whether freehand or stereotactic, without the ability to three-dimensionally see as through our tumorscope, can lead to serious neurological complications and death if bleeding occurs following a procedure.

Metastatic brain tumors can be localized and removed much more accurately and innocuously from this computer-derived digital data. Using a smaller tumorscope which is now in prototype form, these lesions can be approached very specifically and vaporized with a CO_2 laser and the residual tumor base treated with the appropriate adjuvant therapy such as photoradiation therapy. All these sophisticated systems of localization and tumor removal serve only to remove the tumor burden to a degree that allows a specific adjuvant therapy to be successful.

ADJUVANT THERAPY.

Adjuvant therapy at this time can be divided into an immediate and late form, where the immediate phase serves to gain time which is utilized to develop <u>in vitro</u> testing of immunological, chemical or other antitumor measures that specifically react against residual tumor cells. We have approached this with a new hypothermic implantable device that inhibits local enzymatic function. This should theoretically insure a period of "zero growth" for the residual tumor cells. If zero growth is not maintained during the immediate postoperative period, the regrowth of the subsisting tumor cells will again rapidly result in a tumor volume that approaches that at the initial surgery and additionally overwhelms locally applied therapeutic measures.

HYPOTHERMIA.

Adequate temperature reduction to 20° C. at the site of the residual tumor cells is accomplished by a prototype conductive implant - a hypothermic device. This is presently in the form of a Peltier unit. Residual tumor cells are kept inactive with the hypothermia, but viable during this period so that with return to normal temperature they can absorb and metabolize the locally applied antitumor therapy. The specific adjuvant therapy is determined by reactivity of cultured tumor cells to a variety of antitumor chemicals and immunological agents. This in essence is called clonogenic assay and is similar to the selection of a proper antibiotic for a given bacterial infection. The local treatment with chemotherapy directly applied to the area of tumor removal provides greater _in vivo_ efficiency with less systemic reaction than parenteral administration.

IMMUNOTHERAPY.

The immune system, although being the body's greatest defense system, incompletely protects the brain because of problems with the blood-brain barrier. The efficiency of this system is also easily overwhelmed by a massive tumor burden and the blood-brain barrier may prevent early access to the tumor of those humoral or cellular systems which might otherwise be effective. Our investigations to date favor a systemic multiclonal sensitization based on nature's method of controlling splenic metastases in patients with generalized carcinomatosis. Localized tumor implants into the spleen are done for T or killer cell sensitization with the effluent of the thoracic duct then directed back to the tumor cavity.

Monoclonal antibody techniques offer greater specificity but the cytotoxicity may be inefficient as a simple system. However, tagging of amino acid sequences of the monoclonal antibody with drugs or photosensitizing agents may be most effective against specific tumor cells, clones or target organs.

PHOTORADIATION THERAPY.

Photoradiation therapy with hematoporphyrin derivative could become a primary cytotoxic system of great value independent of monoclonal antibodies or other chemotherapeutic modalities in the treatment of small central nervous system neoplasms. We are actively involved in this method and hope to attain a suitable combination of maximum absorption of the dye by tumor cells with a high specificity and least laser damage to surrounding tissue. This work is further documented in the chapter by Dr. Donald Rounds, et al. Presently, photoradiation therapy in the red range has the ability to penetrate and destroy residual tumor cells at a depth of 1-2 cm. away from the bed of the tumor with least damage to the surrounding tissue from the effects of the red light.

In summary, hopefully future brain tumor research will excel in cell culture facilities and techniques to coordinate the tumor cell's individual susceptibility to destructive agents; i.e., the genetic aspects of tumor biology by hybridization of mitochondrial suborganelle proliferation and uncoupling and inhibition.

FUTURE DEVELOPMENTS.

The stereotactic surgical system and its inherent ability to localize and correlate structure with function, particularly now with the marriage of the NMR scanner to the technique, provides a wealth of possibilities for neurosurgeons in the future. We must remember that CT scanning only offers the ability to visualize changes in structure. Nuclear magnetic resonance will eventually allow one to very accurately localize regions of cerebral metabolic dysfunction by the exciting new technique of topical nuclear magnetic resonance spectroscopy. The same algorithms which have been developed for use with our CT scanner and stereotactic system are obviously applicable to any mode of digital information output from whatever the source. We are presently investigating both proton NMR and topical magnetic resonance spectroscopy to develop a spatially oriented representation of the data for mating with our CT scan technique. The future of neurosurgical stereotaxis will soon include the ability to detect structural and metabolic defects almost at the single cell level and, therefore, make the stereotactic approach mandatory. We are now on the verge of the most exciting era in neurosurgery in which investigations and approaches will be at the molecular level.

REFERENCES.

Jacques, S, Shelden, CH, McCann, G, Freshwater, DB, Rand, RW (1980). Computerized three-dimensional stereotaxic removal of small central nervous system lesions in patients. J. Neurosurg 53:816.

Jacques, S, Shelden, CH, McCann, G, Linn, S (1980). A microstereotactic approach to small CNS lesions. Part I. Development of CT localization and 3-D reconstruction techniques. Neurol. Surg, Tokyo 8:527.

Shelden, CH, Jacques, S, McCann, G (1982). The Shelden CT-based microneurosurgical stereotactis system: Its application to CNS pathology. Appl Neurophysiol 45:341-346.

Shelden, CH, McCann, GD, Jacques, S, Lutes, HR, Frazier, RE, Katz, R (1980). Development of a computerized microstereotaxic method for localization and removal of minute CNS lesions under direct 3-D vision. J Neurosurg 52:21.

Porphyrin Localization and Treatment of Tumors, pages 727–745

HPD PHOTODYNAMIC THERAPY FOR OBSTRUCTING LUNG CANCER

Oscar J. Balchum, M.D., Ph.D.;
Daniel R. Doiron, Ph.D.; and
Gerald C. Huth, Ph.D.

Los Angeles County-University of Southern California Medical Center, Los Angeles, California 90033

Abstract

Twenty-two patients with endobronchial cancer of the lungs have been treated with photoradiation therapy (PRT) employing a set protocol of 3.0 mg/kg hematoporphyrin derivative (HpD) administered intravenously 72 hours prior to red light (630 nm) illumination via a bronchoscope.

Twenty patients showed a complete response. Of these, nineteen had large obstructing endobronchial tumors; one had a small mucosal lesion. Only one patient with an obstructing tumor showed a partial response. The one showing no response had an endobronchial mass that consisted of fibrous tissue, not tumor, as shown on three separate bronchoscopic biopsies.

Supported in part by contracts with the National Cancer Institute, N01-CM-27843 and the Division of Biological Environmental Research, Department of Energy, EY-76-5-03-0013 with the Institute of Physics and Imaging Science, USC, and by a Grant (GCRC-RR-43) from the General Clinical Research Centers, Programs of the Division of Research Resources, National Institutes of Health.

Extensive endobronchial tumor was usually abolished by one treatment using one or more of several types of light applicators and delivery techniques.

No immediate tumor destruction nor coagulation by heat nor complications were noted. Tumor debris was removed during subsegment bronchoscopy 2 to 3 days after PRT, opening the bronchus to its full extent.

Introduction

Photoradiation therapy (PRT) of endobronchial cancer depends upon the tumor localizing property and photosensitizing capacity of hematoporphyrin derivative (HpD) in vivo. This tumor localizing property of HpD is presumably due to its long retention time in tumors (7 days or more), compared to normal tissues from which it is largely cleared at 24 hours. The exact mechanism responsible for preferential retention is not known. We have found all histologic types of bronchogenic lung cancer and metastatic tumors to the bronchi to retain HpD and to fluoresce when activated by violet light. Lipson, in the 1960's, first injected humans intravenously with HpD and showed that bronchial and esophageal cancer fluoresced (emitted red light) when illuminated with violet light (405 nm) from an arc lamp via a rigid bronchoscope (Lipson et al 1961). Since then an imaging fluorescence fiberoptic method and instrumentation (Profio et al 1979) have been used to localize small bronchial cancer including carcinoma _in situ_ (CIS) lesions by their fluorescence.

The potential of PRT was first demonstrated by Dougherty et al (1978) in tumors metastatic to the skin from breast cancer, and extended to treatment of endobronchial lung cancer in animals (Dougherty et al 1975) and in man (Kessel, Dougherty 1983).

PRT employs red light (630 nm) and the photosensitizing property of HpD. The HpD retained in tumor tissue absorbs the red light producing an excited state from which photodynamic reactions can be incited. The cytotoxic agent produced by this photodynamic action is apparently singlet oxygen (Weisshaupt et al 1976), formed by energy transfer from the triplet state of excited HpD.

Presumably the action of singlet oxygen is to oxidize biological components and thereby impair cell membrane function and integrity causing tumor cell death. At the light dosages we use, there is no immediate visible change in the appearance of endobronchial tumors but tumor tissue necrosis or debris is seen within 24 hours. Normal tissue (containing little HpD) is not irreversibly damaged. PRT effects in the regimen we use do not result from the immediate destructive action of heat as occurs with coagulation, charring, or vaporization with Nd-YAG or argon laser treatment.

We have now employed PRT in over 60 patients with endobronchial tumors. Various dosages of HpD and red light, and techniques of light application have been tried. This report gives the methods and results in the last 22 patients, employing a fairly well-standardized protocol based on the experience gained in the earlier trials.

Methods

Hematoporphyrin derivative was obtained in sterile injectable solution (5 mg/ml) from Oncology Research and Development, Inc., Cheektowaga, N.Y. under the trade name Photofrin I. It is administered to humans after signed informed consent under IND 12678 held by TJ Dougherty, Roswell Park Memorial Institute, Buffalo, N.Y. Administered dosage in this group of patients was standardized at 3.0 mg/kg body weight, given as a single intravenous injection via a freely flowing IV line (0.5 N Saline in 5% D/W) at 72 hours prior to PRT.

Red light of 630 nm ± 1 nm was used in all cases. It was obtained from a continuous argon pumped dye laser (Spectra Physics Model 164-09 argon laser and Model 375 dye laser with a Model

373 stabilizer) containing rhodamine-B. Delivery of the dye laser output through a flexible fiber-optic bronchoscope (Olympus BF-2T or BF-4B2) was via single step index quartz fibers (Quartz Products, QSF-125, QSF-200, or QSF-400). Application of the light to the endobronchial lesion was either by surface exposure or by direct insertion of the fiber into the tissue.

Surface Exposure: Surface illumination within the bronchus was achieved either by launching the fiber light output into a forward direction or by an isotropic distribution. Forward surface (FS) illumination was obtained by mode scrambling the quartz fiber to provide or attain the desired divergence (maximum of 40 degrees) and sweeping the desired area to be treated. The isotropic distribution was obtained by coupling the light out of a bare fiber into a scattering matrix. This formed a cylindrical source which provided light over 360 degrees along the length of the stripped fiber. We termed this treatment cylindrical surface (CS) illumination when the fiber was within an open bronchus. Typical cylindrical lengths used were 0.5 cm, 1.0 cm, and 1.5 cm.

Insertion Exposure: In order to illuminate a large volume of tumor at depth, the fiber was directly inserted into the tumor mass. This was done either with a forward launching fiber, point insertion (PI) or by cylindrical insertion (CI). When a tissue mass was too firm to allow direct insertion of a fiber, the later was inserted via a 22 guage 1.0 cm flexible-sheath bronchoscopic needle (QSF-125 fiber), PI-Needle.

It was visually determined both during surface and insertion exposure that adequate tumor light exposure and light diffusion, respectively, were taking place during the entire period of illumination with red light.

Bronchoscopic procedures were done under topical anesthesia (1% lidocaine) through the adaptor of a bronchoscopic endotracheal tube (Carden) attached to a Bennett MA-1 ventilator utilizing

100% oxygen. Premedication was standard in type and adjusted individually for each patient.

Fluorescence bronchoscopic examination of the area to be treated was performed prior to light application, to assess HpD uptake and tumor extent. It was performed using a custom intensified imaging fluorescence bronchoscope system described elsewhere in these proceedings (Balchum et al 1983).

Patients were re-bronchoscoped two to four days following PRT, and the loose or firm exudate (membrane or plug) and tumor debris were removed by forceps and the area lavaged with saline and cleaned up. The area of the bronchus treated as well as the distal bronchi were carefully inspected for residual tumor and distal tumor. When necessary, a second treatment was carried out, with appropriate illumination by red light 72 hours after a second administration of HpD (3.0 mg/kg IV), again followed in two to four days by clean-up bronchoscopy. At each clean-up bronchoscopy, the response to treatment was carefully evaluated visually. Bronchial washings and material removed by forceps were cytopathologically evaluated.

Patients: The 22 patients reported here included 4 women and 18 men, age 40 to 71 years (Table 1). The location of the endobronchial tumors was in the lower trachea, main carina, right and left main bronchi, the bronchus intermedius, lobar bronchi, and segmental and sub-segmental bronchi. Twelve patients had squamous cell carcinoma, one large cell carcinoma, five adenocarcinoma, one carcinoid, one metastatic thyroid cancer obstructing the left lower lobe bronchus, one malignant fibrous histiocytoma, and one localized fibrous mass in the right main bronchus occurring one year after right upper lobectomy for squamous cell carcinoma followed by radiation. Two separate fiberoptic bronchoscopic biopsies because of hemoptysis revealed fibrous tissue. After PRT, rigid bronchoscopic biopsy yielded the same tissue diagnosis. All biopsies, bronchial washings, brushings and a number of sputum cytologies over several months were negative for tumor cells.

Table 1. Photoradiation Therapy in Twenty-two Patients With Obstructing Endobronchial Cancer

Patient No., Sex, Age	Location	Histol. Type	Light Applicators Length in cm.	Light Dosages
1. M, 55	MC, RMS, LMS	SC Ca.	CI - 0.5, CI - 0.5	200J/cm, 255 J/cm
2. M, 54	RUL	SC Ca.	CI - 0.5	202 J/cm
3. M. 53	RMS (CIS)	SC Ca.	CS - 1	50 J/cm^2
4. M, 54	MC, RMS	Carcinoid	CI - 0.5	200 J/cm
			CI - 0.5, CI - 0.75	202 J/cm, 198 J/cm
5. F, 41	RMS, BI	Adenoca	CI - 1, Submucosal	200 J/cm
6. M, 60	BI, RMS	SC Ca.	CI - 1; PI; Submuc.	202 J/cm, 182 J
7. M, 44	LMS	Adeno-large cell	CI - 1	200 J/cm
8. M, 67	RLL-6	Adenoca	CI - 0.5	200 J/cm
			CI - 0.5	200 J/cm
9. M, 59	RMS, RUL	SC Ca.	CS - 1; CI - 1	50 J/cm^2, 200 J/cm
10. M, 49	LMS	SC Ca.	CI - 0.75	264 J/cm
11. M, 68	MC, LMS, RMS	SC Ca.	CI - 0.5	418 J/cm (Poor Diffusion)
12. M, 57	LMS, MC, RMS, RLL	SC Ca.	CI - 1.5	225 J/cm
			CS - 1.5, CS - 1.5, FS	55 J/cm^2, 89 J/cm^2, 94 J/cm
13. M, 68	BI, LUL	SC Ca.	CI - 1.5; CI - 0.5	200 J/cm, 216 J/cm
			CS - 1.5, CS - 1.5	60 J/cm^2, 80 J/cm^2
14. F, 53	LMS, MC, Trachea	SC Ca.	PI. CI - 0.5	55 J, 330 J/cm
15. F, 53	LUL, LMS	Adenoca	CI - 1; CI - 0.5	144 J/cm, 144 J/cm
16. M, 71	Lingula, LLL	Adenoca	CI - 0.5; CI - 0.5	296 J/cm, 270 J/cm
			CS - 0.5; FS	86 J/cm^2, 51 J/cm^2
17. M, 69	LUL	SC Ca.	PI	202 J
18. M, 45	RMS	Benign	PI	243 J
19. M, 52	LMS, MC, BI	SC Ca.	PI; PI; PI	220 J, 210 J, 126 J
20. M, 51	LMS	SC Ca.	CS - 1.0	126 J/cm^2
21. M, 36	LLL	Met. Thyroid Ca.	CS - 1.0	76 J/cm^2
22. F, 64	LUL	Malig. Fibrous Histiocytoma	PI, PI	147 J, 149 J

MC-Main Carina
RMS-Right Main Bronchus
LMS-Left Main Bronchus
BI-Bronchus Intermedius
RUL-Right Upper Lobe Bronchus
RLL-Right Lower Lobe Bronchus
RLL-6 Sup Seg RLL
LUL - Left Upper Lobe
LLL - Left Lower Lobe
SC Ca - Squamous Cell Ca.

Total delivered light dosages (Table 1), ranged from 51 to 125 J/cm^2 for surface application, 144 to 438 J/cm for cylindrical insertion therapy (CI) and 55 to 349 J for point insertion therapy (PI). Except for the few cases of poor light diffusion in the tissue for insertion therapy (No. 11, 14, 21), the goal for delivered light was 100 $\pm$ 20 J/cm^2 for surface therapy, 200 $\pm$ 20 J/cm for cylindrical insertion, and 200 $\pm$ 20 J of point insertion therapy. These values are adjusted when multiple treatments in closely related areas are carried out, if the light transmission characteristics of the tissue are low.

Patients were admitted to the Clinical Research Center, and evaluated by complete history, particularly for evidence of shortness of breath with various physical activities, and for debility in the form of weight loss, weakness, poor appetite and food intake, and by physical examination. Chest X-rays in four views and CT scans of the chest were evaluated, as was pulmonary function by spirometry, lung volumes, and arterial blood gas. Krypton 81 m ventilation and Technecium 99 m (99-TC-MAA) perfusion lung scans were useful to assess impairment of the lung containing the bronchus to be treated, and also to determine the gas exchange in the opposite lung. This evaluation was important in order to estimate the risk of temporary impairment of ventilation and gas exchange that might occur after bronchoscopy and following PRT. Cardiac evaluation was by history, EKG, physical examination, and several daily pulse rates at rest and on walking exercise.

Results

Twenty of the twenty-two patients had a complete response to PRT. Nineteen had bronchi obstructed 40 to 90% by tumor; one (No. 3) had CIS or a small superficial mucosal tumor (5 mm diameter). Complete response in the obstructing endobronchial tumors was noted with the opening up of the lumen of the bronchus to its full extent, with no visible residual tumor seen on re-bronchoscopy. In the patient with CIS, complete response was determined when no tumor was visible and bronchial brushings and washings were negative for tumor cells on re-bronchoscopy.

One patient with endobronchial squamous cell carcinoma showed an incomplete response (No. 17). PRT had opened up the bronchus to 50% normal diameter. A residual, hard mass remained that could not be penetrated by fiber or needle, nor forceps biopsied, and apparently was calcified. A second treatment was not indicated for these reasons. In addition, red light transmission and diffusion had been poor. The one patient that showed little or no response (No. 18) had a firm rubbery mass obstructing the right main bronchus 50%, covered by normal appearing mucosa, that showed only fibrous tissue on biospy at two previous separate fiberoptic bronchoscopies because of hemoptysis. He had had right upper lobectomy for squamous cell carcinoma followed by radiation one year previously. Upon re-bronchoscopy after PRT only a small area around the insertion site (PI-needle) showed necrosis and debris. Open tube bronchoscopy and large forceps biopsy a week later again showed fibrous tissue and no tumor cells. All other bronchoscopic specimens and a number of sputum cytologies over several months were negative for tumor cells.

A complete response was achieved by one PRT treatment in 15 of the 21 patients with tumors while 6 required two treatments. The latter either had tumors in more than one bronchus in one lung (four), or the extent of tumor along the bronchus was too long to be adequately illuminated with red light during the first PRT (two). In the latter, after removal of tumor debris from the proximal extent of bronchus after the first PRT, the distal tumor could then be adequately visualized and illuminated with red light during the second PRT. In only one patient of the 19 with large obstructing endobronchial tumors was the lesion discrete and localized, or exophytic (No. 22). This patient had had a right pneumonectomy two years previously for malignant fibrous histicytoma which recurred in the left upper lobe bronchus blocking it by 80%. One PRT removed this lesion down to the level of the bronchial mucosa.

In only one patient was the endobronchial tumor superficial or in-situ (No. 3). In this case a 5 mm diameter area of irregular mucosa in the right main bronchus showed squamous cell carcinoma on biopsy. This lesion was considered to be confined to the bronchus because of negative chest X-rays and CT scan of the chest. The mediastinum, hila, and area around the right main bronchus were free of any density. One week after PRT the mucosa appeared smooth, intact, with only two to three patches of erythema. Bronchial washings and daily sputum cytologies were negative for tumor cells. Presumably this tumor is a carcinoma in situ with or without microinvasion but entirely confined to the bronchus.

All of the lesions showed bright true positive fluorescence, except the fibrous mass (No. 18), the carcinoid (No. 4), and the metastatic thyroid carcinoma (No. 21). The fibrous mass did not fluoresce nor respond to PRT. The carcinoid and thyroid carcinoma showed low-intensity fluorescence and poor red light transmission or diffusion, but did respond to PRT.

In patient No. 21, left lower lobectomy was planned 24 hours after PRT. Metastatic thyroid carcinoma obstructed the left lower lobe bronchus by 60%. Disease elsewhere was stable. Serial thin gross sections of the surgical specimen showed a 4 cm mass. The area of response to PRT was 2.5 cm long and 2 cm in diameter, and appeared hemorrhagic and necrotic. Complete necrosis extended for a radius of 3 mm and partial necrosis out to 8 mm around the needle insertion site (PI-needle), microscopically consisted of red blood cells, mostly polymorphonuclear leucocytes, and tissue debris. All of the partial and complete necrosis was regarded by the pulmonary pathologist (Dr. Michael Koss) to be due to PRT, since no other areas of the tumor were necrotic except the above volume of tumor surrounding the needle insertion site. A greater light dose (349 J) was used because of poor light diffusion noted visually during PI-needle insertion PRT.

Eleven of the twenty patients with obstructing endobronchial tumors had central lesions in the lower trachea, main carina, and/or right and/or left main bronchi. Four of these required two treatments. In some, several areas were treated at the initial and only PRT. Three of the patients with central tumors also had lesions in a lobar bronchus. In all three, both central and lobar lesions were treated at one PRT (Table 1).

Nine patients had lobar lesions. Of these, two required two treatments. Patient No. 13 had two PRT treatments during each of which tumors in the bronchus intermedius and left upper lobe were treated. Patient No. 8 had tumor obstructing the superior segment of the right lower lobe. The first treatment entirely cleared this segmental bronchus of tumor. Distally, a small tumor remnant was visible in a sub-segmental bronchus that required a second treatment once the tumor was accessible.

It has been our aim to resolve or abolish only the tumor within the bronchus, by limiting placement of the fiber in the tumor within the tracheal or bronchial wall, so that during clean-up bronchoscopy removal of tumor debris opens up the bronchus to its full lumen or extent, but not beyond the level of the bronchial mucosa. However, in two patients (Nos. 5,6), much of the bronchial obstruction was due to submucosal tumor with relatively thin mucosal tumor lesions over these areas, in the posterior wall of the bronchus intermedius (No. 5), and in two locations, the posterior wall of bronchus intermedius and right main bronchus (No. 6). In patient No. 5, a 1 cm CI fiber was inserted through the wall of the bronchus into the submucosal mass. Short-term response on clean-up bronchoscopy three days later showed a fair response with increase in lumen size of 20% and absence of mucosal lesion. Patient No. 6 had a 1 cm diffusing CI fiber inserted through the wall of the bronchus intermedius, and PI-needle insertion PRT through the wall of the right main bronchus. Response was good (opening up of 40%) for the bronchus intermedius, and fair for the right main bronchus PRT (increase in lumen of 15%). These two patients will be re-bronchoscoped in 2 or 3 months to determine whether there will be further increase in bronchial lumen diameter over this time from further shrinkage of submucosal tumor volume.

The light dosages and techniques of illumination used (Table 1) resulted in very little bronchial inflammation (erythema, edema). There was no significant bleeding during insertion PRT. Occasionally a slight amount of blood accumulated and clotted around the insertion site. No visible reaction occured on the tumor surface nor around the insertion site such as smoke, visible coagulation, or charring of tumor tissue. Withdrawal of the fiber after PRT showed no coating of carbonaceous or other material. Fibers were clean with only occasional adherence of a small clot of blood. Complications or accidents such as ignition of fiber tips and damage to the bronchoscope that occur with PRT in 100% oxygen were avoided using the following technique: a small amount of light, just enough to see the tip, was applied before insertion. The power was turned up to give the calculated light dose only after insertion was complete and was turned off completely before fiber withdrawal.

Treated endobronchial tumor at the time of rebronchoscopy usually appeared intact, very similar to its appearance before PRT. However, removal by forceps showed that it was debris. Removal resulted in no bleeding indicating that the tissue was necrotic and this was confirmed by microscopic examination. Tumor debris was removed only down to the level of the bronchial mucosa. Bleeding was a signal of unaffected tumor remaining and need for repeat PRT.

A reaction that did commonly occur in the bronchus following PRT and was observed during clean-up bronchoscopy was the accumulation of thick mucus. Often this was loose but at times it was firm to the point of forming a membrane over the surface or even a plug. This reponse was generally only seen in patients receiving higher doses, and occured from stimulation of bronchial and mucosal mucus glands.

Complications: There were no deaths or complications during PRT or clean-up bronchoscopies. Three patients (Nos. 13, 19, 20) developed staphylococcal pneumonia 7 to 10 days after PRT that responded readily to antibiotic therapy. Two (Nos. 7, 13) had left pneumothorax that occured 5 days and 24 hours, respectively, after PRT. PRT in No. 7 was to a lesion in the left main bronchus, and in No. 13 to the left upper lobe and bronchus intermedius. No instrumentation was employed distal to these sites. Coughing was considered the main factor causing pneumothorax in these patients who had significant chronic bronchitis and emphysema (COPD). Cough was moderately severe in the several days following PRT in No. 7. No. 13 had had carcinoma of the pharynx and right mandible previously resected and followed by radiation, and a paralyzed left vocal cord (after radical neck dissection). He aspirated jello, had bouts of coughing, and developed pneumothorax 18 hours following PRT. In both patients chest tube drainage achieved full lung expansion, with removal of the tube by 7 days, without complications.

Patient Outcome: Fifteen of these twenty-two patients remain alive. Ten are working, one at one year, the longest time of follow-up to date. Another was working and died at one year.

Two patients (Nos. 14, 22) died 3 months, and one patient (No. 20) died one year after PRT due to local extension and/or metastasis of tumor. Another (No. 13) died of pneumonia having severe COPD, 3 months after PRT. Another patient (No. 17) died of pulmonary and cardiac failure 3 weeks after COPD, during which he was receiving radiation therapy. Hypoxia, bronchial secretions, and airways obstruction due to severe COPD responded poorly to intensive therapy. He finally had congestive heart failure. Autopsy showed no myocardial infarction. Patient No. 12 died 3.5 weeks after PRT. He was found dead of cause unexplained even after autopsy. The treated left main bronchus was 50% patent with no distal secretions or pneumonia; as it had been 24 hours earlier on bronchoscopy and lavage. He had been discharged without dyspnea on oxygen 10 days after PRT, when he had one radiation treatment. Dyspnea developed 24 hours later and progressed. On re-bronchoscopy, the left main bronchus was extrinsically compressed to 75% of the diameter after PRT, and the distal lobar bronchi were filled with mucus. Bronchoscopic lavage had to be carried out several times to remove mucus, which alleviated his dyspnea. Patient No. 11 took an overdose of sedatives 3 weeks after discharge. He was a 69 year-old male apparently despondent since he was not able to resume full independent physical and business activity.

Discussion

This report outlines the successful palliative treatment of extensive and large obstructing endobronchial cancers by photoradiation therapy (PRT), employing the photodynamic action of HpD activated by red light (630 nm) from a laser light source. The purpose of these clinical trials was to treat endobronchial tumor, both to control its local growth and extension, and to open up bronchi for ventilation in order to decrease shortness of breath and to prevent the retention of secretions and bacteria that cause pneumonia.

Patients were selected whose lung cancer was determined stable by chest X-rays, and who were in reasonable physical condition, ambulatory, and carrying on activities of daily living. None were near-terminal nor bed-ridden.

Comparison with other reports of the effectiveness of PRT is diffucult because of the differences in the extent of bronchial cancer and particularly in the methods and techniques employed. Hayata et al (1982) reported the opening up of bronchi in 4 or 8 patients with obstructing endobronchial tumors. In the remainder of their 14 patients, the size of the endobronchial tumor was small, and 5 had lung resection shortly after PRT. Their protocol and techniques differed substantially from ours, as did criteria for response. Cortese and Kinsey (1982) reported that five large obstructing endobronchial tumors showed a partial response, and four small superficial tumors a complete response. There were substantial differences in light dosage compared to our protocol.

Twenty of the twenty-two patients reported here showing a complete response had large endobronchial cancer masses obstructing bronchi by 40 to 90% at one or more sites. Only one had a localized, exophytic tumor, and a second a superficial carcinoma in situ. Both also showed a complete response.

One patient showed a partial response. Although the first PRT opened up a previously totally obstructed left upper lobe bronchus to 50% of normal diameter, a residual very hard, probably calcified endobronchial mass remained, judged not able to accumulate HpD, transmit light and respond photodynamically. One patient showed no response, having a benign endobronchial fibrous mass in the right main bronchus. Biopsies and other examinations showed no cancer cells. Presumably this mass resulted from previous radiation therapy, did not accumulate HpD, and did not respond to PRT.

Our criteria for complete response was the visual absence of endobronchial tumor at the time of clean-up bronchoscopy within one week following PRT. In eight instances, bronchoscopic washings and biopsies were negative for cancer cells, including the one patient with CIS. In all instances a bronchial lumen of 6 mm or more was achieved, and the bronchus had been opened up to its full extent. Residual narrowing was due to extrinsic tumor and/or fibrosis, since 10 patients previously had had radiation therapy. In one patient, this mechanism of extrinsic compression due to radiation was seen to occur, since he had been bronchoscoped one week before, and then again 72 hours after a single radiation treatment, after which he had had onset of dyspnea. A decrease in lumen size to 75% of its original diameter had occured, presumably due to extrinsic edema from radiation.

The techniques of light application - surface and insertion - varied with the size, extent and location of lesions. Cylinder light application methods (CS, CI) were used for illumination of endobronchial tumors that extended along a considerable length of bronchus, or that had a large volume. Insertion methods (PI, CI) were used for large tumor masses. FS or surface illumination by red light emitted only from the end of a fiber was used to treat discrete areas and small tumor sites.

The light dosage was selected on the basis of the light diffusion characteristics in the tumor tissue, and the degree of anticipated bronchial exudate accumulation. A given site was treated with several light applications using one or more techniques when limited light penetration was noted.

In the previous 40 patients treated with PRT here (to be published), various dosages of HpD were used (2.5 - 5.0 mg/kg) and mainly PI and FS types of illumination with red light. Light dosage was also higher, at 200 J/cm^2 or 400 J. They had similar central, lobar, and segmental obstructing tracheal and bronchial tumors. Clean-up bronchoscopy within a few days after PRT was not consistently done. A second PRT was done when required. Because of this experience, more efficient and varied types of light applicators and methods were developed to effectively illuminate various areas with a sufficient light dose at one or at most two PRT treatments. As a result, rather long areas of endobronchial tumor can now be treated with a complete response, often with only one PRT treatment. And light dosage has been reduced to decrease the degree of bronchial inflammation (redness, edema) around the tumor site, and in an effort toward decreasing thick mucus, tenacious exudate, and membrane formation.

Light dosage delivery was calculated to achieve about 200 J/cm for CI and for PI, and 100 J/cm^2 for FS or CS. The amount of light to be applied to a given tumor mass and the actual dose was determined by estimating the light penetration in the tissue and the extent of tumor mass. There were no visible heat effects such as smoke, charring, or coagulation of tissue around the fiber or on the tumor surface. Visualization during clean-up bronchoscopy several days after PRT, usually at most, showed minimal redness or inflammation and no edema due to PRT, in the surrounding bronchial surface.

A number of patients have lobar or even segmental obstructing endobronchial tumors in small airways. The advantage of PRT as described here is the ability to effectively treat the tumors, since fine, flexible quartz fibers can be maneuvered into place for insertion therapy, as demonstrated in the case of patient No. 8. Here, tumor totally obstructed the superior segmental bronchus, and a fiber was inserted into the length of this bronchial tumor. After clean-up totally opened this bronchus, a small tumor remnant in a subsegmental bronchus could first be visualized. A second PRT effectively treated the tumor in this small, distal sub-segmental bronchus. Because of the rigid and large diameter (2.1 mm) sheath, in Nd-YAG laser destructive therapy, the fiber within the sheath is difficult to maneuver within lobar bronchi, and cannot be aimed within segmental or sub-segmental bronchi.

The more distal the obstructing lesion, that is the smaller the airway, the safer is PRT, in that occlusion of the lumen by loose exudate or firm membrane that can occur in the several days after PRT has little noticeable effect on breathing difficulty. This is not the case in the central airways, the trachea and/or the main bronchus. Here, breathing difficulty may become increased within a day or two after PRT indicating the need for clean-up bronchoscopy. In the 22 patients reported here, breathing difficulties did not create an urgent or emergency situation. The patients were evaluated and monitored for the need for clean-up bronchoscopy, taking into consideration the site of the treated lesion, respiratory symptoms (dyspnea) and the physical condition of the patient.

While no specific test nor method of assessment can predict with certainty an increase in dyspnea after PRT, symptoms of difficulty in breathing, chest ausculation and assessment of the pulse rate help to determine whether clean-up bronchoscopy should be done on the third day following PRT. These patients, many of whom are elderly, all had decreased lung function from COPD, previous radiation, lung resection and lung tumor. In addition to lung function tests and ventilation/ perfusion lung scans, the stamina of the patient was kept in mind, with consideration given to the physical condition, previous record of activity, appetite and food intake, and the apparent degree of debilitation.

Radiographic atelectasis was infrequent, and probably was prevented by prompt clean-up bronchoscopy performed within two or three days and repeated where required. The lower light doses used in these trials resulted in a significant decrease in mucus production thereby preventing subsequent respiratory distress as compared to our experience in several of the 40 previous patients. Of the 22 patients in this report only one (#12) experienced respiratory distress and it occured after radiation therapy.

Patient Outcome:

We judged from follow-up that our patients have benefited from PRT symptomatically. Many returned to work and others maintained reasonably good activity and quality of life, presumably resulting from the successful destruction and local control of endobronchial tumors. The treatment allowed opening of the airways to decrease dyspnea and reduced the retention of secretions with bacteria that can lead to pneumonia. Of our 22 patients, seven are deceased. One patient (#19) was bronchoscoped 8 months after PRT and showed considerable renewal of the bronchial mucosa in the treated bronchus intermedias that previously had been lined with necrotic mucosal tumor. He continues to work full time. Continued follow-up with evaluation bronchoscopies are planned to assess the value of palliative PRT for patients with obstructing endobronchial cancers.

The ultimate effect on life span will depend on the rapidity of the recurrence of endobronchial tumor, local spread in the lung and mediastinum and distant metastasis.

Acknowledgements

The technical assistance and skills of our laser technicians, Nick Razum and Leon Chaput, are greatly appreciated, as was the assistance of Joyce Portugal in records and specimens handling.

References

Balchum OJ, Profio AE, Doiron DR, Huth GC (1983). Imaging fluorescence bronchoscopy for localizing early bronchial cancer and carcinoma in situ. Proceedings Symposium on Porphyrin Localization and Treatment of Tumors, Clayton Foundation, Santa Barbara, California, April 24-28.

Cortese DA, Kinsey JF (1982). Endoscopic management of lung cancer with hematoporphyrin derivative phototherapy. Mayo Clin. Proc 57:543-547.

Cortese DA, Kinsey JK, Woolner LB, Payne WE, Sanderson DR, Fontana RS (1979). Clinical application of a new endoscopic technique for detection of in situ bronchial carcinoma. Mayo Clin Proc 54:635-642.

Dougherty TJ, Grindley GB, Fiel R, Weishaupt KR, Boyle DG (1975). Photoradiation Therapy II. Cure of animal tumors with hematoporphyrin and light. J. Natl. Cancer Inst. 55:115-121.

Dougherty TJ, Kaufman JE, Goldfarb A, Weishaupt KR, Boyle DG, Mittleman A (1978). Photoradiation therapy for the treatment of malignant tumors. Cancer Research 38:2628-2635.

Hayata Y, Kato H, Konaka C, Ono J, Takizaiwa N (1982). Hematoporphyrin derivative and laser photoradiation in the treatment of lung cancer. Chest 81:269-277.

Kessel D, Dougherty TJ, Editors (1983): Porphyrin Sensitization. Vol 160: Advances in Experimental Medicine and Biology, Plenum Press, N.Y.

Lipson RL, Baldes EJ, Olsen AM (1961). The use of a derivative of hematoporphyrin in tumor detection. J. Natl. Cancer Inst. 26:1-10.

Profio AE, Doiron DR, Balchum OJ, Huth GC (1983). Fluorescence bronchoscopy for localization of carcinoma in-situ. Medical Physics 10(1):35-39.

Weishaupt KR, Gomer CJ, Dougherty TJ (1976). Identification of singlet oxygen as the cytotoxic agent in photo-inactivation of a murine tumor. Cancer Research 36:2326-2329.

Porphyrin Localization and Treatment of Tumors, pages 747–758

INDICATIONS OF PHOTORADIATION THERAPY IN EARLY STAGE LUNG CANCER ON THE BASIS OF POST-PRT HISTOLOGIC FINDINGS

Yoshihiro Hayata, M.D., Harubumi Kato, M.D.,
Ryuta Amemiya, M.D. and Jutaro Ono, M.D.
Department of Surgery, Tokyo Medical College
Tokyo 160, Japan

Since Dougherty et al. (1979, 1981) reported the therapeutic effectiveness of photoradiation therapy (PRT) with hematoporphyrin derivative (HpD), increasing attention has been paid not only to the diagnosis or localization of tumors but also treatment of malignant tumors by PRT. We have already reported the valuable results of the method in canine central type lung cancer induced experimentally in dogs (Hayata et al. 1983) and in human central type lung cancer including early stage cases (Hayata et al. 1982; Hayata et al. 1982; Hayata et al. 1982).

Since 1980 we have treated over 167 cases of cancer of various organs by PRT at the Tokyo Medical College Hospital. These consisted of 68 cases of lung cancer, 10 cases of esophageal cancer, 22 cases of gastric cancer, 16 cases of bladder cancer, skin metastases in 14 cases of breast cancer and 37 other cases. Thirty-three of these were early stage cases, including 8 cases of central type lung cancer, 4 cases of esophageal cancer, 12 cases of gastric cancer, one case of vaginal cancer and 8 cancer cases of the uterine cervix (Table 1). In this paper the authors present the therapeutic results of PRT in early stage central type lung cancer and the indications of PRT in these cases.

MATERIALS and METHODS

PRT was performed using HpD (Photofrin) obtained from Oncology Research and Development Inc., (Cheektowaga, N.Y.) and a Spectra-Physics argon dye laser. All cases were treated under local anesthesia. The histologic type of all

	Stage	No. of Cases
Lung ca.	Early stage	8
	Other stages	60
Esophageal ca.	Early stage	4 *
	Other stages	6
Gastric ca.	Early stage	12
	Other stages	10
Bladder ca.		16
Vaginal ca.		7
Ca of vulva	Early stage	1
	Other stages	3
Ca. of uterine cervix	Early stage	8
	Other stages	3
Skin metastasis from breast ca.		14
Skin ca.		7
Ca. of head and neck		8
Total		167

*Including 4 cases from the National Cancer Center

Dec. 1982 TMC

Table 1. Cancer cases treated with PRT and HpD.

8 early stage cases were squamous cell carcinoma, and the age varied from 50 to 76 years, with an average of 64.6 years. The male to female ratio was 7:1. The locations of the tumors were as follows; right upper lobe bronchus in 2, right B_2b and B_8 in one each, left upper lobe bronchus in one, left B_{1+2} in 2 and left B_{10} in one. They were all diagnosed by fiberoptic bronchoscope. Five of the 8 cases were treated with PRT, because 4 were inoperable due to poor pulmonary function and one refused surgery. Surgery was performed in the 3 other cases after PRT and the therapeutic results were examined histologically.

An argon laser (model 171-08, wavelength 457.9 - 514.5 nm, 15 W, Spectra-Physics, Mountain View, CA) pumped a dye laser (model 375-01, wavelength 630-640nm) using rhodamine B dye to obtain red light at a specific wavelength of 630 nm.

The lesion was photoradiated 72 hours after intravenous injection of 2.5-5.0mg/kg body weight HpD with a power of 90-600mW for 10-30 minutes. This is equivalent to 120-240 joules/cm^2. The laser beam was transmitted via a quartz fiber (400 micron, Quartz Products, Plainfield, NJ) inserted through the instrumentation channel of a fiberoptic bronchoscope (Olympus BF-2T). The procedure was performed at a distance of 1-3 cm from the tip of a quartz fiber in superficially invaded tumor or intratumorally by inserting a quartz fiber into protruding tumors.

Tumor response was classified into four grades as follows; complete remission (CR) means no evidence of tumor endoscopically, cytologically or histologically, significant remission (SR) means that more than 60% of the tumor disappeared endoscopically and partial remission means that less than 60% but more than 20% of the tumor disappeared. Decrease of 20% or less was considered as no remission (NR).

RESULTS

CR was obtained in 5 all non-resected cases and they are disease free at 11 to 36 months after PRT and are being followed up carefully endoscopically, cytologically and histologically (Table 2). In 3 resected cases after PRT, CR was obtained in one and SR in two and no recurrence can be observed in all after surgery (Table 3).

Case no.	Age	Sex	Location of tumor	Histologic type	Reason for PRT	Result	Survival	Recurrence
1.	74	M	R.B2b	Sq. ca.	Refused Surgery	CR	33 months	---
2.	76	M	L. B10ai	"	Poor Pulm. func.	CR	23 months	---
3.	59	F	R. u.lobe br.	"	"	CR	22 months	---
4.	70	M	R. u.lobe br.	"	"	CR	16 months	---
5.	62	M	L. B1+2ab	"	"	CR	11 months	---

N.B.; Poor pulm. func. = poor pulmonary function; CR = complete remission

Dec., 1982 TMC

Table 2. Therapeutic results of PRT in non-resected early stage central type lung cancer cases.

Case no.	Age	Sex	Location of tumor	Histologic type	Results	Survival	Recurrence
1.	50	M	L.u.lobe br.	Sq. ca.	SR	25 months	---
2.	54	M	R.B 8	"	SR	14 months	---
3.	74	M	L.B1+2	"	CR	13 months	---

Dec., 1982 TMC

Table 3. Therapeutic results of PRT in resected early stage central type lung cancer cases.

Case No.1. This case was a 74-year old male, chest X-ray negative and sputum cytology positive with cough. A smooth-surfaced tumor, 2.0mm in diameter, was observed endoscopically in right B_2b and the histologic type was squamous cell carcinoma. The case was a good indication for surgery even though the age was 74. PRT was performed, because he completely refused surgery. We performed PRT three times in this case as it was only the second case of PRT in our clinical series. The conditions were as follows; 150mWx20 minutes 72 hours after administration of 2.5mg/kg HpD, 150mWx10 minutes 168 hours and 200mWx10 minutes 240 hours after the first PRT without reinjection of HpD. The tumor disappeared 3 days after the first PRT, and at present he is disease free 36 months later (Fig. 1).

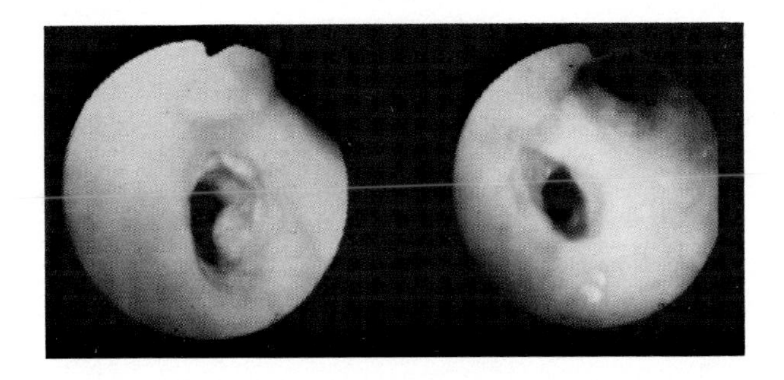

Fig. 1. A tumor, 2.0mm in diameter, can be seen in right B_2b (left) and no tumor can be seen 3 years after PRT (right).

Case No. 2. This 59-year-old female was also chest X-ray negative and sputum cytology positive. An irregular tumor and thickening of the bifurcation of the right upper lobe bronchus was observed endoscopically and the histologic type was squamous cell carcinoma. The patient was inoperable due to her poor pulmonary function and asthma. PRT was performed 72 hours after i.v. injection of 5.0mg/kg HpD with a power of 600mW for 40 minutes. This is equivalent to 360 joules/cm^2. The tumor disappeared after PRT and at present she is disease free 22 months later (Fig. 2).

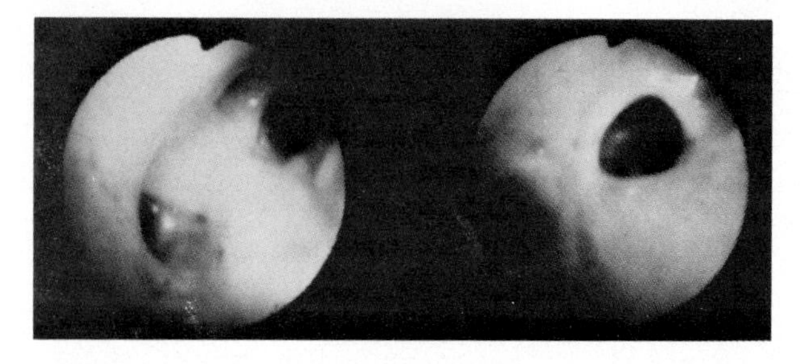

Fig. 2. PRT was performed in the tumor in the right upper lobe bronchus (left) and no tumor was apparent after PRT (right).

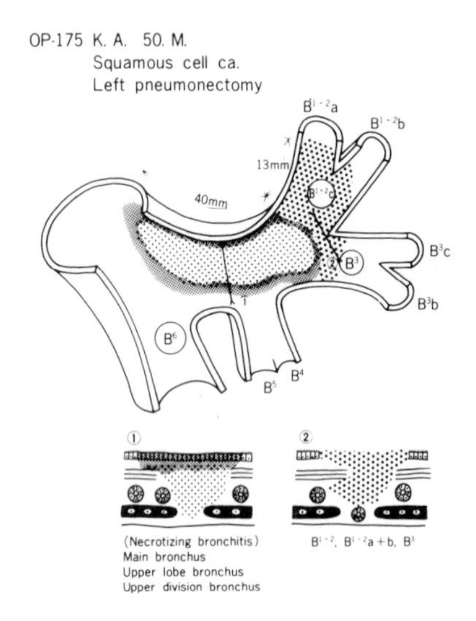

Fig. 3. Schema of case No. 1 of resected group after PRT. Tumor in the upper division bronchus disappeared after PRT and necrotizing bronchitis developed (below left) and tumor remained in the orifice of B_{1+2} (below right).

Case No. 1 of the resected group after PRT. This case was a 50-year-old male, chest X-ray negative and sputum cytology positive. Superficial invading tumor was observed endoscopically in the left upper division bronchus. The histologic type was squamous cell carcinoma. The case was operable, but we performed PRT in this case preoperatively 72 hours after i.v. injection of 5.0mg/kg under the conditions of 300mW for 30 minutes. This is equivalent to 112 joules/cm^2. The tumor disappeared endoscopically after PRT, and pneumonectomy was performed 2 weeks after PRT, because of the danger of recurrence due to the extent of invasion. Histological examinations were performed in the resected specimen and no tumor cells were observed in the upper division bronchus and histologic findings showed features of bronchitis accompanied by necrosis. The reason for these findings was probably that the tumor invasion was extensive and necrotic changes reached up to the extra-muscular layer. However, tumor remained in the orifice of left B_{1+2} and tumor cells were observed in all layers of the bronchus. The reason for this is probably that the laser beam did not reach these areas (Fig. 3).

Case No. 2 of the resected group after PRT was a 54-year-old male with an infiltrative shadow in the right lower lung field. A polypoid tumor was observed in right $B_{8,9}$ and $_{10}$. The histologic type was squamous cell carcinoma. Two weeks before right and lower lobectomy PRT was performed using 80mW power for 40 minutes (e.e.d. 96 joules/cm^2) superficially and 5 minutes intratumorally 48 hours after i.v. injection of 3.0mg/kg HpD. The tumor disappeared endoscopically. The histologic findings of the resected specimen showed features of fibrosis without tumor cells in most areas. However, remaining tumor nests observed in parts in which it was thought the laser beam did not reach perhaps due to the lower power of the argon dye laser (Fig. 4,5).

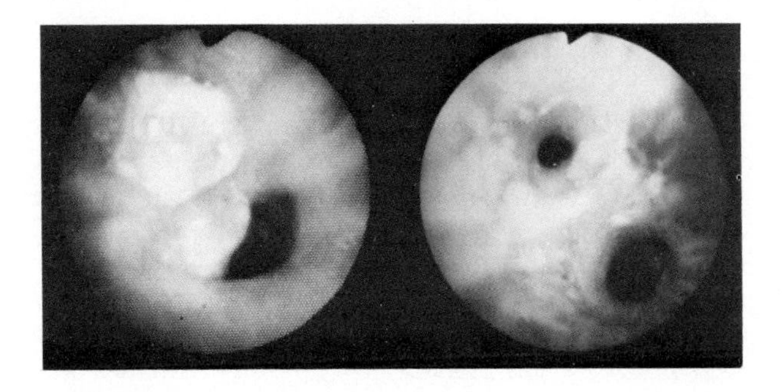

Fig. 4. A polypoid tumor in right $B_{8,9,10}$ (left). Tumor disappeared after PRT (right).

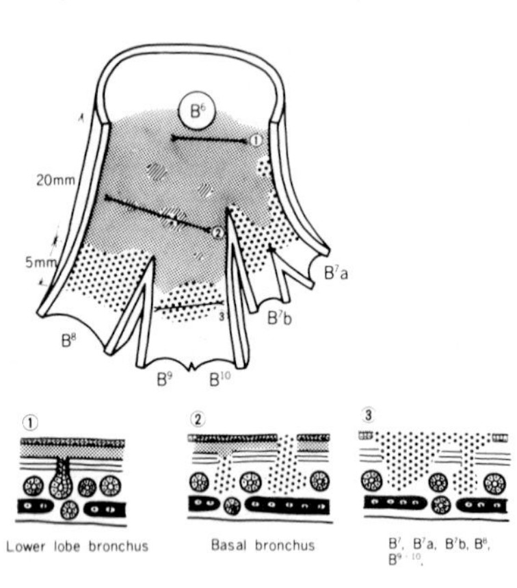

Fig. 5. Schema of case of figure 4.

Tumor disappeared in most areas, but remaining tumor nests observed in parts.

DISCUSSION

Although the activation mechanism of HpD has yet to be completely clarified, it is a fact that PRT has the therapeutic effect in malignant tumor as Dougherty and his colleagues (Dougherty et al., 1975; Dougherty et al., 1978) and this author and his colleagues (Hayata, 1982) reported. The therapeutic effectiveness of PRT in early stage cancer, especially in early stage central type lung cancer, was demonstrated by the author using canine central type lung cancer experimentally induced in dogs by weekly submucosal injections of 20-methylcholanthrene and also clinically in human early stage lung cancer.

Among 8 cases of early stage central type lung cancer 5 cases were treated by PRT only, 4 of which were inoperable due to poor pulmonary function and one refused surgery. The other 3 cases were resected after PRT. Of the 8, CR was obtained in 6 and SR in 2. The fact that CR was obtained in early stage lung cancer by PRT emphasizes the therapeutic value of the method strongly. However, we cannot state that PRT is a difinitively curative therapeutic method at present because of the as yet short period of survival after PRT and presence of some SR cases.

The most important problem in PRT in early stage lung cancer cases is how to diagnose the case as early stage. Ikeda classified early stage central type lung cancer as follows; intrabronchial tumor without lymphatic and hematogenous metastasis (Ikeda 1976). However, these criteria are questionable in terms of selecting candidates for PRT, and classification of early stage central type lung cancer for PRT is necessary. Ono and Amemiya examined the endoscopically viewed pathological findings of central type lung cancer in order to clarify the relationship between the lesion and the mucosal epithelium in relation to histologic type. They divided lesions into the primarily mucosal type and the primarily submucosal type. The primarily mucosal type was classified into three types; superficial infiltrative type, nodular infiltrative type and polypoid type. The primarily submucosal type was also classified into three types; subepithelial type, intramural type and extramural type (Oho and Amemiya 1982). Particularly the mucosal type is found most often in cases of squamous cell carcinoma and cases of this type are indications for PRT (Fig. 6).

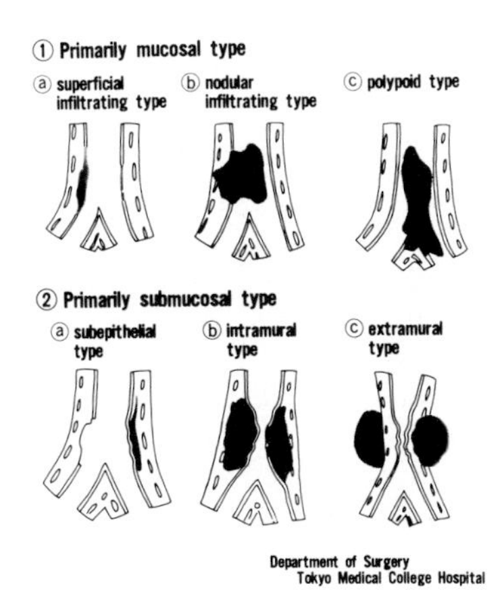

Fig. 6. Classification of endoscopically viewed pathological findings of central type lung cancer.

Among the 8 cases of this series, 6 were superficial type and 2 were polypoid type. However, these are only endoscopical findings and it is impossible to accurately estimate the grade of invasion in the bronchial wall. We would like to suggest the indications of PRT in early stage central type lung cancer tentatively as shown in Figure 7. PRT is effective in lesions consisting of superficial invasion only or superficial lesion plus submucosal invasion as far as the destruction of the muscular layer. However, there is a danger of perforation of the bronchus in cases in which the tumor grew in extramural layer. On the other hand, PRT will not be effective in lesions growing submucosally beyond the normal muscular or cartilage layer. Thus a diagnosis of early stage central type lung cancer is extremely important for PRT, and it should be diagnosed on the basis of roentgenological, histological and endoscopical findings.

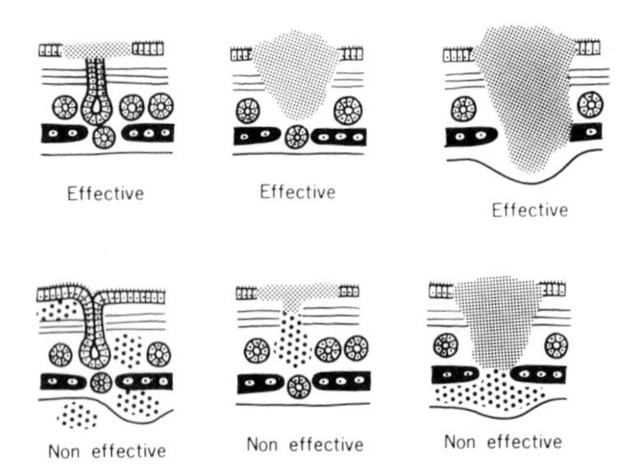

Figure 7. Possible indications of PRT in central type lung cancer.

For histological examination of intrabronchial growth, biopsy technique is also important. Cases of fully exposed lesions or superficial lesions are easily biopsied. However, it is necessary to biopsy deep layers to observe invasion there. Figure 8 shows the biopsy technique in cases of possible lymphatic invasion. Biopsy should be performed at several sites, because a difinitive diagnosis depends to some extent on chance as the amount of tissue is small. Also this technique is necessary in cases in which cancer may have invaded in the extramuscular layer to diagnose the extent of invasion or the relationship between cancer growth and the bronchial structure.

Figure 8. Schema of biopsy technique in case of possible lymphatic invasion or of invasion in deep layer.

Of cases resected after PRT, two showed SR because of limited penetration of the laser beam due to the location of the tumor or insufficient argon dye laser power. In order to prevent failures like this, improvement of light penetration and diagnosis of the intrabronchial extent are necessary in early stage central type lung cancer.

ACKNOWLEDGEMENTS

The authors would like to express their appreciation to Dr. T.J. Dougherty and J.P. Barron for their cooperation. Supported in part by a Grant-in-Aid for Scientific Research from the Ministry of Education and a Cancer Research Grant from the Ministry of Welfare, Japan.

REFERENCES

Dougherty TJ, Grindey GB, Weishaupt KR, Boyle DG (1975). Photoradiation therapy II. Cure of animal tumors with hematoporhyrin and light. J Natl Cancer Inst 55:115.

Dougherty TJ, Kaufman JE, Goldfarb A, Weishaupt KR, Boyle DG, Mittleman A (1978). Photoradiation therapy for the treatment of malignant tumors. Cancer Res 38:2628.

Dougherty TJ, Lawrence G, Kaufman JE, Boyle DG, Weishaupt KR, Goldfarb A (1979). Photoradiation in the treatment of recurrent breast carcinoma. J Natl Cancer Inst 62:231.

Dougherty TJ, Thoma RE, Boyle DG, Weishaupt KR (1981). Photoradiation therapy for treatment of tumors in pet and dogs. Cancer Res 41:401.

Hayata Y, Kato H, Konaka C, Hayashi N, Tahara M, Saito T, Ono J (1983). Fiberoptic bronchoscopic photoradiation in experimentally induced canine lung cancer. Cancer 51:50.

Hayata Y, Kato H, Konaka C, Ono J, Takizawa N (1982). Hematoporphyrin derivative and laser photoradiation in the treatment of lung cancer. Chest 81:269.

Hayata Y, Kato H, Konaka C, Ono J, Matsushima Y, Yoneyama T, Nishimiya K (1982). Fiberoptic bronchoscopic laser photoradiation for tumor localization in lung cancer. Chest 81:10.

Hayata Y, Kato H, Konaka C, Aida M, Ono J, Nishimiya K (1982). Hematoporphyrin derivative and photoradiation for tumor localization and treatment of lung cancer. In Ishikawa, Hayata Y, Suemasu K (eds): "Lung Cancer 1982" Amsterdam-Oxford-Princeton: Excerpta Medica, P 55.

Ikeda S (1976). Atlas of Early Cancer of Major Bronchi. Tokyo: Igaku Shoin Ltd, P 26. (in Japanese)

Oho K, Amemiya R (1982). Practical Fiberoptic Bronchoscopy. Tokyo: Igaku Shoin Ltd, P 74.

Porphyrin Localization and Treatment of Tumors, pages 759–766

PHOTORADIATION THERAPY IN THE TREATMENT OF ADVANCED CARCINOMA OF THE TRACHEA AND BRONCHUS

Ronald Vincent, M.D.

Ellis Fischel State Cancer Center
115 Business Loop 70 W
Columbia, Missouri 65201

Thomas Dougherty, Ph.D., Umo Rao, M.D.
Donn Boyle, William Potter, M.A.

Roswell Park Memorial Institute
666 Elm Street
Buffalo, New York 14263

Photoradiation therapy (PRT) is a new technique being investigated for treatment of a variety of solid malignant tumors. (Dougherty 1978, Dougherty 1979, Proceedings UICC Workshop 1979) It is based upon the tumor localizing ability and efficient photodynamic action of hematoporphyrin derivative (Hpd) resulting in a relatively specific photosensitization of malignant tissue upon exposure to activating visible light in the red region of the spectrum. (Lipson 1961, Lipson 1964, Gregorie 1968, Doiron 1979, Gomer 1979, Weishaupt 1976) It is fortuitous that the properties of tumor localization, photodynamic action, and absorption in the red (near 630 nm) is an essential property since wavelengths in the red portion of the visible spectrum are attenuated least in tissue, dropping to approximately 10% in a distance of 2 cm in an experimental tumor. (Dougherty 1978) The cytotoxic agent appears to be singlet oxygen formed by energy transfer from the porphyrin to endogenous oxygen. (Weishaupt 1976) Oxidation and cross-linking of membrane components followed by cell lysis is the likely mode of cell death. (Dubbelman 1978, Kessel 1977)

A further property of Hpd, useful in tumor localization, is its red fluorescence allowing for its easy identification in situ. This property, known for many years, is currently being utilized for localization of radiologically occult lung tumors in high-risk patients. (Lipson 1961, Lipson 1964, Gregorie 1968, Doiron 1979, Profio 1979)

This paper presents the results of PRT on patients with a carcinoma of the trachea and bronchus.

MATERIALS AND METHODS

Hematoporphyrin derivatives (Photofrin*) was obtained from Oncology Research and Development, Cheektowaga, New York and used under IND 12,678.

Light Sources

A dye-laser system was used which produced up to 4.0 w of power at 635± nm. An 18 w argon laser (Model 171, Spectra-Physics Corp., Mountain View, CA) was used to pump dye laser (Spectra-Physics Model 375) using Rhodamine B dye. The output beam from the dye laser was coupled into a single 400 µm quartz fiber optic (Quartz Silice, Paris France) by means of a modified fiber holder (Oriel Optical Corp., Stamford, CT). Coupling efficiencies of near 80% could be achieved over any desirable length of fiber up to 40 ft.

Patient Selection

Seventeen patients with biopsy proven malignant lesions of the trachea or main stem bronchus were treated by PRT. (Fig. 1) All patients had been treated previously with standard therapy and the disease was regarded to be life threatening due to progressive occlusion of a major airway. Eleven patients had squamous cell carcinoma, five had adenocarcinoma and one large cell carcinoma.

Procedure

Patients received 2.5 to 3.0 mg/kg body weight of Photofrin, intravenously three days prior to light therapy. They were cautioned repeatedly to avoid bright light indoors

* Trade Name

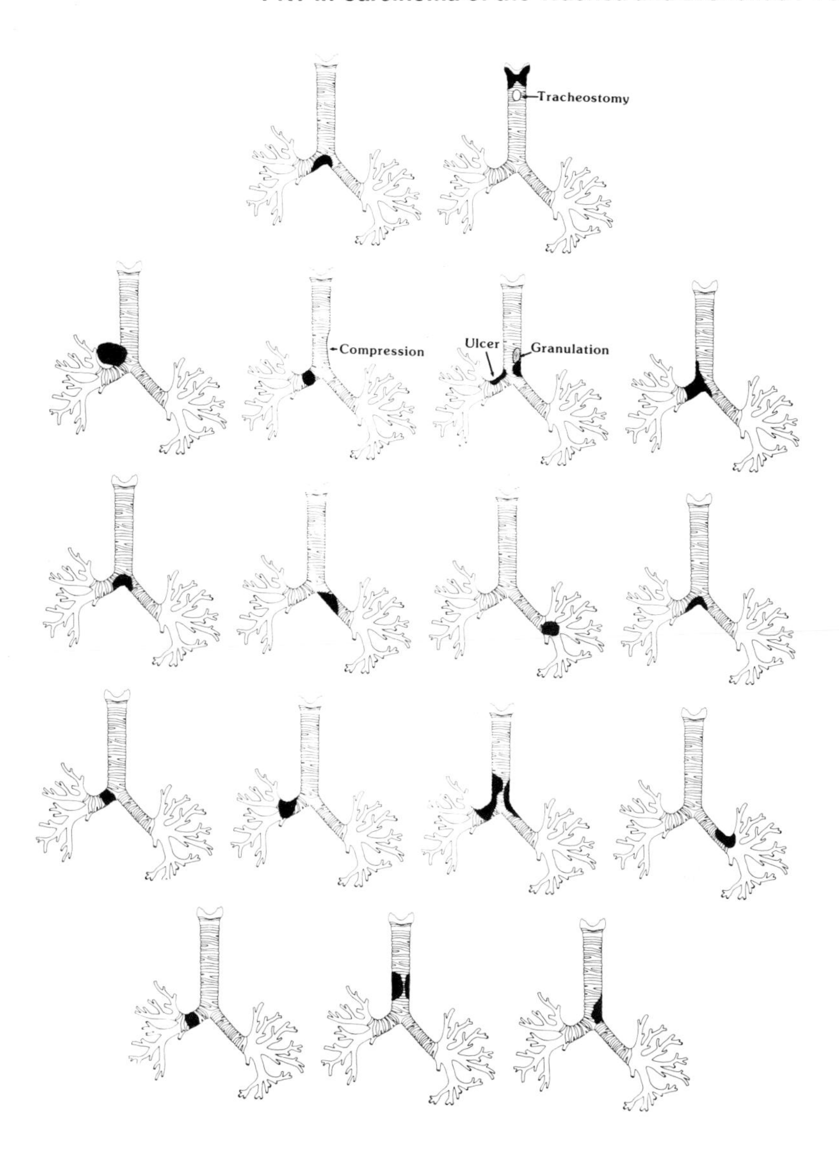

Figure 1. Anatomical location and relative size of seventeen endotracheal and endobronchial lesions treated by photoradiation therapy.

and outdoors particularly sunlight, for thirty days. Patients who disregarded this advice frequently experienced mild to severe sunburn reactions of exposed parts.

The endotracheal and endobronchial lesion was viewed through a 2TB Olympus bronchoscope. The quartz fiber was passed through the large channel of the fiberscope and positioned to achieve maximum exposure of the tumor by light delivered from the laser through the fiber. In order to achieve best exposure of the anatomic profile of the tumor, special fibers were built* with a variety of diffusing tips designed to illuminate in a cylindrical or semi-cylindrical field for up to 3 cm length. Fibers were also designed for direct implantation into the obstructing tumors. Length of treatment was from 10 to 50 minutes generally given every other day for three treatments. Dependent on toxicity and response, the treatments were repeated at three week intervals. The estimated amount of light delivered to the tumor per treatment was 60 to 300 Joules/cm.

RESULTS

In five patients the obstructing lesion was located in the trachea between the larynx and carina. In five patients the lesion was in the carina usually involving trachea and bronchus bilaterally. Seven lesions were primarily of a single bronchus extending to the level of the carina. All lesions but three were extensions of primary lung cancer, one a metastasis from colon cancer, one an extension of nasopharyngeal cancer and one was a primary adenocystic carcinoma of the trachea. All had been under standard treatment for from one to 84 months prior to PRT, (mean 33.5 months). Standard therapy usually included surgical resection of primary lesion followed by radiation therapy, immunotherapy or chemotherapy of the residual disease. Four patients were female and 13 were male with a mean age of 57. Survival ranged from 5 to 210 days after treatment with a median survival of 40 days.

---

* Oncology Research and Development, Cheektowaga, N.Y.

Response

Four patients had no measurable response to the photoradiation. Six patients had necrosis of tumor surface and 7 patients had a greater than 50% estimated reduction in the volume of intraluminal tumor. Two patients had no residual tumor after treatment that could be biopsied within the lumen of the airway.

Complications

All but three patients had significant complications following photoradiation necessitating the placing of 8 patients in the ICU, usually with need of endotracheal intubation to a point beyond the obstructing lesion where possible. Fifty per cent of the patients had symptoms of fever, acute pneumonia, excessive bronchial secretions and mucosal sloughing. Without mechanical removal, the sloughing of necrotic tissue lasted for several weeks. In two patients, the secretion hardened into bronchial casts necessitating mechanical removal. In one instance, the cast could not be removed and resulted in asphyxiation. One patient developed an abscess within the tumor resulting in a significant hemorrhage. Two patients developed severe endotracheal candidiasis.

Autopsy Data

Eight patients came to autopsy and all had significant evidence of severe acute pneumonia with pulmonary edema. Candida was cultured in two patients. Two patients demonstrated extensive tumor necrosis with abscess formation, one of which had bled extensively. Serial pathologic sections were made of the treatment areas of the bronchus and trachea. Uniformly, there was some necrosis of the intraluminal tumor for a depth of 4 to 15 mm. Viable tumor intermixed with necrotic tissue was evident from 5 to 20 mm. Unaffected viable tumor was evident at tissue depths beyond 20 mm from the treatment site.

DISCUSSION

Photoradiation therapy for lesions of the bronchus and trachea is still at an early stage of development. It is evident, however, that in a large percentage of cases, malignant lesions of the bronchus and trachea do take up the photosensitizing drug Photofrin and retain it for at least 3 to 7 days. Necrosis and sloughing tumor tissue up to 1 cm in depth was produced where malignant cells are bound with Photofrin and then exposed to 635±5 nm light. As presently used, there was little effect of photoradiation beyond a tissue depth of 2 cm.

Unfortunately, at this time, the complications are significant and sizeable and frequently life threatening. There is an apparent sequence of events leading to the morbidity that was observed in the 17 cases reported here. Shortly after exposure of the tissues to the photoradiation, there is an issuance of significant amounts of secretions. The volume of the secretions are excessive and continuous, causing the patient to tire after several days of productive coughing to the point that they can no longer manage without frequent tracheal suctioning. These secretions, if not removed, will form a firm hyaline-like cast which mechanically obstructs airways and is removed by instrumentation with considerable difficulty. The secretions may be an excellent media for infection as many of the patients became toxic, developed acute pneumonia and, in some instances, the process culminated in abscess formation and hemorrhage. Attempts to manage this sequela often led to intensive care in specialized units of the hospital with intubation of the airway frequently past the point of obstruction. The intensive measures when taken were often not successful and supportive care and antibiotics did not always favorably effect the pneumonia and associated pulmonary edema.

In several patients relief of obstruction was achieved, resulting in improved respiration for up to 4 months. In general, however, it was not felt that survival was greatly prolonged by this procedure. A considerable amount was learned about light dosimetry, depth of tumor necrosis and toxic morbidity. Preliminary indications are that this procedure may well have significant benefit in early or superficial lesions in the bronchus as reported by Hayata (Hayata 1980) (Hayata 1982), by Cortese (Cortese 1982), and Kato (Kato, In press). The treatment of advanced, large tumor obstructions of airways and bronchus by this method was only of limited success and future progress is dependent upon improvement of the Hpd derivative or a better and more controlled means of giving light exposure.

Other and perhaps more successful methods using the YAG and the carbon dioxide lasers for opening tumor obstructions of major airways have been reported by Dumon (Dumon 1980), (Dumon 1982), McElvein (McElvein 1981), and McDougall (McDougall 1983).

REFERENCES

Cortese DA, Kensey JH: Endoscopic management of lung cancer with hematoporphyrin derivative phototherapy. Mayo Clinic Proc, 57: 543-547, 1982.

Doiron DR, Profio AE, Vincent RG, Dougherty TJ; Fluorescence bronchoscopy for detection of lung cancer. Chest 76: 27-32, 1979.

Dougherty TJ, Kaufman JE, Goldfarb A, Weishaupt KR, Boyle DG, Mittleman A: Photoradiation therapy for the treatment of malignant tumors. Cancer Res 38: 2628-2685, 1978.

Dougherty TJ, Lawrence G, Kaufman JE, Boyle DG, Weishaupt KR, Goldfarb A: Photoradiation in the treatment of recurrent breast carcinoma. J. Natl Cancer Inst 62: 231-237, 1979.

Dumon JF: YAG laser photoirradiation of tracheobronchial lesions. WAB Newsletter 3, 1980.

Dubbelman TMAR, DeGoeij AFOM, van Steveninck J: Protoporphyrin - sensitized photodynamic modification of proteins in isolated human red blood cell membranes. Photochem Photobiol 28: 197-204, 1978.

Dumon JF, Reboud E, Garbe L, Aucomte F, Meric B: Treatment of tracheobronchial lesions by laser photoresection. Chest, 81:3, 278-283, 1982.

Gomer CJ, Dougherty TJ: Determination of ^{3}H and ^{14}C hematoporphyrin derivative distribution in malignant and normal tissue. Cancer Res 39: 146-151, 1979.

Gregorie HB Jr., Horger EO, Ward JL, Green JF, Richards T, Robertson HC Jr., Stevenson TB: Hematoporphyrin derivative fluorescence in malignant neoplasms. Ann Surg 167: 820-828, 1968.

Hayata Y: Basic and clinical studies on the diagnosis and treatment of lung cancer using hematoporphyrin derivative and laser photo-irradiation. Proceedings of the Second World Conference on Lung Cancer, Copenhagen, June 1980.

Hayata Y, Kato H, Konaka C, Ono J, Takizawa N: Hematoporphyrin derivative and laser photoradiation in the treatment of lung cancer. Chest 81:3, March 1982.

Kato E, Konaka C, Ono J, et al: Effectiveness of Hpd and photoradiation therapy in lung cancer. Proceedings of workshop on Porphyrin Photosensitization, Washington D.C., Plenum Press, In Press.

Kessel D: Effects of photo-activated porphyrins in cell surface properties. Biochem Soc Trans 5: 139-140, 1977.

Lipson RL, Baldes EJ, Olsen AM: A further evaluation of the use of hematoporphyrin derivative as a new aid for the endoscopic detection of malignant disease, Dis Chest 46: 676-679, 1964.

Lipson RL, Baldes EJ, Olsen AM: The use of derivative of hematoporphyrin in tumor detection. J. Natl Cancer Inst 26: 1-8, 1961.

McDougall JC, Cortese DA: Neodymium-YAG laser therapy of malignant airway obstruction. Mayo Clin Proc 58: 35-39, 1983.

McElvein RB: Laser endoscopy. Ann Thor Surg, 32:463-467, 1981.

Proceedings of the UICC Workshop on Hematoporphyrin derivative for detection and treatment of malignant tumors, Roswell Park Memorial Institute, Buffalo NY, October 1979.

Profio AE, Doiron DR, King EG, Laser fluorescence bronchoscope for localization of occult lung tumors. Med Phys 6: 523-525, 1979.

Weishaupt KR, Gomer CJ, Dougherty TJ: Identification of singlet oxygen as the cytotoxic agent in photoinactivation of a murine tumor. Cancer Res 36: 2326-2329,1976.

Porphyrin Localization and Treatment of Tumors, pages 767–775

GYNECOLOGIC USES OF PHOTORADIATION THERAPY

M. A. Rettenmaier, M. L. Berman, P.J. Disaia, R. G. Burns, G. D. Weinstein, J.L. McCullough, M. W. Berns

University of California, Irvine, Medical Center
101 City Drive, Orange, CA 92668

INTRODUCTION

Gynecologic malignancies are diagnosed in about 100,000 females annually. Most of these cancers are treated with surgery, radiation, or a combination of these two modalities. A high cure rate is found in early stages of disease in patients so treated. Unfortunately, patients with advanced or recurrent cancer are much less likely to do well since second line therapy does not provide a good chance for cure. Surgical treatment of recurrences in previously irradiated areas in the pelvis usually requires an exenterative procedure for treatment and even then is frequently unsuccessful in prolonging survival. The vigorous nature of this procedure, as well as considerable intraoperative and postoperative morbidity makes this approach unacceptable for many potential candidates. Chemotherapy of pelvic recurrences of cervical, endometrial and ovarian cancer produces responses in only approximately thirty percent of patients.

Premalignant changes of the vagina, vulva and cervix occur in untold thousands of females annually in the United States. For unknown reasons, the number of new cases detected has shown a steady increase in the last ten to fifteen years. Intraepithelial neoplasia (IN) of the genital tract is described by its location (e.g., vaginal intraepithelial neoplasia = VAIN, vulvar intraepithelial neoplasia = VIN, and cervical intraepithelial neoplasia = CIN). Adequate treatment of unifocal, small, low grade IN of the genital tract can usually be performed using a variety of locally destructive techniques (e.g., CO_2 laser vaporization, cryotherapy, electrocautery, local excision). Treatment of multifocal, large or higher grade IN usually required a more extensive surgical procedure (e.g., skinning vulvectomy, partial vaginectomy, or hysterectomy). This is necessary not only because of the known higher failure rate in this situation with local destruction only, but also, especially in multifocal lesions, because of the tendency of these lesions to recur outside of the locally treated area. For this reason, the entire diseased organ must be removed to assure that all microscopic disease is treated.

We have studied the use of photoradiation therapy in the treatment of six patients with recurrent gynecologic malignancies. To minimize prolonged cutaneous phototoxicity or pain associated with the systemic use of hematoporphyrin derivative (HPD), topical therapy was studied in an animal model of premalignant disease. Based on these results topical HPD was evaluated in the treatment of three patients with IN of the genital tract.

MATERIALS AND METHODS:

PHOTOTHERAPY OF MALIGNANT DISEASE

Nine lesions were treated with photoradiation therapy (PRT) in six patients with gynecologic cancers recurring either in the vagina or skin. All patients had previously failed conventional attempts to control their recurrent disease. Informed consent was obtained before PRT was initiated.

Hematoporphyrin derivative (HPD, Oncology Research and Development, Cheektowaga, N.Y.) was given as an intravenous bolus of 3mg/kg. All patients were instructed to avoid exposure to bright light for a four week period after its administration. Seventy-two hours later, the lesion and a 2.0-4.0 cm surrounding clinically uninvolved margin of tissue were exposed to light of 630nm provided by an argon ion pumped due laser (Spectra Physics 375-50). The total light dose per lesion was calculated in Joules/sq. cm. All PRT was given on an outpatient basis and follow-up examinations were performed at twenty-four hours, seventy-hours and weekly thereafter.

PHOTOTHERAPY OF PREMALIGNANT GYNECOLOGIC LESIONS

1. An Animal Model of Intraepithelial Neoplasia

The two-stage method of tumor induction was employed to develop an animal model of IN. Female BDF_1 mice were injected intraperitoneally with 0.6mg of urethane (Sigma Chemical Co., St. Louis, Missouri) per gram of body weight. Seven days after injection, the back of each animal was shaved with electric clippers. Subsequent hair growth was reshaved as needed to permit satisfactory topical therapy. A 0.25ml aliquote of a 0.00125% solution of 12-0 tetradecanoyl phorbol acetate (TPA) in acetone was applied twice weekly to the entire shaved area. Three animals were selected weekly for histologic studies beginning five weeks after injection of urethane. Four separate skin biopsies were taken from the TPA treated area of each animal. Once biopsies had been performed, that animal was not used in any further studies. The skin specimens were embedded in paraffin and ten micron sections were stained with hematoxylin and eosin. Representative sections were also stained by the Feulgen reaction to determine the ratio of nucleic acids between normal and abnormal cells. Fluorescence microscopy of Feulgen stained tissue sections was performed using a laser activated fluorescence microscope. The system consisted of a Zeiss R.A. microscope equipped for standard epi-illumination using a helium-cadium laser (Liconix 4110) as the stimulating radiation source. The variable laser spot size permitted fluorescence of any desired volume within a single cell but the shape and diameter of this spot were kept constant during each set of fluorescence measurements. A spot size which was smaller than the diameter of a normal mouse skin cell nucleus was employed to determine the relative DNA content within a constant nuclear volume. Fluorescence was detected by a Nanospec 10 microfluorometer mounted directly onto the trinocular tube of the Zeiss microscope. This instrument was interfaced with a Tracor Northern TN-170 oscilloscope and TN-43 computer which was programmed to provide multiple 100msec readings of photointensity over time. The fluorescence of 25 dysplastic cells in each tissue section was compared to that of 10 histologically normal epithelial cells in the treated skin adjacent to the lesion being studied to determine the radio of DNA content between the two cell populations.

2. Development of a Topical Formulation of HPD

Two vehicles were investigated for topical application of HPD. Vehicle #1 consisted of Eucerin (Biersdorf Inc., South Norwalk, Conn.). Vehicle #2 consisted of a mixture of N-methylpyrrolidone (43%), isopropyl alcohol (30%), and water (27%). Hematoporphyrin derivative (HPD, Oncology Research and Development, Cheektowaga, N.Y.), was lyophilized and then thoroughly mixed with the two respective vehicles to a final concentration of 1% or 5%. Each topical HPD formulation was applied to the skin of female BDF_1 mice containing intraepithelial neoplasia, as described above. A variety of application schedules were employed and the skin was examined for fluorescence, as described above, at various times after the last application of topical HPD.

3. Human Studies

Three patients with premalignant lesions of the genital tract have been treated to date. All patients had failed conventional attempts at control of their disease previously or were not considered as candidates for standard therapies. Informed consent was obtained before PRT was initiated. Topical HPD (5% HPD in Eucerin) was applied to the treatment area three times a day beginning two days before PRT. A final application was performed on the day of PRT two hours prior to laser exposure. Therefore, each patient had a total of seven topical HPD applications. The treatment area as well as a 2.0-4.0 cm rim of surrounding clinically uninvolved tissue was exposed to light of 630 nm provided by an argon ion pumped dye laser, as described above. The total light dose to each treated lesion was 20-40 joules/cm sq. The patients were studied on an outpatient basis. Follow-up examinations were performed at twenty-four hours, seventy-two hours, and weekly thereafter.

RESULTS

After systemically administered HPD, a cytotoxic response was seen in five of six patients following PRT of their recurrent gynecologic malignancies (Table 1).

Patient	Diagnosis	HPD Dose	Light Dose (J/cm^2)	Response	Toxicity
C.A.	Cervix cancer recurrent to vagina	3mg/kg	40	Partial	First degree burn of hands
B.B.	Vaginal cancer recurrent to skin of back	3mg/kg	20	None	None
N.J.	Bartholins Gland cancer recurrent to perineum and pelvis	3mg/kg	40	Complete	None
D.G.	Cervix cancer recurrent to vagina	3mg/kg	40	Partial	First degree burn of hands
A.H.	Cervix cancer recurrent to vagina	3mg/kg	30	Partial	None
J.J.	Endometrial cancer recurrent to perineum	3mg/kg	30	Complete	First degree burn and edema of face

Table 1. Treatment of recurrent gynecologic malignancies with systemically administered HPD.

Of the nine lesions treated, two showed a complete response and four showed a partial response. No change was seen in three small subcutaneous recurrences of an ovarian cancer in one patient. Toxicity was limited to first degree facial burn and edema in one patient after exposure to a bright artificial light course and first degree burn of the hands in two patients after sun exposure.

Alteration of the two stage method of producing skin tumors in mice using urethane as the inducing agent and TPA as a promoter was successful in producing lesions resembling intraepithelial neoplasia (IN). Normal mouse skin is characterized by keratinizing stratified squamous epithelium approximately 2-3 cell layers deep. Surface keratinization is minimal under which is a single stratum of granular cells and a basal layer of cells with vesicular and slightly irregular nuclei containing an even distribution of chromatin. All skin samples obtained from mice after eleven weeks of TPA treatment showed dysplastic changes. Lesions found after this time were papillary and up to 10 cells thick. Advanced lesions characterized by a higher degree of papillomatosis were seen in specimens obtained after week thirteen of TPA treatment. At this stage, cytologic features of malignancy were present, however, since the base of the membrane appeared intact, the lesions were classified as carcinoma in situ. Fluorescence microscopy of the induced lesions showed a progressive increase in DNA content beginning eight weeks after the initiating dose of urethane. The fluorescence ratio was consistent with a maximum DNA content of 4.86 ± 0.59 N after thirteen weeks. (Fig. 1)

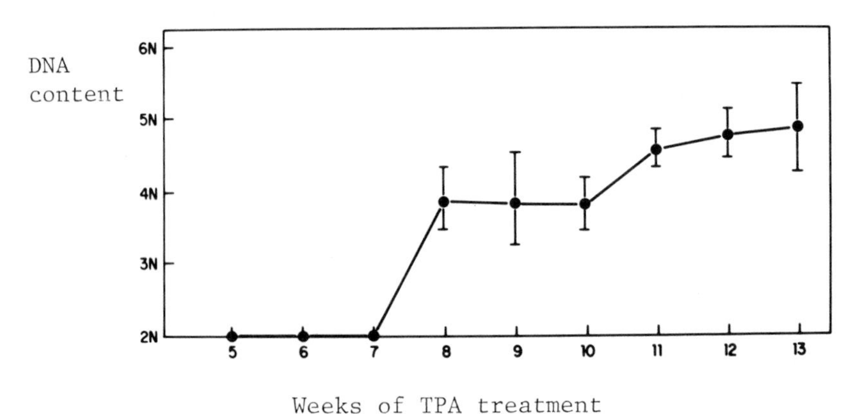

Fig. 1. DNA content of induced intraepithelial neoplasia

Both topical formulations of HPD were applied in a variety of dosage regimens to mice after induction of IN with urethane and thirteen weeks of TPA application. Fluorescence analysis revealed no concentration of HPD in IN compared to adjacent normal skin in animals treated with vehicle #2 (N-methylpyrrolidone, isopropyl alcohol, water). Analysis did reveal a higher HPD concentration in IN compared to surrounding normal skin in the mice treated with vehicle #1 (Eucerin) (Table 2). HPD concentration appears to be related to the number of topical applications and the length of time between application and fluorescence analysis.

Initial prebiopsy application (hrs)	2	24	48	24	72
# applications	1	1	1	3	3
Flourescence Induced IN	3460 +/- 2252	2265 +/- 1813	2448 +/- 2086	7749 +/- 4179	3035 +/- 1728
Normal Skin	857 +/- 304	973 +/- 1034	974 +/- 1025	1241 +/- 794	506 +/- 212
Ratio	4.04	2.33	2.51	6.24	6.00

Table 2. Fluorescence analysis of the skin of BDF_1 mice after topical application of 5% HPD in Eucerin (IN = Intraepithelial Neoplasia)

Phototherapy of premalignant gynecologic lesions (Table 3) after topical application of HPD produced a partial response in one patient with VAIN. This was manifested by a transient return to normal of the vaginal Papanicolaou smear. A biopsy of the treated area two weeks after photoactivation showed no viable evidence of carcinoma in situ. Another patient with VIN has been treated too recently for adequate evaluation at the time of this publication.

Patient	Diagnosis	HPD Formulation	# Applications	Response	Toxicity
J. G.	VAIN	5% in Eucerin	7	Partial	None
S. S.	VIN	5% in Eucerin	7	None	None
C.D.	VIN	5% in Eucerin	7	-	None

Table 3. Treatment of genital tract intraepithelial neoplasia with topically applied HPD (VAIN = Vaginal Intraepithelial Neoplasia; VIN = Vulvar Intraepithelial Neoplasia)

DISCUSSION

Treatment of recurrent gynecologic malignancies following standard first line therapy presents an ongoing challenge to the gynecologic oncologist. Radical surgery and radiotherapy are often unsuitable choices for elderly or infirm women and frequently provide only palliation or little hope for cure. Cytotoxic drugs in all but a few types of tumors have a dismal response rate. Any new tumoricidal therapy in this setting deserves serious study.

Gynecologic tumors often occur in areas easily accessible to activating light sources used with PRT (e.g., cervix, vagina, perineum). Investigation of this modality in this area of oncology, therefore, seems warranted. Because of the small number of patients with gynecologic malignancies treated to date we can draw no conclusion regarding the efficacy of PRT as either a curative or palliative tool. Further study will be needed to overcome the unique problems we have encountered in treating this group of tumors. For instance, determining the degree of pelvic extension of a tumor recurrent to the vagina after prior radiotherapy and resulting pelvic fibrosis can be quite difficult. Treatment of an unsuspected bulky or infiltrative lesion with local methods (e.g., PRT) would be doomed to failure.

Treatment of premalignant lesions of the vagina and vulva is often easily provided by excision or locally destructive measures. However, when these methods have failed or the disease covers a wide geographic area, second line treatment often involves rather extensive surgery (e.g., vulvectomy, vaginectomy). Since the percentage of these lesions which will advance to a frankly malignant state is unknown and may be the minority of instances (DiSaia, P.J., Creasman, W. T. 1981), it is difficult to advocate a treatment with a high incidence of toxic side effects.

The photosensitivity induced after systemic injection of HPD requires that the patient avoid direct sunlight or prolonged contact with bright artificial light. Since areas of intraepithelial neoplasia are usually small and superficial, it seems unreasonable to treat the patient with a systemic medication. To avoid the cutaneous sensitivity of systemic HPD, previous studies have sought to develop topical forms of HPD (McCullough, J.P., Weinstein, G.D., et al., 1983). The HPD formulation used in this study appears to concentrate in areas of intraepithelial neoplasia. Systemic levels of HPD were not measured, however, we would not anticipate significant systemic absorption based on in vitro data (McCullough, J.P., Weinstein, G.D., et al., 1983). No conclusion regarding therapeutic efficacy can be reached because of the small number of patients treated. However, it has been noted in other studies of topical treatment of VAIN and VIN with standard cytotoxic drugs that a lower response is seen after treatment of lesions occurring in keratinized skin (e.g., vulva) as opposed to non-keratinized squamous epithelium (e.g., vagina) (Woodruff, J. D., 1981). This appears to be the case with topically applied HPD. We are currently studying several HPD formulations which may prove to have superior penetrating qualities when applied to keratinized skin.

DiSaia, P.J., Creasman, W. T. (1981). "Clinical Gynecologic Oncology". St. Louis: C. V. Mosby, p.1.

McCullough, J.L., Weinstein, G.D., Lemus, L., Rampone, W., Jenkins, J. (1983). Development of a topical hematoporphyrin derivative formulation: Characterization of photosensitizing effects in vivo. (submitted J. Invest. Derm).

Woodruff, J.D. (1981). Carcinoma in situ of the vagina. Clin Obstet Gynecol 24:485.

Porphyrin Localization and Treatment of Tumors, pages 777–784

PHOTORADIATION OF CHOROIDAL MALIGNANT MELANOMA

Robert A. Bruce Jr., M.D.

Assistant Professor
Department of Ophthalmology
The Ohio State University
Columbus, Ohio 43210

To date the treatment of neoplastic disease has involved radical surgery, radiation therapy, chemotherapy or a combination of all three of these approaches. In all of these modalities, both normal and neoplastic tissue is destroyed. In the mid 1970s, Dr. Thomas Dougherty and his group made a positive breakthrough in a therapy for neoplasia which has been investigated since the early 1900s. This photodynamic process involves pretreatment of tissue with a substance that renders abnormal tissue sensitive to light of a specific wave length. When exposed to the light energy, a photodynamic reaction is initiated that results in the oxidation of cross linkages within the cell wall, causing cell lysis. This process has been named photoradiation therapy (PRT) (Weishaupt 1976; Dougherty 1978; Dougherty 1981; McBride 1979; Pettersen 1979).

Dr. Dougherty's work produced a derivative of bovine hematoporphyrin (HpD). His work has demonstrated a cytotoxic reaction that occurs when tissue pretreated with hematoporphyrin derivative is exposed to red light of a wave length of 630 to 635 nm. To date many types of human cancer have been shown to be responsive to this mode of therapy, both in the laboratory and in the clinical setting (Dougherty 1979).

In 1981 Dr. Francis L'Esperance reported to the American Society of Lasers in Medicine and Surgery the application of the above treatment concept to choroidal malignant melanoma. This report opened a new vista to the treatment of ophthalmic cancer. Prior to this the standard treatment for malignant melanoma was enucleation of the involved eye.

Many investigations of less radical treatment methods for malignant melanoma of the choroid including conventional argon laser and xenon arc photocoagulation, trans-scleral diathermy, local radiation with scleral explants, and proton beam irradiation have been reported. Depending on the size of the tumor, varying degrees of success have been demonstrated with all of these modalities.

Photoradiation therapy joins these treatments as a possible alternative to conventional enucleation. It differs from the above modalities by having no apparent detrimental effects on normal tissue. This report will discuss the early experience with photoradiation treatment in a series of patients with choroidal malignant melanoma evaluated and treated over a 10 month period.

PATIENT EVALUATION

All patients included in this series were initially evaluated at The Ohio State University Department of Ophthalmology and treated at Grant Hospital in Columbus, Ohio with the cooperation of the Laser Medical Research Foundation of Columbus, Ohio.

All patients were referred by a general ophthalmologist to the retinal service at The Ohio State University. Each patient underwent an ophthalmic examination including visual acuity, slit lamp examination, tonometry, and a complete retinal examination to determine the location and size of the mass lesion. This examination was followed by diagnostic fluorescein angiography, photographic documentation, and quantitative ultrasonography to verify the clinical diagnosis and obtain exact measurements of the size of the melanomas being considered for treatment. The ophthalmologic examination has been followed by a medical work-up to evaluate the presence or absence of metastatic disease. Testing includes: a complete physical examination; chest x-ray, liver spleen scan, CT scan, and bone scan; and hematologic evaluations. If all of the testing in this evaluation program reveals no evidence of metastatic disease or other primary neoplasia, photoradiation therapy has been offered to the patient.

TREATMENT REGIMEN

Hematoporphyrin derivative is administered intravenously in a dose of 2.5 mg/kg. Seventy-two hours following the administration of the hematoporphyrin derivative, the tumor mass is exposed to red light generated by a Spectra Physics rhodamine B tunable dye laser. The time of exposure has varied from 10 to 40 minutes. The light has been delivered to the tumor masses transcorneally, trans-sclerally, or via a combination of both of these approaches. Postoperatively, patients' responses have been evaluated by the measurement of visual acuity, fluorescein angiography, ultrasonosgraphy, photography and clinical examination.

POWER LEVELS

The amount of energy delivered to each tumor is outlined in Table 1. The calculation of these energy levels is shown with a sample calculation in Figure 1. The distance measurements in these calculations are approximate values. They reflect the power density delivered to the area of the tumor if the fiberoptic probes were held in the same position throughout the duration of the treatment. In reality, the probes have been hand held, and the beam has been moved deliberately to cover all treated areas of the tumor mass. I believe that this results in lower power densities being delivered to the tumor masses than are reflected in the calculations shown in the table. A more exact method of controlling the distance of the probes from the tumor mass is being developed.

REACTION TO TREATMENT

In all of the patients treated to date, the administration of hematoporphyrin derivative intravenously has not revealed any adverse systemic side effects. The drug is administered intravenously, and the patients are kept in a dark room to eliminate discomfort from sunlight, which results from the generalized photosensitivity caused by the drug. The photosensitivity of the skin appears to be persistent for 8 to 12 weeks after the administration of the dye. This information has been documented by patient experience within our series.

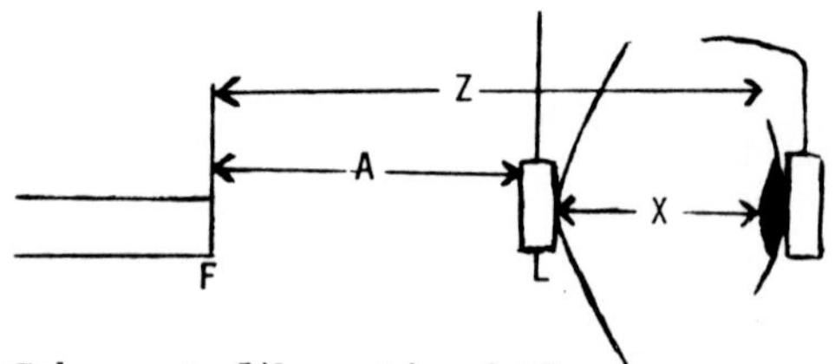

F=Reading, Coherent fiber tip (mW)
A=Distance tip to lens (mm); lens = 2 mm thick
X=Estimated distance surface of eye to tumor (mm)
Z=Distance fiber tip to tumor (mm)
R_Z=Coherent radiometer reading at distance Z with lens in place (mW)
D_Z=Diameter spot size at distance Z (cm)
A_S=Area spot size at distance Z (cm^2)
A_T=Estimated area of tumor (cm^2)

Sample: F=1000 mW, A=12 mm, lens=2 mm, X=10 mm
Z=24 mm, R_Z=850 mW, D_Z=.8 cm, $A_S = \frac{\pi(.8)(.8)}{4} = .5\ cm^2$
A_T=15 mmx20 mm = 3 cm^2
TIME = 30 min

FOR SPOT SIZE: $\frac{R_Z}{A_S} = \frac{850}{.5} = \frac{1700\ mW}{cm^2}$

$$\frac{R_Z\,(mW)}{A_S\,(cm^2)} \times 30\ min \times \frac{60\ sec.}{min.} \times \frac{10^{-3}W}{mW} = \frac{watt\ sec.}{cm^2} = \frac{joules}{cm^2}$$

$$= \frac{1700\ mW}{cm^2} \times 1800\ sec \times \frac{10^{-3}W}{mW} = \frac{3060\ joules}{cm^2}$$

FOR TUMOR SIZE: $\frac{R_Z}{A_T} \times 30 \times 60 \times 10^{-3} = \frac{850mW}{3cm^2} \times 30\ min \times \frac{60sec}{min} \times \frac{10^{-3}}{mW} = \frac{Joules}{cm^2} = \frac{510\ joules}{cm^2}$

RETROBULBAR: $\frac{R_B}{A_B} = \frac{Coherent\ Reading}{Area\ Spot}\left(\frac{mW}{cm^2}\right) = \frac{275}{1.3} = \frac{211\ mW}{cm^2} \times$

$$\frac{275\ mW}{1.3cm^2} \times 30min \times \frac{60\ sec}{min} = \frac{380\ joules}{cm^2}$$

Fig. 1. Sample calculation of dosage.

TABLE 1: AMOUNT OF ENERGY DELIVERED TO EACH TUMOR

PT	AGE	SIZE (mm)	TR.CORN. (watts)	TRSCL. (watts)	TIME (min)	POWER DNSTY (J/cm^2)	PREOP VA	POSTOP VA	TIME FROM Tx (mos)	DECREASED TUMOR SIZE*
EH	59	10.5x6x3	1.00	.500	40	3060	20/25	20/50	10	Total
LM	35	7x7x3.6	1.00	.300	40	6800	20/20	20/20	5	80%
NS	35	12x9x3	1.00	.500	40	2880	20/30	FC	5	50%
RT	53	7.5x7.5x2	1.00	--	40	2600	20/30	20/200	5	Total
LJ	83	6x6x1	1.00	.300	40	1500	20/60	LP	5	Total
ED	55	10x6x1	1.00	.300	40	1920	20/300	HM	5	**
CS	77	10x6x5	1.00	.250	40	3060	LP	LP	3	--
MH	57	7.5x7.5x2	1.00	--	20	614	20/100	HM	1	--
HB	67	12x6x7.5	1.00	.500	40	1700	20/70	20/400	1	--
DB	35	6x7.5x4	1.00	--	10	293	20/70	20/200	1	--
JH	66	12x10x6	1.00	.275	30	3050	FC	HM	1	--

*Approximate measurement by ultrasonography
**Patient has been retreated after 4 months.
LP= Light perception; FC=Finger counting

Changes in the appearance of the tumor mass are noted within minutes after it is exposed to the laser light. These changes include a blanching of the tumor mass, causing it to turn white. Additionally the overlying retina becomes edematous with pinpoint intraretinal hemorrhaging. The adjacent uninvolved retina does not appear to become acutely involved with this edematous/hemorrhagic change in the retina overlying the tumor.

Alterations in the vascular supply to the tumors have been noted in the immediate post-treatment fluorescein angiograms. This is dramatic and has been a persistent change in the studies carried out in the follow-up period of all but one of our patients (Fig 2).

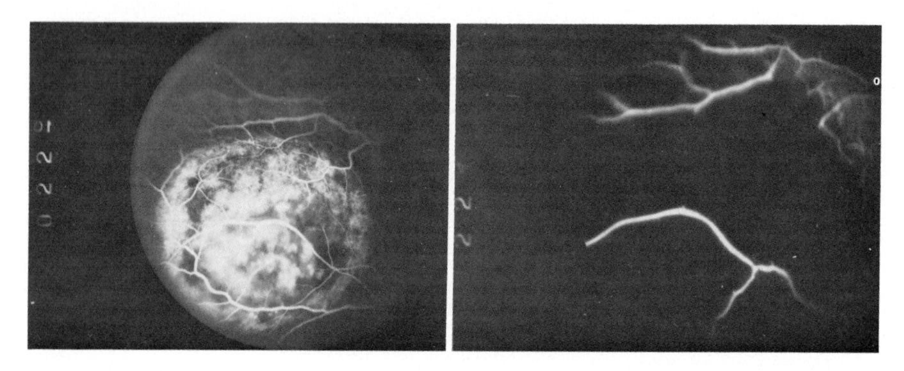

Fig. 2. Pre- and post-treatment fluorescein angiograms showing alterations in the vascular supply to the tumor.

Reduction in tumor mass has been observed in all patients who are five months or longer post-treatment. Shrinkage has been identified initially 10 to 12 weeks after treatment. This clinical observation is one of a gradual change from the initial postoperative white color of the tumor mass to a mottled appearance which is seen in a typical chorioretinal scar. As this occurs, dimpling on the surface of the tumor can be identified. The loss of tumor tissue volume requires several weeks to months to occur. This loss of tumor mass is easily verified by ultrasonography. The final appearance is that of a large chorioretinal scar.

Post-treatment complications in the treated eyes are listed in Table 2. All patients have experienced some degree

TABLE 2: POST-TREATMENT COMPLICATIONS

COMPLICATIONS	NO. CASES (11 TOTAL CASES)
Photosensitivity of skin	11
Iritis	11
Exudative retinal detachment	9
Reduced visual acuity	11
Vitreous reaction	1
Vitreous hemorrhage	2
Chemosis	11
Cataract	1
Choroidal detachment	1

of chemosis, iritis and lid swelling, which have been treated with cycloplegics and corticosteroid drops. The chemosis usually clears within seven days. The iritis lasts up to six weeks. The lid swelling can be eliminated by careful draping of the eyelids with an opaque material.

Exudative retinal detachment has developed in 9 of our 11 cases. This occurs usually within three days of treatment. In those cases in which preoperative detachment existed (3/11), the detachment worsened after treatment. In all but three cases, the detachment of the retina resolved within six weeks of treatment.

Other clinical changes which have been noted include choroidal detachment, cataract, vitreous hemorrhage, vitreous inflammatory reaction, and reduced visual acuity. These changes all appear to be transitory except for the reduction in visual acuity. The final visual acuity appears to be related to the location and the size of the tumor mass. If the tumor has undermined the macular area, central vision is sacrificed. In the case of the larger melanomas, the exudative retinal detachments are more extensive and have resulted in macular involvement. This may be a factor contributing to reduced postoperative visual acuity.

DISCUSSION

Initial short-term results from this series of patients with choroidal malignant melanoma treated with photoradiation therapy are very encouraging. Several aspects of our treatment regimen and results may vary a great deal from the experience of other investigators. The greatest difference is the amount of tumor mass destruction that our series has demonstrated when compared to other series. Dr. L'Esperance (personal communication) found little if any reduction in the size of the tumor mass after 18 months of follow-up. Our results are most likely due to the large power density that we have employed, which may produce a thermal effect in addition to the phototoxic effect of hematcporphyrin derivative plus red light.

Other side effects from the exposure of the retina to the intense light, which may affect the ultimate visual acuity in the involved eyes, have yet to be identified. Electrophysiologic testing pre- and post-treatment may supply some of these answers.

CONCLUSION

Photoradiation therapy as described by Dougherty, et al is a viable alternative to enucleation in the treatment of choroidal malignant melanoma. This conclusion is drawn from our initial impressions from our series of patients. Photoradiation therapy offers a noninvasive form of therapy that may provide a more acceptable mode of therapy to the patient than enucleation. Moreover, it appears to surpass the results obtained with other alternative treatment methods currently being employed.

REFERENCES

Dougherty TJ (1981). Photoradiation therapy for cutaneous and subcutaneous malignancies. J Invest Derm 77:122.
Dougherty TJ, Kaufman JE, Goldfarb A, Weishaupt KR, Boyle D, Mittleman A (1978). Photoradiation therapy for the treatment of malignant tumors. Cancer Research 38:2628.
Dougherty TJ, Lawrence G, Kaufman JH, Boyle D, Weishaupt KR, Goldfarb A (1979). Photoradiation in the treatment of recurrent breast carcinoma. J Natl Cancer Inst 62:231.
McBride G (1979). New treatment for cancer under development. JAMA 242:403.
Moan J, Pettersen EO, Christensen T (1979). The mechanism of photodynamic inactivation of human cells in vitro in the presence of haematoporphyrin. Br J Cancer 39:398.
Weishaupt KR, Gomer CJ, Dougherty TJ (1976). Identification of singlet oxygen as the cytotoxic agent in photo-inactivation of a murine tumor. Cancer Research 36:2326.

Porphyrin Localization and Treatment of Tumors, pages 785–794

PHOTORADIATION THERAPY OF BLADDER TUMORS

Toshimitsu Misaki, Haruo Hisazumi, Norio Miyoshi

Department of Urology, School of Medicine,
Kanazawa University,
Kanazawa, 920, Japan

ABSTRACT

Photoradiation therapy (PRT), in which hematoporphyrin derivative (HpD) is activated by an argon-dye laser, was performed on 46 superficial bladder tumors in 9 patients. It was suggested there is little or no thermal cell killing effect of laser irradiation from a study of the heating effect of the beam intensity. A fluorospectrophotometric study of biopsies from tumors and normal mucosa revealed preferential HpD localization in malignant tissues. In tumors less than 1 cm in diameter treated with 150 mwatts per cm^2 for 5 minutes there were 2 with complete remission (CR) and 5 with no change (NC), whereas with 200 mwatts per cm^2 or more for the same time there were 15 CR and 5 with partial remission (PR). Concerning the accumulated energy intensity of light, in tumors 1 cm in size or less, treatment with 100 - 250 joules per cm^2 obtained CR in 5 of 6 tumors. From these results, it was suggested that the light intensity should be 300 mwatts per cm^2 for 5 - 10 minutes or more and the total light dose should be 100 joules per cm^2 or more in tumors up to 2 cm in size. There was no CR in tumors more than 2 cm in size. No early side efects were seen from administration of HpD. Sensitivity to sunlight was seen in 4 cases.

INTRODUCTION

More than 70% of bladder tumors are superficial tumors (stage Ta - T2) and are multicentric in approximately 30% of cases. Recurrence has been reported in 40 - 70% of cases within 1 - 2 years after initial management and several theories have been advanced to explain this degree of rapid recurrence. Schade and Swinney (1973) biopsied cystoscopically normal areas in patients with tumor evident and reported a 40% incidence of carcinoma in situ in these biopsies, while other more recent studies have revealed atypia in 46%, carcinoma in situ in 14% and carcinoma in 16% of biopsies obtained during a 1-year period from cystoscopically normal-appearing mucosa (Soloway 1978). In such cases of recurrence and multiple occurrence, HpD and PRT may be a convenient and effective approach. Recently, a number of basic and clinical studies regarding HpD and PRT in malignant tumors have been reported. However, in the urological field there is little information on PRT of bladder tumors. In the present study superficial bladder tumors were subjected to PRT and the ability of HpD to photoinduce tumor regression was investigated.

MATERIALS AND METHODS

HpD: Photofrin (Oncology Research and Development Inc., Cheektowaga, N.Y., U.S.A) containing 5.0 mg per ml HpD was stored in the dark at -20°C until immediately before thawing for use.

Light source: A Spectra-Physics Model 375-03 dye laser, pumped by a Model 171-07 argon laser, was employed. The laser beam was transmitted via a 400 μm quartz fiber inserted through a cystourethroscope (UR-27067, Storz, Tuttlingen, FRG). The wavelength of the laser light was 630±1.6 nm.

The heating effect of laser irradiation: A small piece of meat was placed on the bottom of a beaker containing 200 ml physiological saline solution at 25°C and the tip of a probe for temperature measurement of the skin surface was embedded in the center of meat. The fiber tip was kept 1 cm from the meat surface.

Fluorescence spectrum of HpD: Biopsies of tumors and normal mucosa were performed immediately before PRT. The frozen sections were examined by fluorospectrophotometer (Model MPF-4, Hitachi LTD., Tokyo, Japan).

PRT: PRT was performed on 46 superficial bladder tumors in 9 cases. Table 1 shows the details of the individual cases. The patients were injected intravenously with 2.0 - 3.2 mg per kg body weight HpD. After 48 - 72 hours, PRT was performed. The bladder was instilled with 200 ml physiological saline solution. The laser beam

Table 1. Photoradiation therapy of bladder tumors.

Case	Age	Sex	Gross appearance of tumors	T	Grade	No. of tumors	HpD (mg/kg)	Time interval (hr)	Light intensity (mW/cm^2)
1. TN	62	M	papillary sessile	Ta	1	4	3.2	48	200
2. HY	53	M	papillary sessile	Ta	1	7	2.0	72	200
3. HS	68	F	papillary pedunculated	Ta	1	5	2.0	72	150
4. TM	56	M	papillary pedunculated	Ta	2	2	2.0	72	200
5. TI	51	M	papillary pedunculated	Ta	2	3	2.0	48	300
6. KY	69	M	Papillary pedunculated	T1	2	5	2.0	48	300
7. MG	62	M	papillary pedunculated	T1	3	6	2.0	72	150
8. UN	63	M	Papillary pedunculated	T1	3	4	2.5	48	300
9. YI	75	M	papillary sessile	Ta	2	10	3.0	48	300

delivered through a quartz fiber was directed to the tumor through a cystourethroscope. The tip of the fiber was kept 1 - 3 cm from the tumor surface. The light power intensity was 150 - 300 mwatts per cm^2, radiation time 5 - 20 minutes. The effectiveness of treatment was evaluated 3 weeks after treatment by means of cystoscopy and transurethral ultrasonography. Complete disappearance of the tumor was classified as complete remission (CR), less than complete but 50% or more remission as partial remission (PR) and less than 50% as no change (NC).

RESULTS

Laser irradiation resulted in a rapid increase in the meat temperature (1.8 - 6.7°C) within 1 minute after irradiation. Thereafter, 100 - 200 mwatts per cm^2 resulted in little temperature increase. While 300 - 500 mwatts per cm^2 resulted in an increase of 1.7 - 2.3°C. The temperature did not exceed 35°C within 10 minutes after irradiation (Fig. 1).

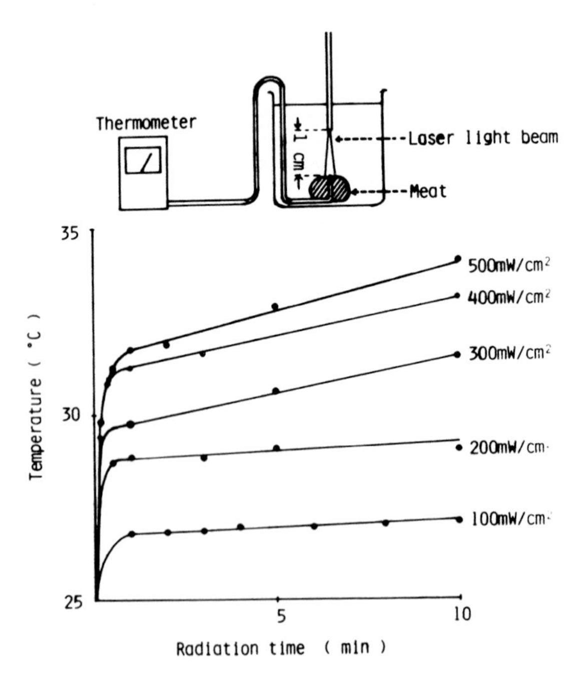

Fig. 1. In vitro temperature increasing effect of photoradiation, wavelength; 630±1.6 nm.

Fluorospectrophotometric studies revealed the fluorescence spectrum responsible for intracellularly incorporated HpD in the biopsy specimens when excited at 400 nm. The major emission spectral peaks were obtained at 630 and 693 nm in the tumor tissue (Curve 1 in Fig. 2) of stage T1, grade 2 bladder tumor that received 3 mg per kg body weight HpD before 48 hours. On the other hand, the fluorescence intensity of HpD in normal mucosa (Curve 2 in Fig. 2) was one fourteenth that of the tumor´s.

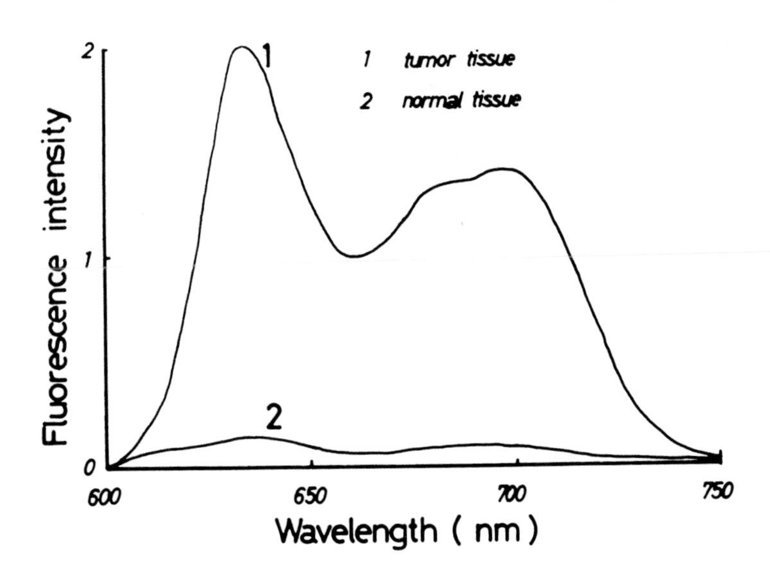

Fig. 2. The fluorescence emission spectrum of HpD.

Immediately after PRT the tumor surface becomes edematous, with necrosis and edema and hyperemia of surrounding mucosa being recognized 4 - 7 days after the PRT. These changes were the most marked at 2 weeks, with appearance of normal findings at 4 weeks. Fig. 3 shows the cystoscopic findings of stage Ta, grade 1 bladder tumor, 0.5

X 0.5 cm in size. PRT was performed for 10 minutes with 200 mwatts per cm^2 72 hours after a single IV injection of 2 mg per kg body weight of HpD. The tumor was observed to have disappeared 1 week after PRT. The histologic changes reflected the degeneration of tumor cells, namely nuclear destruction, cytoplasmic vacuolization with hemorrhage and inflammatory cells (Fig. 4).

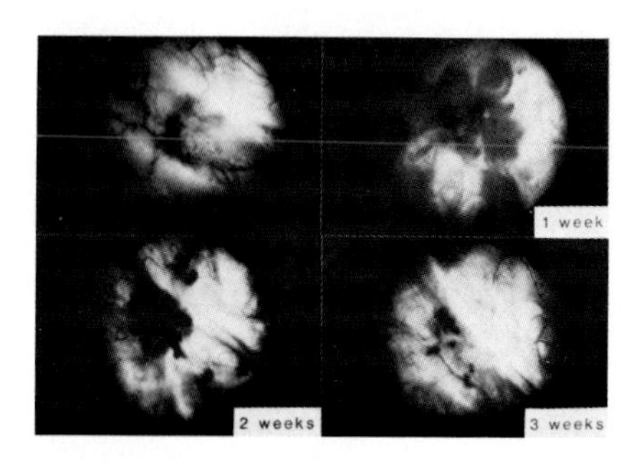

Fig. 3. Cystoscopic changes of bladder tumor after PRT using a dose of 2 mg per kg body weight of HpD and light intensity of 200 mwatts per cm^2 for 10 minutes.

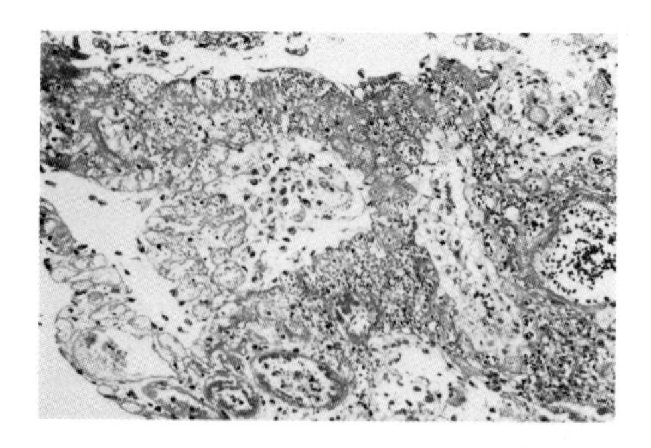

Fig. 4. Histopathology of tumor tissue showed degeneration of tumor cells. Reduced from X 100.

About 1 week after PRT, all cases complained of bladder irritation, including frequent urination, burning on urination and hematuria, in addition to discharge of necrotic materials in the urine. While the effects of PRT were recognized, there were areas in some cases in which the amount of photoradiation was insufficient and viable cells were recognized (Fig. 5).

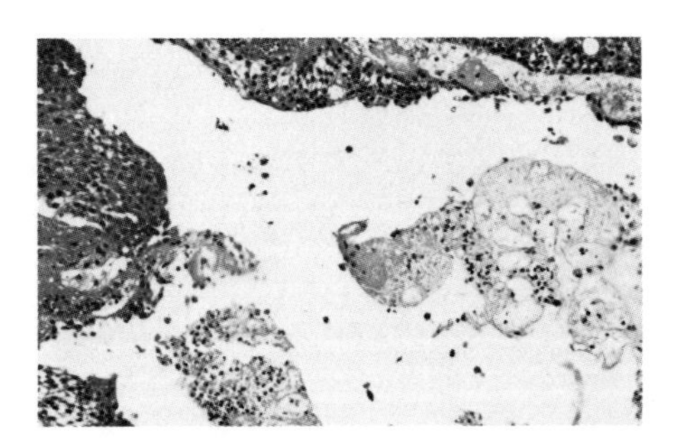

Fig. 5. Microphotograph showing histological changes of a bladder tumor 1 week after PRT using a dose of 2 mg per kg body weight of HpD and light intensity of 300 mwatts per cm^2 for 15 minutes. Note the vacuolar degeneration and detachment of the cells covering the villi, mild edema, slight leukocytic infiltration and bleeding foci in the stroma and debris. Reduced from X 100.

Table 2. Relationship between light intensity and tumor response.

Tumor size	< 1 cm			1 - 2 cm			2 - 3 cm			> 3 cm		
Tumor responce / Total light dose (J/cm^2)	CR	PR	NC	CR	PR	NC	CR	PR	NC	CR	PR	NC
< 50	4	1	5			2						
50 - 100	13	4		2	1	1			1			
100 - 150	3											
150 - 200	1	1		1	1				1			
200 - 250	1			1								
250 - 300								1				1
Total	22	6	5	4	2	3		1	2			1

In 42 tumors measuring less than 2 cm diameter, the following results were obtained according to the light intensity: with 150 mwatts per cm^2, CR 2 and NC 8 tumors; with 200 mwatts per cm^2, CR 10 and PR 2; with 300 mwatts per cm^2, CR 14 and PR 6 tumors. In tumors less than 1 cm in diameter treated with 150 mwatts per cm^2 for 5 minutes there were 2 CR and 5 NC, whereas with 200 mwatts per cm^2 or more for the same time there were 15 CR and 5 PR. When tumor sizes were more than 2 cm, PR and NC were observed in 1 and 3 tumors with 150 - 300 mwatts per cm^2 (Table 2).

In tumors less than 2 cm in diameter, close examination of the relationship between the dosage of light and the tumor response revealed that less than 100 joules per cm^2 resulted in CR, PR, and NC in 19, 6, and 8 tumors, respectively. In cases receiving more than 100 joules per cm^2, CR and PR were observed in 7 and 2 tumors. In tumors 1 cm in size or less, treatment with 100 - 250 joules per cm^2 obtained CR in 5 of 6 tumors, with PR in the other. There was no CR in tumors more than 2 cm in size (Table 3).

Table 3. Relationship between total light dose and tumor response.

Tumor size		< 1 cm			1 - 2 cm			2 - 3 cm			3 cm <		
Light intensity (mW/cm^2)	Exposure time (min.) / Tumor responce	CR	PR	NC	CR	PR	NC	CR	PR	NC	CR	PR	NC
150	< 5	2		5			2						
	5-10									1			
	10-20						1						
200	< 5	4	2										
	5-10	3											
	10-20	2			1					1			
300	< 5	11	3		2	1							
	5-10		1		1	1							
	10-20								1				1
	Total	22	6	5	4	2	3		1	2			1

No early side effects were seen from administration of HpD. Of the 9 patients, 4 developed a slight to moderate reaction upon direct or indirect exposure to bright sunlight 11, 23, 27, and 36 days after PRT. But these photosensitivity reactions disappeared in 5 to 10 days.

DISCUSSION

In this study it was evident that PRT exerts little or no cell killing and HpD is preferentially retained by malignant bladder tissues in a fashion similar to that observed with other malignant tumors. In the urological field, Kelly and associates (1975) have reported that the administration of a HpD, followed 24 hours later by exposure to white light, caused marked destruction of experimental bladder tumors but little or none of normal bladder tissues. HpD or light alone caused no damage to the tumors or normal tissues, suggesting that photodynamic therapy may be applicable in the treatment of superficial transitional cell carcinoma of the bladder. In a subsequent clinical paper (Kelly 1976), preferential HpD localization was observed not only in malignant bladder tissues but also in pre-neoplastic changes occurring in macroscopically normal mucosa. Also, they confirmed that high intensity illumination of an area of HpD sensitized carcinoma causes tumor destruction. From these results, it was suggested that HpD could be used as an aid in the diagnosis and treatment of carcinoma of the bladder.

The results of the present study revealed that PRT possesses the ability to cure bladder tumors. It was suggested that the effectiveness of single PRT sessions involving superficial photoradiation is limited to tumors up to 2 cm in size. The light intensity should be 300 mwatts per cm^2 for 5 to 10 minutes or more and the total dose should be 100 joules per cm^2 or more. When several of the problems that exist at present are solved, such as the improvement of the fiber, the development of a cystoscopy specifically for PRT use and shortening of the time of PRT, this method will hold great hope for the treatment of superficial multicentric tumors and multiple pre-neoplastic urothelial abnormalities.

REFERENCES

Kelly JF, Snell ME, Berenbaum MC (1975). Photodynamic destruction of human bladder carcinoma. Br J Cancer 31:237.

Kelly JF, Snell ME (1976). Hematoporphyrin derivative: A possible aid in the diagnosis and therapy of carcinoma of the bladder. J Urol 115:150.

Schade ROK, Swinney J (1973). The association of urothelial atypism with neoplasia: its importance in treatment and prognosis. J Urol. 109:619.

Soloway MS, Murphy W, Rao MK, Cox C (1978). Serial multiple-site biopsies in patients with bladder cancer. J Urol 120:57.

Porphyrin Localization and Treatment of Tumors, pages 795–804

THE USE OF HEMATOPORPHYRIN DERIVATIVE (HpD) IN THE LOCALIZATION AND TREATMENT OF TRANSITIONAL CELL CARCINOMA (TCC) OF THE BLADDER

Ralph C. Benson, Jr., M. D.

Department of Urology
Mayo Clinic
Rochester, Minnesota 55905

The diagnosis and treatment of transitional cell carcinoma in situ (CIS) of the urinary bladder is often difficult and frequently controversial. The diagnosis may be suggested by the patient's symptoms but it often hinges importantly upon subtle and sometimes imperceptible changes in the urothelium. Any aid in detecting these high grade and potentially lethal lesions would be a welcome addition to the urologist's diagnostic armamentarium.

In most cases, carcinoma in situ is initially treated with conservative measures such as fulguration and intravesical chemotherapy. The majority of patients, however, will experience disease recurrence and the only effective treatment for recurrent resistant carcinoma in situ of the urinary bladder has been cystectomy and urinary diversion (Utz 1980). The necessity for such a radical approach has been demonstrated, yet the use of surgery of such magnitude and consequence for superficial cancer is unfortunate.

Recent studies have indicated that hematoporphyrin derivative (HpD) is preferentially retained in animal and human malignant tissue and that this tissue then emits a characteristic salmon-red fluorescence when irradiated with light of appropriate wavelength (Gray 1967; Gregorie 1968; Leonard 1971; Mossman 1974). It is also known that intracellular HpD can produce relatively selective photosensitization of malignant cells upon exposure to exciting light of appropriate wavelength, resulting in cytotoxicity of these cells (Gregorie 1968; Lipson 1961; Lipson 1960; Kelly 1976; Kinsey 1981; Gomer 1979; Diamond 1972; Dougherty 1975;

Granelli 1975; Dougherty 1976). These observations prompted our investigation of HpD as a localization and treatment aid in transitional cell carcinoma in situ of the urinary bladder.

TUMOR LOCALIZATION

In order to assess the bladder tumor localization potential of HpD three groups of patients who were to undergo cystectomy for invasive carcinoma of the bladder were identified.

In group 1, four patients with invasive transitional cell carcinoma of the bladder were given HpD 2, 3, 24, and 48 hours before they underwent cystectomy and urinary diversion. Immediately after cystectomy, the bladder was opened and exposed to violet light at 400 and 410 nm from a filtered mercury arc lamp. Areas of fluorescence were marked, and then pathologic examination of tissue specimens from these areas were performed. The detection probe from a fluorescence detector previously designed for identification of in situ carcinoma of the bronchus was also scanned across the excised bladder to determine whether a signal could be elicited from any portion of the bladder (Kinsey 1980). Areas emitting positive signals were biopsied and examined pathologically.

In group 2, four patients with invasive transitional cell carcinoma of the bladder received HpD 2 hours before precystectomy cystoscopic examination with the use of our detection device. Sterile water was used as the bladder irrigant. One additional patient had been exposed to a known bladder carcinogen. This patient had suspicious findings on urinary cytologic examination, and only localization cystoscopy and not cystectomy was performed. In all cases, in order to scan the entire bladder, we used the rigid cystoscope to examine the lower hemisphere of the bladder and the flexible bronchoscope or nephroscope-choledochoscope for the upper hemisphere of the bladder. Areas with a positive signal were biopsied. The patients then underwent cystectomy. The opened bladders were exposed to violet light at a wavelength of 400 to 410 nm from a filtered mercury arc lamp, and all fluorescent areas were noted and examined pathologically.

In group 3, eight patients with invasive transitional cell carcinoma or diffuse in situ carcinoma that was refractory to other forms of therapy were given HpD 2 hours before cystectomy. The bladders were opened and pinned to a corkboard, with care being taken not to touch the epithelium. The cystectomy specimens were then exposed to violet-near-ultraviolet light from a General Electric BLB black lamp in a completely darkened room. Color photographs of the fluorescence were taken, and the specimens were immediately fixed in formalin. Pathologic bladder mapping was carried out with systematic semistep blocking and bladder maps were compared with the photographs of fluorescence. In the course of bladder mapping, a minimum of 55 and a maximum of 83 blocks were prepared from the cystectomy specimens. In no case was less than an estimated 70% of the mucosal surface sampled. In transition zones from normal to abnormal mucosa, especially with irregular distribution of lesions, the entire area was usually sampled in continuity.

LOCALIZATION RESULTS

Group 1. Localized fluorescence was noted in all of the bladders removed from patients in this group. Fluorescence was extremely bright and filters were not necessary to observe it. In every case, fluorescent areas were biopsied, and they proved to be severe dysplasia or transitional cell carcinoma. In each case the bladder was then routinely examined pathologically; no additional areas of tumor involvement were noted. Scanning of the excised bladders with our detection device revealed a positive signal only in the areas of fluorescence. Administration of HpD 2 to 3 hours before cystectomy tended to give less background fluorescence than administration at earlier periods before cystectomy. In no case was a positive signal elicited from an area that had proved normal on biopsy.

Group 2. All patients in this group underwent cystoscopic examination with the use of the detection device, as previously described. Areas that emitted a positive signal were not uniformly suspicious visually, but all were biopsied and on pathologic examination all revealed either carcinoma or severe dysplasia. The single patient in this group who did not have a cystectomy proved on biopsy to have severe dysplasia. This area was thoroughly fulgurated, and subsequent urinary cytologic examinations were negative.

All other patients underwent cystectomy. Before routine pathologic examination of the bladder, each specimen was exposed to violet light from a filtered mercury arc lamp. All fluorescent areas had previously been detected at cystoscopic examination and biopsied. Subsequent pathologic examination did not reveal any additional areas of mucosal neoplastic change.

Group 3. All patients in this group had areas in the bladder which showed fluorescence under violet-near-ultraviolet light. Photographs of the bladder fluorescence pattern were then compared with extensive bladder maps. In each case, the areas of dysplasia or carcinoma corresponded closely with the mucosal zones of fluorescence. Faint fluorescence was occasionally observed in regenerative mucosa surrounding healing recent biopsy sites.

TREATMENT

Our demonstration that HpD appears to be an excellent tumor localization aid for carcinoma of the bladder prompted initiation of clinical trials of HpD phototherapy in patients with resistant transitional cell carcinoma in situ of the bladder. Results in the first 5 treated patients will be presented.

Light Irradiation Source and Optical Fiber

The optical guide used was a medical-grade quartz optical fiber 5 m in length and having a core diameter of 400 μm (Quartz and Silice, Paris, France). The exit beam divergence was approximately 20°. The fiber was passed through the open channel of the endoscope so that the distal end protruded from the tip of the instrument. During treatment, the distal end of the fiber was positioned approximately 2 to 2.5 cm from the surface of the tumor whenever the light was directed on the external surface of the tumor. In this case, the diameter of the beam was approximately 2 cm, with a centrally located intense spot of approximately 1 cm. In an attempt to increase the size of the intense spot, the distal tip of the fiber was formed into a microlens by heating in a flame. In this way, the intense spot could be spread over a diameter of 2 cm when the fiber was held 2 to 2.5 cm from the surface of the tumor.

The proximal end of the quartz fiber was attached to an optical positioning device which was coupled to the output of a continuous-wave dye laser. Adjustments were made so that the dye laser beam was focused directly into the optical fiber. Rhodamine B dye was circulated in the dye laser. A continuous-wave argon ion laser was used to pump the dye laser. The argon laser had a maximal power output of 20 W and was capable of producing an output of as much as 4 W at 630 nm from the dye laser. Almost all of the dye laser output could be transmitted through the fiber since the optical losses were minimal. The power output from the optical fiber was measured with a calibrated, flat-response, continuous-wave power meter.

Procedure

For the initial treatment, all patients received HpD intravenously at a dose of 2.5 mg/kg of body weight approximately 2.5 hours before light irradiation. Subsequent treatments were usually 2 to 3 hours after the administration of HpD, but one patient (Pt. 1) was treated 48 hours after injection and another (Pt. 2) was treated 3 and 48 hours after injection. Standard rigid cystoscopes equipped with the Storz laser fiber guide were used to deliver the light when possible. When lesions were located in the dome or at the bladder neck, a flexible fiberoptic bronchoscope or a flexible nephroscope-choledochoscope was employed. Most procedures were done with the patient under general anesthesia. However, if the lesions were easily accessible and the patient requested, treatment was performed under local anesthesia. All areas of in situ carcinoma treated were 3 cm or less in diameter, although in Pts. 1, 2, and 4, the disease was multifocal. Even though the in situ carcinomas were no more than 3 cm in diameter, the diameter of the light irradiation spot was often too small to cover the entire tumor when the probe was held in one position. However, all tumors seemed to be adequately covered by irradiating in two overlapping locations or by moving the probe slowly back and forth over the involved area. In addition, two patients (Pts. 1 and 3) were treated on day 1 and on one succeeding day in order to cover a larger area. Intralesional placement of the fiber was performed in one patient (Pt. 5). Endoscopic photographs were taken before treatment and at each subsequent cystoscopic examination for all patients. Biopsy specimens were taken either several days before the administration of HpD or immediately before the laser irradiation, with careful attention being paid to hemostasis because bleeding would result in diffusion and absorption of the laser light. Light dose approximated 150 joules/cm^2 for each treatment session, and all treatment was performed at a wavelength of 630 nm. Patients were released from the hospital either the evening of the treatment day or the next morning.

As the skin absorbs a quantity of HpD sufficient to produce marked photosensitivity, each patient was carefully and repeatedly warned to avoid exposure to bright light, especially direct and indirect sunlight, for up to 4 weeks.

TREATMENT RESULTS

Four patients with biopsy proven carcinoma in situ who had failed all conservative treatment modalities and who refused cystectomy and one patient with invasive unresectable transitional cell carcinoma were treated.

Cystoscopic examination in all patients 4-8 days after treatment revealed a pronounced exuberant local reaction characterized by edema, hemorrhage, and early necrosis. Biopsy at these times demonstrated mucosal cells which exhibited hemorrhagic and exudative reaction in the submucosal tissue and marked degenerative changes in the overlying urothelium. Thirty to sixty days after treatment, the mucosal surface was populated by essentially normal-appearing urothelial cells.

The pathological data and treatment course for these five patients are shown in Table 1. Each patient with carcinoma in situ had biopsy proven disappearance of his tumor. Multiple recurrences have occurred in 3 patients, both in the initially treated region and also in other areas of the bladder. These recurrences have all been successfully treated--one patient (Pt. 2) awaits biopsy substantiation of his most recent treatment result. The patient with invasive transitional cell carcinoma had treatment because of recurrent hematuria which had necessitated hospitalization on two occasions. Therapy resulted in cessation of his bleeding until the time of his death 15 months later.

Table 1

Pt	Tumor: Stage and grade	Tumor: Present after therapy	Tumor: Recurrence At site	Tumor: Recurrence Elsewhere	Tumor: Present after more Rx	Current status of pt
1	CIS 2-3	No	2	3	No	Tumor-free
2	CIS 3	No	2	1	No	Awaits result of latest Rx
3	CIS 3	No	...	...	...	Tumor-free
4	CIS 3	No	1	1	No	Tumor-free
5*	Gr 3	Yes	...	...	...	Dead--Ca of bladder

*89-year-old man with invasive transitional cell carcinoma.

DISCUSSION

To be clinically useful, we believe that there are three criteria which any bladder tumor localization aid must fulfill: 1) it must be selectively taken up or retained only by dysplastic or neoplastic tissue, 2) it must spare no involved areas, and 3) it must have some feature which makes it detectable during a cystoscopic examination. Fluorescent tumor-localizing drugs such as tetracycline have been extensively investigated as aids in the diagnosis and treatment of various carcinomas. For the most part, their use has been limited either because they are not localized specifically in malignant or premalignant tissue or because their fluorescence cannot be detected by conventional instrumentation.

The three groups of patients employed for tumor localization studies have demonstrated that HpD fulfills our criteria for an effective tumor localization aid. Group 1 and group 3 patients demonstrated that HpD is selectively localized in malignant or severely dysplastic urothelium and that no involved areas were spared. Group 2 patients who underwent cystoscopy prior to cystectomy substantiated that the localized HpD could be detected during a cystoscopic examination. These studies indicate, therefore, that

HpD may be the best tumor localization aid for transitional cell carcinoma of the bladder yet identified.

These preliminary tumor localization findings, plus the knowledge that intracellular HpD can produce photosensitization of malignant cells, indicated that HpD phototherapy might be a clinically useful therapeutic modality for the treatment of bladder cancer.

Our intent was to specifically examine the effect of HpD phototherapy and to exclude other variables that might have an antitumor effect. One such variable when utilizing laser light is heat. It has recently been suggested that much of the antitumor effect of interstitial implantation of the optical fiber for treatment of solid tumors is heat related (Kinsey in press; Svaasand 1981). Although a bladder full of saline should work as a heat sink, we believed that it was necessary to exclude the thermal effect before attempting treatment. Preliminary studies using female dogs with and without HpD administered 2 hours prior to laser treatment indicated that laser light (630 nm) with a power density of 200 mW/cm^2 had no effect on the normal canine bladder. Therefore, the urinary bladder appears to be an organ uniquely well suited for HpD phototherapy because therapy is performed in a liquid medium, thus eliminating local heat effect.

At the present time, we believe that resistant in situ transitional cell carcinoma of the bladder is the tumor stage best treated with HpD phototherapy. We did successfully treat recurrent incapacitating hematuria in one man with inoperable invasive carcinoma. However, our localization studies have indicated that large invasive tumors did not retain HpD as consistently and homogeneously as did the in situ lesions. It is for this reason, plus the possible risk of bladder perforation, that we doubt whether HpD phototherapy as presently administered will be safe or successful in curing large invasive tumors of the bladder. Therefore, despite the apparent suitability of the urinary bladder for HpD phototherapy, it should be used at the present time only in carefully selected patients.

The necessity for delivering light of extremely high energy density into the bladder for effective HpD phototherapy has posed many vexing technical problems. Many patients with carcinoma in situ have diffuse multicentric disease that is impossible to irradiate completely by the use of currently available fibers. Heat treatment of the distal end of the quartz fiber resulting in a convex surface has enlarged the radiation spot, but an even bigger spot size would be helpful for the treatment of large lesions or diffuse disease. The ideal irradiation source would be a point source of light that could be placed in the center of the bladder and that would deliver light of sufficient power to all areas in the bladder. We are actively evaluating several fibers with bulbous tips, but none has given consistent illumination. In addition, improvement in our ability to direct the light beam to the bladder neck and dome is needed--presently available flexible endoscopic instruments can be employed, but only with difficulty.

Several of our patients have experienced tumor recurrences and we, therefore, appear not to be altering the biologic course of the patient's disease with focal phototherapy. Future technical improvements will hopefully allow irradiation of the entire bladder at one time resulting, perhaps, in decreased recurrence rates. We believe that our studies have demonstrated that although HpD phototherapy remains experimental, it is effective in the treatment of transitional cell carcinoma of the bladder.

Diamond I, Granelli SG, McDonagh AF, Nielsen S, Wilson CB, Jaenicke R (1972). Photodynamic therapy of malignant tumours. Lancet 2:1175.

Dougherty TJ, Gomer CJ, Weishaupt KR (1976). Energetics and efficiency of photoinactivation of murine tumor cells containing hematoporphyrin. Cancer Res 36:2330.

Dougherty TJ, Grindey GB, Fiel R, Weishaupt KR, Boyle DG (1975). Photoradiation therapy. II. Cure of animal tumors with hematoporphyrin and light. J Natl Cancer Inst 55:115.

Gomer CJ, Dougherty TJ (1979). Determination of [^{3}H]- and [^{14}C] hematoporphyrin derivative distribution in malignant and normal tissue. Cancer Res 39:146.

Granelli SG, Diamond I, McDonagh AF, Wilson CB, Nielsen SL (1975). Photochemotherapy of glioma cells by visible light and hematoporphyrin. Cancer Res 35:2567.

Gray MJ, Lipson R, Maeck JVS, Parker L, Romeyn D (1967). Use of hematoporphyrin derivative in detection and management of cervical cancer: a preliminary report. Am J Obstet Gynecol 99:766.

Gregorie HB Jr, Horger E, Ward JL, Green JF, Richards T, Robertson HC Jr, Stevenson TB (1968). Hematoporphyrin-derivative fluorescence in malignant neoplasms. Ann Surg 167:820.

Kelly JF, Snell ME (1976). Hematoporphyrin derivative; a possible aid in the diagnosis and therapy of carcinoma of the bladder. J Urol 115:150.

Kinsey JH, Cortese DA (1980). Endoscopic system for simultaneous visual examination and electronic detection of fluorescence. Rev Sci Inst 51:1403.

Kinsey JH, Cortese DA, Moses HL, Ryan RJ, Branum EL (1981). Photodynamic effect of hematoporphyrin derivative as a function of optical spectrum and incident energy density. Cancer Res 41:5020.

Kinsey JH, Cortese DA, Neel HB (in press). Thermal considerations in hematoporphyrin derivative phototherapy. Cancer Res.

Leonard JR, Beck WL (1971). Hematoporphyrin fluorescence: an aid in diagnosis of malignant neoplasms. Laryngoscope 81:365.

Lipson RL, Baldes EJ (1960). The photodynamic properties of a particular hematoporphyrin derivative. Arch Dermatol 82:508.

Lipson R, Baldes EJ, Olsen AM (1961). The use of a derivative of hematoporphyrin in tumor detection. J Natl Cancer Inst 26:1.

Mossman B, Gray MJ, Silberman L, Lipson RL (1974). Identification of neoplastic versus normal cells in human cervical cell culture. Obstet Gynecol 43:635.

Svaasand LO, Doiron DR (1981). On the probability of simultaneous action of hyperthermal and photodynamic effects during photoradiation therapy of malignant tumors. Medical Imaging Sciences Group, University of Southern California and University of Trondheim USC-MISG, p 900.

Utz DC, Farrow GM (1980). Management of carcinoma in situ of the bladder: the case for surgical management. Urol Clin North Am 7:533.

Porphyrin Localization and Treatment of Tumors, pages 805–827

PHOTORADIATION OF MALIGNANT TUMORS PRESENSITIZED WITH HEMATOPORPHYRIN DERIVATIVE

James S. McCaughan Jr., M.D., F.A.C.S.

Director
Laser Medical Research Foundation
300 East Town Street
Columbus, OH 43215

From April 1982 to April 1983, we have treated 55 patients with photoradiation of malignancies following presensitization with hematoporphyrin derivative. The first five patients were treated with Kodak projectors and a #2418 red glass filter for cutaneous and subcutaneous lesions. The remaining 50 patients were treated using the argon dye laser system. These latter include: 18 melanomas of the eye, two squamous cell cancers of the vagina, nine carcinomas involving the esophagus, eight carcinomas involving the endobronchial tree, four cutaneous and subcutaneous breast cancer metastases, three head and neck cancers, recurrent liposarcoma of the arm, melanoma of the esophagus, three basal cell cancers and one squamous cell cancer of the face.

The melanomas of the eye are the subject of a separate report; this report gives a brief description of the tumor types treated and then summarizes our results with 30 of the remaining 32 patients treated with the argon dye laser system. Insufficient data were available on two patients with squamous cell cancer of the vagina.

A dosage of 2.5 to 5.0 mg of hematoporphyrin derivative (Oncology Research & Development, Inc.) per kilogram of body weight is injected as a single intravenous bolus three to seven days prior to photoradiation. Our standard routine now is 2.5 mg of hematoporphyrin derivative (HpD) for ocular lesions and 3.0 mg of hematoporphyrin derivative for all other malignancies; we treat on the third or fourth day following injection.

The light source is generated by a 20 W Spectra Physics model #171 argon laser, which is directed onto rhodamine-B dye circulated in a Spectra Physics tunable dye laser model #375 without the tuning wedge and a pressure of 100 PSI. The 630 nm red beam then is coupled to various beam splitters or directly onto quartz silice fibers. One coupler divides the beam into four separate beams (Oncology Research & Development, Inc.) using 400 micron fibers. This is used for simultaneous interstitial photoradiation through thin-walled hypodermic needles. Another beam splitter (Laser Medical Research Foundation) divides the beam into two for simultaneous photoradiation of anterior and retrobulbar ocular tumors, using a specially designed reflecting probe (Dr. Thomas Dougherty) for the retrobulbar irradiation.

External irradiation and endoscopic irradiation are done using either a 400 micron or a 600 micron fiber. The latter is more rigid and permits direct insertion into intralumenal neoplasms.

A cylinder type fiber delivering the light from its sides radially for a length of 2-3 cm (Oncology Research & Development, Inc., Laser Medical Research Foundation) has been used for esophageal, endobronchial and vaginal neoplasms.

The power delivered by the dye laser and the fibertip point is measured with a Coherent model #210 radiometer. The power delivered by a cylinder type fiber is measured by laying the cylinder flush with the radiometer and using that as the reading. When the single tip fiber is used at a distance from the lesion to deliver a spot size at the target less than the diameter of a Coherent radiometer, the area of the spot size is measured and the power obtained by the Coherent radiometer at this distance is measured. The power density then is calculated as this power divided by the area.

For external irradiation of cutaneous and subcutaneous lesions, the power density is read directly from the Yellow Springs model #65 radiometer.

The rhodamine-B dye solution is prepared by dissolving .976 gm in 10 cc's of reagent grade methanol and diluting it to one liter with reagent grade ethylene glycol. The dye degrades with use, and the power emitted from the dye laser is frequently checked with the radiometer and the wave length with a spectrometer.

All patients have had the benefit of all other known forms of therapy, have been unsuitable for them or have refused them. They all have been advised of the investigational nature of this work in accordance with standardized Investigational Review Board policy.

In over 75 injections of hematoporphyrin derivative, some patients receiving three separate injections, we have noted no adverse effect from the dye per se. However, we have had several patients receive sunburns as much as six weeks after the injection, some forming blisters. Another patient required the ventilator for several days following treatment and developed chemosis of the lower sclera and erythema of the neck and upper chest from artificial light.

ESOPHAGEAL TUMORS

Photoradiation therapy is performed under local anesthesia with a standard flexible gastroscope. Unfortunately, frequently the patients have extensive lesions and are obstructed completely by the time we see them. Even a pediatric esophagoscope cannot be passed through the opening. We have had some extremely satisfying results with the cylinder type fiber passed through small openings - the distance calculated from the CT scans. One patient with adenocarcinoma of the stomach had an esophagus completely obstructed by tumor. Although the tumor is still present, he has been able to eat solid foods for five months following the last of three treatments. His weight remains stable.

A problem with the cylinder fiber, however, is that it is very fragile and breaks easily. Should this occur, the tip might be difficult to retrieve. This is not as serious a problem with esophageal tumors as it is with endobronchial lesions.

We have had three patients that had been dilated previously and no longer could be dilated. They were essentially completely obstructed.

Using a 600 micron fiber and 1000 mW from the tip, we have been able to insert the fiber directly into the tumor. There is an immediate reaction; the tumor becomes swollen, necrotic and soft. At the end of the treatment, patients now can be dilated up to a #40 or #50 Maloney dilator with comparative ease and a #18 sump can be inserted readily into the stomach.

We have had one patient who developed a tracheo-esophageal fistula. However, this was undoubtedly due to his tumor, since he already had recurrent nerve paralysis and tumor involving his trachea. He had been referred because the radiotherapist thought he would develop this kind of fistula if he were irradiated.

Another patient with complete obstruction of the cervical esophagus developed stridor within 24 hours of his therapy. Endoscopy revealed bilateral vocal cord edema, which was not present at the time of his therapy. The esophagoscopy was difficult, and on numerous occasions, the light from the esophagoscope was directed inadvertently onto his vocal cords, but his vocal cords were never mechanically traumatized. The edema subsided spontaneously in three days and the patient was treated with humidified oxygen by mask. We hypothesize that the edema may have been caused by the amount of energy being delivered by the Olympus CLV light source from a GIF-2T Olympus scope. The spectral analysis from the light delivered from a GIF-2T Olympus gastroscope with an Olympus CLV light source is shown in Figure 1. It covers some of the wave lengths that are absorbed by hematoporphyrin derivative.

The maximum light power is obtained with filter A and full power. When held one centimeter from the Coherent model #210 radiometer face, 900 mW were delivered via an Olympus gastroscope to an area 1.76 cm^2 for a power density of 531 mW/cm^2. When the Olympus model BF bronchoscope is used at full power with filter A, 175 mW are delivered over an area of .28 cm^2 for a power density of 625 mW/cm^2.

Although this energy is not completely in the 630 nm range for depth penetration, certainly it is in some of the absorption spectrum of the dye, and when it is used for prolonged periods, it may contribute to the reaction.

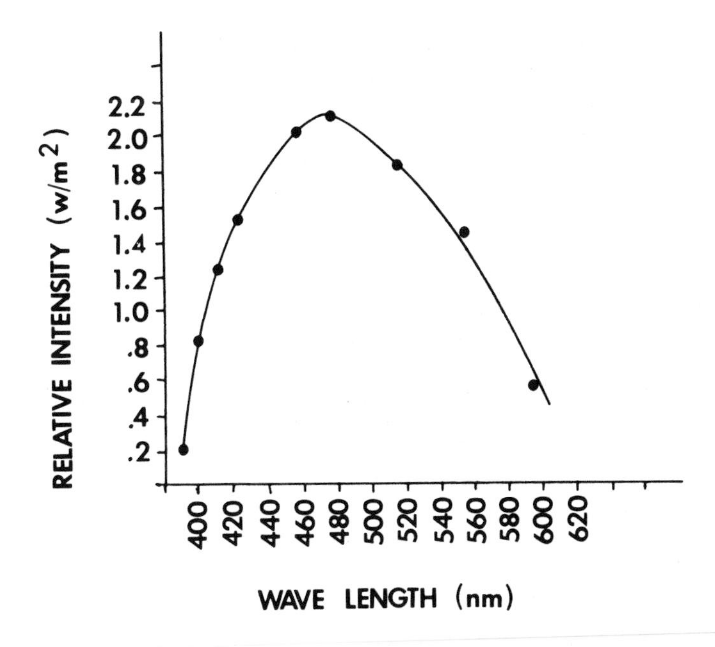

Figure 1. Relative intensity for various wavelengths of Olympus GIF-2T Gastroscope with Olympus light source CLV 228301 K.

ENDOBRONCHIAL LESIONS

Since many of these patients are already in severe respiratory distress, we have found it best to use local anesthesia and a rigid ventilating bronchoscope with assisted ventilation by the anthesthesiologist. We then are able to pass the fiberoptic bronchoscope through the rigid scope, while ventilating the patient. The fiber is inserted through the fiberoptic scope. This gives a much better view of the area being treated than the rigid scope alone. Also, it permits better manipulation of the fibertip, better biopsies, and easier removal of debris through the rigid scope.

We have had one occurence of a flash fire burning the fibertip using 300 mW from the tip of the fiber and 100% oxygen. This apparently occurred from heating the tissue, which was in direct contact with the fibertip. Fortunately, there were no serious sequelae, and the patient recovered. The distal half-inch covering of the fiberoptic bronchoscope did melt, and if there had been a plastic or rubber endotracheal tube in place, the incident might have been more serious. The problem was managed by rapidly withdrawing the fiberoptic bronchoscope. With the rigid scope, we were able to continue ventilating the patient.

Since the tumor does become necrotic and edematous, provision must be made for repeated bronchoscopy to remove debris and even temporary ventilation on the respirator.

EXTERNAL RADIATION OF CUTANEOUS AND SUBCUTANEOUS MALIGNANCIES AND BREAST CANCER METASTASES

There is wide variation in the skin response of different patients. It appears there is a minimal power density required to cause a reaction. However, after this is obtained the main factor now appears to be the total number of joules delivered per cm^2. Separate areas on the same patient treated on the same day that received 25 J/cm^2 to each area with power densities ranging from 30 to 100 mW/cm^2 developed the same amount of skin reaction in all areas treated.

Patients who had skin nodules excised three days prior to photoradiation therapy showed no excessive reaction or different reaction in the excisional sites.

Different patients treated on the same day with the same dose of hematoporphyrin derivative and receiving the same power density and J/cm^2 have exhibited different reactions. Twenty-five J/cm^2 caused blistering on one patient but not on another. As with solar radiation, the type and condition of the skin have some effect on the reaction. We now are spot testing these patients on the third day and treating them on the fourth day.

The skin over inflammatory carcinoma reacts more violently for the same dose per cm^2 than the skin of the same patient that has solitary nodules.

In treating areas 6x7 cm with a single fiber, we have noted the light intensity is not homogeneous. The intensity of light in the center of the spot when read with the Yellow Springs radiometer is more intense than in the periphery.

We have noted recurrences of cutaneous breast cancer in the field of treatment as well as peripherally in patients within two weeks following their therapy, although visible tumors in the field had become necrotic.

Several patients have complained of burning similar to sunburn immediately after treatment. This is relieved somewhat with cold compresses.

HEAD AND NECK CANCER METASTASES

We have had two patients report spontaneously that severe headaches, present before photordiation, had ceased. We do not know if this is from destruction of the sensory nerves or the tumor.

A patient with a T1 lesion of the soft palate has no visible or palpable evidence of tumor six weeks after therapy.

Interstitial photoradiation of large tumors causes marked swelling and edema and pain that lasts for several days and may require narcotics for relief.

Hard intraoral lesions soften while being treated with external radiation from a tip delivering 1000 mW through a 600 micron fiber. After a few minutes, the tip can be inserted directly into the tumor for interstitial treatment.

SQUAMOUS CELL CANCER OF THE VAGINA

A patient with a 2x5 cm tumor of the lateral wall of the vagina underwent two treatments with photoradiation therapy. Six weeks later, the lesion was neither visible nor palpable.

Another patient was treated with interstitial fibers inserted through the vagina into the left paravaginal and cul de sac areas as well as intravaginal external radiation for extensive multiple areas of vaginal squamous cell cancer with cul de sac metastases. Three weeks later, she had developed an orange sized mass in her left cul de sac, which still remained but was approximately lemon sized at six weeks.

CASE SUMMARIES

All treatments and observations were made personally by the author. The treatment regimen for each patient is summarized in the table.

Esophageal Tumors

Patient #1: An 83-year-old white man with primary melanoma of the esophagus was able to swallow only liquids. He was treated with 250 mW from a cylinder-type fiber for ten minutes in each of three areas. Three weeks after treatment, he was able to eat chicken, and the tumor was reduced to one half of the original size.

Patient #2: A 65-year-old white man, with adenocarcinoma of the stomach, had a complete obstruction of the esophagus at 38 cm. Three days after injection of HpD, he was treated with a cylinder fiber that delivered 300 mW for 10 minutes at three successive 3 cm levels. Two days after treatment, gastric juice was seen refluxing through the distal esophagus, and the patient could eat ice cream. At eight days after treatment, there was a 4 mm opening and at 40 days after therapy, a pediatric scope passed easily into the stomach. The patient had been eating solid foods for two weeks, and his weight was stable. Two months after the therapy, however, the tumor had continued to grow, and additional therapy was required. Three days after injection of HpD, the patient was treated with a cylinder fiber that delivered 225 mW for ten minutes at four successive areas. At the end of the procedure, a #16 Levine tube passed easily into the stomach. Four days after treatment, there was a marked inflammatory reaction at 40-34 cm. Fourteen days later, the patient was unable to swallow liquids. The esophagus was blocked with edematous mucoid material but still dilated easily up to a #40 Maloney dilator. At six weeks after treatment, the patient was able to eat solid food and his weight was stable. Five months after the initial treatment; however, additional therapy became necessary. Three days after injection with HpD, a straight fiber tip was inserted into the tumor and 1000 mW were delivered for ten minutes. There was obvious immediate necrosis of the tumor. The esophagus dilated up to a #40 Maloney dilator, and a #18 Salem sump passed without difficulty. One week after treatment the patient was eating pancakes and other solid foods. One month later (six months after the initial treatment), he was eating solid food. The tumor is still present.

Patient #3: A 61-year-old black man with a diagnosis of squamous cell cancer in the distal esophagus was able to swallow only clear liquids, and a pediatric scope could not be passed into the stomach. Three days after injection with HpD, he was treated with a cylinder fiber that delivered 300 mW for 10 minutes in two separate areas and a straight fiber tip that delivered 300 mW for five minutes. Four days later, the esophagus was white and edematous; a pediatric scope could be passed into the stomach. Barium swallow showed a 1.3 cm diameter opening. At one month, an adult esophagoscope (GIF-2T) passed easily into the stomach, and the patient could eat solid foods; his weight remained stable. At ten weeks, the patient's status declined, and he underwent additional treatment. Three days after injection with HpD, he was treated with a cylinder fiber that delivered 300 mW for ten minutes to three separate areas. A large tumor was well demarcated at 39 cm. A pediatric scope could pass into the stomach, and the patient was able to eat steak. Six weeks later, barium swallow showed a 1.3 cm opening. Five months after the first treatment, the patient had difficulty swallowing but the barium still passed through the esophagus. At six months after treatment, the patient requires dilation approximatley every two weeks. The tumor is still present, but no further treatment has been initiated.

Patient #4: a 54-year-old white man with a diagnosis of squamous cell cancer of the cervical esophagus was able to swallow liquids only and was experiencing retrosternal pain. Barium swallow showed a 4 mm opening with aspiration. Three days after injection with HpD, he was treated with a cylinder fiber that delivered 200 mW for 10 minutes to three separate areas. Esophagoscopy performed three days later showed concentric necrosis and an edematous white tumor. The esophagus dilated up to a #16 NG tube. The patient was able to swallow liquids, and his retrosternal pain was gone.

Fourteen days later, however, the patient again was unable to swallow liquids. Three days after a repeat injection with HpD, he was retreated with a cylinder fiber that delivered 200 mW for 10 minutes to five separate areas. Fourteen days after therapy, a 6 mm bronchoscope could pass into the patient's stomach although he still was unable to swallow his saliva. Twenty-eight days after treatment, the patient underwent a gastrostomy and a cervical esophagostomy. He subsequently developed a tracheoesophageal fistula. Five months after the initial treatment, the patient was released from the hospital and was maintaining his weight.

Patient #5: A 67-year-old-white woman was found to have a squamous cell tumor of the esophagus at 30 to 35 cm. A regular gastroscope would not pass through the esophagus, although a pediatric scope would. Three days after injection with HpD, a cylinder type fiber was used to deliver 300 mW for ten minutes in five different areas. Two weeks after treatment, an adult gastroscope could pass into the esophagus without difficulty. Ten weeks after treatment, the patient was eating regular food and had gained ten pounds. Fourteen weeks after treatment, the tumor was still present on esophagoscopy. The patient is still eating without obstruction and has gained two more pounds.

Patient #6: A 75-year-old white man was diagnosed with adenocarcinoma of the stomach involving the esophagus. He had been dilated successfully in the past but now was unable to pass a pediatric scope. He was able to swallow liquids and was experiencing no pain. Three days after injection with HpD, a cylinder fiber was used to deliver 300 mW for 10 minutes to two separate areas and 1000 mW were delivered through a straight fiber tip for 10 minutes to three separate areas for a dose of 600 J. Two days later, the tumor was necrotic, and a pediatric scope passed easily into the stomach. Seven days after another injection with HpD, the patient was retreated with 1000 mW delivered through a straight fiber for 10 minutes to three separate areas for a dose of 600 J. Eighteen days after the initial treatment, an 11 mm esophagoscope passed easily into the stomach. One month after the initial treatment, a #28 and #50 dilator could pass into the stomach. The patient had been eating well but gradually anorexia developed. Thirty-eight days after the initial treatment, cytologic testing of a large left pleural effusion was positive for adenocarcinoma. The patient underwent two treatments with the Nd:YAG laser. Two months after the initial treatment, the patient was on a regular diet but developed progressive anorexia. He died ten weeks after the initial treatment from pneumonia.

Patient #7: A 58-year-old white man diagnosed with adenocarcinoma of the stomach involving the esophagus also was found to have liver metastases. He was able to swallow liquids. Three days after injection with HpD, 900 mW were delivered through a straight fiber for 10 minutes to three separate areas at 540 J. Ten weeks after treatment, the tumor measured 42 to 47 cm and the patient was able to eat solids. The patient was jaundiced, however , due to extensive liver metastases and pedal edema. Twelve weeks after treatment, the patient died from metastatic disease.

Patient #8: A 70-year-old black man with squamous cell of the esophagus was experiencing complete obstruction at 18 cm. Barium swallow showed a 7 mm opening. He was unable to pass a pediatric gastroscope or bronchoscope through the esophagus. Three days after injection of HpD, 1000 mW of external irradiation were delivered through a straight fiber tip for 10 minutes and an additional 1000 mW were delivered through a straight fiber tip inserted into the tumor for 13 minutes. At the end of the procedure, the patient dilated easily through a #48 French Maloney dilator and a #18 Salem sump was inserted into the stomach without difficulty. Twenty-four hours after treatment the patient was experiencing severe stridor due to vocal cord edema. Bronchoscopy showed a small mound of tissue which was found to be a squamous cell tumor. Eleven days after treatment, the patient dilated easily to a #52 Maloney. The tumor was beefy red and necrotic. A large gastroscope passed to 25 cm, and a #18 Salem sump could be inserted without difficulty. Eighteen days after treatment the patient could swallow his own saliva.

Patient #9: A 57-year-old white man with a squamous cell tumor of the esophagus starting at 28 cm could not pass a pediatric gastroscope and could eat only liquids. Three days after injection with HpD, 1000 mW were delivered through a straight fiber tip inserted into the tumor for 10 minutes to three separate areas. Four days after treatment, the patient suffered a saddle pulmonary embolus. Post mortem examination revealed no perforations. The treated tumor was necrotic and liver metastases were present.

Patient #10: A 70-year-old white woman with severe Parkinson's disease and Alzheimer's disease had an adenocarcinoma of the stomach which was obstructing the esophagus. The patient could not be dilated. Three days after injection with HpD, 200 mW were delivered with a cylinder fiber for 10 minutes to two separate areas, and 1000 mW were delivered with a straight fiber tip inserted into the tumor for 10 minutes. Immediate edema and necrosis of the tumor were noted. The patient easily dilated to a #48 Maloney at the end of the treatment, and a gastroscope could be passed into the stomach. Two days after treatment, the patient aspirated the gastrostomy tube feedings and developed pneumonia. She died three weeks after treatment due to an empyema.

Endobronchial Tumors

Patient #11: A 60-year-old white woman had recurrent adenocarcinoma of the lung that completely obstructed the right main bronchus. Three days after injection with HpD, a straight fiber tip was used to deliver 300 mW for nine minutes. One day later, the tumor was necrotic, and two days after treatment, the right main bronchi still were obstructed by tumor. The patient was retreated with a straight fiber tip to deliver 300 mW for ten minutes and 300 mW for 20 minutes above the tumor. Fifteen days after the initial treatment, the right lung aerated spontaneously.

Patient #12: A 45-year-old white woman with a cylindroma was characterized by a severly obstructed trachea and both main bronchi. She suffered from severe dyspnea. She had undergone previous treatment with the Nd:YAG laser. Three days after injection with HpD, a straight fiber tip was used to deliver 300 mW for ten minutes to the trachea and 300 mW for ten minutes to two separate openings in the left main bronchus. One day later, the trachea and left main bronchus were characterized by necrotic tumor, while the right main bronchus was edematous. Five weeks later, the left main bronchus was ulcerous but almost completely open. The patient then underwent treatment to the right main bronchus, and three days after injection with HpD, a cylinder type fiber was used to deliver 150 mW for ten minutes to three separate areas. Four weeks later, the trachea and both main bronchi were open; the patient no longer suffered from dyspnea. Two months after the initial HpD treatment, the patient underwent CO_2 laser therapy to remove nodules from her vocal cords. Three months after initial treatment, the tumor was observed to be growing in the right bronchus; the left bronchus remained partially open. The patient was treated again with HpD and three days later received 200 mW with a cylinder fiber for 10 minutes to three separate areas in the right main and lower bronchi. Two weeks later, the patient had a stridor over the right lung while the left lung was clear. The patient did not complain of dyspnea.

Patient #13: A 66-year-old white man was seen with a recurrent squamous cell cancer in a bronchial stump and was characterized by hemoptysis. A straight fiber tip was used to deliver 1000 mW for 20 minutes three days after injection with HpD. One day later, the area was white and edematous. An additional 1000 mW were delivered via a straight fiber tip for 20 minutes. Several bronchonscopies were required to clear the diffuse edematous white necrotic material that was present. Twelve weeks after treatment, there was no evidence of hemoptysis, and both bronchoscopic biopsies and washings were negative. Four months after treatment, the patient experienced neither hemoptysis nor dyspnea.

Patient #14: A 54-year-old white man had a squamous cell cancer _in situ_ lung. He experienced daily hemoptysis and severe cardiopulmonary impairment. Three days after injection with HpD, a straight fiber tip was used to deliver 250 mW for ten minutes to three separate areas. Two days later, a pearly white reaction was noted. One week after treatment, the hemoptysis had stopped, but five weeks after treatment, an inflammatory reaction still was noted. Biopsies were negative. Twelve weeks after treatment, the lung had a normal appearance, and the patient had no hemoptysis. At six months, the bronchoscopy was normal. Current biopsies show chronic inflammation, but the patient does not exhibit hemoptysis.

Patient #15: A 60-year-old white woman had a squamous cell tumor of the trachea and lumen obstructed by 50%. She previously had undergone X-ray and Nd:YAG laser therapy and when examined was on a respirator. Three days after injection with HpD, a #7 rigid bronchoscope would not pass into the distal trachea. Three hundred mW were delivered with a straight fiber tip and 150 mW were delivered with a cylinder fiber tip into the tumor for 10 minutes. The patient was receiving 100% oxygen during the treatment. As the fiber was withdrawn from the tumor, the fibertip caught fire and burned off. The distal inch of the fiberoptic bronchoscope plastic covering melted. One week after treatment, the patient exhibited chemosis of the lower sclera and erythema of the neck. Two weeks after treatment, she was taken off the respirator. At three weeks, a rigid #9 bronchoscope passed easily to the carina, and both main bronchi were open. Chest X-ray, laminogram and arterial blood gases all were normal.

Patient #16: A 70-year-old white man who was bedridden with severe chronic obstructive pulmonary disease had a diagnosis of squamous cell lung cancer. Three days after injection of HpD, 300 mW were delivered through a cylinder fiber tip for 30 minutes. Three days after treatment, the patient developed pneumonitis and septicemia. He died two weeks after treatment due to renal failure although his lungs were clear.

Patient #17: A 35-year-old white woman was treated for recurrent adenocarcinoma of the lung. On examination, her left lung was completely opque with the tumor extending to the left upper lobe. She was bedridden and suffered from severe dyspnea. Three days after injection of HpD, 300 mW were delivered with a cylinder fiber tip for 30 minutes. Four days later, a tracheostomy revealed a white tumor and the lower lobe was still open. The patient died two weeks after treatment due to disease.

Patient # 18: A 50-year-old white woman had a metastatic lesion to the lung from adenomcarcinoma of the colon. The tumor was obstructing the left upper lobe. Three days after injection with HpD, a 600 micron straight fiber inserted into the tumor was used to deliver 300 mW for ten minutes for a dosage of 180 J, and an additional 300 mW were delivered onto the tumor with a straight fiber tip for 10 minutes for a dosage of 180 J. Four days later, the tumor was pearly white and edematous. A 2x1 cm necrotic tumor mass was removed. Two weeks after tretment, the patient reportedly was breathing better although an X-ray showed that the left upper lobe was still opaque.

Skin Tumors

Patient #19: A 56-year-old white man was characterized by recurrent basal cell cancer of the head. The outer table of the skull was eroded by an ulcer measuring 9 cm in diameter. The patient had suffered from severe, constant headaches for six months. Three days after injection with HpD, 30 mW/cm^2 were delivered for 30 minutes for 54 J/cm^2. Two weeks after treatment, the patient's headaches had stopped. After four treatments three fourths of the circumference of the ulcer was biopsy free of tumor, and the bone was covered with granulation tissue.

Patient #20: A 79-year-old white man was seen with a squamous cell tumor of the eyelid, which was ulcerated and measured 2.0x1.1x.6 cm. Three days after injection with HpD 50 mW/cm^2 were delivered for 8.5 minutess for a dose of 25 J/cm^2. Three days after treatment, the tumor was black and slightly edematous. The eye and adjacent skin were normal. Seven weeks after treatment, the ulcer was healed, and several 2 mm hard nodules could be seen at the periphery. Three months after the initial tratment, a 1.0x1.5 cm area was retreated. Two weeks later, the tumor had shrunk by 50% and there was minimal reaction to the surrounding eyelid.

Patient #21: An 89-year-old white woman, who had undergone Moh's treatment and skin graft for a basal cell tumor of the face, was seen with a recurrent lesion measuring 3.2 cm. Three days after injection with HpD, 17 mW/cm^2 were delivered for 25 minutes for a dose of 25 J/cm^2. Four weeks later, the tumor was black and necrotic, with an ulcer measuring 3x3 cm, but the adjacent skin showed no reaction. Eight weeks after treatment, the ulcerated area had healed, but five months after treatment, there were ulcerations over the previosuly untreated areas. Four separate areas fluoresced salmon color when tested with a Wood's lamp. Three days after another injection of HpD, 55, 60, 40, and 50 mW/cm^2 for a dose of 20 J/cm^2 were delivered to four 3x4 cm areas. One week later, there was no necrosis of the skin, and the ulcerated areas were black.

Patient #22: An 86-year-old white woman with recurrent basal cell cancer of the face was unable to open her mouth to ingest liquids due to pain. After injection of HpD, four different sites were treated. A 3x5 cm ulcer on the chin was treated with 7.5 mW/cm^2 for 29 minutes for a dose of 13 J/cm^2. Three weeks later, the patient was eating soft food. The ulcer was clean, and its raised edges now were flat and well demarcated. Eight weeks later, an ulcer measuring 4x2 cm was treated with 140 mW/cm^2 for 3.5 minutes for a dose of

30 J/cm^2. An angle of ulcer fluoresced when tested with a Wood's lamp, and biopsies were positive. Twelve weeks after initial treatment, a 3 mm mucocutaneous fistula was noted, the ulcer was clean, and the patient was eating.

A second treated site was a 5 cm diameter cutaneous metastasis at the angle of the jaw. Three hundred mW were delivered interstitially and simultaneously through each of four needles for 14 minutes for a dose of 270 joules per fiber. Four days later, there was a slight necrosis of the skin over the tumor. The area was retreated with 50 mW for nine minutes for a dose of 25 J/cm^2. Three weeks later, the tumor had shrunk by 50%. Eight weeks later, a 1.5x2.5 cm lesion was retreated with 400 mW delivered interstitially and simultaneously through each of four needles for 13 minutes for a does of 312 J/F and externally 200 mW/cm^2 for four minutes for a does of 48 J/cm^2.

Site 3 was a 20 mm diameter subcutaneous metastatic lesion on the neck. Three days after injection with HpD, 400 mW were delivered interstitially and simultaneously through each of four needles for 20 minutes for a dose of 480 joules per fiber (J/F). Three weeks later, the tumor had shrunk by 50%.

Site 4 was a 14 mm diameter subcutaneous metastatic lesion on the neck, which was treated with 400 mW delivered interstitially and simultaneously through each of four needles for 20 minutes for a dose of 480 J/F. Three weeks later, the tumor had shrunk by 50%. None of the treated lesions exhibited blistering or necrosis of the adjacent skin.

Liposarcoma of Arm

Patient #23: An 80-year-old white man was seen with recurrent liposarcoma of the arm. Three weeks prior to referral, he had undergone X-ray therapy and re-excision. At that time, tumor was found at the base of the specimen. Three days after injection of HpD, three adjacent areas measuring 6x7 cm were treated. Site 1 received 32 mW/cm^2 for 15 minutes for a dose of 29 J/cm^2. Site 2 received 28 mW/cm^2 for 18 minutes for a dose of 30 J/cm^2. Site 3 received 30 mW for 22 minutes for a dose of 40 J/cm^2. Two days after treatment, blisters were noted at all three sites. Two weeks after treatment, the sites were characterized by black, necrotic eschars. Eleven weeks after therapy, the wounds were granulating.

Breast Tumors

Patient #24: A 60-year-old white woman was seen with a cutaneous and subcutaneous metastatic nodular breast cancer covering approximately a 15x15 cm area. Several large nodules were excised four days prior to treatment, and incisions were made in the treatment fields. Three days after injection with HpD, six contiguous areas, measuring approximately 6x7 cm, were treated with 21 to 36 mW for a dose of 25 J/cm^2 to each area. One day later, the tumors were black; the adjacent skin and incision sites were red and painful but were not necrotic. Twelve days after treatment, the tumors were gone, and areas were black or reddish-brown and flat. By nineteen days after treatment, a black eschar had formed at the tumor area, and the rest of the skin was normal. Eight weeks after treatment, the eschar was separating, and at 14 weeks after treatment, wound healing had occurred with no evidence of tumor recurrence.

Patient #25: A 58-year-old white woman was seen with metastatic cutaneous and subcutaneous lesions in the breast. An ulcerated area was located down to the ribs approximately 7x5 cm in diameter and containing multiple nodules. Several large nodules were excised four days prior to treatment, and incisions were made in the treatment fields. Three days after injection with HpD, eight adjacent sites, ranging in size from 5x6 cm to 9x4 cm, were treated with 32 to 70 mW per site for a dose of 27 to 39 J/cm^2. One day later, the area of tumor around the ribs was black, and the rest of the skin including the incisions, was red and painful. One week after treatment, a black, gangrenous area extended 5 cm laterally to the original ulcer on the ribs. Three weeks after treatment, the wound was granulating, and the rest of the skin returned to its pretreatment status. Seven weeks after treatment, the granulation was healing, but a biopsy of the medial angle was positive. Ten weeks after the original treatment, the patient was retreated at five adjacent sites over the ulcer and skin. Thirty-two to 40 mW were delivered for a dose of 25 J/cm^2 to each area. One day later, the adjacent skin was blistered, and the ulcerated area had blackened. Ten days after the retreatment, the adjacent skin had returned to its pretreatment status. Yellow necrotic material was noted in the ulcer.

Patient #26: A 54-year-old white woman was seen with extensive metastatic nodules in the anterior chest wall and side from inflammatory cancer of the left breast. She previously had undergone X-ray therapy. The right breast was reddened with a peau d'orange effect resembling inflammatory cancer. Three, five and six days after HpD and three days after mutliple excisions were performed with the CO_2 laser, 14 adjacent sites were treated. Twenty-four to 32 mW/cm^2 for a dose of 20, 25 and 30 J/cm^2 were delivered to the remaining 12 sites. One day later, erythematous areas over the right breast were white and other nodules were black and edematous. There was no increased reaction at the incision sites. Five days later, the white areas had become black and gangrenous. The inflammatory areas reacted much more violently. Two weeks after treatment, new nodules were present outside the treated fields as well as in previously treated sites where treated nodules had become necrotic.

Patient #27: A 73-year-old white woman was seen with metastatic nodular cutaneous and subcutaneous breast cancer. Three days after injection of HpD, nine sites were treated. Site 1, measuring 8x5 cm, was treated with 100 mW/cm^2 for 4.4 minutes for 25 J/cm^2. Site 2, measuring 8x4 cm, was treated with 35 mW/cm^2 for 12 minutes for 25 J/cm^2. Site 3, a 2 cm nodule, was treated with 300 mW delivered interstitially and simultaneously through each of four needles for 20 minutes for 360 J/F. Site 4, a 1.5 cm nodule, was treated with 250 mW delivered interstitially and simultaneously through each of four needles for 20 minutes for 300 J/F. Site 5, measuring 12x8 cm, was treated with 35 mW/cm^2 for 12 minutes for 25 J/cm^2. Site 6, measuring 10x8 cm, was treated with 36 mW/cm^2 for 25 J/cm^2. Site 7, measuring 8x9 cm, was treated with 75 mW/cm^2 for six minutes for 27 J/cm^2. Site 8, measuring 8x9 cm, was treated with 30 mW/cm^2 for 14 minutes for 25 J/cm^2. Site 9, measuring 8x10 cm, was treated with 70 mW/cm^2 for six minutes for 25 J/cm^2. One day after treatment, at both sites 1 and 2, the tumor was edematous and well demarcated, and the adjacent skin was reddened. There was no observable reaction at sites 3 and 4. Six days after treatment, the tumor areas in sites 6,7, and 8 were blackened, and the adjacent skin was red and painful. Five weeks after treatment all of the treated areas were sloughing and necrotic, while the adjacent skin was normal. The lesions in sites 3 and 4 were reduced in size by one half. There was no ulceration and no new nodules outside the treatment fields.

Head and Neck Tumors

Patient #28: A 54-year-old white man with a 5 cm metastatic squamous cell tumor of the larynx to the left post auricular area was experiencing constant pain over the post auricular and temporal areas. Three days after injection of HpD, 200 mW were delivered interstitially and simultaneously through each of four needles for 10 minutes at three depths for a does of 120 J/F. One day after treatment, marked swelling at the angle of the jaw was noted. Two days after treatment, there was no pain, and the swelling was subsiding. At one month after treatment, the patient still reported no pain, but a 1 cm nodule was present on the side of the neck.

Patient #29: A 58-year-old white man was treated for a 1 cm squamous cell tumor of the soft palate. Three days after injection of HpD, 87 mW were delivered for eight minutes for a dose of 42 J/cm^2. One day after treatment, the treated area was blackened. At one week after treatment, yellow slough was present. At five weeks, the treated area appeared normal both by palpation and observation. Two months after treatment, the area remains normal.

Patient #30: A 62-year-old white man was seen with massive metastatic tonsil cancer to the neck, requiring the constant administration of intravenous narcotics. Three days after injection of HpD, 300 mW were delivered interstitially and simultaneously through each of four needles at two depths for 20 minutes for 860 J/F. Six hours later, extensive edema of the angle of the jaw developed, which subsided over the next six days. The patient subsequently required a frontal lobotomy for pain.

SUMMARY OF TREATMENT DATA: ENDOSCOPIC TUMORS

PATIENT	AGE	Dx/SITE	NO. TREATED AREAS	DAY AFTER HpD	mW	TYPE FIBER	DURATION (MIN)
1	83	melanoma/ esophageal	3	3	250	cylinder	10 ea
2	65	adenocarcinoma/ stomach	3	3	300	cylinder	10 ea
					(2 months later)		
			4	3	225	cylinder	10 ea
					(3 months later)		
			1	3	1000	cylinder	10
3	61	squamous cell/ esophagus	2	3	300	cylinder	10 ea
			1	3	300	cylinder	5
					(10 weeks later)		
				3	300	cylinder	10 ea
4	54	squamous cell/ cervical esophagus	3	3	200	cylinder	10 ea
					(2 weeks later)		
			5	3	200	cylinder	10 ea
5	67	squamous cell/ esophagus	5	3	300	cylinder	10 ea
6	75	adenocarcinoma/ stomach,esophagus	2	3	300	cylinder	10 ea
			3	3	1000	straight	10 ea
			3	7	1000	straight	10 ea
7	58	adenocarcinoma/ stomach,esophagus	3	3	900	straight	10 ea
8	70	squamous cell/ esophagus	1	3	1000	straight	10
			1		1000	straight	13
9	57	squamous cell/ esophagus	3	3	1000	straight	10 ea

SUMMARY OF TREATMENT DATA: ENDOSCOPIC TUMORS

PATIENT	AGE	Dx/SITE	NO. TREATED AREAS	DAY AFTER HpD	mW	TYPE FIBER	DURATION (MIN)
10	70	adenocarcinoma/	2	3	200	cylinder	10 ea
		stomach/esophagus	1		1000	straight	10
11	60	adenocarcinoma/		3	300	straight	29
		lung			300	straight	10
					300	straight	20
12	45	cylindroma/	1	3	300	straight	10
		trachea,bronchi	2	3	300	straight	10 ea
					(5	weeks later)	
			3	3	150	cylinder	10 ea
					(2	months later)	
			3	3	200	cylinder	10 ea
13	66	squamous cell/		3	1000	straight	20
		bronchi			(1	day later)	
				4	1000	straight	20
14	54	squamous cell/ lung	3	3	250	straight	10 ea
15	60	squamous cell/	1	3	300	straight	10
		trachea			150	cylinder	
16	70	squamous cell/ lung	1	3	300	cylinder	30
17	35	adenocarcinoma/ lung	1	3	300	cylinder	30
18	50	adenocarcinoma/	1	3	300	straight	10
		colon; mets to lung	1		300	straight	10

SUMMARY OF TREATMENT DATA: SURFACE TUMORS

PATIENT	AGE	Dx/SITE	NO. AREAS TREATED	SIZE LESION	DAY AFTER HpD	POWER	DURATION (MIN)	DOSAGE
19	56	basal cell/ head	1	9 cm	3	30mW/cm^2	30	54J/cm^2
20	79	squamous cell/ eyelid	1	2x1.1x.6cm	3	50mW/cm^2	8.5	25J/cm^2
21	89	basal cell/ face	2	3.2cm	3	17mW/cm^2	25	25J/cm^2
				3x4 cm		55,60,40 50 mW/cm^2		20J/cm^2
22	86	recurrent basal cell/ face	4	3x5,4x2cm chin	3	7.5mW/cm^2	29	13 J/cm^2
						140mW/cm^2	3.5	30 J/cm^2
				5cm;jaw		300mW	14	270J/F*
						50mW/cm^2	9	25J/cm^2
						400mW	13	312J/F*
						200mW/cm^2	4	48J/cm^2
				20mm;neck		400mW	20	480J/F*
				14mm;neck		400mW	20	480J/F*
23	80	liposarcoma/	3	6x7cm	3	32mW/cm^2	15	29J/cm^2
						28mW/cm^2	18	30J/cm^2
						30mW/cm^2	22	40J/cm^2
24	60	cutaneous, subcutaneous/ breast	6	6x7cm	3	21-36mW/cm^2		25J/cm^2
25	58	cutaneous, subcutaneous/ breast	8	5x6-9x4cm	3	32-70mW/cm^2		27-39J/cm^2
				(10	weeks	later)		
					3	32-40mW/cm^2		25J/cm^2

SUMMARY OF TREATMENT DATA: SURFACE TUMORS

PATIENT	AGE	Dx/SITE	NO. AREAS TREATED	SIZE LESION	DAY AFTER HpD	POWER	DURATION (MIN)	DOSAGE
26	54	mets chest wall/breast ca	14		3,5,6	24-32mW/cm^2		20,25,30 J/cm^2
27	73	cutaneous, subcutaneous breast	9	8x5cm	3	100mW/cm^2	4.4	25J/cm^2
				8x4cm	3	35mW/cm^2	12	25J/cm^2
				2cm	3	300mW	20	360J/F*
				1.5cm	3	250mW	20	300J/F*
				12x8cm	3	35mW/cm^2	12	25J/cm^2
				10x8cm	3	36mW/cm^2	11	25J/cm^2
				8x9cm	3	75mW/cm^2	6	27J/cm^2
				8x9cm	3	30mW/cm^2	14	25J/cm^2
				8x10cm	3	70mW/cm^2	6	25J/cm^2
28	54	squamous cell/ larynx	1	5cm	3	200mW	10	120J/F
29	58	squamous cell/ soft palate	1	1cm	3	87mW/cm^2	8	42J/cm^2
30	62	metastatic tonsil ca to neck	1		3	300mW	20	360J/F

*J/F=Joules per fiber inserted interstitially.

Porphyrin Localization and Treatment of Tumors, pages 829–840

PHOTORADIATION THERAPY FOR CUTANEOUS AND SUBCUTANEOUS MALIGNANT TUMORS USING HEMATOPORPHYRIN

L. Tomio, F. Calzavara, P.L. Zorat,
L. Corti, C. Polico, E. Reddi, G. Jori,
G. Mandoliti

Divisione di Radioterapia,
Civil Hospital of Padova

Istituto Biologia Animale,
University of Padova, Italy

Recently, increasing attention has been paid to the potential use of hematoporphyrin (Hp) and its derivative (HpD) in combination with red light for the treatment of cancer (Dougherty et al. 1978).

Hp is known to be a powerful photosensitizing agent of several biological substrates (biomolecules, cells, tissues) to visible light (Fowlks 1959; Spikes 1975). Hp absorbs essentially all wavelengths of the visible spectrum.

Moreover, Hp accumulates in larger amounts and for longer times in malignant tissues than in normal tissues (Tsutsui et al. 1975). In the first systematic study of this phenemenon, it was demonstrated that the ability of Hp to be taken up and retained by malignant tissues is a general property: various kinds of induced and transplanted tumors in mice possess a red fluorescence after administration of Hp, while the other tissues, with the exception of lymph nodes, traumatized tissues and embryonic tissues, show no fluorescence (Figge et al. 1948). On these bases Hp was used for diagnostic purposes to recognize and circumscribe neoplastic lesions in humans.

Eleven patients with various types of cancer received 300-1000 mg Hp intravenously: within a few hours all patients (with the exception of one prostatic cancer) showed the typical red fluorescence in correspondence with the tumor lesions (Rassmussen et al. 1955). No fluorescence was detectable in the surrounding normal tissues. Subsequent quantitative studies on Hp distribution in model animals confirmed the preferential accumulation of Hp in the tumor area; the porphyrin was absent in control tissue, such as muscle, at 24 h after administration of Hp doses as high as 80 mg/kg (Winkelman, Rassmussen-Taxdal 1960).

The interest in the tumor-localizing properties of Hp decreased during the subsequent ten-year period in spite of the prospects opened for tumor photodiagnosis by the preparation of HpD; the synthesis of the latter derivative is performed by treating Hp with an acetic acid-sulfuric acid (Lipson, Baldes 1960; Lipson et al. 1961; Gregorie et al. 1968).

The cytotoxic action of the Hp-red light combination had been anticipated in the forties (Figge, Weiland 1949); however the therapeutic benefits of this technique have been exploited only during the last few years as a consequence of the ascertained lethality of the Hp-red light treatment for cultured glioma cells and subcutaneously transplanted gliomas to rats (Diamond et al. 1972; Granelli et al. 1975). At present, however, relatively little information is available as regards the mechanisms governing Hp accumulation in tumors. As a contribution to the elucidation of the above outlined problems, we studied the time-dependence of Hp distribution in Wistar rats, both normal or bearing AH-130 Yoshida hepatoma, as well as in C57Bl/6 mice affected by MBL-2 lymphoma. Different doses of Hp have been administered either i.p. or i.v. and the porphyrin content of selected tissues was estimated spectrophotofluorimetrically. In general, out of the normal tissues examined by us (liver, spleen, lung, kidneys, skin and muscle) the highest Hp levels were observed in liver, although significant amounts of Hp were recovered from all tissues at 1-3 h after injection. In any case, significantly higher Hp concentrations were found in the tumor; the porphyrin uptake was maximal at 1-3 h and remained essentially constant up to 9-12 h. When most Hp had been eliminated from normal tissues, still large Hp amounts persisted in tumor cells at concentrations undergoing only a minor decrease up to 72 h. This behaviour is clearly demonstrated by the time-dependence of Hp concentration in liver and tumor (table 1). Hp accumulated in the liver is eliminated via a bile-involving pathway with no apparent chemical modification, as suggested by Hp determination in rat feces (Tomio et al. 1982). Typical Hp recovery data are given in table 2. The subcellular localization of Hp in hepatocytes and tumor has been studied by Cozzani et al. (1981), suggesting that high-affinity Hp receptors are located in the plasma membrane.

Table 1
Tumor-to-liver ratio of Hp concentration after i.p. injection of 10 mg/kg (Wistar rats) or 2.5 mg/kg (C57 mice) of porphyrin.

Time after injection (h)	[a] Hepatoma AH-130 Solid Type	Ascites Type	[b] Lymphoma MBL-2
1	0.89	0.45	4.00
3	2.11	0.57	----
24	3.57	7.92	5.46
48	4.90	----	----
72	4.07	6.00	6.59

[a] Wistar rats.
[b] C57 mice.

MATERIALS AND METHODS

Drug

Hp was obtained from Porphyrin Products (Logan, Utah, USA) and appeared to be about 97% pure by HPLC. An injectable solution of Hp was prepared by mixing 1 part of Hp with 50 parts by volume of 0.1 M NaOH and stirring at room temperature for 1 h. The pH was brought to 7.2-7.4 by addition of 0.1 M HCl (approximately 15 parts), made isotonic by addition of NaCl, and finally brought to 200 parts total with 0.9% NaCl. The final solution containing 5 mg/ml Hp was sterilized by Millipore filtration and tested for sterility and pirogenicity (Monico, Venezia, Italy). The solution was stored in the dark at -18° until used.

Table 2[a]
Recovery of Hp from feces of normal and tumor-bearing rats at selected times after i.p. or i.v. injection.

Type of rat	Mode of injection	Hp recovered (% of totally administered Hp)	
		24 h	84 h
Normal	i.p.	70.0	81.9
	i.v.	60.0	89.5
Tumor-bearing (ascites AH-130)	i.p.	11.5	58.2
	i.v.	7.7	55.7
Tumor-bearing (solid AH-130)	i.v.	22.0	54.7

[a] from Tomio et al. (1982).

Light sources

The light source was an Osram 4,000 W Xenon arc-lamp endowed with continuous emission in the UV and visible spectral region. Light in the red region of the spectrum (590-690 nm) was isolated by inserting a chemical filter and a heat-reflecting glass filter in the light beam. The beam was focused by a parabolic reflector and a condensing lens placed between the filter and the patient. This system produced a circular light spot having a diameter of 5 cm. The intensity was 25 mW/cm^2 at the level of the patient. In some cases, a He/Ne laser source (Valfivre, Firenze, Italy) was used. The laser produced a monochromatic red beam of light (wavelength, 632.8 nm) which was focused into a 1100-μm quartz optical fiber by means of an alignment device. Output power at the end of the optical fiber was 25 mW. Coupling five sources the intensity can be brought to a maximum of 125 mW.

Procedure

Twelve patients with several types of cutaneous and subcutaneous malignant tumors were included in the clinical study. In many cases these were recurrent or metastatic lesions which had not responded to conventional modes of therapy. There were 5 patients with basal cell carcinoma (recurrent in 3 cases), 3 patients with metastatic breast carcinoma on the chest wall, 2 patients with malignant melanoma and 1 patient each with squamous cell carcinoma of skin and Kaposi's sarcoma (table 3). All patients had biopsy-proven malignancy. Fully informed consent for phototreatment was obtained in all cases.

A total number of 21 lesions were treated. The lesions can be subdivided into two groups. One group (table 4) included single or multiple lesions, which were often exophytic or infiltrating and had an external diameter larger than 2 cm, as well as areas containing multiple lesions

Table 3
Histology of tumors treated by Hp-red light combination

Histology	Number of patients
Basal cell carcinoma	5
Metastatic breast carcinoma	3
Melanoma pigmented	1
Melanoma unpigmented	1
Squamous cell carcinoma	1
Kaposi's sarcoma	1
TOTAL	12

Table 4
Details of first group of lesions.

Histology	Number of lesions	Dimension (cm)	Light Source	Number of treatment courses
Malignant melanoma	3	2-6	Xenon	1
Kaposi's sarcoma	2	2-3	Xenon	2
Squamous cell carcinoma	1	6	Xenon	2
Breast carcinoma	7	Ø.5-5	Xenon Laser He/Ne	1 1

of small diameter. The second group (table 5) included single superficial lesions with an approximate diameter of 1 cm.

Table 5
Details of second group of lesions.

Histology	Number of lesions	Dimension (cm)	Light Source	Number of treatment courses
Basal cell carcinoma	7	1	He/Ne laser	1[a]
Kaposi's sarcoma	1	0.5	He/Ne laser	1

[a] the phototreatment was repeated in 1 case.

Phototherapy with Xenon arc-lamp was performed only for lesions of the former group: the treatment with red light was given twice for 60 min. at 24 h and 48 h after i.v. administration of 5 mg/kg Hp. The total light dose delivered to the patients was about 20 J/cm^2 in each session. In three cases, the patients were given a second dose of Hp at 15-20 days after the first phototreatment and reexposed to light according to the above outlined protocol.

On the other hand laser phototreatment was applied for all lesions in the second group, as well as for a few lesions in the first group. In this case, the total light dose delivered ranged between 30 and 70 J/cm^2 depending on the dimensions of the lesion and the distance between the lesion and the end of the optical fiber.

The tumor response was assessed by following the variations in the size of the tumor and/or by clinical photographs. We report our results in the following way: complete and partial responses, regression, stable and progression. In addition to clinical examination, complete blood tests (liver, renal function), chest roentgenogram and ECG were performed in all patients. Before treatment and during follow-up period, blood and urine samples were taken for Hp analysis.

RESULTS

All twelve patients completed at least one course of phototreatment. The results hitherto obtained are shown in table 6. Out of the 7 lesions which had dimensions larger than 2 cm and were treated with Xenon lamp, two gave a complete response and one underwent regression; no modification was observed upon treatment of two lesions from pigmented malignant melanoma and one large metastatic ulcera from breast carcinoma. A promising response was obtained in one recently treated lesion from amelanotic malignant melanoma. Moreover, no response was given by the three areas containing multiple lesions.

The phototreatment with He/Ne laser was unsuccessful in the case of single lesions from breast cancer, while a complete or partial response was noticed in all the superficial lesions from basal cell carcinoma or Kaposi's sarcoma.

In exophytic or small superficial lesions responses were apparent as early as 24 h after phototreatment, usually leading to the onset of vesicles, erythema and edema. Within one week after phototreatment, the lesions became dark brown; in parallel we noticed the formation of a crust with a hard consistency. On the other hand, no early response could be detected in the case of infiltrating lesions. In general, a complete response was observed within 4-6 weeks after phototreatment of the neoplastic lesions. Generalized cutaneous photosensitivity is the major side effect in phototherapy. Two patients experienced moderate erythema and edema of the face and hands, occurring within the first ten days after Hp injection and clearing within a few days.

Table 6
Conditions and results of Hp + red light-treated patients.

Pts	Histology	Number of lesions	Dimension (cm)	Light Source	Response
1	Squamous cell carcinoma	1	6	Xenon	R
2	Kaposi's sarcoma	3	0.5-3	Xenon He/Ne	CR
3	Malignant melanoma	2	2	Xenon	S
4	Malignant melanoma	1	6	Xenon	NV
5	Basal cell carcinoma	2	1	He/Ne	CR
6	Basal cell carcinoma	2	1	He/Ne	PR
7	Basal cell carcinoma	1	1.5	He/Ne	R
8	Basal cell carcinoma	1	0.5	He/Ne	CR
9	Basal cell carcinoma	1	1	He/Ne	CR
10	Breast carcinoma	2	0.5 (mult.)	Xenon	S
11	Breast carcinoma	3	2-5	Xenon He/Ne	S
12	Breast carcinoma	2	1-2	Xenon He/Ne	S

CR = complete response; PR = partial response; R = regression; S = stable; NV = no value.

The analysis of Hp concentration in serum and urine as a function of time showed that Hp is eliminated by a biphasic process: an initial rapid stage was observed up to about 48 h from i.v. administration and was followed by a remarkably slower elimination phase which led to a nearly stationary value of Hp concentration, at a substantially higher level than that typical of endogenous porphyrins. Moreover, our findings confirmed the previous conclusions drawn by experiments with normal and tumor-bearing rats showing that the main route for Hp elimination occurs via biliary tract. Actually, the urinary concentration of Hp never exceeded 1% of the fecal concentration.

DISCUSSION

Our results demonstrate that the Hp-red light combination can be effective in destroying malignant tumors. Furthermore, the responses depend on tumor histology, morphology and mainly on total light doses delivered. In fact, while a 100% favorable response rate was observed in the treatment of basal cell carcinoma and Kaposi's sarcoma with He/Ne laser when the total light doses ranged between 30 and 70 J/cm^2, no response was obtained in small lesions from breast cancer treated with Xenon lamp, at a dose of 20 J/cm^2. Perhaps, this lack of response may be related to prior high doses of radiation therapy in the PRT field.

Our initial data suggest that Hp phototherapy is clearly more effective in superficial small lesions while in selected cases it can be successfully used for large exophytic tumors.

In any case further studies are required in order to assess more complete data on the pharmacokinetics of Hp in humans, the optimal Hp dose, the total light dose delivered and the most convenient interval between Hp administration and light exposure of the patients.

Acknowledgements

This work was supported by Consiglio Nazionale delle Ricerche under the Progetto Finalizzato "Controllo della Crescita Neoplastica", contract n° 82.00252.96.

REFERENCES

Cozzani I, Jori G, Reddi E, Fortunato A, Granati B, Felice M, Tomio L, Zorat PL (1981). Distribution of endogenous and injected porphyrins at the subcellular level in rat hepatocytes and in ascites hepatoma. Chem Biol Interactions 37:67.

Diamond I, Granelli SG, McDonagh AF, Nielsen S, Wilson CB, Jaenicke R (1972). Photodynamic therapy of malignant tumors. Lancet 2:1175.

Dougherty TJ, Kaufman JE, Goldfarb A, Weishaupt KR, Boyle D, Mittleman A (1978). Photoradiation therapy for the treatment of malignant tumors. Cancer Res. 38:2628.

Figge FHJ, Weiland GS, Manganiello LOJ (1948). Cancer detection and therapy. Affinity of neoplastic, embryonic and traumatized regenerating tissues for porphyrins and metalloporphyrins. Proc Soc Exper Biol Med 68:640.

Figge FHJ, Weiland GS (1949). Studies on cancer detection and therapy: the affinity of neoplastic, embryonic and traumatized tissues for porphyrins and metalloporphyrins. Cancer Res 9:549.

Fowlks WL (1959). The mechanisms of the photodynamic effect. J Invest Dermatol 32:233.

Granelli SG, Diamond I, McDonagh AF, Wilson CB, Nielsen SL (1975). Photochemotherapy of glioma cells by visible light and hematoporphyrin. Cancer Res 35:2567.

Gregorie HB, Horger EO, Ward JL, Green JF, Richards T, Robertson HC, Stevenson TB (1968). Hematoporphyrin derivative fluorescence in malignant neoplasms. Ann Surg 167:820.

Lipson RL, Blodes EJ (1960) The photodynamic properties of a particular hematoporphyrin derivative. Arch Dermatol 82:508.

Lipson RL, Boldes EJ, Olsen A (1961). The use of a derivative of hematoporphyrin in tumor detection. J Natl Cancer Inst 26:1.
Rassmussen-Taxdal DS, Ward GE, Figge FHJ (1955). Fluorescence of human lymphatic and cancer tissues following high doses of hematoporphyrin. Cancer 8:78.
Spikes JD (1975). Porphyrins and related compounds as photodynamic sensitizers. Ann N Y Acad Sci 244:496.
Tomio L, Zorat PL, Jori G, Reddi E, Salvato B, Corti L, Calzavara F (1982). Elimination pathway of hematoporphyrin from normal and tumor-bearing rats. Tumori 68:283.
Tsutsui M, Carrano C, Tsutsui EA (1975). Tumor localizers: porphyrins and related compounds (unusual metalloporphyrins XXIII). Ann N Y Acad Sci 244:674.
Winkelman J, Rassmussen-Taxdal DS (1960). Quantitative determination of porphyrins uptake by tumor tissue following parenteral administration. Bull John Hopkins Hosp 107:228.

Porphyrin Localization and Treatment of Tumors, pages 841–845

A TEACHING PROGRAM FOR ONCOLOGY PATIENTS CONSIDERING AND RECEIVING HEMATOPORPHYRIN SENSITIZATION AND PHOTORADIATION

Elizabeth Z. Olson, R.N., M.S.

Grant Hospital Nursing Service
111 South Grant Ave.
Columbus, Ohio 43215

Written information in lay language on this investigational treatment does not exist. Information-seeking is a healthy coping mechanism and an appropriate step in the decision-making process. However, the necessity of trying to absorb totally new information, frequently presented verbally under time limitations, can create additional anxiety for a patient already under the threat of disease and possibly previous treatment failure. This anxiety can render the decision to accept investigational treatment even more difficult. Utilizing adult learning principles and a combination of audio and visual teaching methods, a three-part instructional program was developed for persons considering and/or receiving hematoporphyrin sensitization and photoradiation. The program can be initiated in the physician's office or elsewhere when the possibility of treatment is first suggested; it continues through self-care instruction after discharge from the hospital. Information from the printed portion of the program has also been useful to nurses in other institutions and in the community in providing follow-up care for these patients.

Why develop a teaching program? Patients have a right to be informed. They must make serious decisions for or against treatment and must have complete, accurate information in order to make those decisions truly "informed". The fact that investigational treatment is not covered by third party payment adds a personal financial element to cost-benefit decisions. The trend toward shorter hospital stays and increased out-patient care intensifies the need for self-care instruction for both patients and family members. Facing the unknown increases a sense of anxiety, loss of control, and helplessness for many people. Lawsuits have been instituted regarding a lack of information or instruction in self-care which resulted in harm to the patient. And finally, the Joint Commission on Accreditation of Hospitals requires patient education and documentation thereof.

Who else needs to be informed? The patient does not exist in a vacuum; he or she is a central figure in a constellation of family and friends who constitute his support system. The less able a patient is to care for himself, the more these people will need that information. Nursing and allied medical staff also need education: first, in order to provide knowledgeable care, and second, in order to appropriately reinforce patient and family understanding.

It is worth noting that the general public is unable to comprehend health-care information written above a sixth-grade reading level. The "fog" index, developed by Robert Gunning, can be used to determine the reading level of a given article. This is a useful tool in evaluating or developing information for patient education. A mild degree of anxiety stimulates motivation to learn; however a high degree of anxiety increases difficulty in absorbing, integrating and retaining information. Patients who are experiencing recurrence of their cancer, who have failed to respond to other forms of treatment, and who now face investigational treatment are likely to be extremely anxious. Therefore, patient education information, in order to be effective, must be relatively simple, presented in lay terms, must be somewhat repetitive, and is best presented through a variety of media.

With these constraints in mind, Grant Hospital supported the development of a teaching program for patients receiving hematoporphyrin sensitization and photoradiation treatment for cancer. There are three components to the program: pamphlets explaining the treatment itself and providing self-care instruction, a slide-tape reviewing the same information, and individualized verbal discussion and instruction.

Requirements for using the program include a knowledgeable nursing staff; a portable self-contained slide-tape projector with screen; software including the literature, slides and tape; secure, convenient storage space; and viewing space free from distraction. During discussion or verbal instruction by the physician or nurse, patients and families should be made as comfortable as possible. Naturally, questions are invited and dealt with honestly.

The program is initiated upon first contact in the hospital or the physician's office, as soon as the possibility of photoradiation is considered. It is helpful for the physician to offer a brief, simple explanation of the treatment and why it should be considered. The patient may wish to take the pamphlets home and discuss the decision with family members. The slide-tape program may be used during the initial visit or not; however when the patient is admitted to the hospital, it should be viewed the evening of admission, if possible. This may raise further questions which can be answered by the nursing staff or referred, if necessary, to the physician. Because individual diagnoses and prognoses vary considerably, the opportunity for individual discussion and instruction is essential in providing complete, accurate information.

Prior to discharge from the hospital, instructions for self-care should be thoroughly reviewed, again individually, with information about this particular patient's activities and physical surroundings taken into account, primarily in regard to protection from sunlight. Some questions to be considered might be: Is this patient fair-skinned or is there a history of sensitivity to sunlight? Are there any areas of circulatory difficulty, such as pedal edema or post-mastectomy lymphedema, which might inhibit normal elimination of hematoporphyrin and require special protection? Must this patient participate in activities which will require exposure to sunlight? If so, for how long, and how can the exposure be minimized? How much light exposure is afforded in the house? Can this be minimized satisfactorily? It is extremely helpful to maintain communication with the staff in the physician's office for feed-back on any difficulty patients are encountering. Preventive teaching can then be included for subsequent patients.

Advantages of this program include the provision of information in lay terms at an appropriate general level of understanding. Patients with higher levels of understanding can be given supplementary, somewhat more technical, information if they desire. The program also provides opportunity for repetition of information: use of a variety of sensory stimuli increases comprehension and retention. The use of standardized materials also provides complete, consistent core information on the topic which is used as a foundation for the individualized discussion and instruction. Use of such an instructional package tends to conserve physician and nursing time, permitting more opportunity to discuss application of the information to the individual. And finally, the flexibility of such a program facilitates up-dating and revision of the contents when necessary, particularly in the slide-tape program as opposed to re-printing booklets or re-filming video-tapes.

Thus far, the two noteworthy disadvantages of the program have been the transport of heavy "portable" equipment and the provision of opportunity for every patient to participate. Storing the slide-tape projector on a rolling cart, with space for secure storage for software below, may eliminate this problem. As increasing numbers of patients are admitted for treatment and it becomes correspondingly difficult to view the slides individually, we may institute group sessions for viewing and discussion. Such a format also has advantages in stimulating questions and learning from others.

In order for any patient-education program to be successful, teaching must be viewed primarily as a responsibility of the staff nurses. No one person can be sufficiently available to provide all the necessary instruction and answer all the questions which are asked twenty-four hours a day, seven days a week. Nurses on every shift must be knowledgeable about the treatment itself, the nursing care required, and pertinent self-care measures; they must be able to provide their patients with consistent, accurate information. They must have opportunity to learn themselves, and must be provided with up-to-date information as it accumulates. We have found that observation of photoradiation treatment in progress is a helpful experience for the nursing staff.

For many reasons, then, it is useful to have one nurse designated as a major resource person but not as the sole teacher: to act as liaison in relaying information as it is developed from the literature, from other institutions and from our own clinical experience. This nurse may have responsibility for developing the teaching materials, for conducting inservices, for leading group patient-teaching sessions, and for contributing to the development of related nursing policies and procedures.

In summary, patients receiving any treatment, but particularly those receiving hematoporphyrin sensitization and photoradiation as an investigational treatment, need and are entitled to adequate information in terms they can comprehend. This responsibility can be time-consuming for the health professionals involved. The three-part teaching program developed at Grant Hospital is proving to be an effective and efficient approach to meeting these needs.

REFERENCES

George, G (1982). If Patient Teaching Tries Your Patience, Try This Plan. Nursing 82 12:5.

Kunkel, J (1981). Patient Teaching: Getting Your Message Through. Nursing Life 1:36.

Taylor, P (1982). Patient Teaching: Keys to More Success More Often. Nursing Life 2:25.

Redman, B (1980). "The Process of Patient Teaching in Nursing", 4th edition. St. Louis: CV Mosby.

Porphyrin Localization and Treatment of Tumors, pages 847–861

IMAGING FLUORESCENCE BRONCHOSCOPY FOR LOCALIZING EARLY BRONCHIAL CANCER AND CARCINOMA IN SITU

Oscar J. Balchum, M.D., Ph.D.; A. Edward Profio, Ph.D.; Daniel R. Doiron, Ph.D.; and Gerald C. Huth, Ph.D.

Los Angeles County-University of Southern California Medical Center, Los Angeles, California 90033

Abstract

A system of imaging fluorescence bronchoscopy instrumentation and methods has been devised that has succeeded in localizing very small (1 x 2 mm) areas of bronchial mucosal cancer, in individuals with radiologically occult lung cancer (positive sputum cytology for malignant cells and a negative chest X-ray). These areas were located solely by their fluorescence, and were visibly normal on white-light examination. The detection of lung cancer in individuals with radiologically occult lung cancer depends upon adequate methods of sputum collection and processing. Proving that fluorescing areas show "true positive" fluorescence depends upon accurate brush and forceps biopsies, providing adequate cytological and biopsy material. The entire system of the diagnosis and localization of early or pre-invasive lung cancer (while still confined to a bronchus) rests on skilled cytopathology methods and interpretation, not only skilled fluorescence bronchoscopic examination and adequate

Supported in part by a grant from the National Cancer Institute, CA25582-04 and the Division of Biological Environmental Research, Department of Energy, EY-76-5-03-0013 with the Institute of Physics and Imaging Science, USC, and by a Grant (GCRC-RR-43) from the General Clinical Research Centers, Programs of the Division of Research Resources, National Institutes of Health.

instrumentation.

Introduction

The detection of early lung cancer, when it is still localized or confined to a bronchus when the chest X-ray is negative, results in a 90 to 100% cure rate by surgical resection. Twenty-five patients with early lung cancer reported by Martini (1983) had a 100% 20-year survival. The high cure rate of early, pre-invasive lung cancer is well-documented (Martini 1980; Melamed et al 1981). It is clear that survival depends upon the extent of disease at the time of initial diagnosis, and not upon the histologic type (Williams et al 1981).

Detection of lung cancer by chest X-ray or respiratory symptoms usually occurs too late, since in the majority of instances the extent of the disease will then be too far advanced for curative resection.

Every one of the 120,000 new cases of lung cancer each year passes through this early or pre-invasive phase, averaging 5 years, during which malignant cells are exfoliated into the sputum. Individuals are asymptomatic and have a negative chest X-ray (radiologically occult lung cancer). For early diagnosis, individuals at high risk for lung cancer should be tested by sputum cytology for malignant cells. "High risk" means 15-20 years or more of cigarette smoking (regardless of age, sex, or the number of cigarettes smoked per day). Risk is proportional to the fifth power of smoking duration but only to the first power of the number of cigarettes per day. Other risk factors are certain known occupational carcinogens: asbestos; coke, asphalt, and coal tar; radon (uranium miners), and certain chemicals (in plastics, chemical manufacturing) and metals (chromates, nickel). Men and women at high risk for lung cancer should have sputum cytology regardless of the reason or illness that brings them to a physician.

Sputum, to maximally collect exfoliated cells, must be produced by deep coughs and collected directly into a preservative (50% alcohol-2% Carbowax). A sensitive method of collection is an "integrated" sputum sample. Collection is started from the time of waking up, over the several morning hours, and for three, four, or five days or more in the same jar of preservative. Ultrasonic aerosol-induced sputum collected by deep coughs during the breathing of high-density nebulized water mist for 30-40 minutes is an equally sensitive method. The problem with "routine" sputum cytology is that most samples consist of saliva from the oropharynx, and not sputum from the tracheo-bronchial tree and lungs (Murray, Washington 1975).

Then, in the cytopathology laboratory, methods must be used that adequately retrieve the cells from each sample. Routinely smearing only two or three particles of mucoid material is not good enough. The sputum sample should be blended, centrifuged, and all of the sediment smeared and stained, to give the cytopathologist enough cells for accurate cytological interpretation (Saccomanno 1978). Even then, there is significant variability amongst cytopathologists in their interpretation of the stage of progression of metaplasia. Preferably therefore, there should be double-reading.

Progression of metaplasia over a number of years to finally frank or invasive cancer has been well-documented by Saccomanno (1978), for the several histologic types of lung cancer. Moderate atypia is a premalignant phase; in some individuals it will progress. Several additional monthly sputum samples should be obtained to confirm a shift to or beyond this stage. When marked atypia, carcinoma in situ (CIS), or cancer cells are found on sputum cytology, then bronchoscopy is indicated to localize the site of cancer, which can be in the oropharynx, or larynx or tracheo-bronchial tree, and in one or more bronchi.

Areas of proliferating marked metaplasia, CIS, or cancer cells are difficult to see, so that routine fiberoptic bronchoscopy is most often negative. Even extensive multiple bronchial brushings and blind biopsies of bronchial bifurcations frequently do not locate the site. And there may be one to several such sites.

This problem of localization has led to the development of a new approach, that is, a more sensitive bronchoscopic method employing hematoporphyrin derivative injected intravenously with bronchoscopy carried out using violet light. Small patches or areas of marked metaplasia, CIS, or cancer fluoresce, so that they can be better located for brush and forceps biopsies. Hematoporphyrin derivative concentrates in and is retained in these bronchial cancer areas for a week or more, and by then is largely washed out from normal tissues. The basis of this method was established in the 1960's by Lipson et al (1964), employing the rigid open-tube bronchoscope and a mercury arc source of violet light. With the development of the fiberoptic bronchoscope, renewed interest has led to the development of a photoelectric fluorescence energy detector that generates a sound signal when the emission of fluorescence is above background and within the visual field of the bronchoscope (Cortese et al 1979). In our hands, an imaging method of visualizing the fluorescence emitted from small areas of bronchial CIS or cancer has been developed (Profio, Doiron 1979; Profio et al 1983) that is highly sensitive in detecting small areas of bronchial carcinoma.

We have previously reported the initial results of imaging fluorescence bronchoscopy (Balchum et al 1982) which demonstrated its high sensitivity in localizing bronchial carcinoma. This report demonstrates that very small (1 x 2 mm) mucosal cancer lesions can be located by their fluorescence even when invisible to the eye under white light examination.

Methods

The hematoporphyrin derivative (HpD) employed was initially obtained from Dr. Thomas J. Dougherty, Roswell Park Memorial Institute, or later as Photofrin I from the Oncology Research & Development Co., Buffalo, N.Y. The 30 ml serum capped bottles (sterile and pyrogen free) were kept frozen until used, and then allowed to liquify at room temperature in the dark, or by gentle warming. The HpD in dosages of 2.0 mg/kg was injected by slow, pulsed push administration over 5 to 10 minutes into the tubing of a fast running intravenous solution, to avoid local extravisation and venous irritation. Patients after this were not exposed to window light or outside light; essentially they were in a room with the shades drawn. No problems were encountered in regard to skin sensitivity to light or sunburn from light from ordinary incandescent or fluorescent light fixtures, but close-up reading lamps, etc. were prohibited.

Bronchoscopy was carried out 72 hours after the single intravenous injection of HpD, employing a double-lumen bronchoscope (Olympus FB-2T. The violet light source was from a krypton continuous ion laser (Spectra Physics Model 164-11/265). Typical output before filtering was 240 mw in 3 lines at 406.7 nm (36%), 413.1 nm (60%) and 415.4 nm (4%). The fluorescence exciting violet light was conducted to the bronchus via a fused quartz step-index optic fiber (0.8 mm OD) inserted to the tip of the bronchoscope via its small channel. A micro-lens was constructed to slip over the tip of the fiber to deliver a diffuse field of fluorescent light, to avoid "hot-spots" that interfere with detecting small areas of positive fluorescence. The violet irradiance of this system (12 nw/cm^2) is insufficient to cause significant heat or photodynamic reaction that might damage normal bronchial tissue or areas of tumor.

An image intensifier with a minimal luminous gain of 30,000 was coupled to the standard ocular of the bronchoscope by means of a transfer lens which gave 2.4 times magnification. The low-level, red fluorescence image was thereby converted to a bright green image more easily visualized by the bronchoscopist (Profio, Doiron 1979). Secondary filters between the bronchoscope ocular and the intensifier blocked the reflected violet light and passed the emitted red light.

Fluorescing sites were viewed first using a broad-band secondary filter (620-730 nm), and then a narrow-band filter (670-720 nm). The former gives a bright image, while the latter has a better signal to background ratio. Photographs were taken by means of a 35 mm single lens reflex camera attached to the image intensifier ocular, allowing direct visualization through the camera during photography.

During the past year, a second type of image intensifier has been developed and used, allowing alternate white-light and fluorescent light examination (Profio et al 1983). Essentially, the white-light images are conducted around the image intensifier by means of a periscope arrangement. A micro-switch attached to the head of the fiberoptic bronchoscope, near the tip-angulation controls, allows alternated switching between white-light and fluorescence light. The white-light source is a modified Olympus CLV unit.

Bronchoscopy is carried out under topical anesthesia employing 1% lidocaine via an endotracheal tube (Carden), connected to Bennett MA-1 ventilator (100% oxygen) by means of an adaptor. The diaphragm port of the adaptor allows easy withdrawal and re-insertion of the fiberoptic bronchoscope. Bronchial brushes are not withdrawn through the channel of the bronchoscope, but only to near its tip, and then the entire unit is withdrawn. The brushes are then quickly inserted into a cytology tube (Saccomanno 1982), for later cytopathological processing.

Results

In vitro, this imaging fluorescence bronchoscopy system is extremely sensitive and capable of visualizing an HpD concentration of 0.1 micrograms/ml in a phantom well, 100 micro-meters (0.1 mm) in depth and 1.5 mm in diameter.

The following two patients demonstrate its potential in vivo, and high sensitivity in localizing small areas (1 x 2 mm) of bronchial cancer that visually under white-light examination appeared normal.

Patient 1 was a 67-year-old man who had been a uranium miner for 28 years, and had smoked 1 pack of cigarettes per day for 50 years. He had been short of breath for 12 years, and had been diagnosed to have chronic bronchitis and emphysema. He had exacerbations of bronchitis requiring antibiotic and respiratory therapy. He had had pneumonia with pleural effusion on the right requiring chest-tube drainage in April 1980.

He was found to have malignant cells in sputum cytology in 1980. Four sputum samples during February to May showed carcinoma in situ cells. The chest X-rays showed over-inflation of the lungs due to emphysema, and a small infiltrate in the right upper lobe posteriorly.

Bronchoscopy in May 1980 showed that the oropharynx and larynx were normal. Visually, upon white-light examination, irregularity of the mucosa was seen in the left upper lobe bronchus, and three smooth whitish areas were seen in the left lower lobe bronchus. Both main bronchi, and the right upper and right middle lobe bronchi were normal. In the right lower lobe, the spur or bifurcation between two segmental bronchi appeared at most perhaps slightly thickened and whitish. (Fig. 1A). None of these areas showed even slight changes in the mucosa that could visually be suspected to be cancer.

Fluorescence bronchoscopy showed positive localized fluorescence only over the base of the spur or bifurcation between segmental bronchi 9 and 10 in the right lower lobe. This area showed bright, localized fluorescence (Fig. 1B). Bronchial brushings were positive for squamous carcinoma cells, thereby confirming the area of positive fluorescence to be "true" positive fluorescence. No other areas of positive fluorescence were seen, and bronchial brushings in other lobes were negative for tumor cells.

Patient 2 was a 66-year-old man who had been a hard rock miner for 23 years, and who had had a short (1 month) previous exposure to asbestos powder. He had smoked 44 years at 1 pack of cigarettes per day. Sputum cytology showed CIS and then squamous cancer cells. The chest X-ray showed no localized densities or infiltrates, but did show bilateral interstitial disease consistent with pulmonary fibrosis due to asbestos.

Bronchoscopy under white-light examination showed no mucosal abnormalities or even subtle changes that might have been suspected to be mucosal cancer. However, fluorescence examination showed an area of bright, discrete or localized fluorescence over the base of the spur between the anterior and apico-posterior segments of the right upper lobe (Fig. 2). No other areas showed positive fluorescence. Re-examination under white-light confirmed that this area had the appearance of entirely normal bronchial mucosa. Brushings of this area showed malignant cells. The patient had a second careful white-light and fluorescence bronchoscopy again a week later. The spur in the right upper lobe at the bifurcation into segmental bronchi again showed positive, localized fluorescence, and brushings again were positive for cancer cells. All other bronchi again appeared normal and did not show localized fluorescence. Selective bronchial brushings elsewhere in other lobes again were negative.

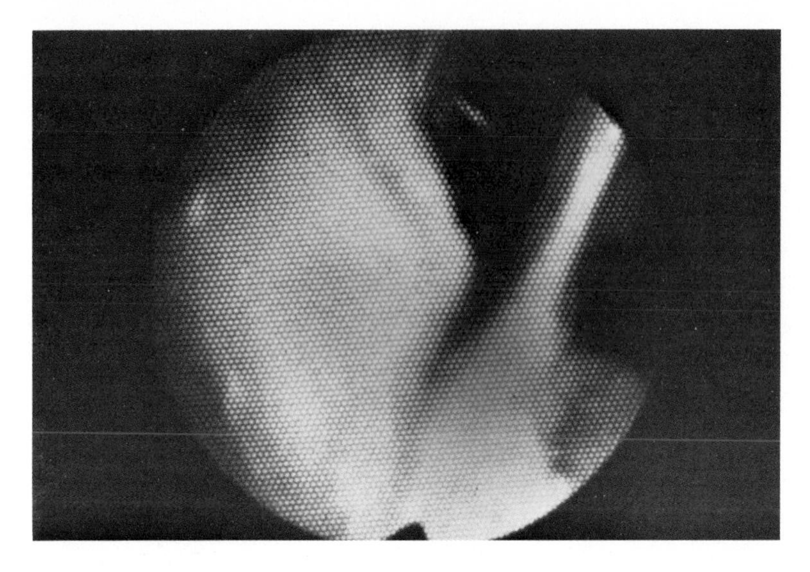

Figure 1: Patient No. 1
A. Spur between segmental bronchi right lower lobe, under white light inspection is normal.

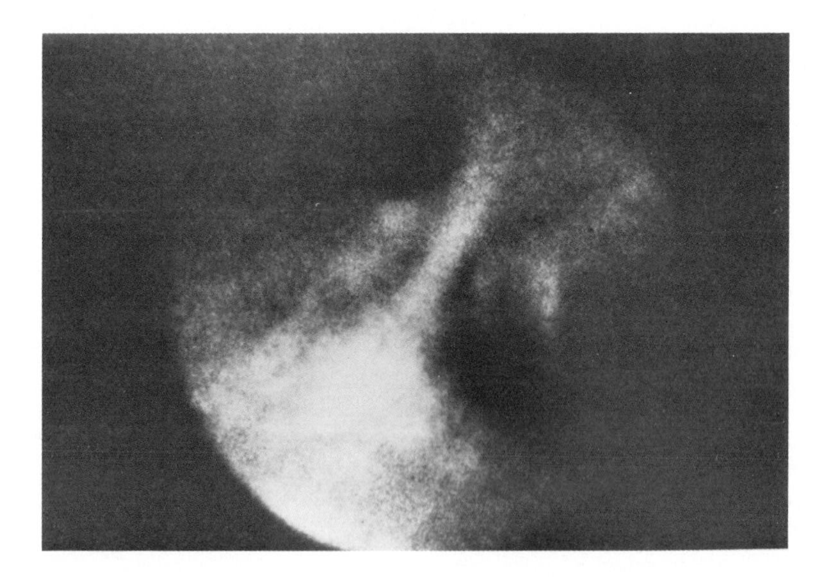

B. Positive Localized Fluorescence over base of spur under fluorescence inspection.

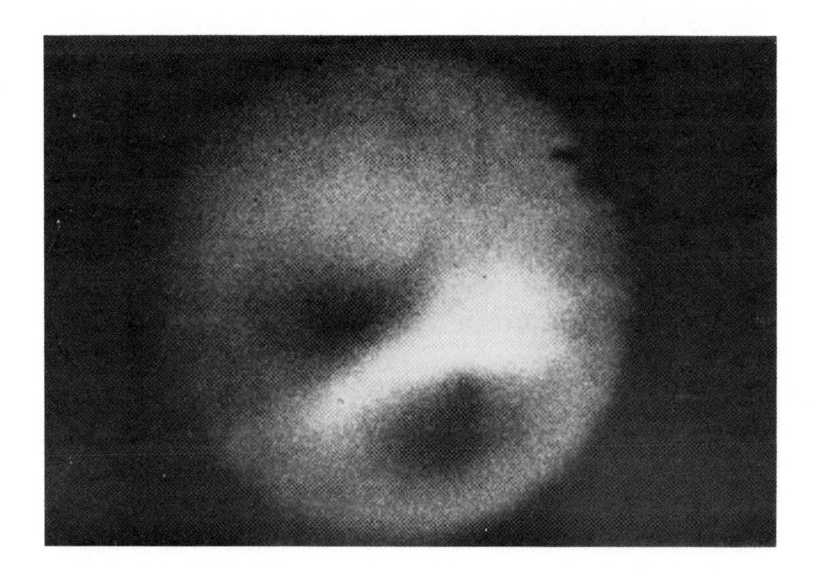

Fig. 2. Patient 2. Positive localized fluorescence base of spur between anterior and apico-posterior segments of right upper lobe. This area was entirely negative under white-light inspection.

Discussion

In these two patients, the potential of imaging fluorescence bronchoscopy has been demonstrated, in that very small (1 x 2 mm) areas of bronchial mucosal cancer have been located solely by their fluorescence. These areas appeared normal under white-light examination, even though search was made for subtle visible changes suggesting cancer, such as mucosal irregularity or small, elevated areas.

The sensitivity of imaging fluorescence bronchoscopy for visible mucosal cancers is greater than 90%, when ranging in size from easily seen to very small (Balchum et al 1982; Doiron et al 1979; King et al 1982). The next task at hand is to examine a large series of patients with radiologically occult lung cancer, to establish the degree of sensitivity of imaging fluorescence bronchoscopy for bronchial mucosal cancer lesions which are located only by their fluorescence, and which show normal mucosa or no more than subtle, non-suggestive mucosal abnormality under white-light examination.

The clinical importance of fluorescence bronchoscopy is that pre-invasive cancer, i.e. that still confined to the bronchus, can have a 100% cure-rate by surgical resection. Routine bronchoscopy is commonly negative. Multiple selective bronchial brushings to attempt to locate the cancerous area in this way is not efficient. This takes several hours, under general anesthesia, and may have to be repeated two or more times.

However, the spectrum of early lung cancer, as detected by sputum cytology positive for malignant cells and a normal chest X-ray, is very wide. Carcinoma in situ can be present in the oropharynx or larynx, and the lower respiratory tract may be normal or may also be involved. The fiberoptic imaging fluorescence bronchoscopic instrumentation system described here is employed for careful fluorescence examination of the large surface area of the mouth, pharynx, and larynx, not only the tracheo-bronchial tree.

Bronchial cancers often reside in small segmental or sub-segmental bronchi, yet within range of the fiberoptic bronchoscope (Olympus FB-2T or FB-4B2). These distal small airways must and can be carefully examined during fluorescence bronchoscopy. A comfortable bronchoscopy for the patient, under topical anesthesia, may take up to 2 hours and averages 1-1/2 hours. This includes careful white-light and fluorescence examination down to the sub-segmental level, brush and bite biopsies of fluorescing and non-fluorescing areas (for study), and photography.

Mucosal cancer may reside in more than one area of the trachea or bronchi. We have encountered several patients having multiple sites; one had up to three other small cancer areas beside the main visible site, each in a different lobar bronchus. Should the lobe with the visible lesion have been resected, the sputum would remain positive for cancer cells, with subsequent recurrence. We have encountered two such patients with positive sputum cytology having had previous lobectomy.

Fluorescence bronchoscopy is also useful for examining the bronchus through which transection will be done at intended lobectomy or pneumonectomy, to be more confident that it is free of tumor pre-operatively.

Cancer may also reside in the lung parenchyma, so that fluorescence bronchoscopy is negative yet the sputum positive for malignant cells. One such patient, with pulmonary fibrosis due to asbestos, has consistently shown adenocarcinoma cells in sputum and bronchial washings, with two fluorescence bronchoscopies that showed no areas of localized or even suspect fluorescence. Diagnosis will be attempted by selective transbronchial lung biopsies from each lobe and segment under fluoroscopic guidance. It is likely that parenchymal bronchioalveolar cell carcinoma is present.

The present instrumentation and system have proven clinically efficient and effective in diagnosis. The bronchoscopist must learn how to adjust the fiber to obtain a uniform field of fluorescent light, and to carry out bronchoscopy through the image intensifier under fluorescent light. Anatomical landmarks under fluorescent light come to be recognized. The new alternate white and fluorescent light image intensifier (Profio et al 1983) makes orientation possible more quickly as one bronchoscopes, and reduces bronchoscopy time. The bronchoscopist must learn to discriminate between an area of localized, discrete positive fluorescence, and a "hot-spot", or areas of diffuse fluorescence or reflection.

Further research here is aimed at developing instruments and methods to improve the accuracy and specificity of localized positive fluorescence detection. One approach is by real-time digital computer analysis to enhance contrast above background and to sharpen the edges of fluorescing areas to improve recognition. Another is the analysis of fluorescence spectra in vivo with a rapid scanning spectrometer to determine whether small areas of mucosal CIS result in characteristic spectra which would aid in specific diagnosis.

Bronchoscopic diagnosis of bronchial cancer rests on good specimens and skilled cytopathologic interpretation. The accurate clinical detection of radiologically occult lung cancer in high risk individuals depends upon adequate sputum sampling to collect exfoliated cells, adequate processing of these samples to maximally retrieve cells from each sample, and upon skilled cytopathologic interpretation. Next, proving that an area of positive localized fluorescence is "true positive" depends upon the methods, skills, and instruments in bronchial brushing and biopsy. Normal or near-normal bronchial mucosa is difficult to brush and biopsy to obtain adequate samples. Areas of early lung cancer that positively fluoresce and can be located by their fluorescence are very small, from 1 x 2 mm to 4 x 5 mm, and may not appear to be raised above the level of the mucosa. These are brushed and biopsied directly under fluorescent light, in order to accurately sample the fluorescing area and not mistakingly obtain samples from adjacent areas.

Brushes must be immediately inserted into cytology tubes (Saccomanno 1982), for prompt fixation to prevent drying and distortion of cells. Biopsies are immediately immersed in 50% alcohol-2% Carbowax in a vial with a conical bottom. All particles must be sectioned and examined. After double reading, we have a final cytopathological interpretation by a reference cytopathologist, since there is significant variability between pathologists. The question that must be answered - whether true positive fluorescence is present, i.e. that the fluorescing area is in fact cancerous - depends upon skills in accurate brushing and biopsying a very small area of bronchial mucosa, obtaining adequate specimens, and upon their cytopathological interpretation.

Hematoporphyrin derivative accumulates and is retained in all histologic types of lung cancer as well as cancer metastatic to the lungs, all of which fluoresce when excited by violet light after hematoporphyrin derivative administration, in our experience.

The use of HpD which is retained in endobronchial cancer and bronchial CIS lesions has led to the development of photoradiation therapy (PRT). This is based upon the photodynamic reaction, in that HpD in tumor cells is activated by red light (630 nm) to produce singlet oxygen, so that cell membranes are impaired and tumor cells slowly die by one to several days. In palliative PRT for endobronchial cancer obstructing the trachea or bronchi, these areas are illuminated with red light 72 hours after the administration of HpD, by means of various light applicators and techniques (Balchum et al 1983). There is no immediate visible reaction in PRT, such as coagulation, charring or vaporization as in Nd-YAG laser therapy. Upon re-bronchoscopy a few days later, the tumor debris is cleared from the bronchi by forceps.

Palliative PRT is for the purposes of abolishing endobronchial tumors obstructing bronchi, to open up the lung or lobes for ventilation to decrease shortness of breath, and to permit the coughing up of secretions and bacteria that cause pneumonia, as well as to control endobronchial tumor growth locally.

In early lung cancer and CIS, bronchial mucosal lesions are thin and PRT has the potential to be curative. Red light penetrates 8-10 mm and PRT may be lethal to cancer cells, with potential for cure. It is an alternative method to lung resection, and can be used in patients who are inoperable because of high risk for resection from significant cardio-pulmonary or other disease. Hayata et al (1982) has employed PRT pre-operatively, before lobectomy, where the intended bronchial transection site is involved by tumor, to attempt to prevent recurrence. There have been good one-year results.

Acknowledgements

The technical assistance and skills of our laser technicians, Nick Razum and Leon Chaput, are greatly appreciated, as was the assistance of Joyce Portugal in records and specimens handling.

References

Balchum OJ, Doiron DR, Profio AE, Huth GC (1982). Fluorescence bronchoscopy for localizing early bronchial cancer and carcinoma in situ. Results in Cancer Research 82:97-120.

Balchum OJ, Doiron DR, Huth GC (1983). Photoradiation

therapy of obstructing endobronchial lung cancer, employing the photodynamic action of hematoporphyrin derivative. Proc. Symp. on Porphyrin Localization and Treatment of Tumors, Clayton Foundation, Santa Barbara, California, 4/24-4/28.

Chalon J, Tang CK, Klein GS, Ramanathan S, Patel C, Turndorf H (1981). Routine cytodiagnosis of pulmonary malignancies. Arch Pathol Lab Med 105:11-14.

Cortese DA, Kinsey JH, Woolner LB, Payne WS, Sanderson DR, Fontana RS (1979). Clinical application of a new endoscopic technique for detection of in situ bronchial carcinoma. Mayo Clin Proc 54:636-642.

Doiron DR, Profio AE, Vincent RG, Dougherty TJ (1979). Fluorescence bronchoscopy for detection of lung cancer. Chest 76:27-32.

Hyata Y, Kato H, Konaka C, Ono J, Takizaiwa N (1982). Hematoporphyrin derivative and laser photoradiation in the treatment of lung cancer. Chest 81:269-277.

King EG, Man G, Le Riche R, Amy R, Profio AE, Doiron DR (1982). Fluorescence bronchoscopy in the localization of bronchogenic carcinoma. Cancer 49:777-782.

Lipson RL, Baldes EJ, Olson AM (1964). Further evaluation of the use of hematoporphyrin derivative as a new aid in endoscopic detection of malignant disease. Dis Chest 46:676-679.

Martini N (1983). Lung mapping. Symp. Advances in Bronchoscopy, ACCP, March 17-19, San Diego, California.

Martini N, Melamed MR (1980). Occult carcinoma of the lungs. Ann Thorac Surg 30:215-223.

Melamed MR, Flehinger BJ, Muhammad BZ, Heeland RT, Hallerman ET, Martini N (1981). Detection of true pathologic stage I lung cancer in a screening program and the effect on survival. Cancer 47:1182-1187.

Murray PR, Washington JA (1975). Microscopic and bacteriologic analysis of expectorated sputum. Mayo Clin Proc 50:339-344.

Profio AE, Doiron DR (1979). Laser fluorescence bronchoscopy for localization of occult lung tumors. Med Phys 6:523-525.

Profio AE, Doiron DR, Balchum OJ, Huth GC (1983). Fluorescence bronchoscopy for localization of carcinoma in situ. Med Phys 10(1):35-39.

Saccomanno G (1978). Diagnostic pulmonary cytology. Amer Soc Clin Pathologists, Chicago, Illinois.

Saccomanno G (1982). Carcinoma in situ of the lung: Its development, detection and treatment. Seminars in Resp Med 4:156-160.

Williams DE, Pairolero PC, Davis CS, Bernatz PE, Payne WS, Taylor WH, Uhlenhopp MA, Fontana RS (1981). Survival of patients surgically treated for stage I lung cancer. J Thor Cardiovasc Surg 82:70-76.

Index